Powell's   Snakes of the Americas: Checklist
$124.00 / 9.98   NDJ
Science & Natural History   117542

Snakes of the Americas

Snakes
of the
Americas
Checklist and Lexicon

Bob L. Tipton

KRIEGER PUBLISHING COMPANY
Malabar, Florida
2005

Original Edition 2005

Printed and Published by
**KRIEGER PUBLISHING COMPANY
KRIEGER DRIVE
MALABAR, FLORIDA 32950**

Copyright © 2005 by Bob L. Tipton

All rights reserved. No part of this book may be reproduced in any form or by any means, electronic or mechanical, including information storage and retrieval systems without permission in writing from the publisher.
No liability is assumed with respect to the use of the information contained herein.
Printed in the United States of America.

FROM A DECLARATION OF PRINCIPLES JOINTLY ADOPTED
BY A COMMITTEE OF THE AMERICAN BAR ASSOCIATION
AND A COMMITTEE OF PUBLISHERS:

This publication is designed to provide accurate and authoritative information inregard to the subject matter covered. It is sold with the understanding that the publisher is not engaged in rendering legal, accounting, or other professional service. If legal advice or other expert assistance is required, the services of a competent professional person should be sought.

Library of Congress Cataloging-in-Publication Data

Tipton, Bob L.
　Snakes of the Americas : checklist and lexicon / Bob L. Tipton. — Original ed.
　　　p. cm.
　Includes bibliographical references (p.).
　ISBN 1-57524-215-X (alk. paper)
　1. Snakes—Latin America. 2. Snakes—North America. 3. Snakes—Latin America—Nomenclature (Popular) 4. Snakes—North America—Nomenclature (Popular) I. Title.
　　QL666.O6T575 2004
　　597.96′097—dc22
　　　　　　　　　　　　　　　　　　　　　　　2003060319

10　9　8　7　6　5　4　3　2

Contents

Foreword by Jonathan A. Campbell vii

Preface ix

Acknowledgments xi

User's Guide to the CD xiii

Introduction 1

Checklist Overview 8

The Primitive Group—Infraorder Scolecophidia Cope, 1864 13

    Superfamily Typhlopoidea Gray, 1825 15

        Family Anomalepididae Taylor, 1939 15

        Family Leptotyphlopidae Stejneger, 1892 19

        Family Typhlopidae Merrem, 1820 28

The Modern Group—Infraorder Alethinophidia Nopsca, 1923 35

    Family Aniliidae Stejneger, 1907 (Fitzinger, 1826) 37

    Primitive (or Basal) Alethinophidia—Macrostomata Müller, [1831] 1832 38

        Family Boidae Gray, 1825 38

        Family Loxocemidae Cope, 1861 52

        Family Tropidophiidae Brongersma, 1951 (Cope, 1894) 52

Advanced Snakes—The Caenophidia Oppel, 1811 58

Family Colubridae Oppel, 1811 58

Family Elapidae Boie, 1827 286

Family Hydrophiidae Boie, 1827 304

Family Viperidae Oppel, 1811 305

Incertae Sedis and Other Questions. 341

Bibliography 351

Index. 433

Foreword

Snakes of the Americas: Checklist and Lexicon provides a current and up-to-date reference to snakes of the western world. One of the more unusual features of this work is the comprehensive synthesis of over 21,000 vernacular names in many different languages. In conversation, such names are sometimes more important than scientific epithets. Most Guatemalans would have no idea what a *Bothrops asper* is, but mention the name barba amarilla, and almost everyone will have a story to tell. However, the situation with common names is often confusing owing to the use of many different names for widespread species. Common names are also important simply because most people find them easier to use than scientific names.

Tipton has ferreted out an amazing number of different names during his extensive travels throughout the Americas and through consulting myriad published works. For those persons conducting field work in the Americas, those who are interested in folklore, or those who enjoy learning the sometimes colorful names of a group of animals firmly rooted in mythology and mystery, Tipton's book provides an interesting and useful reference.

Jonathan A. Campbell
June 2004

Preface

As an amateur, my original goal at the Texas Cooperative Wildlife Collection of Texas A&M University was to learn more about Latin American snakes by researching literature, and when necessary, to look at a few specimens. James R. Dixon encouraged me to expand this project and make the information I had collected over years in the field, or was finding in literature, available to everyone. The information I developed is reflected in this work.

In my travels throughout the world, I continually came across snakes that were difficult to identify. Not being a professional herpetologist, I often did not know where to begin the search for clues to help identify the snakes that I encountered in Latin America, Europe, Scandinavia, Africa, and the Middle East. Those snakes from Latin America seemed to be particularly difficult to identify. This work was organized to help others such as myself, who are not professionals, and who experience problems when they enter the identification maze for Latin American material. In the field, a good place for people like us to begin is to find the local common name and then work backwards to the checklist. The focus of the work, as the title indicates, is the snakes of the Americas. And although the information herein is comprehensive for the Americas, one of my objectives was to identify useful information for snakes of Latin America. Information about snakes found in Canada and the United States is readily available and usually little effort is required to quickly identify what the reader encounters there. Flores-Villela (1993) and Liner (1994) have good checklists for Mexico and their work backs up Smith and Taylor (1966); the Caribbean, although it contains a complex mix of species and subspecies, is fairly straightforward and has been well covered by such authorities as Rivero, Schwartz, Henderson, Crother, and others. Mainland Latin America has a large and confusing selection of snake species, and many are very similar in characteristics. Information for this area is just beginning to evolve from the "dark ages." I am hopeful that the lexicon and checklist combined will offer the reader more tools to apply in identifying some of the more complex snake species and subspecies. Although my original objective was to focus on the Latin American snakes, the checklist and lexicon of common names have been developed to show information for snakes in all of the Americas including the Caribbean, the Galapagos Islands, Easter Islands (a seasnake), and the Hawaiian Islands.

As noted in the beginning of this section, the lexicon was the original objective of the work and should be beneficial to a wide range of readers in various disciplines. Ironically, neither the lexicon nor the checklist evolved into the finished product that I originally envisioned for them. The lexicon was to be a simple work with only about 3000 to 4000 common names cross-referenced to scientific names. The finished lexicon now contains over 21,000 common names in a wide variety of languages. The checklist I foresaw was to be only 25 to 50 pages of genera and species names without accompanying information or subspecies. The checklist as completed is detailed to the subspecies level in about 350 pages. The checklist was only to be a support tool for the lexicon. Ironically, it was the most difficult part of the research, and independently turned out to be a useful stand-alone reference. Much of the final product is due to encouragement from people like James R. Dixon, Bill Lamar, Jonathan Campbell, Douglas Rossman, Ronaldo Fernandes, and Larry David Wilson to improve my investigation of subject matter, become more critical of content, better communicate distribution, and to use more primary references. If there are any criticisms or failings in this work, it is due to my lack of attention, knowledge, or experience, not because of the people who have encouraged and supported me in this project. As the work matured, it became evident that details were required to develop the checklist into a usable tool for the reader. Many of the people who supported me further encouraged me to organize the checklist into a format that could be used by professionals. At the same time, even though the current trend is to move away from subspecies, I decided to attempt to list subspecies to help others such as myself, who were continually confused by what they found in general literature. Although I visited various museums and libraries in the United States, Europe, and Latin America to research this project, the bulk of work was done at the Texas Cooperative Wildlife Collection of Texas A&M University. This project is really the result of my being allowed access to the comprehensive collection of literature and specimens found there, and the generosity of Drs. James R. Dixon and Kathryn Vaughan, who gave their time when I needed some clarification of my findings or direction for my investigations.

As a serious amateur, I have been fortunate to be able to combine my hobby of herpetology with the domestic and international travel required for my work. The checklist and lexicon evolved from a need for information when searching for a particular kind of snake, or trying to identify what some snake might be from a local name I had heard during these travels. The information I encountered has been organized for those people, such as myself, who consider themselves serious novices, and who would like to have a resource to more quickly bracket snakes by geography and common name, in order to identify what they have encountered, and to reference the names in the field and in the literature. The first volume is an effort to provide a checklist of the snakes of the Americas to the subspecies level. The second volume (on compact disk) addresses common names that one finds in literature or in the field. Fortunately, most of the professionals from whom I had asked assistance have gone out of their way to ensure I was given proper directions about how to find information or to guide me through difficult information. This work would not have been completed without their support. Because of the guidance I received, both beginners and professionals who use this book should benefit from the checklist and lexicon.

The reader should be aware of the purpose of this work. As noted in the previous paragraph, it is written for serious novices, such as myself, who are exposed to a wide spectrum of species and must navigate through a confusing maze of literature to learn about the snakes that they might encounter. Because of the simplicity I desired, the professional or academician may have various criticisms about the lack of primary information in this work, but I believe they should recognize the value of the information contained in both the checklist and lexicon.

In putting the checklist together I seem unwittingly to have become somewhat of a taxonomist. During the education period I noticed there is a general interest in common names by zoo people, anthropologists, environmentalists, herpetoculturists, and medical staff, etc., but many of the professional taxonomists do not accept the value of common names. These taxonomists take the position that the scientific name is really the only name that should be accepted for technical work. Without question they are correct about the scientific name being the focus for technical work, but access to information about common names does have value. I believe that Alberto Yanosky expressed the reason very well in a seminar he gave on October 1, 2002, at Texas A&M University. Dr. Yanosky, who lives in Paraguay, noted that around Asunción everyone understands Spanish. In the less populated areas, however, there are fewer people who understand Spanish and knowledge of Guaraní becomes increasingly necessary to communicate with the general population. This shows that common names become important if a researcher is trying to find a particular animal or plant in the field, or to learn something about it. With this in mind, I believe the common names list has a purpose for taxonomists, as well as the rest of the world.

Acknowledgments

I would like to thank the following people: John Aberle, Federico Achával, Héctor F. Aguilar, Manuel Andueza, Luis Anonsí, Johnnie Banks, Janet Barnes, George Baumgardner, Fernando Benitez-Zapata, Marcus Di-Bernardo, Hans Boos, Jeff J. Boundy, Janalee Caldwell, Jeffery D. Camper, Mark A. Carey, Jr., Carlos Castaño, David Chiszar, Monika Cisch, Hugo Classen, Guarino R. Colli, Roger C. Conant, Jesus Cordova, Jaime Crosby Ruso, Xavier Crosby Ruso, Brian I. Crother, Gustavo Cruz-Diaz, Rudolfo Darbelles, Guadeloupe DeSoto, A. Diaz Olivera, Tiffany M. Doan, Carlos Escobedo-Pachero, Daniel Fernandes, Ronaldo Fernandes, Lee Fitzgerald, Cliff Fontenot, Michael J. Forstner, Francisco Franco, Walter García, Alejandro R. Giraudo, Harry W. Greene, Lee L. Grismer, Rodney Guerra, Saul Gutierrez, S. Blair Hedges, Robert W. Henderson, Toby J. Hibbits, Troy D. Hibbits, Ryan Hill, Peter Holm, Marinus S. Hoogmoed, Germán Ibarra, Jorge Ibarra, Javier Icochea, Ivan Ineich, Kelly Irwin, Jerry D. Johnson, Vanessa Kruth Verdade, Karen Kuhl, Travis J. LaDuc, Lionel Landry, Dominic I. Lannutti, R. K. Lawrence, David Lazcano-Villarreal, Joseph Lazlo, Thales de Lema, Paul Longshaw, Julio Cesar Maura Leite, Carl S. Lieb, Luis Martins, José Mendez, Claudia Mercolli, Joseph C. Mitchell, Celso Morato, Robert W. Murphy, Sheldon Parks, Carlos Armando Perez Medina, Victor Petrucci, Louis Porras, Hugh Quinn, Joseph Ramspott, Susan Reed, Gilson Rivas, Carlos Rivero-Blanco, Javier A. Rodrigues-Robles, Douglas A. Rossman, Janis A. Roze, Ethan Russo, Mahmood Sasa Marín, Ivan Sazima, Gordon Schuett, Norman J. Scott, Jr., Rainer Schulte, Juan Silva-Haad, Joseph B. Slowinski, Gerard T. Solmon, James Marvin Thibodeaux, Robert A. Thomas, Angelita Tipton, Santiago Tobon, Jean Marc Touzet, Avi Tuschman, Al Twiss, Peter Uetz, Manfred Velt, Sven Velt, Thomas G. Vermersch, Ramón Villeda-Bermúdez, Laurie J. Vitt, Robert G. Webb, John E. Werler, Martin J. Whiting, Larry David Wilson, and Hussam Zaher.

Jonathan A. Campbell, William W. Lamar, and Van Wallach were kind enough to do initial reviews of at least part of the information so roughly assembled that it could not properly be called a draft. They pointed out the many weaknesses and made important recommendations that allowed me to better organize the work. Various people helped with detailed reviews of individual genera or parts of genera. I appreciate all of their efforts and each reviewer has been acknowledged in the notes at the beginning of each genus section.

The final reviews were undertaken by James R. Dixon, and Alberto Yanosky-Farrán. They were able to comment on much of the checklist manuscript. With me still in the learning process, they sometimes still had to work with crude, improperly prepared manuscripts.

The common names offered special challenges. Lutz Dirksen kindly reviewed the German common names. Alberto Yanosky-Farrán's review of the Guaraní and Spanish names in the common names was important to information on the snakes of the Southern Cone. Dante Martins Teixeira contributed his expertise in Portuguese, Tupi, Guaraní and many Brazilian indigeneous common names. I would like to point out a special note of appreciation to Ernest A. Liner, whose loan of difficult to obtain literature and valuable guidance with comments on the common names and reference sections made the project much easier. Dr. Liner's patience in the tedious review of the reference section improved the accuracy of my work in that section.

Although the lexicon is accessible in a common database program, Microsoft-Excel®, the massive amount of information contained in the files made it rather difficult to manage. Andrew Grant, Paul King, and Brandy Tipton made valuable contributions with their experiences as software engineers in making the database more accessible and user friendly.

I would like to give special thanks to the family of the late Noel Kempff Mercado in Bolivia. They were kind enough to send me a copy of his book via my friend Carlos Armando Perez Medina in La Paz, Bolivia. Kempff Mercado's book covers the snakes of Bolivia, and the copy I received contained the handwritten notes he was making to update the work before his untimely death.

The support of James R. Dixon and R. Kathryn Vaughan at the Texas Cooperative Wildlife Collection of Texas A&M University was invaluable. They were kind enough to give me access to literature and specimens that made preparation of the checklist much easier than it could have been, and their encouragement kept me on the path to completing the project. It took them several years but they also seemed to finally teach me some of the basics of taxonomy. My appreciation for the Texas A&M group also extends to Mary Dixon. Mary seems to adopt any stray animals, graduate students, and visiting herpetologists (professional and amateur alike) that wander into the Collection.

Any amateur attempting to organize a task of this nature finds himself or herself rambling down many dark alleys of conflicting and out-of-date information. It is important to point out that any errors found in the work are mine alone in spite of the many efforts of the above mentioned people to illuminate the often dimly lit path this amateur herpetologist chose to follow.

User's Guide to the CD

System Requirements

The minimum recommended system requirements are:
- Windows 98 or higher
- Microsoft Excel 2000 or higher
- Microsoft Word 98 or higher
- Pentium II (200 MHz)
- 32 MB RAM
- 8x CD-ROM

Although the above are the *minimum* recommended requirements, to obtain the maximum performance from the lexicon, I recommend using a system as follows:
- Windows XP or higher
- Excel 2002 or higher
- Word 2002 or higher
- Pentium 4 (2.53 GHz)
- 512 MB RAM
- 8x CD-ROM

It works efficiently on a system with these parameters, and naturally will perform even faster on upgraded systems.

Opening the CD

Opening the CD is a relatively easy operation. Simply place the CD in the computer's CD or DVD drive. The drive will read the CD, and the following folder will appear:
- Snakes of the Americas: Lexicon of Common Names

Double-click on the folder. Five icons will appear:
- Checklist Distribution (MS Excel worksheet)
- Common Names (folder)
- Contents (MS Word document)
- References (MS Word document)
- User's Guide (MS Word document)

Note: The files can be opened in a number of other ways, including opening directly from the application in which they are written, whether Excel or Word.

Checklist Distribution

The Checklist Distribution tells where a snake may be encountered (by region).

Common Names

Double-clicking on the Common Names folder will reveal three more folders:
- Complete Lists of Common Names
- Language Sort
- Recommended Common Names of Genera, Species, and Subspecies

Complete Lists of Common Names

These files are quite large and may take some time to open. Once open, they also take time to manipulate. For these reasons, the files are presorted into three separate files. The first is sorted alphabetically by common name. The second is sorted by language. The third is sorted by scientific name. All contain the same data.

Several thousand pages of information are found in these files. Choose the subject based on interest. The files are quite extensive and require some minutes to open, so patience is required.

Language Sort

This sort contains a separate file for each major language.

Recommended Common Names of Genera, Species, and Subspecies

These are what I recommend to be the general use common names for all of the snakes in the Americas at genus, species, and subspecies level.

This information is used to update the total common names lists of snakes. The Crother et al. reference is still considered to be the authority on common names in the United States and Canada in English, and Liner's work is the authority for the common names of snakes in English and Spanish in Mexico. This part of the lexicon extends to all of the Americas in several languages and updates and completes the work of Frank and Ramus.

Contents, References, User's Guide

These documents are self-explanatory. The references are those used to generate the lexicon.

Using the Files

The complete lists of common names are quite extensive and require some minutes to open, so patience is required. The other files open relatively quickly.

Once a file is open, use the FIND function in the EDIT pull-down menu to search for a common name, language, or scientific name. These files are Excel worksheets and as such can be sorted by clicking on the down arrows at the bottom of the header columns.

To exit the program, click on the X at the top right of the window.

Introduction

Often a reference in herpetology limits the use of one common name per genus, species, or subspecies, and many times a common name is not included. Rarely does a reference use more than one language or show a common name in a language other than the language in which the reference is written. The checklist gives one name in English for each genus, species, and subspecies as the preferred common name, with a few exceptions as discussed later. The lexicon lists *all* of the names I have encountered for a snake, and it is common for a snake to have 10 or 20 common names in various languages. In several cases more than 100 common names were found for a particular snake. Even if names were encountered that were questionable or obviously not correctly applied, these may be listed. I attempted to find a reference for all of the names I heard in the field. Several names are listed where a reference could not be found. This happened when I could not find a literature cross-reference for that particular name versus the species to which it was linked. In a few cases I developed and recommended a common name in English.

The citations noted in this checklist may not be the "final word," or primary research, but will give the reader points to investigate when using the lexicon. To use the checklist effectively, it is important to have a good working knowledge of the geography of the Americas. If the reader is not proficient in the geography of a certain area, then he or she should use the checklist in conjunction with an atlas to better understand distribution. In many cases there are opposing opinions, and where this occurred I attempted to note opinions in a remark. The checklist is based on what I *perceive* to be the correct listing of scientific names for the snakes of the Americas. As presented here, the Americas encompass all of the Western Hemisphere; the regions are defined in the checklist as follows:

- North America includes Canada, the United States, Mexico, and attendant islands and states. Although not part of the continent, the Pacific chain of islands of Hawaii are a state in the United States and for purposes of this checklist are considered to be part of North America.
- Latin America includes all of the mainland and attendant islands south of the United States in part of North America (Mexico), Central America, and South America.
- The Caribbean encompasses all of the islands and nations in that region including Trinidad and Tobago. The country of Trinidad and Tobago is often listed independently when reference is made to it. It has been separated from the Caribbean locations in the distribution chart.
- Central America is the extreme southern part of North America present as a narrow strip of land which connects it to continental South America and its attendant islands. Technically, Central America is usually considered to extend from the north at the Isthmus of Tehuantepec in southern Mexico, southward to the Isthmus of Panama (not including the southern part of Panama which is considered to be part of South America.) In this checklist, for convenience Central America is considered to be composed of all of the countries of Belize, Guatemala, El Salvador, Honduras, Nicaragua, Costa Rica, and Panama. None of Mexico is considered to be Central America here.
- South America includes the countries of Colombia,

Venezuela, the Guianas (Guyana, Suriname, and French Guiana), Brazil, Ecuador, Peru, Bolivia, Chile, Argentina, Uruguay, Paraguay, and the offshore islands. Even though the nation of Trinidad and Tobago is made up of islands in the Caribbean, its fauna is often grouped with that of mainland South America because of its close proximity to the eastern coast of Venezuela. No part of Panama is considered to be part of South America in this checklist.

- Middle America is used for more diverse geography and encompasses part of southern Mexico and Central America with attendant islands.

The checklist was designed to support the common names lexicon. As such, I have made an effort to list common names in English to all genera, species, and subspecies. Some names are assigned from a wide selection of names. In instances where a common name did not exist in English or did not make sense for that application, I assigned a name based on the meaning of the scientific classification, the author, a physical characteristic, some aspect of its distribution, or other available information. The checklist is the result of the desire to support the lexicon list and was developed around the common names collected over the years in various parts of the Americas, from lists, and most recently the internet. Literature about snakes from Canada, the United States, Mexico, and the Caribbean is relatively accessible, so this checklist will probably be more valuable for those searching for information concerning mainland Latin American snakes.

Not everyone will agree with the genera, species, and subspecies that I have selected, and part of the goal is to encourage the reader to think about arguments to accept or reject genera and species, but hopefully, all should agree that the checklist and/or lexicon have been a worthwhile effort. What I perceive to be the most recently accepted genera are listed and referenced to what appears to me to be the easily accessible and more common citations. The checklist covers citations on genera, species, or subspecies through the end of 2000, although information through 2003 is included if I had access to the literature to review and if it may have contained an important addition or change. If there is a lack of literature for the citation an effort was usually made to avoid citing the original literature so that more current literature could be referenced for the reader.

The checklist is not a panacea for identification, but it should be a valuable tool in the process of elimination. The checklist allows the reader to reference information quickly by focusing on citations that help identify an individual species. Some genera are especially challenging for the readers, but the checklist gives various options as a basis to begin the identification process.

Over the years, even as a novice, I realized that in literature the genera, species, and subspecies must be tied to geographic and physical characteristics supported by photographs, drawings, written descriptions, and if possible, a preserved or live reference specimen. Where difficult-to-identify or unfamiliar individuals are concerned, if the specimen can be compared to a catalogued museum specimen, a basis can be established to give the quickest path for identification.

The references cited should help with these points, and an attempt was made to include references with applicable keys and distribution. The user must be aware that keys do not function universally, especially with the Colubrid snakes. Whenever possible, two or more keys should be compared. This discrepancy does not necessarily occur because an author designed a key poorly, but rather because different keys have different objectives and reference points. The user should understand that a key designed to compare snakes with the same distribution may not match with one designed to compare similar colors or, species in a genera, or form, or habitat.

There are many excellent comprehensive references for snakes of the Americas. Until recently many of these focused on venomous snakes or restricted their subject matter to those found in the United States. This is not to say that good comprehensive references for outside of the United States do not exist. These will not be detailed here, but many of these are listed in the references. Beginning in the late 1970s more references became available. Since then, excellent comprehensive references covering various regions supplemented those few works that had been published previously. Now excellent comprehensive references on snakes of various regions are more accessible. There are new books being published that help focus more technical attention on areas outside of the United States. Hopefully, this checklist will fill some of the voids that may exist in general information and serve as support information for existing and future references. Depending on individual objectives, it may be important that the reader become familiar with some of these.

This checklist is designed to be used in conjunction with the technical literature to develop understanding of a snake in a specific region. As well as the older, well known geographic and local references, more recently there have appeared several references covering regional amphibians and reptiles of lesser known areas. These too can be found in the reference section.

The checklist is relatively comprehensive in information, but the reader is advised to use it alongside other, more definitive references. Although Peters and Orejas-Miranda (1970) is now over 30 years old, it is still a valuable reference and should be considered the "Bible" for beginning any investigation of South American serpents. Vanzolini (1986) supplements their work. Smith and Taylor (1945, 1966) and Smith and Smith (1973, 1976a, 1976b, 1993) are the general references for Mexico. The checklist found in this work used a combination of Peters and Orejas-Miranda (1970), Wilson and Meyer (1985), Vanzolini (1986), Villa et al. (1988), Smith and Taylor (1966), Smith and Smith (1993 and others), Wilson and Meyer (1985), Dixon and Soini (1986), Cunha and Nascimento (1985), Campbell and Lamar (1989), Schwartz and Henderson (1991), and McDiarmid et al. (1999). I recommend the reader obtain copies of the McDiarmid et al. series as the authors complete their work with future volumes. I also recommend the reader find a copy of Campbell and Lamar's update of the Campbell and Lamar (1989) work on venomous snakes of Latin America when it is available.

For those who may be familiar with American snakes, it is easy to confuse the various colubrid genera, and even those who are authorities on American snakes do not agree on classifications. The references cited in the checklist contain reliable guides for identification of these difficult-to-key snakes; however, the references cited do not agree among themselves. This is because of the large number of kinds of snakes and the huge diversity of characteristics, environments, etc.

Ernst and Zug (1996) noted that we are now reasonably certain that all modern snakes belong to major lineages. One lineage is the Blindsnakes classified as Scolecophidians. The second lineage, the Alethinophidia, has been more successful; it consists of over 2000 species and a multitude of lifestyles. The second lineage is broken down further as the Haenophidia, called primitive snakes or basal Alethinophidia, which are the Pythons and Boas, and the Caenophidia, which are sometimes called the advanced or modern snakes. The reader will find hundreds of scientific names on the checklist relating to these lineages. Often, acceptance of a classification is dependent on individual preference, the academic and/or zoological circles in which the reader is active, and/or the herpetocultural groups with which the reader may be associated. Academic pressures of publishing versus the time needed for study and analysis are often in conflict, and frequently do not allow a group of specimens to be properly studied before a new name or group emerges into print. This is not necessarily due to poor habits on the part of the authority but rather the pressure to publish demanded by the job. It is important to point out that there are always some differences of opinion between authorities on snakes or groups of snakes. Like oil and water, time constraints and scholarship to not mix and the best work is always difficult to achieve properly with the time allotted to the academicians who must "publish or die." Even the most esteemed of authorities make errors because of these pressures. This makes the literature research interesting because rarely is there total agreement among herpetologists about the genera, species, and subspecies selected.

To paraphrase James Lazell (1992), just because something has a scientific name attached to it does not mean that name or information is correct. It is always necessary to confirm the information that is presented. This has been exactly what my mentors have expressed to me. They wanted me to form my own opinions, even if in opposition to the norm. In most cases, selection of genus and species was a fairly simple task, but there were several situations where I had to give an opinion and not everyone will be pleased. The original intent of this work was to be a simple literature search. James R. Dixon told me early in the project that I would be obligated to make decisions about many items where various authorities are in conflict. As with most information Dr. Dixon shared with me, he was correct, and I hope that my inexperience in certain areas of this discipline is not too obvious in some of the decisions I made.

As noted in the previous paragraph, the objective of this work is "usable simplicity." There are, however, some valid criticisms of this attempt at simplicity. The citations note several references that are checklists.

Professionals as well as amateurs should find at least some of the checklists cited as valuable for reference purposes. One such example is Smith and Smith (1993). It is basically a detailed checklist of Mexico, and it cites many excellent references. Liner (1994) also covers Mexico and gives common names in English and Spanish. Liner's efforts complement the goals of this work and he is working to update his common names list and checklist of the amphibians and reptiles of Mexico. The most valuable citation I used was the Peters and Orejas-Miranda (1970) checklist. Until this work it was the only comprehensive reference for all South American snakes. Recently, McDiarmid et al. (1999) updated some of the Peters and Orejas-Miranda (1970) checklist. McDiarmid et al. (1999) is a global checklist but remains incomplete for now.

Professionals are not always comfortable with common names. For example, Marinus Hoogmoed (pers. comm. 5 October 2000), responding to a question about a snake I was researching, wrote "as to the use of a common English name, I am completely opposed to the use of such artificially created common names. It is much more logical to use the Latin name if one is a technical person and desires to have a high degree of accuracy in the work." Unquestionably, Dr. Hoogmoed is correct, and accuracy is dependent upon the technical references and primary research, not an artificially created common name. At the same time I believe that in the field and with the general public, the use of common names does have its place. Like the SSAR's common names list of amphibians and reptiles prepared by Crother et al. for Canada and the United States, and Liner's list for Mexico, this is an effort to standardize the common names for Latin American snakes. A particular problem is giving the names in English. In most of Latin America, the name should be in Spanish; in Brazil, Portuguese; in Canada, St. Lucia, and French Guiana, French; etc. Only in Belize, Canada, the United States, Guyana, and some Caribbean islands is English appropriate. Unfortunately, in the checklist the amount of information that can be included is subject to space limitations, therefore names tend to be limited to English.

Although I did look at specimens, the checklist is based mainly on literature searches. In many cases I attempted to supplement this with independent reviews with the support of authorities. There are several genera that have not had an independent peer review. In these cases the information is supported through literature searches only. These are noted for the reader's reference so that he or she will know that there are outstanding questions about a particular genus. An excellent example is the genus *Rhadinaea*. This is a very large, complex group of snakes that is sorely in need of a champion to review the genus and bring the information up to date. Myers (1974) did some very important work on this genus. I was told that at the time of Myers's work there existed only about 2000 specimens in collections that he could review. Since then many new specimens have been added to collections and many scientific names have been assigned to new species.

It is important to discuss introduced species. For years, snakes have shown up in import cargos, such as bananas, into the United States, but rarely do they become established. This also happens in places like Guayaquil, Ecuador, and Manaus, Brazil, where there is high shipping traffic. Sometimes snakes escape their cages or are released on purpose. Lately, several introduced snakes seem to have been successfully establishing themselves in their new environments in the Americas. Those that have been introduced from outside of the Americas with a confirmation or possibility of becoming established are noted with the following symbol: ‡. One introduced snake that is definitely established from outside of the Americas is *Ramphotyphlops braminus*, as confirmed by McKeon (1978), Smith and Smith (1993), Conant and Collins (1998), Censky and Kaiser *in* Crother (1999), Wallach (1999), and Dixon (2000a). This introduced blind snake has been found thriving in Hawaii and several locations in the continental United States and Mexico, and other locations in Latin America and the Caribbean because it is being introduced in the flowerpot mulch of imported plants. Another strong candidate that may have been introduced and may become established is *Boiga irregularis*, as confirmed by Borg (1990), Yoon (1992), Fritts (1994), and McCoid et al. (1994). So far, this snake has shown up in Hawaii and Guam, and has been reported from Corpus Christi, Texas. Hawaii has implemented an aggressive program that is designed to prevent this snake from becoming established there, as it did on Guam. The brown tree snake has been responsible for driving to extinction many of the native species of birds on Guam, and Hawaii is trying to avert a similar disaster.

Other snakes from within the Americas may have the proper environment to establish themselves in other localities, such as the *Boa constrictor*. Several Boas have been caught in the Everglades in Florida, but they have not yet been confirmed as established there. *Elaphe guttata* appears to have established itself in the Cayman Islands. *Diadophis punctatus* has also shown up in the West Indies and *Nerodia* seems have made its way successfully to Cuba. In all of these cases, these species are introduced from within the Americas. The notation ‡ is only used for those snakes that have been introduced from outside of the Americas. Brown (1997) discussed the various foreign snakes that have been found in California but went on to note that none seem to have established themselves in the region. I had an unclaimed Ball Python (*Python regis*) that was picked up roaming a neighborhood in Houston and have heard of other similar situations with exotics. Rivero (1998) comments about finding *Natrix* (*Nerodia*) *sipedon*, *Thamnophis sirtalis*, and *Python regis* in Puerto Rico. Marques et al. (2001) discuss the problem with invading species in Brazil and note that *Elaphe*, *Lampropeltis*, and *Python* are present because they are lost or released by the import trade and buyers. They also pointed out that such native snakes as *Corallus caninus* and *Epicrates cenchria* are sometimes found in places such as Sao Paulo, far from their natural ranges in the Amazon.

Serpentes is divided into two major groups, Scolecophidia (the Blindsnakes) and Alethinophidia (all other living snakes.) Several characters of the eye, skull structure, and soft anatomy support the monophyly of Scolecophidia. Alethinophidia is a much more diverse group and contains the remaining families of snakes. Characters of the skull, nervous system, and axial muscles suggest Alethinophidia relationships among the various families and are presently being investigated by several rearchers.

The currently recognized genera list is comprised of six groups: Scolecophidia, or blind snakes (7 genera); the Haenophidia, which are the primitive part of the advanced snakes (12 genera); generalized snakes in the advanced group (a.k.a. Colubrids) (143 genera); Elapids (3 genera); Sea Snakes (2 genera); and Pit Vipers (12 genera). Another 33 taxa that are undefined or poorly defined are listed in *Incertae Sedis*. The generalized snakes in the advanced group are those that belong (or belonged) to the Colubridae and more recently are being broken into several families. Some of the genera may have more than one accepted common name, depending on the specific group of snakes and their names in common use. Examples of these include Kingsnake and Milksnake, from *Lampropeltis*, Massasauga and Pygmy Rattlesnake, as used with *Sistrurus*, Cribo and Indigo Snake, with *Drymarchon*, and Gophersnake, Pinesnake, and Bullsnake, as used with *Pituophis*.

The following pages contain 179 genera with 1197 species listed for the Americas, excluding those in the *Incertae Sedis* section. I have collected over 21,000 common names for these snakes. The references listed were those for both the checklist and the lexicon. The final results of this effort is the first complete checklist for the snakes of the Americas and lexicon of common names.

The modern trend in taxonomy is moving away from "splitting" subspecies and toward "lumping" into species. Because of this trend, several people advised me to list the accounts to the species level only. After several years of pondering this good advice, I knew that this would certainly make the project much easier but decided to maintain the information as much as possible to the subspecies level. There are several reasons for this: 1) in many cases subspecies are still recognized and often discussed in technical literature; 2) I could not ignore information on variations where solid technical information exists, and this is the best method to present that information in a simple format; 3) including subspecies information would give another tool for understanding the taxonomy in complex situations (such as can be with the *Liophis miliaris intermedius* confusion); and 4) those of us who are serious novices still can review subspecies information to better understand color vs. range vs. comparisons to other subspecies, species, and genera. I hope that my insistence upon including subspecies does not disrupt the flow of productive information, and over time the user should recognize why the inclusion of subspecies information is important, even if there is a move toward lumping.

A Matter of Personal Opinion

Taxonomy is an artificial system that is constantly evolving in an attempt for scientists to better understand and describe various living, supposedly living (like viruses), or once living (fossils) things. Taxonomy is always in a dynamic state of evolution and usually follows trends of the day. H. G. Cogger (1995) as quoted in Golay et al. (1993) states "Taxonomy is a matter of personal opinion..."

Science and technology now utilize a wide variety of tools to apply in attempting to classify a plant or animal. More recently the technology seems to have evolved much faster than the scientists' experience and understanding of how to intererpret results of the technology. This has sometimes been with a disregard for conventional methods that may support the more modern techniques. In an effort to produce "hard" results, I have attempted to include verifiable classifications. Even with this conservative approach, there are still many questions left unanswered and opposing opinions exist that are just as valid as my opinions.

A simple example is my list of introduced species that have established themselves here in the Americas. "Introduced species" is a very loose term. There are those species that are introduced within the Americas outside of captivity which include such snakes as *Elaphe* (in the Cayman Islands) and *Nerodia* (Puerto Rico). I choose to use the term introduced species as those introduced from outside of the Americas and encountered outside of captivity such as the *Boiga* or *Ramphotyphlops*. I use the term established to be a genera, species, or subspecies as populations large enough to produce breeding populations that are present for the long term in the Americas not in captivity. I do not believe that accidental release or releases should be considered established. My list is very restricted, and in reality the list of established introduced species is probably more diverse but I have not been able find any concrete evidence that they have established themselves as long term immigrants. It is very likely that *Boa* and *Python* are established in the southern United States but I did not identify any reliable proof that they exist with viable breeding populations.

Questionable species or subspecies are listed for various reasons. An excellent example is *Agkistrodon bilineatus lemasespinali* Smith & Chiszar (2001). This subspecies of cantil was described from a single specimen reported to have been captured near Palma Sola, Veracruz, Mexico. Its description is based on minor differences in morphology within the *Agkistrodon bilineatus* complex which include aspects of head scalation (fragmented prefrontals), having a shorter tail, having a longer snout, and aspects of the pattern. When combined the information seems to be a rather weak justification for the subspecies. The reason I included this subspecies is that the new classification fills a gap in the distribution of the species complex along the eastern coast of Mexico.

Other questionable species or subspecies may not be listed. An example here is Grismer (1999) where he reclassified the Baja California rattlesnakes. His proposed changes include: *Crotalus enyo cerralvensis* to *Crotalus enyo*, *Crotalus mitchelli angelensis* to *Crotalus angelensis*, *Crotalus mitchelli muertensis* to *Crotalus muertensis*, *Crotalus molossus estebanensis* to *Crotalus estebanensis*, and *Crotalus ruber lorenzoensis* to *Crotalus lorenzoensis*. Although Grismer's evolutionary species concepts may be valid, unfortunately his paper did not include data to support his proposal for independently evolving lineages. As such, his provocative classifications were not included in this work.

With my academic background in biophysics, I often find the biochemical work in taxonomy more appealing than the morphological aspects. It is only more recently that I have learned to appreciate the results of the mundane tasks and hard work in determining the components of morphology. Behavior and ecology combined with morphology and genetic information helps develop a more complete portrait to examine in an effort to more accurately classify a snake. Unquestionably the DNA information is valuable to include in the classification work but is important that the researcher not allow it to become a handicap. This might occur if the researcher uses genetic information without a complete reference to the other components of the taxonomy organism. Recently some species/subspecies assessments seem to have been made based mainly on the genetics and geography of the serpent with limited reference to behavior, ecology and morphology. I did not use some current classifications based on phylogeography or phylogenetics pending more work on these genera, species and subspecies.

The Utiger et al. (2002) paper on the phylogenetic relationships of the Holarctic ratsnakes (*Elaphe*, Auct.), was based on information inferred from protions of two mitochondrial genes, 12-S rRNA and COI. They used the analysis to make assigments of *Elaphe* in the Americas to *Pseudelaphe*, *Pantherophis*, *Senticolis*, etc. Their justifications for the markers they chose may be valid but I chose the more conventional classifications for these snakes. More work in the future may support their assignments.

The *Crotalus viridis* complex has recently received a lot of attention from Pook et al. (2000), Ashton & de Queiroz (2001) and Douglas et al. (2002). Their work seems to have started with an interest in H. R. Quim's 1987 PhD work at the University of Houston. Their assessments appear to be valid but again the results seem to be inferred, therefore I maintained the conventional classifications of subspecies for this large complex.

Pook et al. (2000) note that recognition of the subspecies category is controversal . . . since many applications of the concept have involved the arbitrary subdivision of geographically variable populations, without congruence between characters or the causes of the variations. They also said the use of subspecies rank for differentiated populations is likely the result in misrepresentation of population phylogeny. They believed the phylogeography seeks to reveal patterns of geographic variations relative to the distribution of extant genealogies, permitting further historical phylogenetics, and ecogenetic interpretation. This approach, they state, overcomes the limitations associated with use of morphological character system alone. Although identifying differences, they stopped short of any reclassification of the *Crotalus viridis* complex.

Ashton & de Queiroz (2001) took the Pook et al. (2000) work a step further by proposing a division of the *Crotalus viridis* complex to a group east and another group west of the Rocky Mountains. Based on their analysis of mtDNA, they separated *C. viridis* (including subspecies of *viridis* and *nuntis*) from *C. oreganus* (including other traditional subspecies). Ashton & de Queiroz split the species but retained the traditional subspecies. At the same time, they acknowledged the need for more work on the systematics of *C. oreganus*.

Douglas et al. (2002) reviewed the systematics of the *Crotalus viridis* complex using a mtDNA phylogenetic approach and focused on the those snakes found on the Colorado Plateau. In this work, they recognized seven species within the *Crotalus viridis* complex as: *C. viridis* (including the subspecies *viridis* and *nuntius*, with the later being considered a synonym of *C. viridis*), *C. abyssus*, *C. cereberus*, *C. concolor*, *C. helleri* (including the conventional subspecies *caliginis*, which is considered a synonym of *helleri*), *C. lutosus*, and *C. oreganus*. Douglas et al. (2002) also recognized that the *Crotalus viridis* complex might need more research.

Considerable effort has gone into the work with the *Crotalus viridis* complex but discussions on morphology, etc., were very limited and I decided to maintain the traditional classifications for this group. I am not sure that enough snake genome is sequenced, and mitochondrial DNA (mtDNA) is one of the more commonly used techniques for genetics work in snakes. The segments of the gene such as 12-S, 16-S, ND-4, COI, Lucine, and others, are used by various herpetologists and geneticists to determine evolutionary concepts. The mtDNA gives the female side of the DNA passed through the generations and tells a lot about inheritance from generation to generation. Contrary to those who use this information on a daily basis, and are much more expert than me, I feel it may not be a good tool to define genera (which is an artificial designation anyway). Nuclear genes are slowly evolving genes that may give a more complete story. Genetic work is not universally standardized as a tool for defining genera, species, or subspecies and most authorities will quickly tell the listener that this is not possible under the way the process works. Because of this, at least for now, I prefer to question the deviations from the traditional classifications when other parameters are not included in the analysis. I feel that inclusion of morphology, environment, habits, etc., enhance the authorities' position. I also feel that by relying strictly on the biochemistry there is a move away from what may be important field and laboratory information and voucher specimens of the complete snake may not be properly stored for future reference. On the other hand, Campbell and Smith (2000) based their work for defining *Bothriechis thalassinus* on genetics combined with morphological characteristics. I have listed their classifications because in addition to the generic work, Campbell and Smith included detailed morphological information and summarized comparisons to similar species. This was a more complete presentation using all tools available. I recommend that the reader research all the above-mentioned references to better understand the various researchers' points of view to develop their own opinions.

Checklist Overview

The following list of currently accepted scientific names is designed to indicate classifications of snakes in the Americas (Western Hemisphere) to the subspecies level and is intended to be current through the end of 2000, although some new scientific classifications can be found through 2003. As much as possible, I have attempted to avoid designating or giving an opinion on genera, species, or subspecies, and most of what the reader finds here is drawn mainly from literature. Because of this attempt at a neutral position, more species or subspecies may be reported here than actually are accepted by some authorities.

It is possible that the list indicates fewer species than are generally accepted. There are several reasons why this may occur: 1) I did not have complete information for a particular snake, so names may have been listed (or left out) due to lack of available technical information; 2) authorities may disagree on classification of genus, species, or subspecies; or 3) it is due to my inexperience with a particular subject. In most cases, this list has been simplified by not listing tribes, subfamilies, or other taxonomic categories, as with the generalized list. The exceptions to this are the Blindsnakes and the boids. There are situations where authorities may have opposite views on a classification. In these cases, I gave an opinion and listed a specific genus, species, or subspecies, and citations for each differing point of view. The opinion was not usually made to favor one authority or the other, but rather to ensure that a complete list of accepted names is presented.

An effort has been made to keep the checklist as simple as possible by restricting the technical information to six types: 1) accepted scientific name; 2) author; 3) date original description was published; 4) citations that support the classification in publications and remarks that may be applicable; 5) abbreviated description of the range in which the snake is thought to be found; and 6) remarks as they pertain to a) the accepted technical name, b) author and/or date, or rarely c) distribution. A seventh piece of information, the common name, is also listed but is not critical for scientific purposes.

This checklist covers a very broad spectrum of names and subspecies. Some subspecies in particular will be questioned by many people reading this work. An example is the debate over *Opheodrys*. Some authorities claim four subspecies for *O. aestivus*, while most of the herpetological community does not accept more than two subspecies. In this case an attendant remark is made in some detail. Nothing new has been created in the checklist other than a few common names; it is only a reflection of information from available literature and personal communications. Several of the genera offered particularly interesting challenges. These include *Lichanura/Charina/Calabaria, Urotheca/Pliocercus, Dryadophis/Mastigodryas*, and the coral snakes.

The checklist covers seven major groups of snakes in the Americas. It is divided into two sections: the Blindsnakes, or Scolecophidia, and the modern snakes, or Alethinophidia. These latter are further divided into Macrostomata (the basal Alethinophia) which include the Pipesnakes, Boas, and Pythons; and, the advanced snakes, or Caenophidia (sometimes spelled Xaenophidia). In the Americas the Caenophidia include four families.

Put more simply the checklist contains seven sections of valid genera, species, and subspecies:
1. Blindsnakes and Wormsnakes;
2. Pipesnakes;

3. Boas and Pythons;
4. Colubrids;
5. Elapids;
6. Sea Snakes; and
7. Pit Vipers.

Finally, *Incertae Sedis* and Other Questions lists those taxa with serious unresolved questions.

Genus

A genus brief is included in the introduction of each genus and is not intended to give any taxonomic details; it introduces the checklist for that genus that provides needed basic information on where to locate literature for the genus. The list of species is given in alphabetical order by the "currently recognized" names by genus, subgenus (where applicable), species, and subspecies. When referring to subspecies, the user should keep in mind that this is an artificial division of a species and is a system whose validity is questioned by many experts. In many cases, geographic origin is helpful, or may be necessary to identify a subspecies or species accurately.

For about 250 years, zoologists have applied a binomial nomenclature to plants and animals based on a system developed by Carl von Linné (Linnaeus). Each species is known by a unique combination of generic and specific names usually derived from Greek or Latinized words, often from other languages. The system will not be discussed here because there are so many good references for details of the binomial scientific name system. Common names bring with them very special problems and the binomial system of scientific names is the only designation that should be used to technically reference a particular snake. Greene (1997c) gives a useful overview of both scientific names and common names and I recommend that the reader reads his discussion on the subject.

Species

This section is an alphabetical list of the species and attendant subspecies that are usually considered valid for a particular genus. In the species list, the genus is simply shown with an abbreviated capital letter. Examples of this include:

Genus: *Masticophis* = *Masticophis*
Species: *M. bilineatus* = *Masticophis bilineatus*
Subspecies: *M. b. bilineatus* = *Masticophis bilineatus bilineatus*
Subspecies: *M. b. lineolatus* = *Masticophis bilineatus lineolatus*
Subspecies: *M. b. slevini* = *Masticophis bilineatus slevini*

Subspecies are usually considered to be artificial names based on color forms or some minor difference in morphology that may be associated with geography. Crosses between species are called hybrids. Crosses between subspecies are called intergrades and are very common in which the range of one contacts and overlaps that of another. Subspecies are usually distinguished based on variation of a few characteristics such as scale or ventral count, color, pattern, etc. There are some subspecies that cannot be identified with any confidence if the location where the specimen has been collected is not reported. In older literature, subspecies may have been listed as a variation.

Nomenclature was often accepted from literature at face value without an attempt to evaluate the type specimens. In a few cases, proposed changes have not been used. This is usually due to outstanding questions that still require resolution.

Author(s) and Date

The original author (or authors) credited for naming the genus, species, or subspecies is given with the scientific name of the snake. The date indicates what is believed to be the earliest valid publication that first recognized the classification of the snake. If the generic name changed since the author's original publication, the authors' name and date of publication are listed in brackets. The change of a species to a subspecies has no effect on the use of brackets with the author and date, only the change in generic name. A common name, usually in English, is included under this information. Normally there is one recommended common name. Rarely may two be shown, in cases where there does not seem to be agreement or an option may be in order.

Citations and Remarks

Citations are literature in which more information about a snake may be found. This column may also provide information about references that may be more complete in terms of description, keys, photos, etc. Although there was an attempt to list the more recent literature available for the subject snake, the reader should recognize that, in some cases, older names may be found in the citation, for example, *Natrix* instead of *Nerodia*. Sometimes, references give opposing information. In these cases, an attempt was made to list two or more references. For example, with *Leptomicrurus*, both Roze (1996), and Campbell and Lamar (1989) are listed, even though the benchmark work in this list is the former. Campbell and Lamar (1989) did not consider *Leptomicrurus* valid for current use but their work should be reviewed for complete referencing in the case of this collection of species.

Remarks were sometimes necessary to give the reader information to better understand classifications. For example, if subspecies were listed, but there is not general agreement on the subspecies that are valid, a remark would be made. *Opheodrys aestivus* is an example of using a remark. Sometimes, the snake is listed by another scientific name or has synonyms that sometimes cause confusion. These were listed to give the reader a broader base of information. When researching quoted or other literature, the reader should always watch for species and subspecies names, since the generic names have variable stability.

Some of the citations are checklists. These include such work as Peters and Orejas-Miranda (1970); Smith and Smith (1993); Liner (1994); and Crother et al. (2000). The checklists add information in some form such as geography, keys, reference sections, etc. One checklist that expands references for the amateur is Freiberg (1982) which gives a key and many photos of the snakes of South America.

The particular combination given may not always be obvious in a publication cited, but the information referring to a particular type of snake can be found there.

Abbreviated Range

As genera, species, and subspecies names are changed in updated literature, or information that relates to them, their published ranges frequently change. These changes are often reflected in a single paragraph in such publications as *Herpetological Review*. Although this is not necessarily considered by some to be peer reviewed primary literature, it is valuable information. Cross-check range information to compare ranges as given by various authorities to ensure out-of-date information is not being used. In other cases, accessibility to data may be limited. This is a real problem with snakes in the Amazon. As noted in Roze (1996), some of the snake ranges listed for northern South America had incomplete or unknown range limits for the southern part of their range, which reaches into the less inhabited Amazon basin. In other cases, development is quickly eradicating ranges, and even though a range noted might be valid today, tomorrow it may be invalid. When range is an important issue, the reader should attempt to cross-reference all range information to verify that it is current. Since range data is constantly changing, the distribution in the table is titled "Abbreviated Range."

Abbreviations are used in the checklist range section to note geographic direction; for example, c = central, n = north or northern, s = south or southern, e = east or eastern, and w = west or western. These may be combined, such as ec = eastern central, se = southeastern, nw = northwestern, and so on. As an example: "w slopes of" means "western slopes of"; or "se Venezuela" means "southeastern Venezuela."

References

The references listed in the bibliography are for both the checklist and lexicon.

Reliability Index

Peters and Orejas-Miranda (1970) recognized that all technical publications are not equal in their reliability or scope. They were concerned that the users of their work would regard all information presented as of equal value and felt the need to communicate comparative reliability of the information given from one genus to another. They endeavored to communicate this problem as it pertained to their work with a code to grade the reliability factor based on a number of stars assigned to a genus. One star (★) was considered to be the least reliable information, and a genus classification with four stars (★★★★) was considered to be the most reliable information available, with two grades in between. I have attempted a similar scale to indicate a reliability factor. In Peters and Orejas-Mirada (1970), if information was researched by an outside source and given back to them, it was usually considered to have four stars. In the case of this work, I attempted to develop all of the information from basic literature searches myself. In many cases it was then turned over to other authorities for review. Unlike Peters and Orejas-Mirada (1970), these did not always generate a four-star rating. If questions still remained, these genera may have been coded with three, two, or even one star. These codings recognize that the genus still requires in-depth investigation. In this checklist, the reliability factor is given as follows:

One Star: ★
Low Reliability. There are many outstanding questions or unresolved issues. This may either be in terms of taxonomy, validity of the presence of a species or subspecies in a particular genus, lack of comparative work, need for revisions, limited knowledge of a genus, etc. Available literature may have insufficient information, and there is the possibility, in some cases, that a genus is not valid. It is recommended that the reader review varying opinions in the citations.

Two Stars: ★★
Moderate reliability. In this case, more information is available or accepted, but there remains serious taxonomic controversy over species or genera that is unresolved. Again, it is recommended that the reader review the literature for opinions about these taxa.

Three Stars: ★★★
Good reliability. Most issues have been resolved and there seems to be general acceptance by the herpetological community in published work that does not vary to any great degree in information, at least for those taxa found in the Americas.

Four Stars: ★★★★
High reliability. There seems to be a very high level of acceptance by the herpetological community in published work that does not vary to any great degree in information, at least for those taxa found in the Americas.

Summary of Checklist

The following pages contain as valid:

	Genera	Species
Scolecophidia (Blindsnakes)	7	103
Aniliidae and Boids	12	50
Generalized snakes in the advanced group (Colubrids)	143	870
Elapidae (Coral Snakes)	3	71
Hydrophiidae (Sea Snakes)	2	2
Viperidae/Crotalinae (Pit Vipers)	12	101
Total	179	1197

This checklist describes 179 genera and 1197 species in the Americas. There are another 33 taxa listed in *Incertae Sedis* and Other Questions.

The Primitive Group
Infraorder Scolecophidia

The Primitive Group
Infraorder Scolecophidia Cope, 1864

Superfamily TYPHLOPOIDEA Gray, 1825

These are commonly referred to as Burrowing Snakes or Minute Snakes. They include the Blindsnakes, Wormsnakes, and Threadsnakes.

The Scolecophidians are small snakes with small eyes that lie hidden beneath scales. They have highly polished, round scales that are of uniform size and shape throughout the body. This group includes the families Anomalepididae, Leptotyphlopidae, and Typhlopidae. They are found worldwide and range in size from about 6 inches to 3 feet (15 cm to 1 m) in length. The group is completely harmless to humans and they tend to be restricted to tropical and warm temperate climates.

The Scolecophidians are identified by the lack of differentiation between the head and trunk, obtuse shape of the tail, and belly scales that are the same size as the body scales. These characteristics give the snakes a very close resemblance (visually) to worms. The name comes from Greek and means "worm snake." The Scolecophidian uses its head for burrowing, aided by the rostral scale at the front of the head. The rostral is very well developed and the front of the skull is very rigid. The eye of the Scolecophidian is either hidden by, or barely visible beneath head scales. These snakes have very small mouths, with jaws that are shorter than the skull. The halves of the mandible are joined to each other at the front, which is not the case in Alethinophidians. Primarily, their prey consists of ants, termites, and similar food sources.

CONTENT: The Scolecophidians in the Americas have 7 genera with 103 species.

Family ANOMALEPIDIDAE Taylor, 1939
(previously included in the family TYPHLOPIDAE)

The Anomalepidae are generally differentiated from other scolecophidians by the presence of teeth in both the upper and lower jaws. These snakes are found in a very disjunct range from southern Central America into northwestern South America. Although some authorities claim they are found in Mexico their presence is highly suspect there. The Anomalepididae closely resemble the leptotyphlopids in appearace, habits, and many anatomical features and differ from them by the lack of pelvis vestiges and several other important skeletal features. Although the maxillae are toothed and movable, the frontal bones are shorter and the prefrontal bones, with which the maxillae articulate, extend posteriorly over the orbits. This condition is not found in any other snake group. They are differentiated from the other two scolecophidian families in that they have a long, slender lower jaw, which is hinged and rarely bears more than one small tooth. In the Americas this family is composed of four genera, the *Anomalepis*, the *Helmintophis*, the *Liotyphlops*, and the *Typhlophis*.

Collectively the San Diego Museum of Natural History website (2001) calls these the Dawn Blind Snakes.

GENUS: *Anomalepis* Jan, 1860, is reported in a disjunct range which includes Nicaragua, Costa Rica, Panama, Colombia, Ecuador, Peru, and possibly Mexico. See Peters & Orejas-Miranda (1970); Hahn (1980a); Savage & Villa (1986); Kofron (1988); Smith & Smith (1993); Auth (1994); Wallach & Guenther (1997); Kornacker (1999); McDiarmid et al. (1999) and Köhler (2001).

Review assistance and important contributions for this section were provide by Van Wallach.

CONTENT: Four species. ★★★★

Anomalepis: Dawn Blindsnake. Frank and Ramus (1995) also call this the Dawn Blindsnake.

Species-level Taxa	Citations and Remarks	Abbreviated Range
A. aspinosus Taylor, 1939 Peruvian Dawn or Taylor's Peru Blindsnake	Peters & Orejas-Miranda (1970); Hahn (1980a); Kofron (1988); McDiarmid et al. (1999)	Only known from the type locality in Perico in Departamento Cajamarca, Peru.
A. colombia Marx, 1953 Colombian Dawn Blindsnake	Peters & Orejas-Miranda (1970); Hahn (1980a); Kofron (1988); McDiarmid et al. (1999)	Only known from the type locality in Departamento Caldas in Colombia.
A. flavapices Peters, 1957 Ecuadorian Dawn Blindsnake	Peters (1957b); Peters & Orejas-Miranda (1970); Hahn (1980a); Kofron (1988); Pérez-Santos & Moreno (1991); McDiarmid et al. (1999)	Lowlands of Provencia Esmeraldas and Manabi in nw Ecuador.
A. mexicanus Jan, 1861 Central American Dawn Blindsnake	Amaral (1927h); Peters & Orejas-Miranda (1970); Villa et al. (1988); Kofron (1988); Smith & Smith (1993); Auth (1994); McDiarmid et al. (1999); Köhler (2001)	Disjunct range in Costa Rica, Panama, and questionably Nicaragua and Mexico; Vanzolini (1986) noted that it is not found again until nw Peru in Amazonas and Cajamarca; the type is reported to be from Mexico, but the Mexican location is highly suspect.

GENUS: *Helmintophis* Peters, 1860, is a tropical species that ranges from Costa Rica southward into northern South America to Colombia and Venezuela. Hahn (1980a) reported it may have been introduced into Mauritius in the Indian Ocean. See Orejas-Miranda *in* Peters & Orejas-Miranda (1970); Hahn (1980a); Savage & Villa (1986); Auth (1994); Kornacker (1999); McDiarmid et al. (1999) and Köhler (2001).

Review assistance and important contributions for this section was provided by Van Wallach.

CONTENT: Three species. ★★★★

Helmintophis: Vermiculate Blindsnake or Northern South American Blindsnake. Frank and Ramus (1995) call this the Greater Blindsnake.

Species-level Taxa	Citations and Remarks	Abbreviated Range
H. flavoterminatus (W. Peters, 1857) Yellow-tailed Blindsnake	Amaral (1924a); Roze (1966); Orejas-Miranda *in* Peters & Orejas-Miranda (1970); Savage & Villa (1986); Lancini & Kornacker (1989); Wallach & Guenther (1997); McDiarmid et al. (1999); Kornacker (1999)	Colombia and Venezuela, it may have been introduced into Mauritius fide Hahn (1980a).
H. frontalis (W. Peters, 1860) Pink-headed Blindsnake	Amaral (1924a); Orejas-Miranda *in* Peters & Orejas-Miranda (1970); Villa et al. (1988); Auth (1994); McDiarmid et al. (1999); Köhler (2001)	Costa Rica, Panama, and possibly into n South America in Colombia
H. praeocularis Amaral, 1924 Inter-andean Blindsnake	Amaral (1924a); Orejas-Miranda *in* Peters & Orejas-Miranda (1970); Hahn (1980a); McDiarmid et al. (1999)	Interandean areas of n Colombia in Tolimba, Santander, and Norte de Santander from 200 to 1200 m.

GENUS: *Liotyphlops* W. Peters, 1881, [originally the genus was named *Rhinotyphlops* W. Peters, 1857, but this name was found to be preoccupied by *Rhinotyphlops* Fitzinger], is found from lower Central America in Costa Rica southward into lower South America to as far as Paraguay. Taylor (1951) suggested that the reported records of *Liotyphlops* occurring in Costa Rica are probably in error. Dixon & Kofron (1983) pointed out that no additional specimens have been taken from Costa Rica since Taylor's (1951) summary of its presence there. See Peters & Orejas-Miranda (1970); Hahn (1980a); Dixon & Kofron (1983); Savage & Villa (1986); Starace (1998); Kornacker (1999); McDiarmid et al. (1999) and Köhler (2001).

Review assistance and important contributions for this section were provided by Van Wallach and James R. Dixon.

CONTENT: Seven species in the Americas. ★★★

Liotyphlops: Slender Blindsnake. Frank and Ramus (1995) call this the Lesser Blindsnake.

Species-level Taxa	Citations and Remarks	Abbreviated Range
*L. albirostris** (W. Peters, 1857) Central American Slender Blindsnake	Amaral (1924a); Peters & Orejas-Miranda (1970); Dixon & Kofron (1983); Villa et al. (1988); Lancini & Kornacker (1989); Auth (1994); Coborn (1994); McDiarmid et al. (1999); Esqueda & La Marca (1999); Renjifo & Lundberg (1999); Kornacker (1999); Köhler (2001) *Remark 1: As reported in Villa et al. (1988) fide Dixon & Kofron (1983), *L. rowani* Smith & Grant (1958) is a synonym. *Remark 2: *Liotyphlops petersi* (Boulenger, 1889) and *Liotyphlops rowani* Smith & Grant (1958) from the Pacific shoreline in Ft. Clayton Reservation, Panama Canal Zone are synonyms according to McDiarmid et al. (1999).	Costa Rica southward into Venezuela, Colombia and Ecuador, including the island of Curaçao in the Netherland Antilles off the coast of Venezuela.

Species-level Taxa	Citations and Remarks	Abbreviated Range
	*Remark 3: *Liotyphlops caracasensis* Roze (1952) is a synonym according to Dixon & Kofron (1983).	
	*Remark 4: *Liotyphlops cucutae* Dunn, 1944 is a synonym fide Dixon & Kofron (1983).	
	*Remark 5: It is recommended that the reader also review Peters & Orejas-Miranda (1970) to see their treatment of *Liotyphlops canellei* and *Liotyphlops bordensis* as synonyms; and, *Liotyphlops petersi*, *Liotyphlops cucutae*, *Liotyphlops caracasensis* and *Liotyphlops rowani* as distinct species.	
*L. anops** (Cope, 1899) Cope's Slender Blindsnake	Amaral (1924a); Vanzolini (1948); Peters & Orejas-Miranda (1970); Dixon & Kofron (1983); Vanzolini (1986); McDiarmid et al. (1999)	Departamentos Meta, Santander and Cundimarca in Colombia.
	*Remark: James R. Dixon (pers. comm. 30 June 2000) told me that he and Kofron examined 11 topotypes of *L. metae* Dunn, 1944, but were unable to secure the holotype. The syntypes have been lost. Dunn's original description of *L. metae* is very brief and he noted the head is similar to that of *L. anops*. Cope's drawing is exactly the way Dunn described *L. metae*. Other characteristics also suggest it is *L. anops* (see Dixon & Kofron, 1983).	
L. argaleus Dixon & Kofron, 1983 [1984] Meta Slender Blindsnake	Dixon & Kofron (1983); Vanzolini (1986); McDiarmid et al. (1999).	Cundinamarca in Colombia.
*L. beui** (Amaral, 1924) Dark Slender Blindsnake	Amaral (1924a); Dixon & Kofron (1983); Vanzolini (1986); Cei (1993); Lema (1994a); McDiarmid et al. (1999); Marques et al. (2001)	SE Brazil in Mato Grosso into Paraná and to the border of, and into e Paraguay, Corrientes in Argentina, and possibly Misiones.
	*Remark: Dixon & Kofron (1983) listed *L. caracasensis* Roze, 1952, as a synonym.	
L. schubarti Vanzolini, 1948 São Paulo Slender Blindsnake	Vanzolini (1948); Peters & Orejas-Miranda (1970); Hahn (1980a); Dixon & Kofron (1983); McDiarmid et al. (1999)	State of São Paulo in Brazil; known only from the type locality in Cachoeira de Emas and Sapucai.
*L. ternetzii** (Boulenger, 1896) Ternetz's Slender Blindsnake	Amaral (1924a, 1954, 1976); Peters & Orejas-Miranda (1970); Hahn (1980a); Dixon & Kofron (1983); Vanzolini (1986); Cei (1993); Starace (1998); Yanosky (1998); McDiarmid et al. (1999)	States of Mato Grosso, São Paulo, Brazila, Pará, Goiás, Paraná, and Distrito Federal in Brazil; n Argentina and Paraguay; its presence in Suriname is questionable, although it is found in French Guiana.

	*Remark 1: Fide Dixon & Kofron (1983), *L. incertus* (Amaral, 1924) is a synonym. However, Starace (1998) differentiated the two snakes by the following:
	    L. ternetzi has 4 supralabials
	    L. incertus has 3 supralabials	

Species-level Taxa	Citations and Remarks	Abbreviated Range
	I feel that this is insufficient to be used as a characteristic to differentiate the two, so they remain as synonyms here. Dixon & Kofron (1983) used four supralabials as diagnostic for *Liotyphlops*.	
	*Remark 2: Dixon & Kofron (1983) placed *Liotyphlops incertus* in synonymy.	
*L. wilderi** (Garman, 1883) Wilder's Slender Blindsnake	Amaral (1924a); Vanzolini (1948); Peters & Orejas-Miranda (1970); Hahn (1980a); Dixon & Kofron (1983); Vanzolini (1986); Yanosky (1998); McDiarmid et al. (1999)	State of Minas Geraís and Rio de Janeiro, Brazil into Paraguay.
	*Remark: According to Dixon & Kofron (1983), *L. guentheri* (Boulenger, 1889) is a synonym.	

GENUS: *Typhlophis* Fitzinger, 1843. This genus has been recorded only from northeastern South America and Trinidad. The head is covered with small scales that are indistinguishable from the body scales and there are no teeth in the lower jaw. Found in South America along the Atlantic Versant, this snake's range includes Trinidad. See Peters & Orejas-Miranda (1970); Hahn (1980a); Savage & Villa (1986); Schwartz & Henderson (1991); Starace (1998); Kornacker (1999); and McDiarmid et al. (1999).

Review assistance and important contributions for this section were provided by Van Wallach.

CONTENT: Two species in the Americas. ★★★★

Typhlophis: Similar-scaled Blindsnake

Species-level Taxa	Citations and Remarks	Abbreviated Range
T. ayarzaguenai Señaris, 1998 Ayarzagüena's Similar-scaled Snake	Señaris (1998); Kornacker (1999)	State of Bolívar in Venezuela.
T. squamosus (Schlegel, 1839) Squamos Similar-scaled Snake	Peters & Orejas-Miranda (1970); Amaral (1976); Starace (1998); McDiarmid et al. (1999)	The Guianas and Coastal Brazil in ne South America; it might also eventually be found throughout the Guianas, definitely French Guiana, and possibly Suriname and Guyana; it is closely associated with the northern Atlantic coast of northern South America; according to Amaral (1976) it is abundant all over the Amazon valley. It is not present on Trinidad fide Murphy (1997).

Family LEPTOTYPHLOPIDAE Stejneger, 1892

These are commonly referred to as Threadsnakes and are sometimes called Slender Wormsnakes.

The Leptotyphlopidae are very similar to the Typhlophidae; however, the upper jaw is toothless and the hind-limb vestiges are well developed. These snakes retain a comparatively well-developed pelvic girdle and sometimes vestigal hind limbs, which may show as horny protrusions. Except for the lower jaw, the skull lacks flexibility. The upper jawbones are rigidly attached to each other and to the bones of the snout, coronoid bones are present, supratemporal bones are absent, and all of the bones of the upper jaw and palate have lost their teeth. There is an enlarged scale in front of the vent, and only one lung and oviduct are present and are found on the right side. The eye lacks a brille.

CONTENT: This is a global family with one genus and 49 species in the Americas. The San Diego Museum of Natural History website calls this family collectively the Slender Blind Snakes.

GENUS: *Leptotyphlops* Fitzinger, 1843, [genus originally *Stenostoma* Wagler *in* Spix, 1824, but preoccupied by *Stenostoma* Latreille, 1810] have a global distribution. *Leptotyphlops* is found in southwestern Asia, Africa, and the Americas. In the Americas, this genus ranges from the southwestern United States southward through the Caribbean, Mexico, and Central America into central South America to as far as Uruguay and southern coastal Peru. It has not been found in the higher elevations of the Andes. In the Caribbean the genus is reported from the Lesser Antilles and Watling Island in the Bahamas. See Klauber (1940c); Smith & Taylor (1966); Orejas-Miranda (1967); Orejas-Miranda *in* Peters and Orejas-Miranda (1970); Hahn (1979a, 1980a); Cei (1986, 1993); Savage & Villa (1986); Schwartz & Henderson (1991); Smith & Smith (1993); Starace (1998); Kornacker (1999); McDiarmid et al. (1999); Werler & Dixon (2000) and Köhler (2001).

Review assistance and important contributions for this section were provided by Van Wallach.

CONTENT: A global genus with 49 species in the Americas. ★★★

Leptotyphlops: Threadsnake (sometimes called Slender Wormsnake)

Species-level Taxa	Citations and Remarks	Abbreviated Range
L. affinis (Boulenger, 1884) Venezuelan Threadsnake	Roze (1966); Orejas-Miranda *in* Peters & Orejas-Miranda (1970); Hahn (1980a); Lancini & Kornacker (1989); McDiarmid et al. (1999); Kornacker (1999)	States of Merida and Táchira, Venezuela.
L. albifrons* (Wagler *in* Spix, 1824) White-fronted Threadsnake	Boulenger (1893); Roze (1966); Orejas-Miranda (1967); Amaral (1976); Hahn (1980a); Vanzolini et al. (1980); Hoogmoed & Gruber (1983); Cei (1986, 1993); Lancini & Kornacker (1989); Smith & Smith (1993); Murphy (1997); Kornacker (1999); McDiarmid et al. (1999); Boos (2001) *Remark 1: Vanzolini (1986) quoted Hoogmoed & Gruber (1983) and noted that *L. tenella* Klauber, 1939 is a synonym of this snake. *Remark 2: *L. undecimstriatus* (Schlegel, 1839) can be found as a valid species in Orejas-Miranda *in* Peters & Orejas-Miranda (1970); Hahn (1980a); McDiarmid et al. (1999). It is known only from the type locality in Santa Cruz de la Sierra in Bolivia. I researched all of the literature history to compare descriptions. It appears that after Schlegel, 1839, a complete description was given by Gray (1845:140)	Venezuela into the states of Pará and Rio Grande do Norte in Brazil; and into Colombia, nc Peru, Trinidad and Tobago, Argentina, Paraguay, Uruguay; Bolivia and possibly Ecuador; generally found in humid lowlands.

Superfamily Typhlopoidea

Species-level Taxa	Citations and Remarks	Abbreviated Range
	as *Epicta undecimstriata*. And the next, and what appears to be the last time a complete description was given, was in Boulenger (1893) who synomyzed the species *Epicta undecimstriata* from Gray (1845) (what now might be called *Leptotyphlops undecimestriatus*) with *Glauconia albifrons* (now *Leptotyphlops albifrons*). Orejas-Miranda *in* Peters & Orejas-Miranda (1970) resurrected the species but did not give a reasons for his classifications. In addition, the species was not included in the key so there is nothing to compare. Orejas-Miranda listed only Schlegel, 1839 in his literature and did not list Gray (1845) or Boulenger (1983) and may not have realized that it had been synomyzed. It seems that later literature picked up *L. undecimstriata* without identifying Orejas-Miranda's error, but again without discussion. Van Wallach's (1998) Ph.D. dissertation covers blindsnakes and wormsnakes. Unfortunately, due to the rarity of *L. undecimstriata*, Wallach did not have a specimen to inspect because the type has been lost. He probably could have resolved this question easily because his research is quite detailed on most of the species. I have synomyzed *L. undecimstriatus* as found in Orejas-Miranda *in* Peters & Orejas-Miranda (1970); Hahn (1980a) and McDiarmid et al. (1999), with *L. albifrons* based on Boulenger (1893). *Remark 3: This species is classified with no subspecies in more current literature, there are two subspecies according to Roze (1966). He listed them as *L. a. albifrons* (Wagler *in* Spix, 1821) distributed from the states of Pará and Rio Grande do Norte in Brazil northward to Venezuela; and, *L. a. margaritae* (Roze, 1952) from Margarita Island.	
L. albipunctus (Jan, 1861) Tucumán Threadsnake	Orejas-Miranda *in* Peters & Orejas-Miranda (1970); Hahn (1980a); Laurent (1984); Cei (1993); McDiarmid et al. (1999)	Salta and possibly Tucumán, Argentina, and Tarija, Bolivia.
L. anthracinus Bailey, 1946 Ecuadorian Lowland Threadsnake	Peters (1960b); Orejas-Miranda (1967); Orejas-Miranda *in* Peters & Orejas-Miranda (1970); Hahn (1980a); McDiarmid et al. (1999)	NE Ecuadorian lowlands, with a record from w Ecuador.
L. asbolepis (Thomas, McDiarmid & Thompson, 1985) Martin García Threadsnake	Thomas et al. (1985); Schwartz & Henderson (1991); McDiarmid et al. (1999); Powell et al. *in* Crother (1999)	W Slopes of Lomade Aguacate in Dominican Republic.
L. australis Freiberg & Orejas-Miranda, 1968 Black-ringed Threadsnake	Freiberg (1968): Orejas-Miranda *in* Peters & Orejas-Miranda (1970); Hahn (1980a); Cei (1986, 1993); Scrocchi (1990); Lema (1994a); Yanosky (1998); McDiarmid et al. (1999)	Río Negro to Córdoba in Argentina and into Paraguay.

Species-level Taxa	Citations and Remarks	Abbreviated Range
L. bilineatus (Schlegel, 1839) Martinique Threadsnake	Schwartz & Henderson (1985, 1991); McDiarmid et al. (1999); Censky & Kaiser *in* Crother (1999)	Lesser Antilles in Martinique, St.Lucia, Barbados, and possibly Guadelupe.
L. borapeliotes Vanzolini, 1996 Caatinga Threadsnake	Vanzolini (1996); McDiarmid et al. (1999)	NE Brazil from Paraiba, most likely from Rio Grande do Norte to the middle of São Francisco, associated with caatinga and coast.
L. borrichianus (Degerbøl, 1923) Western Argentina Threadsnake	Freiberg (1951); Orejas-Miranda *in* Peters & Orejas-Miranda (1970); Hahn (1980a); Cei (1986); McDiarmid et al. (1999)	Discontinuous range from w Argentinia in Mendoza into Río Negro northward into La Roja Province. It is presumed this snake will eventually be found in La Pampa.
L. brasiliensis Laurent, 1949 Bahia Threadsnake	Orejas-Miranda *in* Peters & Orejas-Miranda (1970); Vanzolini et al. (1980); Rodrigues & Puorto (1994); Wallach (1996); McDiarmid et al. (1999)	Originally known only from holotype from a poorly defined locality called "Brésil"; two more specimens have now been listed from Barreiros, w Bahia in e Brazil.
L. bressoni Taylor, 1939 Michoacán Threadsnake	Taylor (1939); Smith & Taylor (1966); Hahn (1979a, 1980a); Smith & Smith (1993); McDiarmid et al. (1999)	Known only from the type locality in sw Mexico in El Sabino, Uruapan, Michoacán; and also reported from Airo, Apatzinpan, Michoacán.
L. brevissimus Shreve, 1964 Caqueta Threadsnake	Orejas-Miranda (1967); Orejas-Miranda *in* Peters & Orejas-Miranda (1970); Hahn (1980a); McDiarmid et al. (1999)	Known only from the type locality in Caquetá, Colombia.
L. calypso (Thomas, McDiarmid & Thompson, 1985) Samana Threadsnake	Thomas et al. (1985); Schwartz & Henderson (1991); McDiarmid et al. (1999); Powell et al. *in* Crother (1999)	Hispanola in Samana Province in Dominican Republic.
L. collaris Hoogmoed, 1977 Guianas Threadsnake	Hoogmoed (1977); Hoogmoed (1982 [1983]); Vanzolini (1986); Starace (1998); McDiarmid et al. (1999)	N Suriname and French Guiana; associated with primary forests.
L. columbi Klauber, 1939 Bahamian Threadsnake	Klauber (1939); Thomas (1965c); Schwartz & Henderson (1985, 1991); McDiarmid et al. (1999)	Bahamas on San Salvador Island; known only from the type locality.
L. cupinensis Bailey & Carvalho, 1946 Lely Mountains Threadsnake	Bailey & Carvalho (1946); Orejas-Miranda (1966a, 1967); Orejas-Miranda *in* Peters & Orejas-Miranda; (1970); Hoogmoed (1977); Vanzolini (1986); Starace (1998); McDiarmid et al. (1999)	Suriname in the Lely Mountains and the state of Mato Grosso, Amapá and the territory of Amazonas in Brazil and possibly into French Guiana.
L. diaplocius Orejas-Miranda, 1969 Amazonian Threadsnake	Orejas-Miranda (1969); Orejas-Miranda *in* Peters & Orejas-Miranda (1970); McDiarmid et al. (1999)	NE Peru in the lower Río Ucayali and Huallaga Valleys and the Ouro Preto d'Oeste in Brazil.
L. dimidiatus (Jan, 1861) Rupununi Savanna Threadsnake or Dainty Blindsnake	Orejas-Miranda (1967); Orejas-Miranda *in* Peters & Orejas-Miranda (1970); Amaral (1976); Hoogmoed (1977); Lancini & Kornacker (1989); Starace (1998); McDiarmid et al. (1999); Kornacker (1999)	SE Venezuela, ne Brazil and the Guianas; it is found in the Rupununi Savanna and associated areas; usually only slightly elevated coastal areas; it may eventually be located in French Guiana.

Species-level Taxa	Citations and Remarks	Abbreviated Range
L. dugandi Dunn, 1944 Junamina Threadsnake	Dunn (1944b); Peters & Orejas-Miranda (1970); McDiarmid et al. (1999)	Known only from the Juanamina in Departamento Atlantico, Colombia.
L. dulcis (Baird & Girard, 1853) Plains Threadsnake	Klauber (1940c); Smith & Taylor (1966); Conant (1975); Hahn (1979a, b, 1980a); Stebbins (1985); Smith & Smith (1993); Tennant (1998); Conant & Collins (1998); McDiarmid et al. (1999); Werler & Dixon (2000); Dixon (2000a)	The United States from sc Kansas, c Oklahoma and the panhandle in c Texas west into s New Mexico, se Arizona southward into Mexico in Tamaulipas, Nuevo León, San Luis Potosí, Veracruz, Querrétaro, Hidalgo, Coahuila, and Chihuahua.
L. d. dulcis (Baird & Girard, 1853) Texas Threadsnake	Conant (1975); Hahn (1979b; 1980a); Smith & Smith (1993); Tennant (1998); Conant & Collins (1998); McDiarmid et al. (1999); Werler & Dixon (2000); Tennant & Bartlett (2000); Dixon (2000a)	SW United States in s Oklahoma southward through Texas into Mexico to as far as n Tamaulipas and c Nuevo León; it is associated with the High Plains and n Chihuahua Desert.
L. d. dissectus (Cope, 1896) New Mexico Threadsnake	Conant (1975); Hahn (1970b, 1980a); Smith & Smith (1993); Tennant (1998); Conant & Collins (1998); McDiarmid et al. (1999); Werler & Dixon (2000); Bartlett & Tennant (2000); Tennant & Bartlett (2000); Dixon (2000a)	A very disjunct distribution that ranges from extreme se Colorado, s Kansas and w and c Oklahoma, westward through the panhandle of nw Texas southward into the Trans-Pecos region, and westward though s New Mexico and se Arizona, southward into ne Mexico in Chihuahua, with isolated populations reported in c New Mexico in the United States, and s Coahuila in Mexico.
L. d. iversoni Smith, van Breukelen, Auth & Chiszar, 1998 Tamaulipan Threadsnake	Smith et al. (1998)	SW Tamaulipas, Mexico.
L. d. myopicus* (Garman, 1883) Tampico Threadsnake	Klauber (1940c); Smith & Taylor (1966); Hahn (1979b, 1980a); Pérez-Higareda & Smith (1991); Smith & Smith (1993); Liner (1994); McDiarmid et al. (1999) *Remark: To Smith & Taylor (1966), *myopicus* did not appear to be a race of *L. dulcis*, based on a paper by Klauber (1940), therefore they kept this group independent as its own species of *L. myopicus myopicus* and with subspecies of *L. m. myopicus* (Garman, 1883) and *L. m. dissectus* (Cope, 1896).	N Veracruz and s San Luis Potosí northward across Tamaulipas and into c Nuevo León, Mexico.
L. d. supraocularis* Tanner, 1985 Tanner's Threadsnake	Tanner (1985); Smith & Smith (1993) *Remark: Not mentioned in McDiarmid et al. (1999).	W Chihuahua in Mexico.
L. goudoti (Dumeril, & Bibron, 1844) Black Threadsnake	Orejas-Miranda *in* Peters & Orejas-Miranda (1970); Wilson & Hahn (1973); Hahn (1979a, 1980a); Villa et al. (1988); Schwartz & Henderson (1991); Smith & Smith (1993); Lee (1996, 2000); Campbell (1998b); Kornacker (1999); McDiarmid et al. (1999); Köhler (2001)	SE Mexico in Oaxaca, Colima, Michoacán and through Central America into n South America to Caribbean Colombia and Venezuela, and offshore islands including Isla de Guanaja, Isla Roatán, Isla de Utila, Swan Islands, Utilas, Swan Islands, San Andrés, and Providencia Islands, Suma Islands, Bonaire, and Margarita islands.

Species-level Taxa	Citations and Remarks	Abbreviated Range
L. g. goudoti (Dumeril & Bibron, 1844) Goudoti's Threadsnake	Orejas-Miranda *in* Peters & Orejas-Miranda (1970); Wilson & Hahn (1973); Hahn (1980a); Villa et al. (1988); Kornacker (1999); Lee (1996, 2000); McDiarmid et al. (1999); Mijares-Urrutia & Arends R. (2000)	SE Mexico in Colima, Michoacán, Oaxaca, and Veracruz southward through Central Ameica into Colombia and eastward into Venezuela and associated islands off the coast of the mainland. It is not present on Trinidad fide Murphy (1997).
L. g. ater Taylor, 1939 Taylor's Threadsnake	Taylor (1939); Orejas-Miranda *in* Peters & Orejas-Miranda (1970); Hahn (1980a); McDiarmid et al. (1999)	Nicaragua, Costa Rica, and possibly Honduras and s El Salvador.
L. g. bakewelli Oliver, 1937 Bakewell's Threadsnake	Oliver (1937); Smith & Taylor (1966); Orejas-Miranda *in* Peters & Orejas-Miranda (1970); Smith & Smith (1993); McDiarmid et al. (1999)	SE Mexico in Colima and Michoacán southward into Guatemala.
L. g. magnamaculata Taylor, 1940 Bay Islands Threadsnake	Taylor (1940); Hahn (1980a); Schwartz & Henderson (1991); McDiarmid et al. (1999)	Bay Islands of Honduras, San Andrés Island, Swan Island, and Providence Island.
L. g. phenops (Cope, 1876) Shiny Threadsnake	Orejas-Miranda *in* Peters & Orejas-Miranda (1970); Wilson & Hahn (1973); Wilson & Meyer (1985); Hahn (1980a); Pérez-Higareda & Smith (1991); Smith & Smith (1993); McDiarmid et al. (1999), Lee (2000)	Mexico in the coastal and foothills region of Pacific slopes of Mexico from Colima and Michoacán, then from Veracruz and Yucatán on the Caribbean side, southward through Guatemala, Honduras, El Salvador, Nicaragua, and Swan Island.
L. guayaquilensis Orejas-Miranda & G. Peters, 1970 Guayaquil Threadsnake	Orejas-Miranda & G. Peters (1970); Hahn (1980a); Vanzolini (1986); McDiarmid et al. (1999)	State of Guayas and the Pacific coast of Ecuador.
L. humilis (Baird & Girard, 1853) Western Threadsnake	Klauber (1940c); Cochran (1961); Smith & Taylor (1966); Hahn (1979a, c, 1980a); Stebbins (1985); Smith & Smith (1993); García & Ceballos (1994); Brown (1997); Conant & Collins (1998); Grismer (1999); McDiarmid et al. (1999); McPeak (2000); Dixon (2000a)	SW and Trans-Pecos Texas through sw and c Arizona, s Nevada and s California into Mexico in Sonora, Colima, Chihuahua, Durango, Sinaloa, Jalisco, Tamaulipas and San Luis Potosí.
L. h. humilis (Baird & Girard, 1853) Southwestern Threadsnake	Brattstrom (1953); Smith & Taylor (1966); Hahn (1979c, 1980a); Smith & Smith (1993); Brown (1997); McDiarmid et al. (1997); Bartlett & Tennant (2000)	The United States in coastal and cis-montane s California in the Mojave Desert eastward into the Sierras, from Death Valley region through the southern tip of Nevada to c and se Arizona. It ranges in Mexico into n and c Baja California from Santa Barbara into San Ignacio and is found in Cedros Island.
L. h. boettgeri* (Werner, 1899) Baja California Cape Threadsnake	Smith & Larson (1974a); Hahn (1979c, 1980a); Wallach & Smith (1992); Smith & Smith (1993); McDiarmid et al. (1997) *Remark: Fide Smith & Smith (1993), *L. h. slevini* is a synonym.	Cape region of Baja California Sur and Cerravlo Island.
L. h. cahuilae Klauber, 1931 Desert Threadsnake	Klauber (1931); Smith & Taylor (1966) Hahn (1979c, 1980a); Smith & Smith (1993); Brown (1997); McDiarmid et al. (1999); Bartlett & Tennant (2000)	Yuma and Colorado deserts of se California to sw Arizona southward into Baja California and extreme nw Sonora. It is also reported from the Vizcaí Desert in c Baja California.

Superfamily Typhlopoidea

Species-level Taxa	Citations and Remarks	Abbreviated Range
*L. h. chihuahuaensis** Tanner, 1985 Chihuahua Threadsnake	Tanner (1985); Smith & Smith (1993); Liner (1994) *Remark: Not shown in McDiarmid et al. (1999)	N Mexico.
L. h. dugesi (Bocourt, 1881) Mexican Threadsnake	Klauber (1940c); Smith & Taylor (1966); Hahn (1979c, 1980a); Smith & Smith (1993); Ramírez-Bautista (1994), McDiarmid et al. (1999)	Mexico in Colima, Jalisco, Sinaloa and Guanajuato and probably into some of the surrounding states.
L. h. levitoni Murphy, 1975 Santa Catalina Island Threadsnake	Murphy (1975); Hahn (1979c, 1980a); Smith & Smith (1993); Grismer (1999); McDiarmid et al. (1999) *Remark: Fide Grismer (1999), this should be a junior synonym of *L. humilis*, but his justification for this recommendation is incomplete and a subspecies/species relationship is not clearly defined.	Known only from the type locality of Isla Santa Catalina, Gulf of California.
*L. h. lindsayi** Murphy, 1975 Carmen Island Threadsnake	Murphy (1975); Hahn (1979c, 1980a); Smith & Smith (1993); Grismer (1999); McDiarmid et al. (1999) *Remark: According to Grismer (1999), this should be a junior synonym of *L. humilis*, but his justification for this recommendation is incomplete and a subspecies/species relationship is not clearly defined.	Known only from the type locality of Isla Carmen, Gulf of California.
L. h. segregus Klauber, 1939 Trans-Pecos Threadsnake	Klauber (1939); Smith & Taylor (1966); Conant (1975); Hahn (1979c, 1980a); Smith & Smith (1993); Tennant (1998); Conant & Collins (1998); McDiarmid et al. (1999); Werler & Dixon (2000); Bartlett & Tennant (2000); Tennant & Bartlett (2000); Dixon (2000a)	The United States in se Arizona to sw New Mexico and into Trans-Pecos Texas and Mexico into Coahuila, n Chihuahua and Durango; fide Dixon (2000) this snake is more widespread in w Texas than formally believed.
L. h. tenuiculus (Garman, 1883) Potosí Threadsnake	Klauber (1940c); Smith & Taylor (1966) Hahn (1979c); Smith & Smith (1993); McDiarmid et al. (1999)	San Luis Potosí and Tamaulipas in Mexico.
L. h. utahensis Tanner, 1938 Utah Threadsnake	V. M. Tanner (1938); Klauber (1940); W. W. Tanner (1970); Hahn (1979c); McDiarmid et al. (1999); Bartlett & Tennant (2000)	SE Nevada and extreme sw Utah in Washington County.
L. joshuai Dunn, 1944 Jericó Threadsnake	Orejas-Miranda *in* Peters & Orejas-Miranda (1970); Orejas-Miranda & G. Peters (1970); McDiarmid et al. (1999)	C and w Andes of Colombia in the Provinces of Antioquia and Caldas; the type locality is Jericó, Antioquia.
L. koppesi Amaral, 1955 Terenos Threadsnake	Amaral (1955); Orejas-Miranda *in* Peters & Orejas-Miranda (1970); Amaral (1976); McDiarmid et al. (1999); Campas-Nogueira (2001)	Previously known only from the type locality of Terenos in state of Mato Grosso in Brazil. Recently reported from Minas Gerais.
L. leptepileptus Thomas, McDiarmid & Thompson, 1985 Haitian Border Threadsnake	Thomas et al. (1985); Schwartz & Henderson (1991); McDiarmid et al. (1999); Powell et al. *in* Crother (1999)	Type locality only from the Département de l'Ouest in Haiti.

Species-level Taxa	Citations and Remarks	Abbreviated Range
L. macrolepis (W. Peters, 1857) Big-eyed or Big-scaled Threadsnake	Roze (1966); Orejas-Miranda (1966a, 1967); Orejas-Miranda *in* Peters & Orejas-Miranda (1970); Amaral (1976); Hoogmoed (1977); Hahn (1979a, 1980a); Villa et al. (1988); Lancini & Kornacker (1989); Starace (1998); McDiarmid et al. (1999); Renjifo & Lundberg (1999); Kornacker (1999); Freitas (1999); Köhler (2001)	Panama southward into Colombia, Venezuela, the Guianas, and n Brazil; associated with rainforests from sea level to 1800 m.
L. maximus Loveridge, 1932 Guerrero Threadsnake	Klauber (1940); Davis & Smith (1953); Duellman (1956); Smith & Taylor (1966); Hahn (1979a, 1980a, b); Smith & Smith (1993); McDiarmid et al. (1999)	Mexico in the Upper Balsas Basin and the Upper Río Papaloapan drainage, from low elevations to over 1860 m in desert to broadleaf forests.
L. melanotermus (Cope, 1862) South American Threadsnake	Orejas-Miranda *in* Peters & Orejas-Miranda (1970); Hahn (1980a); Laurent (1984); Cei (1993); McDiarmid et al. (1999); Leynaud & Bucher (1999)	Extreme s Peru, Bolivia, n Argentina in Catamarca, Chaco, Córdoba, Corrientes, Jujuy, Salta, Santa Fé, Santiago del Estero, Tucumán, and possibly in c, w and s Paraguay and sw Brazil.
L. melanurus Schmidt & Walker, 1943 Chiclín Threadsnake	Schmidt & Walker (1943); Orejas-Miranda *in* Peters & Orejas-Miranda (1970); Hahn (1980a); Cei (1993); McDiarmid et al. (1999)	Known only from the type locality of Chiclín, Departamento Libertad in Peru.
L. munoai Orejas-Miranda, 1961 Rio Grande do Sul Threadsnake	Orejas-Miranda (1962); Orejas-Miranda *in* Peters & Orejas-Miranda (1970); Hahn (1980a); Achaval (1987); Cei (1993); Lema (1994a); Yanosky (1998); Giraudo (1999); McDiarmid et al. (1999)	State of Rio Grande do Sul, Brazil, and Uruguay into n Argentina in Buenos Aires, Corrientes, and La Pampa.
L. nasalis Taylor, 1940 Managua Threadsnake	Orejas-Miranda *in* Peters & Orejas-Miranda (1970); Hahn (1980a); Villa et al. (1988); Villa (1990b); McDiarmid et al. (1999); Köhler (2001)	Managua, Nicaragua in the Pacific xeric tropical forests; according to McDiarmid et al. (1999), *L. nasalis* is known only from the type locality.
L. nicefori Dunn, 1946 Mogotes Threadsnake	Orejas-Miranda *in* Peters & Orejas-Miranda (1970); Hahn (1980a); McDiarmid et al. (1999)	Known only from the type locality of Mogotes in Provencia Santander in Colombia.
L. peruvianus Orejas-Miranda, 1969 Peruvian Threadsnake	Orejas-Miranda (1969); Orejas-Miranda *in* Peters & Orejas-Miranda (1970); Hahn (1980a); Henle & Ehrl (1991b); McDiarmid et al. (1999)	Known only from the type locality of Chanchamayo, Departamento Junín in Peru.
L. pyrites Thomas, 1965 Barahona Threadsnake	Thomas (1965); Hahn (1980a); Thomas et al. (1985); Schwartz & Henderson (1991); McDiarmid et al. (1999); Powell et al. *in* Crother (1999)	Hispaniola se coastal plains in Haiti into the sw Dominican Republic in w Barahona Peninsula and Valle de Neiba.
L. rubrolineatus (Werner, 1901) Red-striped or Red-lined Threadsnake	Orejas-Miranda & Zug (1974); Orejas-Miranda *in* Peters & Orejas-Miranda (1970); Vanzolini (1986); McDiarmid et al. (1999)	Peru around Lima; only from the type locality.
L. rufidorsus Taylor, 1940 Huatupilla Threadsnake	Schmidt & Walker (1943b); Orejas-Miranda (1964, 1969); Orejas-Miranda *in* Peters & Orejas-Miranda (1970); Hahn (1980a); Orejas-Miranda & Zug (1974); Vanzolini (1986); McDiarmid et al. (1999)	Known only from Lima and Chiclín in Departamento Libertad in Peru; the distribution seems to be very disjunct.
L. salgueiroi Amaral, 1955 Itá Threadsnake	Amaral (1955, 1976); Orejas-Miranda *in* Peters & Orejas-Miranda (1970); Hahn (1980a); McDiarmid et al. (1999)	Known only from the type locality of Itá in the state of Espirito Santo in Brazil.

Superfamily Typhlopoidea

Species-level Taxa	Citations and Remarks	Abbreviated Range
L. septemstriatus (Schneider, 1801) Seven-rayed or Seven-striped Threadsnake	Orejas-Miranda (1967); Orejas-Miranda *in* Peters & Orejas-Miranda (1970); Amaral (1976); Hoogmoed (1977, [1982] 1983); Hahn (1980a); Cunha et al. (1985); Vanzolini (1986); Nascimento et al. (1987); Lancini & Kornacker (1989); Starace (1998); McDiarmid et al. (1999); Kornacker (1999)	SE Venezuela, the Guianas and n Brazil in Pará and Rondônia; closely associated with rainforest of the Amazon.
L. signatus* (Jan, 1861) Northern Amazon Threadsnake	Orejas-Miranda (1969); Hahn (1979a, 1980a); Lancini & Kornacker (1989); McDiarmid et al. (1999); Kornacker (1999) *Remark: Hahn (1979a) listed *L. amazonicus* Orejas-Miranda (1969) as a synonym.	Venezuela from Bolivar into Amazonas and possibly Colombia into Ecuador and Peru; associated with the n region of the Amazon.
L. striatulus Smith & Laufe, 1945 Yungas Threadsnake	Laurent (1964, 1984); Cei (1993); Vanzolini (1986); McDiarmid et al. (1999)	Yungas in Bolivia and wilderness areas of nw Argentina. The type is from Yaracachi, Sur de Yungas in Bolivia.
L. subcrotillus Klauber, 1939 Klauber's Threadsnake	Klauber (1939); Schmidt & Walker (1943b); Orejas-Miranda *in* Peters & Orejas-Miranda (1970); Hahn (1980a); McDiarmid et al. (1999)	Pacific lowlands of sw Ecuador into Chiclín, Departamento Libertad in Peru.
L. teaguei Orejas-Miranda, 1964 Río Chotano Threadsnake	Orejas-Miranda (1964, 1967); Orejas-Miranda & Zug (1974); Hahn (1980a); McDiarmid et al. (1999)	Only from the type locality in n Río Chotano in n Peru at 2350 m. The type locality is between Chota and Cutervo.
L. tenellus* Klauber, 1939 Yellow-headed Threadsnake or Neotropical Threadsnake	Roze (1966); Orejas-Miranda (1964, 1967); Orejas-Miranda & Zug (1974); Hoogmoed (1977); Moonen et al. (1979); Lancini (1979); Lancini & Kornacker (1989); Schwartz & Henderson (1991); Starace (1998); McDiarmid et al. (1999); Censky & Kaiser *in* Crother (1999) *Remark: Fide McDiarmid et al. (1999), this is a synonym of *L. albifrons* (Wagler, 1824), based on Hoogmoed and Gruber (1983). They also noted that *L. albifrons* may be a complex of closely related species. "*L. tenellus* is very different (externally and internally) from *L. albifrons* and should be recognized as a valid species" (citing Van Wallach pers. comm. 1999). As such, They are maintained as separate species in this work.	N South America in Guyana, Suriname, French Guiana, Trinidad, and se Venezuela. It may eventually be found in Ecuador, Peru and Colombia. One specimen, possibly introduced, is reported from the island of Antigua in the Lesser Antilles; Starace (1998) includes Ecuador and Amazonia Brazil in its range.
L. tesselatus (Tschudi, 1845) Cecilla Threadsnake	Schmidt & Walker (1943b); Orejas-Miranda (1964); Orejas-Miranda & Zug (1974); Hahn (1980a); McDiarmid et al. (1999)	Known only from the type locality in Lima, Peru.
L. tricolor Orejas-Miranda & Zug, 1974 Tricolor Threadsnake	Orejas-Miranda & Zug (1974); Zug (1977); Hahn (1980a); Henle & Ehrl (1991b); McDiarmid et al. (1999)	Peru, from Departamentos of Cajamarca into Ancash and found from 2700 to 3300 m.

Species-level Taxa	Citations and Remarks	Abbreviated Range
L. unguirostris (Boulenger, 1902) Southern Threadsnake	Orejas-Miranda *in* Peters & Orejas-Miranda (1970); Hahn (1980a); Cei (1986, 1993); Kretzschmar (1996); Yanosky (1998); McDiarmid et al. (1999); Leynaud & Bucher (1999)	Discontinuous range in s Paraguay into Córdoba, San Juan to Santiago del Estero in Argentina; including Buenos Aires, Catamarca, La Pampa, and San Luis; Leynaud & Bucher (1999), believed that this snake probably will be found in the Chaco of Bolivia.
L. vellardi Laurent, 1984 Vellard's Threadsnake	Laurent (1984); Vanzolini (1986); Cei (1993); McDiarmid et al. (1999); Leynaud & Bucher (1999)	Formosa and Chaco in Argentina.
L. weyrauchi Orejas-Miranda, 1964 Weyrauch's Threadsnake	Orejas-Miranda (1964); Orejas-Miranda *in* Peters & Orejas-Miranda (1970); Hahn (1980a); Laurent (1984); Cei (1993); Scrocchi (1990); McDiarmid et al. (1999); Leynaud & Bucher (1999)	Argentina in Provincias Tucumán, Santiago del Estero, Formosa, Chaco, and Córdoba.

FAMILY TYPHLOPIDAE Merrem, 1820

The Typhlopidae are also small wormlike snakes that are found in both the Old World and New World. The dentition in this snake is very reduced, the palate and lower jaw are quite or almost devoid of teeth, and are they are rigidly attached to the snout. Vestiges of hind-limb girdles are present. The maxillae are toothed and loosely articulated to the skull, so that they are capable of a certain amount of independent movement. The lower jaw is rigid and toothless. These snakes have a tracheal lung and no enlarged scale in front of the vent.

The San Diego Museum of Natural History website (2001) calls this group collectively the Blindsnakes.

GENUS: *Ramphotyphlops* Fitzinger, 1843‡, is sometimes classified as *Typhlina*, *Typhlops*, or *Tychlina* and has a global distribution. It is an introduced species into the Americas. At times the genus may be seen as *Rhamphotyphlops* (SDMNH website 2001). See Smith & Taylor (1966); Hahn (1980a); Smith & Smith (1993); Tennant (1997b); McDiarmid et al. (1999) and Köhler (2001).

Review assistance and important contributions for this section were provided by Van Wallach.

CONTENT: One introduced species in the Americas. ★★★

Ramphotyphlops: Long-tailed Blindsnake; Crother et al. (2000) call this the Australasian Blindsnake. Frank and Ramus (1995) call this the Common Blindsnake.

Species-Level Taxa	Citations and Remarks	Abbreviated Range
R. braminus (Daudin, 1803) Brahminy's Blindsnake	Smith & Taylor (1966); Dixon & Hendricks (1979); Wilson & Porras (1983); Villa et al. (1988); Smith & Smith (1993); Thomas (1994); Tennant (1997b); Conant & Collins (1998); McDiarmid et al. (1999); Censky & Kaiser *in* Crother (1999); Wallach (1999); Dixon (2000a); Bartlett & Tennant (2000); Tennant & Bartlett (2000); Köhler (2001)	Worldwide. In the Americas, it has been confirmed in at least Florida, Massachusetts; Hawaii, Texas, Mexico, Guatemala; the Lesser Antilles and northeastern South America; fide Dixon (2000), quoting Bartlet (1998), three specimens of this snake have been found in the Rio Grande Valley of Texas; fide Stafford & Meyer (2000) there are unconfirmed reports that it has been found in Belize.

Superfamily Typhlopoidea

GENUS: *Typhlops* Oppel, 1811, has a global distribution. In the Americas, it is found in the United States, Mexico, Central America, the West Indies, and South America. See Peters & Orejas-Miranda (1970); Dixon & Hendricks (1979); Hahn (1980a); Savage & Villa (1986); Schwartz & Henderson (1991); Smith & Smith (1993); Wallach (1998); Kornacker (1999); McDiarmid et al. (1999) and Köhler (2001).

Review assistance and important contributions for this section were provided by Van Wallach and James R. Dixon.

Note: *T. multilineatus* has not been included in this list. Refer to Smith & Smith (1993: 675) for their comments on this species.

CONTENT: Thirty-seven species in the Americas. ★★★

Typhlops: Blindsnake, sometimes called the Earth or Wormsnake. Liner (1994) called it the Common Blind Wormsnake. Frank and Ramus (1995) also call this the Common Wormsnake.

Species-level Taxa	Citations and Remarks	Abbreviated Range
T. annae Breuil, 1999 St. Barts Blindsnake	Breuil (1999)	Island of St. Barthelemy in the Lesser Antilles.
T. biminiensis Richmond, 1955 Bimini Blindsnake	Richmond (1955); Thomas (1968); Schwartz & Henderson (1985, 1991); McDiarmid et al. (1999); Estrada & Ruibal *in* Crother (1999)	W Great Bahama Bank, Cayman Brac, Cay Sul Bank, Cuba, Great Inagua, Andros, Ragged Islands, Ferry Island, New Providence Island and North and South Bimini Islands.
T. b. biminiensis Richmond, 1955 Bimini Blindsnake	Richmond (1955); Schwartz & Henderson (1985, 1991); McDiarmid et al. (1999); Estrada & Ruibal *in* Crother (1999)	Bahamas in n and s Bimini, Andros Island, New Providence, Elbow Cay (Cay Sul Bank) and Little Ragged Islands; on Cuba in Rancho Luna near Cienfuegos and the e side of the Bahia de Guantanamo.
T. b. paradoxus Thomas, 1968 Great Inauga Island Blindsnake	Thomas (1968); Schwartz & Henderson (1985, 1991); McDiarmid et al. (1999)	SE Bahamas on Great Inagua Island.
T. brongersmianus* Vanzolini, 1976 Black Blindsnake or Brongersma's Blindsnake	Dixon & Hendricks (1979); Vanzolini (1972, 1986); Cei (1993); Dixon & Soini (1986); Lema (1994a); Murphy (1997); Yanosky (1998); McDiarmid et al. (1999); Kornacker (1999); Freitas (1999) *Remark: Vanzolini (1986) notes that this is *nomen novum* for *T. brongersmai* Vanzolini, 1972.	Various literature reports that there are two centers of distribution; one in Peru, c Brazil, Bolivia, Paraguay, and Argentina (Río Paraguay Basin region); and, Trinidad, Venezuela, Guyana, Suriname, ne Brazil and extreme coastal Brazil as far south as Rio Grande do Sul; one specimen is reported outside of these ranges from Estancia Breyer, Patquia, La Rioja, Argentina and is thought to be in error, according to Dixon & Hendricks (1979).
T. capitulatus Richmond, 1964 Haitian Pale-lipped Blindsnake	Richmond (1964); Schwartz & Thomas (1975, 1989); Hahn (1980a); Henderson & Schwartz (1984); Schwartz & Henderson (1991); Thomas & Powell (1995a); McDiarmid et al. (1999); Powell et al. *in* Crother (1999)	Hispaniola in mainland Haiti from the type locality in Cul de Sac Plain in the vicinity of Pétionville west along the Tiburon Peninsula to the Miragoâne and on the south coast around Jacmel.

Species-level Taxa	Citations and Remarks	Abbreviated Range
*T. catapontus** Thomas, 1966 British Virgins Blindsnake	Thomas (1966); Hahn (1980a) [as *T. capapontus*]; Hedges & Thomas (1991); Schwartz & Henderson (1991); McDiarmid et al. (1999); Thomas *in* Crother (1999) *Remark: If *T. richardi naugus* Thomas [1965] 1966 is considered a synonym, the range extends to Virgin Gorda according to Hedges & Thomas (1991).	Puerto Rico bank with the type locality in the vicinity of the Settlement, Anegada, British Virgin Islands; also found on Virgin Gorda.
T. caymanensis Sackett, 1940 Grand Cayman Blindsnake	Sackett (1940); Thomas (1968); Schwartz & Henderson (1985, 1991); McDiarmid et al. (1999)	Grand Cayman Island in the West Indies.
T. costaricensis Jiménez & Savage, 1962 Costa Rican Blindsnake	Jiménez & Savage (1962); Peters & Orejas-Miranda (1970); Dixon & Hendricks (1979); Savage & Villa (1986); Villa (1988); Villa et al. (1988); McDiarmid et al. (1999); Köhler (2001)	Central America from s Honduras and Nicaragua into Monteverde, Costa Rica in montane habitats from about 1100 to 1500 m.
T. dominicanus Stejneger, 1904 Dominica Blindsnake	Richmond (1966); Hahn (1980a); Schwartz & Henderson (1985, 1991); Wallach (1998); McDiarmid et al. (1999); Censky & Kaiser *in* Crother (1999)	Lesser Antilles in Dominica and Guadeloupe.
T. d. dominicanus Stejneger, 1904 Dominica Blindsnake	Richmond (1966); Schwartz & Henderson (1991); McDiarmid et al. (1999)	Dominica in the Lesser Antilles.
T. d. guadeloupe Richmond, 1966 Guadeloupe Blindsnake	Richmond (1966); Schwartz & Henderson (1991); McDiarmid et al. (1999)	Guadeloupe Island and probably also on Basse-Terre.
T. epactius Thomas, 1968 Cayman Brac Blindsnake	Thomas (1968, 1989); Schwartz & Henderson (1985, 1991); Wallach (1998); McDiarmid et al. (1999)	Cayman Brac and expected to be found on Little Cayman Island; in the West Indies.
T. gonavensis Richmond, 1964 Gonâve Blindsnake	Richmond (1964); Thomas (1966, 1989); Schwartz & Thomas (1975); Hahn (1980a); Schwartz & Henderson (1991); Thomas & Powell (1995a); McDiarmid et al. (1999); Powell et al. (1999)	Gonâve Island in Haiti; southern part of the island.
T. granti Ruthven & Gaige, 1935 Grant's Blindsnake	Schwartz & Henderson (1985, 1991); Rivero (1998); McDiarmid et al. (1999); Thomas *in* Crother (1999)	SW Puerto Rico from Parguera east into the vicinity of Guánica and also found on Isla Caja de Muertos.
T. hectus Thomas, 1974 La Hotte Blindsnake	Thomas (1974b, 1989); Henderson & Schwartz (1984); Schwartz & Henderson (1991); Thomas & Powell (1992); McDiarmid et al. (1999); Powell et al. *in* Crother (1999)	SW Hispaniola south of the Cul de Sac-Valle de Neiba Plain (except for the Barahona Peninsula) and that portion of Hispaniola north of Cul de Sac-Valle de Neiba Plain and Île Grande Cayemite; sea level to 800 m.
T. hypomethes Hedges & Thomas, 1991 Coastal Dwarf Blindsnake	Hedges & Thomas (1991); Rivero (1998); McDiarmid et al. (1999); Thomas *in* Crother (1999)	Puerto Rico bank with the type locality listed as Río Piedras, San Juan, Puerto Rico; the range is considered to be the eastern periphery of Puerto Rico.

Superfamily Typhlopoidea

Species-level Taxa	Citations and Remarks	Abbreviated Range
T. jamaicensis (Shaw, 1802) Jamaican Blindsnake	Thomas (1966); Schwartz & Henderson (1985, 1991); McDiarmid et al. (1999); Crombie *in* Crother (1999)	Throughout Jamaica below 2000 feet.
T. lehneri Roux, 1926 Falcón Blindsnake	Roze (1966); Peters & Orejas-Miranda (1970); Dixon & Hendricks (1979); Lancini & Kornacker (1989); McDiarmid et al. (1999); Kornacker (1999); Mijares-Urrutia & Arends R. (2000)	Falcón State in Venezuela; type locality is El Pozon.
T. lumbricalis (Linnaeus, 1758) Earthworm Blindsnake	Cochran (1941); Peters & Orejas-Miranda (1970); Schwartz & Henderson (1985, 1991); McDiarmid et al. (1999); Estrada & Ruibal *in* Crother (1999)	Bahamas on Grand Bahama, Eleuthera, Exuma Cays, Cat Island, Long Island into Haiti and found throughout Cuba and its satellites including Isla de la Juventud and Isla de Pinos. It may have been introduced into Guyana. Thomas restricted the holotype to New Providence Island, Bahamas and believed that Bebee's report for the specimen from Karatabo, Guyana is in error and the example actually came from Port-au-Prince region of Haiti since it is similar to those from the Haitian Cul-de-Sac.
T. microstomus Cope, 1866 Yucatán Blindsnake	Smith & Taylor (1966); Dixon & Hendricks (1979); Vanzolini (1986); Villa et al. (1988); Smith & Smith (1993); Lee (1996, 2000); Campbell (1998b); McDiarmid et al. (1999); Stafford & Meyer (2000); Köhler (2001)	A savanna species from the n half of Yucatán peninsula in Merida in southern Mexico southward into El Paso, Guatemala and Belize; it seems to occur in disjunct populations.
T. minuisquamus Dixon and Hendricks, 1979 Amazon Basin Blindsnake or White-nosed Blindsnake	Dixon & Hendricks (1979); Vanzolini (1986); Dixon & Soini (1986); Lancini & Kornacker (1989); McDiarmid et al. (1999); Kornacker (1999)	Amazon Basin from Mishana and Iquitos, Peru eastward toward Manaus, Brazil and northward into Vaupes, Colombia and Morao, Venezuela and into the northern edge of the Guyana Shield in Kartabo and Tacoba, Guyana.
T. monastus Thomas, 1966 Monserrat Blindsnake	Thomas (1966); Hahn (1980a); Schwartz & Henderson (1985, 1991); McDiarmid et al. (1999)	Montserrat and surrounding islands in the West Indies, including Barbuda, Antigua, and Great Bird Island, St. Christopher, and Nevis.
T. m. monastus Thomas, 1966 Monserrat Blindsnake	Thomas (1966); Hahn (1980a); Schwartz & Henderson (1991); McDiarmid et al. (1999)	Island of Montserrat.
T. m. geotomus Thomas, 1966 Leeward Blindsnake	Thomas (1966); Schwartz & Henderson (1985, 1991); Wallach (1998); McDiarmid et al. (1999)	Barabuda, Antigua, St.Christopher, Nevis and Great Bird Island.
T. monensis Schmidt, 1926 Mona Island Blindsnake	Schmidt (1926); Schwartz & Henderson (1991); Rivero (1998); McDiarmid et al. (1999); Censky & Kaiser *in* Crother (1999)	Isla Mona in the West Indies.
T. paucisquamus Dixon & Henricks, 1979 Pernambuco Blindsnake	Dixon & Hendricks (1979); Vanzolini (1986); McDiarmid et al. (1999)	Known only from Pernambuco, Brazil.

Species-level Taxa	Citations and Remarks	Abbreviated Range
T. platycephalus Dumeril & Bibron, 1844 Puerto Rican Forest Blindsnake	Thomas (1989); Hedges & Thomas (1991); Rivero (1998); McDiarmid et al. (1999); Thomas *in* Crother (1999)	Puerto Rico bank on Puerto Rico and its satellites including Caja de Muertos, Vieques and Culebra Island, Cayo Palominitos, Cayo Diablo and Cayo de Tierra as well as St. Croix.
T. pusillus Barbour, 1914 Hispaniolan Common Blindsnake	Cochran (1941); Thomas (1966); Henderson & Schwartz (1984); Schwartz & Henderson (1991); Thomas & Powell (1994); McDiarmid et al. (1999); Powell et al. *in* Crother (1999)	Throughout Hispanola and its satellites including Île Grande Cayemite, Île de Gonâve, Île de Tortue, Isla Catalina and Isla Saone, and now beleived to be introduced into Miami, Florida in the United States; it is found from sea level to about 2400 feet.
T. reticulatus (Linnaeus, 1758) Giant or Reticulated Blindsnake	Roze (1966); Peters & Orejas-Miranda (1970); Vanzolini (1976); Amaral (1976); Dixon & Hendricks (1979); Cunha et al. (1985); Dixon & Soini (1986); Nascimento et al. (1987); Lancini & Kornacker (1989); Starace (1998); McDiarmid et al. (1999); Kornacker (1999); Mijares-Urrutia & Arends R. (2000)	Tropical South America east of the Andes between 12°N and 14°S with one record from coastal Ecuador; its range includes the Guianas, Brazil, e Peru, n Bolivia, Colombia and the Venezuelan states of Falcón, Carabobo, Monagas, and Territorio Federal de Amazonas.
*T. richardi** Dumeril & Bibron, 1844 Richard's Blindsnake	Thomas (1966); Hahn (1980a); Schwartz & Henderson (1985, 1991); Mayer & Lazelle (1988); Hedges & Thomas (1991); Rivero (1998); McDiarmid et al. (1999); Thomas *in* Crother (1999) *Remark: According to McDiarmid et al. (1999), previously recognized subspecies of *T. richardi* were elevated to species status and recommended the reader see *T. catapontus* and *T. platycephalus* in their work.	Virgin Islands, Puerto Rico and its satellites including Caja de Muertos, Vieques and Culebra islands, Caicos Islands, Anegada, Turks Islands, St. Croix, St. Thomas, St. Johns, Tortola, Prickly Pear Island and Virgin Gorda; McDiarmid et al. (1999) noted the range as the eastern islands of the West Indies.
T. rostellatus Stejneger, 1904 Víbora de Pico or Puerto Rican Brown Blindsnake	Thomas (1989); Schwartz & Henderson (1985, 1991); Rivero (1998); McDiarmid et al. (1999); Thomas *in* Crother (1999)	Throughout Puerto Rico but tends to be restricted to the mesic environments and apparently absent from s Puerto Rico, but it does extend into Reserva Forestral de Susúa.
T. schwartzi Thomas *in* Woods, 1989 Eastern Short-tailed Blindsnake	Thomas (1989); Schwartz & Henderson (1991); Thomas & Powell (1994a); McDiarmid et al. (1999); Powell et al. *in* Crother (1999)	E Hispaniola in the eastern two-thirds of the Dominican Republic.
T. stadelmani Schmidt, 1936 Stadelman's Blindsnake	Schmidt (1936); Holdridge (1967); Dixon & Hendricks (1979); Wilson & Meyer (1985); McCranie & Wilson (2001)	Moderate elevations (850 to 1370 m) in the premontane wet forest and lower montane wet forest formations from nw Copán and sw Yoro, Honduras.
T. sulcatus Cope, 1868 Southern Bicolored Blindsnake	Cochran (1941); Richmond (1964); Thomas (1966); Henderson & Schwartz (1984); Schwartz & Henderson (1991); McDiarmid et al. (1999); Powell et al. *in* Crother (1999)	SW Hispanola and its satellites including the Tiburon Peninsula of Haiti west into the Norme Dubois Peninsula east of Aquin, the Cul de Sac-Valle de Neiba Plain, n of 10 km se Montrouis, Peninsula de Barahona; Isla Alto Velo, Île de Gonâve, Île Grande Cayemite, and Navassa Island.
T. syntherus Thomas, 1965 Barahona Blindsnake	Henderson & Schwartz (1984); Schwartz & Henderson (1991); White et al. (1992); McDiarmid et al. (1999); Powell et al. *in* Crother (1999)	Known only from low elevations at the type locality of Barahona Peninsula in the Dominican Republic.

Superfamily Typhlopoidea

Species-level Taxa	Citations and Remarks	Abbreviated Range
T. tasymicris Thomas, 1974 Grenada Blindsnake	Taylor (1940); Thomas (1974); Schwartz & Henderson (1985, 1991); McDiarmid et al. (1999); Censky & Keiser *in* Crother (1999)	Lesser Antilles on Grenada Island only from type locality in St. David Parish and Barbados.
T. tenuis* Salvin, 1860 Coffee Blindsnake	Peters & Orejas-Miranda (1970); Dixon & Hendricks (1979); Wilson & Meyer (1985); Villa et al. (1988); Pérez-Higareda & Smith (1991); Smith & Smith (1993); McDiarmid et al. (1999); Köhler (2001) *Remark: According to Dixon & Hendricks (1979), *T. stadelmani* Schmidt, 1936 is a synonym. However, Wilson (pers. comm. 2 January 2001) advised that McCranie and Wilson will be resurrecting the species (in press).	S Mexico on the Gulf of Mexico versant into Guatemala.
T. tetrathyreus Thomas *in* Woods, 1989 Cul de Sac Blindsnake	Thomas (1989); Schwartz & Henderson (1991); Thomas & Powell (1994c); McDiarmid et al. (1999); Powell et al. *in* Crother (1999)	Hispaniola, primarily in the Plaine Cul-de-Sac of se Haiti with a few records to the north (on the southern slopes of the Montagnes de Trou d'Eau) and southward (from the north slope of the La Selle).
T. titanops Thomas, 1989 Big-eyed Blindsnake	Thomas (1989); Schwartz & Henderson (1991); Thomas & Powell (1995c); McDiarmid et al. (1999); Powell et al. *in* Crother (1999)	Hispaniola in the Massif de la Selle-Sierra de Baoruco Mountain chain in se Haiti into sw Dominican Republic; from 792 to 2400 feet.
T. trinitatus Richmond, 1965 Trinidad Blindsnake	Richmond (1965); Peters & Orejas-Miranda (1970); Hahn (1980a); Murphy (1997); McDiarmid et al. (1999); Boos (2001)	Trinidad and Tobago; on Trinidad it is reported only from the type locality in St. George County; but it is said to be widespread on Tobago.
T. unilineatus* (Dumeril & Bibron, 1844) Lined Blindsnake	Peters & Orejas-Miranda (1970); Dixon & Hendricks (1979); Hahn (1980a); Vanzolini (1986); McDiarmid et al. (1999) *Remark: Dixon & Hendericks (1979) doubted that the type of *T. unilineatus* is actually from South America and suggested that it is probably related to the *Typhlops diardi* group of the Indonesian Archipelago. Starace (1998) did not list this snake as found in French Guiana. Peters & Orejas-Miranda (1970) listed it as a valid species from French Guiana and Suriname, but it seems to be known only from the type locality, therefore, at the very least, Suriname is questionable.	There is some confusion about the distribution of this snake; thought to be from Suriname and French Guiana; however, it is doubtful that it exists in Suriname and is more likely to be known from Cayenne, French Guiana if it really is American.
T. yonenagae Rodrigues, 1991 San Francisco Blindsnake	Rodrigues (1991); McDiarmid et al. (1999)	Rio São Francisco drainage of Bahia, Brazil; associated with sandy soil areas. The type came from Santo Inácio in the state of Bahia from a sandy soil area.

The Modern Group
Infraorder Alethinophidia

The Modern Group
Infraorder ALETHINOPHIDIA Nopsca, 1923

This is a group of often large snakes that vary in size full grown from 8 inches to over 35 feet (20 cm to over 10 m). Alethinophidia includes snakes that can be terrestrial, burrowing, arboreal, semi-aquatic, and aquatic; and represents the "modern snakes." In these snakes, the head is usually clearly differentiated from the body, although in true burrowing types this may be less obvious. The skull has a mouth that is mobile. Eyes are quite visible and, except for a few primitive species, the eye is covered by a specialized scale called a brille, which is clear and fitted to the eye. These snakes are characterized by the ability to swallow prey whose diameter is larger than their own. The mouth is wide with jaws that are as long as or longer than the skull. Various adaptions are used to allow the snake to open their mouths very wide. The mandibles are independent of each other and can open to hyperextend the jaws. However, in some of the burrowing species, the size of the mouth may be reduced. The upper jaw and the mandible both have teeth. In some of the "advanced" Alethinophidia, or Caenophidia, some of the maxillary teeth in the upper jawbone are transformed into fangs, but none of the "basal" Alethinophidians, or Henophidia, have fangs. The ventral area is covered with broad scales in most. The four present-day superfamilies in this group are generally considered to be Anilioids, the Booids, the Acrochordoids (extralimital to the Americas), and the Colubroids.

The Alethinophidia are composed of two groups often referred to as "primitive snakes" and "advanced, or modern snakes." The Henophidian, or Haenophidia, are the Boas and Pythons. These are considered to be the more primitive, or "basal," of the group. All other snakes in the Alethinophidia are considered to be Caenophidia and are considered to be the advanced, or modern snakes.

Primitive (or BASAL) ALETHINOPHIDIA
These include Pipesnakes, Boas, Pythons, and other basal groups. The primitive group does not include the Caenophidia.

This group contains 12 genera with 48 species in the Americas.

Family ANILIIDAE Stejneger, 1907 (Fitzinger, 1826)
(Formerly Ilysiidae)

The Aniliidae in the Americas are commonly called Pipe Snakes or False Coral Snakes. Genera included in this group are *Anomochilus*, *Anilius*, *Xenopeltis*, *Uropeltis*, and *Cylindrophis*. *Anilius* is the only genus in this group found in the Americas and is considered a Basal Alethinophidian.

CONTENT: One genus in the Americas.

GENUS: *Anilius* (Oken, 1816), [genus originally *Tortrix* Oppel, 1811, but found to be preoccupied by *Tortrix* Denis & Schiffermüller, 1775] is found only in northern South America east of the Andes. See Roze (1958a, 1966); Peters & Orejas-Miranda (1970); Starace (1998); Kornacker (1999); and McDiarmid et al. (1999).

Review assistance and/or important contributions for this section were taken from literature based on Roze (1958a, 1966); Peters & Orejas-Miranda (1970); Hahn (1980a) and Lancini & Kornacker (1989).

CONTENT: One species with two subspecies. ★★★★

Anilius: American Pipesnake or Coral Pipesnake.

Species-level Taxa	Citations and Remarks	Abbreviated Range
A. scytale (Linnaeus, 1758) Scarlet Pipesnake	Roze (1958b, 1966); Peters & Orejas-Miranda (1970); Duellman (1978); Moonen et al. (1979); Lancini & Kornacker (1989); Starace (1998); McDiarmid et al. (1999); Kornacker (1999)	N South America, east of the Andes in Colombia, s Venezuela, the Guianas, n Brazil, Peru, and e Ecuador; it is closely associated with the Amazon.
A. s. scytale (Linnaeus, 1758) Coral Pipesnake	Roze (1958b, 1966); Peters & Orejas-Miranda (1970); Amaral (1976); Dixon & Soini (1986); Starace (1989); McDiarmid et al. (1999); Kornacker (1999); Boos (2001)	Venezuela and the Guianas through n Brazil and the Amazon drainage of Colombia, Ecuador, Bolivia, and Peru. According to Murphy (1997) it is not present on Trinidad. However, it is listed by Boos (2001)
A. s. phelpsorum* Roze, 1958 Phelps's Coral Pipesnake	Roze (1958b, 1966); Peters & Orejas-Miranda (1970); Lancini & Kornacker (1989); McDiarmid et al. (1999); Kornacker (1999) *Remark: There is a problem with Lancini & Kornacker (1989) distribution information on this snake.	Roze (1966) confirms this snake is found in the state of Bolivar, Venezuela with the type locality being the Auyan-Tepuí. His map insinuates that it is found in se Venezuela, Guyana (at this time British Guyana), and ne Brazil. However, more recent literature restricts this snake to s and e Venezuela with the type locality listed as above.

MACROSTOMATA Müller, [1831] 1832

The Macrostomata is comprised of the more modern advanced snakes, including the Basal Macrostomatans (boids) and the Caenophidia. The Basal Macrostomatans are true Bolyeriidae, Tropidophiidae, Boidae, and Pythonidae. They have moderately enlarged ventral scales and ligher, more mobile jaw elements than do Basal Alethinophidians.

Family BOIDAE Gray, 1825

Some authors place species listed under Boidae in families such as Loxocemidae and Tropidophidae, as can be referenced in Seigel et al., 1987. They prefered to use Boidae *auctorum* for the time being. This group of snakes is found in southeastern Europe into Asia Minor, northern, central and eastern Africa, Madagascar, Reunión Islands, the Arabian Peninsula through central and southwestern Asia to India and Sri Lanka, and the Moluccas and New Guinea through Melanesia to Somoa, western North America, Central America, South America, and the West Indies.

Subfamily BOINAE Gray, 1825

These are commonly referred to as Boas and Pythons.

CONTENT: Five genera in the Americas.

GENUS: *Boa* Linnaeus, 1758, is found from the Antilles and Mexico southward through Latin America into Argentina. The recent consolidation of *Acrantophis* and *Sanzinia* into *Boa* by Kluge (1991, 1993) extended its range into Madagascar and the Réunion Islands. *Boa constrictor* has been introduced into the Everglades in Florida, United States, and has been caught there several times. However, it has not been proven it is established. See Stull (1935); Smith & Taylor (1966) [as *Constrictor constrictor*]; Peters & Orejas-Miranda (1970), Kempff-Mercado (1975), Savage & Villa (1986); Pérez-Santos & Moreno (1988); Tolson & Henderson (1993), Smith & Smith (1993); Wolff & Ronne (1997); Kornacker (1999); McDiarmid et al. (1999); Dirksen & Auliya (2001); and Köhler (2001).

Review assistance and/or important contributions for this section were provided by Michael J. Forstner.

CONTENT: Three species globally, according to Kluge (1991, 1993), with one species containing eight to eleven subspecies (*Boa constrictor*) in the Americas. Ten subspecies are shown here. Kluge (1991, 1993) synonymized *Acrantophis* and *Sanzinia* with *Boa*. ★★★

Boa: Boa Constrictor (in the Americas)

Species-level Taxa	Citations and Remarks	Abbreviated Range
B. constrictor Linnaeus, 1758 Boa Constrictor	Stull (1932); Roze (1966); Peters & Orejas-Miranda (1970); Villa et al. (1988); Tolson & Henderson (1993); Smith & Smith (1993); Campbell (1998b), Vences et al. (1998); McDiarmid et al. (1999); Dirksen & Auliya (2001); Köhler (2001)	Americas from the Antilles and Mexico from north of the Tropic of Cancer southward into southern South America as far as Argentina and Paraguay and islands associated with the mainland.
*B. c. constrictor** Linnaeus, 1758 Red-tailed Boa Constrictor or Common Boa Constrictor	Roze (1966); Stimson (1969); Peters & Orejas-Miranda (1970); Amaral (1976); Duellman (1978); Dixon & Soini (1986); Böckeler (1988); Lancini & Kornacker (1989); Smith & Smith (1993); Wolff & Ronne (1997); Starace (1998); McDiarmid et al. (1999); Kornacker (1999); Dirksen & Auliya (2001); Boos (2001) *Remark: *Boa constrictor mexicana* is simply a color variation of *Boa constrictor constrictor*, according to some experts; however, see remark under *B. c. imperator*.	Lesser Antilles southward through Trinidad and into the Amazon Basin into Argentina, Bolivia, and Paraguay.
B. c. amarali Stull, 1932 Bolivian Boa Constrictor or Amaral's Boa Constrictor	Stull (1932); Forcart (1951); Stimson (1969); Peters & Orejas-Miranda (1970) Kempff-Mercado (1975); Amaral (1976); McDiarmid et al. (1999); Leynaud & Bucher (1999); Dirksen & Auliya (2001)	Paraguay, se Bolivia, and into Brazil in Mato Grosso and Goiás.
*B. c. imperator** Daudin, 1803 Central American Boa Constrictor	Smith & Taylor (1966); Peters & Orejas-Miranda (1970); Alvarez del Toro (1982); Wilson & Meyer (1985); Schwartz & Henderson (1991); Tolson & Henderson (1993); Smith & Smith (1993); Ramírez-Bautista (1994); Lee (1996, 2000); McDiarmid et al. (1999); Dirksen & Auliya (2001); Knight et al. (1992) *Remark: *Boa constrictor mexicana* is simply a color variation of *Boa constrictor imperator*; according to Peters & Orejas-Miranda (1970), it was noted as *Boa diviniloquax* var. *mexicana* Jan, 1863 and shown as a junior synonym of *B. c. imperator*. Smith & Smith (1993) agree. *B. c. sigma* Smith, 1943 from the Tres	From Tamaulipas in e Mexico and Sonora in w Mexico southward and eastward through Central America into nw South American as far as nw Peru. It is found in the West Indies on San Andrés Island, Isla de Providencia, and Isla Santa Catalina; this includes the Santa Catalina pigmented individuals from off the coast of Honduras called "Hogg Island" and are sometimes referred to as Hogg Island Boas.

Species-level Taxa	Citations and Remarks	Abbreviated Range
	Marías Islands off the west coast of Mexico on María Madre Island is a synonym of *B. c. imperator* (see Peters & Oreajs-Miranda, 1970) based on Stimson (1969). Reference McDiarmid et al. (1999) for more history on this subject.	
*B. c. longicauda** Price & Russo, 1991 Peruvian Long-tailed Boa Constrictor	Price & Russo (1991); Wolff & Ronne (1997); McDiarmid et al. (1999) *Remark: Not listed by Dirksen & Auliya (2001).	Known only from Tumbes Province, nw Peru in the only coastal tropical wet forest in Peru.
*B. c. melanogaster** Langhammer, 1983 Ecuadorian Boa Constrictor	Langhammer (1983); Pérez-Santos & Moreno (1991); Wolff & Ronne (1997); McDiarmid et al. (1999) *Remark 1: Price & Russo (1991) considered *B. c. melanogaster* to be a *nomen dubium*. *Remark 2: Not listed by Dirksen & Auliya (2001).	Ecuador.
*B. c. nebulosus** Lazell, 1964 Dominican Boa Constrictor	Lazell (1964); Stimson (1969); Schwartz & Henderson (1985, 1991); Tolson & Henderson (1993); McDiarmid et al. (1999); Censky & Kaiser *in* Crother (1999); Dirksen & Auliya (2001) *Remark: Price & Russo (1991) elevated this snake to full species status as *B. nebulosa*. There does not seem to be evidence in their discussion to justify this change, therefore, it has been maintained as a subspecies here.	Dominica and the Lesser Antilles in the West Indies.
B. c. occidentalis Philippi, 1873 Argentine Boa Constrictor	Forcart (1951); Stimson (1969); Peters & Orejas-Miranda (1970); Kempff-Mercado (1975); Cei (1986, 1993); McDiarmid et al. (1999); Leynaud & Bucher (1999); Dirksen & Auliya (2001)	N Argentina, n and e Paraguay into w Santa Cruz de la Sierra, Bolivia.
*B. c. orophias** Linneaeus, 1758 St. Lucia Boa Constrictor	Stimson (1969); Schwartz & Henderson (1985, 1991); Price & Russo (1991); Rahak (1995); McDiarmid et al. (1999); Dirksen & Auliya (2001) *Remark: Price & Russo (1991) elevated this snake to full species status as *B. orophias*. There is not sufficient evidence in their discussion to support this change, therefore, it has been maintained as a subspecies here.	St. Lucia Island in the Caribbean Lesser Antilles in the West Indies.
B. c. ortonii Cope, [1877]1878 Peruvian Boa Constrictor	Stimson (1969); Peters & Orejas-Miranda (1970); McDiarmid et al. (1999); Dirksen & Auliya (2001)	NW Peru west of Andes, found from Piura southward to La Libertad.
B. c. sabogae (Barbour, 1906) Saboga Island Boa Constrictor	Forcart (1951); Stimson (1969); Vanzolini (1986); McDiarmid et al. (1999); Dirksen & Auliya (2001)	Pearl Island group on Saboga Island of Panama in Central America.

GENUS: *Corallus* Daudin, 1803, ranges on the mainland from Central America in Guatemala southward to into Peru, Bolivia, and se Brazil in South America, to just south of the Tropic of Capricorn. It is found into the islands of the Caribbean including the islands off the Atlantic and Pacific coasts of Panamá, Isla Margarita, Trinidad and Tobago and the Windward Islands of the Caribbean including St. Vincent, several of the Grenadines, and Grenada. *Corallus* is found off the southeastern coast of Brazil on the Ilha Grande. Campbell (1998b) noted that it is probably in Belize. Fide Kluge (1991), *Xenoboa* has been synonymized with this genus. Note that Henderson (1993) has reevaluated the group and those shown in Peters & Orejas-Miranda (1970) must be referenced to Henderson (1993); Kluge (1991). See Stimson (1969); Peters & Orejas-Miranda (1970); Henderson (1993a, 1997); Stafford & Henderson (1996); Starace (1998); Kornacker (1999); McDiarmid et al. (1999); Dirksen & Auliya (2001) and Köhler (2001).

Review assistance and important contributions for this section were provided by R. W. Henderson.

CONTENT: Seven species. ★★★★

Corallus: Neotropical Tree Boa or South American Tree Boa.

Species-level Taxa	Citations and Remarks	Abbreviated Range
*C. annulatus** (Cope, 1875 [1876]) Ringed or Annulated Tree Boa	Peters & Orejas-Miranda (1970); Savage & Villa (1986); Villa et al. (1988); Henderson (1993a, b); Stafford & Henderson (1996); McDiarmid et al. (1996); Smith & Avecedo (1997); Campbell (1998b); McDiarmid et al. (1999); Dirksen & Auliya (2001); Wirz (2001); Köhler (2001) *Remark 1: Fide Stafford & Henderson (1996), some specimens from Colombia and Ecuador exhibit differences in jawbone structure that suggests an undescribed species. *Remark 2: Peters (1957), Peters & Orejas-Miranda (1970), and Henderson (1993b) all question the validity of subspecies, but listed them anyway. The following demonstrates what Henderson (1993b) considered to be poorly defined subspecies for the species.	Guatemala and n Honduras in Central American southward into South America to as far as s Ecuador west of the Andes; the subspecies are separated into very disjunct ranges.
*C. a. annulatus** (Cope, 1875 [1876]) Common Ringed Tree Boa	Peters & Orejas-Miranda (1970); Pérez-Santos & Moreno (1988); Henderson (1993b); Stafford & Henderson (1996); Campbell (1998b) McDiarmid et al. (1996); Dirksen & Auliya (2001); Wirz (2001) *Remark: Fide Stafford & Henderson (1996), the subspecies of this species are based on small and insubstantial differences in scalation.	Guatemala southward through Central America into nw Colombia; Campbell (1998a) noted that *C. annulatus* is known in Guatemala only from a small portion of ne Izabal, but it probably occurs in neighboring Belize and nearby Petén. It is also known from the Sierra del Merendón in Honduras.
C. a. blombergi (Rendahl & Vestergren, 1941) Blomberg's Annulated Tree Boa	Peters & Orejas-Miranda (1970); Pérez-Santos & Moreno (1991); Henderson (1993b); Stafford & Henderson (1996); McDiarmid et al. (1999); Dirksen & Auliya (2001); Wirz (2001)	Río Zamora just east of the Andes in eastern Ecuador.

Species-level Taxa	Citations and Remarks	Abbreviated Range
C. a. colombianus (Rendahl & Vestergren, 1940) Colombian Annulated Tree Boa	Peters & Orejas-Miranda (1970); Pérez-Santos & Moreno (1988, 1991); Henderson (1993b); Stafford & Henderson (1996); McDiarmid et al. (1999); Dirksen & Auliya (2001); Wirz (2001)	Pacific lowlands of the c Colombia coast region; 2 disjunct populations in c Colombia and the border with Venezuela.
C. caninus (Linnaeus, 1758) Emerald Tree Boa	Carrillo de Espinoza (1966); Peters & Orejas-Miranda (1970); Amaral (1976); Duellman (1978); Moonen et al. (1979); Dixon & Soini (1986); Lancini & Kornacker (1989); Henderson (1993a, 1993c, 1997); Stafford & Henderson (1996); Starace (1998); McDiarmid et al. (1999); Renjifo & Lundberg (1999); Kornacker (1999); Dirksen & Auliya (2001); Wirz (2001)	Amazon Basin of South America in Colombia, Venezuela, Brazil, Ecuador, Peru, the Guianas and presumably Bolivia; normally associated with tropical lowland rainforests east of the Andes, and restricted to Guianan and Amazonian South America.
C. cooki Gray, 1842 Cook's Tree Boa	Henderson (1997); McDiarmid et al. (1999); Dirksen & Auliya (2001); Wirz (2001) [as *C. hortulanus cooki*]	Restricted to St. Vincent Island in the Windward Islands of the Caribbean.
*C. cropanii** Hoge, 1953 Brazilian or Cropani's Boa	Kluge (1991); Amaral (1976) [as *Xenoboa cropanii* Hoge, 1953]; Henderson (1993a, 1993d, 1997); Henderson & Puorto (1993); Stafford & Henderson (1996); McDiarmid et al. (1999); Dirksen & Auliya (2001); Marques et al. (2001) *Remark: According to Stafford & Henderson (1996), this species is probably represented by no more than 3 specimens in museum collections at the time of their paper.	State of São Paulo in Brazil, limited to an area of about 600 km^2 on the coastal plains of se Brazil in what was the Atlantic rainforest.
C. grenadensis (Barbour, 1914) Grenada Tree Boa	Henderson (1997); McDiarmid et al. (1999); Dirksen & Auliya (2001); Wirz (2001) [as *C. hortulanus grenadensis*]	Islands of the Grenada Bank on Bequia, Île Quatre, Baliceaux, Mustique, Canouan, Mayreau, Union, Carriacou and Grenada and probably other Islands. The type is from St. Georgés, Grenada Island, West Indies.
*C. hortulanus** (Linnaeus, 1758) Amazon Tree Boa or Common Tree Boa	Moonen et al. (1979); Stafford & Henderson (1996); McDiarmid et al. (1996); Henderson (1997b); Castro Astor et al. (1998); Starace (1998); McDiarmid et al. (1999); Kornacker (1999); Freitas (1999); Dirksen & Auliya (2001); Wirz (2001) [as *C. hortulanus hortulanus*]; Marques et al. (2001) *Remark: Fide McDiarmid et al. (1996), *C. enydris* is the junior synonym of *C. hortulanus*.	Considered to be found in the Guianas, Amazonia in s Colombia s Venezuela, Ecuador, Peru, Bolivia, and Brazil. In Brazil, it is also found in the Cerrado, Caatinga and Atlantic Rainforests to about 26°08′ S latitude, and the Ilha Grande off se Brazil.
C. ruschenbergeri (Cope, 1875 [1876]) Ruschenberger's Tree Boa	Henderson (1997); McDiarmid et al. (1999); Renjifo & Lundberg (1999); Kornacker (1999); Mijares-Urrutia & Arends R. (2000); Dirksen & Auliya (2001); Wirz (2001) [as *C. hortulanus ruschenbergeri*]; Boos (2001); Köhler (2001)	From s Costa Rica south of 10°N latitude through Panama including offshore Islands of Isla del Rey, Isla Contadora, Isla de Cébaco and Isla Suseantupa, east of the Andes to about north of the Cordilleras Central and Oriental and in Venezuela to north of the Cordillera de Mérida and the Río Orinoco, Isla Margarita and north and west of the Guiana Shield, as well as Trinidad and Tobago.

GENUS: *Epicrates* Wagler, 1830, is found from the Caribbean islands, from the Bahamas southward through Trinidad and Tobago into South America to as far south as Argentina. See Peters & Orejas-Miranda (1970), Cei (1986); Obst et al. (1988); Tolson & Henderson (1993); Starace (1998); Kornacker (1999); McDiarmid et al. (1999) and Köhler (2001).

Review assistance and important contributions for this section were provided by Michael J. Forstner.

CONTENT: Ten species. ★★★

Epicrates: Slender Boa or Rainbow Boa (in South America).

Species-level Taxa	Citations and Remarks	Abbreviated Range
E. angulifer G. Bibron *in* Cocteau & Bibron, 1840 *in* de la Sagra, 1838–1843 Cuban Boa or Majá	Stafford (1986); Schwartz & Henderson (1985, 1991); Tolson & Henderson (1993); McDiarmid et al. (1999); Estrada & Ruibal *in* Crother (1999); Dirksen & Auliya (2001)	Cuba, the Isle of Pines, Isla de la Juventud and the Cayo Cantiles into neighboring islands.
E. cenchria (Linnaeus, 1758) Rainbow Boa	Roze (1966); Peters & Orejas-Miranda (1970); Amaral (1986); Stafford (1986); Savage & Villa (1986); Villa et al. (1988); Lancini & Kornacker (1989); McDiarmid et al. (1999); Dirksen & Auliya (2001); Köhler (2001) *Remark: Peters & Orejas-Miranda (1970) pointed out that some of the authors did not attempt to allocate the early synonyms of *E. cenchria* when describing subspecies and it is possible that some of the older names that have not been taken into consideration have priority over names that are currently used for subspecies.	Caribbean in Trinidad and Tobago and Central America from Costa Rica southward into South America into Argentina and Paraguay.
E. c. cenchria (Linnaeus, 1758) Brazilian Rainbow Boa	Amaral (1955, 1976); Roze (1966); Peters & Orejas-Miranda (1970); Duellman (1978); Fugler & Walls (1978); Lancini & Kornacker (1989); Murphy (1997); Starace (1998); McDiarmid et al. (1999); Kornacker (1999); Dirksen & Auliya (2001)	S Venezuela and the Guianas southward through the Amazon Basin.
E. c. alvarezi Abalos, Baez & Nader, 1964 Argentine Rainbow Boa	Peters & Orejas-Miranda (1970); Lancini (1979); Cei (1986, 1993); Waller & Buongemini P. (1998); McDiarmid et al. (1999); Leynaud & Bucher (1999); Dirksen & Auliya (2001)	N Argentina and frontier zones of Paraguay and Bolivia in Santa Cruz.
*E. c. assisi** Machado, 1945 Caatingas Rainbow Boa	Peters & Orejas-Miranda (1970); Vanzolini et al. (1980); McDiarmid et al. (1999); Freitas (1999) [as *E. c. xerophilus*]; Dirksen & Auliya (2001) *Remark: Peters & Orejas-Miranda (1970) placed *E. c. xerophilus* Amaral (1954) 1955 from ne Brazil as a synonym [see McDiarmid et al. (1999)]. However, *E. c. xerophilus* was listed as a valid subspecies by Dirksen & Auliya (2001).	State of Piauí to n Bahia in the Caatinga region of Brazil.

Species-level Taxa	Citations and Remarks	Abbreviated Range
E. c. barbouri Stull, 1938 Marajó Island Rainbow Boa	Amaral (1955); Peters & Orejas-Miranda (1970); Lema et. al. (1980); Starace (1998); McDiarmid et al. (1999); Dirksen & Auliya (2001)	Known only from the type locality of State of Pará into n Brazil only on Marajó Island.
E. c. crassus Cope, 1862 Paraguayan Rainbow Boa	Amaral (1955, 1976); Peters & Orejas-Miranda (1970); Cei (1993); McDiarmid et al. (1999); Leynaud & Bucher (1999); Dirksen & Auliya (2001)	S and c Brazil, Paraguay, and n and ne Argentina.
E. c. gaigei Stull, 1938 Peruvian Rainbow Boa	Stull (1938); Amaral (1955); Peters & Orejas-Miranda (1970); Dixon & Soini (1986); Cei (1993); McDiarmid et al. (1999); Dirksen & Auliya (2001)	Eastern lowlands of c and s Peru and Bolivia.
E. c. hygrophilus Amaral, 1955 Wetlands Rainbow Boa	Amaral (1955); Peters & Orejas-Miranda (1970); McDiarmid et al. (1999); Freitas (1999); Dirksen & Auliya (2001)	States of Amazonas and Espirito Santo in Brazil and according to Amaral, is restricted to the west by Serra de Espinhaço.
E. c. maurus Gray, 1849 Colombian Rainbow Boa	Stull (1935); Amaral (1955); Roze (1966); Peters & Orejas-Miranda (1970); Chippaux [1986] 1987; Lancini & Kornacker (1989); Starace (1998); McDiarmid et al. (1999); Renjifo & Lundberg (1999); Kornacker (1999); Mijares-Urrutia & Arends R. (2000); Dirksen & Auliya (2001); Boos (2001); Köhler (2001)	Costa Rica south through Colombia into Venezuela, the Guianas and the adjacent Margerita and Trinidad and Tobago.
E. c. polylepis Amaral, 1935 Río Pandeiro Rainbow Boa	Amaral (1955); Peters & Orejas-Miranda (1970); McDiarmid et al. (1999); Dirksen & Auliya (2001)	States of w Bahia, e Goiás and Distrito Federal in e Brazil; in the regions of Rio Canabrava and Rio Pandeiro.
E. chrysogaster (Cope, 1871) Turks and Caicos Island Boa	Sheplan & Schwartz (1974); Schwartz & Henderson (1985, 1991); Tolson & Henderson (1993); McDiarmid et al. (1999)	Caribbean in the Bahamas on Turks Islands, Caicos Islands, Great Inagua, Acklin's Island and Crooked Island.
E. c. chrysogaster (Cope, 1871) Turks and Caicos Island Boa	Sheplan & Schwartz (1974); Schwartz & Henderson (1985, 1991); Tolson & Henderson (1993); McDiarmid et al. (1999)	Turks and Caicos Islands in the Bahamas; including Caicos Islands, Middle Caicos Island, North Caicos Island, Big Ambergris Cay, Long Cay, and presumably on other islands and islets; possibly on Grand Turk Island.
E. c. relicquus Barbour & Shreve, 1935 Great Inagua Island Boa	Sheplan & Schwartz (1974); Schwartz & Henderson (1985, 1991); Tolson & Henderson (1993); McDiarmid et al. (1999)	Great Inagua Island and Sheep Cay in the Bahamas.
E. c. schwartzi Buden, 1975 Schwartz's Boa	Buden (1975); Schwartz & Henderson (1985, 1991); Tolson & Henderson (1993); McDiarmid et al. (1999)	Acklin's Island and Crooked Island in the Bahamas.
E. exsul Netting & Goin, 1944 Abaco Island Boa	Netting & Goin (1944); Schwartz & Henderson (1985, 1991); Tolson & Henderson (1993); McDiarmid et al. (1999); Dirksen & Auliya (2001)	Abaco Islands and Grand Bahama Island in the Bahamas; including Great Abaco Island, Elbow Cay, and Little Abaco Island.

Primitive Alethinophidia—Macrostomata

Species-level Taxa	Citations and Remarks	Abbreviated Range
E. fordi (Günther, 1861) Haitian Ground or Ford's Boa	Schwartz (1979); Henderson & Schwartz (1984); Schwartz & Henderson (1991); Tolson & Henderson (1993); McDiarmid et al. (1999); Powell et al. *in* Crother (1999); Dirksen & Auliya (2001)	Hispaniola and its satellites; ranging from sea level to 305 m.
E. f. fordi (Günther, 1861) Dominican Republic Ground Boa	Schwartz (1979); Henderson & Schwartz (1984); Schwartz & Henderson (1991); Tolson & Henderson (1993); McDiarmid et al. (1999); Powell et al. *in* Crother (1999); Dirksen & Auliya (2001)	Island of Hispaniola including the Valle de Niebal, Plaina de Cul de Sac and Ile de la Gonâve.
E. f. agametus Sheplan & Schwartz, 1974 Haitian Ground Boa	Sheplan & Schwartz (1974); Schwartz (1979); Henderson & Schwartz (1984); Schwartz & Henderson (1988, 1991); Tolson & Henderson (1993); McDiarmid et al. (1999); Powell et al. *in* Crother (1999); Dirksen & Auliya (2001)	Island of Hispaniola; known only from the type locality of Môle St. Nicolas, Départment du Nord-Ouest, Haiti.
E. f. manototus Schwartz, 1979 Cabrit Island Ground Boa	Henderson & Schwartz (1984); Schwartz & Henderson (1988, 1991); Tolson & Henderson (1993); McDiarmid et al. (1999); Powell et al. *in* Crother (1999); Dirksen & Auliya (2001)	Île ù Cabrit, Haiti in Départment de l'Ouest.
E. gracilis (Fischer, 1888) Hispaniolan Gracile Boa or Hispaniolan Boa	Sheplan & Schwartz (1974); Henderson & Schwartz (1984); Schwartz & Henderson (1991); Tolson & Henderson (1993); McDiarmid et al. (1999); Powell et al. *in* Crother (1999); Dirksen & Auliya (2001)	Hispaniola, from sea level to 175 m.
E. g. gracilis (Fischer, 1888) Haitian Gracile Boa	Sheplan & Schwartz (1974); Henderson & Schwartz (1984); Schwartz & Henderson (1991); Tolson & Henderson (1993); McDiarmid et al. (1999); Powell et al. *in* Crother (1999); Dirksen & Auliya (2001)	Hispaniola north of the Plaine de Cul de Sac-Valle de Neiba in scattered localities.
E. g. hapalus (Sheplan & Schwartz, 1974) Hispaniolan Gracile Boa	Sheplan & Schwartz (1974); Henderson & Schwartz (1984); Schwartz & Henderson (1991); Tolson & Henderson (1993); McDiarmid et al. (1999); Powell et al. *in* Crother (1999); Dirksen & Auliya (2001)	Hispaniola in Tiburon Peninsula east to Port-au-Prince and Jacmel.
E. inornatus* (Reinhardt, 1843) Puerto Rican Boa	Sheplan & Schwartz (1974); Schwartz & Henderson (1985); Smith & Wallach (1992); Wallach & Smith (1992); Tolson & Henderson (1993); Rivero (1998); McDiarmid et al. (1999); Thomas *in* Crother (1999); Dirksen & Auliya (2001); Weldon et al. (1992) *Remark: According to Wallach & Smith (1992), *Boella tenella* Smith & Chiszar (1992), is a junior synonym and is not from Oaxaca, Mexico as originally described, but from Puerto Rico in the Caribbean.	Puerto Rico in the Caribbean; from sea level to about 1500 m.
E. monensis Zenneck, 1898 Mona Island Slender Boa	Schwartz & Henderson (1985, 1988, 1991); Tolson & Henderson (1993); Rivero (1998); McDiarmid et al. (1999); Thomas *in* Crother (1999); Dirksen & Auliya (2001)	Isla Mona in the West Indies, on St. Thomas and Tortula in the Virgin Islands as well as Cayo Diablo near Puerto Rico.

Species-level Taxa	Citations and Remarks	Abbreviated Range
E. m. monensis Zenneck, 1898 Mona Island Slender Boa	Schwartz & Henderson (1988, 1991); Tolson & Henderson (1993); Rivero (1998); McDiarmid et al. (1999); Dirksen & Auliya (2001)	Isla Mona in the West Indies.
E. m. granti Stull, 1933 Virgin Islands Boa	Nellis et al. (1983); Henderson & Schwartz (1985, 1988, 1991); Mayer & Lazell (1988); Tolson & Henderson (1993); Rivero (1998); McDiarmid et al. (1999); Dirksen & Auliya (2001)	U.S. and British Virgin Islands in the West Indies, known from St. Thomas and Tortula but thought to be throughout the archipelago. There is one report from Guana Island in Puerto Rico.
E. striatus (Fischer, 1876) Haitian Boa	Cochran (1941); Sheplan & Schwartz (1974); Henderson & Schwartz (1984); Stafford (1986); Schwartz & Henderson (1985, 1991); Tolson & Henderson (1993); McDiarmid et al. (1999); Powell et al. *in* Crother (1999)	Bahamas and Hispaniola including nearby islands.
E. s. striatus (Fischer, 1876) Haitian Boa	Sheplan & Schwartz (1974); Henderson & Schwartz (1984); Schwartz & Henderson (1991); Tolson & Henderson (1993); McDiarmid (1999); Powell et al. *in* Crother (1999); Dirksen & Auliya (2001)	Hispaniola including Île de la Gonâve and Isla Soana and possibly the Bahamas.
E. s. ailurus Sheplan & Schwartz, 1974 Cat Island Slender Boa	Sheplan & Schwartz (1974); Schwartz & Henderson (1985, 1991); Tolson & Henderson (1993); McDiarmid et al. (1999); Dirksen & Auliya (2001)	Cat Island and Alligator Cay in the Bahamas.
E. s. exagistus Sheplan & Schwartz, 1974 Tiburon Island Boa	Sheplan & Schwartz (1974); Schwartz & Henderson (1991); Tolson & Henderson (1993); McDiarmid et al. (1999); Dirksen & Auliya (2001)	Hispaniola in distal portions of the Tiburon Peninsula east to Les Basses, and presumably eastward.
E. s. fosteri Barbour, 1941 Bimini Boa	Sheplan & Schwartz (1974); Schwartz & Henderson (1985, 1991); Tolson & Henderson (1993); McDiarmid et al. (1999); Dirksen & Auliya (2001)	Bimini Islands of North Bimini Island, South Bimini Island, East Bimini Island, Sleeping Snake and Easter Cay in the Bahamas; on Hispaniola, it is found north of, and in (La Descubierta) the Plaine de Cul de Sac-Valle de Neiba, the Sierra de Baoruco, and associated mesic foothills near Oveido.
E. s. fowleri Sheplan & Schwartz, 1974 Fowler's Boa	Sheplan & Schwartz (1974); Schwartz & Henderson (1985, 1991); Tolson & Henderson (1993); McDiarmid et al. (1999); Dirksen & Auliya (2001)	Andros Island and Berry Island in the Bahamas.
E. s. mccraniei Sheplan & Schwartz, 1974 Ragged Island Boa	Sheplan & Schwartz (1974); Schwartz & Henderson (1985, 1991); Tolson & Henderson (1993); McDiarmid et al. (1999); Dirksen & Auliya (2001)	Ragged Islands in the Bahamas.
E. s. strigilatus Cope, 1862 Nassau Boa	Sheplan & Schwartz (1974); Schwartz & Henderson (1985, 1991); Tolson & Henderson (1993); McDiarmid et al. (1999); Dirksen & Auliya (2001)	Bahamas on New Providence, Rose Island, Eleuthera Island, Long Island, and Exuma Cays.
E. s. warreni Sheplan & Schwartz, 1974 Turtle Island Boa	Sheplan & Schwartz (1974); Henderson & Schwartz (1984); Schwartz & Henderson (1991); Tolson & Henderson (1993); McDiarmid et al. (1999); Powell et al. *in* Crother (1999); Dirksen & Auliya (2001)	Île de la Tortue, Haiti.

Species-level Taxa	Citations and Remarks	Abbreviated Range
E. subflavus Stejneger, 1901 Jamaican Boa	Stafford (1986); Schwartz & Henderson (1985, 1991); Tolson & Henderson (1993); McDiarmid et al. (1999); Crombie *in* Crother (1999); Dirksen & Auliya (2001)	Jamaica, including Goat Island.

GENUS: *Eunectes* Wagler, 1830, is from tropical South America. It ranges from Colombia and Venezuela southward into Argentina. See Roze (1966); Peters & Orejas-Miranda (1970); Lancini (1979), Lancini & Kornacker (1989); Dirksen & Böhme (1998a, 1998b); McDiarmid et al. (1999); and Dirksen & Auliya (2001).

Review assistance and/or important contributions for this section provided by Lutz Dirksen.

CONTENT: Three species. ★★★

Eunectes: Anaconda

Species-level Taxa	Citations and Remarks	Abbreviated Range
E. deschauenseei* Dunn & Conant, 1936 Marajó Island Anaconda or Dark-Spotted Anaconda	Dunn & Conant (1936); Peters & Orejas-Miranda (1970); Vanzolini (1986); Strimple et al. (1997); Starace (1998); Dirksen & Böhme (1998a); McDiarmid et al. (1999); Dirksen & Auliya (2001) *Remark: Kluge (1991) treated this snake as a junior synonym of *E. notaeus*.	Marajó Island and Pará throughout ne coastal Brazil into coastal French Guiana.
E. murinus (Linnaeus, 1758) Green Anaconda	Peters & Orejas-Miranda (1970); Duellman (1978); Moonen et al. (1979); Trutnau (1982); Dixon & Soini (1986); Lancini & Kornacker (1989); Strimple et al. (1997); Dirksen & Böhme (1998a); Leynaud & Bucher (1999); McDiarmid et al. (1999); Kornacker (1999); Dirksen & Auliya (2001); Boos (2001) [as *E. murinus gigas*] *Remark 1: Strimple et al. (1997), treated *E. barbouri* (of Pará and the island of Marajó in Brazil), as a synonym of *E. murinus*. Dirksen & Böhme (1998b) supported this. *Remark 2: McDiarmid et al. (1999), quoted Peters & Orejas-Miranda (1970), as showing two subspecies: *E. m. murinus* (Linnaeus, 1758), ranging from northern South America and closely associated wtih the Amazon drainage; and, *E. m. gigas* (Latreille *in* Sonini & Latreille, 1802), found in Colombia, Venezuela, Trinidad and the Guianas. Dirksen & Böhme (1998b) did not recognize these subspecies of *E. murinus*. This work follows Dirksen & Böhme (1998a) as the most recent revisors.	N South America in the Amazon and Orinoco drainages from Venezuela and Colombia into Bolivia, Argentina and Paraguay.
E. notaeus* Cope, 1862 Southern Anaconda or Yellow Anaconda	Peters & Orejas-Miranda (1970); Amaral (1976); Cei (1993); Lema (1994a); Dirksen & Böhme (1998a); Yanosky (1998); McDiarmid et al. (1999); Leynaud & Bucher (1999); Dirksen & Auliya (2001)	S South America in e Bolivia, Paraguay, Uruguay, ne Argentina and w and s Brazil in Rio Grande do Sul and Paraná.

Species-level Taxa	Citations and Remarks	Abbreviated Range
	*Remark: Although Peters & Orejas-Miranda (1970), listed this snake as being from Uruguay, McDiarmid et al. (1999), were unable to verify this distribution.	

Subfamily ERYCINAE Bonaparte, 1831

These are commonly referred to as Sand Boas and Rubber Boas.

This subfamily is found in southern and southeastern Europe, Asia Minor, Africa, Asia, India, southwestern Canada, the western United States and northwestern Mexico.

CONTENT: Two genera in the Americas; sometimes considered to be only one genus in the Americas (see McDiarmid et al. 1999).

GENUS: *Charina* Gray, 1849, is from western North America and found from western Canada southward through the western United States. Kluge (1993) merged the Rosy Boa (*Lichanura*), the Rubber Boa (*Charina*), and the Calabar Burrowing Python (*Calabaria*) into *Charina*. Although Kluge presented a good argument for the consolidation, I still believe that at least the *Charina* and *Lichanura* should be separated and have maintained them separate here. There are valid taxonomic reasons for this separation which are based on Holman (2000), who segregates *Lichanura* and *Charina*. In addition to Holman (2000), the two genera are highly different in that *Charina* is a dweller of wet forests and *Lichanura* is a desert species. Holman noted that *Lichanura* "looks more like a normal boa than *Charina* does." Michael Forstner (pers. comm. 31 October 2000) confirmed that *Charina* and *Lichanura* are molecularly monophyletic groups separate from *Calabaria*. Crother et al. (2000) also maintained them as separate. Kluge (1993) only reviewed the hard structural morphology and did not consider the soft morphology that separates the two. See Klauber (1943b); Stebbins (1966, 1985); Bogert (1968); Stewart (1977); Kluge (1993); Smith & Smith (1993); Uetz (1998); McDiarmid et al. (1999); Holman (2000); and Dirksen & Auliya (2001).

Review assistance and important contributions for this section provided by Michale J. Forstner.

CONTENT: One species containing three subspecies but it is sometimes listed containing up the three species in North America and Africa. ★★★

Charina: Rubber Boa

Species-level Taxa	Citations and Remarks	Abbreviated Range
C. bottae (Blainville, 1835) Rubber Boa	Klauber (1943b); Cunningham (1966); Stebbins (1985); Stewart (1977); Kluge (1993); Brown (1997); McDiarmid et al. (1999); Holman (2000); Dirksen & Auliya (2001)	S Canada to the w United States, from s British Columbia in the vicinity of Quesuel in Canada southward into Wyoming, s Utah, c Nevada and s California; from sea level to about 2800 m.
C. b. bottae (Blainville, 1835) Northern Rubber Boa	Klauber (1943b); Stebbins (1966, 1985); Stewart (1977); Brown (1997); McDiarmid et al. (1999); Bartlett & Tennant (2000); Dirksen & Auliya (2001)	S Canada southward to the United States into sc California and eastward to Utah, Wyoming, Nevada and Montana.

Species-level Taxa	Citations and Remarks	Abbreviated Range
C. b. umbratica Klauber, 1943 Southern Rubber Boa	Klauber (1943b); Stebbins (1966, 1985); Stewart (1977); Brown (1997); McDiarmid et al. (1999); Bartlett & Tennant (2000); Dirksen & Auliya (2001)	S California, United States only in three small disjunct ranges; the Tehachapi, the San Bernardino, and the San Jacinto Mountains.
*C. b. utahensis** Van Denburgh, 1920 Rocky Mountain Rubber Boa	Klauber (1943b); Stebbins (1966, 1985); Stewart (1977); Yingling (1982); Brown (1997); McDiarmid et al. (1997); Bartlett & Tennant (2000); Dirksen & Auliya (2001)	S British Columbia in Canada south into the w United States to as far as Utah.
	*Remark: Stewart (1977) suggested that this subspecies is not valid. It is listed here because it was shown in Stebbins (1985); Brown (1997), and in Bartlett & Tennant (2000).	

GENUS: *Lichanura* Cope, 1861, is found in North America from the southwestern United States southward into western Mexico, including offshore islands. As noted above, Kluge (1993) merged the Rubber Boa (*Lichanura*), the Rosy Boa (*Charina*), and the Calabar Burrowing Python (*Calabaria*) as synonyms of *Charina*, for reasons of taxonomic simplification. I have maintained *Lichanura* independent from *Charina* and suggest *Lichanura* is a group that requires further study, even though Spiteri (1987) has addressed it recently. There are valid taxonomic reasons for this separation which are based on Holman (2000), who segregates *Lichanura* and *Charina*. See the discussion under *Charina*. See Schmidt (1953); Smith & Taylor (1966), Bogert (1968); Yingling (1982); Spiteri (1993), Merker & Merker (1998), Kluge (1993); Smith & Smith (1993); McDiarmid et al. (1999) [as *Charina*]; Holman (2000); and Dirksen & Auliya (2001).

Review assistance and important contributions for this section were provided by Michael J. Forstner.

Note 1: I agree with Smith & Smith (1993) that the taxonomic analyses of *Lichanura* have been confusing (in their words "met with conflicting resolution") in recent years and more investigation and study of the genus is necessary. See Smith & Smith (1993: 470) for an overview of the problem with *Lichanura* nomenclature. I also recommend that Merker & Merker (1998) be reviewed along with Smith & Smith (1993). There are some problems with the Spiteri (1987) *in* Spiteri (1993) taxonomy versus traditional taxonomy, and the subspecies within the species.

Note 2: Troy Hibbits (pers. comm. 15 August 2000) considers the northern Baja orange-striped, clean bellied snakes "intergrades" between *L. roseofusca* and *L. trivirgata*. He feels that clearly they are not intergrades and these snakes need more research. Michael Forstner (pers. comm. 31 October 2000) communicated that if these are intergrades, then they are not differentiated.

Note 3: The reader should attempt to find copies of the Spiteri works to review. After reading Spiteri's information, the reader may decide that the Spiteri designations, for example *L. t. myriolepis* (Cope, 1868), are valid over the subspecies listed below.

Holman (2000) reviewed the fossil snakes of North America. He noted that Kluge (1993) synonymzed *Lichanura* and *Charina*, but recognized what Holman called the <u>very distinct</u> North American genus *Charina* and *Lichanura*. He stated that isolated vertebrae of *Charina* may usually be separated from those of the other North American erycine genus *Lichanura* on the basis of: 1) the longer neural spine, 2) the non-U-shaped zygosphene in dorsal view, 3) the strongly concave (as opposed to flat or moderately concave) zygosphene in anterior view, 4) the more depressed neural arch, and 5) the more deeply incised posterior edge of the neural arch.

CONTENT: One species with about four to seven subspecies. Six subspecies are shown below. The reader should approach the following information very cautiously, since this genus' subspecies designations do not seem to currently be receiving general agreement among herpetologists. Subspecies designations and geographies are confused at the moment and those listed below are my "best guess," based on conversations with people who are much more familiar with these snakes than I. ★

Lichanura: Rosy Boa

Species-level Taxa	Citations and Remarks	Abbreviated Range
*L. trivirgata** Cope, 1861 Rosy Boa	Smith & Taylor (1966); Yingling (1982); Stebbins (1985); Spiteri (1987) *in* Spiteri (1993), Smith & Smith (1993); Kluge (1993); Brown (1997); Merker & Merker (1998); McDiarmid et al. (1999); Holman (2000); Dirksen & Auliya (2001) *Remark: Smith & Smith (1993) consider *L. trivirgata intermedia* as a nomen nudum.	SW United States southward into western Mexico to as far as w Sonora and Baja California.
L. t. trivirgata Cope, 1861 Mexican Rosy Boa	Yingling (1982); Spiteri (1991); Spiteri (1987) *in* Spiteri (1993); Smith & Smith (1993); Merker & Merker (1998); McDiarmid et al. (1999); McPeak (2000); Bartlett & Tennant (2000); Dirksen & Auliya (2001)	In the United States only in s Arizona into Mexico in the Baja California and Sonora.
*L. t. arizona** Spiteri, 1993 Arizona Rosy Boa	Spiteri (1987) *in* Spiteri (1993); Merker & Merker (1998); Dirksen & Auliya (2001) *Remark: This subspecies is not listed by McDiarmid et al (1999) and is noted here based on Spiteri's 1993 work. The reader should consider McDiarmid et al. (1999) the definitive overview for this snake. Bartlett & Tennant (2000) mentioned this subspecies only briefly and did not consider it to be valid.	The United States in s California into adjacent sw Arizona.
*L. t. bostici** (Ottley, 1978) Cedros Island Rosy Boa	Spiteri (1987) *in* Spiteri (1993); Smith & Smith (1993); Merker & Merker (1998); McDiarmid et al. (1999); Dirksen & Auliya (2001) *Remark: Some authorities consider this to be *L. t. trivirgata*. This subspecies was not listed as valid in Dirksen & Auliya (2001).	Isla Cedros, Baja California del Norte, Mexico.
*L. t. gracia** Klauber, 1931 Desert Rosy Boa	Yingling (1982); Spiteri (1987) *in* Spiteri (1993); Smith & Smith (1993); Brown (1997); Merker & Merker (1998); McDiarmid et al. (1999); Bartlett & Tennant (2000); Dirksen & Auliya (2001) *Remark 1: According to Bartlett & Tennant (2000), Spiteri has argued convincingly against the validity of the subspecies *gracia* and the use of the subspecies *roseofusca* for California snakes. They noted that Spiteri believes both to be the subspecies *myriolepis*, which is a term now used by hobbyists to identify	SC and se California into sw and c Arizona in the United States.

Species-level Taxa	Citations and Remarks	Abbreviated Range
	the brightly colored rosy boas of c Baja Peninsula. His findings were ignored, according to the authors, "based more on a technicality rather than due to erroneous conclusions on Spiteri's part." I have difficulty with Spiteri's lumping of of *L. gracia* and *L. roseofusca* into *L. myriolepis*, and have left out *L. myriolepis* and considered *L. gracia* and *L. roseofusca* as separate and valid in this work. Troy Hibbits (pers. comm. 15 August 2000) also has difficulty with Spiteri's lumping of *L. gracia* and *L. roseofusca* into *L. myriolepis*.	
	*Remark 2: Brown (1997) noted Spiteri's decision that only one subspecies of *Lichanura trivirgata* is found in California and Spiteri called it *L. t. myriolepis*. However, Brown maintained the two subspecies of *L. t. gracia* and *L. t. roseofusca* for California and did not list *L. t. myriolepis* as being found there.	
	*Remark 3: Troy Hibbits (pers. comm. 15 August 2000) has seen a large sample of living Rosy Boas and he feels that the Arizona highland boas (Harquahala, Harcuvar, Hualupai, and Cerbat Mountain populations, as well as those populations near Bagdad and Hillside) are something distinct from *gracia*.	
*L. t. roseofusca** Cope, 1868 California Rosy Boa or Coastal Rosy Boa	Yingling (1982); Spiteri (1987) *in* Spiteri (1993); Smith & Smith (1993); Brown (1997); Meker & Merker (1998); McDiarmid et al. (1999); Bartlett & Tennant (2000); Dirksen & Auliya (2001) *Remark: see comments under *L. t. gracia*.	North American in extreme sw California, and Arizona in the United States southward into Mexico on the Baja California; Spiteri restricted *roseofusca* to Ensanada and n Baja California.
*L. t. saslowi** Spiteri *in* Bartlett, 1987 [1988] Mid-Baja California Rosy Boa	Spiteri (1987) *in* Spiteri (1993); Smith & Smith (1993); Merker & Merker (1998); McDiarmid et al. (1999); Dirksen & Auliya (2001) *Remark 1: Smith & Smith (1993) stated that *L. trivirgata saskelli* is a synonym. *Remark 2: See Smith & Smith (1993) for comments on the validation of *L. t. saslowi*. Bartlett (1987) first published the name, based on Spiter's unpublished masters thesis, which was not formally published until 1988. *Remark 3: Walls (1994) suggested that more than one subspecies or type may be recognized within *L. saslowi*. *Remark 4: McDiarmid et al. (1999) did not include the Spiteri literature in their dialogue on *Charina trivirgata* except for his discussions on *L. t. myriolepsis*.	Bahia de los Angeles, Baja California, Mexico.

Family LOXOCEMIDAE Cope, 1861

These are commonly referred to as Mexican Pythons.

Found in Mexico and Central America, these snakes are sometimes shown in the family listed along with *Xenopeltis* in the Xenopeltidae.

CONTENT: One species

GENUS: *Loxocemus* Cope, 1861, is found in Mexico and Central America. See Smith & Taylor (1966); Nelson & Meyer (1967); Peters & Orejas-Miranda (1970), Savage (1980), Smith & Smith (1993); McDiarmid et al. (1999) and Köhler (2001).

Review assistance and important contributions for this section were provided by Michael J. Forstner.

CONTENT: One species. ★★★★

Loxocemus: La Chatilla, Mexican Burrowing Python, Neotropical Python.

Species-level Taxa	Citations and Remarks	Abbreviated Range
L. bicolor Cope, 1861 La Chatilla or Mexican Burrowing Python	Peters & Orejas-Miranda (1970); Nelson & Meyer (1971); Savage (1980); Alvarez del Toro (1982); Wilson & Meyer (1985); Savage & Villa (1986); Villa et al. (1988); Smith & Smith (1993); Ramírez-Baustita (1994); Carcía & Ceballos (1994); Vences et al. (1998); McDiarmid et al. (1999); Köhler (2001)	From Nayarit, Mexico along the Pacific Coast of Central America into n Costa Rica.

Family TROPIDOPHIIDAE Brongersma, 1951 (Cope, 1894)

These are commonly referred to as Dwarf Boas or Eyelash Boas.

These snakes are found from southern Mexico, the West Indies, and Central America into northwestern Pacific South America. See McDiarmid et al. (1999:214) about how the name and dates relate to Ungualiidae Cope, 1894.

Note: Fide Pough et al. (1998), the genera of *Exiliboa* and *Ungaliophis* belong in the Boiidae.

CONTENT: Four genera

GENUS: *Exiliboa* Bogert, 1968, is restricted to southern Mexico. See Bogert (1968); Williams & Wallach (1989); Campbell & Camarillo R. (1992); Smith & Smith (1993) and McDiarmid et al. (1999). Pough et al. (1998) includes *Exiliboa* as a boid.

Review assistance and important contributions for this section were provided by Michael J. Forstner.

CONTENT: One species. ★★★★

Exiliboa: Mexican Dwarf Boa

Primitive Alethinophidia—Macrostomata

Species-level Taxa	Citations and Remarks	Abbreviated Range
E. placata Bogert, 1968 Oaxacan Dwarf Boa	Bogert (1968, 1969); Williams & Wallach (1989); Campbell & Camarillo R. (1992); Smith & Smith (1993); McDiarmid et al. (1999)	Mexico in and around Oaxaca at high altitudes in the Sierra de Juárez and near Totontepec in the Sierra Mixe; associated with cloud forests.

GENUS: *Trachyboa* Peters, 1860, ranges from Panama in Central America southward into northwestern South America to as far south as Colombia, Ecuador, Peru and possibly Amazonian Brazil. See Peters & Orejas-Miranda (1970); McDiarmid et al. (1999) and Köhler (2001).

Review assistance and important contributions for this section were provided by Michael J. Forstner.

CONTENT: Two species. ★★★★

Trachyboa: Eyelash Dwarf Boa, Rough-scaled Boa, or Spiny Boa.

Species-level Taxa	Citations and Remarks	Abbreviated Range
T. boulengeri Peracca, 1910 Northern Eyelash Boa or Rough-scaled Boa	Amaral (1927f, g; 1931b); Dunn & Bailey (1939); Peters (1960a); Peters & Orejas-Miranda (1970); Lehmann (1970); Villa et al. (1988); Pérez-Santos & Moreno (1988, 1991); McDiarmid et al. (1999); Köhler (2001)	C and nw South America from Panama through Colombia into w Ecuador. They are closely associated with the rainforests in the Pacific versant.
T. gularis Peters, 1860 Southern Eyelash Boa or Spiny Boa	W. Peters (1860); Amaral (1927g); Peters (1960b); Peters & Orejas-Miranda (1970); Pérez-Santos & Moreno (1991); McDiarmid et al. (1999)	Coastal w Ecuador; associated with the dry areas of the coast. Peters & Orejas-Miranda (1970) also include Brazil in their range but I have not been able to verify this as valid.

GENUS: *Tropidophis* Bibron *in* de la Sagra, 1840, is found in a widely disjunct distribution in the West Indies, Ecuador, Peru, and Brazil. See Henderson & Schwartz (1984); Tolson and Henderson (1993); Peters & Orejas-Miranda (1970); Schwartz & Henderson (1985); and McDiarmid et al. (1999).

Review assistance and important contributions for this section were provided by Michael J. Forstner.

CONTENT: Twenty species. ★★★

Tropidophis: Caribbean Dwarf Boa or Wood Snake. (Although I like Blair Hedges' common name for this snake of "Trope," it has not yet been commonly used and does not seem to have gained any acceptance in the herpetology community, therefore, I have tended to remain with the more traditional nomenclature.)

Species-level Taxa	Citations and Remarks	Abbreviated Range
T. battersbyi Laurent, 1949 Battersby's Dwarf Boa	Laurent (1949); Peters & Orejas-Miranda (1970); Pérez-Santos & Moreno (1991); Tolson & Henderson (1993); McDiarmid et al. (1999)	Ecuador, known only from the holotype which has no definite locality data.
T. canus (Cope, 1868) Great Inagua Dwarf Boa	Schwartz & Henderson (1985, 1991); Tolson & Henderson (1993); McDiarmid et al. (1999)	Bahamas on Great Inagua, Andros, New Providence, and closely associated islands.

Species-level Taxa	Citations and Remarks	Abbreviated Range
T. c. canus (Cope, 1868) Great Inagua Dwarf Boa	Schwartz & Marsh (1960); Schwartz & Henderson (1991); Tolson & Henderson (1993); McDiarmid et al. (1999)	Great Inagua Island, Bahamas.
T. c. androsi Stull, 1928 Andros Island Dwarf Boa	Stull (1928); Schwartz & Marsh (1960); Schwartz & Henderson (1985, 1991); Tolson & Henderson (1993); McDiarmid et al. (1999)	Andros Island, Bahamas.
T. c. barbouri (Bailey, 1937) Eluthera Island Dwarf Boa	Schwartz & Marsh (1960); Schwartz & Henderson (1985, 1991); Tolson & Henderson (1993); McDiarmid et al. (1999)	Eluthra Island, Eleuthra Cays (Royal Island), Long Island, Cat Island, Exuma Cays, and Great Ragged Island, Bahamas.
T. c. curtus (Garman, 1887) Bimini Island Dwarf Boa	Schwartz & Marsh (1960); Schwartz & Henderson (1985, 1991); Tolson & Henderson (1993); McDiarmid et al. (1999)	New Providence Island, Bimini Islands, and Cay Sal Bank in the Bahamas.
T. caymanensis Battersby, 1938 Cayman Islands Dwarf Boa	Battersby (1938); Schwartz & Henderson (1985, 1991); Tolson & Henderson (1993); McDiarmid et al. (1999)	Cayman Islands on Grand Cayman, Cayman Brac, and Little Cayman islands in the Caribbean.
T. c. caymanensis Battersby, 1938 Grand Cayman Dwarf Boa	Battersby (1938); Schwartz & Henderson (1985, 1991); Tolson & Henderson (1993); McDiarmid et al. (1999)	Grand Cayman Island in the Caribbean.
T. c. parkeri Grant, 1941 Little Cayman Dwarf Boa	Thomas (1963); Schwartz & Henderson (1985, 1991); Tolson & Henderson (1993); McDiarmid et al. (1999)	Little Cayman Island in the Caribbean.
T. c. schwartzi Thomas, 1963 Cayman Brac Dwarf Boa	Thomas (1963); Schwartz & Henderson (1985, 1991); Tolson & Henderson (1993); McDiarmid et al. (1999)	Cayman Brac Island in the Caribbean.
T. celiae Hedges, Estrada & Diaz, 1999 Canasi Dwarf Boa	Hedges et al. (1999)	W Cuba, only known from the type locality which is about 57 km e of La Habana and 33 km w of Matanzas in the Aluras de la Habana.
T. feicki Schwartz, 1957 Broad-banded Dwarf Boa or Feick's Dwarf Boa	Schwartz (1957); Schwart & Henderson (1985, 1991); Tolson & Henderson (1993); McDiarmid et al. (1999); Estrada & Ruibal *in* Crother (1999)	W Cuba from Pedrera de Mendoza and Guane, Pinar del Río Provencia eastward into Pan de Matanzas in Mantanzas Province.
T. fuscus Hedges & Garrido, 1992 Cuban Dusky Dwarf Boa	Hedges & Garrido (1992); Tolson & Henderson (1993); McDiarmid et al. (1999); Estrada & Rubial *in* Crother (1999)	Cuba; known only from two locations, Cruzata and Minas Amorea in ne Guantánamo Province; it is associated with pine woods on red, lateritic soil.
T. greenwayi Barbour & Shreve, 1935 Bahamian Dwarf Boa	Barbour & Shreve (1935); Schwartz & Henderson (1985, 1991); Tolson & Henderson (1993); McDiarmid et al. (1999)	Turks and Caicos Islands and adjacent islands in the West Indies.

Species-level Taxa	Citations and Remarks	Abbreviated Range
T. g. greenwayi Barbour & Shreve, 1935 Ambergris Cay Dwarf Boa	Barbour & Shreve (1935); Schwartz (1963); Schwartz & Henderson (1985, 1991); Tolson & Henderson (1993); McDiarmid et al. (1999)	Ambergris Cay, Caicos Islands in the Bahamas; known only from the type locality.
T. g. lanthanus Schwartz, 1963 Turks & Caicos Islands Dwarf Boa	Schwartz (1963); Schwartz & Henderson (1985, 1991); Tolson & Henderson (1993); McDiarmid et al. (1999)	Caicos Islands, s Caicos, Long Cay, Middleton Cay, North Caicos, Middle Caicos, and probably Providenciales.
T. haetianus (Cope, 1879) Haitian Dwarf Boa	Schwartz (1963); Henderson & Schwartz (1984); Schwartz & Henderson (1991); Tolson & Henderson (1993); McDiarmid et al. (1999); Powell et al. *in* Crother (1999); Estrada & Ruibal *in* Crother (1999)	Hispaniola on Île de la Gonâve and Île de la Tortue; and Cuba in Guardalavaca in Oriente Province, and Jamaica, altitudinal distribution ranges from sea level to 2700 ft.
T. h. haetianus (Cope, 1879) Haitian Dwarf Boa	Schwartz (1963); Henderson & Schwartz (1984); Schwartz & Henderson (1985, 1991); Tolson & Henderson (1993); McDiarmid et al. (1999); Powell et al. *in* Crother (1999); Estrada & Ruibal *in* Crother (1999)	Throughout much of Hispaniola and and e Cuba; its range includes Île de la Gonâve and Île de la Tortue.
T. h. hermerus Schwartz, 1975 Dominican Republic Dwarf Boa	Schwartz (1963); Henderson & Schwartz (1984); Schwartz & Henderson (1991); Tolson & Henderson (1993); McDiarmid et al. (1999); Powell et al. *in* Crother (1999)	Extreme e Dominican Republic in Hispaniola.
T. h. jamaicensis* Stull, 1928 Jamaican Dwarf Boa	Stull (1928); Schwartz (1963); Schwartz & Henderson (1985, 1991); Tolson & Henderson (1993); McDiarmid et al. (1999); Crombie *in* Crother (1999) *Remark: Crombie *in* Crother (1999) lists this snake as the full species *T. jamaicensis*.	S Jamaica from Malvern, St. Elizabeth Parish east into Blue Mountain Estate, St. James Parish (excluding the Portland Peninsula).
T. h. stejnegeri* Grant, 1940 Stejneger's Dwarf Boa	Schwartz (1963); Schwartz & Henderson (1985, 1991); Tolson & Henderson (1993); McDiarmid et al. (1999); Crombie *in* Crother (1999) *Remark: Fide Crombie *in* Crother (1999), this is a full species as *T. stejnegeri*.	N Jamaica from Montego Bay, Mt. Horeb and Plum Park, St. James Parish and Bluefields, Westmoreland Parish, Protland Parish, and eastward into Balaclava, St. Elizabeth Parish.
T. h. stullae* Grant, 1940 Portland Ridge Dwarf Boa	Schwartz (1963); Schwartz & Henderson (1985, 1991); Tolson & Henderson (1993); McDiarmid et al. (1999); Crombie *in* Crother (1999) *Remark: Crombie *in* Crother (1999) state this is *T. stullae*.	Known only from the type locality in Portland Point, Claredon Parish, Jamaica.
T. h. tiburonensis Schwartz, 1975 Tiburon Dwarf Boa	Schwartz (1963); Schwartz & Henderson (1991); Tolson & Henderson (1993); McDiarmid et al. (1999); Powell et al. *in* Crother (1999)	Tiburon Peninsula in Haiti; known from very few areas which include Camp Perrin, Dépt. Du Sud, Haiti (the type locality); vicinity of Jérémie; and near Marceline.
T. hendersoni Hedges & Garrido, 2002 Henderson's Dwarf Boa	Hedges & Garrido (2002)	Known only from the type locality in Holguín Province, located on the northern coast of e Cuba, to the nw of Banes.

Species-level Taxa	Citations and Remarks	Abbreviated Range
T. maculatus (Bibron *in* de la Sagra, 1840) Spotted Dwarf Boa	Schwartz & Henderson (1985, 1991); Tolson & Henderson (1993); McDiarmid et al. (1999); Estrada & Rubial *in* Crother (1999)	W Cuba from Pinar del Río Province east into Matanzas Province, and the Island la Juventud.
T. melanurus (Schlegel, 1837) Cuban Dwarf or Black-tailed Boa	Schwartz & Henderson (1985, 1991); Tolson & Henderson (1993); McDiarmid et al. (1999); Estrada & Rubial *in* Crother (1999)	Cuba and and nearby satellites including Navasa Island, Isla de la Juventud, and Cayos de San Felipe (on Cayo Real).
T. m. melanurus (Schlegel, 1837) Cuban Black-tailed Dwarf Boa	Schwartz & Henderson (1985, 1991); Tolson & Henderson (1993); McDiarmid et al. (1999); Estrada & Ruibal *in* Crother (1999)	Throughout Cuba except where *T. melanurus dysodes* is found (1 km north of La Coloma).
T. m. bucculentus (Cope, 1868) Navassa Island Dwarf Boa	Thomas (1966); Schwartz & Henderson (1985, 1991); Tolson & Henderson (1993); McDiarmid et al. (1999)	Navassa Island west of Haiti in the w Caribbean.
T. m. dysodes Schwartz & Thomas, 1960 Pinar del Rio Dwarf Boa	Schwartz & Thomas (1960); Schwartz & Henderson (1985); Tolson & Henderson (1993); McDiarmid et al. (1999); Estrada & Ruibal *in* Crother (1999)	Known only at the type locality 1 km north of La Coloma in Pinar del Rio Province, Cuba.
T. m. ericksoni Schwartz & Thomas, 1960 Pine Island Dwarf Boa	Schwartz & Thomas (1960); Schwartz & Henderson (1985, 1991); Tolson & Henderson (1993); McDiarmid et al. (1999); Estrada & Ruibal *in* Crother (1999)	Isla de Pinos, Cuba.
T. morenoi Hedges, Garrido & Díaz, 2001 Moreno's Dwarf Boa	Hedges et al. (2001)	Known only from the type locality in ne Cuba, between the large lake Presa Alcranes and the coast at Bahía de Carahatas. The authors think it may inhabit other areas along the ne coast, such as the alturas del Noreste and Llanura del Norte de Las Villas.
T. nigriventris Bailey, 1937 Dark-bellied Trope or Black-bellied Dwarf Boa	Bailey (1937); Schwartz & Henderson (1985, 1991); Tolson & Henderson (1993); McDiarmid et al. (1999); Estrada & Ruibal *in* Crother (1999)	Cuba in e Camagüey Proinve, s Cienfuegos Province and Sancti Spíritus Province.
T. n. nigriventris Bailey, 1937 Black-bellied Dwarf Boa	Bailey (1937); Schwartz & Garrido (1975); Schwartz & Henderson (1985, 1991); Tolson & Henderson (1993); McDiarmid et al. (1999); Estrada & Ruibal *in* Crother (1999)	Cuba, known only from the type locality in e Camagüey Province (6 miles east of Marti); and, 24 km sw of the city of Camagüey.
T. n. hardyi* Schwartz & Garrido, 1975 Hardy's Dwarf Boa	Schwartz & Garrido (1975); Schwartz & Henderson (1985, 1991); Tolson & Henderson (1993); McDiarmid et al. (1999); Estrada & Ruibal *in* Crother (1999) *Remark: Schwartz & Garrido (1975) suggested this may be a distinct species.	S Las Villas Province, Cuba from Soledad to about Trinidad; and s Cienfuegos Province.

Species-level Taxa	Citations and Remarks	Abbreviated Range
T. pardalis (Gundlach, 1840) Leopard Dwarf Boa	Schwartz & Henderson (1985, 1991); Tolson & Henderson (1993); McDiarmid et al. (1999); Estrada & Ruibal *in* Crother (1999)	Throughout Cuba and nearby islands, Isla de la Juventud and the Archipelago de Sebana-Cumagüey; it is less commonly encountered in the east.
T. paucisquamis (Müller *in* Schenkel, [1900] 1901) Peruvian or Brazilian Dwarf Boa	Amaral (1930g); Bogert (1969); Peters & Orejas-Miranda (1970); Amaral (1976); McDiarmid et al. (1999); Marques et al. (2001)	NE Peru into the states of Espirito Santo, Rio de Janeiro, and São Paulo in se Brazil.
T. pilsbryi Bailey, 1937 Cuban White-necked Dwarf Boa or Pilsbry's Dwarf Boa	Bailey (1937); Schwartz & Henderson (1985, 1991); Tolson & Henderson (1993); McDiarmid et al. (1999); Estrada & Ruibal *in* Crother (1999)	E and c Cuba.
T. p. pilsbryi Bailey, 1937 Pilsbry's Dwarf Boa	Schwartz & Garrido (1975); Schwartz & Henderson (1985, 1991); Tolson & Henderson (1993); McDiarmid et al. (1999); Estrada & Ruibal *in* Crother (1999)	E Cuba from the type locality in Santiago de Cuba Province, of Santa Faz near San Vicente, and Guantanamo in Oriente Province.
T. p. galacelidus Schwartz & Garrido, 1975 La Villas Province Dwarf Boa	Schwartz & Garrido (1975); Schwartz & Henderson (1985, 1991); Tolson & Henderson (1993); McDiarmid et al. (1999); Estrada & Ruibal *in* Crother (1999)	Cuba, s Cienfuegos and Sancti Spíritus Provinces, Las Villas Province, and in and adjacent to the Sierra de Trinidad.
T. semicinctus (Gundlach & W.Peters *in* W. Peters [1864] 1865) Yellow-banded Dwarf Boa	Arrlington (1927); Schwartz & Henderson (1985, 1991); Tolson & Henderson (1993); McDiarmid et al. (1999); Estrada & Ruibal *in* Crother (1999)	W and c Cuba in Pinar del Río Province eastward to Sancti Spíritus Province, northward to Matanzas Province, and to near Sagua la Grande, Villa Clara Province in the northeast.
T. spiritus Hedges & Garrido, 1999 Sancti Spíritus Dwarf Boa	Hedges & Garrido (1999)	C Cuba.
T. taczanowskyi (Steindachner, 1880) Amazon Dwarf Boa	Stull (1928); Peters & Orejas-Miranda (1970); Pérez-Santos & Moreno (1991) Tolson & Henderson (1993); McDiarmid et al. (1999)	Amazonian nw South America including Peru, Ecuador and possibly Amazonian Brazil; however, McDiarmid et al. (1999) could not find any records from Brazil and did not include the country in the range for this snake.
T. wrighti Stull, 1938 Wright's Dwarf Boa	Stull (1928); Schwartz & Henderson (1985, 1988); Tolson & Henderson (1993); McDiarmid et al. (1999); Estrada & Ruibal *in* Crother (1999)	EC Cuba in Camagüey Province eastward into Santiago de Cuba Province; Schwartz & Henderson (1988) noted that two specimens from the dolines at the cuevas de Caguanes, Villa Clara Province appear to be this snake.

GENUS: *Ungaliophis* Müller, 1880, is found from southern Mexico through the Pacific versant of Central America into northern South America in Colombia. It can be found from sea level to about 2300 m. See Smith & Taylor (1966); Bogert (1968); Peters & Orejas-Miranda (1970); Corn (1975); Villa et al. (1988); Williams & Wallach (1989); Villa & Wilson (1990); Smith & Smith (1993); and McDiarmid et al. (1999). Pough et al. (1998) included *Ungaliophis* as a boid.

Review assistance and important contributions for this section were provided by Michael J. Forstner.

CONTENT: Two species. ★★★

Ungaliophis: Central American Dwarf Boa or Banana Boa

Species-level Taxa	Citations and Remarks	Abbreviated Range
U. continentalis Müller, 1880 Isthmian Dwarf Boa or Boilla	Conant (1966); Smith & Taylor (1966); Bogert (1969); Peters & Orejas-Miranda (1970); Savage (1980); Alvarez del Toro (1982); Wilson & Meyer (1985); Villa et al. (1988); Villa & Wilson (1990); Smith & Smith (1993); Burger (1996); McDiarmid et al. (1999); Köhler (1997, 2001)	Extreme s Mexico in the Pacific coastal plain and Meseta Central of Chiapas southward along the Pacific versant into Guatemala and Honduras to as far south as Matagalpa, Nicaragua; from low to intermediate elevations to about 2100 m.
U. panamensis* Schmidt, 1933 Panamanian Dwarf Boa	Schmidt (1933); Bogert (1969); Peters & Orejas-Miranda (1970); Corn (1975); Savage & Villa (1986); Ville et al. (1988); Villa & Wilson (1990); McDiarmid et al. (1999); Köhler (2001)	S Nicaragua into Panama and w Colombia into the Andean foothills; from sea level to about 2100 m.
	*Remark: Peters & Orejas-Miranda (1970) stated *U. danieli* Prado, 1940 is a synonym.	

Advanced Snakes—The CAENOPHIDIA Oppel, 1811

As noted earlier, the Alethinophidia are divided into two groups, the Haenophidia and the Caenophidia. The Caenophidia are composed of the groups usually referred to as Colubrids, Elapids, and Vipers (including the Pit Vipers) and are considered to be the advanced or modern snakes. The Colubrids are listed here as the generalized snakes in the advanced group.

Family COLUBRIDAE Oppel, 1811

The descriptive, generalized snakes in the advanced group is used because there is a generally accepted lack of, or cloudy, subfamily classification. The reader should think about this group as the group of genera listed in the family Colubridae. This is now the large, complex group of snakes with a worldwide distribution, that in the Americas, excludes the Blindsnakes, basal Alethinophidians, Elapidae and Pit Vipers (Viperidae or Crotalinae). Relegation of the genera in this group to family or subfamily status has not been attempted because of the number of species contained in the group, and the extremely large variation in morphological differences (such as genetics, soft morphology, habits, environment, physical traits, etc.). I was not prepared to undertake the monumental task of verifying these components for this reference at this time. Ernst & Zug (1996) noted that the Colubridae is sometimes called the "trash-can group," because many unrelated subgroups are thrown together and biologists cannot decide where else to classify them.

CONTENT: In the Americas this group is composed of 143 genera and 870 species.

GENUS: *Adelophis* Dugès *in* Cope, 1879, is restricted to Mexico. See Taylor (1942); Smith & Taylor (1945, 1966); Rossman & Blaney (1968); Rossman & Wallach (1987); Williams & Wallach (1989); and Smith & Smith (1993).

Review assistance for this section was provided by Douglas A. Rossman.

The Caenophidia

Note: I was concerned with the wide gap between the ranges of the two species. Douglas Rossman (pers. comm. 5 June, 2000) confirmed that there is a large geographic discontinuity between the ranges of *A. copei* and *A. foxi*. Dr. Rossman believes that future fieldwork may or may not serve to reduce the gap in distribution.

CONTENT: Two species. ★★★★

Adelophis: Mountain Meadow Snake

Species-level Taxa	Citations and Remarks	Abbreviated Range
A. copei Dugès *in* Cope, 1879 Cope's Mountain Meadow Snake	Smith & Taylor (1945, 1966); Rossman & Blaney (1968); Rossman (1985); Rossman & Wallach (1987); Smith (1987); Smith & Smith (1993)	Jalisco, Guanajuato, Michoacán, and Morelos in Mexico.
A. foxi Rossman & Blaney, 1968 Fox's Mountain Meadow Snake	Rossman & Blaney (1968); Wilson & McCranie (1979); Rossman & Wallach (1987); McCranie & Wilson (1990); Smith & Smith (1993)	Mexico, only from the type locality near El Salto, sw Durango; associated with a meadow in pine forests.

GENUS: *Adelphicos* Jan, 1862, is distributed in a disjunct range from se Mexico into the se Guatemalan highlands, Belize, and Honduras. See Schmidt (1953); Smith & Taylor (1966); Peters & Orejas-Miranda (1970); Campbell & Ford (1982), Wilson & Meyer (1982, 1985); Campbell & Brodie (1988); Smith & Smith (1993); Campbell (1998a); Smith et al. (2000); Lee (2000); and Köhler (2001). Note that according to La Duc (1995), *Adelphicos* is neuter, but the more traditional nomenclature gender was followed by Campbell (1998a). This work follows La Duc (1995) nomenclature recommendations.

Review assistance for this section was provided by Travis J. LaDuc.

CONTENT: Six species. There does not seem to be universal agreement among authorities about the taxonomy of this genus. ★★★

Adelphicos: Middle American Burrowing Snake. Frank and Ramus (1995) call this the American Burrowing Snake.

Species-level Taxa	Citations and Remarks	Abbreviated Range
A. daryi Campbell & Ford, 1982 Highland's Burrowing Snake	Campbell & Ford (1982); Campbell & Brodie (1988); Villa et al. (1988); Köhler (2001)	SE Guatemalan highlands from 1830 to 2135 m.
A. ibarrorum Campbell & Brodie, 1988 Ibarra's Burrowing Snake	Campbell & Brodie (1988); Campbell & Vannini (1989); Smith & Smith (1993)	SC Guatemalan highlands from 2000 to 2100 m.
A. latifasciatum Lynch & Smith, 1966 Oaxacan Burrowing Snake	Lynch & Smith (1966); Peters & Orejas-Miranda (1970); Campbell (1984); Campbell & Brodie (1988); Campbell & Ford (1982); Villa et al. (1988); Johnson (1989); Smith & Smith (1993); La Duc (1995)	SE Oaxaca in Mexico. It is associated with the Sierra Madre de Chiapas from 1500 to 2000 m.

Species-level Taxa	Citations and Remarks	Abbreviated Range
A. nigrilatum Smith, 1942 La Octera	Smith (1942d); Smith & Taylor (1966); Peters & Orejas-Miranda (1970); Campbell & Ford (1982); Alvarez del Toro (1982); Villa et al. (1988); Campbell & Brodie (1988); Johnson (1989); La Duc (1995)	Meseta Central of s Mexico.
A. quadrivirgatum* Jan, 1862 Four-lined Burrowing Snake or Zacatera	Smith & Taylor (1966); Wilson & Meyer (1982, 1985); Alvarez del Toro (1982); Villa et al. (1988); Campbell & Brodie (1988); Johnson (1989); La Duc (1995); Campbell (1998a); Stafford & Meyer (2000); Smith et al. (2000)	SE Mexico in c Veracruz and e Chiapas southward through Belize and Guatemala into n Honduras.
	*Remark: Wilson & Meyer (1985), rejected subspecies because they concluded that a restudy of the species "will demonstrate a complex picture that cannot be adequately handled within the bounds of the subspecies concept." Subspecies are listed here because they are still used in literature. Smith et al. (2001) reconfirmed subepcies of *A. q. quadrivirgattus* and *A. q. newmanorum* with *A. sargii* and *A. visioninus* as full species. Their justification was not detailed so I have maintained all as subspecies in this work.	
A. q. quadrivirgatum Jan, 1862 Jan's or Four-lined Burrowing Snake	Smith & Taylor (1966); Peters & Orejas-Miranda (1970); Wilson & Meyer (1982, 1985); Alvarez del Toro (1982); Campbell & Ford (1982); Johnson (1989, 1990); Pérez-Higareda & Smith (1989, 1991); Mendelson (1990); La Duc (1995); Lee (1996); Smith et al. (2001)	SE Mexico, Yucatán, Belize, Guatemala and Honduras.
A. q. newmanorum Taylor, 1950 Zacatera Roja or Newman's Burrowing Snake	Taylor (1950); Martin (1955); Smith & Taylor (1966); Wilson & Meyer (1982, 1985); Contreras-Arquieta (1989); Smith & Smith (1993); La Duc (1995)	A restricted distribution in Mexico that includes Tamaulipas, Queretaro, San Luis Potosí, and a specimen outside of this range in Nuevo León. David Lazcano V. (pers. comm. 23 June 2000) reported that the specimen outside of Monterrey came from the south side of the municipal park La Estanzuela, located beside the Sierra Madre, south of Monterrey.
A. q. sargii (Fischer, 1885) Sargi's Burrowing Snake	Slevin (1939, 1942d); Davis & Smith (1953); Smith & Taylor (1966); Peters & Orejas-Miranda (1970); Alvarez del Toro (1982); Wilson & Meyer (1985); Smith & Smith (1993); La Duc (1995); Smith et al. (2001)	Foothills on the Pacific slopes from s Chiapas in Mexico and along the Pacific versant of Guatemala.
A. q. visoninum (Cope, 1866) Cope's Burrowing Snake	Schmidt (1941, 1942d); Smith & Taylor (1966); Peters & Orejas-Miranda (1970); Alvarez del Toro (1982); Campbell & Ford (1982); Wilson & Meyer (1982, 1985); McCranie & Wilson (1991a); La Duc (1995); Lee (1996, 2000); Campbell (1998a); Smith et al. (2001)	Foothills on the Atlantic slopes from Tabasco in Mexico southward into Honduras; its range includes the southern part of the Yucatán Peninsula. It has recently been reported from Oaxaca, Mexico.
A. veraepacis Stuart, 1941 Verapaz Burrowing Snake	Stuart (1941a, 1942d); Peters & Orejas-Miranda (1970); Campbell & Ford (1982); Campbell & Brodie (1988); Villa et al. (1988); Campbell & Vannini (1989); Campbell (1998a)	Highlands in s Mexico into Alta Verapaz in Guatemala. It is assoicated with Montañas de Cuilco in the Sierra de las Minas from 1200 to 2200 m.

GENUS: *Alsophis* Cope, 1862, is wide ranging in the Caribbean, western South America, and the Galapagos Islands. *Alsophis* is distinguished from *Arrhyton* by having 2 pits on the outer margins of the dorsal scales. I consider scale pits to be a weak characteristic to use for separting the two genera when other characteristics are not used since other genera such as *Liophis* and *Hydrodynastes* ignore scale pits and synonymize two more genera. Mayer & Lazell (1988) decided that *Liophis* = *Alsophis* and *Liophis* = *Arrhyton*. They did not say why they synonmyzed *Alsophis* and *Arrhyton* with *Liophis*. Fide Vanzolini (1986) *sensu* Peters & Orejas-Miranda (1970); *Dromicus* Bibron, 1843 = *Alsophis* Cope, 1862. Fide Thomas (1977), many of the South American *Alsophis* are referred to the genus *Philodryas*. See Maglio (1970); Peters & Orejas-Miranda (1970); Henderson & Schwartz (1984); Schwartz & Henderson (1985, 1991); Thomas (1997); and Powell & Henderson (1998).

Review assistance for this section was provided by Robert A. Thomas

See also genera *Arrhyton* and *Philodryas*.

CONTENT: Thirteen species. Many may have become, or are becoming extinct because of their restricted ranges on small islands. ★★★

Alsophis: Neotropical Racer

Species-level Taxa	Citations and Remarks	Abbreviated Range
A. anomalus (W. Peters, 1863) Hispaniolan Brown Racer	Cochran (1941); Maglio (1970); Henderson & Schwartz (1984); Schwartz & Henderson (1991); Powell & Henderson (1998); Powell et al. *in* Crother (1999)	Hispaniola in both Haiti and the w Dominican Republic, Île de la Tortue, and Isla Beata.
A. antiguae (Parker, 1933) Antiguan Racer	Parker (1933); Lazell (1967); Schwartz & Henderson (1991); Henderson et al. (1996); Censky & Kaiser *in* Crother (1999)	Antigua and satellite islands, Great Bird Island; Schwartz & Henderson (1991) noted that it is now only known from the 0.2 km^2 islet of Great Bird Island, only 2.5 miles from mongoose infested Antigua; in the Lesser Antilles.
A. a. antiguae (Parker, 1933) Antiguan Racer	Parker (1933); Underwood (1962); Maglio (1970); Henderson (1990); Schwartz & Henderson (1991); Henderson et al. (1996)	Antigua in the Lesser Antilles of the Caribbean; these snakes may be extinct.
A. a. sajdaki Henderson, 1990 Great Bird Island Racer	Henderson (1990); Schwartz & Henderson (1991); Henderson et al. (1996)	Great Bird Island near Antigua in the Lesser Antilles.
A. antillensis (Schlegel, 1837) Leeward Racer	Barbour (1930); Maglio (1970); Schwartz & Henderson (1985, 1991); Censky & Kaiser *in* Crother (1999)	Caribbean in the Lesser Antilles on Guadeloupe, Marie-Galante Guadeloupe, Marie-Galante, Antigua, Les Saintes, Montserrat, and Dominica.
A. a. antillensis (Schlegel, 1837) Guadeloupe Racer	Schlegel, 1837; Maglio (1970); Schwartz & Henderson (1985,1991)	Guadeloupe and Marie-Galante in the Lesser Antilles.
A. a. danforthi Cochran, 1938 Danforth's Racer	Cochran (1938); Schwartz & Henderson (1985, 1991)	Terre-de-Bas and Îles des Saintes in Haiti.

Species-level Taxa	Citations and Remarks	Abbreviated Range
A. a. manselli Parker, 1933 Montserrat Racer	Parker (1933); Underwood (1962); Maglio (1970); Schwartz & Henderson (1985, 1991)	Montserrat in the Lesser Antilles.
A. a. sanctorum Barbour, 1915 Îles des Saintes Racer	Barbour (1930); Underwood (1962); Maglio (1970); Schwartz & Henderson (1985, 1991)	Known only from the type locality on Terre-de-Haut and Îles des Saintes in Haiti.
A. a. sibonius Cope, 1879 Dominica Racer	Cope (1879); Underwood (1962); Maglio (1970); Schwartz & Henderson (1985, 1991)	Dominica in the Lesser Antilles.
A. ater* (Gosse, 1851) Jamaican Tree Snake or Jamaican Racer	Cochran (1940); Maglio (1970); Schwartz & Henderson (1985, 1991); Henderson & Powell (1996a); Crombie *in* Crother (1999) *Remark: Thought to be extinct since 1960 and only two specimens are known in collections. Henderson & Powell (1996a) noted that although this species apparently was once common on Jamaica (Gosse, 1851), it is likely extinct.	Jamaica in the western Caribbean.
A. biserialis (Günther, 1860) Galápagos Racer	Mertens (1960); Maglio (1970); Thomas (1997); Swash & Still (2000)	In the Galápagos islands of Ecuador on the islands of Baltra, Bertólome, Champion, Fernandina, Gardner (near Floreana), Isabela, Rábida, San Cristóbal, Santiago, Santa Cruz, Santa Fé, Floreana, and Tortuga.
A. b. biserialis (Günther, 1860) Eastern Galápagos Racer	Mertens (1960) [as *Dromicus b. eibli*]; Thomas (1997); Swash & Still (2000) [as *Dromicus b. eibli*]	Found on Champion, Gardner (near Floreana), Floreana, and San Cristóbal in the Galápagos Islands.
A. b. dorsalis (Steindachner, 1876) Central Galápagos Racer	Beebe (1924); Mertens (1960); Maglio (1970); Thomas (1997); Swash & Still (2000)	Found on Baltra, Bertólome, Rábida, Santiago, Santa Cruz, and Santa Fé in the Galápagos islands.
A. b. occidentalis (Van Denburgh, 1912) Western Galápagos Racer	Mertens (1960) [as *dorsalis* and *helleri*]; Thomas (1997); Swash & Still (2000)	Found on Fernandina, Isabela, and Tortuga in the Galápagos Islands.
A. cantherigerus (Bibron, 1843) Cuban Racer	Grant (1941); Maglio (1970); Schwartz & Henderson (1985, 1991); Kliment (1993); Estrada & Ruibal *in* Crother (1999)	Cuba and its satellites into the Cayman Islands including Isla de la Juventud, Grand Cayman, Cayman Brac, and Little Cayman.
A. c. cantherigerus (Bibron, 1843) Western Cuba Racer	Maglio (1970); Schwartz & Henderson (1985, 1991); Estrada & Ruibal *in* Crother (1999)	Western Cuba and some satellites including Isla de la Juventud, Archipiélago de los Canarreos, Archipiélago de Sabana-Camagüey, and Cayos de San Felipe.
A. c. adspersus (Gundlach & Peters, 1864) Eastern Cuba Racer	Maglio (1970); Schwartz & Henderson (1985, 1991); Estrada & Ruibal *in* Crother (1999)	Extreme e Cuba.

Species-level Taxa	Citations and Remarks	Abbreviated Range
A. c. brooksi (Barbour, 1914) Swan Island Racer	Barbour (1914); Maglio (1970); Schwartz & Henderson (1985, 1991)	Little Swan Island and Swan Island in the w Caribbean; these islands belong to Honduras, and are east of Belize and north of Honduras.
A. c. caymanus (Garman, 1887) Grand Cayman Racer	Maglio (1970); Schwartz & Henderson (1985, 1991)	Grand Cayman Island in the w Caribbean, south of Cuba.
A. c. fuscicaudus (Garman, 1887) Cayman Brac Racer	Maglio (1970); Schwartz & Henderson (1985, 1991)	Cayman Brac in the Caribbean, south of Cuba.
A. c. pepei (Schwartz & Thomas, 1960) Northern Cuba Racer	Schwartz & Thomas (1960); Maglio (1970); Schwartz & Henderson (1985, 1991); Estrada & Ruibal *in* Crother (1999)	N Cuba north of the mesic coast of Guantánamo and Holguín provinces.
A. c. ruttyi (Grant, 1941) Little Cayman Racer	Grant (1941); Maglio (1970); Schwartz & Henderson (1985, 1991)	Little Cayman in the Caribbean, south of Cuba.
A. c. schwartzi (Lando & Williams, 1969) Schwartz's Racer	Lando & Williams (1969); Maglio (1970); Schwartz & Henderson (1985, 1991); Estrada & Ruibal *in* Crother (1999)	Cuba from sc Sancti Spíritus Province eastward through Camagüey Province, Holguín, Granma, and Santiago de Cuba provinces, to Santiago de Cuba, and to Felicidad in the interior mountains of Guantánamo Province.
A. elegans* Western South American Neotropical Racer or Andean Racer	Donoso-Barros (1966); Peters & Orejas-Miranda (1970); Vanzolini (1986); Thomas & Ineich (1999) *Remark 1: Fide Thomas (1977b), this was reclassified from *Philodryas elegans* (Tschudi) to *Alsophis elegans* (Tschudi). *Remark 2: Fide Vanzolini (1986), and, Myers & Hoogmoed (1974); *Tachymenis surinamensis* Dunn, 1922 is a junior synonym. *Remark 3: Fide Thomas (1999 pers. comm.), *T. elongata* (Despax, 1910) as known only from the type locality in Tablaza de Payta in Peru and is a junior synonym fide Myers & Hoogmoed (1974).	W South America in Ecuador, Peru and Chile.
A. e. elegans (Tschudi, 1845) Elegant Green Racer	Donoso-Barros (1966); Peters & Orejas-Miranda (1970)	Rimac Valley in s Peru southward into n Chile.
A. e. rufidorsatus* (Günther, 1858) Guayaquil Racer	Donoso-Barros (1966); Peters & Orejas-Miranda (1970); Vanzolini (1986); Pérez-Santos & Moreno (1991) *Remark: According to Vanzolini (1986), this is *P. e. rufodorsatus* not *P. e. rufidorsatus* as found in Peters & Orejas-Miranda (1970).	Guayaquil in s Ecuador southward into Departamento Libertad in n Peru.

Species-level Taxa	Citations and Remarks	Abbreviated Range
*A. melanichnus** Cope, 1862 La Vega Racer	Cochran (1941); Maglio (1970); Henderson & Schwartz (1984); Schwartz & Henderson (1991); Powell & Henderson (1998); Powell et al. *in* Crother (1999) *Remark: Thought to be extremely rare or extinct. A literature search suggests that its last appearance was 1910.	Hispaniola where it is rare in both Haiti and Dominican Republic; it is known only from the type locality of near Jérémie in Haiti and La Vega Province in the Dominican Republic.
A. portoricensis (Reinhardt & Lütken, 1863) Puerto Rican Racer	Schwartz & Henderson (1985, 1991); Mayer & Lazell (1988); Rivero (1998); Thomas *in* Crother (1999)	Puerto Rico and Virgin Islands along with their satellites including Andros, Vieques, Buck Island, Isla Desecheo, St. Thomas, and Mona Island; it is found in both the U.S. and British Virgin Islands.
A. p. portoricensis (Reinhardt & Lütken, 1863) Puerto Rican Racer	Maglio (1970); Schwartz & Henderson (1985, 1991); Rivero (1998)	Widely distributed in Puerto Rico including Cayo Santiago; it is not found in the southern third, or extreme western part of Puerto Rico.
A. p. anegadae Barbour, 1917 British Virgin Islands Racer	Maglio (1970); Schwartz & Henderson (1985, 1991); Mayer & Lazell (1988)	British Virgin Islands on Guana Island, Nackas Island, Virgin Gorda, and Anegada, it is probably also found on Tortola.
*A. p. aphantus** Schwartz, 1966 Vieques Island Racer	Schwartz (1966); Maglio (1970); Schwartz & Henderson (1985, 1991); Rivero (1998) *Remark: This subspecies may be extinct.	Isla Vieques near Puerto Rico in the Caribbean.
A. p. nicholsi Grant, 1937 Buck Island Racer	Grant (1937); Maglio (1970); Schwartz & Henderson (1985, 1991); Kliment (1993)	Buck Island of the Capella Islands south of St. Thomas in the U.S. Virgin Islands.
A. p. prymnus Schwartz, 1966 Southern Puerto Rico Racer	Schwartz (1966); Maglio (1970); Schwartz & Henderson (1985, 1991); Rivero (1998)	S Puerto Rico including Caja de Muertos.
A. p. richardi Grant, 1946 Richard's Racer	Grant (1946); Brongersma (1959); Maglio (1970); Schwartz & Henderson (1985, 1991); Rivero (1998)	Isla Culebra, St.Thomas and its satellites, Lovango Cay, Peter Island, and Salt Island in the U.S. Virgin Islands.
A. p. variegatus (Schmidt, 1926) Mona Island Racer	Maglio (1970); Schwartz & Henderson (1985, 1991); Rivero (1998)	Isla Mona and probably Isla Desecheo near Puerto Rico.
A. rijersmai (Cope, 1870) Anguilla Bank Racer	Barbour (1930); Brongersma (1959); Underwood (1962); Maglio (1970); Schwartz & Henderson (1985, 1991); Censky & Kaiser *in* Crother (1999); Townsend et al. (2000)	Anguilla, St. Martin, St. Barthélémy, and Scrub Island in the Lesser Antilles.
A. rufiventris (Dumeril, Bibron & Dumeril, 1854) Saba Racer	Barbour (1930); Brongersma (1959); Underwood (1962); Maglio (1970); Henderson & Sadjak (1986); Schwartz & Henderson (1985, 1991); Censky & Kaiser *in* Crother (1999)	Endemic to the Anguilla Bank where it is reported from Saba, St. Eustatius, St. Christofer, St. Kits, and Nevis in the Lesser Antilles.

Species-level Taxa	Citations and Remarks	Abbreviated Range
A. sanctaecrucis* (Cope, 1862) St. Croix Racer	Barbour (1930); Underwood (1962); Schwartz (1966); Maglio (1970); Schwartz & Henderson (1985, 1991); Henderson & Powell (1996); Thomas *in* Crother (1999) *Remark: Thought to be extinct since 1950.	St. Croix and the satellite islands in the U.S. Virgin Islands; it may also occur on Green Cay.
A. vudii (Cope, 1862) West Indies Racer	Barbour & Shreve (1935); Maglio (1970); Schwartz & Henderson (1985, 1991)	West Indies on New Providence, Eleuthera, Cat Island, Long Island, Exuma, Cays, Andros, Grand Bahama, Great Abaco, Bimini Islands, Crooked Island, Acklin's Island, and Great Inagua.
A. v. vudii (Cope, 1862) West Indies Racer	Barbour & Shreve (1935); Maglio (1970); Schwartz & Henderson (1985, 1991); Kliment (1993)	West Indies in the Bahama Islands including New Providence, Paradise Cay, Eleuthera Island, Cat Island, Long Island, Exuma Cays, Green Cay, Berry Island, Little Ragged Island, and Andros Island; associated with the Great Bahama Bank but not found on Bimini Island, Cat Island or Little San Salvador Island.
A. v. aterrimus Barbour & Shreve, 1935 Grand Bahamas Racer	Barbour & Shreve (1935); Maglio (1970); Schwartz & Henderson (1985, 1991)	Grand Bahama Island and Great Abaco Island in the West Indies.
A. v. picticeps Conant, 1937 Bimini Racer	Conant (1937); Maglio (1970); Schwartz & Henderson (1985, 1991)	Bimini in the Bahama Islands on North, South, and East Bimini.
A. v. raineyi Barbour & Shreve, 1935 Crooked Island Racer	Barbour & Shreve (1935); Maglio (1970); Schwartz & Henderson (1985, 1991)	Bahamas in Crooked Island and Acklin's Island in the West Indies.
A. v. utowanae* Barbour & Shreve, 1935 Great Inagua Racer	Maglio (1970); Schwartz & Henderson (1985, 1991) *Remark: Maglio (1970) suggested this may be a distinct full species as *A. utowanae*.	Great Inagua in the Bahamas, including Sheep Cay in the West Indies.

GENUS: *Amastridium* Cope, 1860 (1861), is a rare snake found in southern Mexico from Nuevo León and Tamaulipas in the Atlantic versant, and Oaxaca through Chiapas in the Pacific versant, southward into Central America to n Guatemala and Honduras and into Panama. See Smith & Taylor (1966); Wilson & Meyer (1969, 1985); Wilson *in* Peters & Orejas-Miranda (1970); Wilson & Robinson (1971); Henderson & Hoevers (1975); Savage & Villa (1986); Wilson (1988a); Smith & Smith (1993); Liner (1994); Lee (1996, 2000); and Köhler (2001). See Wilson (1988a) for a detailed discussion of the species nomenclature of this genus.

CONTENT: It seems that most authorities accept a single species for this genus with no subspecies. However, this does not seem to be universally accepted. The work follows the single species philosophy of Wilson (1988a). ★★★

Amastridium: Rusty-headed Snake

Species-level Taxa	Citations and Remarks	Abbreviated Range
A. veliferum* Cope, 1861 Rusty-headed Snake	Wilson & Meyer (1969, 1985); Wilson *in* Peters & Orejas-Miranda (1970); Wilson & Robinson (1970); Wilson et al. (1986); Wilson (1988a); Villa et al.	S Mexico from Nuevo León southward into Central America to as far as n Panama.

Species-level Taxa	Citations and Remarks	Abbreviated Range
	(1988); Pérez-Higareda & Smith (1991); Lee (1996, 2000); Campbell (1998a); Stafford & Meyer (2000); Köhler (2001) [as both *A. sapperi* and *A. veliferum*]	
	*Remark: The "southern form" is sometimes shown as the subspecies *A.v. veliferum*. Fide Larry David Wilson *in* Peters & Orejas-Miranda (1970), *A. veliferum sapperi* (Werner, 1903), taxon is a variant of *A. veliferum*; or this is sometimes considered to be the "northern form." However, Wilson & Meyer (1985), referred back to Wilson & Meyer (1969), where they demonstrated that characteristics used to distinguish between *sapperi* and *veliferum* are discordant and that there are no additional characteristics that will serve to distinguish the two. Lee (2000) recently showed subspecies but qualified the comment with "northern and southern segments sometimes considered subspecifically distinct,. . . . ."	

GENUS: *Antillophis* Maglio, 1970, is from the Caribbean. See Maglio (1970); Henderson & Schwartz (1984); and Schwartz & Henderson (1985, 1991).

CONTENT: Four species. ★★★

Antillophis: Antillean Racer. Frank and Ramus (1995) call this the Antillean Snake.

Species-level Taxa	Citations and Remarks	Abbreviated Range
A. andreae (Reinhardt & Lütken, 1863) Black and White Racer	Maglio (1970); Schwartz & Henderson (1985, 1991); Estrada & Ruibal *in* Crother (1999)	In and around Cuba including Isla de la Juventud.
A. a. andreae (Reinhardt & Lütken, 1862) Black and White Racer	Maglio (1970); Schwartz & Henderson (1985, 1991); Estrada & Ruibal *in* Crother (1999)	W and c Cuba to extreme now Camagüey Province.
A. a. melopyrrha (Thomas & Garrido, 1967) Cayo Cantiles Racer	Maglio (1970); Schwartz & Henderson (1985, 1991); Estrada & Ruibal *in* Crother (1999)	Cayo Cantiles in Cuba.
A. a. morenoi Garrido, 1973 Moreno's Racer	Schwartz & Henderson (1985, 1991); Estrada & Ruibal *in* Crother (1999)	Cayo Santa Maria and Cayo Guahaba in Cuba.
A. a. nebulatus (Barbour, 1916) Isla de la Juventud Racer	Schwartz & Thomas (1960); Maglio (1970); Schwartz & Henderson (1985, 1991); Estrada & Ruibal *in* Crother (1999)	Isla de la Juventude in Cuba.

The Caenophidia

Species-level Taxa	Citations and Remarks	Abbreviated Range
A. a. orientalis (Barbour & Ramsden, 1919) Oriente Racer	Schwartz & Thomas (1960); Maglio (1970); Schwartz & Henderson (1985, 1991); Estrada & Ruibal *in* Crother (1999)	Throughout Oriente Province in w Cuba.
A. a. peninsulae (Schwartz & Thoms, 1960) Peninsular Racer	Schwartz & Thomas (1960); Maglio (1970); Schwartz & Henderson (1985, 1991); Estrada & Ruibal *in* Crother (1999)	Peninsula of Guanahacabibes in Pinar del Rio Province in c Cuba.
A. parvifrons (Cope, 1862) Hispaniolan Black Racer	Maglio (1970); Henderson & Schwartz (1984); Schwartz & Henderson (1985, 1991); Powell et al. *in* Crother (1999)	In and around Hispaniola including Long Island, and into the Bahamas (status uncertain).
A. p. parvifrons (Cope, 1862) Common Hispaniolan Black Racer	Maglio (1970); Henderson & Schwartz (1984); Schwartz & Henderson (1991); Powell et al. *in* Crother (1999)	Hispaniola in Haiti from the Tiburon Peninsula eastward into about Baradères, Île Grande Cayemite, and Grosse Cay.
A. p. alleni (Dunn, 1920) Allen's Racer	Maglio (1970); Henderson & Schwartz (1984); Schwartz & Henderson (1991); Powell et al. *in* Crother (1999)	Île de la Gonâve and Île Petite Gonâve in Haiti.
A. p. lincolni (Cochran, 1931) Lincoln's Racer	Maglio (1970); Henderson & Schwartz (1984); Schwartz & Henderson (1991); Powell et al. *in* Crother (1999)	From Sierra de Baoruco, Dominican Republic to Jacmel, Haiti; including Isla Beata and Peninsula de Barahona.
A. p. niger (Dunn, 1920) Peninsula de Samaná Racer	Maglio (1970); Henderson & Schwartz (1984); Schwartz & Henderson (1991); Powell et al. *in* Crother (1999)	Peninsula de Samaná, Dominican Republic.
A. p. paraniger (Thomas & Schwartz, 1965) Dominican Republic Racer	Maglio (1970); Henderson & Schwartz (1984); Schwartz & Henderson (1991); Powell et al. *in* Crother (1999)	SE Dominican Republic and Isla Catalina.
A. p. protenus (Jan, 1867) Common Hispaniolan Racer	Maglio (1970); Henderson & Schwartz (1984); Schwartz & Henderson (1991); Powell et al. *in* Crother (1999)	Throughout Hispaniola except where the subspecies *parvifrons, licolni, niger, lincolni* and *paraniger* are found.
A. p. rosamondae (Cochran, 1934) Île-à-Vache Racer	Maglio (1970); Henderson & Schwartz (1984); Schwartz & Henderson (1991); Powell et al. *in* Crother (1999)	Île-à-Vache in Haiti.
A. p. stygius (Thomas & Schwartz, 1965) Isla Saóna Racer	Maglio (1970); Henderson & Schwartz (1984); Schwartz & Henderson (1991); Powell et al. *in* Crother (1999)	Isla Saona in the Dominican Republic.
A. p. tortuganus (Dunn, 1920) Tortuga Island Racer	Maglio (1970); Henderson & Schwartz (1984); Schwartz & Henderson (1991); Powell et al. *in* Crother (1999)	Îles de la Tortue, Haiti.

Species-level Taxa	Citations and Remarks	Abbreviated Range
A. slevini (Van Denburgh, 1912) Banded Galápagos Antillophis	Mertens (1960); Maglio (1970); Thomas (1997); Swash & Still (2000) [as *Alsophis s. slevini*]	Duncan (Pinzón), Fernandina, and Isabella Islands in the Galápagos Islands of Ecuador.
A. steindachneri (Van Denburgh, 1912) Striped Galápagos Antillophis	Mertens (1960); Maglio (1970); Thomas (1997); Swash & Still (2000) [as *Alsophis slevini steindachneri*]	Indefatigable (Santa Cruz), Balta, Rábida, and Santiago (San Salvador) Islands of the Galápagos Islands.

GENUS: *Apostolepis* Cope, 1862, is found in South America from Guyana and eastern Peru into the Chaco of Argentina, Paraguay and Bolivia. See Peters & Orejas-Miranda (1970); Lema (1978c, 1998); Lema & Renner (1998); Cei (1993), Giraudo & Scrocchi (1998); and Harvey (1999).

Review assistance and important contributions for this section were provided by Robert A. Thomas and Thales de Lema.

See also *Elapomorphus* and *Phalotris*.

CONTENT: Twenty-three species. ★★

Apostolepis: South American Wormsnake. Frank and Ramus (1995) call this the Burrowing Snake.

Species-level Taxa	Citations and Remarks	Abbreviated Range
A. ambiniger (W. Peters, 1869) Southern South American Wormsnake	Peters & Orejas-Miranda (1970); Lema (1978c, 1997b); Savage & Slowinski (1992); Yanosky (1998); Harvey et al. (2001)	Argentina, w Brazil, Bolivia, and Paraguay.
A. arenarius Rodrigues, [1992] 1993 Bahia Wormsnake	Rodrigues (1993); Lema (1997b)	Bahia, Brazil.
A. assimilis (Reinhardt, 1861) Reinhardt's Wormsnake	Hoge (1952a); Peters & Orejas-Miranda (1970); Amaral (1976); Cei (1993); Lema (1978c, 1997b); Savage & Slowinski (1992); Lema & Fernandes (1997); Giraudo & Scrocchi (1998)	C and sw Brazil and into the n Argentina Chaco and possibly Paraguay.
A. breviceps Harvey, Gonzales A. & Scrocchi, 2001 Short-headed Wormsnake	Harvey et al. (2001)	Izozog in the Gran Chaco of Santa Cruz, Bolivia.
A. cearensis Gomes, 1915 Gomes's Wormsnake	Peters & Orejas-Miranda (1970); Amaral (1976); Lema (1978c, 1997b); Savage & Slowinski (1992); Zamprogno et al. (1998)	NE Brazil in Piauí, Rio Grande do Norte, Bahia and Tocantins from the caatinga into the cerrado of central Brazil.
A. dimidiata* (Jan, 1862) Jan's Wormsnake	Cei (1993); Lema (1993, 1997b); Romano-Martinez (1996); Giraudo & Scrocchi (1998); Yanosky (1998) *Remark 1: Williams & Wallach (1989), and, Peters	Paraguay, n Argentina in s Misiones, ne Corrientes Formosa, and into c and s Brazil.

The Caenophidia

Species-level Taxa	Citations and Remarks	Abbreviated Range
	& Orejas-Miranda (1970) have this listed as *Elapomojus dimidiatus* Jan, 1862.	
	*Remark 2: Lema (1993, 1997b) included *A. borellii*, *A. barrioi* (Lema, 1978) from Paraguay; *A. ventrimaculata* Lema, 1978 from Paraguay, and, *A. villaricae* (Lema, 1978) from Paraguay all as synonyms. Yanosky (1998) listed all of them as valid in his checklist.	
	*Remark 3: Fide Lema (1997b), *A. ventrimaculata* Lema, 1978 from Paraguay, is a synonym.	
A. dorbignyi (Schlegel, 1837) Dorbigny's Wormsnake	Boulenger (1896); Peters & Orejas-Miranda (1970); Lema (1978c, 1997b); Savage & Slowinski (1992); Yanosky (1998); Harvey (1999); Harvey et al. (2001)	SW Brazil, Paraguay, Peru, and Bolivian Chaco.
A. flavotorquata* (Dumeril, Bibron & Dumeril, 1854) Central Wormsnake	Peters & Orejas-Miranda (1970); Lema (1978c, 1997b); Savage & Slowinski (1992); Lema & Fernandes (1997) *Remark: Lema & Renner (1998) classified the *A. flavotorquata* (in part) as described in Hoge (1958) to be included as a synonym in *A. pymi* Boulenger, 1903. This classification was assigned to those snakes in Suriname; Amazonas and Mato Grosso, Brazil.	C Brazil.
A. gaboi Rodrigues, [1992] 1993 San Francisco Wormsnake	Rodrigues (1993); Lema (1997b)	State of Bahia in Brazil.
A. goiasensis* Prado, 1942 Goiás Wormsnake	Prado (1943b); Peters & Orejas-Miranda (1970); Lema (1978c, 1997b); Harvey et al. (2001) *Remark: Sometimes seen as *A. goyasensis* as can be found in Lema (1997b).	State of Goiás, Brazil.
A. intermedia Koslowsky, 1898 Mato Grosso Wormsnake	Koslowsky (1898); Peters & Orejas-Miranda (1970); Lema (1997b)	State of Mato Grosso, Brazil and into Bolivia.
A. longicaudata Gomes *in* Amaral, 1921 Santa Philomena Wormsnake	Peters & Orejas-Miranda (1970); Amaral (1976); Lema (1978c, 1997b)	State of Piauí, Brazil, known only from the type locality of the municipality of Santa Philomena.
A. multicincta Harvey, 1999 Pampa Grande Wormsnake	Harvey (1999); Harvey et al. (2001)	The type was collected in the vicinity of Pampagrande, Provincia Florida, Departamento Santa Cruz, Bolivia.
A. niceforoi Amaral, 1935 Amazon Wormsnake	Nicéforo María (1942); Peters & Orejas-Miranda (1970); Vanzolini (1986); Pérez-Santos & Moreno (1988); Lema (1978c, 1997b)	Amazonian Colombia and Brazil.

Species-level Taxa	Citations and Remarks	Abbreviated Range
*A. nigrolineata** (W. Peters, 1869) Rainforest Wormsnake	Lema (1997a, 1997b); Lema & Renner (1998) *Remark: Upon re-examination of the holotype of both *Apostolepis nigrolineata* (Peters, 1869) and *Apostolepis pymi* Boulenger, 1903, Lema (1997a) synonymized *A. pymi* with *A. nigrolineata*. His work revealed that *nigrolineata* is the juvenile stage of *pymi*. According to the priority rule in nomenclature, *nigrolineata* has the priority. However, there is some confusion because Lema & Renner (1998) listed the status for *A. pymi* as valid but did not mention *A. nigrolineata*. As part of the validation, Lema & Renner (1998) listed *A. pymi*; *A. coronata* (Amaral 1929, 1936), Peters & Orejas-Miranda, 1970—in part) [from Teresópolis, Rio de Janiero, Brazil]; *A. quinquelineata* (from Hoge, 1958, Lema 1978, and Cunha & Nascimento, 1978—in part) [from Pará oriental, Brazil], and *A. quinquelineata* (in Nascimento & Lima-Verde, 1989) [from Serra do Baturité, Ceará]; and *A. flavotorquata* (Lema, 1878—in part) [from Suriname, Amazonas and Mato Grosso, Brazil] as synonyms. Dr. Lema tells me (pers. comm. 19 September 2001) that the Lema (1997a) was delayed with the publisher and it should be considered the latest information. This, he tells me, causes confusion because the Lema & Renner (1998) is actually older 1997 information than Lema (1997a).	Tropical rainforests in n Brazil, from e Amazonas to Ceará.
A. nigroterminata Boulenger, 1896 Peruvian Wormsnake	Peters & Orejas-Miranda (1970); Lema (1978c, 1997a); Harvey (1999)	E Peru into Bolivia and w Brazil.
A. phillipsi Harvey, 1999 Phillips's Wormsnake	Harvey (1999)	Type location is Estancia El Refugio, Provincia Velasco, Departamento Santa Cruz, Bolivia.
*A. polylepis** Amaral, 1922 Piuaí Wormsnake	Peters & Orejas-Miranda (1970); Amaral (1976); Williams & Wallach (1989); Lema (1997b) *Remark: This snake was listed as *Parapostolepis polylepis* Amaral (1921) fide Peters & Orejas-Miranda (1970) and Amaral (1976). Williams & Wallach (1989) reported that Cadle (1982) replaced *Parapostolepis* Amaral, 1930 in *Apostolepis* and they concurred with him.	State of Piuaí, Brazil, known only from the type locality of Município de Santa Filomena.
*A. quinquelineata** Boulenger, 1896 Guyana Wormsnake	Hoge (1952b, 1959a); Peters & Orejas-Miranda (1970); Cunha et al. (1985); Lema (1978c, 1997b, 1998); Lema & Renner (1998); Vidal et al. (1998) *Remark: Lema & Renner (1998) synonymized *A. rondoni* Amaral, 1925 and *A. coronata* Sauvage, 1877 (in part) with this snake. The *A. coronata* (in part) included those snakes from Guianas; Brazil in Pará and Mato Grosso.	The type is from Demerara, Guyana; however it is also associated with e Amazonas; Lema & Renner (1998) restrict it to w Amazonia in the w Pará, to Amazonas and Rondônia states, Brazil, and toward the north in the Guyanas. It was recently reported from French Guyana.

The Caenophidia

Species-level Taxa	Citations and Remarks	Abbreviated Range
A. quirogai Giraudo & Scrocchi, 1998 Misiones Wormsnake	Giraudo & Scrocchi (1998)	NE Argentina in s Misiones.
A. sanctaeritae Werner, 1924 Santa Rita Wormsnake	Lema & Fernandes (1997); Lema (1997b)	Santa Rita, Brazil [species is known only from the holotype and Werner gave no information about geographically where Santa Rita is located].
A. tenuis Ruthven, 1927 Ruthven's Wormsnake	Lema (1978c, 1997b); Vanzolini (1986); Harvey (1999); Harvey et al. (2001)	Bolivia into ne Brazil.
A. vittata (Cope, 1887) Cope's Wormsnake	Lema (1978c, 1997b); Vanzolini (1986); Harvey (1999); Harvey, et al. (2001)	State of Mato Grosso, Brazil and Departamento Santa Cruz in Bolivia.

GENUS: *Arizona* Kennicott *in* Baird, 1859, is from North America in the United States and Mexico. Some authorities now consider this genus to have as many as three species (*A. elegans*, *A. occidnetalis* and *A. pacta*.) Collins (1991, 1997) and Liner (1994) considered *A. occidentalis* Blanchard, 1924, valid. I have listed *Arizona* as monotypic, with all subspecies classified under *A. elegans* and followed Dixon & Fleet (1976) because Collins (1991) did not discuss details of his taxonomic diagnosis. See Klauber (1946c); Dixon & Fleet (1976); Conant (1975); Stebbins (1985); Williams & Wallach (1989); Collins (1991, 1997); Wright & Wright ([1957]1994); Degenhardt et al. (1996); Tennant (1998); and Dixon (2000a).

Review assistance for this section were provided by James R. Dixon.

CONTENT: One species listed here; with possibly as many as three species. ★★★

Arizona: Glossy Snake

Species-level Taxa	Citations and Remarks	Abbreviated Range
A. elegans Kennicott *in* Baird, 1859 Glossy Snake	Klauber (1946c); Dixon & Fleet (1976); Stebbins (1985); Degenhardt et al. (1996); Brown (1997); Dixon (2000a)	C and w United States into Mexico.
A. e. elegans Kennicott *in* Baird, 1859 Kansas Glossy Snake	Klauber (1946c); Smith & Taylor (1966); Conant (1975); Dixon & Fleet (1976); Stebbins (1985); Tennant (1998); Conant & Collins (1998); Werler & Dixon (2000); Bartlett & Tennant (2000); Tennant & Bartlett (2000); Dixon (2000a)	C United States into Mexico from extreme sw Nebraska and c Kansas through w Texas southeastward into Tamaulipas in n Mexico.
A. e. arenicola Dixon, 1960 Texas Glossy Snake	Conant (1975); Dixon & Fleet (1976); Tennant (1998); Conant & Collins (1998); Dixon (2000a); Werler & Dixon (2000); Tennant & Bartlett (2000)	The United States in sw Texas to the Edwards Plateau and probably into the Tamaulipas biotic province.
A. e. candida Klauber, 1946 Mojave Glossy Snake	Klauber (1946c); Dixon & Fleet (1976); Stebbins (1985); Brown (1997); Bartlett & Tennant (2000)	Mojave Desert region of the United States.

Species-level Taxa	Citations and Remarks	Abbreviated Range
A. e. eburnata Klauber, 1946 Desert Glossy Snake	Klauber (1946c); Dixon & Fleet (1976); Stebbins (1985); Smith & Smith (1993); Brown (1997); Bartlett & Tennant (2000)	W United States.
A. e. expolita Klauber, 1946 Chihuahuan Glossy Snake	Klauber (1946c); Dixon & Fleet (1976); Tanner (1985); Smith & Smith (1993) *Remark: *A. elegans australis* Williams et al. (1961) is a synonym of *A. e. expolita* (see Dixon, 1962).	Distributed from north to south in c Mexico.
A. e. noctivaga Klauber, 1946 Arizona Glossy Snake	Klauber (1946c); Hardy & McDiarmid (1969); Dixon Bartlett & Tennant (2000)	The United States in w Arizona southward into c Sinaloa, Mexico.
A. e. occidentalis Blanchard, 1924 California Glossy Snake	Klauber (1946c); Smith & Taylor (1966); Dixon & Fleet (1976); Stebbins (1985); Smith & Smith (1993); Brown (1997); McPeak (2000); Bartlett & Tennant (2000); Bartlett & Tennant (2000)	San Joaquin Valley in the United States southward into n Baja California, Mexico.
A. e. pacata Klauber 1946 Peninsular Glossy Snake	Klauber (1946c); Dixon & Fleet (1976); Stebbins (1985); Smith & Smith (1993); McPeak (2000)	Baja California, Mexico; endemic to Baja California Sur from near Guerrero Negro to just north of La Paz.
A. e. philipi Klauber, 1946 Painted Desert Glossy Snake	Klauber (1946c); Conant (1975); Dixon & Fleet (1976); Stebbins (1985); Smith & Smith (1993); Liner (1994); Degenhardt et al. (1996); Tennant (1998); Conant & Collins (1998); Werler & Dixon (2000); Bartlett & Tennant (2000); Tennant & Bartlett (2000); Dixon (2000a)	SW United States from Utah, New Mexico, and Arizona through w Texas into extreme n Chihuahua in Mexico.

GENUS: *Arrhyton* Günther, 1858, is a Caribbean snake. The genus *Arrhyton* is distinguished from *Alsophis* by having only one or no pits on each dorsal scales; *Alsophis* has 2 pits per dorsal scale. I consider scale pits to be a weak characteristic to use for separating the two genera when other characteristics are not used since other genera such as *Liophis* and *Hydrodynastes* ignore scale pits and synonymize two more genera. Mayer & Lazell (1988) decided that *Liophis* = *Alsophis* and *Liophis* = *Arrhyton*. They did not say why they synonmyzed *Alsophis* and *Arrhyton* with *Liophis*. Crother (1989) recommended that the Jamaican *Arrhyton* species (*calliliaemus*, *funereum*, and *polylepis*) be included in the genus *Darlingtonia*. Schwartz and Henderson (1991) recognized this recommendation, but maintained these snakes as *Arrhyton* pending more work. This was noted in Hedges & Garrido (1992). Based on Schwartz & Henderson (1991) and Hedges & Garrido (1992), *Arrhyton* has been conserved for this work. See Maglio (1970); Schwartz & Garrido (1981); Schwartz & Henderson (1985, 1991); Mayer & Lazell (1988); Crother (1989); and Hedges & Garrido (1992).

Review assistance for this section provided by James R. Dixon.

See also genera *Alsophis* and *Philodryas*.

CONTENT: Twelve species. ★★

Arrhyton: Caribbean Groundsnake. Frank and Ramus (1995) call this the Island Racer.

The Caenophidia

Species-level Taxa	Citations and Remarks	Abbreviated Range
A. ainictum Schwartz & Garrido, 1981 Las Tunas Groundsnake	Schwartz & Garrido (1981); Schwartz & Henderson (1985, 1991); Estrada & Ruibal *in* Crother (1999)	Known only from the type locality of Francisco, Camgüey Province in Cuba, but presumed to occur in the Sierra de Najasa in Cuba.
A. callilaemum (Gosse, 1851) Jamaican Red Groundsnake	Buden (1966); Maglio (1970); Schwartz & Henderson (1985, 1991); Crother (1989); Crombie *in* Crother (1999)	Jamaica from sea level to about 3000 feet.
A. dolichurum Werner, 1909 Havana Groundsnake	Maglio (1970); Schwartz & Garrido (1981); Schwartz & Henderson (1985, 1991); Estrada & Ruibal *in* Crother (1999)	Cuba, reported from the provinces of Pinar del Río and La Habana.
A. exiguum (Cope, 1862) Puerto Rican Groundsnake	Maglio (1970); Schwartz & Thomas (1975); Rivero (1978, 1998); Schwartz & Henderson (1985, 1991); Thomas *in* Crother (1999)	W Caribbean in the U.S. and British Virgin Islands and Puerto Rico.
A. e. exiguum (Cope, 1862) Common Puerto Rican Groundsnake	Maglio (1970); Rivero (1978, 1998); Schwartz & Henderson (1985, 1991)	W Caribbean in St.Thomas, Culebra Island of Puerto Rico, Hansel Island, Tortola, Peter Island, and Virgin Gorda; U.S. and British Virgin Islands.
A. e. stahli (Stejneger, 1904) Northern Puerto Rican Groundsnake	Maglio (1970); Rivero (1978, 1998); Schwartz & Henderson (1985, 1991)	N Puerto Rico north of a line connecting Mayagüez, Los Rábanos, Aibonito, and Patillas; found from sea level to at least 1800 feet.
A. e. subspadix (Schwartz, 1967) Southwestern Puerto Rican Groundsnake	Maglio (1970); Rivero (1978, 1998); Schwartz & Henderson (1985, 1991)	Found mostly in the dry sw portion of Puerto Rico from Parguera east into Playa de Arroyo and specimens may have been collected in San Germán.
A. funereum (Cope, 1862) Jamaican Black Groundsnake	Buden (1966); Maglio (1970); Crother (1989); Schwartz & Henderson (1985, 1991); Crombie *in* Crother (1999)	Jamaica from sea level to 2000 feet.
A. landoi Schwartz, 1965 Oriente Brown-capped Groundsnake	Schwartz (1965); Maglio (1970); Schwartz & Garrido (1981); Schwartz & Henderson (1985, 1991); Estrada & Ruibal *in* Crother (1999)	S Oriente Province in Cuba from Pilón eastward into the type locality in the mountains north of Imías, Oriente Province.
A. polylepis (Buden, 1966) Jamaican Long-tailed Groundsnake	Buden (1966); Maglio (1970); Crother (1989); Schwartz & Henderson (1985, 1991); Crombie *in* Crother (1999)	E Jamaica; type locality is Port Antonio, Portland Parish.
A. procerum Hedges & Garrido, 1992 Cuban Slender Groundsnake	Hedges & Garrido (1992); Estrada & Ruibal *in* Crother (1999)	Known only from the type locality at 11.4 km ese of Playa Girón, Mantanzas Province in Cuba.

Species-level Taxa	Citations and Remarks	Abbreviated Range
A. supernum Hedges & Garrido, 1992 Oriente Black Groundsnake	Hedges & Garrido (1992); Estrada & Ruibal *in* Crother (1999)	Cuba, known only from the type from El Yunque de Barracoa, and Monte Libano, both from Guantánamo Province.
A. taeniatum Günther, 1858 Broad-striped Groundsnake	Maglio (1970); Schwartz & Garrido (1981); Schwartz & Henderson (1985, 1991); Estrada & Ruibal *in* Crother (1999)	Cuba and Isla de Juventud.
A. tanyplectum Schwartz & Garrido, 1981 Guaniguanico Groundsnake	Schwartz & Garrido (1981); Schwartz & Henderson (1985, 1991); Estrada & Ruibal *in* Crother (1999)	Cuba; the type specimen is from the cliffs at San Vincente, Pinar del Río Province, Cuba.
A. vittatum* (Gundlach *in* Peters, 1861) Common Cuban Groundsnake	Maglio (1970); Schwartz & Henderson (1985, 1991); Estrada & Ruibal *in* Crother (1999) *Remark: Hedges & Garrido (1992) considered *A. bivittatum* to be a synonym.	Cuba and Isla de Juventud.

GENUS: *Atractus* Wagler, 1828, is a wide ranging South American snake that is found from Panama southward into South America in the southern Amazon and northern Andes with its range extending to as far south as Bolivia and Argentina. Although found west of the Andes in northwestern Ecuador, most of its range is east of the Andes. See Savage (1960a); Peters & Orejas-Miranda (1970); Thomas & Greene (1976); Hoogmoed (1980); Cunha & Nascimento (1983); Pérez-Santos & Moreno (1988); Cei (1993); Fernandes (1995b); Kornacker (1999); Starace (1999); Köhler (2001).

Review assistance and important contributions for this section were provided by Ronaldo Fernandes and Van Wallach.

Savage (1960) made the first list of *Atractus*. Peters & Orejas-Miranda (1970), who were heavily influenced by Savage (1960), were the first reviewers to organize all of the known *Atractus* species information into a usable format. They presented a matrix that should be used with the geographic information. The key, although out of date, is an excellent place to begin investigating identification of specimens of *Atractus*. Peters & Orejas-Miranda (1970) listed seventy-three species with two species included in incertae sedis. The list was updated by Vanzolini (1986). The most recent work which includes this genus is that of Fernandes (1995b), but unfortunately it remains an unpublished dissertation. The genus *Atractus* is a very complex group in need of detailed study.

CONTENT: Ninety-two species, with several more possibly included in the *incertae sedis* section. Even though Kornacker (1999) listed *A. vittatus* Boulenger as valid, it is included in *incertae sedis* in this work because questions raised by Peters & Orejas-Miranda (1970) do not seem to have been answered by researchers since their publication. ★

See also the genus *Geophis*.

Atractus: Tellurian Snake. I chose this name because "tellurian" reflects the habits of the snake. The name means "dweller of the earth." Frank and Ramus (1995) call this the Groundsnake.

The Caenophidia

Species-level Taxa	Citations and Remarks	Abbreviated Range
A. albuquerquei Cunha & Nascimento, 1983 Albuquerque Tellurian Snake	Cunha & Nascimento (1983); Vanzolini (1986); Fernandes (1995b)	E Pará and Rondônia in Brazil.
A. alphonsehogei Cunha & Nascimento, 1983 Striped Tellurian Snake	Cunha & Nascimento (1983); Vanzolini (1986); Martins & Oliveria (1993); Fernandes (1995b)	E Pará into ne Maranhão in Brazil.
A. andinus Prado, 1944 Antioquia Tellurian Snake	Prado (1946b); Savage (1960a); Peters & Orejas-Miranda (1970); Pérez-Santos & Moreno (1988); Fernandes (1995b)	Only from the type locality of Antioquia in Colombia.
A. arangoi Prado, 1940 Puerto Asís Tellurian Snake	Prado (1940); Nicéforo María (1942); Savage (1960a); Peters & Orejas-Miranda (1970); Pérez-Santos & Moreno (1988); Fernandes (1995b)	Known only from Puerto Asís in Putumayo in Colombia.
*A. badius** (F. Boie, 1827) Tigrita Tellurian Snake	Savage (1960a); Roze (1966); Peters & Orejas-Miranda (1970); Amaral (1976); Cei (1976); Hoogmoed (1980); Vanzolini (1986); Pérez-Santos & Moreno (1988); Lancini & Kornacker (1989); Starace (1999); Kornacker (1999) *Remark 1: *A. subbcinctum* (Jan, 1862) from Suriname and French Guyana as listed by Peters & Orejas-Miranda (1970), was placed into this classification by Vanzolini (1986). Remark 2: This species was reviewed by Hoogmoed (1980) and he noted that non-Amazonian citations are likely to be in error. *Remark 3: Although *A. micheli* Mocquard, 1904 was listed as valid by Peters & Orejas-Miranda (1970), it is known only from the type specimen from French Guiana. It was not noted in Vanzolini (1986). Fide Cei (1993), is a synonym. *Remark 4: Fide Peters & Orejas-Miranda (1970); Vanzolini (1986); and Dixon & Soini (1986), *A. microrhynchus* (Cope, 1868) reported to be found from Guayaquil, Ecuador into Pebas, Peru is a synonym. *Remark 5: This is not *Atractus microrhynchus* (Cope) as noted in Peters & Orejas-Miranda (1970), which is from the lower Andean slopes in nw Ecuador. Or, from which Vanzolini (1986), referenced Dixon & Soini (1977) who listed it as *Atractus badius* (Boie). However, it was listed in the Lamar (1999) checklist.	N South America east of the Andes in Colombia southward into n Argentina in Chaco; its range includes Colombia, Venezuela, the Guianas, Peru, Brazil, Bolivia, and reported to be found in Argentina [Fernandes (1995c) states reports of this snake from Argentina are probably erroneous.]

Species-level Taxa	Citations and Remarks	Abbreviated Range
	*Remark 6: Argentine citation is probably *A. canedii* as found in Scrocchi & Cei (1991).	
A. balzani Boulenger, 1898 Misiones Mosetenes or Bolivian Tellurian Snake	Boulenger (1898); Savage (1960a); Peters & Orejas-Miranda (1970); Cei (1993); Fernandes (1995b)	Known only from the type locality of Misiones Mosetenes in nw Bolivia.
A. biseriatus Prado, (1940) 1941 Manzinaies or Two-lined Tellurian Snake	Prado (1941c); Nicéforo María (1942); Savage (1960a); Peters & Orejas-Miranda (1970); Pérez-Santos & Moreno (1988); Henle & Ehrl (1991b); Fernandes (1995b)	Known only from the type locality of Manizales in the Departamento de Caldas in Colombia.
A. bocourti Boulenger, 1894 Bocourt's Tellurian Snake	Boulenger (1894); Savage (1960a); Peters & Orejas-Miranda (1970); Pérez-Santos & Moreno (1991); Fernandes (1995b)	Ecuador and ne Peru.
A. boettgeri Boulenger, 1896 Boettger's Tellurian Snake	Boulenger (1896); Savage (1960a); Peters & Orejas-Miranda (1970); Vanzolini (1986); Cei (1993); Fernandes (1995b)	Known only from the type locality in Sierra de las Yungas in Departamento de Cochabamba in Bolivia.
A. boulengeri Peracca, 1896 Boulenger's Tellurian Snake	Peracca (1896); Savage (1960a); Peters & Orejas-Miranda (1970); Fernandes (1995b)	Peters & Orejas-Miranda (1970) wrote that this snake is only known from the type locality that is shown as from South America, and distribution is unknown.
A. canedii Scrocchi & Cei, 1991 Argentine Tellurian Snake	Scrocchi & Cei (1991); Cei (1993); Fernandes (1995c); Leynaud & Bucher (1999)	Salta and Jujuy in Argentina, and presumed will be found in Bolivia.
A. carrioni Parker, 1930 Intermontane Valley or Loja Tellurian Snake	Peters (1960b); Savage (1960a); Peters & Orejas-Miranda (1970); Pérez-Santos & Moreno (1991); Fernandes (1995b)	Intermontane valleys in Loja, Ecuador. A high montane species
A. clarki Dunn & Bailey, 1939 Clark's Tellurian Snake	Dunn & Bailey (1939); Savage (1960a); Peters & Orejas-Miranda (1970); Villa et al. (1988); Auth (1994); Fernandes (1995b); Köhler (2001); Myers (2003)	The type locality is Darién Province at Santa Cruz de Cana Mine in Panama at 500 m. Also from the Chocó in Colombia.
A. collaris Peracca, 1897 Ring-necked Tellurian Snake	Peters (1960b); Savage (1960a); Peters & Orejas-Miranda (1970); Dixon & Soini (1986); Fernandes (1995b)	Amazonian e Ecuador and ne Peru; between 300 to 600 ft. (100–200 m). Considered to be a lowland form.
A. crassicaudatus (Dumeril, Bibron & Dumeril, 1854) Thick-tailed Tellurian Snake	Amaral (1931b, 1932); Peracca (1914); Nicéforo María (1942) Savage (1960a); Peters & Orejas-Miranda (1970); Pérez-Santos & Moreno (1988); Auth (1994); Fernandes (1995b); Köhler (2001) [as *Atractus* cf. *crassicaudatus*]	Panama southward into the highlands of Colombia and Venezuela.

The Caenophidia

Species-level Taxa	Citations and Remarks	Abbreviated Range
A. darienensis Myers, 2003 Darien Tellurian Snake	Myers (2003)	The holotype is from 500 m on the north end of the Serranía de Pirre, Darién, Panama.
A. depressiocellus Myers 2003 Little-eyed Tellurian Snake	Myers (2003)	The type is from Cerro Azul [Cerro Jefe] region in the province of Panama, Panama.
A. duidensis Roze, 1961 Duida or Venezuelan Tellurian Snake	Roze (1961, 1966); Peters & Orejas-Miranda (1970); Hoogmoed (1983 [1982]); Lancini & Kornacker (1989); Fernandes (1995b); Kornacker (1999)	Territorio Federal Amazonas in the region of Cerro Duida in Venezuela.
A. dunni* Savage, 1955 Dunn's Tellurian Snake	Savage (1955, 1960a); Peters (1960b); Peters & Orejas-Miranda (1970); Pérez-Santos & Moreno (1991); Henle & Ehrl (1991b); Fernandes (1995b) *Remark: Originally this snake was called *Rabdosoma maculatum* Bocourt, 1883, but the name was preoccupied by *maculata* Günther, 1858.	Ecuador and Peru; this is a montane form found at higher elevations along the eastern slope of the Andes.
A. ecuadorensis Savage, 1955 Ecuadorian Tellurian Snake	Savage (1955, 1960a); Peters (1960b); Peters & Orejas-Miranda (1970); Pérez-Santos & Moreno (1991); Fernandes (1995b)	Prov. Tunguruhua in Ecuador, it is known only from the type locality which is listed as "Llangante area," probably Llangante Range. Considered a moderate altitude form and found at higher elevations along the eastern slope of the Andes.
A. elaps (Günther, 1858) Ornate Tellurian Snake	Peters (1960b); Savage (1960a); Roze (1966); Peters & Orejas-Miranda (1970); Duellman (1978); Dixon et al. (1976); Hoogmoed (1980); Dixon & Soini (1986); Vanzolini (1986); Pérez-Santos & Moreno (1988, 1991); Savage & Slowinski (1992); Fernandes (1995b); Kornacker (1999)	E Colombia, Venezuela, and Suriname southward into the interandean highlands of Ecuador, n Peru, and Amazonas in Brazil, it may be present in Bolivia. Found from 300 to 3500 ft. (100–1100 m). A lowland form.
A. emigdioi Gonzalez-Sponga, 1971 Emigdio's Tellurian Snake	Gonzalez-Sponga (1971); Vanzolini (1986); Lancini (1979); Lancini & Kornacker (1989); Fernandes (1995b); Kornacker (1999)	Known only from the type locality in Trujillo in Venezuela.
A. emmeli (Boettger, 1888) Emmel's Tellurian Snake	Boulenger (1894); Hoge (1952b); Savage (1960a); Peters & Orejas-Miranda (1970); McCoy (1971); Vanzolini (1986); Fernandes (1995b)	Bolivia and Peru. The type locality is reported to be Río Mapiri, Departamento de La Paz, Bolivia.
A. erythromelas Boulenger, 1903 Red-Black Tellurian Snake	Savage (1960a); Roze (1966); Peters & Orejas-Miranda (1970); Pérez-Santos & Moreno (1988); Lancini & Kornacker (1989) Savage & Slowinski (1992); Fernandes (1995b); Kornacker (1999)	Venezuelan Andes in the state of Mérida above 1000 m; Pérez-Santos & Moreno (1988), and Kornacker (1999) both listed this snake as present in Colombia.
A. favae (Filippi, 1840) Suriname Tellurian Snake	Hoogmoed (1980); Vanzolini (1986); Fernandes (1995b)	The Guiana Shield in Guyana and n Suriname.

Species-level Taxa	Citations and Remarks	Abbreviated Range
A. flammingerus* (F. Boie, 1827) Blotched Tellurian Snake	Hoogmoed (1980, [1982]1993); Cunha & Nascimento (1983); Vanzolini (1986); Dixon & Soini (1986); Fernandes (1995b); Starace (1998)	Brazil, Peru and the Guianas.
	*Remark 1: Vanzolini (1986), listed this taxon with subspecies of A. flammingerus flammingerus (Boie, 1827) and A. fammingerus snethlageae Cunha & Nascimento, 1983. Martins & Oliveria (1993), elevated the subspecies to full species. Note that Peters & Orejas-Miranda (1970) listed A. flammingerus as A. badius.	
	*Remark 2: Vanzolini (1986), reported that Hoogmoed (1980), considered Geophis alasukai Gasc & Rodrigues, 1979 from French Guiana as found in Chippaux ([1986] 1987) to be a synonym. Starace (1998) also listed it as a synonym.	
A. fuliginosus (Hallowell, 1845) Dark Tellurian Snake	Roze (1958c, 1966); Savage (1960a); Peters & Orejas-Miranda (1970); Lancini & Kornacker (1989); Fernandes (1995b); Kornacker (1999)	N Venezuela.
A. gaigeae Savage, 1955 Gaige's Tellurian Snake	Savage (1955, 1960a); Leviton & Banta (1956); Peters (1960b); Peters & Orejas-Miranda (1970); Vanzolini (1986); Pérez-Santos & Moreno (1991); Fernandes (1995b)	Amazonian lowlands of Ecuador where it is restricted to the tropical forests of the upper Amazon basin between 600–2000 ft. (200–600 m). A lowland form.
A. guentheri (Wucherer, 1861) Günther's Tellurian Snake	Savage (1960a); Peters & Orejas-Miranda (1970); Vanzolini (1986); Amaral (1976); Pérez-Santos & Moreno (1988); Savage & Slowinski (1992); Fernandes & Puorto (1993); Lema (1994b) [as c.f. guentheri]; Fernandes (1995b); Fernandes & Argôlo (1999)	Atlantic forests of southern Bahia in Brazil.
A. hostilitractus Myers, 2003 Highly Variable Tellurian Snake	Myers (2003)	The holotype is from 100 to 200 m from Río Mortí, province Darién, Panama.
A. imperfectus Myers, 2003 Imperfect Tellurian Snake	Myers (2003)	The holotype is from the Piedras–Pacora Ridge in c Panama.
A. indistinctus Prado, 1940 Indistinct Tellurian Snake	Prado (1940); Nicéforo María (1942); Savage (1960a); Peters & Orejas-Miranda (1970); Pérez-Santos & Moreno (1988); Fernandes (1995b)	Departamento de Norte de Santander in Colombia; known only from the type locality of Ocaña.
A. insipidus Roze, 1961 Border Tellurian Snake	Roze (1961, 1966); Peters & Orejas-Miranda (1970); Cunha & Nascimento (1983); Lancini & Kornacker (1989); Fernandes (1995b); Kornacker (1999)	In the Amazon at the border between Brazil and Venezuela.
A. iridescens Peracca, 1896 Iridescent Tellurian Snake	Amaral (1931b); Peracca (1896); Nicéforo María (1942); Savage (1960a); Peters & Orejas-Miranda (1970); Pérez-Santos & Moreno (1988); Fernandes (1995b)	Peracca listed the type as South America with distribution unknown.

Species-level Taxa	Citations and Remarks	Abbreviated Range
A. lancinii Roze, 1961 Lancini's Tellurian Snake	Roze (1961, 1966); Peters & Orejas-Miranda (1970); Lancini & Kornacker (1989); Fernandes (1995b) Kornacker (1999)	Distrito Federal in Venezuela, Peters & Orejas-Miranda noted that it is only known from the type locality of El Junquito, Distrito Federal, Venezuela at 1900 m.
A. lasallei Amaral, 1931 Highland Tellurian Snake	Amaral (1931a); Nicéforo María (1942); Savage (1960a); Peters & Orejas-Miranda (1970); Pérez-Santos & Moreno (1988); Fernandes (1995b)	Highlands of n Colombia. The type locality is reported to be Sampedro, south of Medellín, Departamento de Antioquia, Colombia.
A. latifrons (Günther, 1868) Wedge-tailed Tellurian Snake	Savage (1960a); Hoogmoed (1980); Chippaux ([1986] 1987); Cunha & Nascimento (1983); Vanzolini (1986); Dixon & Soini (1986); Nascimento et al. (1987); Pérez-Santos & Moreno (1988); Savage & Slowinski (1992); Martin & Oliveria (1993); Fernandes (1995b); Starace (1998)	E Peru, w Brazil, e Colombia, Suriname, and French Guiana.
A. lehmanni Boettger, 1898 Lehmann's Tellurian Snake	Peters (1960b); Savage (1960a); Peters & Orejas-Miranda (1970); Pérez-Santos & Moreno (1991); Fernandes (1995b)	Known only from the type locality of the Hoyo de Cuenca, Provencia Azuay in Ecuador. This is a river valley draining into the Amazon basin. The type is from about 8500 ft. (2600 m) and it is considered a montane form from the eastern slope of the Andes.
A. limitaneus (Amaral, 1935) La Pedrera Tellurian Snake	Nicéforo María (1942) [as *Leptocalamus limitaneus*]; Savage (1960a); Peters & Orejas-Miranda (1970); Pérez-Santos & Moreno (1988)	Known only from the type locality of La Pedrera, Río Caquetá in Comisaria de Amazonas in Colombia.
A. loveridgei Amaral, 1930 Loveridge's Tellurian Snake	Amaral (1930a; 1931a, b); Nicéforo María (1942); Savage (1960a); Peters & Orejas-Miranda (1970); Pérez-Santos & Moreno (1988); Fernandes (1995b)	Highlands of n Colombia. The type locality is reported to be Jericó, Departamento de Antioquia, Colombia.
A. maculatus (Günther, 1858) Spotted Tellurian Snake	Boulenger (1894); Savage (1960a); Peters & Orejas-Miranda (1970); Fernandes (1995a, c, 2000); Fernandes & Argôlo (1999); Fernandes et al. (2000)	Fernandes et al. (2000) dramatically reduced the range of this snake in their description to surviving in a few dense patches of Atlantic rain forests close to the coast in the State of Algoas, Brazil.
A. major Boulenger, 1893 Giant Tellurian Snake	Peters (1960b); Savage (1960a); Roze (1966); Peters & Orejas-Miranda (1970); Duellman (1978); Dixon & Soini (1986); Pérez-Santos & Moreno (1988, 1991); Lancini & Kornacker (1989); Martins & Oliveria (1993); Fernandes (1995b); Kornacker (1999)	Amazon slopes of Colombia, Brazil, e Ecuador and Venezuela; it may eventually be found in Peru. Associated with forested lowlands.
A. manizalesensis Prado, 1940 Villa María Tellurian Snake	Prado (1940); Nicéforo María (1942); Dunn *in* Savage (1960a); Savage (1960a); Peters & Orejas-Miranda (1970); Pérez-Santos & Moreno (1988); Fernandes (1995b)	Known only from the vicinity of the type location of Villa María in Departamento de Caldas in Colombia.
A. mariselae Lancini, 1969 Marisela's Tellurian Snake	Lancini (1969); Vanzolini (1986); Lancini & Kornacker (1989); Fernandes (1995b); Kornacker (1999)	Trujillo in Venezuela, the type is from Bocono.

Species-level Taxa	Citations and Remarks	Abbreviated Range
A. melanogaster Werner, 1916 Black-bellied Tellurian Snake	Werner (1916); Nicéforo María (1942); Savage (1960a); Peters & Orejas-Miranda (1970); Pérez-Santos & Moreno (1988); Fernandes (1995b)	Departamentos de Tolima and Antioquia in Colombia.
A. melas Boulenger, 1908 Los Mangos Tellurian Snake	Boulenger (1908); Amaral (1931b); Nicéforo María (1942); Savage (1960a); Peters & Orejas-Miranda (1970); Pérez-Santos & Moreno (1988); Fernandes (1995b)	Known only from the type locality in Los Mangos in Departamento Valles in Colombia.
A. modestus* Boulenger, 1894 Modest Tellurian Snake	Savage (1960a); Peters (1960b); Peters & Orejas-Miranda (1970); Pérez-Santos & Moreno (1991); Fernandes (1995b) *Remark: Peters & Orejas-Miranda (1970), listed *Atractus bocki* Werner (1909), known only from the type locality of Cochabamba, Departamento de Cochabamba in Bolivia, and commented that Amaral (1929), indicated that *A. bocki* was a synonym of *A. modestus*, it is included as a synonym here.	W Ecuador and possibly Peru; if *A. bocki* is included as a synonym, then the range extends to Cochabamba in Bolivia.
A. multicinctus (Jan *in* Jan & Sordelli, (1865 [1860–1866]) Many-banded Tellurian Snake	Boulenger (1898); Savage (1960a); Peters (1960b); Peters & Orejas-Miranda (1970); Pérez-Santos & Moreno (1988, 1991); Fernandes (1995b)	NW Ecuador in dense lowland tropical forests northward into the Chocó of Colombia. Found from 100 to 2500 ft.
A. nebularis Bernal-Carlo & Roze, 1997 Cloud Forest Tellurian Snake	Bernal-Carlo & Roze, (1997)	Sierra Nevada de Santa Marta in Colombia; it is associated with cloud forest in the nw part of the Sierra Nevada de Santa Marta, including Serranía de San Lorenzo and Cuchilla Yerbabuena, between 1700 to 2000 m in cloud forest and foggy areas; there is one questionable report from 600 m at Minca.
A. nicefori Amaral, 1930 Niceforo's Tellurian Snake	Amaral (1930a, 1931b); Nicéforo-María (1942); Savage (1960a); Peters & Orejas-Miranda (1970); Pérez-Santos & Moreno (1988); Fernandes (1995b)	Highlands of n Colombia. The type locality is reported to be Jericío, Departamento de Antioquia, Colombia.
A. nigricaudus Schmidt & Walker, 1943 Black-tailed Tellurian Snake	Schmidt & Walker (1943a, 1943c); Savage (1960a); Peters & Orejas-Miranda (1970); Fernandes (1995b)	Known only from the type locality in Huanchon, east of Cerro de Pasco in Departamento de Junín in Peru at 10,000 feet.
A. nigriventris Amaral, 1933 Chita Tellurian Snake	Amaral (1933a); Nicéforo-María (1942); Savage (1960a); Peters & Orejas-Miranda (1970); Pérez-Santos & Moreno (1988); Fernandes (1995b)	Known only from the type locality of Chita, se of San Gil in Departamento de Santander in Colombia.
A. obesus Marx, 1960 Santa Bárbara Tellurian Snake	Marx (1960); Peters & Orejas-Miranda (1970); Vanzolini (1986); Pérez-Santos & Moreno (1988); Savage & Slowinski (1992); Fernandes (1995b)	Known only from the type locality of Santa Bárbara at the base of Cerro Frontimo in the Upper Río Urrao (a tributary of the Río Penserisco in the Cordillera Occidental in Antioquia and El Robala, Río Pichinadé (a tributary of the Río Cali) in Valle de Cauca in Colombia.

Species-level Taxa	Citations and Remarks	Abbreviated Range
A. obtusirostris Werner, 1916 Big-nosed Tellurian Snake	Werner (1916); Amaral (1931a, b); Nicéforo-María (1942); Savage (1960a); Peters & Orejas-Miranda (1970); Pérez-Santos & Moreno (1988); Fernandes (1995b)	Departamento de Tolima and Pensilvania in Colombia.
A. occidentalis Savage, 1955 Andean Tellurian Snake	Savage (1955, 1960a); Peters (1960b); Peters & Orejas-Miranda (1970); Duellman (1978); Pérez-Santos & Moreno (1991); Fernandes (1995b)	Lower Pacific slopes of the Andes in nw Ecuador from 800 to 1200 m. It is found from about Quito to the Hoyo de Chimbo from 2500–4100 ft. (800–1200 m).
A. occipitoalbus (Jan, 1862) Gray Tellurian Snake	Savage (1955, 1960a); Peters (1960b); Peters & Orejas-Miranda (1970); Duellman (1978); Vanzolini (1986); Henle & Ehrl (1991b); Fernandes (1995a, 1995b)	Eastern slopes of Ecuador, Peru, and Bolivia in upper Amazon forests between 600–3500 ft. (200–1100 m). A lowland form.
A. oculotemporalis Amaral, 1932 Jericó Tellurian Snake	Amaral (1932); Nicéforo-María (1942); Savage (1960a); Peters & Orejas-Miranda (1970); Pérez-Santos & Moreno (1988); Fernandes (1995a, 1995b)	Known only from the type locality of Jericó in Departamento de Antioquia in Colombia.
A. pamplonensis Amaral, 1937 Pamplona Tellurian Snake	Amaral (1937); Nicéforo-María (1942); Savage (1960a); Peters & Orejas-Miranda (1970); Pérez-Santos & Moreno (1988); Fernandes (1995b)	Departamento de Norte de Santander in Colombia; the type is from Pamplona and recently reported in Venezuela.
A. pantostictus Fernandes & Puorto, 1993 Central Brazil Tellurian Snake	Fernandes & Puorto (1993); Fernandes (1995b, 1995c)	Brazil, from the state of Goais to the state of São Paulo.
A. paraguayensis* Werner, 1924 Paraguayan Tellurian Snake	Savage (1960a); Peters & Orejas-Miranda (1970); Alvarez et al. (1992); Cei (1993); Fernandes (1995b, 1995c) *Remark: Previously listed as A. reticulatus paraguayensis fide Amaral in Peters & Orejas-Miranda (1970), and Cei (1993). This was reclassified to full species status by Fernandes (1995c).	Known only from the holotype found in Paraguay.
A. paravertebralis Henle & Ehrl, 1991 Madre de Dios Tellurian Snake	Henle & Ehrl (1991b)	Peru, the type location is noted as Departamento Madre de Dios; Baja Tambopata; in secondary forest growth.
A. paucidens Mocquard in Despax, 1910 Rainforest Tellurian Snake	Savage (1955, 1960a); Peters (1960b); Peters & Orejas-Miranda (1970); Pérez-Santos & Moreno (1991); Fernandes (1995b)	Tropical rainforests on the slopes of the Andes in nw Ecuador.
A. pauciscutatus Schmidt & Walker, 1943 Small-scaled Tellurian Snake	Schmidt & Walker (1943b, 1943c); Savage (1960a); Peters & Orejas-Miranda (1970); Fernandes (1995b)	Known only from the type locality of Carpapata in the Upper Chanchamayo Valley ne of Tarma in Departamento de Junín, Peru.
A. peruvianus (Jan, 1862) Peruvian Tellurian Snake	Boulenger (1894); Savage (1960a); Peters & Orejas-Miranda (1970); Henle & Ehrl (1991b); Fernandes (1995b)	Known only from the type locality shown only as Peru.

Species-level Taxa	Citations and Remarks	Abbreviated Range
A. poeppigi (Jan, 1862) Black-backed or Amazon Basin Tellurian Snake	Savage (1960a); Dixon et al. (1976); Vanzolini (1986); Dixon & Soini (1986); Pérez-Santos & Moreno (1988); Savage & Slowinski (1992); Martins & Oliveria (1993); Fernandes (1995b)	Amazon Basin in Amazonas. It has been reported from extreme w Amazon Basin from extreme sw Colombia, e Peru, and possibly extreme w Brazil.
A. potschi Fernandes, 1995 Fernandes's Atlantic Forest Tellurian Snake	Fernandes (1995a, 1995b); Freitas (1999)	Atlantic forests of the states of ne Brazil in Sergipe and Alagoas, more recently it has been reported from Bahia.
A. punctiventris Amaral, 1933 Villavicencio Tellurian Snake	Amaral (1933a); Nicéforo-María (1942); Savage (1960a); Peters & Orejas-Miranda (1970); Pérez-Santos & Moreno (1988); Fernandes (1995a, 1995b)	Known only from the type locality of Villavicencio in the Intendencia de Meta in Colombia.
A. resplendens Werner, 1901 Resplendent Tellurian Snake	Savage (1960a); Peters (1960b); Peters & Orejas-Miranda (1970); Pérez-Santos & Moreno (1991); Fernandes (1995b)	Eastern Amazonian slopes of the Andes of Ecuador between 3500–6200 ft. (1100–1900 m) and considered to be a montane form.
*A. reticulatus** (Boulenger, 1885) Common Tellurian Snake	Nicéforo-María (1942); Savage (1960a); Peters & Orejas-Miranda (1970); Amaral (1976); Alvarez et al. (1992); Savage & Slowinski (1992); Cei (1993); Lema (1994b); Fernandes (1995a, 1995b, 1995c), Yanosky (1998) *Remark: Fernandes (1995c) reviewed specimens labeled as *A. reticulatus* from Colombia. He determined that these were in fact, *A. manizalensis*, and considering the highly restricted distributions of the species in this genus and the distance between southern Brazil and Colombia, it is very likely that all previous reports of *A. reticulatus* from Colombia were based on misidentified specimens.	SE South America in s Brazil in the states of Rio Grande do Sul, São Paulo, Paraná, Santa Catarina, and to ne Argentina in Corrientes, and probably Misiones into Paraguay.
A. riveroi Roze, 1961 Rivero's Tellurian Snake	Roze (1961, 1966); Peters & Orejas-Miranda (1970); Lancini & Kornacker (1989); Fernandes (1995b); Kornacker (1999)	Territorio Federal Amazonas in Venezuela.
A. roulei Mocquard *in* Despax, 1910 Roule's Tellurian Snake	Savage (1960a); Peters (1960b); Peters & Orejas-Miranda (1970); Vanzolini (1986); Pérez-Santos & Moreno (1991); Fernandes (1995b)	Western slopes in sw Ecuador; normally from 4000–8600 ft. (1200 to 1600 m); however, the type is reported from 2350 m in Alausí. It probably will turn up in nw Peru.
A. sanctaemartae Dunn, 1946 Sierra Nevada de Santa Marta Tellurian Snake	Dunn (1946); Savage (1960a); Peters & Orejas-Miranda (1970); Pérez-Santos & Moreno (1988); Bernal-Carlo & Roze (1994); Fernandes (1995b)	Sierra Nevada de Santa Marta in Colombia; it is associated with cloud forest in the nw and s part of the Sierra Nevada de Santa Marta from 900 to 2400 m.
A. sanguineus Prado, (1944–1945) 1945 Yarumal Tellurian Snake	Prado (1946b); Savage (1945); Peters & Orejas-Miranda (1970); Pérez-Santos & Moreno (1988); Martins & Oliveria (1993); Fernandes (1995b)	Known only from the type locality of Yarumal in Departamento de Antioquia in Colombia.
A. schach (F. Boie, 1827) Schach's Tellurian Snake	Hoogmoed (1980); Cunha & Nascimento (1983); Vanzolini (1986); Martins & Oliveria (1993); Starace (1998); Fernandes (1995b)	W and c Suriname through the Guianas into Manaos, and e Pará, Brazil.

Species-level Taxa	Citations and Remarks	Abbreviated Range
A. serranus* Amaral, 1930 Serrana or São Paulo Tellurian Snake	Amaral (1931a); Savage (1960a); Peters & Orejas-Miranda (1970); Dixon & Soini (1986); Fernandes (1995a, 1995b, 1995c) *Remark 1: See the remark in *Incertae Sedis* under *Atractus trihedrurus* for comments about its assumed relationship to this snake. Fernandes et al. (2000), *A. serranus* and *A. trihedrurus* are probably sister taxa, sharing the putative synapomorphy of a deeply bilobate hemipene and other characteristics of unknown polarization, such as unusually high number of teeth (9–10) and similar coloration pattern. *Remark 2: Marques et al. (2001:171), in their comentários taxonômicos noted that *A. serranus* is indistinguishable from *A. trihedrurus* and is probably the same species.	States of Rio de Janiero and Sao Paulo, Brazil, according to Amaral (1976), it is found in the Paranàpiacaba Mountains on the coast.
A. snethlageae* Cunha & Nascimento, 1983 Snethlage's or Brown Tellurian Snake	Cunha & Nascimento (1983); Vanzolini (1986); Martins & Oliveria (1993); Fernandes (1995b) *Remark: Vanzolini (1986) listed this as a subspecies *A. flammigerus snethlageas*, but it was later elevated to full species by Martins & Oliveria (1993).	Amazonas and e Pará in Brazil; the type locality is restricted to Colônia Nova, Rio Gurupi, Pará.
A. steyermarki Roze, 1958 Steyermark's Tellurian Snake	Roze (1958, 1966); Savage (1960a); Peters & Orejas-Miranda (1970); Lancini & Kornacker (1989); Williams & Gudynas (1991); Fernandes (1995b); Kornacker (1999)	Savannas in the state of Bolivar, Venezuela, the type locality is Chimantá Tepui.
A. taeniatus* Griffin, 1916 Chaco Tellurian Snake	Savage (1960a); Peters & Orejas-Miranda (1970); Williams & Gudynas (1991); Cei (1993); Lema (1994b); Fernandes (1995b, 1995c) *Remark: Peters & Orejas-Miranda (1970) listed *A. taeniatus*, and later Vanzolini (1986) noted this as a synonym of *A. boettgeri*. McCoy (1971) also listed it as a synonym of *A. boettgeri*. However, it has been revalidated by Williams & Gudynas (1991), Cei (1993), and Lema (1994b) as a recognized species.	Bolivian Chaco boreal into Entre Rios and Misiones, and including Corrientes in Argentina; some authorities believe that the range includes Rio Grande do Sul; it has recently been reported from Paraná in Brazil. The type locality is reported as Santa Cruz, Santa Cruz, Bolivia.
A. taphorni Schargel & García-Pérez, 2002 Taphorn's Tellurian Snake	Schargel & García-Pérez (2002)	Known only from the type locality at 2200 m in the western range of the Cordillera de Mérida in Venezuela.
A. torquatus (Dumeril, Bibron & Dumeril, 1854) Red Tellurian Snake	Savage (1960a); Peters & Orejas-Miranda (1970); Hoogmoed (1980); Vanzolini (1986); Dixon & Soini (1986); Martins & Oliveria (1993); Fernandes (1995b) Barrio et al. (1999); Kornacker (1999)	The Guianas and Amazon Basin in Venezuela, Guyana, French Guiana, Suriname, and southward and eastward into Colombia, Peru, Bolivia, and Amazonas in Brazil.
A. trihedrurus Amaral, 1926 False São Paulo Tellurian Snake	Savage (1960a); Amaral (1976); Fernandes & Puorto (1994); Fernandes (1995a, 1995b, 1995c); Fernandes et al. (2000); Marques et al. (2001)	Forests of s Brazil in Paraná and Santa Catarina.

Species-level Taxa	Citations and Remarks	Abbreviated Range
	*Remark 1: *A. trihedrurus* is listed here because Fernandes (1995a, 1995b); Fernandes & Puorto (1994); and Fernandes et al. (2000) included this species in their papers as valid. However, Peters & Orejas-Miranda (1970) listed it as incertae sedis and commented that "Savage, Misc. Publ. Mus. Zool. Univ. Mich., 112, 1969, 31, indicated that this was probably not an *Atractus*, but that its generic status must be regarded as uncertain until the holotype can be re-examined." Because of their comment, along with my inability to confirm a formal revalidation, it is also listed in incertae sedis along with its presence here.	
	*Remark 2: Fide Fernandes et al. (2000), *A. serranus* and *A. trihedrurus* are probably sister taxa, sharing the putative synapomorphy of a deeply bilobate hemipene and other characteristics of unknown polarization, such as unusually high number of teeth (9–10) and similar coloration pattern.	
A. trilineatus Wagler, 1828 Three-lined Tellurian Snake	Savage (1960a); Underwood (1962); Roze (1966); Peters & Orejas-Miranda (1970); Cunha & Nascimento (1983); Vanzolini (1986); Lancini & Kornacker (1989); Martins & Oliveria (1993); Fernandes (1995a, 1995b); Murphy (1997); Kornacker (1999); Boos (2001)	E Venezuela, Trinidad, the Guianas, and into Brazil in Amazonas and Território de Roraima in the savannas.
A. trivittatus Amaral, 1933 Three-banded Tellurian Snake	Amaral (1933a); Nicéforo-María (1942); Savage (1960a); Peters & Orejas-Miranda (1970); Pérez-Santos & Moreno (1988); Fernandes (1995b)	Known only from the type locality of Chita in the Departamento de Santander, Colombia.
A. univittatus (Jan, 1862) One-banded Tellurian Snake	Roze (1961, 1966); Savage (1960a); Peters & Orejas-Miranda (1970); Hoogmoed (1980); Lancini & Kornacker (1989); Fernandes (1995b); Murphy (1997); Kornacker (1999)	Cordillera de la Costa in Venezuela and Tobago. Island in Trinidad and Tobago. Hoogmoed (1980) does not believe that it occurs in Suriname.
A. variegatus Prado, [1941] 1942 Variegated Tellurian Snake	Prado (1942b); Nicéforo-María (1942); Savage (1960a); Peters & Orejas-Miranda (1970); Pérez-Santos & Moreno (1988); Fernandes (1995b)	Known only from the type locality of La Uvita in Departamento de Boyocá, Colombia.
A. ventrimaculatus Boulenger, 1905 Speckled Tellurian Snake	Roze (1966); Savage (1960a); Peters & Orejas-Miranda (1970); Vanzolini (1986); Lancini & Kornacker (1989); Fernandes (1995c); Kornacker (1999)	Known only from the type locality in Fuqueros in Andean Venezuela in the state of Merida.
A. vertebralis Boulenger, 1904 Vertebral Tellurian Snake	Boulenger (1904); Savage (1960a); Peters & Orejas-Miranda (1970); Henle & Ehrl (1991b); Fernandes (1995b)	Known only from the type locality of Ocaña in the Cordillera de Carabaja in Departamento de Puno in Peru at 6000 feet.
A. vertebrolineatus Prado, 1940 Ocaña Tellurian Snake	Prado (1941c); Nicéforo-María (1942); Savage (1960a); Peters & Orejas-Miranda (1970); Pérez-Santos & Moreno (1988); Fernandes (1995b)	Known only from the type locality of Ocaña in the Departamento de Norte de Santander in Colombia.

The Caenophidia

Species-level Taxa	Citations and Remarks	Abbreviated Range
A. wagleri Prado, 1945 Wagler's Tellurian Snake	Prado (1945b, 1946b); Savage (1960a); Peters & Orejas-Miranda (1970); Pérez-Santos & Moreno (1988); Fernandes (1995b)	Known only from the type locality of Humbo in Departamento de Boyocá, Colombia.
A. werneri Peracca, 1914 Werner's Tellurian Snake	Amaral (1931b); Prado (1940) [as *A. longimaculatus* and *A. colombianus*]; Moreno (1988); Savage (1960a); Peters & Orejas-Miranda (1970); Pérez-Santos & Moreno (1988); Fernandes (1995b)	Highlands of Colombia.
A. zebrinus* (Jan, 1862) Jan's Atlantic Forest Tellurian Snake	Savage (1960a); Fernandes (1995b, 2000); Fernandes et al. (2000); Marques et al. (2001) *Remark: Fernandes et al. (2000), recently removed this snake from synonymy with *A. maculatus*.	Atlantic Rain Forest species that is found in the dense forests of Brazil, in the states of Minas Geraís, Espírito Santo, Rio de Janeiro, São Paulo, Paraná, and Santa Catarina.
A. zidoki Gasc & Rodrigues, 1979 Zidok's Tellurian Snake	Gasc & Rodrigues (1979); Hoogmoed (1980); Cunha & Nascimento (1983); Vanzolini (1986); Fernandes (1995b); Starace (1998)	E and c Suriname, se French Guiana, and, e Pará and Territorio Federal do Ampaná in Brazil.

GENUS: *Bogertophis* Dowling & Price, 1988, is from the southwestern United States and n Mexico. See Smith & Taylor (1966) [as *Elaphe*]; Stebbins (1985) [as *Elaphe*]; Dowling & Price (1988); Price (1990a); Webb (1990); Degenhardt et al. (1996); Schulz (1996); Tennant & Bartlett (2000); Conant & Collins (1998); and Dixon (2000a).

Also see genera *Elaphe* and *Senticolis*.

Note: Schulz (1996) felt that there was not sufficient justification to separate *Bogertophis* from *Elaphe*. It is recommended that the reader refer to this reference for his comments. This reference separates *Bogertophis* from *Elaphe* based on Dowling & Price (1988); Price (1990a, 1990b); and Webb (1990).

CONTENT: Two species. ★★★

Bogertophis: Desert Ratsnake

Species-level Taxa	Citations and Remarks	Abbreviated Range
B. rosaliae (Mocquard, 1899) Baja California or Santa Rosalia Ratsnake	Smith & Taylor (1966); Stebbins (1985); Dowling & Price (1988, 1990a, 1990b, 1996b); Webb (1990); Schulz (1996); Brown (1997); McPeak (2000); Bartlett & Tennant (2000)	C and s Baja California in Mexico; Stebbins (1985) makes note of one of these snakes being discovered in extreme s California in the United States about 1 mile east of Mt. Spring, Imperial Co. It is found on his map 132, at the frontier with Mexico.
B. subocularis (Brown, 1901) Trans-Pecos Ratsnake	Smith & Taylor (1966); Conant (1975); Worthington (1980); Stebbins (1985); Dowling & Price (1988); Price (1990a, 1990b); Webb (1990); Degenhardt et al. (1996); Schulz (1996); Conant & Collins (1998); Tennant (1998); Werler & Dixon (2000); Dixon (2000a)	SW United States into c Mexico in n Chihuahuan Desert habitat and w Texas sometimes to above 5000 ft on mountain islands.
B. s. subocularis (Brown, 1901) Chihuahuan or Trans-Pecos Ratsnake	Conant (1975); Webb (1990); Smith & Smith (1993); Degenhardt et al. (1996); Schulz (1996); Dixon (2000a); Werler & Dixon (2000); Bartlett & Tennant (2000); Tennant & Bartlett (2000)	SW United States in New Mexico and s Texas into s Coahuila and Chihuahua in Mexico.

Species-level Taxa	Citations and Remarks	Abbreviated Range
B. s. amplinotus Webb, 1990 Durango Ratsnake	Webb (1990); Smith & Smith (1993); Schulz (1996); Bartlett & Tennant (2000)	Coahuila and Durango in n Mexico.

GENUS: *Boiga* Fitzinger, 1826‡, is generally associated with the Indo-Australia area. There are now confirmed reports that in addition to Guam and Saipan, *Boiga irregularis* has been seen, and possibly introduced into Hawaii and Texas. It is thought that these snakes were accidentally brought into Guam and other areas on military transport aircraft. Although none have been confirmed to be actually surviving in these areas other than Guam and possibly Saipan, it is feared that these hardy snakes can become adapted very quickly to the new environments. It is well known that the introduction of a foreign species may have severe consequences in areas where the native fauna have not developed defenses against the intruders. In the 1950's, the brown tree snake (*Boiga irregularis*) was introduced to Guam probably as a stowaway in some imported goods or military aircraft. By the late 1970's, many of the small native birds of Guam were extirpated or are rapidly disappearing. Investigation proved that this was due to the brown tree snake, which had no natural enemies on Guam. It is estimated that the population there is now as high as 10,000 snakes per acre in some areas. See Borg (1990); Yoon (1992); Fritts (1994); McCoid et. al. (1994); and Fritts & Campbell (1994).

CONTENT: Distribution worldwide in the Old World through the Indo-Pacific, but one species is thought to possibly be introduced in the Americas. ★★★★

Boiga: Old World Cat-eyed Snake. Frank and Ramus (1995) call this the Mangrove Snake.

Species-level Taxa	Citations and Remarks	Abbreviated Range
B. irregularis (Merrem, 1820) Brown Tree Snake	Ogawa (1991); Cogger (1988); McCoid et al. (1994); O'Shea (1996)	Indonesia into the Solomon Islands, Papua New Guinea, and the northern and eastern coasts of Australia. Introduced into Guam and possibly Hawaii and Corpus Christi, Texas. David Chiszar (pers. comm. 11 September 2001) advised me that a population appears to be established on Saipan.

GENUS: *Boiruna* Zaher, 1996, is a recently described genus found in Brazil, Bolivia, Paraguay, Uruguay, and Argentina. This snake makes up part of the large, dark snakes often referred to locally as Mussuranas. See Peters & Orejas-Miranda (1970) [as *C. occipitolutea*]; Cei (1993) [as *Clelia*]; and Zaher (1996); Sazima (1992).

Review assistance and important contributions for this section was provided by Hussam Zaher.

Note: Fide Zaher (1996), it is difficult to differentiate *Boiruna* from *Clelia* based on external characteristics because of their similarities. For this reason, Zaher added the internal features to his key in order to complement the external character data.

Also see *Clelia*, *Oxyrhopus*, and *Pseudoboa*.

CONTENT: Two species. ★★★★

Boiruna: False Mussurana

Species-level Taxa	Citations and Remarks	Abbreviated Range
*B. maculata** (Boulenger, 1896) Common False Mussurana	Bailey *in* Peters & Orejas-Miranda (1970) [as *Clelia*]; Cei (1986) [as *Clelia*]; Marques & Lema (1992) [as *Clelia*]; Lema (1994b) [as *Clelia*); Zaher (1996); Yanosky (1998) [as *Clelia occipitolutea*]; Leynaud & Bucher (1999); Campos-Mogueira (2001) *Remarks: G. Scrocchi & M. Vinas (1990) questioned the validity of *C. occipitolutea* and synonymized it with *Clelia clelia*. Cei (1993) followed this arrangement and showed *C. occipitolutea* to be a color form of *C. c. clelia*. Zaher (1996) redescribed the snake as *Boiruna maculata* (Boulenger, 1896) as the valid name for the southernmost South American population that had been historically designated as "*Clelia occipitolutea*," and confirmed it as originally being called *Oxyrhopus maculatus* Boulenger, 1896.	S Bolivia into w Mato Grosso do Sul and s Goiás in Brazil and southward through se and s Brazil into n and w Argentina; it has been found in n Uruguay and Paraguay with an isolated specimen from Humaitá, Amazonas, n Brazil. Recently reported from Distrito Federal and Goías.
B. sertaneja Zaher, 1996 Sertão False Mussurana	Zaher (1996); Freitas (1999) [as *Clelia occipitolutea*]	Restricted to the xeric open formations of ne Brazil in the lowlands of the states of Piauí, Ceará, Paraíba, Pernambuco, Alagoas, Bahia, and Minas Geraís.

GENUS: *Calamodontophis* Amaral, 1936, [substitute name for *Calamodon* Amaral, 1935], is an extremely rare snake from Brazil and Uruguay, and few specimens are known. The snake was originally named *Calamodon* in 1935, and it was later discovered that this name was preoccupied for an extinct mammal described by Cope, 1875, so was changed in 1963. For reference see Amaral (1935, 1963, 1974); Peters & Orejas-Miranda (1970); Pagini & Lema (1987); and Lema (1994b).

Review assistance for this section was provided by Robert A. Thomas.

CONTENT: One species. This genus is presently being reviewed by several authorities and the reader should expect new information or changes in status for this snake in the near future. ★

Calamodontophis: Xadrez Snake. Frank and Ramus (1995) call this the Tropical Forest Snake.

Species-level Taxa	Citations and Remarks	Abbreviated Range
C. paucidens (Amaral, 1935) Xadrez Snake	Amaral (1935, 1963, 1976); Bailey (1966); Peters & Orejas-Miranda (1970); Pagini & Lema (1987); Lema (1994b); Franco et al. (2001)	Extreme southern part of Brazil in the state of Río Grande do Sul into Uruguay.

GENUS: *Carphophis* Gervais, 1843, is restricted to the United States. See Schmidt (1953); Conant (1975); Wright & Wright ([1957]1994); Conant & Collins (1998); and Dixon (2000a).

CONTENT: One or two species; two are shown in the following. ★★★

Carphophis: North American Wormsnake

Species-level Taxa	Citations and Remarks	Abbreviated Range
C. amoenus (Say, 1825) Eastern Wormsnake	Schmidt (1953); Conant (1975); Dundee & Rossman (1989); Dixon (2000a)	E and c United States.
C. a. amoenus (Say, 1825) Eastern Wormsnake	Conant (1975; 1985); Conant & Collins (1998); Tennant & Bartlett (2000)	E United States from s New England into South Carolina, c Georgia and c Alabama.
C. a. helenae (Kennicott, 1859) Midwestern Wormsnake	Schmidt (1953); Conant (1975, 1985); Dundee & Rossman (1989); Conant & Collins (1998); Tennant & Bartlett (2000)	The United States in Mississippi River Valley from s Ohio into s Illinois and e Arkansas into the Gulf Coast.
C. vermis (Kennicott, 1859) Western Wormsnake	Conant (1975); Dundee & Rossman (1989); Wright & Wright ([1957] 1994); Tennant (1998); Conant & Collins (1998); Werler & Dixon (2000); Tennant & Bartlett (2000) [as *C. amoenus vermis*]; Dixon (2000a)	The United States west of the Mississippi River from extreme s Iowa, se Nebraska and w Illinois south into nw Louisiana, ne Texas and disjunct populations in sw Wisconsin and se Arkansas; Dixon (2000a) restricts this snake to Bowie, Harrison, and Red River counties in Texas.

GENUS: *Cemophora* Cope, 1860, is found in North America in the United States and Mexico. See Williams & Wilson (1967); Conant (1975); Williams (1985); Smith & Smith (1993); Dixon (2000a); and Tennant & Bartlett (2000).

CONTENT: One species with three subspecies. ★★★★

Cemophora: North American Scarletsnake

Species-level Taxa	Citations and Remarks	Abbreviated Range
C. coccinea (Blumenbach, 1788) Scarletsnake	Williams & Wilson (1967); Conant (1975, 1985); Williams (1985); Dundee & Rossman (1989); Smith & Smith (1993); Conant & Collins (1998); Dixon (2000a)	E United States into Mexico from extreme s Delaware into s Florida and westward into Louisiana, Ohio, extreme e Texas, and with disjunct populations in New Jersey, Texas, and c Missouri.
C. c. coccinea (Blumenbach, 1788) Florida Scarletsnake	Conant (1975, 1985); Ashton & Ashton (1981); Williams (1985); Tennant (1997b); Conant & Collins (1998); Tennant & Bartlett (2000)	The se United States in peninsular Florida.
C. c. copei Jan, 1863 Northern Scarletsnake	Conant (1975, 1985); Ashton & Ashton (1981); Williams (1985); Dundee & Rossman (1989); Tumlison et al. (1992); Tennant (1997b, 1998); Conant & Collins (1998); Werler & Dixon (2000); Tennant & Bartlett (2000); Dixon (2000a)	E United States in all of the range except s Texas and peninsular Florida.
C. c. lineri Williams, Brown & Wilson 1966 Texas Scarletsnake	Conant (1975, 1985); Williams (1985); Smith & Dixon (1988); Smith & Smith (1993); Tennant (1998); Conant & Collins (1998); Werler & Dixon (2000); Tennant & Bartlett (2000)	The United States in extreme s Texas and ne Mexico.

The Caenophidia

GENUS: *Cercophis* Fitzinger, 1843, is found from Suriname into Brazil. See Vanzolini (1986); and Hoogmoed (1983 [1982], 1997).

Review assistance and important contributions for this section were provided by Hugo Classen and M. S. Hoogmoed.

CONTENT: One species. ★★★

Cercophis: Cercal Snake. Frank and Ramus (1995) call this the Golden Snake.

Species-level Taxa	Citations and Remarks	Abbreviated Range
C. auratus* (Schlegel, 1837) Cercal Snake	Hoogmoed (1983 [1982], 1997); Vanzolini (1986); Williams & Wallach (1989) *Remark: Fide Hoogmoed (1983 [1982] "*Dendrophis aurata* Schlegel, 1837 has never been allocated properly until now, possibly because it was confused with Schlegel's *Dryiophis auratus*, a synonym of *Oxybelis aeneus* (Wagler). The species was described on the basis of a single specimen from Suriname. The species served Fitzinger (1843) as type of his genus *Cercophis*. Duméril, Bibron and Duméril (1854) did not know where to place it and did not pursue the matter. As far as Hoogmoed was aware, the species was only cited by Schlegel (1858), it was not mentioned by Günther (1958) or by Boulenger (1893, 1894, 1896) in their Catalogues of Snakes in the British Museum, by Amaral (1930), nor Peters & Orejas-Miranda (1970) in their respective checklists of South-American snakes. Romer (1956:580) considered *Cercophis* a junior synonym of *Oxybelis*. Keiser (1974), acting in accordance with my (Hoogmoed) advice, (Hoogmoed) did not include *Dendrophis aurata* Schlegel, 1837 in the synonymy of *Oxybelis aeneus* (Wagler). Hoogmoed did investigate the type-specimen (RMNH 813) which unfortunately is in a rather poor condition (e.g. the epidermis has largely disappeared) but still good enough to allow taxonomic conclusions. In his opinion this species, described by Schlegel (1937) and made the type of a new genus by Fitzinger (1843), is completely different from any other known South American snake and therefore properly should be called *Cercophis auratus* (Schlegel, 1837)." Hoogmoed (1983 [1982]) then gave identifying characteristics.	Suriname and recently extended into the Porto Real area of Rio de Janiero in Brazil; until recently only known from the Suriname specimen.

GENUS: *Chapinophis* Campbell & Smith, 1998, is a Guatemalan snake from the cloud forests at high elevations. See Campbell & Smith (1998) and Köhler (2001).

CONTENT: One species. ★★★★

Chapinophis: Guatemalan Snake

Species-level Taxa	Citations and Remarks	Abbreviated Range
C. xanthocheilus Campbell & Smith, 1998 Guatemalan Snake	Campbell & Smith (1998); Köhler (2001)	High elevations in cloud forests of the Sierra de Minas in ec Guatemala.

GENUS: *Chersodromus* Reinhardt, 1860, ranges from Mexico into Guatemala. See Peters & Orejas-Miranda (1970), Cadle (1984a, 1984b), and Flores-Villela & Gerez (1993).

CONTENT: Two species. ★★

Chersodromus: Earth Runner

Species-level Taxa	Citations and Remarks	Estimated Range
C. liebmanni Reinhardt, 1860 Liebmann's Earth Runner	Smith & Taylor (1966); Peters & Orejas-Miranda (1970); Cadle (1984b); Pérez-Higareda & Smith (1991); Mancilla-Moreno & Camarillo-Rangel (1998)	Oaxaca in Mexico and into Guatemala.
C. rubiventris (Taylor, 1949) Red-bellied Earth Runner	Taylor (1949); Dixon & Ketchersid (1969); McCoid et al. (1980); Smith & Smith (1993)	E Queretaro and San Luis Potosí, Mexico.

GENUS: *Chilomeniscus* Cope, 1860, is a genus of sandsnake of North America. See Smith & Taylor (1966); Cadle (1984a, 1988); Stebbins (1985); Williams & Wallach (1989); and Flores-Villela & Gerez (1994).

Important contributions for this section were provided by Peter Holm.

CONTENT: Four species are recognized. A fifth species has been identified by Peter Holm (pers. comm. 22 November 2000) for Baja California, Mexico. The technical paper is being prepared for this additional species and he hopes to have it in press soon. ★★★

Chilomeniscus: North American Sandsnake

Species-level Taxa	Citations and Remarks	Abbreviated Range
C. cinctus* Cope, 1861 Banded Sandsnake	Langebartel & Smith (1954); Smith & Taylor (1966); Stebbins (1985); Brown (1997); McPeak (2000); Bartlett & Tennant (2000); Holm (pers. comm. 7 Nov. 2000) *Remark: Stebbins (1966, 1985) does not list this snake as being found in California. Brown (1997) commented on the results of his research showing that it does not occur in California.	SW United States in sw Arizona southward into Sonora, Mexico then southward into Baja California, Mexico; most of Baja California Peninsula except for northern Baja and the Cape region; Holm (pers. comm 7 Nov. 2000) considers *C. cinctus* to be valid for all populations in Arizona, and in Mexico in Sonora, Sinaloa, and Isla Tiburon.

Species-level Taxa	Citations and Remarks	Abbreviated Range
C. punctatissimus Van Denburgh & Slevin, 1921 Espiritu Santo Island or Island Burrowing Sandsnake	Murphy (1983); Murphy & Ottley (1984); Mattison (1989); Smith & Smith (1993); Flores-Villela & Gerez (1994); Holm (pers. comm 7 Nov. 2000)	Baja California del Sur, Mexico on offshore islands of Islas Espiritu Santo and Partida Sur.
C. savagei Cliff, 1954 Cerralvo Island Sandsnake	Cliff (1954); Mattison (1989); Smith & Smith (1993); Flores-Villela & Gerez (1994); McPeak (2000)	Baja California del Sur, Mexico on Cerralvo Island.
*C. stramineus** Cope, 1860 Spotted Sandsnake	Hoard (1939); Smith & Taylor (1966); Stebbins (1985); Smith & Smith (1993); Flores-Villela & Gerez (1994); McPeak (2000); Holm (pers. comm 7 Nov. 2000) *Remark: Holm (pers. comm 7 Nov. 2000) pointed out that there is a questionable record of a *C. fasciatus* from Isla Cerralvo fide Power & Banta (1974), as *C. stramineus*.	Baja California del Sur, Mexico from Loreto and probably the northern end of Magdalena Plain to the tip of Baja California and across the Gulf of California on the c coast of Sonora in mainland of Mexico; Holm (pers. comm 7 Nov. 2000) considers *C. stramineus* to be valid for the populations southeast of the Sierra de la Laguana.
C. s. stramineus Cope, 1860 Bandless Sandsnake	Hoard (1939); Smith & Taylor (1966); Mattison (1989); Smith & Smith (1993)	S Baja California del Sur, Mexico.
C. s. esterensis Hoard, 1939 Estero Salina Sandsnake	Hoard (1939); Smith & Taylor (1966); Mattison (1989); Smith & Smith (1993)	S Baja California, Mexico only from the type locality of Estero Salina.

GENUS: *Chionactis* Cope, 1860, is in North America. See Klauber (1951); Schmidt (1953); Stebbins (1966, 1985); Smith & Taylor (1966); and Wright & Wright ([1957] 1994).

Review assistance and important contributions for this section were provided by Peter Holm and James R. Dixon.

Note: Stebbins (1985) and Collins (July, 1997), recognized *C. saxatilis* Funk, 1967 and called it a "Mountain Shovelnose Snake." In the original literature of Funk (1967), the genus is described as being a mountain population found in extreme sw Arizona and known only from three unnamed peaks in the Gila Mountains. The snake had been collected between 1800 and 2400 feet. Funk (1967) described the type as being taken 2.5 air miles northeast of the Fortune Mine at an elevation of 2300 feet. It was differentiated from other *Chionactis* by decreasing tail-to-total ratio with maturity, and in males having more ventrals than females. I feel that the information supplied in the original paper contains insufficient data for a classification. I have been unable to find any more reference information on this species that reviews the status of this genus compared to other *Chionactis* to validate the species. Therefore, I have not listed the species as valid. Smith & Smith (1993) is a checklist reference but reflects a reasonable assessment of the snake. Bartlett & Tennant (2000) also chose to leave it out of their line-up but did mention "In 1967, *Chionactis saxatilis* (the mountain shovel-nosed snake) was described from the Gila Mountains of southwestern Arizona. Taxonomists have long questioned the validity of this taxon, and it has been virtually unmentioned in the ensuing years."

CONTENT: Two or three species, depending how *C. saxatilis* Funk (1967) is treated. Two species are shown in the following. ★★

Chionactis: Shovel-nosed Snake

Species-level Taxa	Citations and Remarks	Abbreviated Range
C. occipitalis (Hallowell, 1854) Western Shovel-nosed Snake	Klauber (1951); Schmidt (1953); Smith & Taylor (1966); Stebbins (1985); Smith & Smith (1993); Wright & Wright ([1957] 1994); Brown (1997); McPeak (2000)	SW United States into Mexico from sw Arizona and se California, and adjacent Nevada into Sonora and the ne corner of Baja California, southward to about San Felipe.
C. o. occipitalis (Hallowell, 1854) Mojave Shovel-nosed Snake	Klauber (1951); Stebbins (1966, 1985); Smith & Smith (1993); Wright & Wright ([1957] 1994); Brown (1997); Bartlett & Tennant (2000)	SW United States in s California, s Nevada, and w Arizona.
C. o. annulata* (Baird, 1859) Colorado Desert Shovel-nosed Snake	Klauber (1951); Schmidt (1953); Smith & Taylor (1966); Stebbins (1985); Smith & Smith (1993); Wright & Wright ([1957] 1994); Bartlett & Tennant (2000)	Colorado in sw United States outward into Mexico in n Baja California into nw Sonora.
	*Remark: Fide Smith & Smith (1993), *C. saxatilis* Funk, 1967 is a synonym. Bartlett & Tennant (2000) noted *C. saxatilis* has been virtually unmentioned under this taxon.	
C. o. klauberi (Stickel, 1941) Tucson Shovel-nosed Snake	Klauber (1951); Schmidt (1953); Stebbins (1966, 1985); Smith & Smith (1993); Wright & Wright ([1957] 1994); Bartlett & Tennant (2000)	The United States in sc Arizona in the Sonora Desert southward into n Mexico.
C. o. talpina Klauber, 1951 Nevada Shovel-nosed Snake	Klauber (1951); Schmidt (1953); Stebbins (1966, 1985); Wright & Wright ([1957] 1994); Brown (1997); Bartlett & Tennant (2000)	W United States in Nevada.
C. palarostris (Klauber, 1937) Sonora Shovel-nosed Snake	Klauber (1937); Smith & Taylor (1966); Stebbins (1985); Savage & Slowinski (1992); Smith & Smith (1993)	SW United States in sw Arizona southward into adjacent n Mexico in w Sonora.
C. p. palarostris (Klauber, 1937) Sonora Shovel-nosed Snake	Klauber (1937, 1951); Stebbins (1985); Mattison (1989); Smith & Smith (1993)	The United States in sw Arizona southward into n Mexico.
C. p. organica Klauber, 1951 Organ Pipe Shovel-nosed Snake	Klauber (1951); Stebbins (1966, 1985); Mattison (1989); Smith & Smith (1993); Bartlett & Tennant (2000)	The United States in sw Arizona southward into wc Sonora, Mexico.

GENUS: *Chironius* Fitzinger, 1826, is found in Central and South America, ranging from northern Honduras southward into southern Brazil, northeastern Argentina, and Uruguay. See Peters & Orejas-Miranda (1970); Wiest

(1978); Vanzolini (1986); Schwartz & Henderson (1991); Dixon et al. ([1993] 1995); Cei (1993); Lema (1994b); Starace (1998); Kornacker (1999); and Köhler (2001).

Review assistance and valuable comments for this section were provided by James R. Dixon.

Note: Lancini (1979), and Lancini & Kornacker (1989) mentioned "*Chironius cinnamomeus*" (Wagler, 1824) and described it as ranging in the lower Amazon region of Brazil, Peru, and the Guianas. It is not listed as valid in Hoogmoed & Gruber (1983), nor in Dixon et al. ([1993] 1995). It is discussed here because of its common appearance in literature and uncertain allocation. See the incertae sedis section for a discussion on this snake as *Natrix cinnamomea*.

CONTENT: Thirteen species are considered valid for *Chironius*. See *Natrix cinnamomea* in incertae sedis. ★★★

Chironius: Southern Whipsnake or Neotropical Whipsnake. Frank and Ramus (1995) call this the Sipo.

Species-level Taxa	Citations and Remarks	Abbreviated Range
C. bicarinatus (Wied, 1820) Two-Keeled Whipsnake	Bailey (1955); Peters & Orejas-Miranda (1970); Dixon et al. ([1993] 1995); Cei (1993); Lema (1994b); Yanosky (1998); Marques et al. (2001)	Dixon et al. ([1993] 1995) described this snake as ranging from n of Salvador, Bahia, Brazil southward along the Atlantic coast to Pelotas, Rio Grande do Sul; the range extends westward to the interior of Estado São Paulo, Brazil, the northeastern provinces of Argentina, and the western departments of Uruguay. According to them, several islands off the coast of Brazil are also inhabited by this species.
C. carinatus (Linnaeus, 1758) Agouti Snake	Peters (1960b); Roze (1966); Peters & Orejas-Miranda (1970); Duellman (1978); Savage (1980); Dixon & Soini (1986); Savage & Villa (1986); Villa et al. (1988); Dixon et al. ([1993] 1995); Smith & Smith (1993); Esqueda & La Marca (1999); Renjifo & Lundberg (1999); Kornacker (1999); Freitas (1999); Köhler (2001)	Nicaragua southward through s Central America into South America, into Colombia, Venezuela, Ecuador, Peru, ne Brazil, and its range includes Trinidad.
C. c. carinatus (Linnaeus, 1758) Keeled Whipsnake	Dixon et al. ([1993] 1995); Murphy (1997); Kornacker (1999); Boos (2001)	Wide disjunct range from the Lesser Antilles into Trinidad through the Guianas, e Venezuela, into ne Brazil in Pernambuco, Bahia, Sergipe, Ceará, and Alagoas. It is mainly associated with rain forests north of the Amazon River, usually below 1000 m.
C. c. flavopictus (Werner, 1909) Star Snake or Werner's Sipo	Peters (1960b); Pérez-Santos & Moreno (1988, 1991); Dixon et al. ([1993] 1995)	Mainly evergreen tropical rain forests along the Pacific coast of s Costa Rica, Panama, Colombia, and Ecuador.
C. c. spixii (Hallowell, 1845) Spix's Whipsnake	Pérez-Santos & Moreno (1988); Dixon et al. ([1993] 1995); Kornacker (1999); Mijares-Urrutia & Arends R. (2000)	Associated with the Llanos of Colombia and Venezuela, it is also found in other habitats.
C. exoletus (Linnaeus, 1758) Common Southern Whipsnake	Bailey (1955); Hoge et al. (1978); Savage (1980); Dixon & Soini (1986); Savage & Villa (1986); Villa et al. (1988); Cei (1993); Dixon et al. ([1993] 1995); Starace (1998), Kornacker (1999); Freitas (1999); Marques et al. (2001); Köhler (2001)	Montane cloud forests of Costa Rica and Panama into tropical rainforests of Colombia, Ecuador, e Peru, e and se Venezuela; into Amazonian Brazil, and ne Argentina; its range includes all of the Guianas.

Species-level Taxa	Citations and Remarks	Abbreviated Range
C. flavolineatus (Boettger, 1885) Brown-Striped Whipsnake or Boettger's Sipo	Bailey (1955); Peters & Orejas-Miranda (1970); Dixon et al. ([1993] 1995); Lema (1994b); Yanosky (1998); Freitas (1999)	Mainly savannas of c Bolivia, ne Paraguay, and c and w Bahia into São Paulo; including Bahia, Minas Geraís, Paraná, Mato Grosso, and e Ilhe de Marajó in Pará, Brazil.
C. fuscus (Linnaeus, 1758) Olive Whipsnake	Roze (1966); Peters & Orejas-Miranda (1970); Duellman (1978); Moonen et al. (1979); Dixon et al. ([1993] 1995); Starace (1998); Kornacker (1999); Freitas (1999); Marques et al. (2001)	N South America in Colombia, Venezuela, Suriname, French Guiana, e Peru, e Ecuador, n Bolivia, and into c Brazil. There is an isolated population in the se Brazil Atlantic rainforest. Recorded to 3500 m in Peru.
C. f. fuscus (Linnaeus, 1758) Common Olive Whipsnake	Lancini & Kornacker (1989); Dixon et al. ([1993] 1995); Starace (1998); Kornacker (1999)	Equatorial South America including Amazonian Brazil, n Bolivia, e Peru, e Ecuador, se Colombia, and s Venezuela into the Guianas. It is also closely associated with the evergreen tropical rain forests along the coast of Brazil in Bahia, Espirito Santo, Rio de Janiero, Guanabara, and São Paulo and it may eventually prove to be found in Mato Grosso.
C. f. leucometapus Dixon, Wiest & Cei, [1993] 1995 Peruvian Whipsnake	Dixon et al. ([1993] 1995)	Peru in Departaments of Huanuco, Junín, and San Martin; it ranges from 500 to 3500 m.
C. grandisquamis* (W. Peters, 1868) Ecuadorian Whipsnake	Taylor (1951); Peters (1960b); Peters & Orejas-Miranda (1970); Savage (1980); Wilson & Meyer (1985); Savage & Villa (1986); Villa et al. (1988); Dixon et al. ([1993] 1995); Köhler (2001) *Remark: Fide Dixon et al. ([1993] 1995), *C. schlueteri* (Werner, 1899), as the synonym of *C. grandisquamis*.	N Honduras, Nicaragua, Costa Rica, and Panama through Colombia into Amazonian slopes and nw Ecuador; also found on Isla Gorgona off the sw coast of Colombia. Its range incorporates the Caribbean rainforest of Central America and the Chocoan rainforest of Colombia and Ecuador.
C. laevicollis (Wied, 1824) São Paulo Brown Whipsnake or Brazilian Cipó	Bailey (1955); Peters & Orejas-Miranda (1970); Dixon et al. ([1993] 1995); Lema (1994b); Freitas (1999); Marques et al. (2001)	Rainforests along se coast of Brazil from Espirito Santo into Minas Geraís, Santa Catarina, Paraná, Bahia, and São Paulo in Brazil.
C. laurenti Dixon, Wiest & Cei, 1993 Laurent's Whipsnake	Dixon et al. ([1993] 1995)	Departments of Beni, Cochabamba, and Santa Cruz in Bolivia and into Estado Mato Grosso, Brazil and tropical rainforests of Bolivia. It may occur in the deciduous subtropical forests of sw Brazil.
C. monticola Roze, 1952 Andean Whipsnake	Roze (1952, 1966); Peters & Orejas-Miranda (1970); Lancini & Kornacker (1989); Dixon et al. ([1993] 1995); Esqueda & La Marca (1999); Kornacker (1999); Mijares-Urrutia & Arends R. (2000)	Andean cloud forests in Venezuela, Peru, Colombia, Ecuador, and into the departments of Cochabamba and Santa Cruz in Bolivia.
C. multiventris Schmidt & Walker, 1943 South American Sipo	Schmidt & Walker (1943a); Bailey (1955); Duellman (1978); Dixon et al. ([1993] 1995); Starace (1998); Kornacker (1999); Freitas (1999); Marques et al. (2001)	Cloud forests or evergreen tropical rainforests of n Venezuela, Ecuador, Colombia, the Guianas, Peru, and Brazil. It is found on Trinidad.

The Caenophidia 95

Species-level Taxa	Citations and Remarks	Abbreviated Range
C. m. multiventris Schmidt & Walker, 1943 Long-Tailed Whipsnake	Duellman (1978); Dixon et al. ([1993] 1995); Starace (1998); Kornacker (1999)	Rainforests of se Colombia, s Venezuela, Ecuador, e Peru, and nw Brazil. Reported only to 670 m in Peru.
C. m. cochranae Hoge & Romano, 1969 Cochrane's Whipsnake	Hoge & Romano (1970); Peters & Orejas-Miranda (1970); Dixon et al. (]1993)] 1995); Starace (1998); Kornacker (1999)	NE Brazil, e Venezuela, and the Guianas.
C. m. foveatus Bailey, 1955 Coastal Whipsnake	Bailey (1955); Peters & Orejas-Miranda (1970): Beçak et al. (1971); Dixon et al. ([1993] 1995)	Evergreen tropical forests of eastern coastal Brazil from Bahia to Santa Catarina. It is a lowland species.
C. m. septentrionalis Dixon, Wiest & Cei, 1993 Forest Whipsnake	Dixon et al. ([1993] 1995); Kornacker (1999); Boos (2001)	Cloud forests and evergreen rain forests of n Venezuela and Trinidad. In Trinidad it is a lowland species, but in Venezuela occurs typically at higher elevations up to 1200 m and is found in the states of Aragua, Carabobo, Districto Federal, Guarico, Miranda, and Sucre.
*C. quadricarinatus** (Boie, 1827) Central Sipo	Bailey (1955); Peters & Orejas-Miranda (1970); Dixon et al. ([1993] 1995; Yanosky (1998) *Remark: Dixon et al. ([1993] 1995) pointed out that Bailey (1955) mentioned this species has frequently erroneously been called *sexcarinatus* (Wagler, 1824) because of the presence of more than two rows of keeled dorsal scales.	Lowlands of c Brazil, e Bolivia, Paraguay, and ne Argentina. It is a lowland species.
C. q. quadricarinatus (Boie, 1827) Central Sipo	Dixon et al. ([1993] 1995); Leynaud & Bucher (1999)	Savannas and campos cerrados of ec South America in n Mato Grosso to São Paulo, Brazil into c Bolivia, and c Paraguay; one specimen is reported from Departamento Santa Cruz in the savannas of sc Bolivia.
C. q. maculoventris Dixon, Wiest & Cei, 1993 Chaco Whipsnake	Cei (1993); Dixon et al. ([1993] 1995); Leynaud & Bucher (1999)	Dry to semi-dry Chaco forests of Corrientes, Chaco, Formosa, and Santa Fe in ne Argentina into the chaqueña area of w Paraguay and probably into the Bolivian Chaco in extreme s Bolivia; it has recently been reported from Entre Ríos and Corrientes in Argentina.
*C. scurrulus** (Wagler, 1824) Rusty or Wagler's Whipsnake	Hoge (1952b); Duellman (1978); Peters & Orejas-Miranda (1970); Dixon & Soini (1986); Lancini & Kornacker (1989); Dixon et al. ([1993] 1995); Kornacker (1999); Boos (2001) *Remark: Dixon et al. ([1993] 1995) included *C. barrioi* Donoso-Barros, 1969, as a synonym.	Equatorial and evergreen tropical rainforests of the Amazon Basin including e Peru, se Colombia and e Venezuela into Brazil in the states of Minas Geraís, Rondônia, and Pará and in n Bolivia, Ecuador, Trinidad, and the Guianas.
C. vincenti (Boulenger, 1891) Saint Vincent Racer	Henderson et al. (1988); Schwartz & Henderson (1985, 1991); Dixon et al. ([1993] 1995); Henderson & Powell (1996); Censky & Kaiser *in* Crother (1999)	Rainforests of St. Vincent island in the West Indies; distribution probably from 450 to 1000 feet.

GENUS: *Clelia* Fitzinger, 1826, is distributed throughout much of the Neotropical Americas from the Lesser Antilles and tropical Mexico southward through Latin America to as far as 42° south in Argentina and east of the Andes. See Smith & Taylor (1966); Bailey *in* Peters & Orejas-Miranda (1970), Cunha & Nascimento (1978); Alvarez del Toro (1982); Cei (1986, 1993); Dixon & Soini (1986); Obst et al. (1988); Auth (1990); Scrocchi & Viña (1990); Schwartz & Henderson (1991), Marques & Lema (1992); Savage & Slowinski (1992); Strüssmann & Sazima (1993); Underwood (1993); Lema (1994b); Lee (1996); Zaher (1996); Franco et al. (1997); Campbell (1998a); Starace (1998); and Köhler (2001); Greene (1992).

Review assistance and important contributions for this section were provided by Hussam Zaher.

Note: Also reference *Boiruna*, *Oxyrhopus*, and *Pseudoboa*.

CONTENT: Nine species. ★★★

Clelia: Mussurana. Frank and Ramus (1995) also call this the Mussurana.

Species-level Taxa	Citations and Remarks	Abbreviated Range
C. bicolor* (Peracca, 1904) Bicolored Mussurana	Bailey *in* Peters & Orejas-Miranda (1970); Dixon & Soini (1986); Vanzolini (1986); Cei (1993); Franco et al. (1997); Zaher (1996); Yanosky (1998); Leynaud & Bucher (1999)	S Mato Grosso, Brazil southward into Paraguay, and n Argentina and into c Argentina.
	*Remark: Franco et al. (1997) listed this snake as a full species, based on its geographic separation from *C. quimi*. Until specimens are studied from the geographic void, it will be difficult to determine if there is a taxonomic relationship.	
C. clelia* (Daudin, 1803) Common Mussurana	Greer (1965); Roze (1966); Smith & Taylor (1966); Bailey *in* Peters & Orejas-Miranda (1970); Amaral (1976); Duellman (1978); Savage (1980); Alvarez del Toro (1982); Wilson & Meyer (1985); Dixon & Soini (1986); Savage & Villa (1986); Cei (1986, 1993); Martínez & Cerdas (1986); Villa et al. (1988); Lancini & Kornacker (1989); Scrocchi & Viñas (1990); Schwartz & Henderson (1991); Lema (1994c); Lee (1996, 2000); Zaher (1996); Starace (1998); Censky & Kaiser *in* Crother (1999); Renjifo & Lundberg (1999); Kornacker (1999); Stafford & Meyer (2000); Mijares-Urrutia & Arends R. (2000); Boos (2001) [as *Clelia clelia clelia*]; Köhler (2001)	Tropical lowlands of s Mexico southward through Central America into n South America to as far south as sw Peru, west of the Andes, and c Bolivia, east of the the Andes; and eastward through Colombia into Venezuela, the Guianas, and into the Amazon Basin of Brazil; and found on Trinidad and Grenada into the Lesser Antilles.
	*Remark 1: In older literature the taxon can be found as *Cloelia cloelia* or *Pseudoboa cloelia*.	
	*Remark: Fide Zaher (1996): "More recent taxonomic proposals have modified the composition of the genus *Clelia*, now composed of eight species. Underwood (1993) described *C. errabunda* from St. Lucia, Lesser West Indies. Lema (1994c), based on the spineless condition of the hemipenis, rightfully elevated *Clelia plumbea* to the specific level. Pre-	

The Caenophidia

Species-level Taxa	Citations and Remarks	Abbreviated Range
	viously, two other subspecies of *Clelia clelia*—*C. clelia groomei* and *C. clelia immaculata*-were placed in synonymy of *C. clelia clelia* by Bailey (1970) and Dixon et al. (1962), no subspecies are presently recognized in the genus *Clelia*."	
C. equatoriana (Amaral, 1924) Equatorial Mussurana	Amaral (1924b, [1933–1934] 1934); Bailey *in* Peters & Orejas-Miranda (1970); Vanzolini (1986); Pérez-Santos & Moreno (1988, 1991); Villa et al. (1988); Auth (1994); Zaher (1996); Köhler (2001)	N Costa Rica into the Pirri Range in e Panama southward through the Cauca Valley in Colombia into nw Ecuador, and into Bolivia.
*C. errabunda** Underwood, 1993 Saint Lucia Mussurana	Underwood (1993); Zaher (1996); Franco et al. (1997); Censky & Kaiser *in* Crother (1999) *Remark: This snake is in danger of extinction.	St. Lucia in the West Indian Antilles.
*C. montana** Franco, Marques & Puorto, 1997 Highland Mussurana	Franco et al. (1997); Marques et al. (2001) *Remark: Franco et al. (1997) placed this snake as a full species based on its geographic separation from *C. rustica*. Until specimens are studied from the geographic void between species it will be difficult to establish its taxonomic status.	Restricted to the highlands of nw São Paulo and s Minas Geraís in Brazil.
C. plumbea (Wied, 1820) Common Southeastern Mussurana	Bailey *in* Peters & Orejas-Miranda (1970); Müller (1974); Cunha & Nascimento (1983); Lema (1994b); Zaher (1996); Freitas (1999) [as *Clelia clelia*]; Marques et al. (2001)	S of the Amazon River in the Brazilian Amazon Basin, southward through the open formation of the Cerrados of c Brazil into Mato Grosso do Sul and Paraguay and the Atlantic rainforests of ne, se and s Brazil.
*C. quimi** Franco, Marques & Puorto, 1997 Brazilian Mussurana	Franco et al. (1997); Giraudo (1999) *Remark: Franco et al. (1997) placed this snake as a full species based on its geographic separation from *C. bicolor*. Until specimens are studied from the geographic void, it will be difficult to establish its taxonomic status.	S, se and wc Brazil in São Paulo, Paraná, Santa Catarina, and Minas Geraís into Argentina in Misiones.
*C. rustica** (Cope, 1878) Smooth Brown Mussurana	Bailey *in* Peters & Orejas-Miranda (1970); Amaral (1976), Miranda et al. (1982); Cei (1986, 1993); Achaval (1987); Marques & Lema (1992); Lema (1994b) [as *Pseudoboa rustica*]; Zaher (1996); Franco et al. (1997); Leynaud & Bucher (1999) *Remark: Franco et al. (1997) determined this snake as a full species based on its geographic separation from *C. montana*. Until specimens are studied from the geographic void, it will be difficult to establish its taxonomic status.	S Minas Geraís and Rio de Janiero in s Brazil southward into Paraguay, Uruguay, and s Argentina to as far south as into n Chubut, and west into Mendoza.
C. scytalina (Cope, [1866] 1867) Mexican Snake Eater or Mexican Mussurana	Peters (1960b); Bailey *in* Peters & Orejas-Miranda (1970); Savage (1980); Wilson & Meyer (1985); Savage & Villa (1986); Villa et al. (1988); Smith & Pérez-Higareda (1989); Pérez-Higareda & Smith (1991);	Both coasts of Mexico in Jalisco, Veracruz, and Yucatán southward through Belize in a disjunct range into Central America into Costa Rica and Panama; it is also reported from Colombia and Ecuador.

Species-level Taxa	Citations and Remarks	Abbreviated Range
	García & Ceballos (1994); Ramírez-Bautista (1994); Zaher (1996); Stafford & Meyer (2000); Lee (2000); Köhler (2001)	

GENUS: *Clonophis* Cope, [1888] 1889, is restricted to North American. See Conant (1943, 1975); Rossman (1963); Rossman & Powell (1985); Conant & Collins (1998); and Tennant & Bartlett (2000).

Review assistance for this section was provided by Roger Conant.

CONTENT: One species. ★★★★

Clonophis: Kirtland's Snake

Species-level Taxa	Citations and Remarks	Abbreviated Range
C. kirtlandii* (Kennicott, 1856) Kirtland's Snake	Conant (1943, 1975); Rossman (1963); Rossman & Powell (1985); Conant & Collins (1998); Tennant & Bartlett (2000) *Remark: According to Rossman & Powell (1985), it is now rarely encountered.	Midwestern United States in ne Missouri and se Wisconsin through Illinois, s Michigan, most of Indiana, and Ohio to e Pennsylvania, and adjacent New Jersey, and nc Kentucky.

GENUS: *Coluber* Linnaeus, 1758, is found throughout North America in s Canada, much of the United States, and southward through Mexico into the northern part of Central America in n Guatemala and Belize. See Ortenberger (1928); Smith & Taylor (1966); Conant (1975); Wilson (1978); Stebbins (1985); Schätti & Wilson (1986); Williams & Wallach (1989); Lee (1996, 2000); Degenhardt et al. (1996); Conant & Collins (1998); Tennant & Bartlett (2000); Bartlett & Tennant (2000); and Dixon (2000a).

Review assistance and important contributions for this section were provided by Jeffery D. Camper.

CONTENT: Normally considered to be one species with about eleven subspecies in the Americas. Sometimes *C. c. mormon* is considered a separate species, with *C. mormon* (no subspecies) and *C. constrictor* having ten subspecies. One species with eleven subspecies is listed below. ★★★

Coluber: North American Racer

Species-level Taxa	Citations and Remarks	Abbreviated Range
C. constrictor Linnaeus, 1758 Eastern Racer	Peters & Orejas-Miranda (1970); Conant (1975); Wilson (1978); Stebbins (1985); Schätti & Wilson (1986); Dundee & Rossman (1989); Lee (1996, 2000); Degenhardt et al. (1996); Brown (1997); Bartlett & Tennant (2000); Dixon (2000a); Lillywhite & Smits (1992)	S Canada, much of the United States into e Mexico, Belize, and n Guatemala.
C. c. constrictor Linnaeus, 1758 Northern Black Racer	Conant (1975); Wilson (1978); Villa et al. (1988); Conant & Collins (1998); Dixon (2000a); Tennant & Bartlett (2000)	E United States from s Maine into ne Alabama.

Species-level Taxa	Citations and Remarks	Abbreviated Range
C. c. anthicus (Cope, 1862) Buttermilk Racer	Conant (1975); Wilson (1978); Dundee & Rossman (1989); Conant & Collins (1998); Tenant (1998); Dixon (2000a); Werler & Dixon (2000); Tennant & Bartlett (2000)	W Gulf Coast of the United States in s Arkansas, Louisiana, and e Texas.
C. c. etheridgei Wilson, 1970 Tan Racer	Conant (1975); Wilson (1978); Dundee & Rossman (1989); Conant & Collins (1998); Tenant (1998); Werler & Dixon (2000); Tennant & Bartlett (2000); Dixon (2000a)	E Texas into wc Louisiana in the United States.
C. c. flaviventris Say *in* James, 1823 Eastern Yellow-bellied Racer	Conant (1975); Wilson (1978); Stebbins (1985); Dundee & Rossman (1989); Degenhardt et al. (1996); Conant & Collins (1998); Tennant (1998); Werler & Dixon (2000); Bartlett & Tennant (2000); Tennant & Bartlett (2000); Dixon (2000a)	C United States from Montana, sw North Dakota east into Iowa and south into Texas and extreme sw Louisiana. There is a disjunct population in New Mexico.
C. c. foxi (Baird & Girard, 1853) Blue Racer	Conant (1975); Wilson (1978); Tucker & Paukstis (1995); Conant & Collins (1998); Tennant & Bartlett (2000)	North America from extreme s Ontario in Canada southward into the nc United States in nw Ohio into e Iowa and se Minnesota.
C. c. helvigularis Auffenberg, 1955 Brown-chinned Racer	Conant (1975); Wilson (1978); Ashton & Ashton (1981); Tennant (1997b); Conant & Collins (1998); Tennant & Bartlett (2000)	Georgia into the Lower Chipola and Apalachicola River Valleys in the Florida panhandle in the United States.
C. c. latrunculus Wilson, 1970 Black-masked Racer	Conant (1975); Wilson (1978); Dundee & Rossman (1989) Conant & Colllins (1998); Tennant & Bartlett (2000)	SE Louisiana into n Mississippi in the United States.
C. c. mormon* Baird & Girard, 1852 Western Yellow-bellied Racer	Stebbins (1966, 1985); Wilson (1978); Degenhardt et al. (1996); Brown (1997); Bartlett & Tennant (2000); Dixon (2000a) *Remark: This snake is sometimes listed as the species of *C. mormon* (see Liner, 1994; Collins, 1991, 1997; and Degenhardt et al., 1996). Dixon (1987a), and Smith & Smith (1993), listed this snake as *C. c. mormon*.	W North America from sc British Columbia in Canada southward to s California and east into w Montana and w Colorado in the United States.
C. c. oaxaca* Jan, 1863 Mexican Racer	Smith & Taylor (1966); Wilson (1966b, 1978); Conant (1975); Pérez-Higareda & Smith (1991); Campbell (1998a); Conant & Collins (1998); Tennant (1998); Dixon (2000a); Lee (2000); Stafford & Meyer (2000); Tennant & Bartlett (2000); Köhler (2001) *Remark: According to Wilson (1978), *C. c. stejnegerianus* Cope, 1895 is a synonym.	Rio Grande Valley in and w and s Texas into n Mexico in Tamaulipas and c Veracruz, Nuevo León, Coahuila, Durango, Colima, Oaxaca, Chiapas, and into Belize and n Guatemala.
C. c. paludicola Auffenberg & Babbitt, 1953 Everglades Racer	Conant (1975); Wilson (1978); Ashton & Ashton (1981); Tennant (1997b); Conant & Collins (1998); Tennant & Bartlett (2000)	The United States in se Florida from the Everglades and Miami rim rock into the upper Florida Keys and e Florida.

Species-level Taxa	Citations and Remarks	Abbreviated Range
C. c. priapus Dunn & Wood, 1939 Southern Black Racer	Conant (1975); Wilson (1978); Ashton & Ashton (1981); Dundee & Rossman (1989); Tennant (1997b, 1998); Conant & Collins (1998); Dixon (2000a); Tennant & Bartlett (2000); Werler & Dixon (2000)	Mississippi Valley and se United States from s Indiana and se Oklahoma into s Florida.

GENUS: *Coniophanes* Hallowell *in* Cope, 1860, is distributed from extreme south Texas southward through Mexico and Central America into northern South America. See Dunn (1957); Bailey (1937c, 1939); Smith & Taylor (1966), Myers (1969b); Peters & Orejas-Miranda (1970); Campbell (1984, 1998a); Williams & Wallach (1989); and Dixon (2000a); Köhler (2001).

CONTENT: Fifteen species. ★★

Coniophanes: Black-striped Snake

Species-level Taxa	Citations and Remarks	Abbreviated Range
C. alvarezi Campbell, 1989 Chiapan Stripeless Snake	Campbell (1989); Smith & Smith (1993); Köhler (2001)	Highlands of Chiapas, Mexico from 2010 to 2135 m.
C. andresensis Bailey, 1937 San Andres Groundsnake	Bailey (1937c, 1939); Dunn & Saxe (1950); Myers (1969b); Pérez-Santos & Moreno (1988); Schwartz & Henderson (1991)	Isla San Andres, Colombia in the Lesser Antilles about 125 miles off the coast of Nicaragua.
C. bipunctatus* (Günther, 1858) Two-spotted Snake	Bailey (1939); McCoy (1969); Myers (1969b); Peters & Orejas-Miranda (1970); Villa (1971b); Campbell (1984, 1989; 1998a); Wilson & Meyer (1985); Savage & Villa (1986); Villa et al. (1988); Lee (1996, 2000); Stafford & Meyer (2000); Köhler (2001)	S Veracruz and Tehuantepec in s Mexico through Central America into nw Panama; it is found on the Bay Islands of Honduras and there is a disjunct population in nw Yucatán.
	*Remark: Wilson & Meyer (1985), questioned the validity of subspecies. Lee (2000) notes "Two subspecies recognized by some authorities, rejected by others." Lee then referenced that *C. b. biseriatus* would apply to populations in Yucatán Peninsula. Although Wilson & Meyer (1985) are most likely accurate, subspecies are given below following Lee (2000) listing of a subspecies, since it is the more recent reference.	
C. b. bipunctatus (Günther, 1858) Guenther's Two-spotted Snake	Smith & Taylor (1966); Peters & Orejas-Miranda (1970); Henderson & Hoevers (1975); Savage (1980); Wilson & Meyer (1985)	Humid lowlands from s Veracruz in se Mexico through Central America into nw Panama.
C. b. biseriatus Smith, 1940 Smith's Two-spotted Snake	Conant (1965, 1968); Smith & Taylor (1966); Alvarez del Toro (1982); Wilson & Meyer (1985); Pérez-Higareda & Smith (1991); Lee (2000)	Atlantic se Mexico in nc Chiapas and w Campeche.

Species-level Taxa	Citations and Remarks	Abbreviated Range
C. dromiciformis (W. Peters, 1863) Peters's Running Snake	Bailey (1939); Peters (1960b); Myers (1969b); Peters & Orejas-Miranda (1970); Pérez-Santos & Moreno (1991); Cadle (1989); Campbell (1989)	Pacific coastal of sw Ecuador below 100 m near the Gulf of Guayaquil into nw Peru; Cadle (1989) restricted it to Guayaquil, Ecuador.
C. fissidens (Günther, 1858) Yellow-bellied Snake	Bailey (1937c, 1939); McCoy (1969); Myers (1969b); Peters & Orejas-Miranda (1970); Savage (1980); Campbell (1984, 1989, 1998a); Wilson & Meyer (1985); Savage & Villa (1986); Villa et al. (1988); Lee (1996, 2000); Stafford & Meyer (2000); Köhler (2001)	Nayarit and Veracruz in Mexico southward throughout Central America into South America into Colombia, Ecuador, and Peru.
C. f. fissidens (Günther, 1858) Yellow-bellied Snake	Bailey (1937c); Peters (1960b); Peters & Orejas-Miranda (1970); Henderson & Hoevers (1975); Wilson & Meyer (1982, 1985); Pérez-Higareda & Smith (1991); Campbell (1998a); Lee (2000); Köhler (2001)	C Veracruz, n Chiapas, and ne Oaxaca in Mexico southward through Central America in a disjunct range from Belize, Guatemala, n Honduras and Nicaragua; fide Wilson & Meyer (1985), it supposedly occurs in c Panama; it is also reported from Colombia, Ecuador, and Peru.
C. f. convergens Shannon & Smith, 1950 Veracruz Yellow-bellied Snake	Campbell (1984, 1989); Smith (1987); Pérez-Higareda & Smith (1991); Smith & Smith (1993)	San Luis Potosí, Mexico.
C. f. dispersus Smith, 1941 Smith's Yellow-bellied Snake	Peters (1954); Smith & Taylor (1966); Campbell (1989); Smith & Smith (1993)	Mexico, most likely from Nayarit, Oaxaca, Michoacán, and Guerrero.
C. f. obsoletus Minton & Smith, 1960 Puntarenas Yellow-bellied Snake	Minton & Smith (1960); Myers (1969b); Peters & Orejas-Miranda (1970); Köhler (2001)	Puntarenas Province and Turrialba in Costa Rica.
C. f. proterops Cope, 1860 Cope's Yellow-bellied Snake	Smith & Taylor (1966); Pérez-Higareda & Smith (1991); Campbell (1984, 1989)	San Luis Potosí into wc Veracruz, and e Chiapas, Mexico.
C. f. punctigularis Cope, 1860 Spotted-throat Lizard Snake	Davis & Smith (1953); Smith & Taylor (1966); Peters & Orejas-Miranda (1970); Alvarez del Toro (1982); Wilson & Meyer (1985); Köhler (2001)	Pacific Mexico from the Isthmus of Tehuantepec southward into Honduras and Costa Rica.
C. imperialis (Baird & Girard *in* Baird, 1859) Regal Black-striped Snake	Bailey (1939); Smith & Taylor (1966); Myers (1969b); Peters & Orejas-Miranda (1970); Campbell (1984, 1998a); Wilson & Meyer (1985); Dixon (2000a); Villa et al. (1988); Lee (1996); Stafford & Meyer (2000)	Extreme s Texas southward into Caribbean n Honduras.
C. i. imperialis (Baird and Girard *in* Baird, 1859) Tamaulipan Black-striped Snake	Bailey (1939); Smith & Taylor (1966); Conant (1975); Tennant (1998); Campbell (1998a); Conant & Collins (1998); Mejenes-López et al. (1999); Werler & Dixon (2000); Dixon (2000a); Tennant & Bartlett (2000) (2000)	The United States in extreme s Texas in Cameron County southward into n Veracruz and Puebla, Mexico.

Species-level Taxa	Citations and Remarks	Abbreviated Range
C. i. clavatus (W. Peters, 1864) Red-bellied or Spotted Black-striped Snake	Bailey (1937c, 1939); Peters & Orejas-Miranda (1970); Henderson & Hoevers (1975); Alvarez del Toro (1982); Wilson & Meyer (1985); Pérez-Higareda & Smith (1991); Campbell (1998a); Lee (2000); Köhler (2001)	Atlantic versant of s Mexico from Veracruz southward through Guatemala into n Honduras at low elevations. It is associated with the Caribbean versant.
C. i. copei Hartweg & Oliver, 1938 Cope's Black-striped Snake	Bailey (1939); Smith & Taylor (1966); Wilson & Meyer (1985); Smith & Smith (1993)	Pacific Mexico of the Isthmus of Tehuantepec in Oaxaca.
C. joanae Myers, 1966 Panamanian or Joan's Highland Snake	Myers (1966a, 1969b); Peters & Orejas-Miranda (1970); Campbell (1984, 1989); Villa et al. (1988); Köhler (2001)	Highlands of e Panama in mesic montane forests east of the Canal at elevations of 500 to 1400 m.
C. lateritius Cope, 1861 (1862) Cope's Stripeless Snake	Bailey (1939); Wellman (1959); Smith & Taylor (1966); Hardy & McDiarmid (1969); Campbell (1984, 1989); Webb (1984); Ponce-Campos & Smith (2001); García & Ceballos (1994); Ponce-Campos (2001)	Nayarit, Sinaloa, Jalisco and probably extreme sw Sonora. It is found from 16 to 158 m and is associated with semi-arid regions of lowlands and foothills.
C. longinquus Cadle, 1989 Peruvian Running Snake or Remote Black-striped Snake	Cadle (1989)	NW Peru, in the Río Zaña Valley on the w slope of the Andes from 1200 to 1430 m.
C. melanocephalus (W. Peters, 1870) Black-headed Stripeless Snake	Smith & Grant (1958); Wellman (1959); Saldaña (1981, 1987); Smith & Smith (1993); Ponce-Campos & Smith (2001)	Morelos and Puebla, Mexico in the subhumid scrub forests at higher elevations of the upper Balsas Basin.
C. meridanus Schmidt & Andrews, 1936 Peninsular Stripeless Snake	Smith & Taylor (1966); McCoy (1969); Myers (1969b); Campbell (1984, 1989); Villa et al. (1988); Lee (1996, 2000); Köhler (2001)	N Yucatán in n Campeche and Quintana Roo in s Mexico.
C. piceivittis Cope, [1869] 1870 Southern Black-striped Snake	Bailey (1939); Myers (1969b); Peters & Orejas-Miranda (1970); Savage (1980); Alvarez del Toro (1982); Wilson & Meyer (1985) Savage & Villa (1986); Villa et al. (1988); Campbell (1989); Vences et al. (1998); Köhler (2001)	Guerrero in Mexico into Central America to as far as Bededero, Costa Rica.
C. p. piceivittis Cope, (1869) 1870 Southern Black-striped Snake	Hall (1951); Myers (1969); Peters & Orejas-Miranda (1970); Wilson & Meyer (1982, 1985); Alvarez del Toro (1982)	S Mexico from Guerrero into c Oaxaca and southward into Bebedero, Costa Rica in Central America.
C. p. frangivirgatus Peters, 1950 Peters's Black-striped Snake	Peters (1950); Campbell (1984); Wilson & Meyer (1985); Pérez-Higareda & Smith (1991)	Veracruz, Mexico into Belize.

Species-level Taxa	Citations and Remarks	Abbreviated Range
C. p. taylori Hall, 1951 Taylor's Black-striped Snake	Hall (1951); Peters & Orejas-Miranda (1970); Wilson & Meyer (1985)	S Mexico into n Guatemala.
C. quinquevittatus (Dumeril, Bibron & Dumeril, 1854) Five-striped Snake	Bailey (1937c, 1939); Conant (1965); Smith & Taylor (1966); Peters & Orejas-Miranda (1970); Alvarez del Toro (1982); Campbell (1984, 1989, 1998a); Pérez-Higareda & Smith (1991); Lee (2000); Köhler (2001)	Caribbean lowlands of se Mexico of s Veracruz into Guatemala; distribution seems to be disjunct.
C. sarae* Ponce-Campos & Smith, 2001 Michoacán Stripeless Snake	Wellman (1959); Ponce-Campos & Smith (2001) *Remark: Wellman (1959) recognized this snake was different and described this as an unusual *C. latritus*.	Coalcomán region of Michoacán.
C. schmidti* Bailey, 1937 Schmidt's Black-striped Snake or Faded Black-striped Snake	Bailey (1937c, 1939); McCoy (1969); Peters & Orejas-Miranda (1970); Henderson & Hoevers (1975); Campbell (1984, 1989, 1998a); Wilson & Meyer (1985); Villa et al. (1988); Stafford & Meyer (2000); Lee (2000); Köhler (2001) *Remark: Fide Wilson & Meyer (1982, 1985), this snake was treated as the subspecies *C. piceivittis schmidti*. However, Harrison's (1992) thesis confirms this is a valid species. Smith & Smith (1993) treated this taxon as a full species.	Lowlands of Yucatán Peninsula in se Mexico, Belize, and c El Petén in Guatemala.

GENUS: *Conophis* W. Peters, 1860, is from Middle American and ranges from Mexico southward through northern Central America into Costa Rica. It is usually associated with semiarid regions. See Peters et al. (1945); Wellman (1963); Smith & Taylor (1966); Peters & Orejas-Miranda (1970); Savage & Villa (1986); Lee (1996, 2000); Wilson & Meyer (1982, 1985); Campbell (1998a); and Auth et al. (1998).

CONTENT: Three species. ★★

Conophis: Road Guarder

Species-level Taxa	Citations and Remarks	Abbreviated Range
C. lineatus* (Dumeril, Bibron & Dumeril, 1854) Central American Road Guarder	Wellman (1963); Smith & Taylor (1966); Peters & Orejas-Miranda (1970); Savage (1980); Wilson & Meyer (1985); Savage & Villa (1986); Villa et al. (1988); Lee (1996, 2000); Campbell (1998a); Vences et al. (1998); Stafford & Meyer (2000); Köhler (2001) *Remark: Fide Wilson & Meyer (1985), *C. pulcher* was synonymized with *C. lineatus* and they did not show subspecies.	Veracruz and Yucatán in Mexico southward into Costa Rica; there are isolated populations in Guatemala and Belize.

Species-level Taxa	Citations and Remarks	Abbreviated Range
C. l. lineatus (Dumeril, Bibron & Dumeril, 1854) Lined Road Guarder	Barbour & Loveridge (1929a); Wellman (1963); Smith & Taylor (1966); Pérez-Higareda & Smith (1991); Campbell (1998a)	Atlantic versant of Mexico in Veracruz from Tecolutla to Lerdo de Tejada and Piedras Negras; it is associated with semi-arid habitat in coastal plains.
C. l. concolor Cope, 1866(1867) Peninsular Road Guarder	Wellman (1963); Smith & Taylor (1966); Peters & Orejas-Miranda (1970); Henderson & Hoevers (1975); Campbell (1998a); Lee (2000); Stafford & Meyer (2000); Köhler (2001)	Yucatán peninsula in s Mexico from s Campeche into n Belize, n Guatemala, ne Honduras and nw Costa Rica.
C. l. dunni Smith, 1942 Dunn's Road Guarder	Wellman (1963); Peters & Orejas-Miranda (1970); Henderson & Hoevers (1975); Campbell (1998a); Lee (2000); Stafford & Meyer (2000); Köhler (2001)	Disjunct range from s Yucatán, Belize and the Cuilco Valley in w Guatemala; El Petén southward into nw Costa Rica; associated with semiarid areas from sea level to 1000 m. Savage (1949) reported it as found in Honduras.
C. pulcher* Cope, 1869 Beautiful Road Guarder	Wellman (1963); Peters & Orejas-Miranda (1970); Alvarez del Toro (1982); Vanzolini (1986); Villa et al. (1988); Köhler (2001) *Remark 1: Wilson & Meyer (1982, 1985) do not accept subspecies for this taxon, and go so far as to indicate that *C. pulcher* is a synonym of *C. lineatus*. *Remark 2: Subspecies of *C. p. pulcher* Cope, 1869 and *C. p. similus* Bocourt, 1886 are valid fide Smith & Smith (1993). However, in researching the citations from Smith & Smith (1993), the only reference that could be found where *C. p. similus* Bocourt, 1886 was linked in literature was with their quote of using Alvarez del Toro (1960). It was shown in Appendix 2 of that work only in a list. Wellman (1963) is the last formal reviewer and did not accept subspecies.	S Mexico from Chiapas southward into c Central America as far as Tegucigalpa, Honduras and in El Salvador.
C. vittatus* W. Peters, 1860 Striped Road Guarder	Davis & Smith (1953); Peters (1954); Wellman (1963); Smith & Taylor (1966); Villa et al. (1988); García & Ceballos (1994); Köhler (2001) *Remark: No subspecies of *Conophis vittatus* were recognized by Wellman (1963). He regarded the variation from 4 dark stripes (*C. v. vittatus*) to three (*C. v. viduus*) as a continuous geographic color cline. (See Wellman, 1963: 280–281). However, Smith & Smith (1993) feel that a statement exists in Wellman (1963) that indicates a narrow area of transition in Guerrero. Based on that information, Smith & Smith (1993) stated there is a need for a more careful review of the geographic variation of the pattern. They noted Wellman's review gave no indication of geographic variation in other characteristics, hence the subspecies criteria are limited to pattern. The subspecies are usually listed as *C. v. vittatus* W. Peters	Mexico along the Pacific plain from Nayarit to Chiapas and in the Balsas Basin; and, it has been reported from n Jalisco, c Guerrero, ne Oaxaca, and c Chiapas; although sometimes listed as being from Guatemala, it has not found there according to Campbell (1998a).

Species-level Taxa	Citations and Remarks	Abbreviated Range
	1860, found in humid Pacific coastal areas of se Mexico from Nayarit south into Oaxaca and Chiapas [see Peters (1954)]; and, *C. v. viduus* Cope, 1876 found in se Mexico on the Pacific slopes of the Isthmus of Tehuantepec.	

GENUS: *Conopsis* Günther, 1858, is restricted to eastern and central-southern Mexico. Goyenechea Mayer-Goyenechea (1995), in a master's thesis, synonymized *Toluca* and *Conopsis*. Although she gave a convincing argument, her work remained formally unpublished at the time the manuscript for this work was submitted to the publisher. Although I feel her argument is valid for the consolidation of both genera into *Conopsis*, the genera remain separate in this reference because she had not at that time published her results. Later, Goyenechea & Flores-Villela (2002) defined why *Toluca* should be synonymized with *Conopsis*, with *Conopsis* with the principle priority. They noted even though historically the two genera were differentiated by Taylor & Smith (1942) by *Toluca* having a groove on each posterior maxillary tooth; that is lacking in *Conopsis*, Bogert & Olive (1945) did not consider this to be the putative diagnostic character sufficient for recognizing the genus *Toluca*, and Goyenechea & Flores-Villela concur. Unfortunately, they did not formally describe how species should be consolidated as Goyenechea (1995) had done in her unpublished master's thesis. As such, the information remains confused. Even though I agree with Goyenechea and Flores-Villela (1995), because species remain confused I have maintained both *Toluca* and *Conopsis* valid for the moment. See Taylor & Smith (1942); Smith & Taylor (1966); Bogert & Oliver (1945); Tanner (1985); and Goyenechea Mayer-Goyenechea (1995); Goyenechea & Flores-Villela (2000, 2002).

Review assistance and important contributions for this section were provided by Peter Holm.

See also genera *Ficimia*, *Gyalopion*, *Pseudoficimia*, and *Toluca*.

CONTENT: Two species, if Goyenechea Mayer-Goyenechea (1995) is followed then there are five species with *Toluca* and *Conopsis* in synonymy. ★

Conopsis: Mexican Earthsnake

Species-level Taxa	Citations and Remarks	Abbreviated Range
C. biserialis Taylor & Smith, 1942 Two-lined Mexican Earthsnake	Smith & Taylor (1942, 1966); Davis & Smith (1953); Goyenechea Mayer-Goyenechea (1995); Uribe-Peña et al. (1999)	Central Mexican Plateau.
C. nasus Günther, 1858 Large-nosed Mexican Earthsnake	Tanner (1961, 1985); Smith & Taylor (1966), Goyenechea Mayer-Goyenechea (1995); Goyenechea & Flores-Villela (2000)	Central Mexican Plateau in the Sierra Madre Occidental of Chihuahua and probably e Sonora; Holm (pers. comm 7 Nov. 2000) felt that since this is a montane species, he doubted that it extends northwest across the Río Fuerte Basin and into eastern Sonora.
C. n. nasus Günther, 1858 Plateau Mexican Earthsnake	Tanner (1961, 1985); Hardy & McDiarmid (1969); Goyenechea Mayer-Goyenechea (1995); Uribe-Peña et al. (1999)	Mexican Plateau.

Species-level Taxa	Citations and Remarks	Abbreviated Range
C. n. labialis Tanner, 1961 Chihuahua Mexican Earthsnake	Tanner (1961, 1985); Smith & Smith (1993); Goyenechea Mayer-Goyenechea (1995)	N Mexico in the Sierra Madre Occidental of sw Chihuahua.

GENUS: *Contia* Girard, 1852, is restricted to North America in western Canada and the northwestern United States. See Stebbins (1966, 1985); and Leonard & Ovaska (1998).

CONTENT: One species. ★★★★

Contia: Sharp-tailed Snake

Species-level Taxa	Citations and Remarks	Abbreviated Range
C. tenuis (Baird & Girard, 1852) Sharp-tailed Snake	Stebbins (1966; 1985); Brown (1997); Leonard & Ovaska (1998); Bartlett & Tennant (2000)	W Canada southward into nw United States, from n Pacific coast of the southern tip of Vancouver Island in British Columbia in Canada southward through Puget Sound, Willamette Valley, Cascade Mountains, Sierra Nevada Mountains, and the Central Valley to about Toulare County in California; its range includes extreme sw Canada, Oregon, Washington, and California; its distribution is patchy and is divided by the western mountain ranges from Canada into California.

GENUS: *Crisantophis* Villa, 1971, is a Central American snake. See Wellman (1963); Villa (1971a); Myers & Campbell (1981); Vanzolini (1986); Dowling & Jenner (1987); Villa et al. (1988); Williams & Wallach (1989; and Köhler (2001)).

CONTENT: One species. ★★★

Crisantophis: Crisanta's Snake. Frank and Ramus (1995) call this the Dunn's Road Guarder.

Species-level Taxa	Citations and Remarks	Abbreviated Range
C. nevermanni* (Dunn, 1937) Crisanta's Snake	Wellman (1963); Villa (1969b, 1971a, 1988); Savage (1980); Myers & Campbell (1981); Wilson & Meyer (1985); Vanzolini (1986); Savage & Villa (1986); Villa et al. (1988); Williams & Wallach (1989); Köhler (2001) *Remark: This snake is found in older literature as *Conophis nevermanni*, as seen in Peters & Orejas-Miranda (1970).	Central America from the Pacific coast of Guatemala and El Salvador through w Honduras southward into nw Costa Rica.

GENUS: *Cryophis* Bogert & Duellman, 1963, is confined to Oaxaca, Mexico. See Bogert & Duellman (1963); Duellman (1966), Myers & Campbell (1981), Dowling & Jenner (1987); and Williams & Wallach (1989).

CONTENT: One species. ★★★★

Cryophis: Cloud Forest Snake

Species-level Taxa	Citations and Remarks	Abbreviated Range
C. hallbergi Bogert & Duellman, 1963 Hallberg's Cloud Forest Snake	Bogert & Duellman (1963); Duellman (1966); Myers & Campbell (1981)	Cloud forests in Sierra de Juaréz in Oaxaca, Mexico.

GENUS: *Darlingtonia* Cochran, 1935, is isolated to Hispaniola and endemic to Haiti. Fide Hedges & Garrido (1992), this taxon may represent *Arrhyton*. See Maglio (1970); Henderson & Schwartz (1984); Schwartz (1986); Schwartz & Henderson (1991); and Hedges & Garrido (1992).

CONTENT: One species with three subspecies. ★

Darlingtonia: Hispaniolan Groundsnake. Frank and Ramus (1995) call this the Haitian Groundsnake.

Species-level Taxa	Citations and Remarks	Abbreviated Range
D. haetiana Cochran, 1935 Montane or Haitian Groundsnake	Cochran (1935); Henderson & Schwartz (1984); Schwartz (1986); Schwartz & Henderson (1991); Powell et al. *in* Crother (1999)	Haiti and Dominican Republic in the Sierra de Baoruco.
D. h. haetiana Cochran, 1935 Lower Montane Groundsnake	Schwartz & Thomas (1965); Maglio (1970); Henderson & Schwartz (1984); Schwartz (1986); Schwartz & Henderson (1991); Powell et al. *in* Crother (1999)	Haiti in Massif de la Hotte from 1000 to 5000 feet.
D. h. perfector Schwartz & Thomas, 1965 Intermediate Montane Groundsnake	Schwartz & Thomas (1965); Maglio (1970); Henderson & Schwartz (1984); Schwartz (1986); Schwartz & Henderson (1991); Powell et al. *in* Crother (1999)	Haiti in Sierra de Baoruco and southern Massif de la Selle from 2100 to 4400 feet.
D. h. vaticinata Schwartz, 1970 Higher Montane Groundsnake	Schwartz (1970, 1986); Henderson & Schwartz (1984); Schwartz & Henderson (1991); Powell et al. *in* Crother (1999)	Haiti in the Montagne Noire between 5000 to 5600 feet.

GENUS: *Dendrophidion* Fitzinger, 1843, is found in a disjunct distribution from southern Mexico southward through Central America into northern South America. See Maglio (1970); Peters & Orejas-Miranda (1970); Duellman (1978); Savage (1980); Wilson & Meyer (1982, 1985), Savage & Villa (1986); Lieb (1988, 1991); Starace (1998); Kornacker (1999); and Köhler (2001).

Review assistance for this section was provided by Carl S. Lieb.

CONTENT: Eight species. ★★★

Dendrophidion: Forest Racer. Frank and Ramus (1995) also call this the Forest Racer.

Species-level Taxa	Citations and Remarks	Abbreviated Range
D. bivittatum (Dumeril, Bibron & Dumeril, 1854) Colombian Forest Racer	Amaral (1931b); Stuart (1932); Peters (1960b); Peters & Orejas-Miranda (1970); Lieb (1988); Pérez-Santos & Moreno (1988); Renjifo & Lundberg (1999)	Moderate elevations in c Colombia from 500 to 1500 m.
D. boshelli* Dunn, 1944 Dunn's Forest Racer	Dunn (1944c); Peters & Orejas-Miranda (1970); Lieb (1988); Pérez-Santos & Moreno (1988); *Remark: *D. boshelli* is very similar to *D. percarinatum* fide Lieb (1988), and is known from only two specimens. This snake is also listed in incertae sedis because Lieb (1988) could not clarify its taxonomic status.	Known only from the type locality of Volcanes, Município de Caparrapi in Departamento Cundinamarca in Colombia.
D. brunneum (Günther, 1858) Interandean Forest Racer	Amaral (1930f); Peters (1960b); Peters & Orejas-Miranda (1970); Vanzolini (1986); Lieb (1988)	Interandean valleys in Ecuador and into adjacent Departamento Cajamarca in Peru from 1500 to 2100 m.
D. dendrophis* (Schlegel, 1837) Common Forest Racer or Olive Forest Racer	Peters (1960b); Stuart (1932); Peters & Orejas-Miranda (1970); Amaral (1976); Duellman (1978); Nascimento et al. (1987); Dixon & Soini (1986); Lieb (1988); Pérez-Santos & Moreno (1988, 1991); Starace (1998); Kornacker (1999) *Remark: Specimens from the s Amazon Basin of Peru, Bolivia, and Brazil that are usually referred to as *D. dendrophis* appear to represent an unnamed, widespread species (see Lieb, 1988).	N Amazon Basin from the Guianas and n Brazil westward into e Ecuador and Peru from 140 to 700 m; it is associated with forests and woody places in the Paraguay and Amazon river valleys.
D. nuchale* (W. Peters, 1864) Black-naped Forest Racer	Savage (1980); Wilson & Meyer (1985); Savage & Villa (1986); Lieb (1988, 1991a); Villa et al. (1988); Lancini & Kornacker (1989); Campbell (1998a); Kornacker (1999); Stafford & Meyer (2000); Lee (2000); Köhler (2001) *Remark: *D. clarki* Dunn, 1933, is a junior synonym according to Lieb (1988, 1991). However, Wilson & Meyer (1985), used *D. clarki* and noted Leib's work on the subject, and pointed out that there would be changes in the nomenclature. Smith & Smith (1993), considered *D. clarki* Dunn, 1933 to be valid. I have followed Lieb (1988, 1991).	In a discontinuous range, starting from Central America in Belize southward through Central America into South America to as far as the Andean foothills of Ecuador and the Cordillera de la Costa in Zulia state in the northwest of Venezuela; it may eventually show up in s Mexico. It is often associated with rain forest but can be found in other habitats; found from 125 to 1500 m.
D. paucicarinatum (Cope, 1893) Talamanca Forest Racer	Roze (1966); Peters & Orejas-Miranda (1970); Savage (1980); Savage & Villa (1986); Lieb (1988 1991b);Villa et al. (1988); Lancini & Kornacker (1989); Köhler (2001)	Restricted to the Cordillera Central and Cordillera Talamanca uplands in s Costa Rica and w Panama from 1040 to 1500 m.
D. percarinatum (Cope, 1894) Brown Forest Racer	Stuart (1932); Peters & Orejas-Miranda (1970); Wilson & Meyer (1982, 1985); Savage & Villa (1986); Lieb (1988, 1996); Villa et al. (1988); Lancini & Kornacker (1989); Renjifo & Lundberg (1999); Köhler (2001)	Central America in the Caribbean lowland rainforests from n Honduras and e Nicaragua southward into e Panama from sea level to 550 m and, in the Pacific versant in rainforests from sw Costa Rica

Species-level Taxa	Citations and Remarks	Abbreviated Range
		southward into Colombia, and Ecuador from sea level to 1200 m; and, shown as being found in the Cordillera de la Costa in Venezuela.
D. vinitor Smith, 1941 Barred Forest Racer	Wilson (1966a); Smith & Taylor (1966); Peters & Orejas-Miranda (1970); Savage (1980); Alvarez del Toro (1982); Savage & Villa (1986); Lieb (1988, 1991c); Villa et al. (1988); Pérez-Higareda & Smith (1991); Lee (1996, 2000); Campbell (1998a); Stafford & Meyer (2000); Köhler (2001)	In a discontinuous range from Veracruz in s Mexico southward through Central America into e Panama and adjacent Colombia northern Chacó region from 50 to 1300 m.

GENUS: *Diadophis* Baird & Girard, 1853, is from North America and found in Canada, the United States, Mexico, and has recently been introduced into the Cayman Islands. Note: only one species is indicated here; however, some authorities recognize up to 3 species (*D. amabilis*, *D. punctatus* and *D. regalis*). This group needs further study. See Smith & Taylor (1966); Schmidt (1953); Stebbins (1966, 1985); Dixon (2000a); Schwartz & Henderson (1991); and Smith & Smith (1993). Mecham (1956) listed all forms of western and Texas Trans-Pecos varieties as *D. punctatus arnyi*. Wright and Wright ([1957] 1994) proposed that *D. regalis* and *D. punctatus* as a single species. I followed what seems to be the latest trend of using a single species as *D. punctatus*.

Review assistance and important contributions for this section were provided by Troy D. Hibbits.

1) Smith & Smith (1993) showed that Sokolov (1988) listed *Diadophis elinorae*. But, I have been unable to validate this taxon and it has not been included in the list.

2) In a pers. comm. from Troy Hibbits (14 August 2000), he stated that his research supports the multi-species idea for *Diadophis* and several of the races should be elevated to species. However, his research (to date) has not yet given adequate support to publish his thinking. Troy Hibbits (pers. comm. 15 August 2000), noted that most of the west coast forms are poorly differentiated from one another, and fall into two groups. In this reclassification, one group includes *pulchellus* and *occidnetalis*, and they are undistinguishable from one another, but distinguished from the other group by more extensive encroachment of ventral pigmentation onto the dorsum. In the second group, *amabilis*, *modestus*, *vandenbughi*, *similis*, and *anthonyi* more or less undistinguishable from one another, but can be distinguished from the first group.

CONTENT: Usually considered to be one species, but can be found with as many as three species in literature. ★★

Diadophis: Ring-necked Snake

Species-level Taxa	Citations and Remarks	Abbreviated Range
D. punctatus (Linnaeus, 1766) Ring-necked Snake	Stebbins (1985); Dundee & Rossman (1989); Schwartz & Henderson (1991); Smith & Smith (1993); Brown (1997); Dixon (2000a)	Nova Scotia in se Canada through much of the United States into Mexico.
D. p. punctatus (Linnaeus, 1766) Southern Ring-necked Snake	Conant (1975); Ashton & Ashton (1981); Schwartz & Henderson (1991); Tennant (1997b); Conant & Collins (1998); Tennant & Bartlett (2000)	E United States; Schwartz and Henderson (1991) reported that it has been introduced onto Grand Cayman Island.

Species-level Taxa	Citations and Remarks	Abbreviated Range
D. p. acricus Paulson, 1968 Key Ring-necked Snake	Conant (1975); Tennant (1997b); Conant & Collins (1998); Tennant & Bartlett (2000)	Big Pine Key and the lower keys in Florida.
D. p. amabilis Baird & Girard, 1853 Pacific Ring-necked Snake	Smith & Taylor (1966); Conant (1975); Stebbins (1985); Smith & Smith (1993); Wright & Wright ([1957] 1994) [as *D. amabilis amabilis*]; Brown (1997); Bartlett & Tennant (2000)	Foothills of Central Valley of California between lat. 36° to 39° and the San Francisco Bay region.
D. p. anthonyi Van Denburgh & Slevin, 1923 South Todos Santos Island Ring-necked Snake or Anthony's Ring-necked Snake	Van Denburgh & Slevin (1923); Smith & Taylor (1966); Smith & Smith (1993); McPeak (2000)	South Todos Santos Island, Baja California, Mexico.
D. p. arnyi Kennicott, 1859 Prairie Ring-necked Snake	Schmidt (1953); Conant (1975); Stebbins (1985); Dixon (2000a); Smith & Smith (1993); Degenhardt et al. (1996); Tennant (1998); Conant & Collins (1998); Bartlett & Tennant (2000); Tennant & Bartlett (2000); Werler & Dixon (2000a)	C United States from extreme sw Wisconsin and extreme South Dakota into sc Texas and e New Mexico.
D. p. dugesi Villada, 1875 Mexican Ring-necked Snake	Gelbach (1965); Smith & Taylor (1966); Pérez-Higareda & Smith (1991); Smith & Smith (1993); Uribe-Peña et al. (1999)	Around Mexico City in the Distrito Federal, Mexico.
D. p. edwardsi (Merrem, 1820) Northern Ring-necked Snake	Conant (1975); Conant & Collins (1998); Tennant & Bartlett (2000)	C North America north of Mexico from Nova Scotia in Canada into ne Minnesota southward into n Georgia, ne Alabama, and n Mississippi Valley into Illinois.
D. p. modestus Bocourt, 1866 *in* Dumeril, Mocquard & Bocourt, 1870–1909 San Bernardino Ring-necked Snake	Stebbins (1966, 1985) Wright & Wright ([1957] 1994) [as *D. amabilis modestus*]; Brown (1997); Bartlett & Tennant (2000)	C coastal California.
D. p. occidentalis Blanchard, 1923 Northwestern Ring-necked Snake	Schmidt (1953); Stebbins (1966, 1985); Wright & Wright ([1957] 1994) [as *D. amabilis occidentalis*]; Brown (1997); Bartlett & Tennant (2000)	NW United States from Sonoma county in California northward into Washington.
D. p. pulchellus Baird & Girard, 1853 Coral-bellied Ring-necked Snake	Stebbins (1966, 1985); Wright & Wright ([1957] 1994) [as *D. amabilis pulchellus*]; Brown (1997); Bartlett & Tennant (2000)	W slopes of the Sierra Nevada Mountains California.
*D. p. regalis** Baird & Girard, 1853 Regal Ring-necked Snake	Conant (1975); Stebbins (1985); Smith & Smith (1993); Wright & Wright ([1957] 1994) [as *D. regalis regalis*]; Degenhardt et al. (1996); Brown (1997); Tennant (1998); Conant & Collins (1998); Werler &	W United States into Mexico from Idaho and se California? And wc New Mexico into wc Arizona south into Durango and San Luis Potosí.

The Caenophidia

Species-level Taxa	Citations and Remarks	Abbreviated Range
	Dixon (2000); Bartlett & Tennant (2000); Dixon (2000a); Tennant & Bartlett (2000)	
	*Remark 1: Fide McDiarmid et al. (1999), *D. p. blanchardi* Schmidt & Smith, 1944, is a synonym of *D. p. regalis*.	
	*Remark 2: Brown (1997), doubts the presence of this snake in California, and supported the statement with comments from the results of his research.	
D. p. similis Blanchard, 1923 San Diego Ring-necked Snake	Smith & Taylor (1966); Stebbins (1985); Smith & Smith (1993); Brown (1997); McPeak (2000); Bartlett & Tennant (2000)	S California southward into Baja California in Mexico.
*D. p. stictogenys** Cope, 1860 Mississippi Ring-necked Snake	Conant (1975); Dundee & Rossman (1989); Tennant (1998); Conant & Collins (1998); Dixon (2000a); Tennant & Bartlett (2000); Werler & Dixon (2000)	Mississippi Valley from s Illinois to the Gulf Coast of e Texas into w Alabama.
	*Remark: Troy Hibbits (pers. comm. 15 August 2000), felt that *stictogenys* is, at best, poorly differentiated from *arnyi* and *punctatus*, respectively. Western *stictogenys* appear very *arnyi*-like, and eastern individuals very *punctatus*-like. He stated that this indicated *stictogenys* represents an intergrade form between *punctatus* and *arnyi*.	
D. p. vandenburghi Blanchard, 1923 Monterey Ring-necked Snake	Schmidt (1953); Stebbins (1966, 1985); Wright & Wright ([1957] 1994) [as *D. amabilis vandenburghi*]; Brown (1997); Bartlett & Tennant (2000)	California from Ventura County into Santa Cruz County.

GENUS: *Diaphorolepis* Jan, 1863, is found in northwestern South America. See Bogert (1964); Nicéforo-María (1950); Peters & Orejas-Miranda (1970); Hillis (1990); Kornacker (1999); Köhler (2001); and Dowling & Pinou (2003).

CONTENT: Two species. ★★★★

Diaphorolepis: Double-Keeled-Scale or Frog-Eating Snake. Frank and Ramus (1995) call this the Frog-eating Snake.

Species-level Taxa	Citations and Remarks	Abbreviated Range
D. laevis Werner, 1923 Colombian Frog-eating Snake	Werner (1923); Amaral (1931b); Peters & Orejas-Miranda (1970); Pérez-Santos & Moreno (1988); Hillis (1990); Kornacker (1999)	Known only from one specimen and the holotype, which is from Colombia, lacks precise locality data.
D. wagneri Jan, 1863 Ecuadorian Frog-eating Snake	Peters (1960c); Bogert (1964); Peters & Orejas-Miranda (1970); Villa et al. (1988); Pérez-Santos & Moreno (1988, 1991); Kornacker (1999); Köhler (2001)	Darién, Panama into w Colombia and w Ecuador. The report of one from eastern Ecuador is suspect.

GENUS: *Dipsas* Laurenti, 1768, is found from tropical Mexico southward into northern South America. See Peters (1960c); Smith & Taylor (1966); Peters & Orejas-Miranda (1970), Duellman (1978); Savage (1980); Lancini (1979); Williams & Wallach (1989); Lancini & Kornacker (1989); Lema (1994b); Starace (1998); Kornacker (1999); Köhler (2001).

Review assistance and important contributions for this section were provided by Ronaldo Fernandes.

See also genera *Sibon*, *Sibynomorphus*, and *Tropidodipsas*.

CONTENT: About thirty species; thirty species are listed below and *D. brevifaces* has been placed in incertae sedis. ★★

Dipsas: American or Neotropical Snail-eating Snake, Snail-eater or Snail-sucker. It is also sometimes referred to as the Thirst Snake. Frank and Ramus (1995) call this the Snail Eater.

Species-level Taxa	Citations and Remarks	Abbreviated Range
*D. albifrons** (Sauvage, 1884) Sauvage's Snail-eater	Hoge (1951); Peters (1960c); Peters & Orejas-Miranda (1970); Amaral (1976); Sazima (1992); Wallach (1995); Porto & Fernandes (1996); Yanosky (1998); Marques et al. (2001)	S Brazil in Santa Catarina, São Paulo, and Mato Grosso and into Paraguay.
	*Remark 1: Hoge (1951), originally described subspecies as *D. a. albifrons* (Sauvage, 1884) from s Brazil and Paraguay; and, *D. a. cavalheiroi* Hoge, 1950 (1951) restricted to São Paulo, Queimada Grande Island in Brazil. Peters (1960c), concluded that the subspecies Hoge recognized were not possible at that time. Peters & Orejas-Miranda (1970), did not recognize subspecies, and there does not seem to be a formal revalidation since then.	
	*Remark 2: Fide Porto & Fernandes (1996), the single specimen of *Dipsas neivai* examined by Peters (1960c) was erroneously identified as a *D. albifrons*.	
D. articulata (Cope, 1868) Lower Central American Snail-eater	Peters (1960c, 1964b); Peters & Orejas-Miranda (1970); Savage (1980); Savage & Villa (1986); Villa et al. (1988); Savage & Slowinski (1992); Auth (1994); Köhler (2001)	Lower elevations in Costa Rica and Panama.
D. bicolor (Günther, 1895) Bicolored Snail-eater	Peters (1960c); Peters & Orejas-Miranda (1970); Savage (1980); Wilson & Meyer (1985); Savage & Villa (1986); Villa et al. (1988); Savage & Slowinski (1992); Cruz (1997); Köhler (2001)	Pacific versant of Honduras, s Nicaragua, and Costa Rica, southward to Peru.
D. boettgeri (Werner, 1901) Boettger's Snail-eater	Peters (1960c); Peters & Orejas-Miranda (1970); Wallach (1995)	Andean slopes of n Bolivia and s Peru.
D. catesbyi (Sentzen, 1796) Catesby's Snail-eater	Peters (1956, 1960c); Peters & Orejas-Miranda (1970); Duellman (1978); Fulger & Walls (1978); Dixon & Soini (1986); Lancini & Kornacker (1989); Wallach (1995); Starace (1998); Kornacker (1999); Freitas (1999)	Amazon regions of n South America and the Andean slopes of Bolivia, Peru, Ecuador, and Colombia to the coast of Venezuela, the Guianas, and the northern half of Brazil.

Species-level Taxa	Citations and Remarks	Abbreviated Range
D. chaparensis Reynolds & Foster, 1992 Chapare Snail-eater	Reynolds & Foster (1992)	Chapare region of Bolivia; known only from the type locality listed "as the road between Cochabamba to Villa Tunari, Chapare Province, Departamento Cochabamba, Bolivia."
D. copei (Günther, 1872) Cope's Snail-eater	Peters (1960c); Roze (1966); Peters & Orejas-Miranda (1970); Lancini & Kornacker (1989); Starace (1998); Vidal et al. (1998); Kornacker (1999)	S Venezuela and the Guianas into Peru.
D. elegans* (Boulenger, 1896) Elegant Snail-eater	Peters (1960c); Smith & Taylor (1966); Kofron (1982); Vanzolini (1986); Wallach (1995) *Remark: Fide Wallach (1995), this is *D. oreas elegans*.	Known only from the type locality in Oaxaca, Mexico in Tehuantepec.
D. ellipsifera* (Boulenger, 1898) Boulenger's Snail-eater	Peters (1960b, 1960c); Peters & Orejas-Miranda (1970); Kofron (1982); Vanzolini (1986); Smith & Smith (1993); Wallach (1995) *Remark 1: Vanzolini (1986) noted that Kofron (1982) listed this snake as *D. elegans*, but Smith & Smith (1993) listed *D. elegans* and a synonym of *D. ellipsifera*. *Remark 2: Fide Wallach (1995) this is *D. oreas ellipsifera*.	High-andean slopes of w Ecuador.
D. gaigeae (Oliver, 1937) Gaige's Snail-eater	Oliver (1937); Peters (1960c, 1964b); Smith & Taylor (1966); Casas-Andreu (1982); Kofron (1982); Savage & Slowinski (1992); García & Ceballos (1994); Ramírez-Bautista (1994); Wallach (1995)	Mexico, known only from along the Colima and Jalisco coasts.
D. gracilis (Boulenger, 1902) Graceful Snail-eater	Peters (1960b, 1960c); Peters & Orejas-Miranda (1970); Pérez-Santos & Moreno (1991); Savage & Slowinski (1992); Wallach (1995)	NW Ecuador into Peru.
D. incerta* (Jan, 1863) Jan's Snail-eater	Peters (1960c); Peters & Orejas-Miranda (1970); Lema (1994b); Marques et al. (2001) *Remark 1: *D. incerta* is being studied by Ronaldo Fernandes and the readers should expect revisions by him on this species in the near future. *Remark 2: Marques et al. (2001:171) noted that this name is used in describing a snake from French Guiana and may not be the species represented in their work.	Coastal se Brazil from Espirito Santo into Santa Catarina and Rio Grande do Sul.
D. indica Laurenti, 1768 Amazonian Snail-eater	Peters (1960c); Peters & Orejas-Miranda (1970); Amaral (1976); Wallach (1995); Starace (1998); Yanosky (1998); Freitas (1999); Kornacker (1999); Marques et al. (2001); Sazima 1992	Amazonia in South America north of the Tropic of Capricorn.
D. i. indica Laurenti, 1768 Amazonian Big-headed Snail-eater	Peters (1960b, 1960c); Peters & Orejas-Miranda (1970); Dixon & Soini (1986); Starace (1998); Kornacker (1999)	Amazon Basin of Brazil, Colombia, Ecuador, Peru, and the Guianas.

Species-level Taxa	Citations and Remarks	Abbreviated Range
*D. i. bucephala** (Shaw, 1802) Large-headed Snaileater	Peters (1960c) Peters & Orejas-Miranda (1970); Vanzolini (1986); Cei (1993); Sazima (1992)	High Amazon in Ecuador and Peru through Bolivia into the Beni Basin; Misiones in Argentina, and the se Atlantic coast of Brazil.
D. i. cisticeps (Boettger, 1885) Box-headed Snail-eater or False Yarará Snail-eater	Peters (1960c); Peters & Orejas-Miranda (1970); Hoge (1970)	Paraguay and Bolivia.
D. i. ecuadorensis Peters, 1960 Ecuadorian Amazon Snail-eater	Peters (1960b, 1960c); Peters & Orejas-Miranda (1970); Duellman (1978); Pérez-Santos & Moreno (1991)	Amazon drainage of Ecuador and Peru.
D. i. petersi Hoge & Romano, 1976 Peters's Atlantic Snail-sucker	Hoge & Romano (1976); Vanzolini (1986)	Atlantic versant of the coast of Brazil from Paraná into s Bahia.
D. latifasciata (Boulenger, 1913) Striped Snail-eater	Peters (1960b, 1960c); Peters & Orejas-Miranda (1970); Pérez-Santos & Moreno (1991); Wallach (1995)	Amazonian slopes of the Andes in extreme s Ecuador and n Peru.
D. latifrontalis (Boulenger, 1905) Venezuelan Snail-eater	Peters (1960a, 1960c); Roze (1966); Peters & Orejas-Miranda (1970); Lancini & Kornacker (1989); Wallach (1995); Esqueda & La Marca (1999); Kornacker (1999)	Lower Amazonian slopes from Venezuela into s Ecuador.
D. maxillaris (Werner, 1909) Werner's Snail-eater	Peters (1960c); Smith & Taylor (1966); Villa et al. (1988); Smith & Smith (1993); Köhler (2001)	Known only from the type locality in Tabasco, Mexico.
D. neivai Amaral, 1926 Atlantic or Bahia Snail-eater	Peters (1960c); Peters & Orejas-Miranda (1970); Amaral (1976); Vanzolini (1986); Porto & Fernandes (1996); Freitas (1999); Marques et al. (2001)	Closely associated with the Atlantic forests in the Brazilian states of Alagoas, Espírito Santo, Minas Geraís; Paraná, São Paulo, and Santa Catarina; and it also inhabits the semideciduous forest of Itagibá in Bahia.
D. oreas (Cope, 1868) Ecuadorian Montane Snail-eater	Peters (1960b, 1960c) Peters & Orejas-Miranda (1970); Vanzolini (1986); Pérez-Santos & Moreno (1991)	Western slopes of the Andes in Ecuador at higher elevations.
D. pavonina Schlegel, 1837 Ringed Thirst Snake	Peters (1960b, 1960c, 1964b); Roze (1966); Peters & Orejas-Miranda (1970); Duellman (1978); Fugler & Walls (1978); Dixon & Soini (1986); Lancini & Kornacker (1989); Savage & Slowinski (1992); Wallach (1995); Porto & Fernandes (1996); Starace (1998); Kornacker (1999)	N South America from Colombia through Venezuela and the Guianas into Peru, Pará in Brazil, and Bolivia.
D. perijanensis (Alemán, 1953) Perija Snail-eater	Alemán (1953); Peters & Orejas-Miranda (1970); Peters (1970b); Lancini & Kornacker (1989); Kornacker (1999)	The type is from Jamayaujaina in the Sierra de Perijá in the state of Zulia, Venezuela.

Species-level Taxa	Citations and Remarks	Abbreviated Range
D. peruana (Boettger, 1898) Peruvian Snail-eater	Schmidt & Walker (1943a); Peters (1960c); Peters & Orejas-Miranda (1970); Wallach (1995)	Eastern slopes of the Andes in s Peru.
D. polylepis (Boulenger, 1912) Many-scaled Snail-eater	Boulenger (1912); Peters (1960c); Peters & Orejas-Miranda (1970); Reynolds & Foster (1992)	Known only from the type locality of Huancabamba, Peru above 3000 feet.
D. pratti (Boulenger, 1897) Pratt's Snail-eater	Amaral (1931b); Prado (1941a) [as *Dipsas niceforoi*]; Prado (1941c, 1943) [as *Dipsas tolimensis*]; Peters (1960c); Peters & Orejas-Miranda (1970); Pérez-Santos & Moreno (1988)	Cordillera Central in Colombia.
D. sanctijoannis (Boulenger, 1911) Saint Joann Snail-eater	Rendahl & Vestergren (1941); Peters (1960c); Peters & Orejas-Miranda (1970); Pérez-Santos & Moreno (1988); Wallach (1995)	Slopes of Cordillera Occidental and Medellín in Colombia.
D. schunkii (Boulenger, 1908) Schunk's Snail-eater	Boulenger (1908); Peters (1960c); Peters & Orejas-Miranda (1970); Wallach (1995)	Amazonian slopes of the Andes in Peru.
D. temporalis (Werner, 1909) Temporal Snail-eater	Peters (1960b, 1960c, 1964b); Peters & Orejas-Miranda (1970); Villa et al. (1988); Pérez-Santos & Moreno (1988); Savage & Slowinski (1992); Köhler (2001)	Atlantic versant of Panama into the Pacific versants of Colombia and Ecuador.
D. tenuissima Taylor, 1954 Taylor's Snail-eater	Taylor (1954); Peters (1960c); Peters & Orejas-Miranda (1970); Savage (1980); Savage & Villa (1986); Villa et al. (1988); Savage & Slowinski (1992); Auth (1994); Köhler (2001)	Costa Rica and Panama.
D. variegata (Dumeril, Bibron & Dumeril, 1854) Variegated Snail-eater	Peters (1960c); Roze (1966); Peters & Orejas-Miranda (1970); Amaral (1976); Villa et al. (1988); Auth (1994); Wallach (1995); Starace (1998); Kornacker (1999); Köhler (2001)	Panama in s Central America southward into n South America to Colombia, Venezuela, Ecuador, Peru, and possibly Bolivia; it is closely associated with the Amazon.
D. v. variegata (Dumeril, Bibron & Dumeril, 1854) Common Variegated Snail-eater	Peters (1960b, 1960c, 1964b); Peters & Orejas-Miranda (1970); Lancini & Kornacker (1989); Starace (1998); Kornacker (1999)	N South America from Ecuador and Peru northward into Colombia, Venezuela, the Guianas, and Amazonian Brazil.
D. v. nicholsi (Dunn, 1933) Nichols's Snail-eater	Dunn (1933a); Peters (1960b, 1960c); Peters & Orejas-Miranda (1970)	Atlantic versant in e coast of Panama southward through Colombia into nw Ecuador.
D. v. trinitatis Parker, 1926 Trinidad Thirst Snake	Parker (1926); Peters (1960c); Underwood (1962); Peters & Orejas-Miranda (1970); Murphy (1997); Boos (2001)	Confined to Trinidad in the southern Caribbean.
D. vermiculata Peters, 1960 Vermiculate Snail-eater	Peters (1960b, 1960c); Peters & Orejas-Miranda (1970); Wallach (1995)	Amazonian Ecuador and ne Peru.

Species-level Taxa	Citations and Remarks	Abbreviated Range
D. viguieri (Bocourt, 1884) Bocourt's Snail-eater	Breeder (1946); Peters (1960c); Peters & Orejas-Miranda (1970); Villa et al. (1988); Savage & Slowinski (1992); Wallach (1995); Köhler (2001)	Pacific coast of Costa Rica and Panama.

GENUS: *Ditaxodon* Hoge, 1958, is from southeastern Brazil. See Peters & Orejas-Miranda (1970); Lema (1994b); and Thomas et al. (in press).

Review and important contributions for this section were provided by Robert A. Thomas.

CONTENT: One species. ★★★

Ditaxodon: Brown Bush Snake

Species-level Taxa	Citations and Remarks	Abbreviated Range
D. taeniatus (Hensel, 1868) Brown Bush Snake or Hensel's Snake	Hoge (1958); Peters & Orejas-Miranda (1970); Amaral (1976) [as *Conophis taeniatus*]; Lema (1994b); Thomas et al. (in press)	SE Brazil in Rio Grande do Sul, São Paulo and Paraná; it may eventually be found in Mato Grosso do Sul.

GENUS: *Drepanoides* Dunn, 1928, [*Drepanodon* Peracca, 1896 was the original name for the genus and was found to be preoccupied by *Drepanodon* Nesti, 1826; *Drepanoides* Dunn, 1928 is the substitute name for *Drepanodon* Peracca, 1896] is found from northern into central South America and is sometimes called the Egg-Eater. See Bailey *in* Peters & Orejas-Miranda (1970); Vanzolini (1986); and Starace (1998). Peters & Orejas-Miranda (1970) commented that the species *Arrhyton quenselii* Andersson, 1901, may belong here, but it was omitted awaiting additional information on the species.

CONTENT: One species. ★★★★

Drepanoides: American Egg-eating or Black-collared Snake

Species-level Taxa	Citations and Remarks	Abbreviated Range
D. anomalus (Jan, 1863) Amazon Egg-eating Snake or Black-collared Snake	Rendahl & Vestergren (1941); Hoge (1952b); Peters (1960b); Bailey *in* Peters & Orejas-Miranda (1970); Duellman (1978); Fugler & Walls (1978); Hoogmoed (1983 [1982]); Vanzolini (1986); Dixon & Soini (1986); Pérez-Santos & Moreno (1988, 1991); Savage & Slowinski (1992); Starace (1998)	C Bolivia into s Colombia along the Andes, and Brazil in Pará and Rondônia, as well in the Guianas in French Guiana.

GENUS: *Drymarchon* Fitzinger, 1843, is a genus that can be found from the southern United States in Texas southward into South America to as far as northern Argentina. There is an allopatric population in Florida. See Smith & Taylor (1966); Peters & Orejas-Miranda (1970); McCranie (1980c); Savage & Villa (1986); Obst et al. (1988); Smith & Smith (1993); Cei (1993); Starace (1998); Campbell (1998a); Kornacker (1999); and Dixon (2000a).

The Caenophidia

Review assistance and important contributions for this section were provided by James R. Dixon.

CONTENT: One species, with about eight subspecies. Eight subspecies are shown here. ★★★

Drymarchon: Cribo and Indigo Snake

Species-level Taxa	Citations and Remarks	Abbreviated Range
D. corais (Boie, 1827) Indigo Snake or Cribo	Peters & Orejas-Miranda (1970); McCranie (1980c); Freiberg (1982); Wilson & Meyer (1985); Villa et al. (1988); García & Ceballos (1994); La Marca et al. (1995); Lee (1996, 2000); Starace (1998); Campbell (1998a); Kornacker (1999); Dixon (2000a)	SE United States, allopatric in Florida, from Texas and Sonora in Mexico southward through Latin America into n Argentina and Paraguay.
D. c. corais (Boie, 1827) Common Cribo	Roze (1966); Peters & Orejas-Miranda (1970); Amaral (1976); Dixon & Soini (1977); McCranie (1980c); Dixon & Soini (1986); Dixon & Kornacker (1989); Henle & Ehrl (1991b); Cei (1993); Starace (1998); Leynaud & Bucher (1999); Kornacker (1999); Freitas (1999); Boos (2001)	The Caribbean nation of Trinidad and Tobago, and South America from Venezuela southward through the Amazon and Paraguay basins into Chaco and Formosa in n Argentina.
*D. c. couperi** (Holbrook, 1842) Eastern Indigo Snake	Conant (1975); McCranie (1980c); Ashton & Ashton (1981); Tennant (1997b); Conant & Collins (1998); Tennant & Bartlett (2000) *Remark: Collins (1991) elevated this to full species based on allopatry and what I feel is a weak diagnosis, so I have maintained it as a subspecies.	This is an allopatric population in the se United States in Florida.
D. c. erebennus (Cope, 1860) Western Indigo Snake	Smith & Taylor (1966); Conant (1975); McCranie (1980c); Pérez-Higareda & Smith (1991); Tennant (1998); Conant & Collins (1998); Werler & Dixon (2000); Tennant & Bartlett (2000); Dixon (2000a)	North America from s Texas into adjacent nc Mexico.
D. c. margaritae Roze, 1959 Margarita Island Cribo	Roze (1966); Peters & Orejas-Miranda (1970); McCranie (1980c); Lancini & Kornacker (1989); Kornacker (1999)	Restricted to W Margarita Island off the coast of e Venezuela.
D. c. melanurus (Dumeril, Bibron & Dumeril, 1854) Black-tailed Cribo	Peters (1960b); Smith & Taylor (1966); Roze (1966); Peters & Orejas-Miranda (1970); McCranie (1980c); Alvarez del Toro (1982); Wilson & Meyer (1985); Lancini & Kornacker (1989); Pérez-Higareda & Smith (1991); Smith & Smith (1993); Campbell (1998a); Kornacker (1999); Lee (2000); Stafford & Meyer (2000)	S Mexico through Central America into n South America as far as Venezuela and Peru; Degenhardt et al. (1996) question the Bailey (1928) reference that discusses this snake's presence in New Mexico.
D. c. orizabensis (Dugès, 1905) Orizaba Cribo	Smith & Taylor (1966); Ramírez-Bautista (1978); McCranie (1980c); Pérez-Higareda & Smith (1991); Smith & Smith (1993)	S Mexico, known only from a few localities in Veracruz.
*D. c. rubidus** Smith, 1941 Mexican West Coast Cribo	Davis & Smith (1953); Peters (1954); Smith & Taylor (1966); Peters & Orejas-Miranda (1970); McCranie (1980c); Casas-Andreu (1982); Smith & Smith (1993); Ramírez-Bautista (1994)	W Mexico from sw Chihuahua and Sinaloa southward into extreme sw Guatemala.

Species-level Taxa	Citations and Remarks	Abbreviated Range
	*Remark: Although *D. c. cleofae* Brock, 1942, from Tres Marias Islands, Mexico was listed in Smith & Smith (1993), it is included here as a synonym of *D. c. rubidus* based on McCranie (1980c). The classification does not seem to have been formally challenged or supported since that time.	
D. c. unicolor Smith, 1941 Central American Cribo	Smith & Taylor (1966); Peters & Orejas-Miranda (1970); McCranie (1980c); Alvarez del Toro (1982); Wilson & Meyer (1985); Smith & Smith (1993)	Chiapas, Mexico southward into the Pacific versant of Nicaragua; at low and moderate elevations.

GENUS: *Drymobius* Fitzinger, 1843, is distributed from extreme southern Texas southward through Latin America into western South America in Colombia, Venezuela, and Ecuador to as far as southern Peru. See Stuart (1932); Smith & Taylor (1966); Peters & Orejas-Miranda (1970); Wilson (1975a); Savage (1980); Obst et al. (1988); Starace (1998); Kornacker (1999); Dixon (2000a) and Köhler (2001).

CONTENT: Four species based on Wilson (1975a). ★★★

Drymobius: Speckled, Central American, or Neotropical Speckled Racers; Crother et al. (2000) listed this as Neotropical Racers.

Species-level Taxa	Citations and Remarks	Abbreviated Range
D. chloroticus (Cope, 1886) Green Highland Racer	Davis & Smith (1953); Smith & Taylor (1966); Peters & Orejas-Miranda (1970); Wilson (1970, 1975a, 1975b); Alvarez del Toro (1982); Wilson & Meyer (1985); Wilson & Cruz-Diaz (1986); Villa et al. (1988); Pérez-Higareda & Smith (1991); Köhler (2001)	Atlantic versant of Mexico from San Luis Potosí southward into Central America to the Pacific versant of Costa Rica; in moderate and intermediate elevations.
D. margaritiferus (Schlegel, 1837) Speckled Racer	Dunn (1944f); Roze (1966); Peters & Orejas-Miranda (1970); Wilson (1974a, 1975c); Conant (1975); Savage (1980); Wilson & Meyer (1985); Wilson & Cruz-Diaz (1986); Savage & Villa (1986); Villa et al. (1988); García & Ceballos (1994); Lee (1996); Campbell (1998a); Stafford & Meyer (2000); Köhler (2001)	Extreme s Texas in the United States southward into Mexico along the Atlantic coast and in s Sonora on the Pacific versant southward to the Caribbean coast of Colombia in South America; it seems to be absent from the Central Plateau of Mexico.
D. m. margaritiferus (Schlegel, 1837) Northern Speckled Racer	Smith & Taylor (1966); Peters & Orejas-Miranda (1970); Wilson (1974a); Conant (1975); Alvarez del Toro (1982); Campbell (1998a); Tennant (1998); Conant & Collins (1998); Werler & Dixon (2000); Tennant & Bartlett (2000); Lee (2000); Dixon (2000a); Köhler (2001)	North, Central, and n South America from extreme s Texas in the Rio Grande Valley of the United States southward into s Colombia in low and moderate elevations.
D. m. fistulosus Smith, 1942 Central American Speckled Racer	Davis & Smith (1953); Peters (1954); Smith & Taylor (1966); Wilson (1974a, 1975c); Casas-Andreu (1982); Wilson & Meyer (1985); Ramírez-Bautista (1994)	Sonora into Nayarit and Oaxaca in w Mexico.

Species-level Taxa	Citations and Remarks	Abbreviated Range
D. m. maydis Villa, 1968 Great Corn Island Racer	Villa (1968); Peters & Orejas-Miranda (1970); Wilson (1974a, 1975c); Wilson & Meyer (1985); Köhler (2001)	Only from Great Corn Island of Nicaragua.
D. m. occidentalis Bocourt, 1890 *in* Dumeril, Mocquard & Bocourt, 1870–1909 Western Speckled Racer	Smith & Taylor (1966); Peters & Orejas-Miranda (1970); Wilson (1974a, 1975c); Alvarez del Toro (1982); Wilson & Meyer (1985); Köhler (2001)	Pacific versant of Chiapas, Mexico into Guatemala and El Salvador in low and moderate elevations.
D. melanotropis (Cope, 1875) Black Speckled Racer	Peters & Orejas-Miranda (1970); Wilson (1975d); Savage (1980); Wilson & Meyer (1985); Wilson & Cruz-Diaz (1986); Savage & Villa (1986); Villa et al. (1988); Köhler (2001)	Central America from Honduras and Nicaragua into Panama.
D. rhombifer (Günther, 1860) Blotched Racer	Peters (1960b); Roze (1966); Peters & Orejas-Miranda (1970); Wilson (1975a); Duellman (1978); Savage (1980); Vanzolini (1986); Dixon & Soini (1986); Savage & Villa (1986); Villa et al. (1988); Lancini & Kornacker (1989); Starace (1998); Kornacker (1999); Köhler (2001)	Nicaragua in Central America into n South America in Colombia, Peru, Venezuela, Ecuador, and Bolivia; also reported from Ilha de Maraca in Brazil, and the Guianas.

GENUS: *Drymoluber* Amaral, 1929, is distributed in tropical South America from nw Colombia, Ecuador, and Peru into the Guianas and ne Brazil. See Stuart (1932); Peters & Orejas-Miranda (1970); Obst et al. (1988); Lancini & Kornacker (1989); Starace (1998); and Kornacker (1999).

CONTENT: Two species. ★★★

Drymoluber: Glossy Racer. Frank and Ramus (1995) call this the Woodland Racer.

Species-level Taxa	Citations and Remarks	Abbreviated Range
D. brazili (Gomes, 1918) Brazil's Glossy Racer	Gomes (1918); Amaral (1923, 1976); Stuart (1932); Peters & Orejas-Miranda (1970); Campos-Nogueira (2001)	SC Brazil.
D. dichrous (W. Peters, 1863) Common Glossy Racer	Maglio (1932); Peters (1960b); Peters & Orejas-Miranda (1970); Amaral (1976); Duellman (1978); Fugler & Walls (1978); Cunha et al. (1985); Dixon & Soini (1986); Lancini & Kornacker (1989); Henle & Ehrl (1991b); Starace (1998); Freitas (1999); Kornacker (1999)	N South America in Colombia, Ecuador, e Peru, and Amazonian Venezuela into Amazonian n Brazil in Amazônia to Marahão; also found in the Guianas.

GENUS: *Echinanthera* Cope, 1894, is formed from the *Rhadinaea "undulatus"* group and is found from Brazil southward into Argentina. See Peters & Orejas-Mi randa (1970) [*Lygophis, Leimadophis,* and *Liophis*]; Myers (1974); Di-Bernardo & Lema (1987); Di-Bernardo (1992, 1994, 1996); Giraudo et al. (1996); and Marques et al. (2001).

See also the genus *Taeniophallus*. *Taeniophallus* is listed with seven species, but based on Di-Bernardo (1996), only *T. nicagus* belongs to this genus. If this is the case, then the species *affinis* (Günther, 1858), *bilineata* (Fischer, 1885), *brevirostris* (W. Peters, 1863), *occipitalis* (Jan, 1863), *persimillis* (Cope, 1869), and *poecilopogon* (Cope, 1863) may belong in *Echinanthera*. Di-Bernardo (1996) is an abstract of a paper given at the Fourth Latin American Congress of Herpetology in Santiago-Chile. It differentiated *nicagus* (with a simple sulcus) from the other species (that have divided sulcus). I chose to keep this group in *Echinanthera* because separating genera based on this single hemipenal characteristic seems to be the only basic morphological argument between Di-Bernardo (1996) and Myers & Cadle (1994) on what constitutes the genus *Taeniophallus* and *Echinanthera*. To divide the genera based on this single characteristic does not seem to be logical considering the Myers & Cadle (1994) work.

CONTENT: Six species listed here, but as may as 12 species may be involved. ★★

Echinanthera: Brazilian Tropical Snake. Frank and Ramus (1995) also call this the Brazilian Tropical Snake.

Species-level Taxa	Citations and Remarks	Abbreviated Range
E. amonena (Jan, 1863) Brazilian Tropical Snake	Peters & Orejas-Miranda (1970) [as *Lygophis*]; Myers (1974); Di-Bernardo (1992); Marques et al. (2001)	SE Brazil in s Minas Geraís into s Paraná, São Paulo, and Rio de Janiero.
E. cephalomaculata Di-Bernardo, 1994 Di-Bernardo's Tropical Snake	Di-Bernardo (1994, 1996)	SE Brazil.
E. cephalostriata Di-Bernardo, 1996 Espirito Santo Tropical Snake	Di-Bernardo (1992 part; 1996); Marques et al. (2001)	SE Brazil in Espirito Santo into ne Santa Catarina.
E. cyanopleura (Cope, 1885) Cope's Tropical Snake	Di-Bernardo (1992, 1996); Lema (1994b); Giraudo et al. (1996); Marques et al. (2001)	SE and s Brazil east of Espirito Santo along the Atlantic coast into se Rio Grande do Sul, through Minas Geraís, Catarina; Rio de Janiero, São Paulo, Paraná, and Santa and, into Misiones in Argentina.
E. melanostigma (Wagler *in* Spix, 1824) Wagler's Tropical Snake	Peters & Orejas-Miranda (1970) [as *Liophis*]; Amaral (1976); Dixon (1980) [as incertae sedis *Natrix melanostigma*]; Lema (1989a, 1994b); Di-Bernardo (1992); Marques et al. (2001)	Brazil in se Bahia southward through Espirito Santo, se Minas Geraís, Rio de Janiero, and e São Paulo.
E. undulata (Wied, 1824) Wied's Tropical Snake	Amaral (1976); Di-Bernardo (1992); Marques et al. (2001)	SE and s Brazil in s Rio de Janiero, s Minas Geraís, São Paulo, Paraná, and into ne Santa Catarina; a lowland form.

GENUS: *Elaphe* Fitzinger *in* Wagler, 1833, is found from Canada and the United States southward into Central America to Nicaragua. See Smith & Taylor (1966); Stebbins (1966, 1985); Conant (1975); Staszko & Walls (1994), Wright & Wright ([1957] 1994); Degenhardt et al. (1996); Schulz (1996); Vaughan et al. (1997); Campbell (1998a); Conant & Collins (1998), Dixon (2000a) and Köhler (2001).

Review assistance and important contributions for this section were provided by Robert A. Thomas and R. K. Vaughan.

Note: Also reference genera *Bogertophis* Dowling & Price, 1988 and *Senticolis* Dowling & Fries, 1987. See Schulz (1996) for his comments about the relationship of *Bogertophis* and *Senticolis* and his justification for not recognizing these genera and listing them as *Elaphe*.

CONTENT: About six species depending on the classification accepted for *E. flavirufa phaescens*, *E. obsoleta rossalleni*, *E. obsoleta*, *E. guttata*, and *E. vulpina* subspecies. ★★

Elaphe: Common Ratsnake

Species-level Taxa	Citations and Remarks	Abbreviated Range
E. bairdi (Yarrow *in* Cope, 1880) Baird's Ratsnake	Smith & Taylor (1966); Conant (1975); Olson (1977); Staszko & Walls (1994); Schulz (1996); Tennant (1998); Conant & Collins (1998); Werler & Dixon (2000); Tennant & Bartlett (2000); Dixon (2000a)	Very spotty distribution in w, c and s Texas into Mexico in Coahuila and Nuevo León into Tamaulipas.
E. flavirufa (Cope, 1866 [1867]) Central American Ratsnake or Mexican Corn or Nightsnake	Dowling (1952a); Peters & Orejas-Miranda (1970); Wilson & Meyer (1985); Villa et al. (1988); Staszko & Walls (1994); Lee (1996, 2000); Schulz (1996); Campbell (1998a)	S Mexico through c Central America into Nicaragua; some authorities believe it is found into Colombia.
E. f. flavirufa (Cope, 1866[1867]) Tamaulipian Ratsnake or Tamaulipian Nightsnake	P. W. Smith & Darling (1952); Smith & Taylor (1966); Alvarez del Toro (1982); Pérez-Higareda & Smith (1991); Staszko & Walls (1994); Lee (1996, 2000); Schulz (1996); Stafford & Meyer (2000); Köhler (2001)	Coastal distribution in e Mexico from Tamaulipas southward into Belize, and, w and n Guatemala.
E. f. matudai* Smith, 1941 Matuda's Ratsnake or Matuda's Nightsnake	Smith & Taylor (1966); Alvarez del Toro (1982); Staszko & Walls (1994); Schulz (1996); Köhler (2001) *Remark: This snake is known from a single specimen collected over 50 years ago, and fide Staszko & Walls (1994), along with other authorities, suggest that it might be an aberrant *E. f. pardalina*. Nonetheless, it has very distinctive dorsal features that warrants its separation until more information is available.	Salto de Agua, Chiapas in Mexico.
E. f. pardalina (W. Peters, 1868) Honduran Ratsnake or Honduran Nightsnake	Dowling (1952); Peters & Orejas-Miranda (1970); Henderson & Hoevers (1975); Schultz (1993); Staszko & Walls (1994); Schulz (1996); Köhler (2001)	S Chiapas, Mexico into Nicaragua including Corn Island, Nicaragua; its range includes Guatemala and Honduras.
E. f. phaescens* Dowling, 1952 Yucatán Ratsnake or Yucatán Nightsnake	Dowling (1952); Flores-Villela (1993); Lee (1996, 2000); Schulz (1996); Köhler (2001) *Remark: Both Flores-Villela (1993) and Liner (1994), listed this snake as *Elaphe phaescens* Dowling, 1952, and it is often encountered as such in the literature. The specific recognition may have been based on Smith & Smith (1976), who listed *phaescens* as a full species. Smith & Taylor (1966), noted in their work that there was a geographic void between the range of this snake and that of its closest	Yucatán in Mexico.

Species-level Taxa	Citations and Remarks	Abbreviated Range
	relative. They used this for the decision that it was a full species. However, Lee (1996) felt that the only difference with *E. f. flavirufa* and *phaescens*, is that *flavirufa* occurs at the base of the Yucatán peninsula and has more than 31 body blotches. *Elaphe phaescens* is found at the dry outer end of the peninsula and has fewer than 32 body blotches. Based on this information, Lee (1996), felt that *phaescens* should be treated as a subspecies. I follow Lee (1996, 2000), and have listed *phaescens* as a subspecies here.	
E. f. polysticha Smith & Williams, 1966 Roatán Island Ratsnake	Smith & Williams (1966b); Peters & Orejas-Miranda (1970); Smith & Smith (1993); Schulz (1996)	Only from Roatán Island in the Bay Islands of Honduras.
E. gloydi* Conant, 1940 Eastern Foxsnake	Conant (1940; 1975); Powell (1990); Staszko & Walls (1994); Schulz (1996); Conant & Collins (1998); Tennant & Bartlett (2000); Crother et al. (2000) *Remark: Collins (1991, 1997) elevated this snake to full specific status.	S Ontario, Canada southward into the United States into e Michigan and nc Ohio.
E. guttata (Linnaeus, 1766) Cornsnake	Dundee & Rossman (1989); Schwarz & Henderson (1991); Staszko & Walls (1994); Smith et al. (1994); Schulz (1996); Vaughan et al. (1997); Dixon (2000a)	The United States into northern Mexico and introduced onto Grand Cayman Island in the Caribbean.
E. g. guttata* (Linnaeus, 1766) Cornsnake	Thomas (1974); Conant (1975); Ashton & Ashton (1981); Smith et al. (1994); Staszko & Walls (1994); Schulz (1996); Tennant (1997b); Vaughan et al. (1997); Conant & Collins (1998); Werler & Dixon (2000); Tennant & Bartlett (2000); Dixon (2000a) *Remark: Duellman & Schwartz (1958), and Ashton & Ashton (1981), reported that *E. g. rosaeca* (Cope, 1888), is a synonym of *E. g. rosacea* (Cope, 1888), of the Florida Keys. This was later supported by Tennant (1997b), and Smith et al. (1994), but they stated that a reappraisal is in order, particularly in view of the recognition of *E. g. meahllmorum*. Schulz (1996), separated it as a valid subspecies. I have followed what seems to me to be the more current thinking in the United States, as noted by the absence of *E. g. rosacea* in such works as Tennant (1997b), Conant & Collins (1998), and Tennant & Bartlett (2000).	Fide Schwartz and Henderson (1991), it has been introduced onto Grand Cayman Island; its original range is the e United States west and north of the *E. g. emoryi* range to as far south as into the Florida Keys.
E. g. emoryi* (Baird & Girard, 1853) Great Plains Ratsnake	Thomas (1974); Conant (1975); Dundee & Rossman (1989); Pérez-Higareda & Smith (1991); Staszko & Walls (1994); Smith et al. (1994); Schulz (1996); Conant & Collins (1998); Tennant (1998); Dixon (2000a); Werler & Dixon (2000); Bartlett & Tennant (2000); Tennant & Bartlett (2000) *Remark: *E. g. intermontana* Woodbury & Woodbury, 1942, of Utah and Colorado is a synonym fide	Utah southward into n Mexico as far south as San Luis Potosí and northward into the Great Plains of the United States, and as far east as from c Louisiana in the south and sw Illinois [the race with more numerous blotches in a discontinuous range from 29° N latitude, just south of San Antonio in Texas into Utah and Colorado]. Pérez-Higareda & Smith (1991) show this snake to be present in Veracruz.

The Caenophidia

Species-level Taxa	Citations and Remarks	Abbreviated Range
	Thomas (1974) in an unpublished master's thesis. This is also noted by Smith et al. (1994).	
E. g. meahllmorum Smith, Chiszar, Staley & Tepedelen, 1994 Southwestern Ratsnake	Smith et al. (1994); Degenhardt et al. (1996); Vaughan et al. (1997); Dixon (2000a); Werler & Dixon (2000)	Missouri, s Arkansas, Louisiana, and eastward into s Texas and n Mexico.
E. obsoleta (Say *in* James, 1823) Eastern Ratsnake	Dundee & Rossman (1989); Staszko & Walls (1994); Schulz (1996); Dixon (2000a) *Remark 1: *Scotophis laetus* Baird & Girard, 1853 [later *Elaphe laetus*], is a junior synonym fide Smith et al. (1994). *Remark 2: Burbrink et al. (2000), felt that the subspecies of *E. obsoleta* do not represent distinct genetic lineages. Instead, they think the evidence from their work with mitochondrial DNA analysis points to three well-supported mitochondrial DNA clades confined to particular geographic areas in the eastern United States. They intend to discuss taxonomic recommendations elsewhere following evaluations of their morphometric data. They also include *E. o. deckerti* as a subspecies in their checklist. It is not normally a recognized subspecies	S Canada southward through the midwestern and e United States almost to the frontier with Mexico.
E. o. obsoleta (Say *in* James, 1823) Black Ratsnake	Conant (1975); Dundee & Rossman (1989); Staszko & Walls (1994); Schulz (1996); Conant & Collins (1998); Tennant & Bartlett (2000)	S Ontario in Canada south through the ne United States from sw New England south into Georgia, and from sw Wisconsin into Oklahoma and n Louisiana with disjunct populations in e Ontario, New York, and the Oklahoma panhandle.
E. o. lindheimeri (Baird & Girard, 1853) Texas Ratsnake or Lindheimer's Ratsnake	Conant (1975); Dundee & Rossman (1989); Smith & Smith (1993); Staszko & Walls (1994); Schulz (1996); Conant & Collins (1998); Tennant (1998); Werler & Dixon (2000); Tennant & Bartlett (2000); Dixon (2000a)	The United States in bayou and swampland areas of Louisiana, Arkansas, and Texas. From c and s Texas, eastward into the Pearl River and Mississippi River.
E. o. quadrivittata (Holbrook, 1836) Yellow Ratsnake	Conant (1975); Ashton & Ashton (1981); Campbell *in* Wright & Wright ([1957] 1994); Schulz (1996); Tennant (1997b); Collins (1998); Tennant & Bartlett (2000)	SE United States in s North Carolina south into s Florida and Upper Keys.
*E. o. rossalleni** Neill, 1949 Everglades Ratsnake	Ashton & Ashton (1981); Staszko & Walls (1994); Schulz (1996); Tennant (1997b); Conant & Collins (1998); Tennant & Bartlett (2000) *Remark: This is a questionable subspecies.	The United States in the Everglades and Kissimmee Prairie of s Florida.
E. o. spiloides (Dumeril, Bibron & Dumeril, 1854) Gray Ratsnake	Conant (1975); Ashton & Ashton (1981); Staszko & Walls (1994); Schulz (1996); Tennant (1997b); Conant & Collins (1998); Tennant & Bartlett (2000)	In the midwest and southeast in the United States, from sw Georgia into Mississippi and the n Mississippi Valley in e and s Illinois and sw Indiana.

Species-level Taxa	Citations and Remarks	Abbreviated Range
E. vulpina* (Baird & Girard, 1853) Western Foxsnake	Conant (1940, 1975); Powell (1990); Staszko & Walls (1994); Schulz (1996); Conant & Collins (1998); Tennant & Bartlett (2000); Crother et al. (2000) *Remark: Collins (1991, 1997) elevated this snake to full species status.	S Ontario in Canada southward into the United States in Nebraska, and the upper peninsula of Michigan into c Illinois and n Missouri.

GENUS: *Elapomorphus* Wiegmann, 1843, is from tropical South America. Lema (1984) defined *Elapomorphus* as being divided into two subgenera, *Elapomorphus* (*sensu stricto*) and *Phalotris* Cope, 1862. He considered *Elapomorphus* as containing tropical forms found from the Guyanas southward into eastern Brazil in Minas Geraís and Rio de Janeiro. The *Phalotris* group is composed of more species and Lema noted these species were found from the central "planaltos" and southern Brazil into the llanos of Uruguay and Argentina. Recently, *Phalotris* Cope, 1862, was elevated to a full genus in a master's thesis by Ferrarezzi. Ferrarezzi later separated the two as full genera in several papers. Although this separation does not seem to have been formally published with details, the two are separated here, based on Ferrarezzi (1993) and Puorto & Ferrarezzi (1993) general use of the names and descriptions. The reader should note that further analysis may catagorize *Phaltoris* as either a subgenus or genus, and that I do not feel that this question has yet been properly resolved. Lema (pers. comm. 24 July 2000), noted that there are plans to resolve this by publications from Ferrarezzi and Hofstadler-Deiques. See *Phalotris* for further information on this group. Only two species have been left in *Elapomorphus* (*lepidus* and *quinquelineatus*). The rest of the species have been relegated to *Phalotris*. See Peters & Orejas-Miranda (1970); Lema (1975, 1984c, 1994b); Amaral (1976); Cei (1986, 1993); Williams & Wallach (1989); Ferrarezzi (1993b); Puorto & Ferrarezzi (1993), Lema & Hofstadler-Deiques (1995); and Marques et al. (2001).

Review assistance and important contributions for this section were provided by Thales de Lema.

See also *Apostolepis* and *Phalotris*.

CONTENT: Two species; but as many as fifteen may be recognized depending on the authority and if *Phalotris* is considered to be a subgenus of *Elapomorphus*. ★

Elapomorphus: Brazilian Burrowing Snake [Frank & Ramus (1995) refer to these as Diadem Snakes which is appropriate but a name which I believe should be relegated to a group of Old World snakes.]

Species-level Taxa	Citations and Remarks	Abbreviated Range
E. lepidus* Reinhardt, 1861 Black Crown Burrowing Snake	Peters & Orejas-Miranda (1970); Amaral (1976) Lema (1984c); Savage & Slowinski (1992); Lema & Hofstadler-Deiques (1995); Marques et al. (2001) *Remark: *E. (P.) coronatus* Sauvage, 1877, and *E. wuchereri* (part) Günther, 1861, are synonyms according to Lema & Hofstadler-Deiques (1995). The *E. (P.) coronatus* Sauvage, 1877 is the *Apostolepis coronata* (Sauvage, 1877) mentioned in Peters & Orejas-Miranda (1970) and Amaral (1976). Thales de Lema (pers. comm. 19 September 2001) examined the holotype of *Apostolepis coronatus* in Paris and tells me that the Sauvage specimen described as	Atlantic slopes in the states of Bahia, Espirito Santo, se Minas Geraís, and Rio de Janeiro and e São Paulo in Brazil.

Species-level Taxa	Citations and Remarks	Abbreviated Range
	having the internasal fused with the prefrontal plates is definitely an *Elapomorphus*.	
E. quinquelineatus* (Raddi, 1820) Five-lined Burrowing Snake	Hoge (1959bc); Peters & Orejas-Miranda (1970); Lema (1984c, 1994b); Cei (1986); Lema & Hofstadler-Deiques (1995); Marques et al. (2001)	E and c Brazil in Bahia, Minas Gerais, Rio de Janeiro, São Paulo, Santa Catarina, and Rio Grande do Sul.
	*Remark: *E. accedens* Jan, 1862, and *E. wuchereri* (in part) Günther, 1861, are synonyms according to Lema & Hofstadler-Deiques(1995).	

GENUS: *Emmochliophis* Fritts & Smith, 1969, is from Ecuador. These are very rare snakes and the two species are only known from the holotype for each. See Fritts & Smith (1969); Hillis (1990); Sheil (1998); and Dowling and Pinou (2003).

See also *Synophis*.

CONTENT: Two species. ★★★

Emmochilophis: Stiff Snake. Frank and Ramus (1995) call this the Pinchinda Snake.

Species-level Taxa	Citations and Remarks	Abbreviated Range
E. fugleri Fritts & Smith, 1969 Pichinda Snake	Fritts & Smith (1969); Vanzolini (1986); Hillis (1990); Pérez-Santos & Moreno (1991); Sheil & Grant (2001)	W Ecuador, the type locality is Pichincha; known only from the holotype.
E. miops* (Boulenger, 1898) Peramba Snake	Nicéforo-María (1950) [as *Diaphololepis*]; Peters & Orejas-Miranda (1970); Hillis (1990); Sheil (1998); Sheil & Grant (2001)	Paramba in western Ecuador; known only from the holotype.
	*Remark: Sometimes placed in the genus *Symophis*, but the change to *Emmochliophis* was recently confirmed by Sheil (1998).	

GENUS: *Enuliophis* McCranie & Villa, 1993, lives in Latin America and is found from Central America southward to as far as Peru. It has not been reported from Ecuador but I would not be surprised if someday it is found there. See Peters & Orejas-Miranda (1970); McCranie & Villa (1993) and Köhler (2001). This genus was listed in Peters & Orejas-Miranda (1970), as *Enulius sclateri* (Boulenger, 1894).

CONTENT: One species. ★★★★

Enuliophis: White-headed Snake. Frank and Ramus (1995) call this the the Colombia Long-tailed Snake.

Species-level Taxa	Citations and Remarks	Abbreviated Range
E. sclateri (Boulenger, 1894) White-headed Snake	Savage (1980); Savage & Villa (1986); Pérez-Santos & Moreno (1988); McCranie & Villa (1993); Köhler (2001)	S Nicaragua and Costa Rica and Panama southward into Colombia and Peru.

GENUS: *Enulius* Cope, 1871, ranges from Mexico into South America to as far as Colombia. See Dunn (1938); Smith & Taylor (1966); Smith et al. (1967); Peters & Orejas-Miranda (1970); Wilson & Meyer (1985); and Köhler (2001).

CONTENT: Four species. ★★★

Enulius: Middle American Long-tailed Snake. Frank and Ramus (1995) call this the Mexican Long-tailed Snake.

Species-level Taxa	Citations and Remarks	Abbreviated Range
E. bifoveatus McCranie & Köhler, 1999 Guanaja Long-tailed Snake	McCranie & Köhler (1999); Köhler (2001)	Isla Guanaja in the Islas de la Bahía off the north coast of Honduras.
E. flavitorques (Cope, 1869) Pacific Long-tailed Snake	Dunn (1938); Peters & Orejas-Miranda (1970); Savage (1980); Wilson & Meyer (1985); Savage & Villa (1986); Villa et al. (1988); Köhler (2001)	S Jalisco, Mexico into n Colombia.
E. f. flavitorques (Cope, 1869) Pacific Long-tailed Snake	Peters & Orejas-Miranda (1970); Wilson & Meyer (1985); Pérez-Santos & Moreno (1988); Smith & Smith (1993); Köhler (2001)	Mexico through Central America into Colombia.
E. f. sumichrasti Bocourt, 1883 *in* Dumeril, Mocquard & Bocourt, 1870–1909 Sumichrast's Long-tailed Snake	Smith & Taylor (1966); Smith et al. (1967); Alvarez del Toro (1982); Wilson & Meyer (1985); Smith & Smith (1993); Köhler (2001)	Pacific slopes of s Mexico.
E. f. unicolor (Fischer, 1880) Fischer's Long-tailed Snake	Peters (1954); Smith & Taylor (1966); Smith et al. (1967); Alvarez del Toro (1982); Smith & Smith (1993)	S Mexico in Michoacán, Morelos, and c Guerrero south into the Isthmus of Tehuantepec including Chiapas and Oaxaca.
E. oligostichus Smith, Arndt & Sherbrook, 1967 Mexican Long-tailed Snake	Smith et al. (1967)	W Mexico from s Sinaloa southward in Mexico.
E. roatanensis McCranie & Köhler, 1999 Roatán Long-tailed Snake	McCranie & Köhler (1999); Köhler (2001)	Isla de Roatán in the Islas de la Bahía off the northern coast of Honduras.

GENUS: *Eridiphas* Leviton & Tanner, 1960, has a spotty distribution in Baja California, Mexico. See Ottley & Tanner (1978), Stebbins (1985), Dowling & Jenner (1987), Grismer (1999) and McPeak (2000).

Review assistance and important contributions for this section were provided by James R. Dixon.

See also *Hypsiglena* and *Pseudoleptodeira*.

The Caenophidia

CONTENT: One species, with two subspecies. ★★

Eridiphas: Baja California Nightsnake

Species-level Taxa	Citations and Remarks	Abbreviated Range
E. slevini (Tanner, 1943) Baja California Nightsnake	Ottley & Tanner (1978); Stebbins (1985); Smith & Smith (1993); Grismer (1999)	Baja California in Mexico; it has a spotty distribution from the north starting about the vicinity of Bahia de los Angeles southward to the Cape; it is expected to occur further north, probably on the e side of the San Pedro Martír Mountains, on Cerralvo, Panzante and San Marcos islands in the Gulf of California.
E. s. slevini* (Tanner, 1943) Slevin's Nightsnake	Ottley & Tanner (1978); Stebbins (1985); Smith & Smith (1993); Grismer (1999); McPeak (2000) *Remark: Grismer (1999) and McPeak (2000) list this as *E. slevini*. See Remark under *E. s. marcosensis*.	Most of the *E. slevini* range in Baja California and is the mainland variant. According to McPeak (2000), it is found primarily on the e side of the peninsula form about Bahia de de los Angeles to the Cape Region. It is also reported from Islas Coronado, Isla Danzante and Isla Cerralvo in the Gulf of California; and Isla Santa Margarita off the Pacific Coast.
E. s. marcosensis* Ottley & Tanner, 1978 San Marcos Island Nightsnake	Ottley & Tanner (1978); Smith & Smith (1993); Grismer (1999); McPeak (2000) *Remark: This subspecies is known from only two specimens, and Grismer (1999) noted that available data suggested that it be recognized as a full species (*E. marcosensis*). The available data he refers to is not shared in the literature and outside of this remark, nothing seems to be given to justify the change, therefore, I have retained the name as a subspecies here.	Endemic to San Marcos Island in the Gulf of California off the coast of Baja California in Mexico.

GENUS: *Erythrolamprus* Wagler, 1830, ranges from Guatemala and Honduras southward into South America west of the Andes in Peru, southern Brazil, Argentina, and Paraguay, and Pacific nw Ecuador. See Dowling & Gibson (1970); Peters & Orejas-Miranda (1970); Obst et al. (1988); Lancini & Kornacker (1989); Savage & Slowinski (1992); Auth (1994); Hardy & Boos (1995); Starace (1998); Kornacker (1999) and Köhler (2001).

CONTENT: Five species. ★★

Erythrolamprus: South American False Coral Snake. Frank and Ramus (1995) call this the False Coral Snake.

Species-level Taxa	Citations and Remarks	Abbreviated Range
E. aesculapii (Linnaeus, 1758) Aesculapian False Coral Snake	Roze (1966); Peters & Orejas-Miranda (1970); Amaral (1976); Savage & Slowinski (1992); Hardy & Boos (1995); Starace (1998); Kornacker (1999); Freitas (1999); Marques et al. (2001); Boos (2001)	Central America southward into Trinidad and Tobago, and then widely distributed throughout Amazonian South America into c Brazil, Argentina, Bolivia, and Paraguay.
E. a. aesculapii* (Linnaeus, 1766) Common Aesculapian False Coral Snake	Peters & Orejas-Miranda (1970); Duellman (1978); Dixon & Soini (1986); Pérez-Santos & Moreno (1988, 1991); Lancini & Kornacker (1989); Starace (1998); Kornacker (1999)	Amazon Basin in South America.

Species-level Taxa	Citations and Remarks	Abbreviated Range
	*Remark: *Erythrolamprus baupertuisii* Dumeril et al. 1854, as found in Roze (1966), from Venezuela, is a synonym (see Vanzolini, 1986, and Lancini & Kornacker, 1989).	
E. a. monzonus Jan, 1863 Atlantic Forest False Coral Snake	Machado (1945d); Peters & Orejas-Miranda (1970)	State of Bahia southward into Rio de Janeiro, Brazil.
*E. a. ocellatus** Peters, 1868 Tobago False Coral Snake	Underwood (1962); Emsley (1966a, b); Peters & Orejas-Miranda (1970); Hardy & Boos (1995); Murphy (1997); Boos (2001) [as *E. ocellatus*]	Restricted to the island of Trinidad in the s Caribbean.
	*Remark: Murphy (1997) lists this as a full species, *E. ocellatus*. However, Murphy (1997) shows his most current reference for this snake as Peters & Orejas-Miranda (1970). Peters & Orejas-Miranda (1970), use *E. aesculapii ocellatus* and reference Emsley (1966b), who also uses *E. aesculapii ocellatus*. Murphy (1997) may be correct in elevating this snake to full species, but my investigation was unable to identify a formal, peer-reviewed publication that supports him. Therefore, I have retained the subspecies position for this name.	
E. a. tetrazonus Jan, 1863 Bolivian Aesculapian False Coral Snake	Mertens (1956); Peters & Orejas-Miranda (1970)	SW Bolivia.
E. a. venustissimus (Wied, 1821) Southern Aesculapian False Coral Snake	Machado (1945d); Peters & Orejas-Miranda (1970); Cei (1993); Hardy & Boos (1995); Marques et al. (2001)	E Bolivia into Misiones in ne Argentina, and into Brazil in Minas Geraís and Rio de Janeiro.
E. bizonus Jan, 1863 Slender False Coral or Double-banded False Coral Snake	Dunn & Bailey (1939); Roze (1966); Peters & Orejas-Miranda (1970); Savage & Vial (1974); Savage (1980); Savage & Villa (1986); Villa et al. (1988); Pérez-Santos & Moreno (1988); Lancini & Kornacker (1989); Savage & Slowinski (1992); Auth (1994); Hardy & Boos (1995); Murphy (1997); Kornacker (1999); Mijares-Urrutia & Arends R. (2000); Boos (2001); Köhler (2001)	Nicaragua and Costa Rica southward into Colombia and N Venezuela, and possibly into Trinidad; it has been reported from sea level to 1450 m in Costa Rica and 400 to 2000 m in Venezuela; Pérez-Santos & Moreno (1988) reported it from 2500 to 2700 m in Colombia.
E. guentheri Garman, (1883) 1884 Günther's or Pink-naped False Coral Snake	Peters & Orejas-Miranda (1970); Pérez-Santos & Moreno (1991); Savage & Slowinski (1992); Hardy & Boos (1995); Fugler &Walls (1978)	Amazonian slopes of e Ecuador and Peru.
E. mimus (Cope, 1868) Central American or Mimic False Coral Snake	Dunn & Bailey (1939); Peters & Orejas-Miranda (1970); Savage & Vial (1974); Savage (1980); Wilson & Meyer (1985) Savage & Villa (1986); Villa et al. (1988); Savage & Slowinski (1992); Auth (1994); Cruz-Diaz (1997); Köhler (2001)	Honduras and Nicaragua in Central America into South America in w Ecuador and Peru; from 500 to 2000 m.

Species-level Taxa	Citations and Remarks	Abbreviated Range
E. m. mimus (Cope, 1868) Mimic False Coral Snake	Dunn & Bailey (1939); Peters & Orejas-Miranda (1970); Wilson & Meyer (1985); Pérez-Santos & Moreno (1991); Köhler (2001)	Colombia, Ecuador and E Peru.
E. m. impar Schmidt, 1936 Central American False Coral Snake	Dunn & Bailey (1939); Peters & Orejas-Miranda (1970); Wilson & Meyer (1982, 1985); Köhler (2001)	Honduras, Nicaragua and Costa Rica.
E. m. micrurus Dunn & Bailey, 1939 Trans-andean False Coral Snake	Dunn & Bailey (1939); Nicéforo María (1942); Peters & Orejas-Miranda (1970); Wilson & Meyer (1985); Pérez-Santos & Moreno (1988, 1991); Köhler (2001)	Atlantic versant of Panama southward into w Colombia and w Ecuador.
E. pseudocorallus Roze, 1959 Maracaibo False Coral Snake	Roze (1959, 1966); Peters & Orejas-Miranda (1970); Lancini & Kornacker (1989); Savage & Slowinski (1992); Fuentes & Barrio (1999); Kornacker (1999)	Mountainous regions of the Venezuelan states of Zulia, Amazonas, and Perijá.

GENUS: *Farancia* Gray, 1842, contains the semi-aquatic Mud and Rainbow snakes that are endemic to the central, eastern, and southeastern United States. There are disjunct populations in Georgia and Alabama. See Conant (1975); Mitchell (1983a, 1983b); McDaniel & Karges (1983); Tennant & Bartlett (2000); and Dixon (2000a).

CONTENT: Two species. ★★★★

Farancia: North American Mudsnake and Rainbow Snake

Species-level Taxa	Citations and Remarks	Abbreviated Range
F. abacura (Holbrook, 1836) Red-bellied Mudsnake	Conant (1975); Mitchell (1982a, 1994) McDaniel & Karges (1983); Dundee & Rossman (1989); Wright & Wright ([1957] 1994); Savage & Slowinski (1992); Conant & Collins (1998); Dixon ((2000a)	C, e and s United States in lowlands and swamps.
F. a. abacura (Holbrook, 1836) Eastern Mudsnake	Conant (1975); Ashton & Ashton (1981); McDaniel & Karges (1983); Mitchell (1994); Tennant (1997b); Conant & Collins (1998); Tennant & Bartlett (2000)	SE United States from se Virginia into s Florida.
F. a. reinwardti (Schlegel, 1837) Western Mudsnake	Smith (1938); Conant (1975); McDaniel & Karges (1983); Dundee & Rossman (1989); Tennant (1997b, 1989); Tennant (1997b, 1998); Conant & Collins (1998); Werler & Dixon (2000); Tennant & Bartlett (2000); Dixon (2000a)	SC United States from s Illinois and sw Indiana south into e Texas to Alabama.
F. erythrogramma Palisot de Beauvois *in* Sonnini & Latreille, 1801 North American Rainbow Snake	Neill (1964a); Conant (1975); Mitchell (1982b); Dundee & Rossman (1989); Conant & Collins (1991, 1998); Savage & Slowinski (1992); Mitchell (1994)	E United States from s Maryland into n Florida, and westward into e Louisiana.

Species-level Taxa	Citations and Remarks	Abbreviated Range
F. e. erythrogramma Palisot de Beauvois in Sonnini & Latreille, 1801 Common Rainbow Snake	Neill (1964a); Conant (1975); Ashton & Ashton (1981); Mitchell (1982, 1994); Dundee & Rossman (1989); Tennant (1997b); Conant & Collins (1998); Tennant & Bartlett (2000)	Mississippi Valley and e United States.
F. e. seminola* Neill, 1964 Southern Florida Rainbow Snake	Neill (1964a); Conant (1975); Ashton & Ashton (1981); Mitchell (1982b, 1994); Tennant (1997b); Conant & Collins (1991, 1998); Tennant & Bartlett (2000)	S Florida.
	*Remark: Tennant (1997b) stated that F. e. seminola is the rarest snake in North America and is known from only 3 specimens. He suggested that it may be extinct.	

GENUS: *Ficimia* Gray, 1849, ranges from the United States in southern Texas southward through Mexico into northern Central America. See Slevin (1934); Smith & Taylor (1966); Peters & Orejas-Miranda (1970); Hardy (1970, 1975a, 1975b, 1990); Mendoza-Quijano & Smith (1993); Smith & Smith (1993); Tennant & Bartlett (2000); Dixon (2000a) and Köhler (2001).

Important contributions for this section were provided by Peter Holm.

See also genera *Conopsis*, *Gyalopion*, *Pseudoficimia*, and *Toluca*.

CONTENT: Seven species. ★★★

Ficimia: Eastern Hook-nosed Snake

Species-level Taxa	Citations and Remarks	Abbreviated Range
F. hardyi Mendoza-Quijano & Smith, 1993 Hidalgo Hook-nosed Snake	Mendoza-Quijano & Smith (1993); Hernández-Ibarra et al. (1999)	1480 to 2100 m in n Hidalgo, Mexico.
F. olivacea Gray, 1849 Brown Hook-nosed Snake	Smith & Taylor (1966); Hardy (1975a, 1978, 1980c, 1990); Pérez-Higareda & Navarro L. (1980); Pérez-Higareda & Smith (1991); Mendoza-Quijano & Smith (1993)	Mexico in Tamaulipas, Nuevo León, and Veracruz; and southward into Central America.
F. publia* Cope, 1866 Blotched Hook-nosed Snake	Smith & Langebartel (1949); Smith & Taylor (1966); Peters & Orejas-Miranda (1970); Hardy (1975a, 1980a, 1990); Alvarez del Toro (1982); Wilson & Meyer (1985); Villa et al. (1988); Pérez-Higareda & Smith (1991); Mendoza-Quijano & Smith (1993);	Atlantic versant from Jalisco and the Isthmus of Tehuantepec in e and s Mexico into Central America to Honduras, and in the Pacific versant from Guerrero, Mexico into s Guatemala.

Species-level Taxa	Citations and Remarks	Abbreviated Range
	Lee (1996, 2000); Campbell (1998a); Stafford & Meyer (2000); Köhler (2001)	
	*Remark: Hardy (1975, 1980a); Smith & Taylor (1966), and Smith & Smith (1993), did not recognize subspecies. When listed, the subspecies are generally shown as *F. p. publia* Cope, 1866, as can be seen in Peters & Orejas-Miranda (1970), and Henderson & Hoevers (1975). The distribution reported to be Atlantic versant of Isthmus of Tehuantepec, Yucatán peninsula, Guatemala, Belize, and, Honduras; and in the Pacific versant from Guerrero, Mexico into Guatemala. *F. p. taylori* Smith, 1947, as listed in Peters & Orejas-Miranda (1970), Hardy (1969, 1975a), with a distribution in s Mexico; *F. p. wolffsohni* Neill, 1965, as shown in Hardy (1969, 1980a); Neill (1965); Henderson & Hoevers (1975), distributed in Belize and Honduras. Subspecies are shown here because Campbell (1998a), referenced *F. p. publia* and *F. p. wolffsohni*.	
F. ramirezi Smith & Langebartel, (1949) 1950 Ramirez's Hook-nosed Snake	Smith & Langebartel (1949); Hardy (1979, 1990); Villa et al. (1988); Mendoza-Quijano & Smith (1993); Köhler (2001)	Associated with the Sierra de Chiapas in the state of Chiapas and se Oaxaca in se Mexico.
F. ruspator Smith & Taylor, 1941 Guerreran Hook-nosed Snake	Smith & Taylor (1945, 1966); Smith & Langebartel (1949); Smith (1943); Hardy (1975a, 1975b, 1976a, 1979, 1980c); Lee (1996); Campbell (1998a); Mendoza-Quijano & Smith (1993)	Mexico in c Guerrero and Morelos.
F. streckeri Taylor, 1931 Tamaulipan Hook-nosed Snake	Smith & Taylor (1941, 1966); Conant (1975); Hardy (1975a, 1976a, 1990); Pérez-Higareda & Navarro L. (1980); Pérez-Higareda & Smith (1991); Mendoza-Quijano & Smith (1993); Tennant (1998); Conant & Collins (1998); Werler & Dixon (2000); Tennant & Bartlett (2000); Dixon (2000a)	S Texas and n Tamaulipas in Mexico southward into the states of Nuevo León, e San Luis Potosí, Hidalgo, and n Veracruz.
F. variegata (Günther, 1858) Tehuantepec Hook-nosed Snake	Smith & Taylor (1941, 1966); Hardy (1980c, 1990); Alvarez del Toro (1982); Pérez-Higareda & Smith (1991); Smith & Smith (1993); Mendoza-Quijano & Smith (1993)	S Mexico in c Isthmus of Tehuantepec and nw Chiapas with limited records from Oaxaca.

GENUS: *Geagras* Cope, 1876, is endemic to the Pacific versant of Oaxaca in Mexico. See Smith & Taylor (1966); Wilson & Meyer (1981); and Wilson (1987b).

CONTENT: One species. ★★★★

Geagras: Tehuantepec Striped Snake. Frank and Ramus (1995) call this the Tehmantepec Striped Snake.

Species-level Taxa	Citations and Remarks	Abbreviated Range
*G. redimitus** Cope, 1876 Tehuantepec Striped Snake	Hartweg & Oliver (1940); Smith (1943); Smith & Taylor (1966); Wilson & Meyer (1981); Wilson (1987b) *Remark: Wilson (1987b) noted that *Tantilla depressa* is a synonym.	Lowland plains of Tehuantepec in the Pacific versant of se Oaxaca, Mexico.

GENUS: *Geophis* Wagler, 1830, is a small Neotropical snake associated with Middle America from Mexico southward into Colombia. [The genus was originally named *Catostoma* Wagler, 1830, and the name was later discovered to be similar to *Catostomus* Lesueur, 1817. According to Smith & Taylor (1966), *Catastoma* Wagler, 1830, and *Geophis* Wagler, 1830, were used in the same article. Either name may be used, according to Wagler, 1830, depending on the choice of first revisor, even though *Catastoma* has page priority]. Downs (1967) is the first revisor and really defined the name of *Geophis* that has been used continuously now for over thirty years. He recognized thirty-three forms. *Geophis* is a complex group of ground snakes that are found mainly in Middle America, but range from Mexico into northern South America to Colombia. See Smith (1941e); Smith & Taylor (1966); Downs (1967); Peters & Orejas-Miranda (1970); Savage (1960a, 1980, 1981); Campbell, et al. (1983); Vanzolini (1986); Williams & Wallach (1989); Smith & Flores-Villela (1993); Smith & Smith (1993); Lips & Savage (1994) and Köhler (2001).

Note: Fide Smith & Smith (1993) *Geophis alasukai* is considered to be a nomen nudum.

Important contributions to this section were provided by Ernest A. Liner.

CONTENT: Forty-four species. ★★★

Geophis: Middle American Earthsnake or Central American Earthsnake. Frank and Ramus (1995) call this the Latin American Earthsnake.

Species-level Taxa	Citations and Remarks	Abbreviated Range
G. anocularis Dunn, 1920 Sierra Mije Earthsnake	Dunn (1920); Smith & Taylor (1966); Downs (1967); Campbell, et al. (1983); Smith & Smith (1993)	Mexico, known only from the type locality in Totontepec, Oaxaca.
G. bellus Myers, 2003 Pretty Earthsnake	Myers (2003)	Panama. Holotype is from 700 m near the community of Altos de Pacora (east of Cerro Jefe), Province of Panama, in e Panama.
G. betaniensis Restrepo & Wright, 1987 Colombian Earthsnake	Cadle (1984b); Restrepo & Wright (1987); Lips & Savage (1994); Wilson et al. (1998)	WC Colombia.
G. bicolor Günther, 1868 Mexican Plateau Earthsnake	Smith & Taylor (1966); Downs (1967); Cadle (1984b); Smith & Flores-Villela (1993); Smith & Smith (1993)	States of México and Michoacán in Mexico.
G. blanchardi Taylor & Smith, [1938] 1939 Blanchard's Earthsnake	Smith & Taylor (1966); Downs (1967); Pérez-Higareda & Smith (1991); Smith & Smith (1993)	C Veracruz, Mexico in the Sierra Madre Oriental.

Species-level Taxa	Citations and Remarks	Abbreviated Range
G. brachycephalus (Cope, 1871) Gray Earthsnake	Downs (1967); Peters & Orejas-Miranda (1970); Savage (1980, 1981); Savage & Villa (1986); Villa et al. (1988); Pérez-Santos & Moreno (1988); Savage & Slowinski (1992); Smith & Smith (1993); Lips & Savage (1994); Wilson et al. (1998); Campbell & Solorzano (1992); Myers (2003)	Cordillera Central in n Costa Rica into w Panamá 250 to 2000 m.
G. cancellatus Smith, 1941 Chiapan Earthsnake	Smith (1941d); Hartweg (1959); Landy et al. (1966); Smith & Taylor (1966); Downs (1967); Peters & Orejas-Miranda (1970); Alvarez del Toro (1982); Vanzolini (1986); Villa et al. (1988); Savage & Slowinski (1992); Smith & Smith (1993)	Chiapas and Oaxaca in se Mexico into Guatemala.
G. carinosus Stuart, 1941 Keeled Earthsnake	Stuart (1941a); Downs (1967); Alvarez del Toro (1982); Villa et al. (1988); Pérez-Higareda & Smith (1991); Smith & Smith (1993); Lee (2000); Köhler (2001)	SE Mexico in Yucatán and Guatemala, from 200 to 2000 m.
G. chalybeus (Wagler, 1830) Veracruz Earthsnake	Smith & Taylor (1966); Downs (1967); Pérez-Higareda & Smith (1991); Smith & Smith (1993); Wilson et al. (1998)	Veracruz in Mexico.
G. championi Boulenger, 1894 Volcán Chiriqui Earthsnake	Downs (1967) Peters & Orejas-Miranda (1970); Savage (1981); Villa et al. (1988); Lips & Savage (1994); Wilson et al. (1998); Köhler (2001)	Known only from the type locality of Chiriqui and Boquete on the eastern slopes of Volcán Chiriqui in Panama.
G. damiani Wilson, McCranie & Williams, 1998 Damian's Earthsnake	Wilson et al. (1998); Köhler (2001)	Known only from the type locality within the limits of the Refugio Silvestre Cerro Texíguat in lower montane wet forest from the Cordillera Nombre de Dios in n Honduras.
G. downsi Savage, 1981 Savage's Earthsnake	Savage (1981); Vanzolini (1986); Savage & Villa (1986); Villa et al. (1988); Lips & Savage (1994); Köhler (2001)	Premontane rainforest of the Cordillera Costeña in Costa Rica from 1100 to 1200 m.
G. dubius (W. Peters, 1861) Mesa del Sur Earthsnake	Smith & Taylor (1966); Bogert & Porter (1966b); Downs (1967); Campbell & Ford (1983); Pérez-Higareda & Smith (1991); Smith & Smith (1993); Wilson et al. (1998)	Uncertain range from Veracruz, Mexico southward into Guatemala. There is a male questionably reported from Puebla, Mexico. Found to 1700 m which is closely associated with pine-oak habitat in mountains.
G. duellmani Smith & Holland, 1969 Sierra Juarez Earthsnake	Smith & Holland (1960); Campbell & Ford (1983); Savage & Slowinski (1992) Smith & Flores-Villela (1993); Smith & Smith (1993)	Oaxaca, Mexico; it seems to be confined to the vicinity of Vista Hermosa.
G. dugesi Bocourt, 1883 *in* Dumeril, Mocquard & Bocourt, 1870–1909 Dugès Earthsnake	Hartweg (1959); Smith & Taylor (1966); Downs (1967); Hardy & McDiarmid (1969); Smith & Smith (1993)	Sinaloa, Durango, and Chihuahua into n Michoacán in Mexico.

Species-level Taxa	Citations and Remarks	Abbreviated Range
G. dugesi dugesi Bocourt, 1883 *in* Dumeril, Mocquard & Bocourt, 1870–1909 Dugès Earthsnake	Hartweg (1959); Webb (1977); Smith & Smith (1993)	Sinaloa and Durango into n Michoacán in Mexico.
G. dugesi aquilonaris Legler, 1959 Chihuahuan Earthsnake	Hartweg (1959); Downs (1967) [as *G. aquilonaris*]; Webb (1977); Tanner (1985); Smith & Smith (1993); Vásquez-Díaz & Quintevo-Díaz (1999a)	Chihuahua and Aguascalientes in Mexico.
G. dunni Schmidt, 1932 Matagalpa Earthsnake	Downs (1967); Peters & Orejas-Miranda (1970); Villa et al. (1988); Lips & Savage (1994); Wilson et al. (1998); Köhler (2001)	Known only from the type locality of Matagalpa in Nicaragua.
G. fulvoguttatus Mertens, 1952 Mertens's Earthsnake	Downs (1967); Peters & Orejas-Miranda (1970); Campbell & Ford (1983); Villa et al. (1988); McCranie & Wilson (1991); Köhler (2001)	Known only from the type locality of Hacienda Monte Cristo in a cloud forest at 2200 m in the Sierra de Metapan in Departamento Santa Ana in El Salvador and into w Honduras.
G. godmani* Boulenger, 1894 Godman's Earthsnake	Peters & Orejas-Miranda (1970); Savage (1980); Savage (1981); Campbell & Ford (1983); Savage & Villa (1986); Villa et al. (1988); Smith & Smith (1993); Lips & Savage (1994); Köhler (2001); Campbell & Solorzano (1992); Myers (2003) *Remark: Fide Smith & Smith (1993), this snake is found in Mexico.	Costa Rica and Panama between 1300 to 2100 m.
G. hoffmanni (W. Peters, 1859) Hoffman's Earthsnake	Downs (1967); Peters & Orejas-Miranda (1970); Savage (1960a, 1980, 1981); Wilson & Meyer (1985); Savage & Villa (1986); Villa et al. (1988); Lips & Savage (1994); Wilson et al. (1998); Köhler (2001); Myers (2003)	Low and moderate elevations from e Honduras southward into w Panamá. Myers (2003) reported one specimen from Colombia.
G. immaculatus Downs, 1967 Downs's Earthsnake	Downs (1967); Peters & Orejas-Miranda (1970); Villa et al. (1988); Smith & Smith (1993); Köhler (2001)	Pacific versant of se Mexico and Guatemala.
G. incomptus Duellman, 1959 Sierra Coalcomán Earthsnake	Duellman (1959); Bogert & Porter (1966a); Downs (1967); Smith & Smith (1993)	Michoacán in Mexico.
G. isthmicus (Boulenger, 1894) Isthmian Earthsnake	Savage (1960a); Smith & Taylor (1966); Downs (1967); Smith & Smith (1993)	Known only from the type locality in Tehuantepec, Mexico.
G. juliai Pérez-Higareda, Smith & López-Luna (2001) Los Tuxtlas Rainforest Earthsnake	Pérez-Higareda et al. (2001)	Tropical rainforests in the Los Tuxtlas area of s Veracruz, Mexico.

Species-level Taxa	Citations and Remarks	Abbreviated Range
*G. laticinctus** Smith & Williams, 1963 Mesa Central Earthsnake	Downs (1967); Johnson (1979); Flores-Villela (1993); Alvarez del Toro (1982); Villa et al. (1988); Savage & Slowinski (1992); Smith & Smith (1993); Köhler (2001); Pérez-Higareda et al. (2001) *Remark: Several authorities still list subspecies as can be found in Smith & Smith (1993); when shown, subspecies are generally as follow: *G. l. laticinctus* Smith & Williams, 1963, which is differentiated only by dark (black) ventral and associated with pine-oak forests at +/−1800 m; and, *G. l. albiventris*, Smith & Holland, 1969 differentiated by light (white) venter and associated with lower montane rainforests. However, Johnson (1979) determined this species is monotypic.	Atlantic versant of Mexico from e Oaxaca into nc Chiapas in Mexico from about 730 to 1800 m; it may eventually be found in e Chiapas and El Petén in Guatemala.
*G. laticollaris** Smith, Lynch & Altig, 1965 Wide Collar Earthsnake	Smith et al. (1965); Smith & Chiszar (1992b); Smith & Smith (1993); Wilson et al. (1998) *Remark: Downs (1967) considered the holotype *G. laticollaris* to be a synonym of *G. sallaei*.	S Mexico; Smith & Chiszar (1992b) noted that it is found on the n slopes of the Sierra Madre del Sur and compare it to the distribution of *G. sallaei*.
G. latifrontalis Garman, 1883 Potosí Earthsnake	Savage (1960a); Smith & Taylor (1966); Downs (1967); Dixon & Thomas (1974); Smith & Smith (1993); Wilson et al. (1998)	S San Luis Potosí, Guerrero, Hidalgo, and possibly in Colima (?), Mexico.
G. l. latifrontalis Garman, 1883 Potosí Earthsnake	Smith & Taylor (1966); Downs (1967); Dixon & Thomas (1974); Smith & Smith (1993)	Mexico; s San Luis Potosí in Mexico; known only from the vicinity of Alvarez.
G. l. semiannulatus Smith, 1941 Mountain Earthsnake or Broken-ringed Earthsnake	Smith & Taylor (1966); Downs (1967); Smith & Smith (1993); Dixon & Thomas (1974)	Known only from the type locality of Colima (possibly in error), Guerrero, and Hidalgo in Mexico.
G. maculiferus Taylor, 1942 Michoacán Earthsnake	Smith & Taylor (1966); Bogert & Porter (1966a); Downs (1967); Smith & Smith (1993)	Michoacán in Mexico.
G. mutitorques (Cope, 1885) Highlands Earthsnake	Downs (1961, 1967); Smith & Taylor (1966); Cadle (1984b); Pérez-Higareda & Smith (1991); Smith & Smith (1993)	C Mexico in s Hidalgo, cw Veracruz and n Puebla.
G. nasalis (Cope, 1868) Coffee Earthsnake	Davis & Smith (1953); Peters (1954); Smith & Taylor (1966); Bogert & Porter (1966b) Downs (1967); Peters & Orejas-Miranda (1970); Campbell & Murphy (1977); Alvarez del Toro (1982); Villa et al. (1988); Smith & Smith (1993); Lips & Savage (1994); Wilson et al. (1998); Köhler (2001)	Pacific versant from Chiapas in se Mexico into adjacent e Guatemala in the se highlands and Guatemalan Plateau, associated with the "coffee zone" from 600 to 1500 m.
G. nigroalbus Boulenger, 1908 Black and White Earthsnake	Savage (1960a); Downs (1967); Peters & Orejas-Miranda (1970); Lips & Savage (1994); Wilson et al. (1998); Myers (2003)	E Panama and Colombia.

Species-level Taxa	Citations and Remarks	Abbreviated Range
G. nigrocinctus Duellman, 1959 Sierra Coalcomán Earthsnake	Duellman (1959); Downs (1967); Campbell & Murphy (1977); Smith & Smith (1993)	Michoacán and Jalisco in Mexico.
G. omiltemanus Günther, 1893 Guerreran Earthsnake	Smith & Taylor (1966); Downs (1967); Smith & Smith (1993); Wilson et al. (1998)	In the Sierra Madre del Sur in c Guerrero, Mexico.
G. petersi Boulenger, 1894 Peters's Earthsnake	Hartweg (1959); Smith & Taylor (1966); Downs (1967); Restrepo & Wright (1987); Smith & Smith (1993); Lips & Savage (1994); Wilson et al. (1998)	Southern edge of the Mexican Plateau in Michoacán and Oaxaca.
G. pyburni Campbell & Murphy, 1977 Pyburn's Earthsnake	Campbell & Murphy (1977); Smith & Smith (1993); Wilson et al. (1998)	Michoacán in Mexico.
G. rhodogaster (Cope, 1868) Rose-bellied Earthsnake	Downs (1967); Peters & Orejas-Miranda (1970); Campbell et al. (1983); Villa et al. (1988); Campbell & Vannini (1988); Smith & Smith (1993); Lips & Savage (1994); N. Smith (1995); Köhler (2001); Campbell & Solorzano (1992)	Yucatán in Mexico southward into El Salvador and the w Guatemalan highlands and se highlands. It is generally found in oak-pine forests between 1500 and 2500 m.
G. rostralis Jan, 1864 Sierra Madre Earthsnake	Smith & Taylor (1966); Bogert & Porter (1966b); Campbell et al. (1983); Restrepo & Wright (1987); Smith & Smith (1993)	S Mexico likely in w Oaxaca and Michoacán.
G. russatus Smith & Williams, 1966 Red Earthsnake	Restrepo & Wright (1987); Savage & Slowinski (1992); Smith & Smith (1993); Flores-Villela (1993); Lips & Savage (1994); Wilson et al. (1998)	S Mexico in Oaxaca.
G. ruthveni Werner, 1925 Ruthven's Earthsnake	Downs (1967); Peters & Orejas-Miranda (1970); Savage (1960a, 1980, 1981); Savage & Villa (1986); Villa et al. (1988); Lips & Savage (1994); Köhler (2001)	Costa Rica in the Cordillera Central and Cordillera Guanacaste between 550 and 1600 m.
*G. sallaei** Boulenger, 1894 Salle's Earthsnake	Hartweg (1959); Smith & Taylor (1966); Downs (1967); Smith & Chiszar (1992b); Smith & Smith (1993); Lips & Savage (1994) *Remark: Smith & Smith (1993) listed the spelling of *G. sallei* based on Smith & Chiszar (1992) who proposed the *G. sallei* spelling.	Oaxaca in Mexico; it is reported by Smith & Chiszar (1992) from two locations, Cafetal Concordia, near Pluma, Hidalgo, Oaxaca and Santa Rosa, near Lachao, Distrito de Juquila, Oaxaca; both locations are near each other in the foothills of the Sierra Madre del Sur.
G. semidoliatas (Dumeril, Bibron & Dumeril, 1854) Coral Earthsnake	Hartweg (1959); Smith & Taylor (1966); Downs (1967); Alvarez del Toro (1982); Pérez-Higareda & Smith (1991); Savage & Slowinski (1992); Smith & Smith (1993); Wilson et al. (1998); Pérez-Higareda et al. (2001)	CE Mexico in the foothills of cw Veracruz and adjacent portions o fthe state of Hidalgo and ne Oaxaca.
G. sieboldi (Jan, 1862) Siebold's Earthsnake	Smith & Taylor (1966); Downs (1967); Campbell & Murphy (1977); Smith & Smith (1993); Lips & Savage (1994); Wilson et al. (1998)	Mexico; the type locality originally was shown as "Mexico" and "Guadelupa." The distribution remains undetermined. There is supposed to be a specimen from Amula (=Almolonga), Guerrero, and one from

Species-level Taxa	Citations and Remarks	Abbreviated Range
		Coalcomán, Michoacán. It seems likely that *G. sieboldi* occurs in moderate elevation along the Pacific versant west of the Isthmus of Tehuantepec.
G. talamancae Lips & Savage, 1994 Talamanca Earthsnake	Lips & Savage (1994); Wilson et al. (1998); Köhler (2001); Myers (2003)	SC Costa Rica in the Cordillera de Talamanca; known only from the type locality of the zona Protectora Las Tablas of the Reserva de la Biosfera la Amistad near the Costa Rican-Panama border at 1800 m.
G. tarascae Hartweg, 1959 Tarasca Earthsnake	Hartweg (1959); Downs (1967); Restrepo & Wright (1987); Smith & Smith (1993)	Michoacán in Mexico.
G. zeledoni Taylor, 1954 Volcán Poás Earthsnake	Taylor (1954); Downs (1967); Peters & Orejas-Miranda (1970); Savage (1980, 1981); Savage & Villa (1986); Villa et al. (1988); Savage & Slowinski (1992); Lips & Savage (1994); Wilson et al. (1998); Köhler (2001)	Slopes of Volcán Poás in the Cordillera Central of c Costa Rica between 1600 and 2000 m.

GENUS: *Gomesophis* Hoge & Mertens, 1959, is from southeastern Brazil. See Peters & Orejas-Miranda (1970); Williams & Wallach (1989); and Lema (1994b).

Important contributions for this section were provided by Robert A. Thomas.

CONTENT: One species. ★★★

Gomesophis: Gomes's Snake, Ball Snake, or Cobra Bola.

Species-level Taxa	Citations and Remarks	Abbreviated Range
G. brasiliensis (Gomes, 1918) Brazilian Ball Snake	Gomes (1918); Peters & Orejas-Miranda (1970); Amaral (1976); Williams & Wallach (1989); Lema (1994a); Marques et al. (2001)	SE Brazil in the states of Minas Geraís, São Paulo, Rio Grande do Sul and Paraná.

GENUS: *Gyalopion* Cope, 1860, is distributed from the sw United States southward into Mexico. See Smith & Taylor (1966); Hardy (1975b, 1976b); Stebbins (1985); Peters & Orejas-Miranda (1970); Smith & Smith (1993); Degenhardt et al. (1996); Tennant & Bartlett (2000); and Dixon (2000a).

Review assistance and important contributions for this section were provided by Peter Holm.

See also genera *Conopsis*, *Ficimia*, *Pseudoficimia*, and *Toluca*.

CONTENT: Two species (unless *G. desertorum* is recognized.) ★★★

Gyalopion: Western Hook-nosed Snake. Frank and Ramus (1995) call this the Desert Hook-nosed Snake.

Species-level Taxa	Citations and Remarks	Abbreviated Range
G. canum Cope, 1860 Chihuahuan Hook-nosed Snake	Leviton & Banta (1961); Smith & Taylor (1966); Conant (1975); Hardy (1975b, 1976b); Stebbins (1985); Degenhardt et al. (1996); Tennant (1998); Conant & Collins (1998); Werler & Dixon (2000); Bartlett & Tennant (2000); Tennant & Bartlett (2000); Dixon (2000a)	SW United States from se Arizona into c New Mexico and w Texas southward into c Mexico in Zacatecas and San Luis Potosí; in Mexico its range includes ne Sonora, Coahuila, San Luis Potosí, Nuevo León, ne Chihuahua, ne Durango and n Zacatecas.
*G. quadrangulare** (Günther, 1893) Thornscrub Hook-nosed Snake	Dixon & Fugler (1959); Woodin (1962); Smith & Taylor (1966); Conant (1975); Hardy (1976b); Stebbins (1985); Campbell & Lamar (1989); Bartlett & Tennant (2000) *Remark: Fide Hardy (1976b), this is a monotypic species. Subspecies are not normally considered valid but when listed, are: *G. q. quadrangulare* (Günther, 1893), and, fide Stebbins (1985), *G. q. desertorum*, which is known only from the type locality 12 km nw of Guyamas, Sonora. This taxon is sometimes shown as *G. desertorum* (see Smith & Taylor, 1966).	SE United States into Mexico from s Arizona through the w coast of Mexico into Sonora, Sinaloa, and Nayarit; usually found below 1200 m.

GENUS: *Helicops* Wagler, 1830, is widely distributed throughout northern South America southward into Argentina, and is absent from the more arid and cooler parts of the continent. This snake tends to be closely associated with river basins. See Rossman *in* Peters & Orejas-Miranda (1970); Vanzolini (1986); Obst et al. (1988); and Kornacker (1999).

Review assistance and important contributions for this section were provided by Douglas A. Rossman.

CONTENT: Sixteen species. ★★★★

Helicops: South American Watersnake, Neotropical Water Snake, or Water Mapanare. Frank and Ramus (1995) call this the Keelback.

Species-level Taxa	Citations and Remarks	Abbreviated Range
H. angulatus (Linnaeus, 1758) Brown-banded Watersnake	Hoge (1952a); Peters (1960b); Underwood (1962); Roze (1966); Rossman *in* Peters & Orejas-Miranda (1970); Amaral (1976); Pérez-Bravo [1976/77] (1978); Duellman (1978); Moonen et al. (1979); Dixon & Soini (1986); Lancini & Kornacker (1989); Savage & Slowinski (1992); Lema (1994b); Murphy (1997); Starace (1998); Kornacker (1999); Freitas (1999); Boos (2001)	Throughout n South America east of the Andes and south to Bolivia and ec Brazil; also found in Trinidad.
*H. carinicaudus** (Wied, 1825) Rough-tailed Watersnake	Lema (1958); Rossman *in* Peters & Orejas-Miranda (1970); Amaral (1976); Vanzolini (1986); Deiques & Cechin (1991); Yuki (1994); Yanosky (1998); Marques et al. (2001) *Remark 1: Yuki (1994) examined specimens from Colombia previously identified as *Helicops carini*-	SE Brazil from Espírito Santo southward into Rio Grande do Sul.

Species-level Taxa	Citations and Remarks	Abbreviated Range
	caudus by Pérez-Santos & Moreno (1988), and determined that these specimens are *H. danieli*.	
	*Remark 2: Deiques & Chechin (1991) determined that *infrataenatus* is a separate species.	
H. danieli Amaral, (1937) 1938 Colombian Watersnake	Nicéforo María (1942); Rossman *in* Peters & Orejas-Miranda (1970); Yuki (1994); Yuki & Castano (1999); Renjifo & Lundberg (1999)	NW Colombia.
H. gomesi Amaral, [1921] 1922 Gomes's or São Paulo Watersnake	Rossman *in* Peters & Orejas-Miranda (1970); Amaral (1976); Vanzolini (1986)	Rio Tieté Basin in Brazilian state of São Paulo.
*H. hagmanni** Roux, 1910 Hagmann's Squint-eyed or Painted Watersnake	Rossman *in* Peters & Orejas-Miranda (1970); Rossman (1975); Vanzolini (1986); Paolillo (1986); Lancini & Kornacker (1989); Kornacker (1999) *Remark: Rossman (pers. comm. 28 March 2000), noted that Lancini & Kornacker (1989), illustrated *H. polylepis* in the figure of their book labeled as *H. hagmanni* [see Lancini & Kornacker (1989:157, abb. 77).	N Brazil, s Colombia, s Venezuela and probably into ne Peru.
*H. hogei** Lancini, 1964 Hoge's Water Mapanare	Pérez-Bravo [1976/77] (1978); Lancini (1964, 1979); Rossman *in* Peters & Orejas-Miranda (1970); Lancini & Kornacker (1989); Kornacker (1999); Rossman (2002) *Remark: Rossman *in* Peters & Orejas-Miranda (1970), lists the type locality as Tamacuro. This is a typographic error and should be Amacuro.	Originally known only from the type locality of Río Autana in Territorio Federal Amazonas and from Territorio Federal de Amacuro in Venezuela. Another specimen identified collected in the Río Cerro La Paloma, Sierra Imataca, Delta Amacuro, Venezuela.
*H. infrataeniatus** Jan, 1865 Brazilian Watersnake	Jan (1965); Miranda et al. (1982); Deiques & Cechin (1991); Cei (1993); Lema (1994b); Yanosky (1998) *Remark: Rossman (2000) indicates that *H. septemvittatus* is a synonym.	Found from the Brazilian state of São Paulo southward and westward into Río Grande do Sul, Uruguay, and ne Argentina.
H. leopardinus (Schlegel, 1837) Spotted Watersnake	Hoge (1952a); Rossman *in* Peters & Orejas-Miranda (1970); Amaral (1976); Hoge et al. ([1978/79] 1981); Miranda et al. (1982); Dixon & Soini (1986); Böckeler (1988); Cei (1993); Starace (1998); Yanosky (1998); Leynaud & Bucher (1999); Freitas (1999); Campos-Nogueira (2001)	NE South America from Amazonian Colombia, Peru, and the Guianas southward through Brazil into Bolivia, Paraguay, and Argentina.
H. modestus Günther, 1861 Olive Watersnake	Rossman *in* Peters & Orejas-Miranda (1970); Amaral (1976); Vanzolini (1986); Yanosky (1998); Campos-Nogueira (2001)	EC and s Brazil.
H. pastazae Shreve, 1934 Pastaza Watersnake	Shreve (1934); Peters (1960b); Rossman *in* Peters & Orejas-Miranda (1970); Rossman (1976); Vanzolini (1986); Dixon & Soini (1986); Lancini & Kornacker (1989); Kornacker (1999)	Amazonian Ecuador, Colombia, Peru, and Venezuela.

Species-level Taxa	Citations and Remarks	Abbreviated Range
H. petersi Rossman, 1976 Peters's Watersnake	Rossman (1976); Duellman (1978); Vanzolini (1986); Pérez-Santos & Moreno (1991)	Eastern Andean foothills of Ecuador.
H. pictiventris Werner, 1897 Werner's Watersnake	Rossman in Peters & Orejas-Miranda (1970); Amaral (1976)	S Brazil where it is found in the river basins between the states of Paraná and Rio Grande do Sul (Italjaí, Iguaçu, and Paraná rivers).
H. polylepis Günther, 1861 Black-chinned Watersnake	Hoge (1952a); Rossman in Peters & Orejas-Miranda (1970); Kempff-Mercado (1975); Amaral (1976); Pérez-Bravo [1976/77] (1978); Hoogmoed ([1982] 1983); Dixon & Soini (1986)	Amazon in Brazil, Colombia, Peru, and Bolivia.
H. scalaris Jan, 1865 Dark Water Mapanare	Jan (1965a); Lancini (1964); Roze (1966); Rossman in Peters & Orejas-Miranda (1970); Pérez-Bravo [1976/77] (1978); Pérez-Santos & Moreno (1988); Lancini & Kornacker (1989); Kornacker (1999); Rossman (2003)	NW Venezuela and adjacent Colombia in drainages that empty into Lake Maracaibo or the Caribbean.
H. trivittatus (Gray, 1849) Equatorial Watersnake	Hoge (1952a); Rossman in Peters & Orejas-Miranda (1970); Amaral (1976)	E equatorial Brazil where it is associated with the eastern section of the Amazon river basin.
H. yacu Rossman & Dixon, 1975 Forest Watersnake	Rossman & Dixon (1975); Rossman & Abe (1979); Dixon & Soini (1986)	Departamento Loreto in Peru and the Brazilian state of Acre.

GENUS: *Heterodon* Latreille, 1802, is distributed from southern central Canada southward through the central and eastern United States into northeastern Mexico. See Smith & Taylor (1966); Conant (1975); Platt (1983); Stebbins (1985); Degenhardt et al. (1996); Tennant (1998); Conant & Collins (1998); Dixon (2000a); and Tennant & Bartlett (2000).

Review assistance and important contributions for this section were provided by James R. Dixon.

CONTENT: Three species. ★★★

Heterodon: North American Hog-nosed Snake

Species-level Taxa	Citations and Remarks	Abbreviated Range
H. nasicus Baird & Girard, 1852 Western Hog-nosed Snake	Smith & Taylor (1966); Conant (1975); Platt (1983); Stebbins (1985); Degenhardt et al. (1996); Eckerman (1996); Conant & Collins (1998); Walley & Eckerman (1999); Dixon (2000a)	S Alberta in Canada southward through the United States from Illinois into se Arizona into c Mexico.
H. n. nasicus* Baird & Girard, 1852 Plains Hog-nosed Snake	Smith & Taylor (1966); Conant (1975); Stebbins (1985); Degenhardt et al. (1996); Tennant (1998); Conant & Collins (1998); Walley & Eckerman (1999); Werler & Dixon (2000); Bartlett & Tennant (2000); Tennant & Bartlett (2000); Dixon (2000a)	Western plains and prairies of Canada and the United States; one individual is reported from Sonora; its range is now considered to include c and e Texas, and e New Mexico through Oklahoma and Kansas to sw Manitoba and se Saskatchewan and in-

Species-level Taxa	Citations and Remarks	Abbreviated Range
	*Remark: Eckerman (1996), in a master's thesis, included *H. n. gloydi* Edgren, 1952, as a junior synonym, based on genetic and morphological data. This was later formalized in Walley & Eckerman (1999). *Heterodon n. gloydi*'s range was considered to be Kansas to Texas and is referenced in Conant (1975); Stebbins (1985); Dixon (2000a); Smith & Smith (1993); Degenhardt et al. (1996); Tennant (1998); and, Conant & Collins (1998). Werler & Dixon (2000), Tennant & Bartlett (2000), and Crother, *et. al.* (2000) listed *H. n. gloydi* as valid. However, this work follows Walley & Eckerman (1999) as the most currently published revisor.	cludes distribution in the prairie peninsula in Illinois, Iowa, and Minnesota.
H. n. kennerlyi Kennicott, 1860 Mexican Hog-nosed Snake	Conant (1975); Smith & Taylor (1966); Stebbins (1985); Degenhardt et al. (1996); Tennant (1998); Conant & Collins (1998); Walley & Eckerman (1999); Werler & Dixon (2000); Bartlett & Tennant (2000); Tennant & Bartlett (2000); Dixon (2000a)	Mexico from Tamaulipas, c San Luis Potosí, and e Agua Calientes northward and westward along the Coahuila Foldel, and enters the U.S. in the extreme s Rio Grande Valley and Trans-Pecos, Texas, se Arizona, and, according to Eckerman (1996), sw New Mexico and se Arizona; Degenhardt *et. al.* (1996) questioned its reported presence in New Mexico, but feel that it may eventually be found there.
*H. platirhinos** Latreille *in* Sonnini and Latrielle, 1801 Eastern Hog-nosed Snake	Blem (1981); Ashton & Ashton (1981); Platt (1983) Conant (1985); Dundee & Rossman (1989); Smith & Smith (1993); Tennant (1997b, 1998); Conant & Collins (1998); Boundy & Burbrink (1998); Werler & Dixon (2000); Tennant & Bartlett (2000); Dixon (2000a) *Remark 1: Historically the species name was *H. platyrhinos*. The present spelling was resurrected in Platt (1985). *Remark 2: Fide Stebbins (1985), a specimen of *H. platirhinos* was found in 1943, 9 miles west of Lamar, Bent Co., Colorado. Since none have been found since, he did not include them in his map for Colorado.	Throughout the eastern half of North America from Massachusetts and the s portions of New York, Pennsylvania, Ohio, Indiana, Illinois, and Missouri southward to the Gulf Coast. Northward, the species occurs locally in s New Hampshire, w Pennsylvania, the peninsula of s Ontario, nw Ohio, the s peninsula of Michigan, n Indiana, much of the s half of Wisconsin ,and c Minnesota. In the sw of its range, it is found in e Texas to Fort Worth, and locally in favorable areas of the Edwards Plateau, with limits to the South Texas Plains, westward it reaches the forest-prairie border and extends westward along river valleys.
H. simus (Linnaeus, 1766) Southern Hog-nosed Snake	Conant (1975); Ashton & Ashton (1981); Platt (1983); Meylan (1985); Tennant (1997b); Conant & Collins (1998); Tennant & Bartlett (2000)	SE United States in se North Carolina southward into sc Florida, s Mississippi, and n Alabama.

GENUS: *Hydrodynastes* Fitzinger, 1843 is found from northern to southern South America east of the Andes. See Peters & Orejas-Miranda (1970); Dowling & Gibson (1970); Lancini & Kornacker (1989); Chippaux ([1986] 1987); Obst et al. (1988) [as *Cyclagras* and *Hydrodynastes*]; Cei (1993); Starace (1998); and Kornacker (1999).

CONTENT: Two species. ★★★★

Hydrodynastes: South American False Water Cobra. Frank and Ramus (1995) call this the Water Cobra.

Species-level Taxa	Citations and Remarks	Abbreviated Range
*H. bicinctus** (Herrmann, 1804) Blotched Amazonian False Water Cobra	Hoge (1958); Neill (1963b); Roze (1966); Peters & Orejas-Miranda (1970); Chippaux ([1986] 1987); Lancini & Kornacker (1989); Savage & Slowinski (1992); Starace (1998); Kornacker (1999) *Remark: Kornacker (1999) showed this species as monotypic without subspecies.	Amazonia in the Guianas, Amazonas in Brazil, Colombia, and Venezuela.
H. b. bicinctus (Herrmann, 1804) Double-banded Watersnake	Hoge (1952b, 1966); Peters & Orejas-Miranda (1970); Beçcak et al. (1971); Lancini & Kornacker (1989)	Amazon Region including Colombia, Venezuela, the Guianas, and Amazonian Brazil.
H. b. schultzi Hoge, 1966 Schultz's False Water Cobra	Hoge (1966); Peters & Orejas-Miranda (1970)	State of São Paulo in Brazil; the range is uncertain with the type locality reported as Presidenté.
*H. gigas** (Dumeril, Bibron & Dumeril, 1854) Giant False Water Cobra or Brazilian False Water Cobra	Hoge (1958, 1966); Peters & Orejas-Miranda (1970); Peters (1970a); Dowling & Gibson (1970); Amaral (1976); Campbell & Murphy (1984); Vanzolini (1986); Böckeler (1988); Cei (1993); Lema (1994b); Starace (1998); Bernarde & Moura-Leite (1999); Leynaud & Bucher (1999) *Remark: According to Dowling & Gibson (1970), except for the presence of apical pits in *C. gigas* and their absence in *H. bicinctus*, the scutellation is virtually identical. They believed that the identity or great similarity of the two in body size and form, color and pattern, scutellation, dentition, hemipenis, and karyotype discounted the importance of apical scale pits in one and their absence in the other. They recommended placing *Cyclagras* as a synonym in the genus *Hydrodynastes*.	Chaco in e Bolivia into Amazonas, Pará, Amapa, Marajo Island, Rio Grande do Sul, Paraná, and Rondônia in Brazil, as well as Paraguay and n Argentina in Misiones, Corrientes, Entre Ríos, Santa Fe, Chaco, and Formosa; its range includes the Guianas and Peru; it is closely associated with water sources such as rivers, marshes, and swamps.

GENUS: *Hydromorphus* W. Peters, 1859, ranges from Guatemala southward through Central America into Panama. See Nelson (1966); Peters & Orejas-Miranda (1970); Villa (1970 [1971], 1990); Wilson & Meyer (1985); Savage & Donnelly (1988); Solórzano et al. (1989); Auth (1994) and Köhler (2001).

CONTENT: Two species. ★★★★

Hydromorphus: Central American Watersnake. Frank and Ramus (1995) call this the Tropical Watersnake.

Species-level Taxa	Citations and Remarks	Abbreviated Range
*H. concolor** W. Peters, 1859	Nelson (1966); Peters & Orejas-Miranda (1970); Villa ([1970] 1971, 1990); Wilson & Meyer (1985);	Guatemala, Honduras, Nicaragua, Costa Rica, and Panama; in Honduras it is reported from about 100

Species-level Taxa	Citations and Remarks	Abbreviated Range
Fresh-water or Costa Rican Watersnake	Villa et al. (1988); Savage & Donnelly (1988); Crother (1989); Solórzano et al. (1989); Auth (1994); Vences et al. (1998); Köhler (2001)	to 1400 m in tropical dry forest and subtropical wet forest.
	*Remark: Savage & Donnelly (1988), and Villa (1990), included *H. clarki* Dunn, 1942 form the lowlands of c Panama, as a synonym.	
H. dunni Slevin, 1942 Dunn's Watersnake	Slevin (1942); Savage & Donnelly (1988); Villa (1990); Auth (1994); Köhler (2001)	Known only from the type locality in the vicinity north of Boquette in Provencia Chiriquí, Panama at about 1250 m. Associated with low and moderate elevations of the Pacific versant of w Panama.

GENUS: *Hydrops* Wagler, 1830, is associated with the Amazon and drainage areas east of the Andes and above 15°S. It is also reported from the island of Trinidad. See Roze (1957, 1966); Peters & Orejas-Miranda (1970); Lancini (1979); Lancini & Kornacker (1989); Starace (1998); and Kornacker (1999).

CONTENT: Two species. ★★★

Hydrops: Neotropical Mudsnake or South American Mudsnake. Frank and Ramus (1995) call this the Amazon Watersnake.

Species-level Taxa	Citations and Remarks	Abbreviated Range
H. martii (Wagler *in* Spix, 1824) Coral Mudsnake or Amazon Watersnake	Nicéforo-María (1956); Roze (1957); Peters & Orejas-Miranda (1970); Savage & Slowinski (1992)	Amazon region in e Peru, Colombia, and Brazil.
H. m. martii (Wagler *in* Spix, 1824) Coral Mudsnake	Roze (1957); Peters & Orejas-Miranda (1970); Amaral (1976); Pérez-Santos & Moreno (1988)	Amazon Basin including Peru, Colombia, and Brazil to Maranhão. According to Amaral (1976), it is associated with the Amazon, Solimões, and Paraguay basins.
H. m. callostictus Günther, 1868 Peruvian Mudsnake	Roze (1957); Peters & Orejas-Miranda (1970); Dixon & Soini (1986)	NE Peru in the drainage of the Río Ucayali and Río Marañón.
H. triangularis (Wagler *in* Spix, 1824) Red and Black-banded False Coral Mud or Triangle Watersnake	Roze (1957); Underwood (1962); Peters & Orejas-Miranda (1970); Lancini (1979); Moonen et al. (1979); Lancini & Kornacker (1989); Savage & Slowinski (1992); Starace (1998); Kornacker (1999)	Trinidad and the Amazon Basin in Venezuela, the Guianas, and, southward into e Pará, Piauí, and Maranhão in Brazil, n Bolivia, Paraguay, and Argentina into Corrientes.
H. t. triangularis (Wagler *in* Spix, 1824) False Coral Mudsnake	Hoge (1952b); Roze (1957, 1966); Peters & Orejas-Miranda (1970); Amaral (1976); Starace (1998)	Amazon Basin of Brazil into Belem, Pará, and possibly into French Guiana.
H. t. bassleri Roze, 1957 Common Mudsnake	Roze (1957); Nicéforo-María (1970); Peters & Orejas-Miranda (1970); Dixon & Soini (1986); Henle & Ehrl (1991b)	NE Peru and Ecuador and is associated with the drainage of the Río Ucayali and Río Marañón.

Species-level Taxa	Citations and Remarks	Abbreviated Range
H. t. bolivianus Roze, 1957 Bolivian Mudsnake	Roze (1957); Peters & Orejas-Miranda (1970); Williams & Couturier (1984); Vanzolini (1986); Williams & Gudynas (1991); Cei (1993)	Beni in ne Bolivia into Corrientes, Argentina; associated with the Mamoré-Madeira river basin of the high Amazon.
H. t. fasciatus (Gray, 1849) Guianan Mudsnake	Roze (1957); Peters & Orejas-Miranda (1970); Abuys (1984)	Essequebo, Guyana into Suriname.
H. t. neglectus Roze, 1957 Red-Sided Watersnake or Red-Sided Mudsnake	Roze (1957); Peters & Orejas-Miranda (1970); Abuys (1984); Murphy (1997); Starace (1998); Boos (2001)	Trinidad and Tobago as well as w Guyana, and possibly French Guiana.
H. t. venezuelensis Roze, 1957 Venezuelan False Coral Water or Venezuelan False Coral Mudsnake	Roze (1957, 1966); Peters & Orejas-Miranda (1970); Lancini (1979); Lancini & Kornacker (1989); Kornacker (1999)	Río Orinoco Basin in Venezuela, and Colombia from the southern llanos into Río Vaupés.

GENUS: *Hypsiglena* Cope, 1860, is distributed in North America where it is found from the western and central United States southward into Mexico. See Tanner (1944); Smith & Taylor (1966), Peters & Orejas-Miranda (1970), Stebbins (1966, 1985); Dixon & Dean (1986); Degenhardt et al. (1996); Tennant (1998); and Dixon (2000a).

Review assistance and important contributions for this section were provided by James R. Dixon.

See also *Eridiphas* and *Pseudoleptodeira*.

CONTENT: Two to three or more species; two species are shown as valid below. ★★

Hypsiglena: Fangless Nightsnake

Species-level Taxa	Citations and Remarks	Abbreviated Range
*H. tanzeri** Dixon & Lieb, 1972 Tanzer's Nightsnake	Dixon & Lieb (1972); Tanner (1981); Dixon & Dean (1986); Smith & Smith (1993) *Remark: Dixon & Dean (1986), recognized this as *H. tanzeri* and because intergradation has not been shown to exist between it and *H. t. jani* in areas of near sympatry, their classification as full species still appears valid to me.	Central Mexico.
H. torquata (Günther, 1860) Nightsnake	Peters & Orejas-Miranda (1970); Stebbins (1985); Dixon & Dean (1986); Smith & Smith (1993); Degenhardt et al. (1996); Brown (2000); McPeak (2000); Dixon (2000a)	W and midwestern United States southward into Baja California, Pacific versant, and the Gulf Coast of Tamaulipas in Mexico.
H. t. torquata (Günther, 1860) Common Nightsnake	Taylor (1938 [1939]); Peters & Orejas-Miranda (1970); Dixon & Dean (1986); Smith & Smith (1993)	Sonora and Nayarit southward into Michoacán in the Pacific lowlands of Mexico.

Species-level Taxa	Citations and Remarks	Abbreviated Range
H. t. affinis Boulenger, 1894 Boulenger's Nightsnake	Smith & Taylor (1966); Dixon & Dean (1986); Smith & Smith (1993)	W Zacatecas and Jalisco in Pacific highlands of Mexico.
H. t. baueri Zweifel, 1958 Bauer's Nightsnake	Zweifel (1958); Tanner (1981); Smith & Smith (1993)	Cedros Islands in the Pacific off Baja California in Mexico.
H. t. catalinae* Tanner, 1966 Santa Catalina Nightsnake	Tanner (1966); Murphy (1983); Murphy & Ottley (1984); Smith & Smith (1993) *Remark: Grismer (1999), stated that this snake should be tentatively considered a synonym of *H. torquata*, but he did not give detail for the change.	Santa Catalina Island off Baja California, Mexico.
H. t. chlorophaea Cope, 1860 Sonoran Nightsnake	Langebartel & Smith (1954); Tanner (1985); Dixon & Dean (1986); Smith & Smith (1993); Crother et al. (2000)	Sonoran desert in n Sinaloa and s Sonora on the western mainland of Mexico, and in the w United States in Arizona.
H. t. deserticola Tanner, 1944 Desert Nightsnake	Stebbins (1966); Tanner (1981, 1985); Dixon & Dean (1986); Smith & Smith (1993); Brown (2000); Bartlett & Tennant (2000)	Deserts of the Great Basin in w United States in Utah, Washington, Oregon, Idaho, Nevada, Arizona, and California into Mexico.
H. t. gularis* Tanner, 1954 Partida Norte Nightsnake	Tanner (1954); Murphy (1983); Murphy & Ottley (1984); Smith & Smith (1993) *Remark: Grismer (1999), stated that this snake should be tentatively considered a distinct species, *H. gularis*, but he did not give details to justify a change.	Mexico on Isla Partida (north) in Baja California at 28°53′ N latitude and 113°3′ W longitude.
H. t. jani* (Dugès, 1866) Texas Nightsnake	Dixon & Dean (1986); Degenhardt et al. (1996); Tennant (1998); Conant & Collins (1998); Werler & Dixon (2000); Bartlett & Tennant (2000); Tennant & Bartlett (2000); Dixon (2000a) *Remark: The junior synonyms *H. torquata texana* Stejneger, 1893 (from Texas) and *H. torquata dunkeli* Taylor, 1938 (from the lowlands of Tamaulipas in Mexico), are still sometimes used, see Smith & Smith (1993), and Lemos-Espinal et al. (1992). These are synonyms according to Dixon & Dean (1986), and more current justifications do not seem to have been detailed since their work. I use Dixon & Dean (1986), and Dixon (2000a) as the most current critically investigated publications to support that these are junior synonyms of *H. t. jani*.	C United States southward into n Mexico from sc Kansas, se Colorado, through New Mexico and Texas into adjacent n Mexico into the Mexican Plateau north of Mexico City, and into Michoacán, San Luis Potosí, and Guanajuato; there is a disjunct population in ne Texas.
H. t. klauberi Tanner, (1944) 1946 San Diego Nightsnake	Stebbins (1966); Tanner (1985); Smith & Smith (1993); Brown (1997); Bartlett & Tennant (2000)	The United States in coastal southern California southward into extreme nw Baja California, Mexico.
H. t. loreala Tanner, 1944 Mesa Verde Nightsnake	Tanner (1944); Stebbins (1966); Bartlett & Tennant (2000)	W United States in Colorado and Utah, it is closely associated with the San Juan River Basin.

Species-level Taxa	Citations and Remarks	Abbreviated Range
H. t. martinensis Tanner & Banta, 1962 San Martin Nightsnake	Tanner & Banta (1962); Smith & Smith (1993)	San Martin Island in the Pacific off the coast of Baja California Norte in Mexico.
H. t. nuchalata Tanner, 1943 California Nightsnake	Stebbins (1966); Smith & Smith (1993); Brown (1997); Bartlett & Tennant (2000)	The United States in coastal California on the w slopes of the Sierra Nevada mountains.
*H. t. ochrorhyncha** Cope, 1860 Spotted Nightsnake	Conant (1975); Tanner (1981, 1985); Dixon & Dean (1986); Smith & Smith (1993); Bartlett & Tennant (2000) *Remark: Tennant (1984) stated this snake is found in far west Texas. Tennant (1998), revised the taxon he had originally listed in far west Texas to *H. t. jani* and did not list *H. t. ochrorhynchus* as found in Texas. Dixon (2000a), listed only *H. t. jani* as being found in Texas and he confirmed in a personal communication (11 June 1999), that this has not changed. See the Crother et al. (2000) comment about this snake.	W United States from Arizona, s lower California and southward through Chihuahua and Baja; in s Baja California, Mexico it seems to intergrade with *H. t. chlorophae* about the central Baja peninsula.
*H. t. tiburonensis** Tanner, 1981 Tiburon Island Nightsnake	Tanner (1981); Murphy (1983); Murphy & Ottley (1984); Smith & Smith (1993); Grismer (1999) *Remark: Grismer (1999) indicated this snake should be tentatively considered a synonym of *H. torquata*, but he did not give details for a change.	Isla Tiburón off the Sonora, Mexico coast in the Gulf of California.
*H. t. tortugaensis** Tanner, 1981 Tortuga Island Nightsnake	Tanner (1981, 1985); Murphy (1983); Murphy & Ottley (1984); Smith & Smith (1993); Grismer (1999) *Remark: Fide Grismer (1999), this snake should be tentatively considered a synonym of *H. torquata*, but he did not give details for the change.	Isla Tortuga in the Gulf of California in Mexico.
H. t. unaocularis Tanner, [1944] 1946 Clarion Island Nightsnake	Tanner (1944); Smith & Smith (1993)	Isla Clarion in the sw Revilla Gigedo Island complex of Mexico off Baja California.
H. t. venusta (Mocquard, 1899) Islands Nightsnake	Tanner (1944, 1981, 1985); Smith & Smith (1993)	Known only from Tortuga Island in Baja California 20 miles ne of Santa Rosalia.

GENUS: *Hypsirhynchus* Günther, 1858, is a Caribbean snake found on Hispaniola. See Maglio (1970); Schwartz (1971); Henderson & Schwartz (1984); Schwartz & Henderson (1985, 1988, 1991); and Sosa et al. (1995).

Important contributions for this section were provided by Robert A. Thomas.

CONTENT: One species with four subspecies. ★★★★

Hypsirhynchus: Hog-nosed Racer or Hispaniolan Cat-eyed Snake.

Species-level Taxa	Citations and Remarks	Abbreviated Range
H. ferox Günther, 1858 Hispaniolan Hog-nosed Racer	Maglio (1970); Schwartz (1971); Henderson & Schwartz (1984); Schwartz & Henderson (1991); Sosa et al. (1995); Powell et al. *in* Crother (1999)	Hispaniola and its satellites from sea level to 5600 feet.
H. f. ferox Günther, 1858 Common Hog-nosed Racer	Schwartz (1971); Henderson & Schwartz (1984); Schwartz & Henderson (1991); Sosa et al. (1995); Powell et al. *in* Crother (1999)	Hispaniola in Haiti and Dominican Republic and adjacent islands; north of, and including Plaine de Cul-de-Sac-Valle de Neiba, s Haiti north and south of Massif de la Selle, Peninsula de Barahona.
H. f. exedrus Schwartz, 1971 Saona Island Hog-nosed Racer	Schwartz (1971); Henderson & Schwartz (1984); Schwartz & Henderson (1991); Powell (1993); Sosa et al. (1995); Powell et al. *in* Crother (1999)	Isla Saona, Dominican Republic on Hispaniola.
H. f. paracrousis Schwartz, 1971 Gonave Hog-nosed Racer	Schwartz (1971); Henderson & Schwartz (1984); Schwartz & Henderson (1991); Powell (1993); Sosa et al. (1995); Powell et al. *in* Crother (1999)	Île de la Gonâve; Haiti on Hispaniola.
H. f. scalaris (Cope, 1863) Tiburon Hog-nosed Racer	Schwartz (1971); Henderson & Schwartz (1984); Schwartz & Henderson (1991); Sosa et al. (1995); Powell et al. *in* Crother (1999)	Haiti in the Tiburon Peninsula east into Diquini and 3.6 miles east of Jacmel in Département de l'Ouest; mainly assoicated with the eastern half of Hispaniola.

GENUS: *Ialtris* Cope, 1862, is a Caribbean snake from Hispaniola. See Maglio (1970); Schwartz (1971, 1980); Schwartz & Rossman (1976); Henderson & Schwartz (1984); Schwartz & Henderson (1991); and Powell & Henderson (1994a).

Review assistance and important contributions for this section were provided by Douglas A. Rossman.

CONTENT: Three species. ★★★★

Ialtris: Hispaniolan Racer. Frank and Ramus (1995) call this the Fanged Snake.

Species-level Taxa	Citations and Remarks	Abbreviated Range
I. agyrtes Schwartz & Rossman, 1976 Southern Desert Racer	Schwartz & Rossman (1976); Schwartz (1980); Henderson & Schwartz (1984); Schwartz & Henderson (1991); Powell & Henderson (1994a, 1994b); Powell et al. *in* Crother (1999)	Hispaniola in the sw Dominican Republic; it is known only from the type locality of Barreras, Azua Province and two localities on the Barahona Peninsula; all are in xeric forests.
I. dorsalis (Günther, 1858) W-headed Racer or Brown Racer	Cochran (1941); Maglio (1970); Schwartz & Rossman (1976); Henderson & Schwartz (1984); Schwartz & Henderson (1991); Zippel et al. (1994); Powell & Henderson (1994a); Powell et al. in Crother (1999)	Widely distributed in Hispaniola but apparently more common in Haiti, in Île-à-Vacha, Île de la Gonâve and Île de Tortue.

Species-level Taxa	Citations and Remarks	Abbreviated Range
I. parishi Cochran, 1932 Tiburon Banded Racer	Cochran (1932, 1941); Maglio (1970); Schwartz & Rossman (1976); Schwartz & Thomas (1975); Henderson & Schwartz (1984); Schwartz & Henderson (1991); Powell & Henderson (1994a, 1994c); Powell et al. *in* Crother (1999)	Hispaniola in Haiti, only known from the type locality 10 miles east of Baradères in Département du Sud and Île de la Tortue.

GENUS: *Imantodes* A. Dumeril, 1853. *Imantodes* is an emendation of *Himantodes*. According to Amaral (1976), the correct spelling of the generic name is *Himantodes*, as based on its Greek etymon which means "thong-like," and was preferred by Cope (1863). However, the spelling with the unaspirated initial "*I*" as used in Erpétologie Générale (1854), has been kept through priority. It is found from Mexico southward into South America to as far as northern Argentina and Paraguay. See Zweifel (1959b); Smith & Taylor (1966); Peters & Orejas-Miranda (1970); Myers (1982); Abuys (1984); Savage & Scott (1985); Vanzolini (1986); Wilson et al. (1986); Dowling & Jenner (1987), Obst et al. (1988); Smith & Smith (1993); Starace (1998); Kornacker (1999) and Köhler (2001).

CONTENT: Six species. ★★

Imantodes: Blunt-headed Tree Snake. Frank and Ramus (1995) call this the Central American Treesnake.

Species-level Taxa	Citations and Remarks	Abbreviated Range
*I. cenchoa** (Linnaeus, 1758) Blunt-headed Tree Snake	Zweifel (1959b); Peters & Orejas-Miranda (1970); Savage (1980); Myers (1982); Savage & Scott (1985); Wilson & Meyer (1985) Vanzolini (1986); Savage & Villa (1986); Lancini & Kornacker (1989); Lee (1996, 2000); Starace (1998); Renjifo & Lundberg (1999); Kornacker (1999); Freitas (1999); Stafford & Meyer (2000); Köhler (2001) *Remark: Myers (1982) synonymized subspecies. Kornacker (1999) showed without subspecies. However, Peters & Orejas-Miranda (1970); Wilson & Meyer (1985), Cei (1993); and Campbell (1998b), discussed subspecies and that is why they are used here.	S Tamaulipas southward through the Isthmus of Tehuantepec in Mexico southward into n Argentina, Paraguay, and Brazil; recently reported from São Paulo and Goías in Brazil; from sea level to 1500 m.
*I. c. cenchoa** (Linnaeus, 1758) Common Blunt-headed Tree Snake	Peters (1960b); Roze (1966); Peters & Orejas-Miranda (1970); Amaral (1976); Duellman (1978); Fugler & Walls (1978); Alvarez del Toro (1982); Abuys (1984); Dixon & Soini (1986); Chippaux ([1986] 1987); Villa et al. (1988); Smith & Smith (1993); Cei (1993); Campbell (1998a); Starace (1998); Esqueda & La Marca (1999); Marques et al. (2001); Boos (2001) *Remark: I follow Cei (1993), who synonymized *I. semifasciatus* (Cope, 1894) with this snake.	Panama southward to Bolivia, Paraguay, Argentina, and Brazil, it is also found in Trinidad.
I. c. leucomelas (Cope, 1861) Cope's Blunt-headed Tree Snake	Smith (1942e); Smith & Taylor (1966); Peters & Orejas-Miranda (1970); Henderson & Hoevers (1975); Wilson & Meyer (1985); Pérez-Higareda & Smith (1991); Smith et al. (1995); Campbell (1998a); Lee (2000)	Veracruz and Chiapas in s Mexico southward into n Honduras; Smith et al. (1995) show the range to be Mexico and n Central America.

Species-level Taxa	Citations and Remarks	Abbreviated Range
I. c. semifasciatus (Cope, 1894) Half-banded Blunt-headed Tree Snake	Peters & Orejas-Miranda (1970); Wilson & Meyer (1985); Smith et al. (1995)	Guatemala? to Panama; Smith et al. (1995) show it to be Nicaragua to Panama in Pacific versant.
I. gemmistratus* (Cope, [1860]1861) Central American Tree Snake or Greater Blunt-headed Tree Snake	Zweifel (1959b); Smith (1942e); Smith & Taylor (1966); Peters & Orejas-Miranda (1970); Myers (1982, 1992); Savage & Scott (1985); Wilson & Meyer (1985); Savage & Villa (1986); Villa et al. (1988); García & Ceballos (1994); Lee (1996, 2000); Ponce-Campos & Huerta-Ortega (1998); Köhler (2001) *Remarks: Myers (1982) discussed that classification of this species was problematic. Wilson & Meyer (1985), discussed the problems with trying to identify subspecies within this species. Subspecies are listed by Peters & Orejas-Miranda (1970), and others. Subspecies are not listed by Smith & Taylor (1966). Campbell (1998a) referred to subspecies. Subspecies have been included here, but the reader should review reference literature before accepting the list. Zweifel (1959b), assigned subspecies based on definable differences in the materials from western Mexico. Also see Smith et al. (1995) about comments on subspecies.	Along both coasts of Mexico from n Veracruz and s Sonora southward through Central America to Colombia into the Magdalena Valley from sea level to 750 m.
I. g. gemmistratus (Cope, [1860]1861) Greater Blunt-headed Tree Snake	Smith & Taylor (1966); Peters & Orejas-Miranda (1970); Alvarez del Toro (1982); Wilson & Meyer (1985); Smith & Smith (1993)	Sonora, Mexico southward into Panama.
I. g. gracillimus* (Günther, 1895) Slender Blunt-headed Tree Snake	Zweifel (1959b); Smith & Taylor (1966); Peters & Orejas-Miranda (1970); Smith & Smith (1993) *Remark: Zweifel (1959b) stated that this snake was known from only seven snakes at the time of his work.	Rare from Mexico, found from coastal Colima into Acapulco, into c Guerrero, and Tres Marías Islands off the coast of Nayarit.
I. g. latistratus (Cope, 1887) Red Blunt-headed Tree Snake	Zweifel (1959b); Peters & Orejas-Miranda (1970); Casas-Andreu (1982); Wilson & Meyer (1985); Smith & Smith (1993); Ramíez-Bautista (1994)	Mexico on the Pacific coast in low jungle from Sonora into Jalisco, possibly ranges from Sinaloa through Nayarit, Morelos, c Guerrero, and Tres Marías Islands; Zweifel (1959b), expected this snake to eventually be found in w Puebla and adjacent Oaxaca.
I. g. luciodorsus Oliver, 1942 Southern Blunt-headed Tree Snake	Zweifel (1959b); Smith & Taylor (1966); Peters & Orejas-Miranda (1970); Pérez-Higareda & Smith (1991); Smith & Smith (1993); Ramíez-Bautista (1994)	SE Mexico from c Veracruz into Guatemala in coastal or lowlands areas of the Atlantic versant from c Veracruz through Campeche and n Chiapas into Guatemala, but is not reported from the northern half of the Yucatán peninsula. It possibly will eventually be found in Oaxaca and Tabasco.
I. g. oliveri Smith, 1942 Oliver's Blunt-headed Tree Snake	Peters (1954); Zweifel (1959b); Smith & Taylor (1966); Peters & Orejas-Miranda (1970); Alvarez del Toro (1982) [as *I. splendidus oliveri*]; Smith & Smith (1993); Smith et al. (1995)	Oaxaca in s Mexico southward into w Guatemala.

Species-level Taxa	Citations and Remarks	Abbreviated Range
I. g. reticulatus* (Müller, 1882) Reticulated Blunt-headed Tree Snake	Peters & Orejas-Miranda (1970); Ponce-Campos & & Huerta-Ortega (1998); Smith et al. (1995) *Remark: Fide Smith & Smith (1993), *I. g. oliveri* Smith, 1942, is a synonym; however, lists *I. g. oliveri* only as valid. It has been included here based on Liner (1994), but it probably is not valid. It is also shown in the incertae sedis section as *Species Inquirenda*.	Fide Peters & Orejas-Miranda (1970) range and type locality cannot be properly identified so is *Species Inquirenda*; however, is listed by Liner (1994) as present in Mexico; based on *I. g. oliveri* the range should be Tehuantepec area of Oaxaca in s Mexico into w Guatemala.
I. g. splendidus* (Günther, 1895) Splendid Blunt-headed Tree Snake	Zweifel (1959b); Smith & Taylor (1966); Peters & Orejas-Miranda (1970); Smith & Smith (1993); Campbell (1998a); Lee (2000) *Remark: Campbell (1998a), noted that *I. gemmistratus* is found in n Yucatán, but was not sure if this was the subspecies that occurs in Petén, Guatemala.	Yucatán in se Mexico.
I. inornatus (Boulenger, 1896) Western Tree Snake or Chocan Blunt-headed Tree Snake	Peters (1960b); Fugler & Walls (1978); Savage (1980); Myers (1982); Savage & Scott (1985); Wilson & Meyer (1985); Wilson et al. (1986); Vanzolini (1986); Savage & Villa (1986); Villa et al. (1988); Köhler (2001)	Honduras and Nicaragua southward to w Colombia and nw Ecuador; generally associated with low tropical dry forest formations.
I. lentiferus (Cope, 1894) Amazon Tree Snake	Peters (1960b); Duellman (1978); Myers (1982); Abuys (1984); Vanzolini (1986); Dixon & Soini (1986); Starace (1998); Kornacker (1999)	Amazon Basin of Brazil, Peru, s Colombia, Ecuador, and into Venezuela and the Guianas.
I. phantasma Myers, 1982 Ghost Blunt-headed Tree Snake	Myers (1982); Vanzolini (1986); Villa et al. (1988); Auth (1994); Köhler (2001)	Provencia Darién in Panama in the Serranía de Pirre range.
I. tenuissimus (Cope, [1866]1867) Yucatán Blunt-headed Tree Snake	Zweifel (1959b) Smith & Taylor (1966); Peters & Orejas-Miranda (1970); Myers (1982); Dundee et al. (1986); Villa et al. (1988); Lee (1996, 2000); Köhler (2001)	Yucatán peninsula of Mexico where it is found in several localities in Campeche, Quintana Roo, and Yucatán; there is an unconfirmed record from Prussia, Campeche.

GENUS: *Lampropeltis* Fitzinger, 1843, is wide ranging in the northern part of the Western Hemisphere. It is found in the north part of its range south of the 48th Parallel in s Ontario and sw Quebec in Canada, westward to s Washington, and southward through the United States, Mexico, and Central America into extreme northwestern South America in Colombia, Ecuador, and into the Cordillera de la Costa of Venezuela. See Peters & Orejas-Miranda (1970); Blaney (1973); Conant (1975); Williams (1978); Markel (1980, 1990); Garstka (1982); Stebbins (1985); Savage & Slowinski (1992) Smith & Smith (1993); Degenhardt et al. (1996); Campbell (1998a); Tennant (1998); Conant & Collins (1998); Kornacker (1999); Bartlett & Tennant (2000); Dixon (2000a); and Tennant & Bartlett (2000).

Important contributions for this section were provided by Javier A. Rodrigues-Robles (for *L. zonata*) and Gerard T. Solman (for *L. alterna*, *L. ruthveni*, and *L. triangulum*).

CONTENT: Eight species. ★★★

Lampropeltis: Kingsnake, Milksnake, and Molesnake

Species-level Taxa	Citations and Remarks	Abbreviated Range
L. alterna* (Brown, 1901) Gray-banded Kingsnake	Smith & Taylor (1966); Gelbach (1967); Conant (1975); Markel (1980) [as *L. mexicana alterna*]; Garstka (1982); Degenhardt et al. (1996); Tennant (1998); Hilken & Schlepper (1998); Conant & Collins (1998); Werler & Dixon (2000); Bartlett & Tennant (2000); Tennant & Bartlett (2000); Dixon (2000a) *Remark: Fide Garstka (1982), a color form of this snake exists that was named *L. blairi* Flury, 1950, and later called *L. mexicana blairi*. When compared with *alterna*, the "*alterna*" form was considered to have bands or rings and the "*blairi*" phase has saddles or blotches. These forms are reviewed in detail in the references. Hilken & Schlepper (1998) stated that *alterna* and *blairi* are subspecies of *L. alterna*. Although they obviously put a lot of work into the paper, I did not find their data convincing therefore retained *blairi* as a color form only. However, their work was recognized by Crother et al. (2000) and subspecies of *L. a. alterna* and *L. a. blairi* were listed as valid in that checklist.	SW Texas in the United States southward into n Mexico in Coahuila, Durango, and extreme w Nuevo León.
L. calligaster (Harlan, 1827) Yellow-bellied Kingsnake	Blaney (1979); Markel (1980, 1990); Price (1987); Dundee & Rossman (1989); Dixon (2000a)	Eastern United States from se Nebraska, s Iowa, Illinois, and extreme w Indiana southward to Texas, and eastward to the Atlantic coast from Maryland and Virginia to n Florida.
L. c. calligaster (Harlan, 1827) Prairie Kingsnake	Conant (1975); Blaney (1979); Markel (1980, 1990); Smith & Smith (1993); Tennant (1998); Conant & Collins (1998); Dixon (2000a); Werler & Dixon (2000); Tennant & Bartlett (2000)	C United States from w Indiana, c Kentucky and nw Mississippi westward through Kansas, Oklahoma, and e Texas.
L. c. occipitolineata Price, 1987 South Florida Mole Kingsnake	Markel (1980, 1990); Price (1987); Tennant (1997b); Conant & Collins (1998); Tennant & Bartlett (2000)	SE United States in c Florida north of Lake Okeechobee to the Atlantic coast.
L. c. rhombomaculata (Holbrook, 1840) Mole Kingsnake	Conant (1975); Blaney (1979); Markel (1980); Ashton & Ashton (1981); Dundee & Rossman (1989); Tennant (1997b); Conant & Collins (1998); Tennant & Bartlett (2000)	E and s United States from Maryland southward into the Florida panhandle and west into c Tennessee and s Mississippi.
L. getula (Linnaeus, 1766) Common Kingsnake	Markel (1980, 1990); Ashton & Ashton (1981); Stebbins (1985); Dundee & Rossman (1989); Savage & Slowinski (1992); Degenhardt et al. (1996); Brown (1997); Tennant (1998); Dixon (2000a)	Coast to coast in the United States below 41° on the Atlantic coast to below 43° on the Pacific coast, southward into most of Baja California and northern Mexico into Zacatecas and San Luis Potosí.
L. g. getula (Linnaeus, 1766) Eastern Kingsnake	Conant (1975); Ashton & Ashton (1981); Markel (1990); Tennant (1997b, 1998); Conant & Collins (1998); Tennant & Bartlett (2000)	E United States.

Species-level Taxa	Citations and Remarks	Abbreviated Range
L. g. californiae * (Blainville, 1835) California Kingsnake	Markel (1980, 1990); Stebbins (1985); Degenhardt et al. (1996); Brown (1997); Bartlett & Tennant (2000) *Remark: Fide Markel (1990), and Brown (1997), *L. g. yumensis* Blanchard, 1919, sometimes called the Yuma Kingsnake, is a synonym.	W United States in California southward into Baja California del Norte, Mexico.
L. g. catalensis * Van Denburgh & Slevin, 1921 Santa Catalina Island Kingsnake	Van Denburgh & Slevin (1921); Smith & Smith (1976); Grismer (1999) *Remark: Not found in Markel (1990) or Liner (1994), and probably not a valid subspecies. However, Grismer (1999) suggested that this is a full species, *L. catalinensis*. Since Grismer (1999) based his decision on color, I decided to leave this snake as *L. getulus catalensis* because physical and/or biochemical parameters are generally used to define taxonomic differences. I did not elevate it to species level because Grismer (1999) did not give details for the justification.	Baja California, Mexico on offshore islands in the Sea of Cortez in the Gulf of California.
L. g. floridana * Blanchard, 1919 Florida Kingsnake	Conant (1975); Ashton & Ashton (1981); Markel (1980, 1990); Tennant (1997b); Conant & Collins (1998); Tennant & Bartlett (2000) *Remark 1: Those individuals found in the Apalachicola region of Florida were once classified as *L. g. goini*. They are now considered intergrades between *L. g. getulus* and *L. g. floridana* (Blaney, 1977). *Remark 2: The highly spotted version was sometimes called *L. g. brooksi* (Conant, 1997).	S Florida in the e United States.
L. g. holbrooki Stejneger, 1902 Speckled Kingsnake	Conant (1975); Markel (1980, 1990); Dundee & Rossman (1989); Tennant (1998); Conant & Collins (1998); Werler & Dixon (2000); Tennant & Bartlett (2000); Dixon (2000a)	Mississippi Valley to the Gulf Coast of the United States.
L. g. niger (Yarrow, 1882) Eastern Black Kingsnake	Conant (1975); Markel (1980, 1990); Conant & Collins (1998) [as *L. getula nigra*]; Tennant & Bartlett (2000) [as *L. getula nigra*]	United States from s Ohio and se Illinois southward into c Alabama nw Georgia.
L. g. nigrita Zweifel & Norris, 1955 Western Desert Kingsnake	Markel (1980, 1990); Stebbins (1985); Bartlett & Tennant (2000)	Deserts of nw Mexico in w Sonora and Sinaloa, it intergrades with *splendida* and *californiae* in se Arizona in the w United Sates at the frontier.
L. g. splendida (Baird & Girard, 1853) Desert Kingsnake	Conant (1975); Markel (1980); Stebbins (1985); Degenhardt et al. (1996); Tennant (1998); Conant & Collins (1998); Werler & Dixon (2000); Bartlett & Tennant (2000); Tennant & Bartlett (2000); Dixon (2000a)	Frontier of sw United States and Mexico from extreme se Arizona into c Texas, and into nc Mexico. Disjunct populations in n New Mexico and s Colorado.

The Caenophidia

Species-level Taxa	Citations and Remarks	Abbreviated Range
*L. g. sticticeps** Barbour & Engles, 1942 Outer Banks Kingsnake	Conant (1975); Conant & Collins (1998); Tennant & Bartlett (2000) *Remark: Fide Tennant & Bartlett (2000), this is probably only a local color phase; and, Crother et al. (2000) did not list it.	Outer Banks of North Carolina from Cape Hatteras to Cape Lookout.
*L. mexicana** (Garman, 1884) San Luis Potosí Kingsnake or Mexican Kingsnake	Gelbach & Baker (1962); Markel (1980, 1990); Garstka (1982); Trutnau (1986); Savage & Slowinski (1992); Smith & Smith (1993); Hilken & Schlepper (1998); Salmon et al. (2001) *Remark 1: Subspecies seen in Gelbach (1967, 1982), and Trutnau (1986) are now considered only color variants. Using Smith & Smith (1993) as a guide, the color phases are often designated as: *L. m.* "greeri" Webb, 1961, found from Trans-Pecos region of Texas southward into n Mexico; *L. m.* "leonis" (Günther, 1893), is rare from Mexico, the type specimen was collected in Nuevo León; and, *L. m.* "thayeri" Loveridge, 1924, with a relatively few specimens from the Miquihauna area of Tamaulipas in Mexico. *Lampropeltis mexicana* is now considered to be monotypic, polymorphic. Hilken & Schlepper (1998) suggested *greeri*, *leonis* and *thayeri* are subspecies of *mexicana*. They put a lot of work into the paper but I did not find that the data was compelling enough to justify separating these snakes and chose to maintain them as color forms in this work. *Remark 2: *L. mexicana* was mentioned under *L. alterna* in Crother et al. (2000), but was not listed as a valid species. *Remark 3: Gerry Salmon (personnal commuication 8 August 2002), along with others have proposed a new common name that more accurately reflects the status of this snake.	Normally found between 3400 and 7500 feet in the Chihuahuan desert which extends from Texas and New Mexico border area southward into c Mexico.
L. pyromelana (Cope, 1866 [1867]) Sonora Mountain Kingsnake	Osborn (1931); Schmidt (1953a); Markel (1980, 1990); Tanner (1953, 1983, 1985); Stebbins (1985); Smith & Slowinski (1992); Smith & Smith (1993); Degenhardt et al. (1996)	W United States southward into c Mexico.
*L. p. pyromelana** (Cope, 1867) Arizona Mountain Kingsnake	Markel (1980, 1990); Tanner (1953, 1983, 1985); Stebbins (1985); Smith & Smith (1993); Degenhardt et al. (1996); Bartlett & Tennant (2000) *Remark: According to Tanner (1983) and Van Devender et al. (1993), *L. p. knoblocki* Taylor, 1940 is a synonym. Markel (1980, 1990) considered *L. p. knoblocki* to be valid and as found from sw Chihuahua and extreme se Sonora possibly into n Durango in Mexico.	C Arizona and New Mexico, and into Mexico in Chihuahua and Sonora; not found in the Huachuca Mountains.

Species-level Taxa	Citations and Remarks	Abbreviated Range
L. p. infralabialis Tanner, 1953 Utah Mountain Kingsnake	Markel (1980, 1990); Tanner (1953, 1983, 1985); Stebbins (1985); Bartlett & Tennant (2000)	United States from the Grand Canyon region northward into Nevada.
L. p. woodini Tanner, 1953 Huachuca Mountain Kingsnake	Markel (1980, 1990); Tanner (1953, 1983); Smith & Smith (1993); Bartlett & Tennant (2000)	S Arizona in the United States southward into adjacent n Mexico; restricted to the Huachuca Mountains.
L. ruthveni Blanchard, 1921 Queretaro Kingsnake	Garstka (1982); Williams (1988); Markel (1990); Hilken & Schlepper (1998); Salmon et al. (2001)	Michoacán, Jalisco, Queretaro and Oaxaca in southern Mexico.
L. triangulum* (Lacépède, 1788) Milksnake	Mittleman (1952); Smith (1952); Peters & Orejas-Miranda (1970); Williams (1974, 1978, 1994); Conant (1975); Markel (1980, 1990); Garstka (1982); Wilson & Meyer (1985); Stebbins (1985); Villa et al. (1988); Dundee & Rossman (1989); Savage & Slowinski (1992); García & Ceballos (1994); Lee (1996, 2000); Degenhardt et al. (1996); Campbell (1998a); Kornacker (1999); Dixon (2000a) *Remark: In older literature, this species is often referred to as *L. doliata* (Linnaeus, 1766), (see Wright & Wright, [1957] 1994).	Wide ranging in the United States southward into northern South America, it has the most widespread distribution of any *Lampropeltis*; found from se Canada through the c and e United States, and from s Sonora in Mexico southward through Central America into Colombia; Ecuador, and into the Cordillera de la Costa in Venezuela.
L. t. triangulum (Lacépèdè, 1788) Eastern Milksnake	Williams (1978, 1988); Markel (1980, 1990); Conant & Collins (1998); Tennant & Bartlett (2000); Dixon (2000a)	E United States from Maine into Minnesota and Ontario, Canada, southward into n Alabama.
L. t. abnorma (Bocourt, 1886 *in* Dumeril, Mocquard & Bocourt, 1870–1909) Guatemalan Milksnake	Klauber (1948); Peters & Orejas-Miranda (1970); Williams (1978, 1988); Markel (1980, 1990); Lee (2000)	S Mexico, Guatemala, and Honduras.
L. t. amaura Cope, 1861 Louisiana Milksnake	Klauber (1948); Conant (1975); Williams (1978, 1988, 1994); Markel (1980, 1990); Dundee & Rossman (1989); Tennant (1998); Conant & Collins (1998); Werler & Dixon (2000); Tennant & Bartlett (2000); Dixon (2000a)	United States in Louisiana, e Texas, sw Arkansas, and se Oklahoma.
L. t. andesiana Williams, 1978 Andean Milksnake	Peters & Orejas-Miranda (1970); Williams (1978, 1988, 1994); Markel (1980, 1990); Vanzolini (1986)	NW Colombian Andes where it is found at 220 to 2700 m in elevation.
L. t. annulata Kennicott, 1861 Tamaulipian or Mexican Milksnake	Klauber (1948); Conant (1975); Williams (1978, 1988, 1994); Markel (1980, 1990); Garstka (1982); Tennant (1998); Conant & Collins (1998); Werler & Dixon (2000); Tennant & Bartlett (2000); Dixon (2000a)	C Texas in the United States into c Mexico to as far as Tamaulipas, and westward to as far as c Nuevo León, and southward and eastward into Coahuila.
L. t. arcifera Werner, 1903 Jalisco Milksnake	Davis & Smith (1953); Williams (1978, 1988, 1994); Markel (1980, 1990); Englemann & Obst (1981); Garstka (1982)	The Sierras in central and western Mexico from 700 to 3500 feet. It is closely associated with the highlands.

The Caenophidia

Species-level Taxa	Citations and Remarks	Abbreviated Range
L. t. blanchardi Stuart, 1935 Blanchard's Milksnake	Dunn (1937); Williams (1978, 1988, 1994); Markel (1980, 1990); Smith & Smith (1993); Lee (2000)	S and se Mexico in the deciduous forest zone in the Yucatán, and the rain forest zone in Quintana Roo and Campeche into n El Petén in Guatemala.
L. t. campbelli Quinn, 1983 Pueblan Milksnake	Quinn (1983); Williams (1988, 1994); Markel (1990); Smith & Smith (1993)	Mexico in s Puebla west into e Morelos, and southward into n Oaxaca.
L. t. celaenops Stejneger, 1903 New Mexico Milksnake	Conant (1975); Williams (1978, 1988, 1994); Markel (1980, 1990); Garstka (1982); Stebbins (1966, 1985); Degenhardt et al. (1996); Tennant (1998); Werler & Dixon (2000); Bartlett & Tennant (2000); Tennant & Bartlett (2000); Dixon (2000a)	New Mexico and Texas in the United States southward into Mexico, with disjunct populations in ec Arizona and Colorado. It seems to prefer gamma grasslands and piñon-juniper woodland at about 6900 feet.
L. t. conanti Williams, 1978 Conant's Milksnake	Williams (1978, 1988, 1994); Markel (1980, 1990); Smith & Smith (1993)	C and s Mexico in tropical lowlands near the coast and inland into tropical highlands.
L. t. dixoni Quinn, 1983 Dixon's Milksnake	Quinn (1983); Williams (1988, 1994); Markel (1990); Smith & Smith (1993)	Mexico in the mountain passes and valleys from s San Luis Potosí into the Jalapan Valley of ne Queretaro.
L. t. elapsoides (Holbrook, 1838) Scarlet Kingsnake	Klauber (1948); Conant (1975); Williams (1978, 1988, 1994); Markel (1980, 1990); Ashton & Ashton (1981); Dundee & Rossman (1989); Tennant (1997b); Conant & Collins (1998); Tennant & Bartlett (2000)	United States from Virginia southward into Florida and inland into Tennessee, s Kentucky, and Mississippi.
L. t. gaigae Dunn, 1937 Giant Central American Milksnake	Peters & Orejas-Miranda (1970); Williams (1978, 1988, 1994); Markel (1980, 1990); Smith & Smith (1993); Frank (1994)	Mountains of e Costa Rica and w Panama in moist to wet areas at relatively high elevations of 4300 to 6500 feet in Panama and 5000 to 7400 feet in Costa Rica.
L. t. gentilis (Baird & Girard, 1853) Central Plains Milksnake	Conant (1975); Williams (1978, 1988, 1994); Markel (1980, 1990); Stebbins (1985); Degenhardt et al. (1996); Tennant (1998); Conant & Collins (1998); Dixon (2000a); Werler & Dixon (2000); Bartlett & Tennant (2000); Tennant & Bartlett (2000)	Central Plains of the United States in Kansas, Oklahoma, and extreme nc Texas into c Colorado.
L. t. hondurensis Williams, 1978 Honduran Milksnake	Peters & Orejas-Miranda (1970); Williams (1978, 1988, 1994); Markel (1980, 1990); Wilson & Meyer (1985); Vanzolini (1986)	Caribbean versant of Guatemala (se El Petén), southward into Honduras, Nicaragua, and Costa Rica in low to moderate elevations.
L. t. micropholis Cope, 1861 Ecuadorian Milksnake	Klauber (1948); Roze (1966); Peters & Orejas-Miranda (1970); Williams (1978, 1988, 1994); Markel (1980, 1990); Lancini & Kornacker (1989); Kornacker (1999)	Costa Rica south into nw and Caribbean Colombia, nw Venezuela, and nw Ecuador, higher elevations of the Andes are avoided.
L. t. multistrata Kennicott, 1861 Pale Milksnake	Conant (1975); Williams (1978, 1988, 1994); Markel (1980, 1990); Stebbins (1966, 1985); Frank (1994); Conant & Collins (1998); Bartlett & Tennant (2000); Tennant & Bartlett (2000)	NW prairies of w United States from c Montana into Nebraska, and is closely associated with the high plains.
*L. t. nelsoni** Blanchard, 1920 Nelson's Milksnake	Williams (1978, 1988, 1994); Markel (1980, 1990); Garstka (1982); Ramírez-Bautista (1994)	Coastal Jalisco and Colima moving into the western edge of the Mesa Central, and southward into Jalisco; there is a population on Tres Marias Island.

Species-level Taxa	Citations and Remarks	Abbreviated Range
	*Remark: Williams (1994), included *L. t. schmidti* Stuart, 1935 from María Madre Island in the Tres María Islands of Mexico as a synonym of this snake. However, Markel (1990), listed it as a valid species.	
L. t. oligozona (Bocourt, 1886 *in* Dumeril, Mocquard & Bocourt, 1870–1909) Central American Milksnake	Klauber (1948); Peters & Orejas-Miranda (1970); Williams (1978, 1988, 1994); Markel (1980); Alvarez del Toro (1982); Smith & Smith (1993)	Pacific versant in the Isthmus of Tehuantepec in s Mexico, southward into s Guatemala and El Salvador.
L. t. polyzona Cope, 1861 Veracruz Milksnake	Klauber (1948); Peters & Orejas-Miranda (1970); Williams (1978, 1988, 1994); Markel (1980, 1990); Alvarez del Toro (1982); Pérez-Higareda & Smith (1991); Smith & Smith (1993); Lee (2000)	Gulf coast of San Luis Potosí, Jalisco and c Veracruz in Mexico southward into Belize, Guatemala, and Nicaragua, and probably also present Honduras? It has been collected at 6000 feet on Volcán San Martin, Veracruz.
L. t. sinaloae Williams, 1978 Sinaloan Milksnake	Williams (1978, 1988, 1994); Markel (1980, 1990); Smith & Smith (1993)	W Mexico around Sinaloa usually below 1000 m along coastal lowlands.
L. t. smithi Williams, 1978 Smith's Milksnake	Williams (1978, 1988, 1994); Markel (1980, 1990); Pérez-Higareda & Smith (1991); Smith & Smith (1993)	Mexico in the Sierra Madre Oriental from se San Luis Potosí south through e Queretaro, Hidalgo, ne Puebla, and into the Jalapa area of Veracruz.
L. t. stuarti Williams, 1978 Stuart's Milksnake	Peters & Orejas-Miranda (1970); Williams (1978, 1988, 1994); Vanzolini (1986); Markel (1980, 1990); Smith & Smith (1993)	Pacific Versant of Central America in El Salvador, Honduras, Nicaragua, and nw Costa Rica; said to prefer dry tropical forests and coastal plains.
L. t. syspila Cope, 1888 Red Milksnake	Conant (1975); Williams (1978, 1988, 1994); Markel (1980, 1990); Dundee & Rossman (1989); Conant & Collins (1998); Tennant & Bartlett (2000)	C United States from s Indiana into nw Mississippi, and westward into extreme se South Dakota and e Oklahoma
L. t. taylori Tanner & Loomis, 1957 Utah Milksnake	Williams (1978, 1988, 1994); Markel (1980, 1990); Stebbins (1966, 1985); Smith & Smith (1993); Bartlett & Tennant (2000)	SW United States in Utah and Colorado.
L. zonata (Blainville, 1835) California Mountain Kingsnake	Klauber (1943); Zweifel (1974); Markel (1980, 1990); Stebbins (1985); Savage & Slowinski (1992); Smith & Smith (1993); Brown (1997)	W United States southward into n Baja California, Mexico from sea level to 8000 feet.
L. z. zonata (Blainville, 1835) St. Helena Mountain Kingsnake	Klauber (1943a); Zweifel (1974); Markel (1980, 1990); Stebbins (1985); Brown (1997); Bartlett & Tennant (2000)	The United States in California, se Oregon, and Washington.
L. z. agalma Van Denburgh & Slevin, 1923 San Pedro Mountain Kingsnake	Klauber (1943a); Zweifel (1974); Markel (1980, 1990); Stebbins (1985); Smith & Smith (1993); McPeak (2000)	Baja California in Mexico; it is endemic to Sierra San Pedro Mártir, and Sierra Juárez from 5200 ft to around 9200 ft (1600 to 2800 m).

The Caenophidia

Species-level Taxa	Citations and Remarks	Abbreviated Range
L. z. herrerae Van Denburgh & Slevin, 1923 Todos Santos Island Kingsnake	Klauber (1943a); Smith & Taylor (1945); Zweifel (1974); Markel (1980, 1990); Smith & Smith (1993); McPeak (2000)	Todos Island, Baja California in Mexico.
L. z. multicincta (Yarrow, 1882) Sierra Mountain Kingsnake	Klauber (1943a), Zweifel (1952, 1974); Markel (1980, 1990); Stebbins (1985); Smith & Smith (1993); Brown (1997); Bartlett & Tennant (2000)	The United States from sw Oregon into California; it has a very disjunct range; in the Sierra Mountains.
L. z. multifasciata (Bocourt, 1886 *in* Dumeril, Mocquard & Bocourt, 1870–1909) Coast Mountain Kingsnake	Klauber (1943a); Zweifel (1974); Markel (1980, 1990); Stebbins (1985); Smith & Smith (1993); Brown (1997); Bartlett & Tennant (2000)	W United States restricted to California; disjunct range along the coastal mountains.
L. z. parvirubra Zweifel, 1952 San Bernardino Mountain Kingsnake	Zweifel (1952, 1974); Markel (1980, 1990); Stebbins (1985); Smith & Smith (1993); Brown (1997); Bartlett & Tennant (2000)	Restricted to California in the w United States; in Los Angeles, San Bernardino, and Riverside counties.
L. z. pulchra Zweifel, 1952 San Diego Mountain Kingsnake	Klauber (1943a); Zweifel (1952,1974); Markel (1980, 1990); Stebbins (1985); Smith & Smith (1993); Brown (1997); Bartlett & Tennant (2000)	W United States restricted to s California; in Los Angeles, Orange, Riverside, and San Diego counties.

GENUS: *Leptodeira* Fitzinger, 1843, ranges from south Texas southward through Middle America through South America to as far south as northern Argentina and Paraguay. See Duellman (1958); Smith & Taylor (1966); Peters & Orejas-Miranda (1970); Lancini (1979); Dowling & Jenner (1987), Auth (1994); Smith & Smith (1993); Kornacker (1999); Dixon (2000a) and Köhler (2001).

Review assistance and important contributions for this section were provided by Travis J. La Duc and James R. Dixon.

Duellman (1958) originally reviewed the *Leptodeira* species and subspecies listing nine species and one in incertae sedis.

Also reference the genera *Tantalophis* Duellman, 1958 and *Pseudoleptodeira*, Taylor, 1938 [1939].

CONTENT: Nine species. ★★★★

Leptodeira: American Cat-eyed Snake

Species-level Taxa	Citations and Remarks	Abbreviated Range
L. annulata (Linnaeus, 1758) Banded Cat-eyed Snake	Amaral, 1934 [1933–1934]; Duellman (1958); Peters & Orejas-Miranda (1970); Wilson & Meyer (1985); Savage & Villa (1986); Villa et al. (1988); Smith &	Mexico southward through the Amazon Basin into n Argentina and Paraguay.

Species-level Taxa	Citations and Remarks	Abbreviated Range
	Smith (1993); Duellman & Mendelson (1995); Lee (1996); Starace (1998); Kornacker (1999); Freitas (1999); Köhler (2001)	
L. a. annulata (Linnaeus, 1758) Common Cat-eyed Snake	Peters (1960b); Duellman (1958, 1978); Roze (1966); Peters & Orejas-Miranda (1970); Amaral (1976); Fugler & Walls (1978); Lancini (1979); Dixon & Soini (1986); Lancini & Kornacker (1989); Starace (1998); Kornacker (1999); Boos (2001)	Amazon Basin in Venezuela, Ecuador, Peru, and Bolivia into São Paulo, Brazil as well as the Guianas; below 1100 m in tropical rain forests.
L. a. ashmeadi (Hallowell, 1845) Ashmead's Cat-eyed Snake	Duellman (1958); Roze (1966); Peters & Orejas-Miranda (1970); Lancini & Kornacker (1989); Kornacker (1999); Mijares-Urrutia & Arends R. (2000)	East of the Santa María Mountains in Colombia and into n Venezuela associated with mountain valleys along the coast southward to the Río Orinoco; its range includes Trinidad and Tobago and Isla Margarita; sea level to 1000 m.
L. a. cussiliris Duellman, 1958 Duellman's Cat-eyed Snake	Duellman (1958); Peters & Orejas-Miranda (1970); Alvarez del Toro (1982); Wilson & Meyer (1982, 1985); Pérez-Higareda & Smith (1991); Smith & Smith (1993); Mendelson (1995); Lee (2000); Köhler (2001)	Mexico in the coastal lowlands and low foothills of the Sierra Madre from s Tamaulipas, southeastward to the Isthmus of Tehuantepec, and westward along the Pacific lowlands to the Río Balsas, southeastward along the Pacific coast to se Chiapas, and through the Grijalva Valley of Chiapas; reported to prefer semi-arid environments from sea level to about 2000 m.
L. a. pulchriceps Duellman, 1958 Southern Banded Cat-eyed Snake	Duellman (1958) Peters & Orejas-Miranda (1970); Cei (1993); Leynaud & Bucher (1999)	Santa Cruz de la Sierra in Bolivia, s Paraguay, Chaco region of n Argentina including Corrientes, Entre Ríos, Santa Fe, Chaco, Formosa, Salta, and Tucumán, and into the s Mato Grosso state in Brazil.
L. a. rhombifera Günther, 1872 Rhombic Cat-eyed Snake	Duellman (1958); Peters & Orejas-Miranda (1970); Wilson & Meyer (1982, 1985); Smith & Smith (1993); Köhler (2001)	C and se Guatemala eastward and southward into Panama and the Archipiélago de las Perlas in Panama in subhumid habitats; reported from sea level to about 1000 m.
L. bakeri Ruthven, 1936 Aruban or Baker's Cat-eyed Snake	Ruthuen (1936); Duellman (1958); Peters & Orejas-Miranda (1970); Mijares-Urrutia & Arends R (1995, 2000); Kornacker (1999)	Aruba Island in the Dutch West Indies, and n Venezuela in Peninsular de Paraguana in the state of Falcon.
L. frenata (Cope *in* Ferrari-Pérez, 1886) Rainforest Cat-eyed Snake	Duellman & Werler (1955); Duellman (1958); Peters & Orejas-Miranda (1970); Henderson & Hoevers (1975); Villa et al. (1988); Smith & Smith (1993); Lee (1996, 2000); Campbell (1998a); Stafford & Meyer (2000); Köhler (2001)	SE Mexico in Veracruz, Tabasco, Chiapas, and Yucatán southward into Belize and into Petén, Guatemala.
L. f. frenata (Cope *in* Ferrari-Pérez, 1886) Cope's Rainforest Cat-eyed Snake	Duellman (1958); Smith & Taylor (1966); Peters & Orejas-Miranda (1970); Pérez-Higareda & Smith (1991); Smith & Smith (1993); Campbell (1998a)	SE Mexico in c Veracruz southward and eastward into the Isthmus of Tehuantepec in scattered locations of tropical evergreen forests.
L. f. malleisi Dunn & Stuart, 1935 Malleis's Cat-eyed Snake	Duellman (1958); Smith & Taylor (1966); Peters & Orejas-Miranda (1970); Henderson & Hoevers (1975); Smith & Smith (1993); Campbell (1998a); Lee (2000); Köhler (2001)	SE Mexico in Campeche, and w Yucatán through n Chiapas into n Belize and Petén in Guatemala in rainforests and tropical savannas.

Species-level Taxa	Citations and Remarks	Abbreviated Range
L. f. yucatanensis (Cope, 1887) Peninsular Cat-eyed Snake	Duellman (1958); Smith & Taylor (1966); Peters & Orejas-Miranda (1970); Alvarez del Toro (1982); Smith & Smith (1993); Campbell (1998a); Lee (2000); Köhler (2001)	S Mexico in n Yucatán and Isla Cozumel into Central America in arid scrub forests and parts of tropical desert forests.
L. maculata* (Hallowell, 1860 [1861]) Southwestern Cat-eyed Snake	P. W. Smith & Darling (1952); Duellman (1958); Smith & Taylor (1966); Casas-Andreu (1982); Flores-Villela (1993); Smith & Smith (1993); Ramírez-Bautista (1994); García & Caballos (1994) *Remark: *L. smithi* Taylor, 1938, from the Río Balsin Basin in Mexico, is a synonym (see Duellman, 1958 and Smith & Smith, 1993).	Mexico in the Pacific coastal lowlands and slopes of the Sierra Madre Occidental from s Sinaloa southwest to the Río Balsas and inland into the Balsas Basin in Guerrero and Michoacán, and the n Yucatán peninsula; it is associated with intermediate and low jungle to about 2000 m.
L. nigrofasciata* Günther, 1868 Black-banded Cat-eyed Snake	Peters & Orejas-Miranda (1970); Savage (1980); Alvarez del Toro (1982); Wilson & Meyer (1985); Savage & Villa (1986); Villa et al. (1988); Smith & Smith (1993); Vences et al. (1998); Köhler (2001) *Remark: The subspecies of *L. n. nigrofasciata* Günther, 1868, and *L. n. mystacina* Cope, 1869, are sometimes listed.	Pacific lowlands of Guerrero in Mexico through Central America into Costa Rica and Panama; from sea level to 1300 m.
L. punctata (W. Peters, 1867) Western Cat-eyed Snake	Duellman (1958); Smith & Taylor (1966); Dowling & Jenner (1987); Smith & Smith (1993)	Mexico in c Sinaloa south into Lago de Chapala in Jalisco; this is a lowland form.
L. rubricata* (Cope, 1893) Costa Rican Cat-eyed Snake	Duellman (1958); Peters & Orejas-Miranda (1970) [as *L. a. rhombifera*]; Savage & Villa (1986); Villa et al. (1988); Savage & Slowinski (1992); Köhler (2001) *Remark: Duellman (1958) listed this snake as a synonym of *L. annulata rhombifera*. However, Savage & Villa (1986) stated this is a valid species. They showed that ventrals overlap, but the body bands do not overlap, and separated it based on this characteristic comparison. This is the only evidence that I can find where anyone has taken it out of synonymy.	Costa Rica.
L. septentrionalis (Kennicott *in* Baird, 1859) Northern Cat-eyed Snake	Duellman (1958); Peters & Orejas-Miranda (1970); Savage (1980); Wilson & Meyer (1985); Savage & Villa (1986); Villa et al. (1988); Smith & Smith (1993); Kornacker (1999); Stafford & Meyer (2000); Lee (2000); Dixon (2000a); Köhler (2001)	Extreme s Texas southward into w South America; Degenhardt et al. (1996) noted that this snake has been reported in New Mexico but question its presence there; generally found from sea level to 2000 m.
L. s. septentrionalis (Kennicott *in* Baird, 1859) Northern Cat-eyed Snake	Duellman (1958); Conant (1975); Smith & Smith (1993); Lee (1996, 2000); Tennant (1998); Conant & Collins (1998); Dixon (2000a); Werler & Dixon (2000); Tennant & Bartlett (2000)	Extreme s Texas lowlands of the Rio Grande Embayment southward through the lowlands of ne Mexico from sea level to about 2000 m.
L. s. larcorum Schmidt & Walker, 1943 Río Marañón Valley Cat-eyed Snake	Peters (1960b); Duellman (1958); Peters & Orejas-Miranda (1970); Pérez-Santos & Moreno (1991)	Coastal and upper Río Marañón Valley in coastal n Peru intergrading in Ecuador.

Species-level Taxa	Citations and Remarks	Abbreviated Range
L. s. ornata (Bocourt, 1884) Ornate Cat-eyed Snake	Peters (1960b); Duellman (1958); Lancini & Kornacker (1989); Kornacker (1999); Köhler (2001)	S Costa Rica southward into n South America into w Venezuela and southward through Colombia into nw Ecuador from the lowlands into about 1000 m.
L. s. polysticta* Günther, 1895 Central American Cat-eyed Snake	Davis & Smith (1953); Duellman (1958); Hardy & McDiarmid (1969); Henderson & Hoevers (1975); Alvarez del Toro (1982); Wilson & Meyer (1982, 1985); Pérez-Higareda & Smith (1991); Campbell (1998); Wallach (personal observation as per pers. comm. 1999); Lee (2000); Köhler (2001) *Remark: Campbell (1998a) lists this taxon as a full species, *L. polysticta*, but most literature maintains it as a subspecies, so it is listed as such here.	S Mexico into Central America from below 2000 m from Nayarit and s Veracruz, southward into El Salvador and c Costa Rica.
L. splendida Günther, 1895 Splendid Cat-eyed Snake	Davis & Smith (1953); Smith & Taylor (1966); Tanner (1985); Smith & Smith (1993)	N and c Mexico.
L. s. splendida Günther, 1895 Guenther's Splendid Cat-eyed Snake	Duellman (1958); Smith & Taylor (1966); Tanner (1985); Smith & Smith (1993)	Morelos, n Guerrero, and sw Puebla in Mexico; associated with the upper Balsas Basin and adjacent slopes of the Mexican Plateau in semiarid regions from 900 to 1700 m.
L. s. bressoni Taylor, 1938 Bresson's Splendid Cat-eyed Snake	Taylor (1938); Peters (1954); Duellman (1958); Smith & Taylor (1966); Smith & Smith (1993)	C and s Mexico from s Sinaloa southward into Colima and s Michoacán; scattered from 500 to 1500 m.
L. s. ephippiata Smith & Tanner, 1944 Saddled Cat-eyed Snake	Duellman (1958); Smith & Taylor (1966); Hardy & McDiarmid (1969); Tanner (1985); Smith & Smith (1993)	Pacific n Mexico from Chihuahua at the Texas frontier southward into Sonora and Sinaloa.

GENUS: *Leptodrymus* Amaral, 1927, is a Central American snake. See Peters & Orejas-Miranda (1970); Wilson & Meyer (1985); Savage & Villa (1986); Smith & Smith (1993); Auth (1994); and Köhler (2001).

CONTENT: Monotypic species. ★★★★

Leptodrymus: Green-headed Racer. Frank & Ramus (1995) refer to this as the Striped Lowland Snake.

Species-level Taxa	Citations and Remarks	Abbreviated Range
L. pulcherrimus (Cope, 1874) Green-headed Racer or Striped Lowland Snake	Bogert (1938); Peters & Orejas-Miranda (1970); Wilson & Meyer (1985); Savage & Villa (1986); Villa et al. (1988); Smith & Smith (1993); Auth (1994); Köhler (2001)	Central America in the Pacific versant from Guatemala into Nicaragua and Atlantic versant from Honduras into Costa Rica; it is associated with coastal and low elevations; found from sea level to 1300 m.

GENUS: *Leptophis* Bell, 1825, is widely distributed throughout Latin America from Mexico southward into South America west of the Andes into Ecuador, and east of the Andes into central Argentina. See Oliver (1942a, 1942b,

1948); Smith & Taylor (1966); Peters & Orejas-Miranda (1970); Mertens (1973); Wilson & Meyer (1982, 1985); Smith & Smith (1993); Starace (1998); Kornacker (1999); and Köhler (2001).

This group was reviewed by Oliver (1948) as the genus *Thalerophis*, Oliver then transferred the classification to *Leptophis* based on the International Commission of Zoological Nomenclature (1958). The list was updated by Peters & Orejas-Miranda (1970). Vanzolini (1986) expanded the Peters & Orejas-Miranda (1970) work with the addition of *L. ahaetulla chacoensis* Oliver = *L. a. urostictus*; and, *L. modestus* (Günther).

CONTENT: Ten species. ★★★

Leptophis: Parrot Snake

Species-level Taxa	Citations and Remarks	Abbreviated Range
L. ahaetulla (Linnaeus, 1758) Common or Green Parrot Snake	Oliver (1942a, 1948); Mertens (1973); Vanzolini (1986); Savage (1980); Wilson & Meyer (1982, 1985); Savage & Villa (1986); Villa et al. (1988); Smith & Smith (1993); Lee (1996, 2000); Campbell (1998a); Starace (1998); Kornacker(1999); Köhler (2001)	Throughout Latin America from Mexico southward through Central America into South America and the Guianas into as far south as Brazil.
L. a. ahaetulla (Linnaeus, 1758) Green Parrot Snake	Oliver (1942a, 1948); Peters & Orejas-Miranda (1970); Wilson & Myers (1982, 1985); Smith & Smith (1993); Starace (1998)	The Guianas southward along the Atlantic coastal regions into about the southern boundary of the state of Bahia, Brazil.
L. a. bocourti Boulenger, 1898 Bocourt's Parrot Snake	Oliver (1942a, 1948); Peters & Orejas-Miranda (1970); Pérez-Santos & Moreno (1988, 1991)	Gorgona Island, Colombia and nw Ecuador.
L. a. bolivianus Oliver, 1942 Bolivian Parrot Snake	Oliver (1942a, 1948); Peters & Orejas-Miranda (1970)	Bolivia in Departamentos Beni and Santa Cruz.
L. a. coeruleodorsus Oliver, 1942 Green Horse-whip Snake	Oliver (1942a, 1948); Roze (1966); Peters & Orejas-Miranda (1970); Lancini & Kornacker (1989); Kornacker (1999); Boos (2001)	Coast of ne central Venezuela, and into Trinidad and Tobago islands.
L. a. copei Oliver, 1942 Cope's Parrot Snake	Oliver (1942a, 1948); Roze (1966); Peters & Orejas-Miranda (1970); Lancini & Kornacker (1989); Kornacker (1999)	Colombia, Brazil and the Rio Orinoco-Negro region of Venezuela. The type specimen is reported from Salto do Huá, Brazil-Venezuela boundary.
L. a. liocercus (Wied, 1824) Atlantic Parrot Snake	Oliver (1942a, 1948); Peters & Orejas-Miranda (1970); Amaral (1976); Hoge et al. ([1978/79] 1981); Freitas (1999)	Atlantic drainage of se Brazil, closely associated with woodlands.
L. a. marginatus (Cope, 1862) Blue Parrot Snake	Oliver (1942a, 1948); Freiberg (1954); Peters & Orejas-Miranda (1970); Duellman (1978); Dixon & Soini (1986); Cei (1993); Meneghel & Achaval (1997); Leynaud & Bucher (1999)	N Argentina into se Bolivia and east into s and w São Paulo, Brazil southward into s Paraguay and n Argentina in Corrientes, Entre Ríos, Chaco, Misiones, and probably Santa Fe and Salta; also found in Uruguay.
L. a. nigromarginatus (Günther, 1866) Black-skinned Parrot Snake	Oliver (1942a, 1947, 1948); Peters & Orejas-Miranda (1970); Amaral (1976); Duellman (1978); Dixon & Soini (1986)	Amazon lowland region of Colombia into e Ecuador, Peru, n Bolivia and possibly to w Brazil; closely associated with woodlands.

Species-level Taxa	Citations and Remarks	Abbreviated Range
L. a. occidentalis (Günther, 1859) Green Tree Snake or Western Parrot Snake	Oliver (1942a, 1948); Dunn (1944f); Roze (1966); Peters & Orejas-Miranda (1970); Wilson & Meyer (1985); Lancini & Kornacker (1989); Bernal-Carlo & Roze (1994); Renjifo & Lundberg (1999); Kornacker (1999); Mijares-Urrutia & Arends R. (2000); Köhler (2001)	SE Mexico into n South America; distribution is usually considered to be from Nicaragua to w Venezuela and to Ecuador.
L. a. ortoni Cope, 1875 Orton's Parrot Snake	Oliver (1942a, 1948); Nicéforo María (1942); Peters & Orejas-Miranda (1970); Pérez-Santos & Moreno (1988); Köhler (2001)	Amazon Basin from se Colombia southward into extreme n Bolivia and eastward into c Brazil.
L. a. praestans (Cope, 1868 [1869]) Northern Green Frogger	Oliver (1942a, 1948); Smith & Taylor (1966); Peters & Orejas-Miranda (1970); Alvarez del Toro (1982); Wilson & Meyer (1985); Pérez-Higareda & Smith (1991); Lee (1996); Campbell (1998a)	S Veracruz in Mexico southward along the Caribbean versant into n Honduras; in low and moderate elevations.
L. a. urostictus* (W. Peters, 1873) Chocó Parrot Snake	W. Peters (1873); Oliver (1948); Mertens (1973); Vanzolini (1986) *Remark: Mertens (1973) listed *L. a. chocoensis* Oliver, 1942, as a synonym.	In and around Bogotá, Colombia (as shown in Oliver 1948 under *Thalerophis richardi occidentalis*.)
L. cupreus (Cope, 1868) Brown Parrot Snake	Andersson (1901); Peters & Orcés-V. (1960); Peters & Orejas-Miranda (1970); Dixon & Soini (1986); Pérez-Santos & Moreno (1991)	Amazonian lowlands in Ecuador and Peru.
L. depressirostris (Cope, 1861) Cope's Central American Parrot Snake	Andersson (1901); Oliver (1942a, 1948); Brattstrom & Howell (1954); Savage (1980); Savage & Villa (1986); Villa et al. (1988); Auth (1994); Köhler (2001)	Atlantic versant of Central America from Nicaragua into n South American in the Pacific versant of Colombia, nw Ecuador, and possibly Peru.
L. diplotrophis (Günther, 1872) Pacific Coast Parrot Snake	Oliver (1942a); Peters (1954); Smith & Taylor (1966); Wilson & McCranie (1979); Alvarez del Torro (1982); Villa et al. (1988); Pérez-Higareda & Smith (1989b); Smith & Smith (1993); García & Ceballos (1994); Köhler (2001)	Mexico from se Chihuahua and s Sonora to Oaxaca along the Pacific plain; it is also found in c Guerrero n Michoacán and se Oaxaca; Pérez-Higareda & Smith (1989b) suggest this is a lowland species whose distribution coincides with the most part in areas of tropical scrub and is a member of the Pacific lowland subhumid assemblage. They say to expect to find it in Durango, Zacatecas, Puebla, and possibly Aguascalientes. It is also found on the Tres Marias Islands.
L. d. diplotrophis (Günther, 1872) Pacific Coast Parrot Snake	Davis & Smith (1953); Smith & Taylor (1966); Alvarez del Toro (1982); Tanner (1985); Smith & Smith (1993); Ramírez-Bautista (1994)	C and s Mexico from Nayarit into the Isthmus of Tehuantepec. It has been reported from Guerrero, Jalisco, Michoacán, Nayarit, Oaxaca, Sinaloa, and Sonora.
L. d. forreri Smith, 1943 Forrer's Parrot Snake	Smith (1943); Smith & Taylor (1966); Smith & Smith (1993)	Tres Marias Islands in Mexico.
L. mexicanus Dumeril, Bibron & Dumeril, 1854 Mexican Parrot Snake	Smith & Taylor (1966); Henderson (1976); Savage (1980); Wilson & Meyer (1982, 1985); Savage & Villa (1986); Villa et al. (1988); Lee (1996, 2000); Campbell (1998a); Köhler (2001); McCoy & Censky (1992)	Tamaulipas in Mexico, excluding the n Yucatán peninsula, into Costa Rica.

Species-level Taxa	Citations and Remarks	Abbreviated Range
L. m. mexicanus Dumeril, Bibron & Dumeril, 1854 Green-headed Parrot Snake	Oliver (1942a, 1948); Smith & Taylor (1966); Henderson & Hoevers (1975); Henderson (1976); Alvarez del Toro (1982); Wilson & Meyer (1985); Pérez-Higareda & Smith (1991); Campbell (1998a); Vences et al. (1998); Stafford & Meyer (2000); Lee (2000)	S Tamaulipas in Mexico southward into Central America in Guatemala in the Pacific versant, and to Costa Rica in the Atlantic versant; low and moderate elevations.
L. m. hoeversi Henderson, 1976 Turneffe Islands Parrot Snake	Henderson (1976, 1976); Wilson & Meyer (1985); Smith & Smith (1993); Stafford & Meyer (2000); Lee (2000); Köhler (2001)	Big Cay Borkel, Belize and is expected to be found in other islands in the Turneffe Islands group.
L. m. septentrionalis Mertens, 1972 Tamaulipan Parrot Snake	Mertens (1972); Wilson & Meyer (1982; 1985); Padilla G. & Mendoza Q. (1996)	Tamaulipas and San Luis Potosí into Queretaro in Mexico.
L. m. yucatanensis Oliver, 1942 Yucatán Parrot Snake	Oliver (1942a); Smith & Taylor (1966); Henderson (1976); Wilson & Meyer (1985); Campbell (1998a); Lee (2000)	N Yucatán and Quintana Roo in s Mexico.
L. modestus (Günther, 1872) Cloud Forest Parrot Snake	Oliver (1942a) [as incertae sedis *Ahaetulla modesta*]; Hoyt (1964); Mertens (1973); Smith & Smith (1973, 1993); Alvarez del Toro (1982); Wilson & Meyer (1985); Wilson et al. (1986); Vanzolini (1986); Villa et al. (1988); McCranie & Wilson (1993); Köhler (2001)	Atlantic versant of sw Mexico in ne Chiapas southward into nw El Salvador and w Honduras; it is found in at disjunct range between 1335–2590 m in lower montane moist forest formation.
L. nebulosus Oliver, 1942 Striped Parrot Snake or Bronze-striped Parrot Snake	Oliver (1942a, 1942b, 1948); Savage (1980); Wilson & Meyer (1985); Savage & Villa (1986); Villa et al. (1988); Köhler (2001)	Central America from Patuca, Honduras southward through Nicaragua to Cariblanco, Costa Rica; in low and moderate elevations.
L. riveti Despax, 1910 Turquoise Parrot Snake	Oliver (1942a, 1948); Nicéforo María (1942); Emsley (1977); Savage (1980); Savage & Villa (1986); Villa et al. (1988); Auth (1994); Murphy (1997); Köhler (2001)	Costa Rica, Panama, n and c Colombia, Peru, and Ecuador; there is supposed to be a single record from Trinidad but that seems to be far out of its normal range. It is usually found at high altitudes to 5000 ft on both sides of the Andes.
L. santamartensis Bernal-Carlo & Roze, 1994 Santa Marta Parrot Snake	Bernal-Carlo & Roze (1994)	Sierra Nevada de Santa Marta in Colombia; it is associated with cloud forest on the n slopes of the Santa Marta Mountains.
*L. stimsoni** Harding, 1995 Mt. Aripo Parrot Snake	Harding (1995); Boos (2001) *Remark: Murphy (1997) did not mention this snake.	Type locality is Mt. Aripo, Trinidad, Boos (2001) lists its range as Trinidad and notes a few more specimens other than the type have now been discovered.

GENUS: *Liochlorophis* Smith & Oldham, 1991, is the North American Smooth Green or Grass Snake that used to be in *Opheodrys* and is still considered to be a synonym by some authorities. See Conant (1975) [as *Opheodrys*]; Stebbins (1985) [as *Opheodrys*]; Oldham & Smith (1991); Grobman (1992); Degenhardt et al. (1996); Conant &

Collins (1998); Dixon (2000a); Bartlett & Tennant (2000). Grobman (1992) is still the most recent formal reviewer of this genus.

Important contributions for this section were provided by James R. Dixon.

Note: This generic designation is still not fully accepted by a some authorities, who continue use *Opheodrys*. Smith & Smith (1993) stated that *Liochlorophis* may be a *nomen nudum*. Mitchell (1994), did not use *Liochlorophis* because he felt that Oldham & Smith (1991) was not peer reviewed. Crother et al. (2000) maintained this in *Opheodrys*.

CONTENT: One species with possibly up to three subspecies. Bartlett & Tennant (2000) noted that difficult-to-differentiate subspecies are occasionally assigned to this snake, but, the current trend seems to consider the smooth green snake a somewhat variable form. Although I agree with Bartlett & Tennant (2000), subspecies are listed here based on Grobman (1992). ★★

Liochlorophis: North American Smooth Green Snake or Grass Snake.

Species-level Taxa	Citations and Remarks	Abbreviated Range
L. vernalis (Harlan, 1827) North American Smooth Green Snake	Conant (1975); Stebbins (1985) [as *Opheodrys*]; Oldham & Smith (1991); Grobman (1992) [as *Opheodrys*]; Degenhardt et al. (1996); Tennant (1998); Conant & Collins (1998) [as *Opheodrys* with no subspecies]; Bartlett & Tennant (2000); Tennant & Bartlett (2000); Dixon (2000a)	Eastern part of the United States and southern e Canada.
L. v. vernalis (Harlan, 1828) Eastern Smooth Green Snake	Conant (1975) [as *Opheodrys*]; Stebbins (1985) [as *Opheodrys*]; Oldham & Smith (1991); Grobman (1992) [as *Opheodrys*]	NE North America.
L. v. blanchardi* (Grobman, 1941) Western Smooth Green Snake	Anderson (1965); Conant (1975) [as *Opheodrys*]; Stebbins (1985) [as *Opheodrys*]; Oldham & Smith (1991); Grobman (1992) [as *Opheodrys*]; Smith & Smith (1993); Degenhardt et al. (1996); Dixon (2000a); Werler & Dixon (2000) *Remark: Dixon (2000a) noted that a living specimen of this snake has not been observed in over 30 years in Texas.	Associated with the northern great plains and prairies west to the Black Hills of North America from s Manitoba in sc Canada, and southward ith a disjunct population reported from Texas. Tennant reports these snakes from counties around the Galveston Bay region of Texas. This is some 500 miles from the main Midwest population. Dixon has been unable to confirm validity of this statement.
L. v. borealis (Grobman, 1992) Northern Smooth Green Snake	Grobman (1992) [as *Opheodrys*]	NE United States and southern e Canada.

GENUS: *Lioheterophis* Amaral, 1934 [1935], is restricted to Brazil. See Peters & Orejas-Miranda (1970); and Amaral (1976).

CONTENT: One species. ★ [The problem with ranking the validity for the information on this snake is the limited number of specimens available, only one.]

Lioheterophis: Ihring's or Jericoa Snake

Species-level Taxa	Citations and Remarks	Abbreviated Range
L. iheringi Amaral, 1934 (1935) Ihering's or Jericoa Snake	Amaral (1934, 1976); Peters & Orejas-Miranda (1970)	Brazil; known only from the type specimen originally found in Campina Grande, ne part of the Estado de Paraiba.

GENUS: *Liophis* Wagler, 1830, is sometimes referred to by a wide variety of names such as the South American Queen Snake or Water Snake, sabanera, tree snake, or swampsnake. There has been some confusion between several genera that are similar to *Liophis*. Some of the genera include *Dromicus, Leimadophis, Lygophis, Philodryas, Alsophis, Antillophis, Urotheca* and *Arrhyton*. The following comments give some history to reference how the group nomenclature has developed. Per Vanzolini (1986), *sensu* Peters & Orejas-Miranda (1970), *Leimadophis* Fitzinger, 1843 = *Liophis* Wagler, 1830, except for *Leimadophis atahuallpae* (Steindachner), *Leimadophis simonsii* (Boulenger) and *Leimadophis pygmaeus* (Cope). Fide Myers (1973), part of the *Lygophis* was *Saphenophis* and fide Dixon (1980), the rest were *sensu Lygophis* Fitzinger, 1843 = *Liophis* Wagler. See Dunn (1957); Dowling & Gibson (1970); Maglio (1970); Peters & Orejas-Miranda (1970), Cunha & Nascimento (1976); Dixon (1980, 1981b, 1983a, 1983b, 1989, 1993); Vanzolini (1986); Cei (1986, 1993); Lancini & Kornacker (1989); Schwartz & Henderson (1991); Lema (1994b); Starace (1998); Kornacker (1999); and Köhler (2001).

Review assistance and important contributions for this section were provided by James R. Dixon.

Dixon first organized the genus in Dixon (1980) and updated the information in Dixon (1989).

Note: *L. leucogaster* (Jan, 1863), as found in Peters & Orejas-Miranda (1970), and said to be known only from the type specimen and type locality as unknown, is not shown by Dixon (1989), in his review of *Liophis* therefore is not referenced here.

CONTENT: A complex group of snakes with forty-one species. ★★★

Liophis: Legion Snake. These ubiquitous snakes fill the niche in South American similar to what the Gartersnakes and Ribbonsnakes (*Thamnophis*) do in North America. Frank & Ramus (1995) called this group the Amazon Groundsnakes; however, this genus is not restricted to the Amazon and often take on the habits of water snakes. I have modified the common name and reflected one more suitable for its use. I call it the Legion Snake because it is very common and widespread geographically throughout South America.

Species-level Taxa	Citations and Remarks	Abbreviated Range
L. almadensis (Wagler *in* Spix, 1824) Almada Legion Snake	Peters & Orejas-Miranda (1970); Amaral (1976); Dixon & Thomas (1985); Vanzolini (1986); Achaval (1987); Dixon (1989, 1991); Cei (1993); Dixon et al. (1993); Lema (1994b); Yanosky (1998); Leynaud & Bucher (1999); Freitas (1999)	Restricted to the south side of the Amazon River from Ilha Marajó along the Atlantic coast to the Brazilian state of Rio Grande do Sul and Maranhão east into the chaqueña of se Bolivia and into the n Paraguyan chaco to the south in Argentina in Formosa, Chaco and Corrientes, and then, into Uruguay and Paraguay; Dixon (1989) noted that published reports of this taxon from the state of Misiones in Argentina have not been verified; according to Amaral (1976) it prefers open fields.

Species-level Taxa	Citations and Remarks	Abbreviated Range
L. andinus Dixon, 1983 Incachaca Legion Snake	Dixon (1983a, 1989); Vanzolini (1986)	Known only from the type locality in Incachaca, Cochabamba in Bolivia at 2500 m.
L. anomalus (Günther, 1858) Red-tailed Legion Snake	Freiberg (1954); Fabian (1971); Amaral (1976); Miranda et al. (1982); Dixon (1985b, 1986, 1989, 1993); Achaval (1987); Savage & Slowinski (1992); Lema (1994b); Yanosky (1998); Leynaud & Bucher (1999)	S South America from Paraguay into Uruguay and Rio Grande do Sul in s Brazil and Argentina in Buenos Aires, Córdoba, Santa Fe, Entre Ríos, Corrientes, Chaco, Formosa, and Salta. It occurs in Pampan and Chacoan plant formations.
L. atraventer Dixon & Thomas, 1985 Black-bellied Legion Snake	Dixon & Thomas (1985); Dixon (1987b, 1989); Marques et al. (2001)	Known only from the type locality of Estação Biologica da Boracéia in São Paulo, Brazil.
L. breviceps Cope, 1860 Red-bellied Legion Snake	Peters (1960b); Vanzolini (1986); Dixon (1983e, f, 1989); Dixon & Soini (1986); Starace (1998); Kornacker (1999)	Amazonia in Bolivia, Brazil, Colombia, Peru, Venezuela, and the Guianas; Dixon (1989) described it as being found on the eastern flanks of the Andes of Ecuador and Perú, eastward to Obidos, Brazil; on the north from c Colombia and n Guyana south to Río Mamoré (Trinidad) Bolivia, and Posto Diuarum, Brazil.
L. b. breviceps Cope, 1860 Amazon Legion Snake	Dixon (1983e, f, 1989); Vanzolini (1986); Dixon & Soini (1986); Lancini & Kornacker (1989); Kornacker (1999)	Amazon Basin in forested regions of Bolivia, Brazil, Peru, Colombia, and into the Guianas and Venezuela.
L. b. canaima Roze, 1956 Roze's Legion Snake	Roze (1956b, 1966); Vanzolini (1986); Dixon (1983f, 1989); Lancini & Kornacker (1989); Kornacker (1999)	S Venezuela, known only from the region of the Río Ugueto in Amazonas.
L. carajasensis Cunha, Nascimento & Avila-Pires, 1985 Serra Norte Legion Snake	Cunha et al. (1985); Dixon (1989, 1991)	Known only from the type locality in the Serra Norte of Pará in Brazil. The type was from Campo Rupestre de Norte 1.
L. ceii Dixon, 1991 Cei's Legion Snake	Dixon (1991); Cei (1993)	From the province of Santa Cruz and Tarija in e Bolivia south into the states of Catamarca, Tucumán, Salta, and Jujuy in Argentina; from 600 to 2000 m.
L. cobellus* (Linnaeus, 1758) Cobel Legion Snake	Peters (1960b); Emsley (1977); Duellman (1978); Amaral (1976); Hoge et al. (1978); Moonen et al. (1979); Cunha & Nascimento (1976, 1978); Dixon (1983e, f, 1989); Starace (1998); Kornacker (1999); Freitas (1999) *Remark: Dixon (1980) shows *L. purpurans* as valid but it is not shown in Dixon (1989). In a personal commuciation (12 September 2002) James R. Dixon told me that *L. purpurans* is simply a juvenile of *L. cobellus*.	N South America from Villavicencio in Colombia southward into Buenavista, Bolivia; and, from Carapito in Venezuela southeast into Bahia, Brazil; its range includes the Guianas.
L. c. cobellus* (Linnaeus, 1758) Common Cobel Legion Snake	Roze (1966); Dixon (1983e, f, 1989); Lancini & Kornacker (1989); Murphy (1997); Kornacker (1999); Boos (2001)	N South America east of the Andes in eastern area of the Guiana Shield exclusive of the Venezuelan tepuí region, it is also found in Trinidad.

The Caenophidia

Species-level Taxa	Citations and Remarks	Abbreviated Range
	*Remark: Dixon (1989), listed *L. serpentinus* (Daudin, 1803) and *L. cenchrus* (Daudin, 1803), as synonyms.	
L. c. dyticus Dixon, 1983 South American Legion Snake	Dixon (1983e, f, 1989); Dixon & Soini (1986); Hurtado & Blanco (1994 [1995])	W Amazon Basin from Loma Linda, Colombia southward into Buenavista, Bolivia then east into Pôrto Velho, Brazil, the type locality is listed as Monte Carmelo, Loreto, Peru; with a recent report that it is found in Peru in the Nublado Forest in the Q'osnipata Valley.
L. c. taeniogaster Jan, 1863 Eastern Legion Snake	Cunha & Nascimento (1976); Dixon (1983e, f, 1989)	NE Brazil from Isla Bananal east into central Bahia and north into Rio Amazonas (the south bank).
*L. c. trebbaui** Roze, 1958 Bolivar Legion Snake	Roze (1958b, 1966); Dixon (1983f, 1989); Vanzolini (1986); Lancini & Kornacker (1989); Kornacker (1999) *Remark: Dixon (1989), included *L. ingeri* Roze, 1958, as a synonym, although it is described in the key as *L. cobellus ingeri*. Kornacker (1999), listed *L. ingeri* as a valid species.	The Auyantepui region of the state of Bolivar in Venezuela, from the Chimantá Tepui, and km marker 144 of the El Dorado Santa Elena highway also in the state of Bolivar in Venezuela.
L. cursor (Lacépède, 1789) Martinique Legion Snake	Barbour (1930); Underwood (1962); Maglio (1970); Schwartz & Thomas (1975); Dixon (1981b, 1989); Schwartz & Henderson (1985, 1991); Censky & Kaiser *in* Crother (1999)	Known only from the island of Martinique in the Lesser Antilles, and its satellite Rocher de Dimant.
L. dilepis (Cope, 1862) Dry Forest Legion Snake	Hoge (1953b); Hoge et al. ([1978/79] 1981); Michaud & Dixon (1987); Dixon (1989); Yanosky et al. (1992); Cei (1993); Yanosky (1998); Leynaud & Bucher (1999)	NE into s Brazil extending into s Bolivia, Paraguay, and n Argentina into Chaco; there is said to be two population centers, one in the ne of Brazil (Caatinga), and the other in the south of Brazil (Cerrado), Bolivia, Paraguay, and ne Argentina; it is closely associated with the Caatinga and Cerrado.
L. elegantissimus (Koslowsky, 1895) Sierra de la Ventana Legion Snake	Miranda et al. (1982); Dixon (1985b, 1989); Cei (1993)	800 to 1000 m in Sierra de la Ventana in the province of Buenos Aires in Argentina, fide Dixon (1989) it is known only from five localities.
L. epinephelus Cope, 1862 Fire-bellied Legion Snake	Peters & Orejas-Miranda (1970); Greene (1976); Dixon (1983b, f, 1989, 1991); Savage & Villa (1986); Villa et al. (1988); Savage & Slowinski (1992); Kornacker (1999); Köhler (2001)	C Costa Rica and western Panama in Central America southward into northern South America into extreme n Ecuador and Peru along the Pacific coast at and above 2200 m; and, in Trans-Andean South America from Venezuela into Peru.
L. e. epinephelus Cope, 1862 Common Fire-bellied Legion Snake	Dunn (1957); Vanzolini (1986); Pérez-Santos & Moreno (1988, 1991); Dixon (1989); Bernal-Carlos (1994); Köhler (2001)	Costa Rica and Panama southward into Colombian lowlands and most inter-Andean valleys below 1500 m, and to the south along the Colombian coast into n Ecuador.
*L. e. albiventris** Jan, 1863 Ecuadorian Legion Snake	Prado (1942a); Dunn (1944c); Peters & Orejas-Miranda (1970); Dixon (1983b, 1989, 1991); Bernal-Carlos (1994)	W Ecuador from sea level to 2600 m.

Species-level Taxa	Citations and Remarks	Abbreviated Range
	*Remark: Dixon (1989) included *L. alticolus* Cope, 1868, *L. ater* Günther, 1872, and *L. quadrilineatus* Jan, 1863, as synonyms.	
*L. e. bimaculatus** Cope, 1899 Twin-spotted Legion Snake	Dunn (1944c, 1957); Roze (1066); Peters & Orejas-Miranda (1970); Dixon (1983b, 1989); Lancini & Kornacker (1989); Kornacker (1999)	Eastern high andean slopes from 2600 to 3300 m of w Venezuela, c Colombia, and southward into n Peru.
	*Remark: Dixon (1980, 1989), included *Leimadophis epinephelus ecuadorensis* (Laurent, 1949), as listed in Peters & Orejas-Miranda (1970), as a synonym. Dixon also included *L. bipraeocularis* Boulenger, 1903, as a synonym.	
L. e. fraseri Boulenger, 1894 Fraser's Legion Snake	Peters (1960b); Dixon (1983b, 1989), Pérez-Santos & Moreno (1991)	Intermediate elevations of the eastern and western slopes of s Ecuador, and southward into c Peru.
L. e. juvenalis Dunn, 1937 Dunn's Legion Snake	Dixon (1983b, 1989); Köhler (2001)	Intermediate slopes of mountains of c Costa Rica into w Panama.
L. e. kogiorum A. Bernal-Carlo, 1994 Kogi Cloud Forest Legion Snake	Bernal-Carlo (1994)	Santa Marta Mountains in Colombia; it is associated with cloud forest in the w and nw part of the Sierra Nevada de Santa Marta, from 1000 to about 1700 m, but might be found as high as above 2000 m.
L. e. lamonae Dunn, 1944 Lamon Legion Snake	Dunn (1957); Peters (1960b); Dixon (1983b, 1989); Pérez-Santos & Moreno (1988 1991)	EC Ecuador into Colombia along the eastern slopes of the Andes at 1500 to 2600 m.
L. e. opisthotaenius Boulenger, 1908 Renita Legion Snake	Roze (1966); Dixon (1983b, 1989); Pérez-Santos & Moreno (1988); Lancini & Kornacker (1989); Bernal-Carlos (1994); Kornacker (1999)	Andes of Venezuela and Colombia, found in Mérida in Venezuela and the Páramo de Tama region of Colombia and Venezuela.
*L. e. pseudocobellus** Peracca, 1914 Peracca's Legion Snake	Dunn (1957); Dixon (1983b, 1989) *Remark: Dixon (1989) listed *L. alticolus* (Amaral, 1931), as a synonym.	Intermediate elevations of the central and western Andes of Colombia southward to the Ecuadorian border.
L. festae (Peracca, 1897) Drab Legion Snake	Peters (1960b); Dixon & Markezich (1979); Dixon (1983f, 1989); Henle & Ehrl (1991b)	Intermediate elevations in s Colombia southward through Ecuador into the intermediate elevations of the Cis-Andean c Peru.
L. flavifrenatus (Cope, 1862) Jararaquinha Listada or Cope's Legion Snake	Peters & Orejas-Miranda (1970) [as *Lygophis*]; Amaral (1976); Cunha & Nascimento (1976); Dixon (1985b, 1989); Achaval (1987); Michaud & Dixon (1987); Cei (1993); Lema (1994b); Yanosky (1998)	S Brazil, c and s Paraguay, Uruguay, ne Argentina, and extreme se Brazil; closely associated with the deciduous mesophytic subtropcial forest that surrounds Araucarian forest, dry to slightly moist forest of the e Chaco, and grass and shrub cover of the hilly Pampas.
L. frenatus (Werner, 1909) Common Swamp Legion Snake	Dixon (1983e, f, 1989); Giraudo (1985, 1999); Savage & Slowinski (1992); Yanosky (1998)	Primavera, Paraguay southward into se Brazil to Guayuvira; recently reported from Misiones and Corrientes in Argentina.

The Caenophidia

Species-level Taxa	Citations and Remarks	Abbreviated Range
L. guentheri Peracca, 1897 Guenther's Legion Snake	Dixon & Thomas (1985); Dixon (1987b, 1989); Cei (1993); Yanosky (1998); Leynaud & Bucher (1999)	Restricted to the dry central Chaco region of Bolivia, Paraguay, and Argentina by Dixon (1987).
L. jaegeri (Günther, 1858) Jaeger's Swamp Legion Snake	Fabian (1971); Amaral (1976); Miranda et al. (1982); Dixon (1983e, 1985, 1987b, 1989); Dixon & Thomas (1985); Achaval (1987); Lema (1994b); Yanosky (1998); Freitas (1999)	C and s Brazil, coastal Uruguay, n Argentina, and Paraguay. It is found from about 19°S latitude in Brazil to about 35°S latitude in Uruguay and Argentina, westward to about 61°W longitude, along the Río Paraguay Basin in Argentina, Paraguay, and Brazil.
L. j. jaegeri* (Günther, 1858) Jaeger's Swamp Legion Snake	Dixon (1987b, 1989); Lema (1994b); Leynaud & Bucher (1999) *Remark: Dixon (1989), listed *L. dorsalis* Peters, 1863, and *L. lineata* (Jensen, 1900), as synonyms.	SE Brazil, coastal Uruguay, Argentina, and Paraguay. It is found east of the Rio Paraná, from São Paulo area of Brazil.
L. j. coralliventris (Boulenger, 1894) Boulenger's Red-bellied Swamp Legion Snake	Peters & Orejas-Miranda (1970 [as *Lygophis*]; Dixon (1987b, 1989); Cei (1993); Yanosky (1998); Leynaud & Bucher (1999)	Argentina in Misiones, Corrientes, Entre Ríos, Santa Fe, and Buenos Aires, and in Brazil into the eastern lowlands into Paraná to the Atlantic coast, in Uruguay, and in Paraguay in the Paraguay River Basin.
L. janaleeae Dixon, 2000 Janalee's Legion Snake	Dixon (2000b)	E slopes of the Peruvian Andes from 854 to 2700 m; the type specimen is from Moyobamba, Peru.
L. juliae (Cope, 1879) Leeward Legion Snake	Barbour (1930); Underwood (1962); Howes (1970); Schwartz & Thomas (1975); Dixon (1980, 1981b); Schwartz & Henderson (1985, 1991) *Remark: Fide Dixon (1980) no subspecies are recognized; Schwartz & Henderson (1985, 1991), listed the subspecies of *L. j. juliae* (Linnaeus, 1758) on Dominica, *L. juliae copeae* Parker, 1936 on Guadeloupe, and *L. juliae mariae* Barbour, 1914 on Marie-Galante. This work follows Dixon (1980).	Lesser Antilles in the West Indies on Dominica, Guadeloupe, and Maire-Galante.
L. lineatus (Linnaeus, 1758) Lined Legion Snake	Peters (1960b); Roze (1966); Cunha & Nascimento (1978); Hoge et al. (1978); Moonen et al. (1979); Dixon (1980); Michaud (1984); Michaud & Dixon (1987); Villa et al. (1988); Lancini & Kornacker (1989); Kornacker (1999); Dixon & Michaud (1992); Yanosky (1998); Starace (1998); Esqueda & La Marca (1999); Renjifo & Lundberg (1999); Mijares-Urrutia & Arends R. (2000); Köhler (2001)	C Panama e into South America Colombia, Venezuela, Guyana, Suriname and French Guiana to north of the Amazon River; Dixon (1989), noted that a few specimens are known from Esmeraldas and Guayaquil, Ecuador, and are probably accidental introductions via the shipping trade. This snake is found in principally three types of major vegetation, grass and bush steppes, dry forest, and coastal savanna. Michaud & Dixon (1987) suggested that it may also occur in the coastal grass steppe of nw Maranhão, Brazil.
L. longiventris Amaral, 1925 Rio Manjura Legion Snake	Amaral (1976); Peters & Orejas-Miranda (1970); Dixon (1983f, 1989); Vanzolini (1986); Cei (1993); Strussman & Carvalho (1998)	Known only from Rio Manjuro, Amazonas in Brazil, and 12°51′S-51°46′W in Mato Grosso, Brazil.

Species-level Taxa	Citations and Remarks	Abbreviated Range
L. maryellenae Dixon, 1985 Mary Ellen's Legion Snake	Dixon (1985, 1987b, 1989); Freitas (1999) [as *Liophis mariahellanae*]	Tablelands of se Brazil in c-se Brazil from Annapolis on the west, to Grão Mogol on the east and from near Barreiras on the north to Itambe do Dentro on the south; its range includes Bahia.
L. melanotus (Shaw, 1802) Yellow Queen Legion Snake	Underwood (1962); Roze (1966); Peters & Orejas-Miranda (1970); Schwartz & Thomas (1975); Emsley (1977); Lancini (1979); Dixon (1980, 1981b); Lancini & Kornacker (1989); Schwartz & Henderson (1985, 1991); Dixon & Michaud (1992); Censky & Kaiser in Crother (1999); Renjifo & Lundberg (1999); Kornacker (1999); Mijares-Urrutia & Arends R. (2000)	N South America in n Colombia and Venezuela including Trinidad and Tobago, and also reported from Grenada, it is found on both sides of the Andes.
L. m. melanotus (Shaw, 1802) Yellow Queen Legion Snake	Dixon (1989); Dixon & Michaud (1992); Kornacker (1999)	Llanos of Venezuela east and south into the llanos of Colombia.
L. m. lamari Dixon & Michaud, 1992 Lamar's Swamp Legion Snake	Dixon & Michaud (1992); Kornacker (1999)	Foothills within inter-Andean Valleys in Colombia and w Venezuela.
L. m. nesos Dixon & Michaud, 1992 Beh Belle Chemin	Dixon & Michaud (1992); Murphy (1997); Boos (2001)	Trinidad & Tobago in the southern Caribbean.
L. meridonalis (Schenkel, 1901) Southern Legion Snake	Hoge (1953b); Dixon (1980, 1989); Michaud & Dixon (1987); Cei (1993); Lema (1994b); Leynaud & Bucher (1999)	C Brazil and n Bolivia south into s Paraguay including all of the Bolivian and Paraguayan Chaco, and the chaqueña in the ne tip of Argentina and se Brazil. It is found in a wide variety of habitats.
*L. miliaris** (Linnaeus, 1758) Common South American Watersnake or Swamp Legion Snake	Gans (1964); Cunha & Nascimento (1970); Fabian (1971); Miranda et al. (1982); Dixon (1983e, 1983f, 1989); Achaval (1987); Lema (1994b); Starace (1998); Marques et al. (2001) *Remark: Henle & Ehrl (1991b), designated a new subspecies of *L. miliaris* as *L. m. intermedius* from Departamento Madre de Dios in Peru. There seem to be two problems with their assessment. 1) The *intermedius* subspecies already exists as *L. poecilogyrus intermedius*, which appears to be a completely different snake; and, 2) the snake shown in the photograph is not a *Liophis miliaris*. James R. Dixon (pers. comm. 23 May 2000), agrees that the statistics do not seem to indicate a *L. miliaris*. According to Dixon, the color and scale counts, along with other information, seem to fit *L. reginae*. Dixon & Tipton (in preparation) will detail the determination of the snake's classification.	Wide ranging in e South America including Brazil, Peru, Colombia, Ecuador, and the Guianas, Uruguay, Paraguay, and Argentina; records are scattered in the Amazon Basin and the Cerrado of Brazil.

Species-level Taxa	Citations and Remarks	Abbreviated Range
*L. m. miliaris** (Linnaeus, 1758) Common South American Swamp Legion Snake	Gans (1964); Amaral (1976); Dixon (1983e, 1983f, 1989) *Remark: Dixon (1989) listed *L. purpurans* (Dumeril, Bibron and Dumeril, 1854), as can be seen in Peters (1960b), *L. orientalis* (Günther, 1864), and *L. collaris* Jan, 1863, as synonyms.	The Guianas in ne South America.
L. m. amazonicus (Dunn, 1922) Southern Amazon Legion Snake	Dixon (1983e, 1983f, 1989); Cunha & Nascimento (1983); Nascimento et al. (1987)	Santarem, Brazil southwest into Río Itenez in Bolivia, and into Mato Grosso in Brazil.
L. m. chrysostomus (Cope, 1868) White-lipped Legion Snake	Dixon (1983e, 1983e, 1989); Dixon & Soini (1986); Pérez-Santos & Moreno (1988, 1991)	Rainforests of Brazil, Colombia, Ecuador, and Peru.
*L. m. merremi** (Wied, 1821) Merrem's Legion Snake	Gans (1964); Dixon (1983e, 1983f, 1989); Marques et al. (2001:171) *Remark: Dixon (1989) considers *L. bicolor* (Reuss, 1834), *L. australis* (Günther, 1858), and *L. dictyodes* (Wied, 1824), as synonyms.	Mainly in Atlantic forests of Brazil from Pernambuco into Rio de Janeiro.
L. m. mossorensis Hoge & Lima-Verge, 1972 Cerrados Legion Snake	Hoge & Lima-Verde (1972); Vanzolini, Ramos-Costa & Vitt (1980); Hoge et al. ([1978/79] 1981); Dixon (1983e, 1983f, 1989)	NE Brazil, primarily in caatinga and dry cerrado.
*L. m. orinus** (Griffin, 1915) Serrana Smooth Legion Snake	Gans (1964); Dixon (1983, 1989); Lema (1994b); Marques et al. (2001:171) *Remark: Dixon (1989), listed *L. natricoides* (Werner, 1926), as a synonym.	SE Brazil from s Minas Geraís southward through São Paulo, Parana, Santa Catarina, and the northern third of Río Grande do Sul.
*L. m. semiaureus** (Cope, 1862) Pampa Smooth Legion Snake	Gans (1964); Dixon (1983e, 1983f, 1989); Cei (1993); Lema (1994b) *Remark: Dixon (1989) listed *L. fuscus* (Cope, 1885) and *L. ornata* Jan, 1863, as synonyms. *Rhadinaea fusca* (Cope, 1865) is a name preoccupied by the var. *semiaureus* Cope, 1862 (Gans, 1964).	Paraguay, s and e Uruguay, s Rio Grande do Sul in Brazil, and in n Argentina in Buenos Aires, Córdoba, Santa Fe, Entre Ríos, Formosa, and Misiones.
L. oligolepis (Boulenger, 1905) False Regal Legion Snake	Amaral (1976); Cunha & Nascimento (1978); Cunha et al. (1985); Dixon & Tipton (in preparation)	This species occurs in low densities from the east of Pará, between Rio Guamá, the Atlantic and the Rio Gurupi. It has been captured in Paruá (west of Maranhão), to the south of the Rio Guamá (Pará) and Serra Norte. Vanzolini (pers. comm. in Cunha et al., 1985) has examples from Santarém, Canindé and Oriximiná (Pará), Porto Velho (Rondonia), Postoí Walter (Acre) and Estirón (Peru).

Species-level Taxa	Citations and Remarks	Abbreviated Range
L. ornatus (Garman, 1887) St. Lucia Legion Snake	Underwood (1962); Schwartz & Thomas (1975); Dixon (1981b, 1989); Swartz & Henderson (1985); Smith et al. (1993); Censky & Kaiser *in* Crother (1999) *Remark: Reported to be last collected in 1973.	St. Lucia and Maria Islands in the Lesser Antilles.
L. paucidens (Hoge, 1953) Goiás Swamp Legion Snake	Hoge (1952a, 1953b); Peters & Orejas-Miranda (1970); Dixon (1980, 1981b, 1989); Michaud & Dixon (1987); Campos & Nogueira (2001)	Known only from the type locality in ec Brazil around Mato Verde in Goiás state. It inhabits the Campo Cerrados and groundwater cover of the Babacu region. Recently reported from Distrito Federal.
L. perfuscus Cope, 1862 Barbados Legion Snake	Barbour (1930); Underwood (1962); Emsley (1963b); Schwartz & Thomas (1975); Dixon (1981b, 1989); Schwartz & Henderson (1985, 1991); Censky & Kaiser *in* Crother (1999)	Barbados in the Lesser Antilles.
L. poecilogyrus* (Wied, 1824) Copeia Legion Snake	Peters (1960b); Hoge et al. (1978); Dixon & Markezich (1979, 1992); Dixon (1980, 1983f, 1989); Vanzolini et al. (1980); Campbell & Murphy (1984); Dixon & Thomas (1985); Savage & Slowinski (1992); Lema (1994b); Fuentes & Barrio (1999); Leynaud & Bucher (1999); Kornacker (1999); Freitas (1999); Marques et al. (2001) *Remark: Because of the color variation and ontogenetic shift in pattern in this snake, Dixon (1989) recommended that subspecies not be used and did not recognize them at that time. However, Dixon & Markezich (1992) recognized four subspecies as valid [*L. p. poecilogyrus* (of the Atlantic rainforests), *L. p. caesius* (of the Chaco), *L. p. sublineatus* (of the Pampas), and *L. p. schotti* (of the Amazon-Cerrado-Caatinga)]. They noted that *L. p. schotti* may represent more than one taxon, but there was not enough material available to resolve the questions. They also noted that *L. p. xerophilus* (Amaral, 1944), from semi-arid regions of Pernambuco into Piauí in ne Brazil, may be resurrected in the future based on several traits. *Remark 2: Dixon & Markezich (1992) found that the names *m-nigrum*, *alternans*, and *forsteri* are senior to the name *poecilogryus*. However, in the spirit of the International Code of Zoological Nomenclature, the name *poecilogyrus* should be preserved because of its usage in 52 primary articles (at that time) as the valid name of the taxon. *Remark 3: *L. p. intermedius* (Amaral, 1944); found in Peters & Orejas-Miranda (1970) [as *Leimadophis*] with distribution in Goiás and Mato Grosso in Brazil is different than *L. miliaris intermedius* fide Henle & Ehrl (1991b).	Associated with the Atlantic coastal rain forests, chaco, caatinga, pampa and campo cerrado region s of South America from Guyana and se Venezuela south-southeast into Bolivia, and southward through the Pantanal and chaco into pampas of Argentina and from Ilhe Marajó at mouth of the Río Amazona southward into Uruguay and Paraguay.

Species-level Taxa	Citations and Remarks	Abbreviated Range
	*Remark 4: *L. p. pictostriatus* (Amaral, 1944), from c and s Rio Grande do Sul to the coastal region of Santa Catarina in Brazil, and into Uruguay was listed by Lema (1994b). It was not considered valid by Dixon & Markezich (1992) so is not included here.	
	*Remark 5: *L. p. platensis* (Amaral, 1944), Paraguay, Uruguay, and much of Brazil from was listed by Cei (1993). It was not considered valid by Dixon & Markezich (1992) so is not included here.	
	*Remark 6: *L. p. subfasciatus* (Cope, 1862), from Paraguay and Entre Ríos, Argentina was listed in Lema (1994b). It is not listed by Dixon anywhere and may not be valid. Since it was not considered valid by Dixon, it is not included here.	
L. p. poecilogyrus (Wied, 1824) Angry Legion Snake or Common Copeia Legion Snake	Amaral (1976); Dixon & Markezich (1992)	Atlantic rainforests of Brazil, w Guyana, and e Venezuela.
L. p. caesius (Cope, 1862) Chaco Legion Snake	Dixon (1980); Dixon & Markezich (1992); Leynaud & Bucher (1999)	Chaco of Bolivia, Paraguay, and n Argentina.
L. p. shottii (Schlegel, 1837) Schott's Legion Snake	Peters & Orejas-Miranda (1970); Beçak et al. (1971); Dixon & Markezich (1992); Cei (1993); Leynaud & Bucher (1999)	Amazon rainforest and cerrado-caatinga of ne Brazil, Paraguay, and Misiones in Argentina.
L. p. sublineatus Cope, 1860 Argentine Legion Snake	Dixon (1989); Dixon & Markezich (1992); Cei (1993); Lema (1994b); Leynaud & Bucher (1999)	Pampas of s Brazil in Porto Alegre, Uruguay, and Argentina.
L. problematicus Myers, 1986 Problematic Legion Snake	Myers (1986); Dixon (1989); Henle & Ehrl (1991b)	Known only from the type locality of San Juan, Río Tambopata in Sandia Province, Puno, Peru. This is the Amazonian side of the Andes about 150 km n of Lake Titicaca in the border region between Peru and Bolivia. The holotype was taken at 1570 m.
L. reginae (Linnaeus, 1758) Regal Legion Snake	Roze (1966); Peters & Orejas-Miranda (1970); Amaral (1976); Emsley (1977); Duellman (1978); Hoge et al. (1978); Moonen et al. (1979); Dixon (1980, 1983a, 1989); Lema (1994b); Murphy (1997); Kornacker (1999); Freitas (1999); Marques et al. (2001)	Associated primarily with lowlands in cis-Andean South America with a wide distribution from Colombia into Argentina and Paraguay.
*L. r. reginae** (Linnaeus, 1758) Common Regal Legion Snake	Peters (1960b); Peters & Orejas-Miranda (1970); Cunha & Nascimento (1978); Hoge et al. (1978); Dixon (1983a, 1989); Abuys (1984) *Remark: Dixon (1989) listed *L. violaceus* (Lacépède, 1789) and *L. graphicus* (Shaw, 1802), as synonyms.	The Guianas in Guyana, Suriname, and French Guiana.

Species-level Taxa	Citations and Remarks	Abbreviated Range
*L. r. macrosomus** (Amaral, 1935) Southern Regal Legion Snake	Peters & Orejas-Miranda (1970); Dixon (1983a, 1989); Cei (1993); Lema (1994b); Leynaud & Bucher (1999) *Remark: Dixon (1989), included *L. r. maculicaudus* (Hoge, 1954), as a synonym.	Chaco boreal and cerrado vegetation of central meridional Brazil, Paraguay, Bolivia, and Argentina in Salta, Formosa, Corrientes, and Misiones.
*L. r. semilineatus** (Wagler *in* Spix, 1824) Stripe-tailed Regal Legion Snake	Dixon (1983a, 1989); Cunha et al. (1985); Dixon & Soini (1986); Lancini & Kornacker (1989); Kornacker (1999); Henle & Ehrl (1991); Marques et al. (2001); Dixon & Tipton (2003) *Remark 1: Dixon (1989) included *Leimadophis oligolepis* (Boulenger, 1905), as a synonym. However, Dixon (pers. comm. June 11, 1998), based on access to a large sample of *oligolepis*, now considers the taxon valid (based on ventral pattern/color and scale row count of 15-15-15). It has not been broken out here because this should be validated in Dixon and Tipton's work overviewing the *Liophis* group that is soon to be published. Remark 2: Henle & Ehrl (1991) named *Liophis miliaris intermedius* from the type locality of Dept. Madre de Dios, Peru. From the description and photos this does not appear to be a *Liophis miliaris*. Based on the location of the type and the name intermedius being occupied, Dixon and Tipton borrowed the type and discovered that this snake is really *L. reginae semilineatus* and is discussed in Dixon & Tipton (2003).	Amazonian Ecuador, Colombia, Peru, Bolivia, Venezuela, Brazil, and the Atlantic forests of Brazil.
L. r. zweifeli (Roze, 1959) Reticulated Regal Legion Snake	Roze (1966); Peters & Orejas-Miranda (1970); Dixon (1983a, 1989); Lancini & Kornacker (1989); Murphy (1997); Kornacker (1999); Mijares-Urrutia & Arends R. (2000); Boos (2001)	N coast of Colombia and montane forest zones of Venezuela and into Trinidad in the southern Caribbean.
L. sagittifer (Jan, 1863) Painted Steppe Legion Snake	Peters & Orejas-Miranda (1970); Amaral (1976); Dixon & Thomas (1982); Dixon (1989); Yanosky (1998)	S South America in Bolivia, nw Argentina into s Brazil, and Paraguay; Dixon & Thomas (1982) indicated the two major plant formations in which it is found are monte and chaco. This is the southernmost ranging form of *Liophis*.
*L. s. sagittifer** (Jan, 1863) Painted or Spotted Steppe Legion Snake	Dixon & Thomas (1982); Cei (1986, 1993); Dixon (1989); Leynaud & Bucher (1999) *Remark: Dixon (1989), listed *L. pulcher* Steindachner, 1867 and *L. argentinus* Bréthès, 1917, as synonyms.	N Argentina from Tucumán, Catamarca, Santiago del Estero, La Rioja, Córdoba, San Luis, San Juan, La Pampa, Mendoza, Neuquen, Rio Negro, and Chubut. It is presumed to be in Rio Grande do Sul, Brazil.
*L. s. modestus** (Koslowsky, 1896) Stripeless Steppe Legion Snake	Dixon & Thomas (1982); Dixon (1989); Cei (1993); Leynaud & Bucher (1999) *Remark: Dixon (1989), listed *L. trifasciatus* Werner, 1899, as a synonym.	Bolivian-Paraguayan chaco in e Bolivia, n Argentina in Chaco, Formosa, Jujuy, Salta, Santiago del Estero, Tucumán, and Catamarca, and into Rio Grande de Sul, Brazil, and w Paraguay; it is associated with dry Chaco-type habitat.

Species-level Taxa	Citations and Remarks	Abbreviated Range
L. taeniurus Tschudi, 1845 Thin Legion Snake	Amaral (1927f); Dunn (1944c); Peters (1960b); Peters & Orejas-Miranda (1970); Dixon & Markezich (1979); Dixon (1983f, 1989, 2000)	Intermediate and upper e Andean slopes of s Ecuador, Peru, and into Cochabamba in Bolivia in the Amazon region.
L. torrenicola Donnelly & Myers, 1991 Cerro Guaiquinima Legion Snake	Donnelly & Myers (1991); Kornacker (1999)	Venezuela, Tepuí on Cerro Guaiquinima.
L. triscalis (Linnaeus, 1758) Curacao Legion Snake	Brongersma (1959); Roze (1966); Peters & Orejas-Miranda (1970); Dixon (1981b, 1989)	Leeward Island of Curaçao in the Caribbean and a questionable record from Venezuela.
*L. typhlus** (Linnaeus, 1758) Queen Legion Snake	Peters (1960b); Dunn (1957); Peters & Orejas-Miranda (1970); Dixon & Markezich (1979); Moonen et al. (1979); Dixon (1980, 1987b, 1989); Dixon & Thomas (1985); Lema (1994b); Starace (1998); Yanosky (1998); Kornacker (1999); Freitas (1999); Marques et al. (2001) *Remark: Dixon (1980) and Vanzolini (1986), did not recognize subspecies; however, Dixon (1989) used the taxon and did recognize subspecies.	South America from the Cis-Andean region southward into n Argentina and Paraguay, associated with rainforests of the Guiana Shield and Amazon Basin, and also the Chaco and Cerrado of Bolivia, Brazil, and Paraguay; Peters & Orejas-Miranda (1970) described the distribution as South America east of the Andes and north of about 35°S latitude.
*L. t. typhlus** (Linnaeus, 1758) Velvety Swamp Legion Snake	Roze (1966); Peters & Orejas-Miranda (1970) [as *Leimadophis*]; Cunha & Nascimento (1978); Abuys (1984); Dixon & Soini (1986); Dixon (1987b, 1989); Lancini & Kornacker (1989); Starace (1998); Kornacker (1999) *Remark: Dixon (1989), listed *L. isolepis* (Cope, 1870), and *L. macrops* Werner, 1925, as synonyms.	South America east of the Andes, the Guianas into the Amazon Basin. It is found throughout the Amazon Basin in rain forest to about 1000 m.
L. t. brachyurus (Cope, 1887) Forest Queen Legion Snake	Peters & Orejas-Miranda (1970); Dixon (1987b, (1989); Marques et al. (2001:172)	Deciduous mesophytic forests of se Brazil and in Campo Cerrado forests of sc Brazil.
L. t. elaeoides Griffin, 1916 Chaco Forest Legion Snake	Griffin (1916); Dixon (1987b, 1989); Leynaud & Bucher (1999)	Chaco forests of se Bolivia, n Paraguay, and w Mato Grosso in Brazil.
L. vanzolinii Dixon, 1985 Vanzolini's Legion Snake	Dixon (1985b, 1989); Cei (1993)	Argentina; limited to the serrano puntado-cordobés system (San Luís, Córdoba) in Argentina, and known from three localities. Is restricted to the dry grass and and brush covered slopes of the e Sierra of Córdoba.
L. viridis Günther, 1862 Brazilian Green Legion Snake	Peters & Orejas-Miranda (1970); Amaral (1976); Vanzolini et al. (1980); Dixon (1985a, 1987b, 1989); Dixon & Thomas (1985); Vanzolini (1986); Yanosky (1998); Freitas (1999)	NE Brazil southward into n Argentina and Paraguay; it is associated with the Caatinga, Agreste, and Atlantic rainforests of Brazil.

Species-level Taxa	Citations and Remarks	Abbreviated Range
L. v. viridis Günther, 1862 Common Brazilian Green Legion Snake	Dixon (1985a, 1987b, 1989)	Atlantic rainforests and agreste forest in Brazil from Recife, Pernambuco, eastward and southward to Salvador, Bahia. It is assoicated with the agreste and Atlantic rainforest.
L. v. prasinus Jan, 1866 *in* Jan & Sordelli, 1866–1870 Caatingas Green Legion Snake	Dixon (1985a, 1987b, 1989); Dixon & Michaud (1992)	Caatinga forests of Brazil form Joao Pereira, Maranhão in ne Brazil, southward to Sao Francisco, Minas Geraís, then ne to about Parauagua, Rio Grande do Norte, Brazil.
L. vitti Dixon, 2000 Vitt's Legion Snake	Dixon (2000b)	Pacific Andean slopes of Ecuador near the Colombian border; the type locality is Carchi, Ecuador at 1410 m.
L. williamsi (Roze, 1958) Williams Legion Snake	Roze (1958a); Dixon (1980, 1983a, 1989); Myers (1986); Vanzolini (1986); Lancini & Kornacker (1989); Kornacker (1999)	Cloud forests of the e Andes of n Venezuela in the Cordillera de la Costa from Rancho Grande, Aragua into Cerro El Avila in the Distrito Federal.

GENUS: *Lystrophis* Cope, 1885, is the South American Hognose Snake that is found in south and southeastern Brazil, Bolivia, Paraguay, Uruguay, and north and central Argentina. See Peters & Orejas-Miranda (1970); Cei (1986, 1993); Obst et al. (1988); Scrocchi & Druz (1993); and Lema (1994b).

Review assistance and important contributions for this section were provided by Thales de Lema.

CONTENT: Five species. ★★★

Lystrophis: South American Hog-nosed Snake

Species-level Taxa	Citations and Remarks	Abbreviated Range
L. dorbignyi (Dumeril, Bibron & Dumeril, 1854) South American Hog-nosed Snake	Orejas-Miranda (1966b); Peters & Orejas-Miranda (1970); Amaral (1976); Miranda et al. (1982); Cei (1986, 1993); Achaval (1987); Yanosky & Chani (1988); Savage & Slowinski (1992); Lema (1994b); Yanosky (1998); Leynaud & Bucher (1999)	East of the Andes in Argentina, Uruguay, s Paraguay, Bolivia, and s Brazil in Rio Grande do Sul, it is present between 25°S to 40°S latitude. Reported in Argentina from Chaco, Misiones, Entre Ríos, Rio Negro, Buenos Aires, La Pampa, Santa Fe, Córdoba, and Santiago del Estero; associated with water sources, such as rivers and streams.
L. d. dorbignyi (Dumeril, Bibron & Dumeril, 1854) Argentine Hog-nosed Snake	Lema (1994b)	E Brazil into s Buenos Aires Province.
L. d. chacoensis Lema, 1994 Chacoan Hog-nosed Snake	Lema (1994b)	Chaco into the n of Argentina and Paraguay southward and eastward into the east southeast of Argentina.

Species-level Taxa	Citations and Remarks	Abbreviated Range
L. d. orientalis Lema, 1994 Dunes Hog-nosed Snake	Lema (1994b)	Extreme se Brazil in Santa Catarina and Rio Grande do Sul.
L. d. uruguayensis Lema, 1994 Uruguayan Hog-nosed Snake	Lema (1994b)	Uruguay into Rio Grande do Sul in Brazil and into ne Argentina.
L. histricus (Jan, 1863) Rayed or Jan's Hog-nosed Snake	Orejas-Miranda (1966b); Peters & Orejas-Miranda (1970); Amaral (1976); Cei (1993); Achaval (1987); Savage & Slowinski (1992); Lema (1994b); Yanosky (1998) *Remark: Vanzolini (1986) showed *L. nattereri* (Steindachner, 1867) from s Brazil as a valid species, Cei (1993) listed it as a synonym of *L. histricus*. This work follows Cei (1993).	Hoge notes it is from ne Argentina, Paraguay, ne Uruguay and s and se Brazil from between 15°S to 34°S latitude, and also notes it as being found in the Amazon. Cei claimed it is found in Misiones, and Corrientes in Argentina; Lema (1994b) believes it is found in se Brazil into Argentina, Paraguay, and Uruguay. It is also reported from Chaco in Argentina.
L. matogrossensis Scrocchi & Cruz, 1993 Mato Grosso Hog-nosed Snake	Scrocchi & Cruz (1993); Giraudo (1997)	Mato Grosso in Brazil west into Estancia Cascavel, Sandoval in Santa Cruz, Bolivia.
*L. pulcher** (Jan, 1863) Hog-nosed False Coral Snake	Scrocchi & Cruz (1993); Leynaud & Bucher (1999) *Remark: Sometimes considered a synonym of *L. semicinctus* (see Peters & Orejas-Miranda, 1970 and Cei, 1993). However, Scrocchi & Cruz (1993) showed this to be a valid species and it was also listed as valid by Leynaud & Bucher (1999).	SW Brazil into s and e Bolivia, s Paraguay and in Argentina in Jujuy, Salta, Chaco, Catamarca, Tucuman, Santiago del Estero, Formosa, Cordobá, San Luis, Entre Rios, and Corrientes.
L. semicinctus (Dumeril, Bibron & Dumeril, 1854) Banded or Tricolor Hog-nosed Snake	Orejas-Miranda (1966b); Peters & Orejas-Miranda (1970); Amaral (1976); Miranda et al. (1982); Cei (1986, 1993); Savage & Slowinski (1992); Scrocchi & Cruz (1993); Yanosky (1998) *Remark: Cei (1993) showed the ranges of various color forms of this snake and called them the forms "*L. s. aequicinctus*", "*L. s. multicinctus*", and "*L. s. nigrocinctus*".	C and n Argentina, sw Brazil, s Bolivia and probably n Paraguay; from between 20°S to 30°S latitude. Boulenger cites it from Uruguay. Orejas-Miranda questions its presence there. Cei (1993) claims it is associated with the Bolivian and Paraguayan chaco and widespread in ne part of Argentina.

GENUS: *Manolepis* Jan, 1863, is endemic to Mexico. See Smith & Taylor (1966); Smith & Smith (1976, 1993); Underwood (1979); Williams & Wallach (1989) and Köhler (2001).

Review assistance and important contributions for this section were provided by Robert A. Thomas.

CONTENT: One species. ★★★★

Manolepis: Ridge-headed Snake

Species-level Taxa	Citations and Remarks	Abbreviated Range
M. putnami (Jan, 1863) Ridge-headed Snake	Peters (1954); Smith & Taylor (1966); Johnson (1978); Villa et al. (1988); Casas-Andreu (1982); Smith & Smith (1993); Ramírez-Bautista (1994); Köhler (2001)	W Mexico from Nayarit into Tehuantepec on the Pacific coast into Chiapas.

GENUS: *Masticophis* Baird & Girard, 1853, is the genus referred to as the Common Whip Snakes, Striped Racers, and Coachwhips of the Americas. It ranges from the northwestern United States southward through Mexico and Central America into northern South America in northern Colombia and Venezuela. Distribution becomes disjunct in the southern part of this taxon's range in southern Mexico, Panama, Colombia, and Venezuela; and, in the northeastern part of its range in the United States. See Ortenberger (1928); Smith & Taylor (1966), Peters & Orejas-Miranda (1970); Wilson (1973a); Conant (1975); J. D. Johnson (1977); Stebbins (1985); Smith & Smith (1993); Dixon & Camper (1994); Degenhardt et al. (1996); Tennant (1998); Kornacker (1999); Dixon (2000a); Tennant & Bartlett (2000) and Köhler (2001); Lillywhite & Smits (1992).

Review assistance and important contributions for this section were provided by Jeffery D. Camper.

Ortenberger (1923, 1928) were the first reviews of the North American whipsnakes.

CONTENT: Eight species. ★★★

Masticophis: North American Whipsnake or Coachwhip

Species-level Taxa	Citations and Remarks	Estimates Range
M. anthonyi (Stejneger, 1901) Clarion Island Whipsnake	Stejneger (1901); Brattstrom (1955); Smith & Taylor (1966); Wilson (1973a); Smith & Smith (1993)	Clarion island in the Revillagigedo Archipelago of Mexico approximately 400 miles sw of Cape San Lucas in Lower Baja California.
M. aurigulus (Cope, 1861) Baja California Striped Whipsnake or Cape Whipsnake	Smith & Taylor (1966); Wilson (1973a); Stebbins (1985); Grismer (1990a, 1999); Smith & Smith (1993)	Mexico in the Cape region of Baja California Sur and satellite islands in the Gulf of California.
M. a. aurigulus (Cope, 1861) Cape San Lucas Racer	Ortenburger (1923); Savage (1960b); Smith & Taylor (1966); Grismer (1990a, 1990b); Smith & Smith (1993); Smith & Grismer (1999); McPeak (2000)	Cape region of Baja California Sur, Mexico where where it is restricted to the upper elevations of the Sierra de la Laguna and its eastern foothills.
M. a. barbouri (Van Denburgh & Slevin, 1921) Espiritu Santo Striped Whipsnake	Ortenburger (1923); Soulé & Sloan (1966); Smith & Taylor (1966); Wilson (1973a); Stebbins (1985); Grismer (1990a, 1990b, 1999); Smith & Smith (1993); McPeak (2000)	Islands of lower Baja California in the Gulf of California in Mexico including Espíritu Santo Island and Partida Sur off the east coast of La Paz.
M. bilineatus* Jan, 1863 Sonoran Whipsnake	Ortenberger (1923, 1928); Hensley (1950); Lowe & Norris (1955); Smith & Taylor (1966); Wilson (1973a); Stebbins (1985); Smith & Smith (1993); Camper & Dixon (1994); Degenhardt et al. (1996); Camper (1996a); McPeak (2000)	W United States into Mexico; from s Arizona and extreme sw New Mexico south through the w coast of Mexico into Jalisco and Colima with distribution into Zacatecas through canyons that bisect the western Sierra Madre Occidental. It is found on both Isla Tiburón and Isla San Esteban in the Gulf of California.

Species-level Taxa	Citations and Remarks	Abbreviated Range
	*Remark 1: This species is sometimes shown with subspecies, as can be seen in Stebbins (1985), as *M. b. bilineatus* Jan, 1863, from the sw United States into Mexico and called the Sonoran Mountain Whipsnake; *M. b. lineolatus* Hensley, 1950, from s and se Arizona southward into Mexico and called the Ajo Mountain Whipsnake; and, *M. b. slevini* Lowe & Norris, 1955, from San Esteban Island in Mexico and sometimes called the San Esteban Island Whipsnake. However, Camper & Dixon (1994) did not recognize subspecies. *M. lineolatus* Hensley, 1950, is sometimes listed as a full species (Liner, 1994). Bartlett & Tennant (2000), listed the subspecies *M. b. bilineatus* and *M. b. lineolatus*.	
	*Remark 2: Grismer (1999), elevated *sleveni* to full (evolutionary) species status without detail for the change.	
M. flagellum (Shaw, 1802) Coachwhip	Ortenberger (1928); Wilson (1970d, 1973a, 1973b); Conant (1975); Stebbins (1985); Dundee & Rossman (1989); Smith & Smith (1993); Degenhardt et al. (1996); Brown (1997); Tennant & Bartlett (2000); Dixon (2000a); Lillywhite & Smits (1992)	Across the s United States from California to the east coast, and south into Mexico on the entire Baja California and to as far south as n Veracruz and Querrétaro in the east, and s Durango and c Sinaloa on the western mainland.
M. f. flagellum (Shaw, 1802) Eastern Coachwhip	Ortenberger (1923); Wilson (1973b); Conant (1975); Ashton & Ashton (1981); Dundee & Rossman (1989); Tennant (1997b, 1998); Conant & Collins (1998); Dixon (2000a); Werler & Dixon (2000)	E United States from North Carolina into s Florida and west into Texas, Oklahoma, and se Kansas. An isolated population in sc Kentucky. It seems to be absent from much of the Mississippi River Valley.
M. f. cingulum Lowe & Woodin, 1954 Sonoran Coachwhip	Lowe & Woodin (1954); Stebbins (1966, 1985); Wilson (1973b); Smith & Smith (1993); Bartlett & Tennant (2000)	The sw United States in Arizona southward into Mexico.
M. f. fulginosus (Cope, 1895) Baja California Coachwhip	Wilson (1970d, 1973a, 1973b); Stebbins (1985); Smith & Smith (1993); Brown (1997); Bartlett & Tennant (2000)	The United States in s California southward into Baja California, Mexico.
M. f. lineatulus Smith, 1941 Lined Coachwhip	Smith & Taylor (1966); Conant (1975); Tanner (1985); Stebbins (1985); Conant & Collins (1998); Bartlett & Tennant (2000)	W United States in extreme se Arizona to sw New Mexico, and into n and c Mexico to as far as San Luis Potosí.
*M. f. piceus** (Cope, 1892) Red Racer or Red Coachwhip	Ortenberger (1923); Klauber (1942b); Smith & Taylor (1966); Tanner (1985); Stebbins (1985); Degenhardt et al. (1996); Brown (1997); Bartlett & Tennant (2000)	W North America from southern Nevada into Mexico in Baja California and Sonora.
	*Remark: According to Wilson (1970d), *M. f. frenatus* is simply the red phase of this snake, and is synonym.	
M. f. ruddocki Brattstrom & Warren, 1953 San Joaquin Coachwhip	Brattstrom & Warren (1953); Stebbins (1966, 1985); Wilson (1973b); Brown (1997); Bartlett & Tennant (2000)	The western United States in e and s California.

Species-level Taxa	Citations and Remarks	Abbreviated Range
M. f. testaceus* (Say *in* James, 1823) Western Coachwhip	Maslin (1953); Smith & Taylor (1966); Conant (1975); Tanner (1985); Stebbins (1985); Degenhardt et al. (1996); Tennant (1998); Conant & Collins (1998); Werler & Dixon (2000); Bartlett & Tennant (2000); Tennant & Bartlett (2000); Dixon (2000a) *Remark 1: According to Wilson (1970d), *M. f. flavigularis* is a synonym of this snake and is the brown eastern phase. *Remark 2: Conant & Collins (1998), recorded reddish-colored individuals in some populations in Trans-Pecos Texas, ec New Mexico, and se Colorado.	C North America from s Nebraska, e Colorado, and w Kansas into ne Mexico in San Luis Potosí and ne Sonora. Pérez-Higareda & Smith (1991) listed this subspecies as being present in Veracruz, Mexico.
M. lateralis (Hallowell, 1853) Striped Racer	Ortenberger (1923, 1928); Riemer (1954); Smith & Taylor (1966); Wilson (1973a); Jennings (1983); Stebbins (1985); Brown (1997); Grismer (1990b)	SW California in the w United States southward into Mexico in Baja California Norte with another population in Baja California del Sur; it is absent from the central mountain areas in the middle of California.
M. l. lateralis (Hallowell, 1853) California Striped Racer	Ortenberger (1923, 1928); Riemer (1954); Stebbins (1966, 1985); Jennings (1983); Brown (1997); McPeak (2000); Bartlett & Tennant (2000)	North America in w California south into Baja California in Mexico; this subspecies is found in most of the range including the disjunct population in Baja California del Sur.
M. l. euryxanthus Riemer, 1954 Alameda Striped Racer	Riemer (1954); Riemer (1954); Stebbins (1966, 1985); Jennings (1983); Brown (1997); Bartlett & Tennant (2000)	Restricted to the San Francisco Bay area of California.
M. menotvarius (Dumeril, Bibron & Dumeril, 1854) Neotropical Whipsnake	Ortenburger (1923); Smith (1942f); Roze (1953b); Smith & Taylor (1966); Peters & Orejas-Miranda (1970); Wilson (1973a); Johnson (1977, 1982); Alvarez del Toro (1982); Wilson & Meyer (1985); Villa et al. (1988); García & Ceballos (1994); Lee (1996, 2000); Campbell (1998b); Kornacker (1999); Köhler (2001)	Mexico in s San Luis Potosí to n coast of Veracruz, and from the west coast in s Sonora and Guerrero along both coast of Mexico southward into n South America, the populations are disjunct in s Pacific Panama, ne Colombia, and Venezuela; with an isolated record from Chihuahua in n Mexico.
M. m. mentovarius (Dumeril, Bibron & Dumeril, 1854) Common Neotropical Whipsnake	Roze (1953b); Fulger (1960); Smith & Taylor (1966); Peters & Orejas-Miranda (1970); Johnson (1977, 1982); Pérez-Higareda & Smith (1991); Campbell (1998a); Stafford & Meyer (2000); Lee (2000); Köhler (2001)	Tropical Mexico southward into n Central America in low and moderate elevations form San Luis Potosí, Mexico, to Honduras on the Caribbean slopes and from Guerrero, Mexico, to Costa Rica along the Pacific slopes.
M. m. centralis (Roze, 1953) Colombian Whipsnake	Roze (1953b, 1966); Peters & Orejas-Miranda (1970); Lancini (1979); Johnson (1982); Lancini & Kornacker (1989); Kornacker (1999); Köhler (2001)	Panama into n Colombia and nw Venezuela.
M. m. striolatus* (Mertens, 1934) Mexican Whipsnake	Davis & Smith (1953); Zweifel & Norris (1955); Hardy & McDiarmid (1969); Wilson (1973a); Johnson (1977, 1982); Smith & Smith (1993); Ramírez-Bautista (1994) Remark: *M. f. lineatus* (Bocourt, 1890) is a synonym of *M. flagellum striolatus* (Mertens), (see Smith & Taylor, 1966, and Smith & Smith, 1993	Mexico in s Sonora into the Balsas Basin in Colima, Guerrero, Jalisco, Michoacán, Morelos, Nayarit, and Sinaloa.

Species-level Taxa	Citations and Remarks	Abbreviated Range
M. m. suborbitalis (W. Peters, 1868) Venezuelan Whipsnake	Roze (1953b, 1966); Peters & Orejas-Miranda (1970); Johnson (1982); Lancini & Kornacker (1989); Kornacker (1999)	N Venezuela including Isla Margarita.
M. m. variolosus Smith, 1943 Tres Marias Islands Whipsnake	Johnson (1977, 1982); Jennings (1983); Smith & Smith (1993); Lee (1996)	Only from María Magdalena Island in the Tres Marías islands of Mexico.
M. schotti Baird & Girard, 1853 Schott's Whipsnake	Ortenberger (1928); Gloyd & Conant (1934b); Smith & Taylor (1966); Smith & Smith (1976); Parker (1982); Camper & Dixon (1994); Camper (1996b); Tennant (1998); Dixon (2000a); Tennant & Bartlett (2000)	C Texas south of the Balcones Escarpment west and south on the Gulf Coastal Plain into c Mexico to c Veracruz; found in arid brush, grasslands, desert, and montane forest to about 2300 m.
M. s. schotti Baird & Girard, 1853 Schott's Whipsnake	Conant (1975); Smith & Smith (1993); Camper & Dixon (1994); Camper (1996b); Conant & Collins (1998) [as *M. taeniatus schotti*]; Dixon (2000a); Werler & Dixon (2000)	C Texas south of the Balcones Escarpment and west and south into Coahuila, Mexico.
*M. s. ruthveni** Ortenburger, 1923 Ruthven's Whipsnake	Ortenberger (1923); Smith & Taylor (1966); Conant (1975); Parker (1982); Pérez-Higareda & Smith (1991); Camper & Dixon (1994); Camper (1996b); Tennant (1998); Conant & Collins (1998) [as *M. taeniatus ruthveni*]; Dixon (2000a); Werler & Dixon (2000); Tennant & Bartlett (2000) [as *M. ruthveni*] *Remark: Pérez-Higareda & Smith (1991), as well as Smith & Smith (1993), listed *M. t. australis* as valid However, *M. t. australis* Smith, 1941, which is found in the Mexican Plateau from s Guanajuato and San Luis Potosí to as far south as the state of Michoacán, is a synonym. According to Camper & Dixon (1994), the holotype is a juvenile of *M. s. ruthveni*.	Extreme s Texas in the Rio Grande Valley into Mexico in Tamaulipas, Nuevo León, San Luis Potosí, and Veracruz.
*M. taeniatus** (Hallowell, 1852) Striped Whipsnake	Ortenberger (1923, 1928); Smith & Taylor (1966); Wilson (1973a); Parker & Brown (1980); Parker (1982); Stebbins (1985); Camper & Dixon (1994); Camper (1996c); Brown (1997); Dixon (2000a); Martin (1992) *Remark: Fide Camper (1996), the eastern and southern range limits in Mexico are poorly understood.	W United States in the Great Basin south through c Texas into c Mexico to as far south as ne Jalisco; found in arid grassland, desert, woodland, and montane habitats.
*M. t. taeniatus** (Hallowell, 1852) Desert Striped Whipsnake	Ortenberger (1923); Smith & Taylor (1966); Conant (1975); Stebbins (1985); Camper & Dixon (1994); Camper (1996c); Tennant (1998); Conant & Collins (1998); Brown (1997); Bartlett & Tennant (2000); Tennant & Bartlett (2000); Dixon (2000a) *Remark: Conant & Collins (1998) and Stebbins (1966, 1985), showed range maps for *M. t. taeniatus* that included extreme western Texas and adjacent Mexico. These maps seem to have been copied by	W United States from California and Washington southward into s New Mexico; this snake is not found in Mexico or Texas. It is restricted to e of the Continental Divide.

Species-level Taxa	Citations and Remarks	Abbreviated Range
	Bartlett & Tennant (2000), and included in their book. Camper & Dixon (1994) confirmed that this snake <u>does not</u> occur in Texas or Mexico.	
*M. t. girardi** (Stejneger & Barbour, 1917) Central Texas Whipsnake	Ortenberger (1923); Tanner (1985); Camper & Dixon (1994); Camper (1996c); Tennant (1998); Conant & Collins (1998); Werler & Dixon (2000); Tennant & Bartlett (2000); Dixon (2000a) *Remark: *M. t. ornatus* Baird & Girard, 1853, is a synonym (Parker, 1982). *M. t. ornatus* is not listed in Dixon (2000a), but is listed in Crother et al. (2000) and given the common name of Central Texas Whipsnake, but they do not list *M. t. girardi*.	W and c Texas southward through the Mexican Plateau into Guanajuato and ne Jalisco. It is closely associated with the Chihuahuan Desert-western Mexican Plateau into the Edwards Plateau of Texas.

GENUS: *Mastigodryas* Amaral, 1935 (1933–1934), is another tropical racer that in recent history was restricted to Colombia and usually considered to contain a single species. Dixon & Tipton (in press) recently verified that *Dryadophis* Stuart, 1939 is a synonym of *Mastigodryas* and the genus now is considered to range from Mexico southward into Argentina. Peters & Orejas-Miranda (1970) showed that Romer (1956) did not use *Mastigodryas*, but did indicate that *Mastigodryas* and *Dryadophis* were synonyms; and Vanzolini (1986), used *Dryadophis* as a junior synonym of *Mastigodryas*. Dixon & Tipton were unable to locate Peters' notes so reviewed the actual characteristics of the holotype to come to the conclusion that *Dryadophis* a junior synonym of *Mastigodryas*. However, over the years the two genera were separated as distinct or synonymized depending on the authority. Stuart's name *Dryadophis*, 1939 is a substitute for *Eudryas* Fitzinger, 1843, which is preoccupied by *Eudryas* Boivduval, 1836. Before Dixon & Tipton (in press), Peters is the last critical reviewer of the genus.

For more information on this genus (and *Dryadophis*) see Amaral, 1935 (1933–1934, 1964), Stuart (1938, 1941b), Smith & Larsen (1947), Romer (1956), Smith & Taylor (1966), Peters *in* Peters & Orejas-Miranda (1970), Smith & Larson (1974b), Wilson & Meyer (1985), Smith & Pérez-Higareda (1986), Vanzolini (1986), Pérez-Santos & Moreno (1988), Williams & Wallach (1989), Flores-Villela (1993), Smith & Smith (1993), Liner (1994), Lee (1996), Murphy (1997), Campbell (1998a), Starace (1998), Köhler (2001), Dixon & Tipton (in press).

Mastigodryas: Neotropical Lizard-eating Snake or Lizard-eater.

CONTENT: Twelve species. ★★

Species-level Taxa	Citations and Remarks	Abbreviated Range
M. amarali (Stuart, 1938) Amaral's Lizard-eater	Stuart (1941b); Roze (1966); Peters & Orejas-Miranda (1970); Lancini & Kornacker (1989); Kornacker (1999) [as *Mastigodryas*]	NE Venezuela ands Isla Margarita; it seems to be associated with arid regions. It is not found on Trinidad fide Murphy (1997).
*M. bifossatus** (Raddi, 1820) Brown-lined Lizard-eater	Hardy (1963); Peters & Orejas-Miranda (1970); Smith & Pérez-Higareda (1986); Lancini & Kornacker (1989); Starace (1998); Kornacker (1999) [as *Mastigodryas*]; Freitas (1999 as *Mastigodryas*); Marques et al. (2001) *Remark: Vanzolini (1986) noted subspecies do not	Widespread throughout South America from Venezuela and Colombia into s Brazil, Bolivia, Paraguay and n Argentina; its range includes the Guianas.

Species-level Taxa	Citations and Remarks	Abbreviated Range
	seem to be biologically sound. Lancini (1979) also seems to support the idea of no subspecies. Subspecies are included here because later works recognize subspecies, including Cei (1993), Starace (1998), and others.	
M. b. bifossatus (Raddi, 1820) Common Brown-lined Lizard-eater	Stuart (1941b); Peters & Orejas-Miranda (1970); Amaral (1976); Cunha & Nascimento (1978); Cei (1993); Lema (1994b); Starace (1998)	Rio Grande do Sul into Rio de Janeiro, and Minas Gerais in Brazil, and Uruguay; and possibly the Guianas; associated with the more humid areas.
M. b. lacerdai Cunha & Nascimento, 1978 Lacerda's Lizard-eater	Cunha & Nascimento (1978); Hoge et al. ([1978/79] 1981); Cunha et al. (1985); Starace (1998)	E Pará into w Maranhão in Brazil; and possibly the Guianas.
M. b. striatus (Amaral, 1931) Savanna Striped Lizard-eater	Amaral (1931a, 1932); Stuart (1941b); Nicéforo María (1942); Roze (1966); Peters & Orejas-Miranda (1970); Pérez-Santos (1986c); Pérez-Santos & Moreno (1988); Kornacker (1999) [as *Mastigodryas*]	E Colombia, s Venezuela, n Brazil, and the Guianas; widespread in savannas.
M. b. triseriatus (Amaral, 1931) Gray-bellied Lizard-eater	Amaral (1931a, 1976); Stuart (1941b); Peters & Orejas-Miranda (1970); Cunha & Nascimento (1978); Hoge Hoge et al. ([1978/79] 1981) Cei (1993); Leynaud & Bucher (1999)	Chaqueña region of n Argentina, Paraguay, and Bolivian chaco and ne into c (Mato Grosso) Brazil. In Argentina, it is found in Salta, Jujuy, Formosa, Chaco, Tucumán, Corrientes, Misiones, Entre Ríos, and Santa Fe.
M. b. villelai (Hoge, 1952) Villela's Lizard-eater	Hoge (1952a); Peters & Orejas-Miranda (1970); Cunha & Nascimento (1978)	Goais, Pará, Mato Grosso, and Banol Island in Brazil.
M. boddaerti (Sentzen, 1796) Boddaert's Lizard-eater or Neotropical Lizard-eater	Stuart (1933, 1941b); Roze (1966); Peters & Orejas-Miranda (1970); Chippaux ([1986] 1987); Pérez-Santos (1986c); Renjifo & Lundberg (1999); Kornacker (1999) [as *Mastigodryas*]	Found in northern South America and Trinidad and Tobago. It is reported in Colombia, Venezuela, Ecuador, Brazil, Bolivia, Peru (?), and the Guianas.
M. b. boddaerti (Sentzen, 1796) Tan Lizard-eater	Stuart (1933, 1941b); Peters (1960b); Peters & Orejas-Miranda (1970); Kempff-Mercado (1975); Emsley (1977); Amaral (1976); Pérez-Santos (1986c); Lancini & Kornacker (1989); Murphy (1997); [as *Mastigodryas*]; Kornacker (1999) [as *Mastigodryas*]; Boos (2001) [as *Mastigodryas*]	N South America in Colombia, Venezuela, Ecuador, Bolivia, the Guianas. Stuart (1941) reports an isolated colony in Bahía, Brazil and his range description may include Peru.
M. b. dunni (Stuart, 1933) Dunn's Lizard-eater	Stuart (1933, 1941b); Peters & Orejas-Miranda (1970); Emsley (1977); Murphy (1997); Boos (2001)	Tobago Island in the southern Caribbean.
M. b. ruthveni (Stuart, 1933) Ruthven's Lizard-eater	Stuart (1933, 1941b); Nicéforo María (1942); Peters & Orejas-Miranda (1970); Pérez-Santos (1986c); Pérez-Santos & Moreno (1988)	Sierra Nevada of Santa Marta in Colombia above 2200 m.
M. bruesi (Barbour, 1914) Windward Tree Racer	Stuart (1941b); Schwartz & Henderson (1985, 1991); Censky & Kaiser *in* Crother (1999)	St. Vincent, the Grenadines, and Grenada in the Lesser Antilles.

Species-level Taxa	Citations and Remarks	Abbreviated Range
M. cliftoni (Hardy, 1964) Clifton's Lizard-eater	Hardy (1963 [as *D. fasciatus*], 1964); Nickerson & Heringhi (1966); Hardy & McDiarmid (1969); Ponce-Campos & Huerta-Ortega (1998)	Mexico in Nayarit, Sinaloa, Sonora, Durango, and Jalisco.
M. danieli Amaral, 1935 Colombian Racer	Amaral ([1933–34] 1935), 1935c, 1935d); Stuart (1939); Peters & Orejas-Miranda (1970); Pérez-Santos (1986c); Pérez-Santos & Moreno (1988); Dixon & Tipton (in press)	Pérez-Santos (1986) included a map that indicates a range in the lowlands of the Caribbean versant of Colombia.
M. dorsalis* (Bocourt, 1890 *in* Dumeril, Mocquard & Bocourt, 1870–1909) Striped Lizard-eater	Stuart (1941b); Peters & Orejas-Miranda (1970); Villa (1971b); Wilson & Meyer (1985); Villa et al. (1988); Köhler (2001) *Remark 1: Sometimes listed as *M. melanolomus dorsalis* as seen in Lynch & Smith (1966). *Remark 2: Smith & Smith (1993) and Liner (1994), showed this snake as being found in Mexico.	Highlands of Guatemala, Honduras, Nicaragua, Mexico, and possibly in Belize.
M. heathi (Cope, 1875) Heath's Lizard-eater	Stuart (1941b); Peters & Orejas-Miranda (1970); Henle & Ehrl (1991b)	Stuart (1941b) showed this as being restricted from very arid coastal desert from Ecuador to possibly as far south as Chile; however, Peters & Orejas-Miranda (1970) restrict this snake to coastal nw Peru.
M. melanolomus* (Cope, 1868) Orange-bellied Lizard-eater	Smith & Taylor (1966); Peters & Orejas-Miranda (1970); Wilson & Meyer (1985); Pérez-Santos (1986c); Villa et al. (1988); Censky & McCoy (1988); García & Ceballos (1994); Lee (1996, 2000); Monroy-Ibarra et al. (1996); Campbell (1998a); Stafford & Meyer (2000); Köhler (2001); Savage (2002) *Remark: Savage (200) shows *M. sanguiventris* (Taylor, 1951) from the Golfo Dulce area of Costa Rica to be a synonym of *M. melanolomus*.	Mexico along both coasts from Tamaulipas in the Atlantic versant and Nayarit and Jalisco in the Pacific versant through Central America into as far south as Panama.
M. m. melanolomus (Cope, 1868) Common Lizard-eater	Stuart (1941b); Smith & Taylor (1966); Peters & Orejas-Miranda (1970); Wilson & Meyer (1985); Campbell (1998a); Lee (2000); Köhler (2001)	Lowlands of ne Yucatán in Mexico southward into c El Petén in Guatemala.
M. m. alternatus (Bocourt, 1884) Central American Lizard-eater	Stuart (1933, 1941b); Peters & Orejas-Miranda (1970); Wilson & Meyer (1985); Smith & Pérez-Higareda (1986); Pérez-Santos (1986c); Köhler (2001)	Honduras southward into Panama and islands off the coast.
M. m. laevis (Fischer, 1881) Guatemalan Lizard-eater	Stuart (1941b); Peters & Orejas-Miranda (1970); Wilson & Meyer (1985); Campbell (1998a); Lee (2000); Köhler (2001)	Foothills of the mountains of Alta Verapaz and Izabal in Guatemala; at low and moderate elevations.
M. m. slevini (Stuart, 1933) Slevin's Lizard-eater	Stuart (1933, 1941b); Hardy (1963); Smith & Taylor (1966); Wilson & Meyer (1985)	Colima, Mexico southward along the Pacific versant lowlands to Guatemala; and on Tres Marías Islands.

Species-level Taxa	Citations and Remarks	Abbreviated Range
M. m. stuarti Smith, 1943 Stuart's Lizard-eater	Stuart (1941b); Peters (1954); Hardy (1963); Smith & Taylor (1966); Hardy & McDiarmid (1969); Casas-Andreu (1982); Alvarez del Toro (1982); Wilson & Meyer (1985); Ramírez-Bautista (1994)	Mexico in the Pacific slopes from Colima into Chiapas (excluding the Isthmus of Tehuantepec), including Jalisco. It is assoiciated with lowlands.
M. m. tehuanae (Smith, 1943) Tehuana Lizard-eater	Smith (1943); Peters & Orejas-Miranda (1970); Wilson & Meyer (1985); Smith & Smith (1993); Köhler (2001)	Nayarit in Mexico and the Pacific slopes of the Isthmus of Tehuantepec southward into w Guatemala.
M. m. veraecrucis (Stuart, 1941) Olive Lizard-eater	Stuart (1941b); Smith & Taylor (1966); Alvarez del Toro (1982); Wilson & Meyer (1985); Smith & Pérez-Higareda (1986); Pérez-Higareda & Smith (1991)	Mexico in the mainland Atlantic coast slopes from s Tamaulipas into Tabasco and Hidalgo.
M. pleei (Dumeril, Bibron & Dumeril, 1854) Plee's Lizard-eater	Stuart (1941b); Dunn (1944f); Roze (1966); Peters & Orejas-Miranda (1970); Hoogmoed (1983 [1982]); Pérez-Santos (1986c); Smith & Pérez-Higareda (1986); Villa et al. (1988); Lancini & Kornacker (1989); Auth (1994); Esqueda & La Marca (1999); Kornacker (1999) [as *Mastigodryas*]; Köhler (2001)	Panama into n Colombia and nw Venezuela including the islands of Margarita and Testigos. It is associated with the more arid regions.
M. pulchriceps (Cope, 1868) Cope's Lizard-eater	Stuart (1941b); Peters (1960b); Hardy (1963); Peters & Orejas-Miranda (1970); Smith & Pérez-Higareda (1986); Pérez-Santos (1986c); Pérez-Santos & Moreno (1991)	WC Ecuador and Peru; Stuart (1941b) showed this snake to be restricted to the Guyanas Basin and more humid habitats of wc Ecuador.
M. sanguiventris* (Taylor, 1954) Costa Rican Lizard-eater	Taylor (1954); Minton & Smith (1960); Peters & Orejas-Miranda (1970) *Remark: Not in Savage & Villa (1986) or Köhler (2001); however, does not seem to be invalidated.	Costa Rica.

GENUS: *Nerodia* Baird & Girard, 1853, was once included in *Natrix* Laurenti, 1768, the genus *Natrix* which is now restricted to the Old World. *Nerodia* ranges from southern Canada southward with wide distribution throughout the United States and into Mexico, and the *Nerodia* have now been introduced into several Caribbean islands. See Smith & Taylor (1966); Schmidt (1953), Conant (1969, 1975); McCranie (1990); Degenhardt et al. (1996); Tennant (1998); Conant & Collins (1998); Dixon (2000a); Tennant & Bartlett (2000) and Köhler (2001); Reinert (1992); B. Savitzky (1992).

Review assistance and important contributions for this section were provided by Roger Conant.

Conant (1958) was the first complete review of *Nerodia* (at the time called *Natrix*). He later increased the scope of his work to include Mexican watersnakes in Conant (1969).

As found in Dundee (1989) and Rossman et al. (1996), the *Thamnophis* are similar to *Nerodia*, and are distinguished by the presence of a single anal plate or cloacal scute. *Nerodia* has a divided anal plate. However, according to Rossman et al. (1996), the alternate character-state does occur in fairly low frequencies and/or concentrated in localized populations of some species of *Thamnophis*. Rossman (1995), identified a second characteristic to separate *Nerodia* from *Thamnophis*. The second characteristic is that in *Nerodia*, a typically less enlarged first dorsal row (DSR-1) is present.

CONTENT: Ten species. ★★

See also *Clonophis* and *Regina*.

Nerodia: North American Watersnake

Species-level Taxa	Citations and Remarks	Abbreviated Range
N. clarki (Baird & Girard, 1853) Saltmarsh Snake	Ashton & Ashton (1981); Lawson (1987); Dundee & Rossman (1989); Schwartz & Henderson (1991); Dixon (2000a)	The United States Gulf Coast into the s Atlantic Coast from se Texas into s Florida and the Florida Keys; also found on the n coast of Cuba.
N. c. clarki (Baird & Girard, 1853) Gulf Saltmarsh Snake	Conant (1975); Ashton & Ashton (1981); Dundee & Rossman (1989); Wright & Wright ([1957] 1994); Tennant (1997b, 1998); Conant & Collins (1998); Estrada & Ruibal *in* Crother (1999) [as *N. clarki*]; Werler & Dixon (2000); Tennant & Bartlett (2000); Dixon (2000a)	The United States Gulf Coast from s Texas into wc Florida.
N. c. compressicauda Kennicott, 1860 Mangrove Saltmarsh Snake	Conant (1975); Dundee & Rossman (1989); Schwartz & Henderson (1985, 1991); Tennant (1997b), Conant & Collins (1998); Estrada & Ruibal *in* Crother (1999); Tennant & Bartlett (2000)	Mangrove swamps from Tampa, Florida southward into n Cuba; in Florida it is found along the Gulf and Atlantic coasts of the United States into the Keys.
N. c. taeniata (Cope, 1895) Atlantic Saltmarsh Snake	Conant (1975); Campbell *in* Wright & Wright ([1957] 1994); Tennant (1997b); Conant & Collins (1998); Tennant & Bartlett (2000)	The United States in coastal Florida; restricted to the counties of Brevard, Volusia, and Indian River.
N. cyclopion* (Dumeril, Bibron & Dumeril, 1854) Green or Mississippi Green Watersnake	Conant (1975); Sanderson (1993); Dundee & Rossman (1989); Tennant (1997b, 1998); Conant & Collins (1998); Werler & Dixon (2000); Tennant & Bartlett (2000); Dixon (2000a) *Remark: Dundee & Rossman (1989) indicate that no subspecies of this snake are currently recognized.	C and s United States in the Mississippi Valley from extreme s Illinois into the Gulf Coast from se Texas into Florida.
N. erythrogaster (Forster *in* Bossu, 1771) Plain-bellied Watersnake	Clay (1938); Conant (1969, 1975); Lawson (1987); Dundee & Rossman (1989); McCranie (1990); Wright & Wright ([1957] 1994); Rossman et al. (1996); Dixon (2000a)	C and s United States south into Mexico; with disjunct populations outside of the main range for the species.
N. e. erythrogaster (Forster, 1771) Red-bellied Watersnake	Conant (1975); Ashton & Ashton (1981); McCranie (1990); Smith & Smith (1993); Tennant (1997b); Conant & Collins (1998); Tennant & Bartlett (2000)	Extreme e United States from se Virginia into se Alabama and n Florida, and another population in s Delaware and Maryland.
N. e. alta (Conant, 1963) Aguanaval Watersnake	Conant (1963a, 1969); McCranie (1990); Smith & Smith (1993)	Río Aguanaval drainage in Zacatecas, Mexico in the n Mexican Altiplano.
N. e. bogerti (Conant, 1953) Nazas Watersnake	Conant (1953, 1969); Fitch (1981); McCranie (1990); Smith & Smith (1976, 1993)	Río Nazas drainage in Durango, Mexico in the n Mexican Altiplano.

Species-level Taxa	Citations and Remarks	Abbreviated Range
N. e. flavigaster (Conant, 1949) Yellow-bellied Watersnake	Conant (1949, 1975); Ashton & Ashton (1981); Rossman (1989); McCranie (1990); Tennant (1997b, 1998); Conant & Collins (1998); Werler & Dixon (2000); Tennant & Bartlett (2000); Dixon (2000a)	C United States from w Illinois into nc Georgia and e Texas, and the Gulf Coast. Isolated colonies in se Iowa, nw Illinois, and sc Missouri.
N. e. neglecta (Conant, 1949) Copper-bellied Watersnake	Conant (1949, 1975); McCranie (1990); Conant & Collins (1998); Tennant & Bartlett (2000)	The United States in w Kentucky, adjacent Illinois, and sw Indiana, with disjunct and isolated populations in Ohio, Indiana, Michigan, ne Texas, and Kentucky.
N. e. transversa (Hallowell, 1852) Blotched Watersnake	Smith & Taylor (1966); Conant (1969, 1975); Dundee & Rossman (1989); McCranie (1990); Degenhardt et al. (1996); Tennant (1998); Conant & Collins & (1998); Werler & Dixon (2000); Bartlett & Tennant (2000); Tennant & Bartlett (2000); Dixon (2000a)	C and sw United States in Oklahoma, Texas, and New Mexico southward into n Mexico into Coahuila, Nuevo León, and Tamaulipas.
N. fasciata (Linnaeus, 1766) Southern Watersnake	Conant (1963b, 1969, 1975); Lawson (1987); Dundee & Rossman (1989); Savage & Slowinski (1992); Reinert (1992) Dixon (2000a)	C, e and s United States, Cuba and possibly extreme ne Mexico.
N. f. fasciata (Linnaeus, 1766) Banded Watersnake	Conant (1975); Ashton & Ashton (1981); Tennant (1997b); Conant & Collins (1998); Tennant & Bartlett (2000)	SE United States in the Coastal Plain from North Carolina into s Alabama.
N. f. confluens (Blanchard, 1923) Broad-banded Watersnake	Conant (1975); Dundee & Rossman (1989); Tennant (1998); Conant & Collins (1998); Werler & Dixon (2000); Tennant & Bartlett (2000); Dixon (2000a)	The United States where it is centered around the Mississippi Valley in lowlands from extreme s Illinois into c Texas along the Gulf Coast.
N. f. pictiventris (Cope, 1895) Florida Watersnake	Conant (1975); Ashton & Ashton (1981); Smith & Smith (1993); Tennant (1997b, 1998); Conant & Collins (1998); Werler & Dixon (2000); Tennant & Bartlett (2000); Dixon (2000a)	SE United States; it has also been introduced into Brownsville, Texas and is possibly now found in extreme ne Mexico.
*N. floridana** (Goff, 1936) Florida Green Watersnake	Schmidt (1953); Conant (1975); Ashton & Ashton (1981); Lawson (1987); Sanderson (1993); Tennant (1997b); Conant & Collins (1998); Tennant & Bartlett (2000) *Remark: The older name of *N. cyclopion floridana* may prove to be valid.	SE United States in s South Carolina, extreme e Georgia, and another population in s Georgia south into Florida.
N. harteri (Trapido, 1941) Brazos River Watersnake	Trapido (1941); Conant (1975); Mecham (1983); Rossman et al. (1996); Conant & Collins (1998); Dixon (2000a); Werler & Dixon (2000); Tennant & Bartlett (2000)	Only found in c Texas in the Palo Pinto region of the upper Brazos River.
*N. paucimaculata** (Tinkle & Conant, 1961) Concho Watersnake	Tinkle & Conant (1961); Conant (1975); Mecham (1983); Densmore et al. (1992), Tennant (1998); Conant & Collins (1998); Werler & Dixon (2000); Tennant & Bartlett (2000); Dixon (2000a), Crother et al. (2000)	Isolated to the Colorado River drainage from the Coke and Runnels county line to near Bend in San Saba county and Concho River drainage from San Angelo in Tom Green county into the confluence with the Colorado River; its range covers only about

Species-level Taxa	Citations and Remarks	Abbreviated Range
	*Remark: Densmore et al. (1992), developed molecular data on this snake, and Roxe & Selcer (1989) suggested elevating this to specific status based on the information. Werler & Dixon (2000) and Dixon (2000a) continue to list as *N. harteri paucimaculata*.	276 river miles of the Colorado and Concho River basins of west Texas.
N. rhombifer (Hallowell, 1852) Diamond-backed Watersnake	Clay (1938); Cliburn (1956); Conant (1969); Rossman & Eberle (1977); McAllister (1985); Lawson (1987); Villa et al. (1988); Dundee & Rossman (1989); Smith & Smith (1993); Lee (1996, 2000); Rossman et al. (1996); Dixon (2000a); Köhler (2001); B. Savitzky (1992)	C United States southward to the Gulf Coast and along the Gulf Coastal Plain into Mexico to as far south as sw Campeche; there are several records from outside the continuous range.
N. r. rhombifer (Hallowell, 1852) Northern Diamond-backed Watersnake	Clay (1938); Smith & Taylor (1966); Conant (1969) (1975); Rossman & Eberle (1977); McAllister (1985); Dundee & Rossman (1989); Smith & Smith (1993); Tennant (1998); Conant & Collins (1998); Werler & Dixon (2000); Tennant & Bartlett (2000); Dixon (2000a)	Mississippi Valley and the Gulf Coast Gulf Coast of the United States southward into Mexico in Coahuila, Tamaulipas, and Nuevo León. Smith & Taylor (1966) speculated that this snake could reach as far south as Guatemala, but this seems unlikely.
N. r. blanchardi (Clay, 1938) Tampico Diamond-backed Watersnake	Clay (1938); Smith & Taylor (1966); Conant (1969); Rossman & Eberle (1977); McAllister (1985); Pérez-Higareda & Smith (1991); Smith & Smith (1993)	Tamaulipas, Mexico in the Río Pánuco-Río Tamesí drainage complex and a few streams to the n and s.
N. r. werleri (Conant, 1953) Tabasco Diamond-backed Watersnake	Conant (1953, 1969); Rossman & Eberle (1977); McAllister (1985); Pérez-Higareda & Smith (1991); Smith & Smith (1993); Lee (2000); Köhler (2001)	S and s Veracruz to Tabasco into Campeche, Chiapas and Oaxaca, Mexico; Campbell (1998a) theorizes that it might show up in the far south. It is known from the Río Usumacinta and its tributaries in Tabasco and n Chiapas, Mexico, it is possible that further exploration of the Río Usumacinta will reveal the presence of this snake along the w border of Petén.
N. sipedon (Linnaeus, 1758) Northern Watersnake	Clay (1938); Conant & Clay (1937); Conant (1975); Schwaner & Mount (1976); Lawson (1987); Dundee & Rossman (1989); Bartlett & Tennant (2000); Tennant & Bartlett (2000)	S Canada southward through the c United States.
N. s. sipedon (Linnaeus, 1758) Common Watersnake	Conant & Clay (1937); Conant (1975); Conant & Collins (1998)	S Canada in Quebec into e and c United States in extreme ne North Carolina westward into Colorado.
N. s. insularum (Conant & Clay, 1937) Lake Erie Watersnake	Conant & Clay (1937); Conant (1975); Conant & Collins (1998); Tennant & Bartlett (2000)	NC United States where it is confined to the islands of Lake Erie's Put-in-Bay; it is found in the large group of limestone islands in w Lake Erie and on the associated mainland peninsula between Catawba and Marblehead in Ohio and Pointe Pelee in Ontario.
N. s. pleuralis (Cope, 1892) Midland Watersnake	Conant (1975); Ashton & Ashton (1981); Dundee & Rossman (1989);Tennant (1997b); Conant & Collins (1998); Dixon (2000a); Werler & Dixon (2000); Tennant & Bartlett (2000)	C and se United States from Indiana into ne Arkansas to the Gulf of Mexico and into the mountains of the Carolinas; fide Dixon (2000a) the first verification in Texas is from two specimens taken 5 miles north of Sherman in Grayson County, Texas.

The Caenophidia

Species-level Taxa	Citations and Remarks	Abbreviated Range
N. s. williamengelsi* (Conant & Lazell, 1973) Carolina Watersnake	Conant & Lazell (1973); Conant (1975); Conant & Collins (1998); Tennant & Bartlett (2000) *Remark: Formerly *N. sipedon engelsi* Barbour, 1943, and *N. fasciatus engelsi* Barbour, 1943.	Outer bank isles and the mainland coast of North Carolina in the e United States.
N. taxispilota (Holbrook, 1838) Brown Watersnake	Clay (1938); Cagle (1952); Cliburn (1956); Conant (1975); Ashton & Ashton (1981); McCranie (1983c); Lawson (1987); Rossman et al. (1996); Tennant (1997b); Conant & Collins (1998); Tennant & Bartlett (2000)	The United States, mainly in the coastal plains but can be found up the Piedmont streams in se Virginia and sw Alabama south into the southern tip of Florida; its range includes Virginia, Alabama, e North Carolina, South Carolina, Georgia, and Florida.

GENUS: *Ninia* Baird & Girard, 1853, is found from Mexico southward into South America in Venezuela, Colombia, and Ecuador. See Burger & Werler (1954); Smith & Taylor (1966); Peters & Orejas-Miranda (1970); Savage & Lahanas (1991); McCranie & Wilson (1995); Kornacker (1999); and Köhler (2001)

CONTENT: Nine species. ★★

Ninia: Coffee Snake. Frank and Ramus (1995) call this the Common Coffee Snake.

Species-level Taxa	Citations and Remarks	Abbreviated Range
N. atrata (Hallowell, 1845) Hallowell's Coffee Snake	Peters (1960b); Underwood (1962); Roze (1966); Peters & Orejas-Miranda (1970); Wilson & Meyer (1985); Savage & Villa (1986); Villa et al. (1988); Lancini & Kornacker (1989); Savage & Lahanas (1991); Murphy (1997); Kornacker (1999); Boos (2001); Köhler (2001)	E Panama southward to Ecuador on the Pacific versant and n Colombia and Venezuela on the Atlantic versant, as well as Trinidad and Tobago.
N. celata McCraine & Wilson, 1995 Costa Rican Coffee Snake	McCranie & Wilson (1995); Köhler (2001)	Atlantic slopes of Volcán Poás in nc Costa Rica and the n slopes of Cerro Pando in w Panama.
N. diademata* Baird & Girard, 1853 Ring-necked Coffee Snake	Burger & Werler (1954); Smith & Taylor (1966); Peters & Orejas-Miranda (1970); Wilson & Meyer (1985); Villa et al. (1988); Lee (1996); Campbell (1998a); Stafford & Meyer (2000); Köhler (2001) *Remark: Wilson & Meyer (1985) questioned the validity of subspecies in this species.	Mexico southward through Central America into Honduras.
N. d. diademata Baird & Girard, 1853 Veracruz Ring-necked Coffee Snake	Burger & Werler (1954); Smith & Taylor (1966); Peters & Orejas-Miranda (1970); Alvarez del Toro (1982); Pérez-Higareda & Smith (1991); Campbell (1998a)	Foothills and mountains of Caribbean s Mexico from c Veracruz southward into c Central America.
N. d. labiosa (Bocourt, 1883 *in*	Peters & Orejas-Miranda (1970); Alvarez del Toro (1982); Wilson & Meyer (1985); Smith & Smith (1983)	Pacific slopes in the coffee zones of Oaxaca in s Mexico into se Guatemala; found in moderate elevations.

Species-level Taxa	Citations and Remarks	Abbreviated Range
Dumeril, Mocquard & Bocourt, 1870–1909) Pacific Ring-necked Coffee Snake		
N. d. nietoi Burger & Werler, 1954 Caribbean Ring-necked Coffee Snake	Berger & Werler (1954); Peters & Orejas-Miranda (1970); Pérez-Higareda & Smith (1991); Campbell (1998a); Lee (2000)	Caribbean slopes of s Veracruz in Mexico through Guatemala into ne Honduras.
N. d. plorator Smith, 1942 Hidalgo Ring-necked Coffee Snake	Smith & Taylor (1966); Smith & Smith (1993); Peters & Orejas-Miranda(1970); Köhler (2001)	Montane subspecies from Mexico in ne Hidalgo and the southern part of San Luis Potosí.
N. espinali McCranie & Wilson, 1995 Honduran Coffee Snake	Wilson & McCranie (1995); Köhler (2001)	Restricted to the hardwood cloud forest in w Honduras and n El Salvador; from 1590 to 2242 m.
N. hudsoni Parker, 1940 Amazon Coffee Snake	Parker (1940); Peters (1960b); Vanzolini (1986); Burger & Werler (1954); Peters & Orejas-Miranda (1970); Duellman (1978); Lehr & Fernandez (2000)	Lower elevations of Amazonian e Ecuador and the Guianas into Acre and Rondônia in Brazil; recently reported from Departamento Pasco in Peru.
N. maculata (Peters, 1940) Banded Coffee Snake	Peters & Orejas-Miranda (1970); Villa et al. (1988); Savage & Villa (1986); Savage & Lahanas (1991); Smith & Campbell (1996); Köhler (2001); McCranie et al. (2001)	Central America from Alta Verapaz, Guatemala into Darién, Panama.
N. m. maculata (Peters, 1940) Pacific Banded Coffee Snake	Stuart (1948); Peters & Orejas-Miranda (1970)	W and c Costa Rica southward into Darien and the Canal Zone of Panama.
N. m. tessellata Cope, 1876 Caribbean Banded Coffee Snake	Stuart (1948); Burger (1954); Peters & Orejas-Miranda (1970)	Atlantic versant of s Nicaragua, Costa Rica and probably Panama, in c Costa Rica it is found in the Cordillera Central in subtropical zones above approximately 4000 feet. Probably Guatemala.
N. pavimentata (Bocourt, 1883) Red-bellied Coffee Snake	Stuart (1948); Peters & Orejas-Miranda (1970); Smith & Campbell (1996); Köhler (2001)	Mountains of in Guatemala from Finca Chichén in Departamento Alta Verapaz; at moderate elevations.
N. psephota* (Cope, 1876) Highland Coffee Snake	Dunn (1935); Peters & Orejas-Miranda (1970); Savage & Lahanas (1991); Savage & Villa (1986); Villa, et al. (1988); Köhler (2001) *Remark: Villa et al. (1988), included *N. cerroensis* Taylor, 1954 and *N. oxynota* (Werner, 1910), as synonyms.	Higher elevations in the subtropical zone above approximately 4000 feet in Volcán de Chiriquí in w Panama and adjacent mountains of the Cordillera de Talamanca in e Costa Rica.
N. sebae* (Dumeril, Bibron & Dumeril, 1854)	Dunn (1935); Schmidt & Rand (1957); Peters & Orejas-Miranda (1970); Wilson & Meyer (1985); Savage & Villa (1986); Villa et al. (1988); Savage &	Oaxaca and Veracruz in Mexico southward into Central America into Costa Rica; found mainly in coffee zones, it may prove to be in Chiriquí in w Panama.

Species-level Taxa	Citations and Remarks	Abbreviated Range
Red-backed Coffee Snake	Lahanas (1991); Savage & Slowinski (1992); Lee (1996, 2000); Cruz (1997); Campbell (1998a); Stafford & Meyer (2000); Köhler (2001)	They are associated with low, moderate, and intermediate elevations from sea level to 1900 m.
	*Remark: Wilson & Meyer (1985) questioned the validity of *N. sebae* subspecies. They preferred not to recognize subspecies as described by Schmidt & Rand (1957) because it appeared to them that the whole species consists of a mosaic of only slightly differentiated populations. However, they also noted that data on the Honduran material augments and confirms the trends in subcaudal variation shown by Schmidt & Rand (1957) who differentiated more subspecies mainly on the presence or absence of pattern plus subcaudal and ventral count. Although Wilson & Meyer (1985) presented a good argument, I believe more work needs to be performed to confirm the status of the group with respect to subspecies.	
N. s. sebae (Dumeril, Bibron & Dumeril, 1854) Common Red-backed Coffee Snake	Schmidt & Andrews (1936); Smith & Taylor (1966); Peters & Orejas-Miranda (1970); Alvarez del Toro (1982); Wilson & Meyer (1985); Pérez-Higareda & Smith (1991); Campbell (1998a); Lee (2000)	Coffee zones of s Mexico, Belize, Guatemala, Honduras, and El Salvador; recently reported from Campeche, Mexico.
N. s. immaculata Schmidt & Rand, 1957 Immaculate Red-backed Coffee Snake	Schmidt & Rand (1957); Peters & Orejas-Miranda (1970); Wilson & Meyer (1985)	SC Honduras, Nicaragua, and Costa Rica.
N. s. morleyi Schmidt & Andrews, 1936 Yucatán Red-backed Coffee Snake	Schmidt & Andrews (1936); Smith & Taylor (1966); Peters & Orejas-Miranda (1970); Wilson & Meyer (1985); Smith & Smith (1993); Campbell (1998a); Lee (2000)	Dryer areas of the Yucatán peninsula in Yucatán and n Campeche into Quintana Roo and south into Belize and El Petén, Guatemala.
N. s. punctulata (Bocourt, 1883) Bocourt's Red-backed Coffee Snake	Schmidt & Rand (1957); Peters & Orejas-Miranda (1970); Alvarez del Toro (1982); Wilson & Meyer (1985)	Pacific versant of Chiapas, Mexico and n Guatemala between 500 and 2000 m.

GENUS: *Nothopsis* Cope, 1871, is found from Honduras southward into Colombia and Ecuador. Note: at times this genus has been noted as *Nothopis*. See Taylor (1951); Dunn & Dowling (1957); Peters & Orejas-Miranda (1970); Savage & Villa (1986); Köhler (2001); and Dowling & Pinou (2003).

CONTENT: One species. ★★★★

Nothopsis: Mongrel or Rough Coffee Snake. Frank and Ramus (1995) call this the Rough Coffee Snake.

Species-level Taxa	Citations and Remarks	Abbreviated Range
N. rugosus Cope, 1871 Mongrel or Rough Coffee Snake	Taylor (1951); Dunn & Dowling (1957); Peters (1960b); Peters & Orejas-Miranda (1970); Savage & Villa (1986); Villa et al. (1988); Pérez-Santos & Moreno (1988, 1991); Köhler (2001)	Atlantic versant of coastal Honduras and Nicaragua southward through Central America into the Pacific versant of w Colombia and Ecuador.

GENUS: *Opheodrys* Fitzinger, 1843 is confined to North America, where it is found from Canada southward into extreme northern Mexico. See *Liochlorophis* for comparison to the Smooth Green Snake. See Smith & Taylor (1966); Conant (1975); Grobman (1984, 1992, 1994); Dixon (2000a); and Walley & Plummer (2000).

Review assistance and important contributions for this section was provided by James R. Dixon.

CONTENT: One species, with zero to four subspecies, depending on the authority. ★★★

Opheodrys: Rough Green Snake or Rough Grass Snake

Species-level Taxa	Citations and Remarks	Abbreviated Range
O. aestivus* (Linnaeus, 1766) Rough Green Snake	Conant (1975); Ashton & Ashton (1981); Grobman (1984); Dundee & Rossman (1989); Oldham & Smith (1991); Pérez-Higareda & Smith (1991); Degenhardt et al. (1996); Tennant (1997b, 1998); Conant & Collins (1998); Werler & Dixon (2000); Tennant & Bartlett (2000); Dixon (2000a); Walley & Plummer (2000) *Remark 1: Oldham & Smith (1991) divided the two species of North American green snakes into separate genera. They relegated *O. vernalis* to a monotypic genus, now *Liochlorophis*. *Remark 2: Up to four subspecies can be found in literature. They can be listed as *O. a. aestivus* (Linnaeus, 1766), *O. a. carinatus* Grobman, 1984, *O. a. conanti* Grobman, 1984, and *O. a. majalis* (Baird & Girard, 1853), (see Grobman, 1984, Liner, 1994, Mitchell, 1994, and Dixon, 2000a). Walley & Plummer (2000) reported *Opheodrys* to be a monotypic species with no subspecies. Crother et al. (2000) listed the subspecies *O. a. aestivus* and *O. a. carinatus*.	E North America from Connecticut southward into extreme n Mexico into Tamaulipas and Nuevo León; Degenhardt et al. (1996) pointed out that some authorities think that this snake could show up in New Mexico; its range is broken in the thornbush area of south Texas northward and along the Texas side of the Rio Grande River. Pérez-Higareda & Smith (1991) show *O. a. majalis* to be found in Veracruz, Mexico.

GENUS: *Oxybelis* Wagler, 1830, is wide ranging from extreme sw United States in Arizona southward into northern South America to as far as eastern Peru, Bolivia, and Amazonian Brazil. Originally the generic name was *Dryophis* Fitzinger, 1826, but it is preoccupied by *Dryophis* Dalhamn, 1823. See Smith & Taylor (1966); Stebbins (1966); Alvarez del Toro (1982); Keiser *in* Peters & Orejas-Miranda (1970); Keiser (1982); Lancini & Kornacker (1989); Campbell (1998a); Starace (1998); Kornacker (1999 and Köhler (2001)).

Review assistance and important contributions for this section were provided by Hussam Zaher.

Note: *O. argenteus* (Daudin, 1803) and *O. boulengeri* (Procter, 1923), are found from Trinidad and Venezuela into the Guianas, Amazonian Brazil, e Peru, Bolivia, Argentina, and Paraguay have been placed into the genus *Xenoxybelis*.

CONTENT: Four species. ★★★

Oxybelis: Neotropical or American Vinesnake

Species-level Taxa	Citations and Remarks	Abbreviated Range
O. aeneus* (Wagler *in* Spix, 1824) Brown Vinesnake	Peters (1960b); Keiser (1974, 1982); Henderson & Hoevers (1975); Kempff-Mercado (1975); Amaral (1976); Lancini (1979); Moonen et al. (1979); Alvarez del Toro (1982); Wilson & Meyer (1985); Stebbins (1985); Savage & Villa (1986); Dixon & Soini (1986); Villa et al. (1988); Lancini & Kornacker (1989); Pérez-Higareda & Smith (1991); Wright & Wright ([1957] 1994); García & Ceballos (1994); Ramírez-Bautista (1994); Villa & McCranie (1995); Lee (1996, 2000); Campbell (1998a); Starace (1998); Vasquez-Diaz et al. (1998); Kornacker (1999); Freitas (1999); Stafford & Meyer (2000); Bartlett & Tennant (2000); Marques et al. (2001); Boos (2001); Köhler (2001) *Remark: Keiser (1974, 1982), Keiser *in* Peters & Orejas-Miranda (1970), Wilson & Meyer (1982); Smith & Smith (1993), and Campbell (1998a) did not recognize subspecies.	S Arizona in the sw United States southward into Latin America to as far as Peru, Bolivia, Ecuador; Brazil, and the Guianas. Its range includes Trinidad.
O. brevirostris (Cope, 1861) Short-nosed or Cope's Vinesnake	Andersson (1916); Taylor (1951); Peters (1960b); Keiser *in* Peters & Orejas-Miranda (1970); Wilson & Meyer (1985); Cruz & Wilson (1986); Savage & Villa (1986); Villa et al. (1988); Villa & McCranie (1995); Köhler (2001)	Atlantic versant in lowlands from Honduras and Nicaragua southward into nw South America; in the Pacific versant of Colombia and Ecuador.
O. fulgidus (Daudin, 1803) Green Vinesnake	Peters (1960b); Smith & Taylor (1966); Roze (1966); Keiser *in* Peters & Orejas-Miranda (1970); Alvarez del Toro (1982); Wilson & Meyer (1985); Savage & Villa (1986); Dixon & Soini (1986); Villa et al. (1988); Lancini & Kornacker (1989); Pérez-Higareda & Smith (1991); Flores-Villela (1993); Smith & Smith (1993); Villa & McCranie (1995); Lee (1996, 2000); Campbell (1998a); Starace (1998); Strussman & Carvalho (1998); Stafford & Meyer (2000); Kornacker (1999); Köhler (2001)	S Mexico southward into tropical South America east of the Andes into Mato Grosso in Brazil and into Bolivia; *Oxybelis* is not listed in Cei (1993). It is not present on Trinidad and Tobago according to Murphy (1997).
O. wilsoni Villa & McCranie, 1995 Roatán Island Vinesnake	Villa & McCranie (1995); Groves (1995); McCranie (1999); Köhler (2001)	Honduras on Isla Roatán in the Bay Islands.

GENUS: *Oxyrhopus* Wagler, 1830, is another type of American false coral snake. It is found from southern Mexico southward into South America to just south of 35°S east of the Andes in Argentina, and to about Lima, Peru west of the Andes. This is another group of snakes that seems to have confusing taxonomy and requires further study. See Smith & Taylor (1966); Bailey *in* Peters & Orejas-Miranda (1970); Cei (1986, 1993); Vanzolini (1986); Obst et al. (1988); Zaher & Caramaschi (1992); Smith & Smith (1993); Starace (1998) and Köhler (2001).

Also see *Boiruna*, *Clelia*, and *Pseudoboa*.

CONTENT: Thirteen species are listed below. ★★

Oxyrhopus: Calico Snakes (a type of false coral snake). Frank and Ramus (1995) call this the False Coral Snake.

Species-level Taxa	Citations and Remarks	Abbreviated Range
O. clathratus Dumeril, Bibron & Dumeril, 1854 Serrana False Coral Snake	Bailey *in* Peters & Orejas-Miranda (1970); Cei (1993); Lema (1994b); Giraudo (1999); Marques et al. (2001)	Minas Gerais and Rio de Janiero into Rio Grande do Sul and Santa Catarina in se Brazil, and in Misiones, Argentina.
O. doliatus Dumeril, Bibron & Dumeril, 1854 Cuzco Calico Snake	Barbour (1913); Bailey *in* Peters & Orejas-Miranda (1970)	Known only from the province of Cuzco in Peru.
O. fitzingeri (Tschudi, 1845) Fitzinger's False Coral Snake	Schmidt & Walker (1943b); Bailey *in* Peters & Orejas-Miranda (1970); Zaher & Caramaschi (1992)	Coastal areas of Ecuador and Peru.
O. f. fitzingeri (Tschudi, 1845) Fitzinger's False Coral Snake	Schmidt & Walker (1943b); Peters (1960b); Bailey *in* Peters & Orejas-Miranda (1970)	Coast of sw Peru.
O. f. frizzelli Schmidt & Walker, 1943 Pacific Lowland False Coral Snake	Schmidt & Walker (1943b); Peters (1960b); Bailey *in* Peters & Orejas-Miranda (1970), Pérez-Santos & Moreno (1991)	Dry Pacific lowlands of w Ecuador and Peru.
O. formosus* (Wied, 1820) Yellow-headed Calico Snake	Hoge (1952); Peters (1960b); Bailey *in* Peters & Orejas-Miranda (1970); Duellman (1978); Zaher & Caramaschi (1992); Savage & Slowinski (1992) Starace (1998) *Remark: Bailey *in* Peters & Orejas-Miranda (1970) commented that this is a complex of forms and needs further review. Kornacker (1999) continued to list the subspecies *O. t. trigeminus* alluding to *formosus* being a subspecies.	Forests of South America north of 20° latitude. It is thought to be present in at least Colombia, Brazil, and Peru.
O. guibei* Hoge & Romano, 1977 Guibe's False Coral Snake	Hoge & Roamn (1977); Vanzolini (1979); Zaher & Caramaschi (1992); Cei (1993); Yanosky (1998); Freire (1999); Freitas (1999); Marques et al. (2001) *Remark: Previous to Zaher & Caramaschi (1992) *O. guibei*, was often noted as a subspecies of *O.*	Bahia, Minas Geraís, São Paulo, Bahia, Paraná, Alagoas, Mato Grosso and Mato Grosso in Brazil, Paraguay, and the Amazon region of Peru, Bolivia, and into Chaco and Corrientes in Argentina. Zaher & Caramaschi (1992) indicate Bolivia in its range but not Peru.

Species-level Taxa	Citations and Remarks	Abbreviated Range
	trigeminus as *O. trigeminus guibei* (see Vanzolini, 1979 and Lancini & Kornacker, 1989).	
O. leucomelas (Werner, 1916) Werner's Calico Snake	Downs (1961); Kofron (1980); Pérez-Santos & Moreno (1988, 1991); Henle & Ehrl (1991b); Savage & Slowinski (1992)	Amazonian Peru, Ecuador, and Colombia in Río Cauca and Río Magdalena valleys.
O. marcapatae (Boulenger, 1902) Peruvian Highland False Coral Snake	Thomas (1913); Bailey *in* Peters & Orejas-Miranda (1970)	Highlands in Peru in Marcapata and Urubamba valleys.
O. melanogenys (Tschudi, 1845) Tschudi's False Coral Snake	Peters (1960b); Bailey *in* Peters & Orejas-Miranda (1970); Duellman (1978); Dixon & Soini (1986); Savage & Slowinski (1992); Zaher & Caramaschi (1992)	Amazon Basin of Brazil, Bolivia, Ecuador and Peru.
O. m. melanogenys (Tschudi, 1845) Black-headed Calico Snake	Cunha & Nascimento (1983); Starace (1998)	Amazon Basin.
O. m. orientalis Cunha & Nascimento, 1983 Eastern Calico Snake	Cunha & Nascimento (1983); Nascimento et al. (1987)	Brazil in easternmost Pará into adjacent Maranhão.
O. occipitalis (Wagler *in* Spix, 1824) Western Calico Snake	Hoge et al. (1973); Vanzolini (1986); Savage & Slowinski (1992)	Brazil and Suriname.
O. petola* (Linnaeus, 1758) Flame Snake or Calico False Coral Snake	Amaral, 1934 [1933–1934]; Peters (1960b); Bailey *in* Peters & Orejas-Miranda (1970); Amaral (1976) [as *Pseudoboa petola*]; Wilson & Meyer (1985); Savage & Villa (1986); Villa et al. (1988); Savage & Slowinski (1992); Cei (1993); Lee (1996, 2000); Murphy (1997); Campbell (1998a); Starace (1998); Giraudo (1999); Stafford & Meyer (2000); Giraudo (1999); Renjifo & Lundberg (1999); Kornacker (1999); Marques et al. (2001); Köhler (2001) *Remark 1: *O. baileyi* is a synonym (see Villa et al., 1988). *Remark 2: *O. petola* is composed of three subspecies fide Starace (1998), who listed them as *O. p. petola* (Linné, 1758), *O. p. digitalis* (Reuss, 1834), and *O. p. sebae* (Duméril, Bibron and Duméril). He did not include *O. p. semifasciatus* (Tschudi, 1845).	Mexico into n South America as far as nw Ecuador west of the Andes and Amazonian Bolivia, and Brazil east of the Andes to as far south as Misiones in Argentina. It is found in Trinidad and Tobago.
O. p. petola (Linnaeus, 1758) Forest Flame Snake	Roze (1966); Bailey *in* Peters & Orejas-Miranda (1970); Lancini & Kornacker (1989); Murphy (1997); Starace (1998); Boos (2001)	Colombia eastward through Venezuela to the Guianas and Trinidad.

Species-level Taxa	Citations and Remarks	Abbreviated Range
O. p. digitalis* (Reuss, 1834) Banded Calico Snake	Bailey *in* Peters & Orejas-Miranda (1970); Duellman (1978); Fugler & Walls (1978); Dixon & Soini (1986); Lancini & Kornacker (1989); Starace (1998); Kornacker (1999); Freitas (1999); Marques et al. (2001) *Remark: Kornacker (1999) included *O. p. semifasciatus* (Tschudi, 1845) as listed in Roze (1966) and Chippaux ([1986] 1987) from s Venezuela, Peru, Ecuador, and Brazil as a synonym of *O. p. digitalis*.	N and c South America in Amazonian Bolivia, Peru, and Ecuador into forested and coastal c Brazil, Chocó Colombia, and c Panama; may eventually be found in the Guianas.
O. p. sebae* Dumeril, Bibron & Dumeril, 1854 Seba's False Coral Snake	Bailey *in* Peters & Orejas-Miranda (1970); Smith & Smith (1993); Starace (1998); Campbell (1998a); Lee (2000); Köhler (2001) *Remark 1: Fide Bailey *in* Peters & Orejas-Miranda (1970), *Clelia baileyi*, is a synonym of *O. petola sebae*. Smith & Smith (1993) reported that both *O. baileyi* (Smith, 1942), and *O. p. baileyi* (Smith 1942), are synonyms of this snake. *Remark 2: Pérez-Higareda & Smith (1991), and Smith & Smith (1993) listed *O. p. aequifasciatus*, and based on Remark 1, this should be *O. p. sebae*.	Veracruz in s Mexico southward into c Colombia and w Ecuador.
O. rhombifer Dumeril, Bibron & Dumeril, 1854 Argentine False Coral Snake	Hoge (1952a); Bailey *in* Peters & Orejas-Miranda (1970); Amaral (1976); Achaval (1987); Savage & Slowinski (1992) Zaher & Caramaschi (1992); Lema (1994b); Yanosky (1998); Freitas (1999)	Amazon River in Brazil southward into c Argentina, Bolivia, Peru, Uruguay, and Paraguay.
O. r. rhombifer Dumeril, Bibron & Dumeril, 1854 Common Calico Snake	Bailey *in* Peters & Orejas-Miranda (1970); Miranda et al. (1982); Cei (1986); Lema (1994b)	SE Brazil in s Minas Geraís, Paraná and Rio de Janeiro and southward into Argentina in the provinces of Buenos Aires, Córdoba, to the east of Paraná (Brazil) in Entre Rios and Corrientes, and into Uruguay.
O. r. bachmanni (Weyenbergh, 1876) Bachmann's False Coral Snake	Bailey *in* Peters & Orejas-Miranda (1970); Cei (1986, 1993); Leynaud & Bucher (1999)	W, c and n Argentina west of Río Paraná southward into Mendoza and Córdoba including La Pampa, San Juan, Rio Negro, La Rioja, Catamarca, Tucumán; and Santiago del Estero.
O. r. inaequifasciatus Werner, 1909 Chaqueña False Coral Snake	Hoge & Mertens (1955) [as *Pseudoboa ornata*]; Bailey *in* Peters & Orejas-Miranda (1970); Cei (1986, 1993); Rey & Lions (1997); Leynaud & Bucher (1999)	Jujuy, Salta, Formosa, Chaco and Santa Fe in nw Argentina into the Pantanal in sw Mato Grosso, Brazil and into the Chaco in w Bolivia; Paraguay; also reported from Corrientes.
O. r. septentrionalis Vellard, 1943 Amazon False Coral Snake	Vellard (1943); Bailey *in* Peters & Orejas-Miranda (1970)	C planalto excluding the Pantanal in Mato Grosso, Goiás and north to the Amazon River near Santarem in Brazil and into Peru, it is not found in the Pantanal.
O. trigeminus* Dumeril, Bibron & Dumeril, 1854 Brazilian Calico Snake	Bailey *in* Peters & Orejas-Miranda (1970); Amaral (1976); Vanzolini et al. (1980); Hoogmoed ([1982] 1983); Dixon & Soini (1986); Lancini & Kornacker (1989); Savage & Slowinski (1992); Zaher & Cara-	N South America including most of Brazil north of Rio de Janeiro to the Amazon River including Amazonian Peru and westward into Mato Grosso and Marajó Island, its range includes Paraguay and Bolivia.

Species-level Taxa	Citations and Remarks	Abbreviated Range
	maschi (1992); Yanosky (1998); Kornacker (1999); Freitas (1999)	
	*Remark: Kornacker (1999) continued to show the subspecies *O. trigeminus trigeminus*.	
O. venezuelanus Shreve, 1947 Venezuelan Calico Snake	Shreve (1947); Roze (1966); Bailey *in* Peters & Orejas-Miranda (1970); Lancini & Kornacker (1989); Savage & Slowinski (1992); Kornacker (1999); Mijares-Urrutia & Arends R. (2000)	Andes of nc Venezuela; associated with semi-deciduous forest to coastal cloud forest.

GENUS: *Phalotris* Cope, 1862, is restricted to subtropical South America. It is a genus surrounded by much uncertainty, especially its relationship with *Elapomorphus*. Lema (1984) defined *Elapomorphus* into two subgeneric entities: *Elapomorphus* (*sensu stricto*) and *Phalotris* Cope, 1862 as a subgenus. He considered *Elapomorphus* as being tropical forms found from the Guyanas south into eastern Brazil in Minas Geraís and Rio de Janeiro. Lema noted that the *Phalotris* group are from the central "planaltos" and southern Brazil into the llanos of Uruguay and Argentina. Ferrarezzi (1993b) found them from central Brazil and southern Bolivia into Patagonia, and it is reported from Paraguay. Recently *Phalotris* Cope, 1862 is less often treated as a subgenus of *Elapomorphus* and more often as a full genus. Two species remain in *Elapomorphus* as *lepidus* and *quinquelineatus*, the rest are relegated to *Phalotris*. Based on recent literature, this work is treating *Phalotris* as a genus. This is confirmed as valid based on a Lema (pers. comm. July 24, 2000). In an unpublished master's thesis, Ferrarezzi demonstrated the validity of the two subgenera as full genera. Later, Ferrarezzi supported this in publications such as Ferrarezzi (1993), and Puorto & Ferrarezzi (1993). Hoffstadler-Deiques is publishing a dissertation about the skull osteology of several species of elapomorphini (*Elapomorphus*, *Phalotris* and *Apostolepis*) which, I am told, should further support the separation of the genera. See Peters & Orejas-Miranda (1970); Lema (1975); Lema (1984c, 1994b); Amaral (1976); Cei (1986, 1993); Williams & Wallach (1989); Ferrarezzi (1993a, 1993b); Puorto & Ferrarezzi (1993); and Lema & Hofstadler-Deiques (1995). Lema is continuing studies of the group, and will be publishing more on this subject soon.

Review assistance and important contributions for this section were provided by Thales de Lema.

The reader should note that this elevation of *Phaltoris* to full genus from subgenus status is not based on the Ferrarezzi (1993b) unpublished thesis, but rather, on the information from Ferrarezzi (1993a), and Puorto & Ferrarezzi (1993), where parameters for differentiating *Phalotris* were defined from *Elapomorphus*. The reader should be aware that some authorities may consider the status of *Phalotris* to still be in question, but Lema (pers. comm.) expects that this should be resolved soon by Ferrarezzi and Hoffstadler-Deiques in separate works.

See also *Elapomorphus* and *Apostolepis*.

CONTENT: About thirteen species. [If *Phalotris* is considered to be a subgenus of *Elapomorphus*, this contributes the thirteen species to that group.] ★

Phalotris: South American Burrowing Snake. Frank & Ramus (1995) included this group in *Elapomorphus* and called them Diadem Snakes.

Species-level Taxa	Citations and Remarks	Abbreviated Range
*Phalotris bilineatus** (Dumeril, Bibron & Dumeril, 1854) Two-lined Burrowing Snake	Peters & Orejas-Miranda (1970); Lema (1970, 1975 1978f, 1979b, 1984c); Amaral (1976); Achaval (1987); Puorto & Ferrarezzi (1993); Ferrarezzi (1993b); Yanosky (1998); Leynaud & Bucher (1999) *Note: *P. bilineatus* is a hybrid between two taxa and the type is a hybrid (Lema, 1979). As such, the next available name is *P. lemniscatus*. Therefore, all names previously referred to as *P. bilineatus* now are *P. lemniscatus*. Later, Lema (1984) further split the nomenclature of this snake into two separate species, *P. lemniscatus* and *P. spegazzini*. The name is referenced here because it still appears in literature periodically.	S Brazil, Uruguay, Argentina, and Paraguay.
P. concolor Ferrarezzi, 1993 Minas Geraís Burrowing Snake	Ferrarezzi (1993b)	In the ne part of the state of Minas Geraís in Brazil.
P. cuyanus (Cei, 1984) Cei's Burrowing Snake	Cei (1986, 1993); Ferrarezzi (1993b); Leynaud & Bucher (1999)	San Juan and s Mendoza in Argentina.
*P. dimidiatus** (Jan, 1862) Jan's Burrowing Snake	Peters & Orejas-Miranda (1970); Williams & Wallach (1989); Cadle (1982) *Remark: Previously *Elapomojus dimidiatus* fide Peters & Orejas-Miranda (1970).	Paraguay, n. Argentina, and, c and s Brazil.
P. lativittatus Ferrarezzi, 1993 Ferrarezzi's Burrowing Snake	Ferrarezzi (1993b)	Interior of the state of Sao Paulo in Brazil; also reported from the state of Santa Catarina but this needs further confirmation.
*P. lemniscatus** (Dumeril, Bibron & Dumeril, 1854) Dumeril's Burrowing Snake	Lema (1984c, 1985, 1994a); Savage & Slowinski (1992); Cei (1993); Puorto & Ferrarezzi (1993, 1993b) *Remark: See remarks under *P. bilineatus*.	Uruguay, Paraguay, Bolivia, s Brazil in Rio Grande do Sul, and n Argentina in Misiones and Corrientes.
P. l. lemniscatus (Dumeril, Bibron and Dumeril, 1854) Pampa Burrowing Snake	Lema (1984c, 1994b); Cei (1993)	Rio Grande do Sul, Brazil; Uruguay, ne Argentina, s Paraguay, and Bolivia.
P. l. divittatus (Lema, 1984) Lema's Burrowing Snake	Lema (1984c, 1994b)	Brazil in se Rio Grande do Sul into Punta del Este and Montevideo in Uruguay.
P. l. iheringi (Strauch, 1845) Ihering's Burrowing Snake	Lema (1984c, 1994b); Cei (1993)	Brazil in Planalto Meridional from São Paulo to Rio Grande do Sul and into Misiones in Argentina.

The Caenophidia

Species-level Taxa	Citations and Remarks	Abbreviated Range
P. l. trilineatus (Boulenger, 1889) Three-lined Burrowing Snake	Prado (1941b); Lema (1984c, 1994a)	Coastal areas from s Brazil southward into Maldonado, Uruguay.
P. mertensi (Hoge, 1955) Mertens's Burrowing Snake	Peters & Orejas-Miranda (1970); Amaral (1976); Savage & Slowinski (1992); Ferrarezzi (1993b)	S Paraguay and the states of Sao Paulo, Santa Catarina, Minas Gerais, n Paraná, and Mato Grosso in Brazil.
P. multipunctatus Puorto & Ferrarezzi, 1993 Spotted Burrowing Snake	Puorto & Ferrarezzi (1993)	S Brazil in the states of São Paulo and Mato Grosso do Sul below 500 m.
P. nasutus (Gomes, 1915) Gomes's Burrowing Snake	Hoge & Garcia (1949a, 1949b); Peters & Orejas-Miranda (1970); Amaral (1976); Vanzolini (1986); Ferrarezzi (1993b); Lema & Hofstadler-Deques (1995)	States of São Paulo, n Santa Catarina, Minas Geraís, and Mato Grosso do Sul in Brazil as well as Brasilia, D.F.; typical of the cerrados of the Planalto Central.
P. nigrilatus Ferrarezzi, 1993 Paraguayan Burrowing Snake	Ferrarezzi (1993b)	E Paraguay; the type locality is Depto. San Pedro and the range is believed to be between the Paraguay River and Paraná River.
P. punctatus (Lema, 1979) Dotted Burrowing Snake	Lema (1979a); Cei (1986, 1993); Vanzolini (1986); Savage & Slowinski (1992); Ferrarezzi (1993b); Leynaud & Bucher (1999)	NE and nw Argentina and Paraguay.
P. spegazzinii (Boulenger, 1913) Spegazzini's Burrowing Snake	Lema (1978b, 1984c); Cei (1986, 1993)	Paraguay into Argentina excluding Patagonia and the northeast, its range includes Uruguay and s Brazil.
*P. s. spegazzinii** (Boulenger, 1913) Marsh Burrowing Snake	Lema (1978b, 1984c); Cei (1993) *Remark: Fide Lema (1978b), *E. bollei* (Mertens, 1959), is a synonym.	Swampy pampas, mainly Prov. Buenos Aires, Argentina.
P. s. suspectus (Amaral, 1924) Southern Burrowing Snake	Lema (1978b, 1978d, 1984c); Cei (1986, 1993)	Argentina in arid sub-Andes and high Patagonia to as far south as 42°30′ lat.
P. tricolor (Dumeril, Bibron & Dumeril, 1854) Tricolor Burrowing Snake	Peters & Orejas-Miranda (1970); Savage & Slowinski (1992); Cei (1993); Ferrarezzi (1993b); Yanosky (1998); Leynaud & Bucher (1999)	SE South America in the Chaco region of se Bolivia, ne Uruguay, ne Argentina, n Paraguay, and s and w Brazil in São Paulo, Rio Grande do Sul, Paraná, and Mato Grosso.

GENUS: *Philodryas* Wagler, 1830, has a widespread distribution throughout South America. According to Freiberg, in his *Snakes of South America*, the genus *Philodryas*, as used in the checklist, is approximately equivalent to several species. Freiberg and other authorities feel the status of the genera *Alsophis*, *Antillophis*, *Arrhyton*, *Liophis* (including *Dromicus*, *Leimadophis* and *Lygophis*) and *Philodryas* on the South American Mainland is highly uncertain and species often seem to be assigned to genera at random. Some of the genera such as *Dromicus*, *Leimadophis*, and *Lygophis* are placed in the genus *Liophis* (see Dixon, 1980). Therefore, the listings for the above genera are only rough attempts at organization. As noted in Vanzolini (1986), this includes *Alsophis sensu* Maglio (1970), and fide Thomas (1977a), *Dromicus* Bibron, 1843, and South American *Alsophis* are placed in synonymy with *Philodryas* Wagler, 1830. See Dowling & Gibson (1970); Peters & Orejas-Miranda (1970), Thomas (1976) [Ph.D. dissertation]; Vanzolini (1986); Cei (1986, 1993); Obst et al. (1988); Lancini & Kornacker (1989); Thomas & Fernandes (1996); Starace (1998); and Kornacker (1999).

Review assistance and important contributions for this section were provided by Robert A. Thomas.

As a comment, *Philodryas* is one of the few Neotropical colubrids that has been known to cause human fatalities.

Note: *Philodryas oligolepis* Gomes *in* Amaral, 1921 (see Peters & Orejas-Miranda, 1970, and Amaral, 1976) is reported to occur in the state of Minas Geraís, Brazil. According to Starace (1998), it occurs in French Guiana, and possibly Suriname. It is not listed here because it is not a valid species (Thomas, 1999 pers. comm.).

*Remark: GENUS: *Platyinion* Amaral, 1923 is a Latin American snake found in Brazil. Thomas & Fernandes (1996) placed this genus in the synonymy of *Philodryas* (see Peters & Orejas-Miranda, 1970) [as *Platynion*]; and Vanzolini, 1986 [as *Platyinion*].) See Peters & Orejas-Miranda (1970); Amaral (1976); Vanzolini (1986); Cei (1986, 1993); and Kornacker (1999).

CONTENT: A large, complex group with about eighteen species. Eighteen species are shown below. Thomas (1976) (unpublished doctoral dissertation) lists two subspecies that have not been validated through a formal publication. Thomas' (1976) doctoral dissertation is the first and only complete review of this genus since Peters & Orejas-Miranda (1970). The dissertation contains much valuable data and it is unfortunate that the work remains formally unpublished. Bob Thomas tells me these should be published soon. ★

Note: Donnelly and Myers (1991) assigned the name *Philodryas cordata* for a new species they found in the Tepui region of Venezuela. Because they had a assigned a feminine name to as the species, they felt that a footnote was required to explain this action. They noted that historically *Philodryas* has been treated as masculine. The gender of *Philodryas* is determined by its ending. They pointed out that the masculine *Dryas* means the father of Lycurgus, king of Thrace; or one of the Lapithæ; or a party of the Calydonian hunt. They stated that the generic name actually is based on the other *Dryas*, a spirit who is physically and grammatically feminine. They reported that in the original description of *Philodryas* (type species *Coluber Olfersii* Lichtenstein), Wagler (1830:185) and inserted a footnote explaining that the name was derived from Greek. The authors concluded that thus, *Philos* (friend or friendly) + *dryas* (tree nymph) seems to convey the notion of friendly tree snakes-quite inappropriate for somewhat venomous snakes that aggressively bite. Based on the Donnelly & Myers (1991) designation that *Philodryas* is feminie, I have listed the species names in the feminine gender.

Philodryas: South American Racer. Frank and Ramus (1995) call this the Green Racer.

Species-level Taxa	Citations and Remarks	Abbreviated Range
P. aestiva (Dumeril, Bibron & Dumeril, 1854)	Peters & Orejas-Miranda (1970); Amaral (1976); Thomas (1976); Miranda et al. (1982); Lema (1994b); Yanosky (1998); Leynaud & Bucher (1999); Freitas (1999)	S South America in se Brazil, Bolivia in Cochabamba and into the Chaco, Paraguay, Uruguay, and n Argentina.

The Caenophidia

Species-level Taxa	Citations and Remarks	Abbreviated Range
Brazilian Green Tree Snake		
P. a. aestiva (Dumeril, Bibron & Dumeril, 1854) Northern Green Tree Snake	Peters & Orejas-Miranda (1970); Thomas (1976); Lema (1994b); D'Agostini (1998); Marques et al. (2001)	Bolivia into São Paulo in se Brazil, Paraguay, and n Argentina.
*P. a. subcarinata** (Boulenger, 1902) Southern Green Snake or Brazilian Green Racer	Thomas (1976); Cei (1986, 1993); Achaval (1987); Lema (1994b); Leynaud & Bucher (1999) *Remark: *Philodryas a. manegarzoni* Orejas-Miranda, 1959, is a synonym (see Barrio et al. 1977).	Bolivia, Paraguay, Uruguay, the chaqueña and the paranense region of n Argentina in Misiones, Buenos Aires, Córdoba, Corrientes, Entre Ríos, Formosa, Salta, Tucumán, Santiago del Estero, and Santa Fe into Brazil. in Rio Grande do Sul.
P. arnaldoi (Amaral, 1932) Brazilian Speckled or Arnaldo's Green Racer	Lema (1962, 1994b); Peters & Orejas-Miranda (1970); Amaral (1976); Thomas (1976)	SE Brazil in states of São Paulo, Paraná, Rio Grande do Sul, and Santa Catarina in open grassland areas.
P. baroni Berg, 1895 Baron's Racer	Freiberg (1954); Peters & Orejas-Miranda (1970); Thomas (1976); Cei (1986, 1993); Böckeler (1988); Yanosky (1998); Arzamendia (1999); Leynaud & Bucher (1999)	Argentina in Tucumán, Salta, Catamarca, Córdoba, Chaco, Santa Fe, and Santiago del Estero, and recently found in Paraguay and Bolivia.
*P. boliviana** Boulenger, 1896 Bolivian Racer	Thomas (1976), Cei (1993) *Remark: This was a synonym for *P. psammophideus* in Peters & Orejas-Miranda (1970). The species is listed here because Thomas (1976) resurrected the name in his thesis and Cei (1993) does not consider it to be a synonym.	Departamento La Paz in Bolivia where it seems to be closely associated with the puna.
*P. chamissona** (Wiegmann, 1834) Chilean Racer	Donoso-Barros (1974); Thomas (1976, 1977a, 1997); Vanzolini (1986); Cei (1986); Núñez (1992) *Remark: Listed in Donoso-Barros (1966) as *Dromicus chamissonis*, with an excellent description for individuals from Chile.	Chile; Cei (1986) noted that this snake may eventually prove to be found in Argentina because it ranges to the Chile/Argentina border. But since it has not been collected in Argentina, he preferred not to list it as being found there. Thomas (1976) restricted its presence to west of the continental divide and noted that there is a specimen from Mendoza, Argentina (USNM 7327).
P. c. chamissona (Wiegmann, 1834) Chilean Racer	Donoso-Barros (1974); Thomas (1976, other in preparation)	C Chile west of the continental divide between 26°S and 41°S and in Provincia Mendoza, Argentina (fide Thomas, 1976).
P. c. eremicola (Donoso-Barros, 1974) Atacama Racer	Donoso-Barros (1974); Thomas (1976, other in preparation)	Atacama Desert, Chile on the dry slopes between Copiapó and Caldera.
P. cordata Donelly & Myers, 1991 Venezuelan Racer	Donelly & Myers (1991); La Marca (1997); Kornacker (1999)	Venezuela, the type is from the north side of Cerro Guaiquinima (1030 m), Bolívar state.

Species-level Taxa	Citations and Remarks	Abbreviated Range
P. hoodensis (Van Denburgh, 1912) Españolan Galápagos Racer	Mertens (1960); Maglio (1970); Thomas (1997); Swash & Still (2000)	Española and Gardner (near Española) in the Galápagos Archipelago.
*P. livida** Amaral, 1923 Guiacurus Blue Racer	Peters & Orejas-Miranda (1970); Amaral (1976); Vanzolini (1979); Thomas & Fernandes (1996); Valdujo & Nogueira (1999) *Remark: *Platyinion lividum* was a monotypic genus and considered to be a rare snake. It was recently designated as a synonym of *Philodryas* (Thomas & Fernandes, 1996).	States of Paraná, Mato Grosso, São Paulo, and Goiás in Brazil generally associated with Campos Cerrados. Its range is characterized by Hueck & Seiber (1972) as being deciduous forest, alternating with grassland and campos cerrados.
P. mattogrossensis Koslowsky, 1898 Mato Grosso or Two-colored Racer	Peters & Orejas-Miranda (1970); Amaral (1976); Thomas (1976); Cei (1986, 1993); Yanosky (1998); Leynaud & Bucher (1999)	Paraguay, Bolivian chaco, and Mato Grosso in sw Brazil and into Salta, Formosa, and Chaco in Argentina. Thomas (1976) noted that its range involves the chaparral and dry forested chaco of c South America.
P. nattereri Steindachner, 1870 Paraguayan Racer	Hoge (1952a); Peters & Orejas-Miranda (1970); Amaral (1976); Thomas (1976, 1977c); Vanzolini et al. (1980); Hoge et al. ([1978/79] 1981); Böckeler (1988); Yanosky (1998); Freitas (1999)	Paraguay and wc Brazil, it is associated with the dry zones.
P. olfersi (Lichtenstein, 1823) Lichtenstein's Green Racer	Peters & Orejas-Miranda (1970); Thomas & Dixon (1975); Thomas (1976, 1977); Moonen et al. (1979); Vanzolini et al. (1980); Achaval (1987); Lema (1994b); Starace (1998); Yanosky (1998); Kornacker (1999); Freitas (1999); Marques et al. (2001)	Wide ranging in e and s South America including the Guianas.
P. o. olfersi (Lichtenstein, 1823) Southeastern Green Racer	Thomas (1976); Böckeler (1988); Cei (1993); Lema (1994b)	E Peru, Uruguay, Bolivia, Paraguay, and w Brazil.
*P. o. herbea** (Wied, 1825) Green-bellied Palm Snake	Thomas & Dixon (1975); Thomas (1976); Lancini & Kornacker (1989); Kornacker (1999) *Remark: *P. carbonelli* Roze, 1957, is a synonym (see Thomas, 1976).	Venezuela, s Colombia, and Amazonian drainage of Brazil as well as the Guianas.
P. o. latirostris Cope, 1862 Cope's or Southern Green Racer	Thomas (1976); Cei (1986, 1993); Leynaud & Bucher (1999)	Eastern cis-andean part of South America between Peru; non-andean Bolivia; Mato Grosso, Brazil; Paraguay; Uruguay; and nw Argentina.
P. patagoniensis (Girard, 1858) Patagonian Racer	Bianchi (1969); Peters & Orejas-Miranda (1970); Thomas (1976); Miranda et al. (1982); Campbell & Murphy (1984); Cei (1986, 1993); Achaval (1987); Böckeler (1988); Lema (1994b); Thomas & Fernandes (1996); Yanosky (1998); Leynaud & Bucher (1999); Freitas (1999); Marques et al. (2001); Campos-Nogueira (2001)	Mato Grosso Goías, Distrito Federal and Rio Grande do Sul in Brazil; non-andean Bolivia, Uruguay, Paraguay, and almost all of Argentina excluding Santa Cruz, but not south of Chubut. Thomas (1976) describes its range as from about 19°S in Brazil (more northward in the Cochabamba area of Bolivia) to as far south as Provincia Chubut, Argentina. It is found eastward into Chile with a record from Provincia Santiago.

Species-level Taxa	Citations and Remarks	Abbreviated Range
P. psammophidea Günther, 1872 South American Striped Racer	Peters & Orejas-Miranda (1970); Thomas (1976); Cei (1986, 1993); Yanosky (1998); Leynaud & Bucher (1999)	W and s Brazil, e Bolivia, Paraguay, Uruguay, and Argentina.
P. p. psammophidea Günther, 1872 Common Striped Racer	Thomas (1976); Cei (1986, 1993); Leynaud & Bucher (1999)	Paraguayan and Bolivian chaco including Argentina from Río Negro and Neuquén south into Mendoza, San Juan, La Rioja, Catamarca, Tucumán, Salta, Jujuy, Chaco, Formosa, Santiago del Estero, Córdoba, Santa Fe, San Luis, La Pampa and Corrientes.
P. p. lativittata (Cope, 1887) Brazilian Striped Racer	Thomas (1976); Cei (1993)	Mato Grosso in Brazil and the adjacent andean slopes in Bolivia up to Tarija.
P. simonsi* Boulenger, 1900 Simons' Racer	Peters (1960b); Thomas (1976, 1977a, other in press); Vanzolini (1986) *Remark 1: *Dromicus angustilineatus* Schmidt & Walker, 1943, and *Dromicus inca* Schmidt & Walker (1943), are synonyms of this snake (see Peters & Orejas-Miranda, 1970). *Remark 2: *Alsophis angustilineatus* Schmidt & Walker, 1943, and *Alsophis inca* Schmidt & Walker, 1943, are synonyms of this snake (Thomas, 1977a). *Remark 3: *Sensu Incaspis* Donoso-Barros, 1974 is *Philodryas*, and, *Incapis cercostroha* Donoso-Barros, 1974, are *P. simonsii* synonyms, (see Thomas, 1977a).	S Ecuador, Peru and n Chile; Peters & Orejas-Miranda (1970) noted this snake is found to about 10,000 feet. Thomas (1976) noted it is restricted to the western slopes of the Andes. Its habitat is generally very dry scrub forest to virtually barren desert.
P. tachymenoides* (Schmidt & Walker, 1943) Twin-spotted Racer	Donoso-Barros (1966); Peters & Orejas-Miranda (1970); Thomas (1976, 1977a); Vanzolini (1986) *Remark: *Dromicus tachymenoides* Schmidt & Walker (1943) is a synonym (see Donoso-Barros, 1966); and Peters & Orejas-Miranda, 1970).	Restricted to extreme n Chile and s coastal Peru around the frontier from sea level to about 10,000 feet. The habitat is characterized by coastal desert and transitional dry forests.
P. trilineata* (Burmeister, 1861) Argentine Racer	Peters & Orejas-Miranda (1970); Thomas & Johnson (1984); Cei (1985, 1993); Leynaud & Bucher (1999) *Remark: *P. burmeisteri* Jan, 1863 is simply a juvenile of *P. trilinaeatus* (see Thomas & Johnson, 1984); and Cei, 1993).	N Argentina in Patagonia and arid chaqueña forests to semiarid monte treeless dense tangled growth environments to 2000 m in Chubut, Río Negro, Neuquén into Mendoza, San Luis, Salta, La Pampa, La Rioja, Catamarca, San Juan, Tucumán, and probably into the chaco in Bolivia.
P. varia* (Jan, 1863) Jan's or Borelli's Racer	Vanzolini (1986); Thomas & Johnson (1984); Cei (1986, 1993) *Remark: Thomas et al. (1976) elevated *P. borellii*, which is found along the eastern side of the Andes in Argentina and Bolivia, to a full species. Then, Thomas (1977a), Cei (1986), and Thomas & Johnson (1984), placed *P. v. varia* as a synonym.	Puna region in Bolivia in Cochabamba, Tarija, Santa Cruz de la Sierra, and into n Argentina in Catamarca, Tucumán, Salta, and Jujuy.

Species-level Taxa	Citations and Remarks	Abbreviated Range
*P. viridissima** (Linnaeus, 1758) Common Green Racer	Roze (1966); Peters & Orejas-Miranda (1970); Thomas (1976); Amaral (1976); Lancini (1979); Cei (1986, 1993); Dixon & Soini (1986); Lancini & Kornacker (1989); Starace (1998); Yanosky (1998); Kornacker (1999) *Remark: Fide Thomas (1976), there are two subspecies as *P. v. viridissimus* (Linnaeus, 1758) which he called the Northern Green Tree Snake, and *P. v. laticeps* Werner, 1900 which he calls the Bolivian Green Tree Snake. I could not find any further valid references to subspecies in literature. Since Thomas (1976) is a Ph.D. dissertation and has never been formally published, it should remain without subspecies until further work is competed and published.	South America in the Amazon and Paraguay basins in s Venezuela, the Guianas, Amazonas in Brazil, Bolivia, Argentina, and Paraguay; closely associated with the canopy in the forests but is also found in savanna and chaparral, Thomas (1976) presumed this is in gallery forests.

GENUS: *Phimophis* Cope, 1860, [Genus originally listed as *Rhinosimus* by Dumeril, Bibron & Dumeril, 1854 but the name was preoccupied by *Rhinosimus* Latreille, 1802–1803]. This is a wide-ranging Latin American snake. See Roze (1966); Bailey *in* Peters & Orejas-Miranda (1970), Moonen et al. (1979); Cadle (1984); Cei (1986, 1993); Lancini & Kornacker (1989); Starace (1998); Kornacker (1999) and Köhler (2001).

CONTENT: Six species. ★★★★

Phimophis: Miner Snake or Miner. Frank & Ramus (1995) call this the Pampas Snake.

Species-level Taxa	Citations and Remarks	Abbreviated Range
P. chui Rodrigues, 1993 Bahia Miner Snake	Rodrigues (1993)	Bahia in Brazil; associated with the palaeoquaternary sand dunes of the middle São Francisco River.
P. guerini (Dumeril, Bibron and Dumeril, 1854) Common Miner Snake	Hoge (1952a); Bailey *in* Peters & Orejas-Miranda (1970); Cei (1986, 1993); Yanosky (1998); Freitas (1999)	States of Piaui into São Paulo in Brazil southwestward into Argentina in Misiones, Corrientes, Entre Ríos, Santa Fe, Chaco, Tucumán, and Córdoba. Hoogmoed (1982) lists it as being found in the Guianas. Romano-Martínez lists it in Paraguay.
P. guianensis (Troschel, 1848) Guianan Miner Snake	Roze (1966); Bailey *in* Peters & Orejas-Miranda (1970); Amaral (1976); Moonen et al. (1979); Hoogmoed (1983 [1982]); Villa et al. (1988); Lancini & Kornacker (1989); Cei (1993); Starace (1998); Kornacker (1999); Mijares-Urrutia & Arends R. (2000); Köhler (2001)	Panama, n Colombia, Venezuela, and into the Guinas in Suriname and French Guiana.
P. iglesiasi (Gomes, 1915) Gomes's Miner Snake	Bailey (1957); Bailey *in* Peters & Orejas-Miranda (1970); Amaral (1976); Hoge et al. ([1978/79] 1981); Freitas (1999);	Interior of ec Brazil from Pirapora, Minas Geraís into Piauí.
P. scriptorcibatus Rodrigues, 1993 Rodrigues's Miner Snake	Rodrigues (1993)	Bahia in Brazil; associated with the palaeoquaternary sand dunes of the middle São Francisco River.

Species-level Taxa	Citations and Remarks	Abbreviated Range
P. vittatus (Boulenger, 1896) Chaqueña Miner Snake	Bailey *in* Peters & Orejas-Miranda (1970); Cei (1986, 1993); Yanosky (1998); Leynaud & Bucher (1999)	Paraguay into the chaqueña in ne Argentina, and probably into the chaqueña in s Bolivia. In Argentina, it is associated with the western semiarid areas and is found in San Juan, La Rioja, Catamarca, Córdoba, Santiago del Estero, Tucumán, Salta, Chaco, Formosa, Santa Fe, Entre Ríos, and Corrientes.

GENUS: *Phyllorhynchus* Stejneger, 1890, is confined to North America from southern California and Nevada southward into the northern tip of Baja California, Mazatlan, Sinaloa, and various islands in the Gulf of California. See Savage & Cliff (1954); Leviton & Banta (1964); Smith & Taylor (1966); Tanner (1969); Stebbins (1972, 1985); Powers & Banta (1974); Campbell & Christman (1982); Obst et al. (1988); McDiarmid & McCleary (1993); McCleary & McDiarmid (1993).

Note: McDiarmid & McCleary (1993) and McCleary & McDiarmid (1993) listed subspecies but were not convinced that they should be recognized. Crother et al. (2000) did not list subspecies. Subspecies are listed here because they were included in the McDiarmid and McCleary works even though validity was questioned.

CONTENT: Two species. ★★

Phyllorhynchus: Leaf-nosed Snake

Species-level Taxa	Citations and Remarks	Abbreviated Range
P. browni* Stejneger, 1890 Saddled Leaf-nosed Snake	Hardy & McDiarmid (1969); Stebbins (1985); McDiarmid & McCleary (1993) *Remark: McDiarmid & McCleary (1993) questioned the validity of subspecies. They doubted the evolutionary verity and taxonomic distinctiveness, and questioned the utility of recognizing certain of them. According to the authors, most of these subspecies are based on perceived differences in pattern, coloration, ventral/subcaudal counts, and the degree of dorsal scale keeling. The reader should reference McDiarmid & McCleary (1993) for details.	SW United States southward to n Mexico and all of Baja California, Mexico; Degenhardt et al. (1996) did not think this snake is present in New Mexico.
P. b. browni Stejneger, 1890 Pima Leaf-nosed Snake	Smith & Taylor (1966); Stebbins (1985); McDiarmid & McCleary (1993); Bartlett & Tennant (2000)	The United States in s Arizona south into n Sonora in Mexico.
P. b. fortitus Bogert & Oliver, 1945 Sonoran Leaf-nosed Snake	Bogert & Oliver (1945); McDiarmid & McCleary (1993)	Sonora, Mexico.
P. b. klauberi Shannon & Humphry, 1959 Klauber's Leaf-nosed Snake	Shannon & Humpfrey (1959); Hardy & McDiarmid (1969)	Sinaloa, Mexico.

Species-level Taxa	Citations and Remarks	Abbreviated Range
P. b. lucidus Klauber, 1940 Maricopa Leaf-nosed Snake	Stebbins (1985); McDiarmid & McCleary (1993); Wright & Wright ([1957] 1994); Bartlett & Tennant (2000)	SW United States in Maricopa County, Arizona south into adjacent Mexico.
P. decurtatus* (Cope, 1868) Spotted Leaf-nosed Snake	Smith & Langebartel (1951); Hardy & McDiarmid (1969); Tanner (1969); Powers & Banta (1974); Campbell & Christman (1982); Stebbins (1985); McDiarmid & McCleary (1993); Brown (1997); Grossman (1999); McPeak (2000) *Remark: McDiarmid & McCleary (1993) questioned the validity of subspecies. They doubted the evolutionary verity and taxonomic distinctiveness, and questioned the utility of recognizing certain of them. According to the authors, most of these subspecies are based on perceived differences in pattern, coloration, ventral/subcaudal counts, and the degree of dorsal scale keeling.	SW United States in the Colorado desert region southward into n Baja California Norte, Sonora, and Sinaloa in nw Mexico.
P. d. decurtatus (Cope, 1868) Baja California Leaf-nosed Snake	Smith & Langebartel (1951); Smith & Taylor (1966); Hardy & McDiarmid (1969); Tanner (1969); Powers & Banta (1974); Campbell & Christman (1982); Stebbins (1985); McDiarmid & McCleary (1993)	S Baja California Norte in w Mexico to near the southern tip of Baja California Sur.
P. d. arenicola* Savage & Cliff, 1954 Monserrate Leaf-nosed Snake	Savage & Cliff (1954); Smith & Smith (1976); Murphy & Ottley (1980); McDiarmid & McCleary (1993); Wright & Wright ([1957] 1994) *Remark: Grismer (1999), in an examination of material from Isla Monserrate and the adjacent peninsula, indicated that *P. d. arenicola* is not morphologically distinct form *P. d. decurtatus*. He placed it as a synonym of *P. d. decuratus*. However, his detail seemed to be rather sparse so I have retained the separation until further information is available.	Restricted to the Isla Monserrate in the Gulf of California in Mexico.
P. d. norrisi Smith & Langebartel, 1951 Norris' Leaf-nosed Snake	Smith & Langebartel (1951); McDiarmid & McCleary (1993); McCleary & McDiarmid (1993)	NW Mexico along the Gulf of California in Sonora and Sinaloa.
P. d. nubilis Klauber, 1940 Clouded Leaf-nosed Snake	Bogert & Oliver (1945); Stebbins (1985); McCleary & McDiarmid (1993); McDiarmid & McCleary (1993); Wright & Wright ([1957] 1994); Bartlett & Tennant (2000)	SW United States from sc Arizona southward into Sonora, Mexico.
P. d. perkinsi Klauber, 1935 Western Leaf-nosed Snake	Brattstrom (1953); Smith & Taylor (1966); Stebbins (1985); McDiarmid & McCleary (1993); McCleary & McDiarmid (1993); Brown (1997); Bartlett & Tennant (2000)	SW United States from sc California, s Nevada and c Arizona southward into n Mexico in ne Baja California Norte and nw Sonora; it is also reported from Isla Angel de la Guarda in the Gulf of California.

GENUS: *Pituophis* Holbrook, 1842, is a wide ranging genus present from southern Canada southward through the United States, Mexico and into northern Central America as far south as central Guatemala. The taxonomy is inadequately defined and classification positions are constantly being debated and will probably remain unclear for some time. See Stull (1940); Klauber (1946a, 1946d, 1947); Smith & Taylor (1966); Stebbins (1966, 1985); Peters & Orejas-Miranda (1970); Conant (1975); Obst et al. (1988); Sweet & Parker (1990); Mara (1994); Wright & Wright [1957] (1994); Degenhardt et al. (1996), Tennant (1998); Rodríguez-Robles & Jesús-Escobar (2000); Dixon (2000a); and Grismer (2001).

Review assistance and important contributions for this section were provided by Javier A. Rodriguez-Robles and James R. Dixon.

Stull (1940) was the first reviewer to organize the species and subspecies, followed by Klauber (1946d), Sweet & Parker (1990), and Reichling (1995). Others such as Stebbins (1985), list most, if not all, of the *Pituophis* as *P. melanoleucus* subspecies. Sweet & Parker (1990) listed 15 subspecies of *P. melanoleucus*, however, two to three species are recognized in recent publications. Those listed here from that group are *P. catenifer*, *P. melanoleucus*, and *P. vertebralis*, except for *P. ruthveni* which was separated out by Reichling (1995). Rodríguez-Robles & Jesús-Escobar (2000) took the taxonomy a step further with a review of molecular systematics. Using Rodríguez-Robles & Jesús-Escobar (2000) as the basis for my classification, I listed the western subspecies in the United States as *P. catenifer*. The elevation of *P. ruthveni* to full species status gives support to the separation of *P. melanoleucus* as the eastern U.S. subspecies, and *P. cantenifer* as the western U.S. subspecies (see Dixon, 2000a). Mexican species being further separated with subspecies under *P. deppei*. *Pituophis lineaticollis* represents the group found from the Mexican Plateau into Guatemala. The *P. vertebralis* represents a central and lower Baja California group. Grismer (2001) does not agree with the classifications that Rodríguez-Robles & Jesús-Escobar (2000) used. I prefer the Rodríguez-Robles & Jesús-Escobar classification but I also recommend that the reader refer to Grismer (2001) for his well thought out views for the classifications of this genus in the Baja California, Mexico region and attendant offshore islands.

CONTENT: About three to seven species. Six species are listed here. Rodríguez-Robles & Jesús-Escobar (2000), based on molecular systematics, concluded that their phylogenetic analyses indicated that three distinct species in the *melanoleucus* complex, *P. melanoleucus* (sensu stricto), *P. catenifer*, and *P. ruthveni* should be recognized. They did not agree with *vertebralis* as presently defined and noted that *bimaris* and *vertebralis* are considered by some taxonomists as a different species but included them in the *catenifer* clade. All of these were compared to *deppei* and *lineaticollis*. I recommend that interested readers obtain a copy of the Rodríguez-Robles & Jesús-Escobar (2000) paper for details of their conclusions on *Pituophis*. ★★

Pituophis: Bullsnake, Gophersnake, and Pinesnake

Species-level Taxa	Citations and Remarks	Abbreviated Range
P. catenifer (Blainville, 1835) Gophersnake	Smith & Taylor (1966); Wright & Wright ([1957] 1994); Brown (1997); Rodríguez-Robles & Jesús-Escobar (2000); Grismer (2001)	W Canada southward through the w United States into Mexico.
P. c. catenifer (Blainville, 1835)	Klauber (1947); Smith & Kennedy (1951); Stebbins (1985); Sweet & Parker (1990); Wright & Wright	W coast United States in Oregon and California from the Cascade Mountains southward into Califor-

Species-level Taxa	Citations and Remarks	Abbreviated Range
Pacific Gophersnake	([1957] 1994); Brown (1997); Dixon (2000a); Bartlett & Tennant (2000); Rodríguez-Robles & Jesús-Escobar (2000); Grismer (2001)	nia, west of the Sierras into n Santa Barbara county and the Tehachapi Mountains.
P. c. affinis Hallowell, 1852 Sonoran Gophersnake	Klauber (1946b, 1947); Smith & Kennedy (1951); Smith & Taylor (1966); Conant (1975); Stebbins (1985); Sweet & Paker (1990); Wright & Wright ([1957] 1994); Degenhardt et al. (1996); Brown (1997); Tennant (1998); Conant & Collins (1998); Werler & Dixon (2000); Bartlett & Tennant (2000); Tennant & Bartlett (2000); Dixon (2000a); Rodríguez-Robles & Jesús-Escobar (2000)	The United States in se California into w Texas and southward into Mexico in Zacatecas and s Sinaloa.
P. c. annectens Baird & Girard, 1853 San Diego Gophersnake	Klauber (1946b, 1947); Smith & Kennedy (1951); Smith & Taylor (1966); Stebbins (1985); Sweet & Paker (1990); Smith & Smith (1993); Wright & Wright ([1957] 1994); Brown (1997); Bartlett & Tennant (2000); Rodríguez-Robles & Jesús-Escobar (2000); Grismer (2001)	The United States in coastal California southward into nw Baja California, Mexico.
P. c. coronalis Klauber, 1946 Coronado Island Gophersnake	Klauber (1946b, 1947); Smith & Kennedy (1951); Sweet & Paker (1990); Smith & Smith (1993); Rodríguez-Robles & Jesús-Escobar (2000); Grismer (2001)	Coronado Island in the Gulf of California, w Mexico.
P. c. deserticolus Stejneger, 1893 Great Basin Gophersnake	Klauber (1947); Smith & Kennedy (1951); Smith & Taylor (1966); Stebbins (1985); Sweet & Parker (1990); Smith & Smith (1993); Wright & Wright ([1957] 1994); Degenhardt et al. (1996); Bartlett & Tennant (2000); Rodríguez-Robles & Jesús-Escobar (2000); Grismer (2001)	SC British Colombia, Canada southward into the w United States and ne Baja California, Mexico.
P. c. fulginatus Klauber, 1946 San Martin Gophersnake	Klauber (1946b, 1947); Smith & Kennedy (1951); Sweet & Paker (1990); Smith & Smith (1993); McPeak (2000); Rodríguez-Robles & Jesús-Escobar (2000); Grismer (2001)	San Martin Island in the Gulf of California, w Mexico.
P. c. insulanus Klauber, 1946 Cedros Island Gopher snake	Klauber (1946a, 1946b, 1947); Smith & Kennedy (1951); Slevin & Leviton (1956); Sweet & Paker (1990); Smith & Smith (1993); Rodríguez-Robles & Jesús-Escobar (2000); Grismer (2001)	Cedros Island in the Gulf of California, w Mexico.
P. c. pumilus Klauber, 1946 Santa Cruz Gophersnake	Klauber (1947); Schmidt (1953); Stebbins (1966, 1985); Sweet & Paker (1990); Smith & Smith (1993); Wright & Wright ([1957] 1994); Brown (1997); Bartlett & Tennant (2000); Rodríguez-Robles & Jesús-Escobar (2000); Grismer (2001)	The United States only on Santa Cruz and Santa Rosa Islands off the mainland in the Santa Barbara group of islands in Santa Barbara County, California.
P. c. sayi (Schlegel, 1837) Bullsnake	Stull (1940); Klauber (1947); Smith & Kennedy (1951); Stebbins (1966, 1985); Conant (1975); Sweet & Paker (1990); Pérez-Higareda & Smith (1991); Wright & Wright ([1957] 1994); Degenhardt et al. (1996); Tennant (1998); Conant & Collins (1998); Dixon (2000a); Werler & Dixon (2000); Bartlett	Much of the grassy prairie of w Canada through w USA from Texas into ne Mexico.

Species-level Taxa	Citations and Remarks	Abbreviated Range
	& Tennant (2000); Tennant & Bartlett (2000); Rodríguez-Robles & Jesús-Escobar (2000); Grismer (2001)	
P. deppei (Dumeril & Bibron, 1854) Mexican Bullsnake	Stull (1940); Duellman (1960b); Smith & Taylor (1966); Smith & Smith (1993); Rodríguez-Robles & Jesús-Escobar (2000)	S Chihuahua in n Mexico southward through the country.
P. d. deppei (Dumeril & Bibron, 1854) Western Mexican Bullsnake	Stull (1940); Smith & Taylor (1966); Pérez-Higareda & Smith (1991) Smith & Smith (1993); Uribe-Peña (1999); Rodríguez-Robles & Jesús-Escobar (2000)	NC Nuevo León westward into s Chihuahua and Jalisco then through Coahuila, San Luis Potosí, and c Puebla.
P. d. jani (Cope, 1860) Eastern Mexican Bullsnake	Stull (1940); Smith & Taylor (1966); Smith & Smith (1993); Rodríguez-Robles & Jesús-Escobar (2000)	N Hidalgo into se Coahuila in the arid slopes of the Sierra Madre Oriental.
P. lineaticollis (Cope, 1861) Middle American Gophersnake	Stull (1940); Peters & Orejas-Miranda (1970); Villa et al. (1988); Smith & Smith (1993); Rodríguez-Robles & Jesús-Escobar (2000); Köhler (2001)	Mexico from the edge of the Mexican Plateau in c Veracruz south into the highlands of Guatemala.
P. l. lineaticollis (Cope, 1861) Cope's Gophersnake	Duellman (1960); Peters & Orejas-Miranda (1970); Smith & Smith (1993); Rodríguez-Robles & Jesús-Escobar (2000)	Oaxaca in Mexico southward into Guatemala.
P. l. gibsoni Stuart, 1954 Gibson's Gophersnake	Stuart (1954); Duellman (1960); Peters & Orejas-Miranda (1970); Alvarez del Toro (1982); Smith & Smith (1993); Rodríguez-Robles & Jesús-Escobar (2000)	Moderate and intermediate elevations on the Pacific versant of w Guatemala and the Caribbean coast of Sierra de los Cuchumatanes, Guatemala.
P. melanoleucus (Daudin, 1803) Pinesnake	Conant (1975); Dundee & Rossman (1989); Sweet & Parker (1990); Smith & Smith (1993); Wright & Wright ([1957] 1994); Rodríguez-Robles & Jesús-Escobar (2000)	E United States.
P. m. melanoleucus (Daudin, 1803) Northern Pinesnake	Stull (1940); Conant (1975); Sweet & Parker (1990); Wright & Wright ([1957] 1994); Conant & Collins (1998); Tennant & Bartlett (2000); Rodríguez-Robles & Jesús-Escobar (2000)	E United States from New York and New Jersey into Georgia, Tennessee, s Kentucky, and the Carolinas.
P. m. lodingi Blanchard, 1924 Black Pinesnake	Conant (1975); Dundee & Rossman (1989); Sweet & Paker (1990); Wright & Wright ([1957] 1994); Tennant (1997b); Conant & Collins (1998); Tennant & Bartlett (2000); Rodríguez-Robles & Jesús-Escobar (2000)	The United States in extreme e Louisiana into sw Alabama.
P. m. mugitus Barbour, 1921 Florida Pinesnake	Stull (1940); Conant (1975); Ashton & Ashton (1981); Sweet & Paker (1990); Wright & Wright ([1957] 1994); Tennant (1997b); Conant & Collins (1998); Tennant & Bartlett (2000); Rodríguez-Robles & Jesús-Escobar (2000)	SE United States from s South Carolina into Georgia and s Florida.

Species-level Taxa	Citations and Remarks	Abbreviated Range
P. ruthveni Stull, 1929 Louisiana Pinesnake	Conant (1975); Thomas et al. (1976); Dundee & Rossman (1989); Sweet & Paker (1990); Wright & Wright ([1957] 1994); Reichling (1995); Tennant (1998); Conant & Collins (1998) [as *P. melanoleucus ruthveni*]; Werler & Dixon (2000); Tennant & Bartlett (2000); Dixon (2000a); Rodríguez-Robles & Jesús-Escobar (2000); Grismer (2001)	The United States in long-leaf pine forests of e Texas and wc Louisiana.
P. vertebralis* (Blainville, 1835) Cape Gophersnake	Smith & Taylor (1966); Murphy (1975, 1983b, c) Smith & Smith (1993); Grismer (1994, 1997); McPeak (2000); Rodríguez-Robles & Jesús-Escobar (2000); Grismer (2001) *Remark: According to Rodríguez-Robles & Jesús-Escobar (2000), the two subspecies in this group were included in the clade from central and western United States and northern Mexico; i.e., the *P. catenifer* clade, implying it may include *bimaris* and *vertebralis*. Their findings led them to reject the recognition of *Pituophis vertebralis* as defined by Grismer (1994, 1997). The vicariance model developed by Murphy (1975, 1983b, c) which predicted the endemic Baja California *Pituophis* relationship to *P. deppei* and *P. lineaticollis* was not supported by the Rodríguez-Robles & Jesús-Escobar (2000) results.	C and s Baja California, Mexico.
P. v. vertebralis* (Blainville, 1835) Central Baja California Gophersnake	Klauber (1946b, 1947); Smith & Kennedy (1951); Stebbins (1985); Sweet & Paker (1990); Smith & Smith (1993); Mellnick (1995); Rodríguez-Robles & Jesús-Escobar (2000); Grismer (2001) *Remark: Classified as *P. catenifer vertebralis* in Rodríguez-Robles & & Jesús-Escobar (2000).	Extreme s Baja California, Mexico; found from the outskirts of the town of La Paz southward to the tip of the peninsula in the Cape region.
P. v. bimaris* (Klauber, 1946) Baja California Gophersnake	Klauber (1946b, 1947); Smith & Kennedy (1951); Sweet & Paker (1990) Smith & Smith (1993); Mellnick (1995); Rodríguez-Robles & Jesús-Escobar (2000); Grismer (2001) Remark: Classified as *P. catenifer bimaris* in Rodríguez-Robles & & Jesús-Escobar (2000).	C and s Baja California, Mexico.

GENUS: *Pseudablabes* Boulenger, 1896, is present in southern South America. See Peters & Orejas-Miranda (1970); Cei (1993); Lema (1994b).

Review assistance and important contributions for this section were provided by Thales de Lema and James R. Dixon.

CONTENT: One species. ★★★★

Pseudablabes: False Harmless Snake or South American Scorpion Snake. Frank & Ramus (1995) called this the Burrowing Night Snake.

Species-level Taxa	Citations and Remarks	Abbreviated Range
P. agassizii (Jan, 1863) False Harmless Snake or South American Scorpion Snake	Peters & Orejas-Miranda (1970); Amaral (1976); Achaval (1987); Cei (1993); Lema (1994b); Kiefer (1998); Giraudo (1999); Leynaud & Bucher (1999); Campos-Nogueira (2001)	Uruguay, s and sw Brazil in Rio Grande do Sul, Goías and Minas Geraís; and, ne Argentina in Buenos Aires into Ventanía, La Pampa, Santa Fe, Entre Ríos, Chaco, Santiago del Estero, and Misiones.

GENUS: *Pseudoboa* Schneider, 1801, is a genus of Latin American false coral snakes that is associated mainly with the Amazon Basin, although they range from southern Central America into Argentina and are also found on some of the Caribbean islands. See Roze (1966); Bailey *in* Peters & Orejas-Miranda (1970); Lema & Matschula-Ely (1979); Vanzolini (1986); Lancini & Kornacker (1989); Schwartz & Henderson (1991); Cei (1993); Smith & Smith (1993); Starace (1998); Kornacker (1999) and Köhler (2001).

Review assistance and important contributions for this section were provided by Hussam Zaher.

See also *Boiruna*, *Clelia* and *Oxyrhopus*.

CONTENT: Five species. ★★★★

Pseudoboa: South American Scarletsnake and South American False Boa. Frank and Ramus (1995) call this the False Boa.

Species-level Taxa	Citations and Remarks	Abbreviated Range
P. coronata Schneider, 1801 Amazon Scarletsnake	Roze (1966); Bailey *in* Peters & Orejas-Miranda (1970); Duellman (1978); Fugler & Walls (1978); Moonen et al. (1979); Dixon & Soini (1986); Lancini & Kornacker (1989); Savage & Slowinski (1992); Starace (1998); Kornacker (1999)	Amazon Basin in Brazil, Colombia, Ecuador, Peru, Venezuela, Bolivia, and the Guianas. It is not present on Trinidad according to Murphy (1997).
P. haasi (Boettger, 1905) Haas's False Mussurana	Bailey *in* Peters & Orejas-Miranda (1970); Amaral (1976); Vanzolini (1986); Lema & Matschulat-Ely (1979); Savage & Slowinski (1992); Lema (1994b); Giraudo (1999); Marques et al. (2001)	Brazilian states of Paraná, Rio de Janeiro, n Santa Catarina, and Rio Grande do Sul in Brazil, into Misiones in Argentina.
P. neuwiedi (Dumeril, Bibron & Dumeril, 1854) Neuwied's False Boa	Hoge & Lancini (1960); Underwood (1962); Roze (1966); Bailey *in* Peters & Orejas-Miranda (1970); Amaral (1976); Emsley (1977); Lancini (1979); Lancini & Kornacker (1989); Schwartz & Henderson (1985, 1991); Savage & Slowinski (1992); Murphy (1997); Starace (1998); Censky & Kaiser *in* Crother (1999); Esqueda & La Marca (1999); Renjifo & Lundberg (1999); Kornacker (1999); Mijares-Urrutia & Arends R. (2000); Boos (2001); Köhler (2001)	Panama into n South America in Colombia eastward into Venezuela, Suriname, and southward into Brazil to along the Amazon River. Its range includes Trinidad and Tobago, Grenada and the Guianas; Schwartz & Henderson (1991) noted that it may be extirpated from Grenada.

Species-level Taxa	Citations and Remarks	Abbreviated Range
P. nigra (Dumeril, Bibron & Dumeril, 1854) Cerrados Black Snake	Bailey *in* Peters & Orejas-Miranda (1970); Vanzolini et al. (1980); Villa et al. (1988); Cei (1993); Yanosky (1998); Freitas (1999)	Central cerrados into Paraná in s Brazil, Paraguay, c Bolivia, and Misiones in Corrientes in Argentina.
P. serrana Morato, Moura-Leite, Prudente & Bérnils, 1995 Serrana False Boa	Morato et al. (1995)	SE Brazil

GENUS: *Pseudoeryx* Fitzinger, 1826, is a wide ranging South American snake. See Peters & Orejas-Miranda (1970); Lancini & Kornacker (1989); Moonen et al. (1979); Vanzolini (1986); Cei (1993); Starace (1998); and Kornacker (1999).

CONTENT: One species with two subspecies. ★★★★

Pseudoeryx: Eel or Pond Snake. Frank and Ramus (1995) call this the Pond Snake.

Species-level Taxa	Citations and Remarks	Abbreviated Range
P. plicatilis (Linnaeus, 1758) Eel Snake or South American Pond Snake	Peters & Orejas-Miranda (1970); Amaral (1976); Moonen et al. (1979); Dixon & Soini (1986); Starace (1998); Yanosky (1998); Giraudo (1999); Kornacker (1999)	N and c South America in Colombia, Venezuela, and the Guianas south into Bolivia, Paraguay, and n Argentina; lowlands.
P. p. plicatilis (Linnaeus, 1758) Common Eel Snake	Peters & Orejas-Miranda (1970); Lancini & Kornacker (1989); Cei (1993); Starace (1998); Kornacker (1999)	Poorly defined distribution in South America, including Mato Grosso in Brazil into Corrientes, Chaco, and Formosa in Argentina, as well as the Guianas. Kornacker (1999) defines its range as being Argentina, Bolivia, Brazil, Colombia, Ecuador, French Guiana, Guyana, Paraguay, Peru, Suriname, and Venezuela.
P. p. mimeticus* (Cope, 1885) Bolivian Eel Snake	Peters & Orejas-Miranda (1970); Cunha & Nascimento (1978); Dixon & Soini (1986); Henle & Ehrl (1991b); Cei (1993); Smith & Chiszar (1996) *Remark: Dixon & Soini (1986) discussed integradation. In the same paper, they placed *P. p. ecuadorensis* Mertens, 1965, in synonymy with this snake.	Amazonian Bolivia.

GENUS: *Pseudoficimia* Bocourt, 1883, is a Mexican Snake. See Campbell & Simmons (1962); Smith & Taylor (1966); Hardy (1972, 1973, 1975a); and Smith & Smith (1993).

Important contributions for this section were contributed by Peter Holm.

See also genera *Conopsis*, *Ficimia*, *Gyalopion*, and *Toluca*.

CONTENT: One species. ★★★★

Pseudoficimia: False Ficimia

Species-level Taxa	Citations and Remarks	Abbreviated Range
P. frontalis* (Cope, 1864) False Ficimia	Peters (1954); Campbell & Simmons (1962); Hardy (1972, 1973, 1975a); Smith & Smith (1993); Ramírez-Bautista (1994); García & Ceballos (1994) *Remark: Two subspecies have been described and shown as *P. f. frontails* (Cope, 1864), from w Mexico; and, *P. f. hiltoni* (Bogert & Oliver, 1945) from c Sonora into n Sinaloa, Mexico. But, at the moment, are not recognized as synonyms (see Smith & Smith, 1993).	W Mexico; definitely from se Sonora through the Balsas Basin into c Guerrero, Michoacán; Puebla, Morelos, and s Sinaloa; possibly also found in sw Durango and se Chihuahua.

GENUS: *Pseudoleptodeira* Taylor, 1938 [1939], is a Mexican genus. See Duellman (1958); Smith & Taylor (1966); Dowling & Jenner (1987); Smith & Smith (1993) and Ramírez-Bautista (1994).

Also see genera *Rhadinophanes*, *Eridiphas*, *Hypsiglena*, *Leptodeira* and *Tantalophis*.

CONTENT: Two species. ★★★★

Pseudoleptodeira: False Cat-eyed Snake

Species-level Taxa	Citations and Remarks	Abbreviated Range
P. latifasciata (Günther, 1894) False Cat-eyed Snake	Tanner (1944); Duellman (1958); Smith & Taylor (1966); Dowling & Jenner (1987); Ramírez-Bautista & Smith (1992); Smith & Smith (1993)	Mexico along the Pacific coast from Jalisco into the Río Balsas Basin; lowlands from 100 to 1300 m. Duellman (1958) lists it from the states of Colima, Guerrero, Michoacán, Morelos, and Puebla.
P. uribei Ramírez-Bautista & Smith, 1992 Uribe's False Cat-eyed Snake	Ramírez-Bautista & Smith (1992); Ramírez-Bautista (1994); García & Ceballos (1994)	Mexico coast in low forests in Jalisco into Guerrero.

GENUS: *Pseudotomodon* Koslowsky, 1896, is a South American genus. See Bailey *in* Peters & Orejas-Miranda (1970); Cei (1986, 1993); and Obst et al. (1988). This genus is currently the focus of attention by various researchers. I expect that the genus could experience taxonomic changes in the near future.

Review assistance and important contributions for this section were provided by Robert A. Thomas.

CONTENT: One species. ★★

Pseudotomodon: False Yarará or Argentine False Pit Viper. Frank and Ramus (1995) call this the False Tomodon Snake.

Species-level Taxa	Citations and Remarks	Abbreviated Range
*P. trigonatus** (Leybold, 1873) False Yarará or False Tomodon Snake	Bailey *in* Peters & Orejas-Miranda (1970); Miranda et al. (1982); Cei (1986, 1993); Obst et al. (1988); Scrocchi (1997); Thomas (in press); Cruz et al. (1999)	W Argentina in Chubut, Río Negro, Neuquén, La Pampa, Mendoza, San Luis, San Juan, La Rioja, Córdoba, Catamarca, Santiago del Estero, and southward into Chubut.
	*Remark 1: This was considered to be a subspecies of *Tomodon ocellatus* Dumeril, Bibron & Dumeril, but, Bailey *in* Peters & Orejas-Miranda (1970) felt it was distinct enough to remain as a separate species.	
	*Remark 2: Thomas (pers. comm.) feels that this genus should be placed in a separate genus and is working on a paper for publication. This statement was made in his presentation to the Texas Herpetological Society in the annual fall meeting at Texas A & M University on October 23, 1999.	
	*Remark 3: In a personal communication on December 2, 1999, Lema (1999) suggests that *Pseudotomdon* Koslowsky, 1896 should be placed in another genus and is working on a paper for publication.	

GENUS: *Pseustes* Fitzinger, 1843, [Originally *Thamnobius* Fitzinger, 1843 but preoccupied by *Thamnobius* Schenher, 1836)], is found from San Luis Potosí in Mexico southward through Central America into northern and central South America. See Roze (1966); Smith & Taylor (1966); Peters & Orejas-Miranda (1970); Hoogmoed & Gruber (1983); Savage & Villa (1986); Obst et al. (1988); Lancini & Kornacker (1989); Cei (1993); Smith & Smith (1993); Starace (1998); and Kornacker (1999).

Review assistance and important contributions for this section were provided by James R. Dixon.

CONTENT: About four species. Four species are shown here. ★★

Pseustes: Neotropical Bird Snake or Puffing Snake

Species-level Taxa	Citations and Remarks	Abbreviated Range
*P. poecilonotus** (Günther, 1858) Bird-eating Tree Snake or Puffing Snake	Peters & Orejas-Miranda (1970); Wilson & Meyer (1985); Villa et al. (1988); Smith & Smith (1993); Lee (1996); Campbell (1998a); Starace (1998); Kornacker (1999); Stafford & Meyer (2000); Köhler (2001)	Wide ranging from s Mexico the Isthmus of Tehuantepec and the Yucatán peninsula southward through Central America into South America as far as the Amazon drainage of Brazil, Ecuador, Peru, and Bolivia. Its range includes Trinidad.
	*Remark: Subspecies are problematic.	
P. p. poecilonotus (Günther, 1858) Central American Bird Snake	Kofron (1980); Smith & Smith (1993); Campbell (1998a); Lee (2000)	Yucatán in Mexico, Belize, Guatemala, and Honduras.
P. p. argus (Bocourt, 1888 *in* Dumeril, Mocquard & Bocourt, 1870–1909) Arboreal Bird Snake	Smith & Taylor (1966); Peters & Orejas-Miranda; (1970); Alvarez del Toro (1982); Pérez-Higareda & Smith (1991); Smith & Smith (1993); Campbell (1998a)	San Luis Potosí, Mexico southward into Guatemala and Honduras.

Species-level Taxa	Citations and Remarks	Abbreviated Range
P. p. chrysobronchus (Cope, [1875] 1876) Costa Rican Bird Snake	Amaral (1929 [1930c]); Brongersma (1937); Peters & Orejas-Miranda (1970)	Costa Rica and Nicaragua.
P. p. polylepis (W. Peters, 1867) Common Bird-eating Snake	Roze (1966); Amaral (1976); Dixon & Soini (1986); Lancini & Kornacker (1989); Murphy (1997); Starace (1998); Renjifo & Lundberg (1999); Kornacker (1999); Boos (2001)	Amazonian South America southward into Bolivia, Brazil, Ecuador, Peru, Venezuela, the Guianas, and Trinidad and Tobago; associated with woodlands.
P. sexcarinatus* (Wagler *in* Spix, 1824) Amazon Bird-eating Snake	Hoge (1966); Peters & Orejas-Miranda (1970); Amaral (1976) [as *Chironius sexcarinatus*]; Cei (1993); Yanosky (1998) *Remark: Hoge (1964) described this snake in a manner that keyed very similar to *P. poecilonotus* (see Peters & Orejas-Miranda, 1970).	Amazonia in Pará, Brazil into the Province of Misiones, Argentina.
P. shropshirei (Barbour & Amaral, 1924) Shropshire's Puffing Snake	Nicéforo María (1942); Peters (1960b); Savage (1980); Vanzolini (1986); Fuenmayor & Molina (1998); Renjifo & Lundberg (1999); Kornacker (1999); Köhler (2001) [as a synonym of *P. poecilonotus*]	Panama into w Colombia and Ecuador; recently reported from Distrito Federal in Venezuela.
P. sulphureus (Wagler *in* Spix, 1824) Yellow-throated Puffing Snake	Roze (1966); Peters & Orejas-Miranda (1970); Moonen et al. (1979); Vanzolini (1986); Dixon & Soini (1986); Starace (1998); Kornacker (1999); Marques et al. (2001)	N South America in Peru, Ecuador, Brazil, Venezuela, the Guianas, and Trinidad.
P. s. sulphureus (Wagler *in* Spix, 1824) Giant Bird Snake	Roze (1966); Peters & Orejas-Miranda (1970); Amaral (1976); Duellman (1978); Fugler & Walls (1978); Dixon & Soini (1986); Lancini & Kornacker (1989); Murphy (1997); Starace (1998); Kornacker (1999)	Peru, Trinidad, the Guianas, Ecuador, and equatorial Brazil.
P. s. diperkini* (Schlegel, 1837) Guianan Bird Snake	Brongersma (1937); Peters & Orejas-Miranda (1970); Hoge & Romano (1970); Vanzolini (1986) *Remark: Smith & Taylor (1966) placed the taxon as a synonym of *P. s. sulphureus* without further comment; however, Brongersma (1937) listed it as *P. s. sulphurerus* with overview. It is maintained as a subspecies here because it is listed in the later primary literature.	The Guianas.
P. s. poecilostomus (Wied, 1824) Black-tailed Golden Snake	Peters & Orejas-Miranda (1970); Amaral (1976); Freitas (1999)	SE Brazil.

GENUS: *Psomophis* Myers & Cadle, 1994, is a genus based (in part) as the *Rhadinaea "brevirostris"* group of ground snakes that occur in southern South America. See Myers (1974) and Myers & Cadle (1994).

CONTENT: Four species. ★★★★

Psomophis: Chaco Diminutive Snake. Frank & Ramus (1995) have this listed this snake as Spirit Snake.

Species-level Taxa	Citations and Remarks	Abbreviated Range
P. brevirostris (W. Peters, 1863) Amazonian Short-nosed Snake	Peters & Orejas-Miranda (1970); Myers (1974); Lancini (1979); Freiberg (1982); Dixon & Soini (1986); Lancini & Kornacker (1989); Myers & Cadle (1994); Starace (1998)	Amazon Basin from the Venezuela, Guianas, and s Colombia southward to Bolivia and Brazil.
P. genimaculatus (Boettger, 1885) Spirit Diminutive Snake	Meyers & Cadle (1994); Lions & Alvarez (1997); Leynaud & Bucher (1999)	S and e Bolivia in the Llanos of Majos, the Chaco Boreal and sw Brazil in the Pantanal, southward into the Chaco of n Paraguay and n Argentina.
P. joberti (Sauvage, 1884) Jobert's Diminutive Snake	Meyers & Cadle (1994); Yanosky (1998) [as *Liophis joberti* and *Psomophis joberti*]; Freitas (1999) [as *Psomophis joberti*]	Paraguay into e Brazil excluding the Amazon; the range includes Marajo Island.
P. obtusus (Cope, 1863) Red-bellied Diminutive Snake	Myers (1974); Cei (1993) [as *Rhadinaea obtusa* Cope]; Meyers & Cadle (1994); Lema (1994b); Yanosky (1998) [as *Liophis obtusus*]; Leynaud & Bucher (1999)	S Brazil, s Paraguay, Uruguay, and n Argentina in Chaco, and probably in Entre Ríos and Corrientes.

GENUS: *Ptychophis* Gomes, 1915, is a South American snake that is found only in Brazil. See Hoge & Romano (1969); Peters & Orejas-Miranda (1970); Vanzolini (1989); and Lema (1994b).

Important contributions for this section were provided by Robert A. Thomas.

CONTENT: One species. ★★★

Ptychophis: Brazilian Yellow-lined Snake

Species-level Taxa	Citations and Remarks	Abbreviated Range
*P. flavovirgatus** Gomes, 1915 Brazilian Yellow-lined Snake	Hoge & Romano (1969); Peters & Orejas-Miranda (1970); Amaral (1976); Vanzolini (1986), Porto & Caramaschi (1988); Lema (1994b) *Remark: The name *Paraptychophis meyeri* Lema, 1967 is a junior synonym (see Hoge & Romano, 1969).	States of Paraná, Rio Grande do Sul, and Santa Catarina in Brazil.

GENUS: *Regina* Baird & Girard, 1853, was previously classified as *Natrix* and *Liodytes*. It is found in Canada and the United States. See Neill (1963); Auffenberg (1950); Conant (1975); Tennant (1998), Conant & Collins (1998); Dixon (2000a); and Tennant & Bartlett (2000).

Review assistance and important comments for this section was provided by Roger Conant.

CONTENT: Four species. ★★★★

Regina: Crayfish Snake

The Caenophidia

Species-level Taxa	Citations and Remarks	Abbreviated Range
R. alleni* (Garman, 1874) Striped Crayfish Snake	Conant (1975, 1978); Lawson (1987); Ashton & Ashton (1981); Tennant (1997b); Conant & Collins (1998); Tennant & Bartlett (2000)	S Georgia and peninsular Florida.
	*Remark: Subspecies are usually not considered valid. When shown, they are listed as, *R. a. alleni* (Garman, 1874), from the Okefenokee Swamp in Georgia south through peninsular Florida into Lake Okeechobee; and, *R. a. lineapiatus* (Auffenberg, 1950), of s Florida	
R. grahami Baird & Girard, 1853 Graham's Crayfish Snake	Clay (1938); Conant (1975, 1978); Lawson (1987); Dundee & Rossman (1989); Tennant (1998); Conant & Collins (1998); Dixon (2000a); Werler & Dixon (2000); Tennant & Bartlett (2000)	C United States from Iowa and Illinois into Louisiana and Texas.
R. rigida (Say, 1825) Glossy Crayfish Snake	Clay (1938); Conant (1975, 1978); Price (1983); Lawson (1987); Dundee & Rossman (1989); Conant & (1998); Dixon (2000a)	Gulf Coast and Coastal Plain of the United States from North Carolina into nc Florida and west into ec Texas. Isolated populations occur in Virginia and wc Mississippi.
R. r. rigida (Say, 1825) Glossy Crayfish Snake	Huheey (1959); Conant (1975); Ashton & Ashton (1981); Dundee & Rossman (1989); Tennant (1997b); Conant & Collins (1998); Tennant & Bartlett (2000)	Atlantic se coast of United States from North Carolina into nc Florida.
R. r. deltae (Huheey, 1959) Delta Crayfish Snake	Huheey (1959); Conant (1975); Dundee & Rossman (1989); Conant & Collins (1998); Tennant & Bartlett (2000)	Mississippi Delta in se Louisiana and sw Mississippi in the s United States.
R. r. sinicola (Huheey, 1959) Gulf Crayfish Snake	Huheey (1959); Conant (1975); Ashton & Ashton (1981); Lawson (1987); Dundee & Rossman (1989); Tennant (1997b, 1998); Conant & Collins (1998); Werler & Dixon (2000); Tennant & Bartlett (2000); Dixon (2000a)	The United States Gulf Coast from Texas into Georgia and Mississippi and ec Texas and se Oklahoma into s Arkansas.
R. septemvittata* (Say, 1825) North American Queen Snake	Conant (1975); Ashton & Ashton (1981); Lawson (1987); Tennant (1997b); Conant & Collins (1998) [with no subspecies]; Tennant & Bartlett (2000)	Great Lakes region from s Ontario in Canada southward into New York and Pennsylvania into the Gulf Coast of the United States from Florida, Alabama and Mississippi.
	*Remark: Subspecies are not usually considered valid. When shown, they normally appear as, *R. s. septemvittata* (Say, 1825), from the Great Lakes southward nearly to the Gulf of Mexico and westward into Arkansas and Missouri; and, *R. s. mabila* (Neill, 1963), which is restricted to the Gulf Coast in s Alabama and extreme w Florida	

GENUS: *Rhachidelus* Boulenger, 1908, is a South American snake from Brazil and Argentina. See Peters & Orejas-Miranda (1970); Cei (1993); and Lema (1994b).

Review assistance for this section was provded by Thales de Lema.

CONTENT: One species. ★★★

Rhachidelus: South American Black Snake. Frank & Ramus (1995) call this the Brazilian Bird Snake.

Species-level Taxa	Citations and Remarks	Abbreviated Range
R. brazili Boulenger, 1908 South American Black Snake or Brazilian Bird Snake	Peters & Orejas-Miranda (1970); Amaral (1976); Cei (1993); Lema (1994b); Campos-Nogueira (2001)	S Brazil in São Paulo, Rio Grande do Sul, Distrito Federal and Paraná to Misiones in ne Argentina.

GENUS: *Rhadinaea* Cope, 1863, is often found in pine-oak and cloud forests. Historically, the genus *Rhadinaea* (along with *Liophis*) was a "catch-all" genus for some of the American snakes which are often referred to as littersnakes. One allopatric species is found in the United States, which is far north and east of the *Rhadinaea* normal range limits. The rest are widely distributed throughout Latin America. Those snakes originally listed as *Rhadinaea undulatus* group are now found in *Echinanthera* Cope, 1894. Those originally listed in *Rhadinaea brevirostris* group are now in *Taeniophallus* Cope, 1895, but three of the taxa are placed in *Psomophis* Myers & Cadle, 1994. The *Rhadinaea laterstriga* group is now *Urotheca* Bibron, 1843. See Dunn (1957) Smith & Taylor (1966); Peters & Orejas-Miranda (1970); Myers (1969b, 1974); Wilson et al. (1986); Smith & Smith (1993); Smith & Campbell (1994); Kornacker (1999) and Köhler (2001).

The first revision on the genus *Rhadinaea* was made by Myers (1974). At the time of the Myers (1974) publication, he listed *Leimadophis melanostigma* and discussed it under *Rhadinaea affinis*. He also referred to *Liophis undulatus* outside of *Rhadinaea*. Myers (1974), listed *genimaculata* Boettger (= *joberti* Sauvage), *obtusus* Cope, and *steinbachi* Boulenger and noted that they were under investigation. As noted in Myers (1974, the species *genimaculata*, *joberti* and *obtusus*, which were designated as *incertae sedis* by Dixon, were later classified by Myers & Cadle (1994) as genus *Psomophis*. Vanzolini (1986) expanded the Peters & Orejas-Miranda (1970 listing of *Rhadinaea* based on Myers (1974) and the work was later updated into the genera:
- *Echinanthera* Cope, 1894 =species of the *Rhadinaea brevirostris* group and *Echinanthera* (part), see Myers (1974), Di-Bernardo (1992, 1994, 1996); and Myers & Cadle (1994).
- *Taeniophallus* Cope, 1895 = species of the *Rhadinaea brevirostris* group/*Echinanthera* group, see Myers (1974), Myers & Cadle (1994), and Di-Bernardo (1992, 1996).
- *Urotheca* Bibron, 1843 = taken from the *Rhadinaea laterstriga* group, see Savage & Crother (1989), and Savage & Lahanas (1989).
- *Psomophis* Myers & Cadle (1994) = taken from the incertae sedis *Rhadinaea* group, see Myers & Cadle (1994).

CONTENT: A large, complex genus of snakes, with thirty-six species. ★★

Rhadinaea: Graceful Brownsnake or Littersnake. Frank and Ramus (1995) also call this the Graceful Brown Snake

Species-level Taxa	Citations and Remarks	Abbreviated Range
R. anachoreta Smith & Campbell, 1994 Reclusive Forest Snake	Smith & Campbell (1994); Campbell (1998a); Köhler (2001)	Guatemala in the Sierra de Santa Cruz and Sierra de Cara, from 500 to 1180 m.
R. bogertorum Myers, 1974 Oaxacan Graceful Brownsnake	Myers (1974); Smith & Smith (1993)	Known only from the type locality and vicinity of Cerro Pelón, Sierra de Juárez part of the Sierra Madre del Sur of nc Oaxaca in Mexico in cloud forests.

The Caenophidia

Species-level Taxa	Citations and Remarks	Abbreviated Range
R. calligaster (Cope, [1875] 1876) Green Littersnake	Taylor (1951); Myers (1974); Savage (1980); Savage & Villa (1986); Villa et al. (1988); Köhler (2001)	Central America in middle Costa Rica in the Cordillera de Talamanca into extreme w Panama in wet, montane forests, from 1220 to 2440 m.
R. cuneata Myers, 1974 Veracruz Graceful Brownsnake	Myers (1974); Dixon (1981a); Pérez-Higareda & Smith (1991) Smith & Smith (1993)	Known only from the Córdoba region of c Veracruz, on the eastern side of the Sierra Madre de Oaxaca in Mexico.
R. decorata (Günther, 1859) Adorned Graceful Brownsnake	Taylor (1951); Myers (1974); Savage (1980); Alvarez del Toro (1982); Savage & Villa (1986); Villa et al. (1988); Pérez-Higareda & Smith (1991); Smith & Smith (1993); Lee (1996, 2000); Smith & Campbell (1994); Campbell (1998a); Stafford & Meyer (2000); Köhler (2001)	Atlantic versant from San Luis Potosí in Mexico south through the Isthmus of Tehuantepec, briefly crossing into the Pacific versant in a narrow strip, southward through Costa Rica and Panama into nw South America into as far as nw Ecuador.
R. dumerili Bibron *in* Sagra, 1840 Dumeril's Graceful Brownsnake	Myers (1974); Pérez-Santos & Moreno (1988)	N Colombia from the area of Río San Juan in Depto. Chocó in the Pacific versant.
R. flavilata (Cope, 1871) Pine Woods Littersnake	Myers (1974, 1987); Conant (1975); Ashton & Ashton (1981); Dundee & Rossman (1989); Tennant (1997b); Conant & Collins (1998); Walley (1999); Tennant & Bartlett (2000)	Coastal regions of se United States from North Carolina southward into Florida to about Palm Beach Co., and eastward to e Louisiana just east of the Mississippi River; this is an allopatric species whose range leaves a large distributional gap in the genus.
R. forbesi Smith, 1942 Forbes's Graceful Brownsnake	Smith & Taylor (1966); Myers (1974); Pérez-Higareda & Smith (1991); Smith & Smith (1993); Nieto-Montes de Oca & Mendelson III (1997)	Southern end of the Sierra Madre Oriental in c Veracruz in Mexico.
R. fulvivittis Cope, 1875 Ribbon Graceful Brownsnake	Smith & Taylor (1966); Myers (1974); Myers & Campbell (1981); Pérez-Higareda & Smith (1991); Smith & Smith (1993)	Pine-oak forests in the Sierra Madre del Sur of Oaxaca and along the Madre de Oaxaca northward into se Puebla and c Veracruz in Mexico.
R. gaigeae* Bailey, 1937 Gaige's Pine Forest Snake	Bailey (1937b); Smith & Taylor (1966); Myers (1974); Dixon (1981a); Smith & Smith (1993) *Remark: *R. crassa* Smith, 1942 as seen in Myers (1974) and distributed in c Mexico from c Hidalgo into se San Luis Potosí, is a synonym (see Smith & Smith, 1993).	Pine-oak and cloud forests in the Sierra Madre Oriental of s Tamaulipas, e San Luis Potosí, n Hidalgo, and a mountainous region to the west in sc San Luis Potosí, Mexico from about 200 to 2700 m.
R. godmani* (Günther, 1865) Godman's Graceful Brownsnake	Peters & Orejas-Miranda (1970); Myers (1974); Savage (1980); Wilson & Meyer (1985); Savage & Villa (1986); Wilson et al. (1986); Villa et al. (1988); McCranie & Wilson (1992); Smith & Smith (1993); Smith & Campbell (1994); Mendelson III & Kizirian (1995); Köhler (2001) *Remark: Subspecies are not valid according to McCranie & Wilson (1992).	Fragmented distribution throughout the montane areas of Middle America, but the distribution is very poorly defined; it ranges from the se Oaxacan highlands southward through the highlands of Guatemala into Ocotepeque, Honduras and Cerro Montecristo in El Salvador; Myers (1974) notes a distributional gap in Honduras and Nicaragua. Köhler (2001) lists the altitudinal range as 1000 to 2650 m.

Species-level Taxa	Citations and Remarks	Abbreviated Range
*R. hannsteini** (Stuart, 1949) Hannstein's Spot-lipped Snake	Myers (1974); Alvarez del Toro (1982) [as *Trimetopon hannsteini*]; Vanzolini (1986); Villa et al. (1988); Smith & Campbell (1994); Köhler (2001) *Remark: This species is not shown in Campbell (1998a).	Pacific versant of the Sierra Madre of Chiapas, Mexico into Guatemala, found from 500 to 2000 m.
*R. hempsteadae** Stuart & Bailey, 1941 Hempstead's Graceful Brownsnake	Hardy & McDiarmid (1969) Myers (1974); Villa et al. (1988); Smith & Campbell (1994); Mendelson, III & Kizirian (1995); Köhler (2001) *Remark: This species not listed in Campbell (1998a)	W Chiapas, Mexico the highlands in the mountains of Alta Verapaz and Sierra del las Minas in e Guatemala, from 1200 to 3000 m.
*R. hesperia** Bailey, 1940 Western Graceful Brownsnake	Davis & Smith (1953); Hardy & McDiarmid (1969) Myers (1974); Smith & Smith (1993); García & Ceballos (1994); Ramírez-Bautista (1994); Vásquez-Díaz et al. (1999b) *Remark: Myers (1974) noted it would seem simplest to drop subspecies of this taxon and to recognize it as a geographically variable species. When listed, the subspecies are usually shown as: *R. h. hesperia* Bailey, 1940 found from the upper Río Balsas drainage, in the Sierra Madre del Sur of Guerrero and the Cordillera Volcánica of Morelos; *R. h. baileyi* Smith, 1942 Pacific slopes of the Sierra Madre del Sur of Guerrero in Mexico; and, *R. h. hesperiodes* Smith, 1942, (See Peters, 1954), is the northern subspecies of the c Mexican Plateau in Sinaloa and Jalisco	W Mexico from Sinaloa southward into Guerrero and Morelos, in the Sierra Madre Occidental, Cordillera Volcánica, Sierra de Coalcomán, and the Sierra Madre del Sur; it was reported from as far inland as Guanajuato in Mexico and has been reported from Jalisco and the state of México.
R. kanalchutchan Mendelson III & Kizirian, 1995 Kanalchutchan	Mendelson III & Kizirian (1995); Köhler (2001)	Guatemala in and around San Cristóbal de las Casas on the Meseta Central de Chiapas and the leeward valley of the Río Grijalva drainage in the northern highlands (La Selva Negra) of Chiapas in Mexico, from 2300 to 2700 m.
*R. kinkelini** Boettger, 1898 Kinkelini's Graceful Brownsnake	Peters & Orejas-Miranda (1970); Myers (1974); Wilson & Meyer (1985); Villa et al. (1988); McCranie & Wilson (1991a); Smith & Smith (1993); Smith & Campbell (1994); Mendelson III & Kizirian (1995); Vences et al. (1989); Köhler (2001) *Remark: *Rhadinaea pinicola* Mertens is a synonym of this snake (see Köhler & McCranie, 1999).	Highlands of c Guatemala in Alta Verapaz through Honduras and extreme nw El Salvador into n Nicaragua in a disjunct range, from 1300 to 2200 m.
*R. lachrymans** (Cope, [1869] 1870) Tearful Pine-Oak Snake	Smith & Taylor (1966); Myers (1974); Alvarez del Toro (1982); Villa et al. (1988); Smith & Smith (1993); Holm & Cruz D. (1994); Köhler (2001) *Remark: Not shown in Campbell (1998a).	Pacific mountains of extreme s Mexico in Chiapas southward into Guatemala; associated with the Plateau of Guatemala and the Sierra Madre of w Guatemala, from 500 to 3000 m.
R. laureata (Günther, 1868)	Davis & Smith (1953); Myers (1974); Smith & Smith (1993); Auth et al. (1999); Uribe-Peña et al. (1999)	From about 1500 to 3100 m in mountains west and south of the Mexican Plateau in pine-oak forests of

Species-level Taxa	Citations and Remarks	Abbreviated Range
Crowned Graceful Brownsnake		the Sierra Madre Occidental and the Cordillera Volcánica from Durango southward to c Michoacán, and then eastward into México and Morelos.
R. macdougalli Smith & Langebartel, 1949 (1950) MacDougall's Graceful Brownsnake	Myers (1974); Villa et al. (1988); Smith & Smith (1993); Köhler (2001)	Known definitely only from two localities in e Oaxaca, Mexico, in the Sierra Madre de Chiapas and in the Sierra Madre del Sur on the Cerro Zempoaltepec Ridge, east and west of the Plains of Tehuántepec from about 1140 to 1220 m.
R. marcellae Taylor, 1949 Marcella's Graceful Brownsnake	Myers (1974); Hernández G. & Mendoza Q. (1994); Nieto-Montes de Oca & Mendelson, III (1997)	In the Sierra Madre Oriental from Tamaulipas and Luis Potosí south into Oaxaca and in the cloud forest mountains of the Xilitla region s into Hidalgo.
R. montana Smith, 1944 Nuevo León Graceful Brownsnake	Smith (1944); Smith & Taylor (1966); Myers (1974); Smith & Smith (1993), Liner (1996)	Known only from c Nuevo León in the n Sierra Madre Oriental of Mexico.
R. montecristi Mertens, 1952 Monte Cristi Graceful Brownsnake	Myers (1974); Wilson & Meyer (1985); McCranie & Wilson (1991b); Smith & Campbell (1994); Holm & Cruz D. (1994); Köhler (2001)	Central American mountains from sw Honduras into nw El Salvador, and possibly Guatemala in a disjunct distribution.
R. myersi Rossman, 1965 Myers's Graceful Brownsnake	Rossman (1965); Myers (1974); Smith & Smith (1993)	Known only from the Sierra Madre del Sur in the vicinity of Pluma Hidalgo and La Soledad, near the Pacific Coast of c Oaxaca in Mexico.
R. omiltemana (Günther, 1894) Guerreran Pine Woods Snake	Smith & Taylor (1966); Myers (1974); Smith & Smith (1993)	Sierra Madre del Sur in Guerrero, Mexico where it is only known from Omilteme and the vicinity of Chilpancingo.
R. pilonaorum (Stuart, 1954) Stuart's Graceful Brownsnake	Myers (1974); Vanzolini (1986); Villa et al. (1988); Smith & Smith (1993); Köhler (2001)	Known only from the Pacific versant of se Guatemala and w El Salvador, from 670 to 950 m.
R. posadasi (Slevin, 1936) Posadas's Graceful Brownsnake	Slevin (1936); Myers (1974); Vanzolini (1986); Villa et al. (1988); Smith & Smith (1993); Köhler (2001)	Known only from the slopes of the volcanoes in the Pacific versant of the Sierra Madre of sw Guatemala.
R. pulveriventris Boulenger, 1896 Common Graceful Brownsnake	Myers (1974); Savage (1980); Savage & Villa (1986); Villa et al. (1988); Köhler (2001)	C Costa Rica in the Cordillera Central and the Cordillera de Talamanca into extreme w Panama.
R. quinquelineata Cope, 1885 [1886] Pueblan Graceful Brownsnake	Smith & Taylor (1966); Myers (1974); Dixon (1981a); Smith & Smith (1993)	Known definitely only from the type locality of Teziutlán in n Puebla, near the Veracruz border in the Sierra Madre Oriental of Mexico.

Species-level Taxa	Citations and Remarks	Abbreviated Range
R. rogerromani Köhler & McCranie, 1999 Cerro Saslaya Graceful Brownsnake	Köhler & McCranie (1999); Köhler (2001)	Cerro Saslaya in Nicaragua.
R. sargenti Dunn & Bailey, 1939 Sargent's Graceful Brownsnake	Dunn & Bailey (1939); Myers (1974); Villa et al. (1988); Köhler (2001)	Hills east of the Canal Zone in c Panama in the watersheds of the Río Chagres and Río Pequení.
R. schistosa* (Smith, 1941) Broken-collar Graceful Brownsnake	Smith (1941c); Myers (1974); Pérez-Higareda & Smith (1991); Smith & Smith (1993) *Remark: *Rhadinella schitosa* Smith, 1941 is a synonym, (see Smith & Taylor, 1966).	Foothills along the eastern slope of the Sierra Madre de Oaxaca, in Veracruz and n Oaxaca in Mexico.
R. serperaster Cope, 1871 Southern Graceful Brownsnake	Taylor (1951); Taylor (1954); Myers (1974); Savage (1980); Savage & Villa (1986); Villa et al. (1988); Smith & Campbell (1994); Köhler (2001)	Mountains of c Costa Rica in the Cordillera Central and in the n end of the Cordillera de Talamanca and Panama, from 1220 to 1450 m.
R. stadelmani* Stuart & Bailey, 1941 Stadelman's Graceful Brownsnake	Stuart & Bailey (1941); Peters & Orejas-Miranda (1970); Myers (1974); Mendelson, III & Kirzirian (1995); Köhler (2001) *Remark: Myers (1974) listed *R. stadelmani* as a synonym of *R. hempsteade*. Mendelson, III & Kirzirian (1995), however recognized the species.	Intermediate elevations on the eastern and western sides of the Sierra de los Cuchumatanes and the Montañas del Cuilco in extreme w Guatemala.
R. taeniata (W. Peters, 1863) Pine-Oak Snake	W. Peters, 1863; Smith & Taylor (1966); Myers (1974); Smith & Smith (1993)	Pine-oak and possibly fir forests of Mexico in the mountains from c Jalisco into Michoacán.
R. t. taeniata (W. Peters, 1863) Giant Pine-Oak Snake	W. Peters, 1863; Bailey (1940); Myers (1974); Smith & Smith (1993)	Northern populations are found north of the Balsas Basin with its range extending southeastward across the narrow, lower end of the Río Balsas, into the Sierra Madre del Sur of w Guerrero in Michoacán and Jalisco and doubtfully in the vicinity of Mexico City, in the Cordillera Volcánica and the Sierra de Coalcomán at elevations from 1524 to 2835 m.
R. t. aemula Bailey, 1940 Mimic Pine-Oak Snake	Bailey (1940); Myers (1974); Smith & Smith (1993)	Southern populations found mainly south and southeast of the arid plain in the groups of mountains that comprise the Sierra Madre del Sur, with at least one population in the north in the Cordillera Volcánica of Morelos from ec Guerrero in the region of Chilpancingo, and just north of San Vicente de Jesús, both in the Sierra Madre del Sur from c Oaxaca into c Guerrero into Morelos from 1700 to 2439 m.
R. tolpanorum Holm & Cruz, 1994 Texiguat Graceful Brownsnake	Holm & Cruz (1994); Köhler (2001)	N Honduras in a cloud forest of the proposed wildlife refuge (Refugio de Vida Silvestre Texiguat) in the Montaña de Texiguat.

The Caenophidia

Species-level Taxa	Citations and Remarks	Abbreviated Range
R. vermiculaticeps (Cope, 1860) Cope's Graceful Brownsnake	Taylor (1951); Myers (1974); Villa et al. (1988); Köhler (2001)	WC Panama, most likely from humid forests at moderate elevations, from 700 to 900 m.

GENUS: *Rhadinophanes* Myers & Campbell, 1981, is confined to the Mexican highlands. See Myers & Campbell (1981); Dowling & Jenner (1987), Williams & Wallach (1989); and Smith & Smith (1993).

Review assistance and important contributions for this section were provided by Jonathan A. Campbell.

See also *Tantalophis*.

CONTENT: One species. ★★★★

Rhadinophanes: Graceful Mountain Snake

Species-level Taxa	Citations and Remarks	Abbreviated Range
R. monticola Myers & Campbell, 1981 Graceful Mountain Snake	Myers & Campbell (1981); Smith & Smith (1993)	Sierra Madre del Sur of Guerrero, Mexico in the Mexican highlands; from high montane forests in humid pine-oak-fir forest at about 2750 m on Cerro Teótepec.

GENUS: *Rhinobothryum* Wagler, 1830, is a found from Guatemala and Honduras southward into South America in the Amazon and Paraguay basins. Peters & Orejas-Miranda (1970); Moonen et al. (1979); Dixon & Soini (1986); Savage & Villa (1986); Obst et al. (1988); Lancini & Kornacker (1989); Molina R. & Rivas F. (1996); Starace (1998); Kornacker (1999); and Köhler (2001).

CONTENT: Two species. ★★★

Rhinobothryum: Ringed Tree Snake or Banded Tree Snake. Frank and Ramus (1995) call this the Banded Snake.

Species-level Taxa	Citations and Remarks	Abbreviated Range
R. bovallii Andersson, 1916 Costa Rican Banded Tree Snake	Savage & Vial (1980); Wilson & Meyer (1985); Savage & Villa (1986); Villa et al. (1988); Lancini & Kornacker (1989); Savage & Slowinski (1992); Smith & Smith (1993); Cruz (1997); Kornacker (1999); Köhler (2001)	Honduras in Central America southward into n South America into nw Colombia, Ecuador, and nw Venezuela.
R. lentiginosum (Scopoli, 1785) Amazonian Banded Tree Snake	Peters & Orejas-Miranda (1970); Cunha & Nascimento (1975, 1978); Moonen et al. (1979); Dixon & Soini (1986); Chippaux ([1986]1987); Savage & Slowinski (1992); Molina R. & Rivas F. (1996); Starace (1998); Yanosky (1998); Kornacker (1999)	Amazon and Paraguay basins into as far north as Bolívar, Venezuela and including Brazil, Bolivia, Colombia and Paraguay, and to as far as the Guianas in French Guiana and Suriname.

GENUS: *Rhinocheilus* Baird & Girard, 1853 is confined to North America. See Smith & Taylor (1966); Medica (1975); Stebbins (1966, 1985); Savage & Slowinski (1992); Smith & Smith (1993); Degenhardt et al. (1996); Tennant (1998); and Tennant & Bartlett (2000).

CONTENT: One species with four subspecies. ★★★

Rhinocheilus: North American Long-nosed Snake

Species-level Taxa	Citations and Remarks	Abbreviated Range
R. lecontei Baird & Girard, 1853 Long-nosed Snake	Medica (1975); Stebbins (1985); Degenhardt et al. (1996); Grismer (1999); Brown (1997); McPeak (2000); Dixon (2000a)	C and sw United States southward into Mexico.
*R. l. lecontei** Baird & Girard, 1853 Western Long-nosed Snake	Klauber (1941b); Langebartel & Smith (1954); Smith & Taylor (1966); Medica (1975); Stebbins (1985); Smith & Smith (1993); Degenhardt et al. (1996); Brown (1997); Bartlett & Tennant (2000) *Remark: The Desert Long-nosed Snake, *R. l. clarus* Klauber, 1941, is generally considered to be found in sw United States in s California, s Nevada, and w Arizona southward into extreme n Baja California. It was long considered to be a subspecies; however, the designation was dropped when it was discovered that both bicolored and tricolored examples were present in a given population (see Bartlett & Tennant, 2000).	SW United States from s Arizona southward into Baja California and Sonora in Mexico.
R. l. antoni Dugès, 1886 Mexican Long-nosed Snake	Klauber (1941b); Langebartel & Smith (1954); Smith & Taylor (1966); Medica (1975); Smith & Smith (1993); Liner (1994)	Pacific coastal w Mexico from Sonora into Jalisco.
*R. l. etheridgei** Grismer, 1990 Cerralvo Island Long-nosed Snake	Grismer (1990, 1999); McPeak (2000) *Remark: Not shown in Smith & Smith (1993). Grismer (1999) considers this taxon a distinct species as *R. etheridgei* but detail is not given to support the change.	Cerralvo Island in Mexico.
R. l. tessellatus Garman, 1883 Texas Long-nosed Snake	Klauber (1941b); Smith & Taylor (1966); Conant (1975); Stebbins (1985); Degenhardt et al. (1996); Tennant (1998); Conant & Collins (1998); Werler & Dixon (2000); Bartlett & Tennant (2000); Tennant & Bartlett (2000); Dixon (2000a)	The United States in sw Kansas and sw New Mexico southward into extreme n Mexico in ne Sonora, ne Chihuahua, and San Luis Potosí.

GENUS: *Salvadora* Baird & Girard, 1853, is from North America in the United States and Mexico. See Smith & Taylor (1966); Stebbins (1966, 1985); Conant (1975); Smith & Smith (1993); Degenhardt et al. (1996); Tennant (1998); Conant & Collins (1998); Dixon (2000a); and Tennant & Bartlett (2000).

Review assistance and important contributions for this section came from: Dominic I. Lannutti.

The Caenophidia

CONTENT: About seven species. Seven species are listed here. ★★★

Salvadora: Patch-nosed Snake

Species-level Taxa	Citations and Remarks	Abbreviated Range
S. bairdi Jan, 1860 Baird's Patch-nosed Snake	Davis & Smith (1953); Smith & Taylor (1966); Hardy & McDiarmid (1969); Pérez-Higareda & Smith (1991); Smith & Smith (1993); Uribe-Peña et al. (1999)	Mexico in Veracruz and c Hidalgo westward into w Jalisco; it is reported from Aguascalientes, Coahuila, Chihuahua, Guerrero, Guanajuato, Jalisco, México, Oaxaca, Puebla, Querétaro, Sinaloa, Veracruz, Zacatecas, and the Federal District.
S. deserticola* Schmidt, 1940 Big Bend Patch-nosed Snake	Conant (1975); Tanner (1985); Stebbins (1985); Degenhardt et al. (1996); Tennant (1998); Conant & Collins (1998); Werler & Dixon (2000); Bartlett & Tennant (2000); Tennant & Bartlett (2000); Dixon (2000a) *Remark: Crother et al. (2000) list this as *S. hexalepis deserticola* because they feel that Bogert & Degenhardt (1961) or Bogert (1945) did not demonstrate why it had been elevated.	SW United States, nw New Mexico southward into se Arizona, and eastward into Big Bend region of Texas, then southward into s Sinaloa, Sonora, and e and s Chihuahua.
S. grahamiae* Baird & Girard, 1853 Eastern Patch-nosed Snake	Schmidt (1940); Smith & Taylor (1966); Conant (1975); Stebbins (1985); Degenhardt et al. (1996); Dixon (2000a) *Remark: Smith & Smith (1993) proposed the spelling *S. grahamae*, not *S. grahamiae*. Refer to Smith & Smith (1993: 470) for reasons why I retain the traditional spelling as the latter did in their primary list.	SW United States into n Mexico from se Arizona, nc New Mexico, and c Texas southward into n Zacatecas in Mexico.
S. g. grahamiae Baird & Girard, 1853 Mountain Patch-nosed Snake	Schmidt (1940); Conant (1975); Tanner (1985); Stebbins (1985); Degenhardt et al. (1996); Tennant (1998); Conant & Collins (1998); Werler & Dixon (2000); Bartlett & Tennant (2000); Tennant & Bartlett (2000); Dixon (2000a)	SW United States from se Arizona and nc New Mexico eastward into the Trans-Pecos region of Texas with a disjunct population in Chihuahua.
S. g. lineata Schmidt, 1940 Texas Patch-nosed Snake	Schmidt (1940); Tennant (1998); Conant & Collins (1998); Werler & Dixon (2000); Tennant & Bartlett (2000); Dixon (2000a)	The United States in nc Texas southward into n Mexico along the Chihuahua-Durango border, and, Chihuahua and Hidalgo.
S. hexalepis (Cope, 1866 [1867]) Western Patch-nosed Snake	Smith & Taylor (1966); Stebbins (1985); Smith & Smith (1993); Brown (1997); McPeak (2000)	SW United States into Mexico from wc Nevada southward to the tip of Baja California and sw Sonora, and from coastal southern California into sw Utah and c Arizona.
S. h. hexalepis (Cope, 1866 [1867]) Desert Patch-nosed Snake	Smith & Taylor (1966); Stebbins (1985); Smith & Smith (1993); Brown (1997); Bartlett & Tennant (2000)	SW United States southward through nw Mexico into Baja California including the Tiburón Island.
S. h. klauberi Bogert, 1945 Baja California Patch-nosed Snake	Bogert (1945); Stebbins (1985); Smith & Smith (1993); Grismer et al. (1994)	W Mexico in the southern 2/3 of Baja California, and the Vizcaino Peninsula, and on Isla San Geronimo.

Species-level Taxa	Citations and Remarks	Abbreviated Range
S. h. mojavensis Bogert, 1945 Mojave Patch-nosed Snake	Smith & Taylor (1966); Stebbins (1985); Smith & Smith (1993); Brown (1997); Bartlett & Tennant (2000)	SW United States in the Mojave Desert region; ranging from nw Nevada and sc Utah, southward to s California and c Arizona.
S. h. virgultea Bogert, 1935 Coast Patch-nosed Snake	Smith & Taylor (1966); Stebbins (1985); Smith & Smith (1993); Brown (1997); Bartlett & Tennant (2000)	Foothills in sw United States in coastal California southward into nw Baja California, Mexico, west of the coast range and San Pedro Mártir mountains.
S. intermedia Hartweg, 1940 Oaxacan Patch-nosed Snake	Hartweg (1940); Smith & Taylor (1966); Smith & Smith (1993)	C Mexico in the Sierra Madre del Sur in c Guerrero and the upper Río Balsas Basin in Tehuacán, Mexico.
S. lemniscata (Cope, 1895 [1896]) Pacific Patch-nosed Snake	Bogert (1939); Smith & Taylor (1966); Alvarez del Toro (1982); Villa et al. (1988); Smith & Smith (1993); Köhler (2001)	Mexico in coastal c Guerrero, southward into Guatemala.
S. mexicana (Dumeril, Bibron & Dumeril, 1854) Mexican Patch-nosed Snake	Davis & Smith (1953); Peters (1954); Smith & Taylor (1966); Ramírez-Bautista (1994); García & Ceballos (1994)	W into s Mexico from Nayarit and probably s Sinaloa southward in the Pacific versant into c Guerrero and probably w Oaxaca; it has been reported from n Michoacán.

GENUS: *Saphenophis* Myers, 1973, is a South American snake. See Myers (1969a, 1973) and Vanzolini (1986).

CONTENT: Five species. ★★★★

Saphenophis: False Rhadinaea. Frank & Ramus (1995) called this the Saphenophis Snake.

Species-level Taxa	Citations and Remarks	Abbreviated Range
S. antioquiensis (Dunn, 1943) Antioquia False Rhadinaea	Myers (1969a, 1973); Peters & Orejas-Miranda (1970) [as *Lygophis*]; Pérez-Santos & Moreno (1988)	Known only from the type locality of San Pedro in Depto. Antioquia, Colombia, at the northern end of the Cordillera Central.
S. atahuallpae (Steindachner, 1901) Atahuallpa False Rhadinaea	Myers (1969a, 1973); Peters & Orejas-Miranda (1970) [as *Leimadophis*]; Vanzolini (1986); Pérez-Santos & Moreno (1991)	Known only from the type locality of La Palmas in the western spur of the Andes between Babahoyo and Guaranda at 2500 m in Ecuador.
S. boursieri (Jan, 1867 *in* Jan & Sordelli, 1866–1870 Boursier's False Rhadinaea	Myers (1966b, 1969a, 1973); Peters & Orejas-Miranda (1970) [as *Lygophis*]; Pérez-Santos & Moreno (1988, 1991)	SW Colombia and the Río Pastaza region, Amazonian slopes and western slopes above 1000 and Andean Ecuador. It may also be present in Amazonian Peru.
S. sneiderni Myers, 1973 Cauca False Rhadinaea	Myers (1973); Vanzolini (1986); Pérez-Santos & Moreno (1988)	Colombia in Cauca in the Cordillera de Occidental.

Species-level Taxa	Citations and Remarks	Abbreviated Range
S. tristriatus (Rendahl & Vestergren, 1940) Striped False Rhadinaea	Myers (1969a, 1973); Peters & Orejas-Miranda (1970) [as *Lygophis*]; Pérez-Santos & Moreno (1988)	Known only from the type locality in "Cauca," Colombia and vicinity of Gabrielopez, at about 3200 m in the Malvasá Valley on the w slope of the Cordillera Central.

GENUS: *Scaphiodontophis* Taylor & Smith, 1943, is found from southern and eastern Mexico into northern Colombia. See Smith & Taylor (1966); Morgan (1973); Peters & Orejas-Miranda (1970); Savage (1980); Wilson & Meyer (1982); Savage & Slowinski (1992); Smith & Smith (1993); Savage & Crother (1996); Campbell (1998a); Savage & Slowinski (1990, 1992, 1996); and Köhler (2001).

Review assistance and important contributions for this section provided by Brian J. Crother and Joseph B. Slowinski.

CONTENT: This is considered to be a complex polymorphic genus consisting of only one species with no subspecies. ★★★

Scaphiodontophis: Neck-banded Snake or Shovel-toothed Snake

Species-level Taxa	Citations and Remarks	Abbreviated Range
S. annulatus* (Dumeril, Bibron & Dumeril, 1854) Central American Neck-banded Snake	Brattstrom & Adis (1952) [as *S. albonuchalis*] Roze (1969); Alvarez del Toro (1982); Wilson & Meyer (1985); Villa et al. (1988); Pérez-Higareda & Smith (1991); Savage & Slowinski (1992); Smith & Smith (1993) Lee (1996, 2000); Savage (1996); Savage & Crother (1996); Campbell (1998a); Stafford & Meyer (2000); Köhler (2001) *Remark: All of the subspecies were synonymized back into *Scaphiodontophis annulatus* without subspecies by Savage & Slowinski (1992), reconfirmed in Savage & Slowinski (1996), and Savage & Crother (1996). In a communication with J. Slowinski (pers. comm. November 2,1999), he reconfirmed that he and Savage believed as far as *Scaphiodontophis* is concerned, there was evidence which only supported the existence of a single, highly polymorphic species of *S. annulata*. All other names are synonyms. Their decision to recognize only a single species of *Scaphiodontophis* is strongly influenced by the extreme variability in coloration found in every sample. Almost every individual in their samples had one or more anomalous elements in the pattern (Savage & Slowinski, 1996).	S Mexico from the Yucatán through Central America into w Panama, including Guatemala, Belize, Honduras, Nicaragua, and into Colombia.

GENUS: *Scolecophis* Fitzinger, 1843, is restricted to Central America. See Wilson et al. (1986); Savage & Slowinski (1992); Smith & Smith (1993); and Köhler (2001).

CONTENT: One species. ★★★★

Scolecophis: Central American Banded Centipede Snake. Frank & Ramus (1995) calls this the Black-banded Snake.

Species-level Taxa	Citations and Remarks	Abbreviated Range
S. atrocinctus (Schlegel, 1837) Central American Banded Centipede Snake or Black-banded Snake	Wilson & Meyer (1985); Wilson et al. (1986); Savage & Villa (1986); Villa et al. (1988); Cruz (1997); Köhler (2001)	Central America from Guatemala and El Salvador southward into Costa Rica; found in low to moderate elevations.

GENUS: *Seminatrix* Cope, 1895, is restricted to North America in the Untied States. See Dowling (1950); Ashton & Ashton (1981); Wright & Wright ([1957] 1994); Conant & Collins (1998); Dorcas et al. (1998); Tennant & Bartlett (2000).

Review assistance and important contributions for this section were provided by Roger Conant.

CONTENT: One species with three subspecies. ★★★★

Seminatrix: North American Swampsnake

Species-level Taxa	Citations and Remarks	Abbreviated Range
S. pygaea (Cope, 1871) Black Swampsnake	Dowling (1950); Wright & Wright ([1957] 1994); Conant & Collins (1998)	SE United States from coastal North Carolina into s Florida.
S. p. pgyaea (Cope, 1871) Northern Florida Swampsnake	Dowling (1950); Ashton & Ashton (1981) Wright & Wright ([1957] 1994); Conant & Collins (1998); Tennant & Bartlett (2000)	SE United States in s Georgia, s Alabama and n Florida.
S. p. cyclas Dowling, 1950 Southern Florida Swampsnake	Dowling (1950); Ashton & Ashton (1981) Wright & Wright ([1957] 1994); Tennant (1997b); Conant & Collins (1998); Tennant & Bartlett (2000)	The United States in s Florida from c Florida southward to the tip of the peninsula.
S. p. paludis Dowling, 1950 Carolina Swampsnake	Dowling (1950); Wright & Wright ([1957] 1994); Conant & Collins (1998); Tennant & Bartlett (2000)	E United States in the Coastal Plain of South Carolina and e North Carolina.

GENUS: *Senticolis* Dowling & Fries, 1987, ranges from the sw United States in se Arizona and southern Tamaulipas in Mexico, southward into Costa Rica. See Peters & Orejas-Miranda (1970)[as *Elaphe triaspis*]; Smith & Taylor (1966)[as *Elaphe triaspis*]; Price (1991); Smith & Smith (1993); Staszko & Walls (1994); Schulz (1996); Degenhardt et al. (1996); Crother et al. (2000); and Köhler (2001).

Also see *Elaphe* and *Bogertophis*.

Note: Schulz (1996) did not feel there was sufficient evidence for separating *Senticolis* from *Elaphe*. It is recommended that the reader see his comments on the subject. This reference maintains *Senticolis* as separate from *Elaphe* based on Dowling and Fries (1987).

CONTENT: One species with three subspecies. ★★★

Senticolis: Neotropical Ratsnake, Mountain Ratsnake or Green Ratsnake; note that Lee (2000) calls this the Peninsular Rat Snake. [The snake that is commonly called the Green Ratsnake is that snake in the northern part of the range. I have found those in the southern part of the range to have little green or no green on them. Because of this, a move was made away from the commonly used "Green Rat Snake" as can be found in Crother et al. (2000) to a more universal name that covers it over its entire range.]

Species-level Taxa	Citations and Remarks	Abbreviated Range
S. triaspis (Cope, 1866) Neotropical or Green Ratsnake	Smith & Taylor (1966)[as *Elaphe triaspis*]; Peters & Orejas-Miranda (1970) [as *Elaphe triaspis*]; Wilson & Meyer (1985); Dowling & Fries (1987); Villa et al. (1988); Price (1991); Staszko & Walls (1994); García & Ceballos (1994); Degenhardt et al. (1996); Lee (1996, 2000); Schluz (1996); Campbell (1998a); Crother et al. (2000); Köhler (2001)	SW United States from se Arizona southward into s Tamaulipas through Mexico and much of Central America into Costa Rica; it is not reported from the Central Plateau in Mexico but is found on the Pacific coast from Sinaloa to Colima.
S. t. triaspis (Cope, 1866) Guatamalan Neotropical Ratsnake	Dowling (1960b); Peters & Orejas-Miranda (1970) [as *Elaphe t. triaspis*]; Wilson & Meyer (1985); Dowling & Fries (1987); Price (1991); Staszko & Walls (1994); Schulz (1996); Campbell (1998a); Stafford & Meyer (2000); Lee (2000)	Yucatán in Mexico southward into e Guatemala in El Petén; Stafford & Meyer (2000) listed it from s Belize.
S. t. intermedia (Boettger, 1883) Green Ratsnake	Alvarez del Toro (1982); Wilson & Meyer (1985); Stebbins (1985); Dowling & Fries (1987); Price (1991); Pérez-Higareda & Smith (1991); Ramírez-Bautista (1994); Degenhardt et al. (1996); Schulz (1996); Bartlett & Tennant (2000); Lee (2000)	S Arizona and New Mexico in the United States southward through Chiapas in w southern Mexico along the Pacific coast.
S. t. mutabilis (Cope, 1885) Honduran Neotropical Ratsnake	Dowling (1960b); Peters & Orejas-Miranda (1970) [as *Elaphe t. mutabilis*]; Wilson & Meyer (1985); Dowling & Fries (1987); Price (1991); Staszko & Walls (1994); Schulz (1996); Campbell (1998a); Stafford & Meyer (2000); Lee (2000)	Highlands of Guatemala south into Costa Rica to about San José; Stafford & Meyer (2000) listed it from n Belize; Lee (2000) listed it as being found in the Lacandon region of Chiapas.

GENUS: *Sibon* Fitzinger, 1826, was the genus *sensu Tropidodipsas* (Günther, 1858), and was synonymized with *Sibon* Fitzinger, 1826 by Kofron (1985). However, *Tropidodipsas* is separated here because it was resurrected by Wallach (1995). *Sibon* ranges from southern Mexico into northern South America to as far as Ecuador, Trinidad, and Tobago. See Peters (1960c); Smith & Taylor (1966); Peters & Orejas-Miranda (1970); Kofron (1980, 1985); Vanzolini (1986); Wallach (1995); Starace (1998); Kornacker (1999) and Köhler (2001). *Exelencophis* is a synonym of *Sibon* (see Smith & Smith, 1993).

Review assistance and important contributions for this section came from: Ronaldo Fernandes and Van Wallach.

The genus *Tropidodipsis* (*annulifera, fasciata, fischeri, philippi, sartorii,* and *zweifeli*) is closely associated with *Sibon*. See comments about the relationship in the section on *Tropidodipsas*. See also genera *Dipsas* and *Sibynomorphus*.

CONTENT: Nine species or more depending if *Tropidodipsas* species are included in this genus. Ten species are listed here but validity of *S. brevis* is questionable. ★★

Sibon: Middle American Snail-eating Snake or Snail-sucker

Species-level Taxa	Citations and Remarks	Abbreviated Range
S. anthracops (Cope, 1868) Cope's Snail-sucker	Peters (1960c); Smith & Taylor (1966); Peters & Orejas-Miranda (1970); Hidalgo (1979); Savage (1980); Wilson & Meyer (1985); Vanzolini (1986); Savage & Villa (1986); Villa et al. (1988); Savage & McDiarmid (1992); Savage & Slowinski (1992); Wallach (1995); Cruz-D. (1997); Köhler (2001)	Pacific versant of Central America in Honduras, El Salvador, Nicaragua, and Costa Rica.
S. argus (Cope, 1875) Sipurio Snail-eater	Peters (1960c); Peters & Orejas-Miranda (1970); Savage (1980); Savage & Villa (1986); Villa et al. (1988); Savage & McDiarmid (1992); Köhler (2001)	Known only from the type locality listed as Sipurio, Costa Rica.
S. brevis* (Dumeril, Bibron & Dumeril, 1854) Lost Snail-eater	Smith & Taylor (1966); Smith & Smith (1993) *Remark: This taxon is listed as a valid species but there is no way to confirm the validity of the nomenclature against a type or other reference specimen (see Smith & Smith, 1993). This species is also shown in the incertae sedis section.	Type locality is thought to be Mexico, range is unknown and type specimen is lost.
S. carri (Shreve, 1951) Carr's Snail-eater	Peters (1960c) Kofron (1980, 1985); Wilson & Meyer (1985); Villa et al. (1988); Wallach (1995); Köhler (2001)	Pacific versant of Guatemala, s Honduras, and El Salvador.
S. dimidiata* (Günther, 1872) Slender Snail-eater	Peters (1960c) Peters & Orejas-Miranda (1970); Savage (1980); Wilson & Meyer (1985); Savage & Villa (1986); Villa et al. (1988); Kofron (1990); Pérez-Higareda & Smith (1991); Savage & McDiarmid (1992); Wallach (1995); Lee (1996, 2000); Campbell (1998a); Köhler (2001) *Remark: Kofron (1990) and Wilson & Meyer (1985), did not recognize subspecies. However, subspecies were referenced by Campbell (1998a) so are included here.	SE Mexico into Central America.
S. d. dimidiata (Günther, 1872) Coral-backed Snail-eater	Peters (1960c); Peters & Orejas-Miranda (1970); Alvarez del Toro (1982); Campbell (1998a); Stafford & Meyer (2000); Lee (2000)	Atlantic versant from Veracruz in se Mexico into Guatemala and Nicaragua, excluding the Yucatán peninsula.
S. d. grandocula (Müller, 1887) Big-eyed Snail-eater	Peters (1960c); Peters & Orejas-Miranda (1970)	Pacific slopes of Chiapas in Mexico and w Guatemala.
S. dunni Peters, 1957 Dunn's Snail-eater	Peters (1957a, 1960b, 1960c); Peters & Orejas-Miranda (1970); Pérez-Santos & Moreno (1991); Wallach (1995)	Known only from the type locality of Pimanpiro, San Nicholas in Provencia Imabura, Ecuador.

Species-level Taxa	Citations and Remarks	Abbreviated Range
S. linearis Pérez-Higareda, López-Luna & Smith, 2002 Balzapote Snail-eater	Pérez-Higareda et al. (2002)	Los Tuxtlas region of Veracruz, Mexico. Known only from the holotype collected in Balzapote.
S. longifrenis (Stejneger, 1909) Mottled Snail-eater	Peters (1960c, 1964b); Peters & Orejas-Miranda (1970); Savage (1980); Savage & Villa (1986); Villa et al. (1988); Savage & McDiarmid (1992); Köhler (2001); McCranie et al. (2001)	Known only from three localities in the Atlantic versant of Costa Rica and Panama. Recently reported from Honduras.
S. nebulata (Linnaeus, 1758) Cloudy Snail-sucker	Peters (1960c); Peters & Orejas-Miranda (1970); Moonen et al. (1979); Savage (1980); Wilson & Meyer (1985); Villa et al. (1988); Savage & McDiarmid (1992); Ramírez-Bautista (1994); García & Ceballos (1994); Auth (1994); Wallach (1995); Lee (1996, 2000); Campbell (1998a); Starace (1998); Freira (1998); Kornacker (1999); Köhler (2001)	W and se Mexico along both coasts from Nayarit and Veracruz southward through Central America into n South America in Venezuela, Brazil, Ecuador, the Guianas, and Trinidad and Tobago.
S. n. nebulata (Linnaeus, 1758) Cloudy Snail-sucker	Peters (1960c, 1964b); Roze (1966); Peters & Orejas-Miranda (1970); Kofron (1980); Alvarez del Toro (1982); Wilson & Meyer (1985); Lancini & Kornacker (1989); Kofron (1990); Pérez-Higareda & Smith (1991); Murphy (1997); Campbell (1998a); Starace (1998); Stafford & Meyer (2000); Lee (2000); Kornacker (1999); Mijares-Urrutia & Arends R. (2000); Boos (2001); Köhler (2001)	Nayarit and Veracruz in s Mexico southward into n South America north of the Andes chain and Trinidad and Tobago, into as far as the Guianas; there is a disjunct population found below the range of *S. n. leucomelas* in nw Ecuador.
S. n. hartwegi Peters, 1960 Hartweg's Snail-sucker	Peters (1960c); Peters & Orejas-Miranda (1970); Pérez-Santos & Moreno (1988)	Upper reaches of the Río Magdalena and it tributaries, and the Río Porce Valley of Colombia.
S. n. leucomelas (Boulenger, 1896) Coastal Snail-sucker	Peters (1960b, 1960c); Peters & Orejas-Miranda (1970); Smith & Smith (1993); Köhler (2001)	S Panama—w Colombia border through c Colombia into nw Ecuador.
S. n. popayanensis Peters, 1960 Popayán Snail-sucker	Peters (1960c); Peters & Orejas-Miranda (1970); Pérez-Santos & Moreno (1988)	Upper Río Cauca in Colombia.
S. sanniola (Cope, 1866) Yucatán Snail-eater	Peters (1960c, 1964b); Peters & Orejas-Miranda (1970); Villa et al. (1988); Kofron (1990); Lee (1996, 2000); Wallach (1995); Köhler (2001)	S Mexico in the northern and eastern parts of the Yucatán peninsula, southward into Belize.
S. s. sanniola (Cope, 1866) Pygmy Snail-sucker	Peters & Orejas-Miranda (1970); Campbell (1998a); Stafford & Meyer (2000); Lee (2000); Köhler (2001)	Yucatán Peninsula in Mexico and Belize.
S. s. neilli Henderson, Hoevers & Wilson, 1977 Neill's Pygmy Snail-eater	Kofron (1985); Vanzolini (1986); Villa et al. (1988); Campbell (1998a); Stafford & Meyer (2000); Lee (2000); Köhler (2001)	Belize and e El Petén in Guatemala.

GENUS: *Sibynomorphus* Fitzinger, 1843, is found in South America south of the equator. See Peters (1960c); Peters & Orejas-Miranda (1970); Rossman & Thomas (1979); Cunha et al. (1980); Vanzolini (1986); Orcés V. & Almendáriz (1989a); Smith & Smith (1993); Lema (1994b); and Wallach (1995).

Review assistance and important contributions for this section were provided by Ronaldo Fernandes.

See also genera *Dipsas*, *Sibon* and *Tropidodipsas*.

CONTENT: Twelve species. ★★★

Sibynomorphus: South American Snail-eating Snake or South American Tree Snake. Frank and Ramus (1995) also call this the South American Tree Snake.

Species-level Taxa	Citations and Remarks	Abbreviated Range
S. inaequifasciatus (Dumeril, Bibron & Dumeril, 1854) Equator Tree Snake	Peters (1960c); Peters & Orejas-Miranda (1970); Orcés V. & Almendáriz (1989a)	Unknown; type locality listed as "South America doubtfully from Brazil."
S. lavillai Scrocchi, Porto & Rey, 1993 La Villa's Tree Snake	Scrocchi et al. (1993); Leynaud & Bucher (1999)	Santiago del Estero, Jujuy, Salto, Chaco, and Formosa in Argentina into Bolivia and Paraguay.
S. mikani* (Schlegel, 1837) Brazilian Slug-eater	Peters (1960c) Peters & Orejas-Miranda (1970); Hoge et al. [(1978/79] 1981); Cei (1993); Scrocchi et al. (1993); Lema (1994b); Wallach (1995); Yanosky (1998); Orcés V. & Almendáriz (1989a) *Remark: Shown in Cei (1993) without subspecies.	
S. m. mikani (Schlegel, 1837) Serrana Slug-eater	Peters (1960c) Peters & Orejas-Miranda (1970); Amaral (1976); Vanzolini (1986)	Inland se Brazil south of the Amazon Basin in Mato Grosso, Minas Geraís, Paraná, Río Grande do Norte, Río Grande do Sul, and São Paulo.
S. m. septentrionalis Cunha, Nascimento & Hoge, 1980 Maranhão Slug-eater	Cunha et al. (1980); Vanzolini (1986)	Maranhao, Brazil.
S. neuwiedi (Ihering, 1910) Eastern or Neuwied's Snail-eater	Peters (1960c); Hoge et al. [(1978/79] 1981); Vanzolini (1986); Cei (1993); Lema (1994); Freitas (1999) [as *Sibynomorphus neuwiedii*]; Marques et al. (2001)	S and se Brazil coastal area from Bahia into Río Grande do Sul.
S. oligozonatus* Orcés V. & Almendáriz, 1989 Ecuadorian Snail-eater	Orcés V. & Almendáriz (1989a) *Remark: The taxonomic status of this snake is unclear.	Peru and Ecuador.
S. oneilli Rossman & Thomas, 1979 O'Neill's Snail-eater	Rossman & Thomas (1979); Vanzolini (1986); Rossman & Kirzirian (1993); Orcés V. & Almendáriz (1989a)	Departamento Amazonas of nw Peru in semiarid brushland in the Río Marañón valley. The type is from about 1645 m.

Species-level Taxa	Citations and Remarks	Abbreviated Range
S. petersi* Orcés V. & Almendáriz, 1989 Peters's Snail-eater	Orcés V. & Almendáriz (1989) *Remark: The taxonomic status of this snake is unclear.	Ecuador and Peru.
S. turgidus (Cope, 1868) Tigrada Snail-eater	Orejas-Miranda (1958); Peters (1960c); Peters & Orejas-Miranda (1970); Amaral (1976); Campbell & Murphy (1984); Achaval (1987); Böckeler (1988); Orcés V. & Almendáriz (1989a); Cei (1993); Lema (1994b); Wallach (1995); Yanosky (1998); Leynaud & Bucher (1999)	N Paraguay; se Bolivia; the state of Mato Grosso in Brazil and Argentina in Misiones, Corrientes, Chaco, Entre Rios, Santa Fe, Córdoba, Tucumán, Salta, and Jujuy; Orejas-Miranda (1958) listed it as present in Uruguay.
S. vagrans (Dunn, 1923) Bellavista Snail-eater	Peters (1960c); Orcés V. & Almendáriz (1989a); Rossman & Thomas (1979); Kofron (1980); Henle & Ehrl (1991b)	Known only from the type locality in Bellavista, Departamento Cajamarca, Peru in dry forest and cactus-acacia scrub.
S. vagus (Jan, 1863) Huancabamba Snail-eater	Peters (1960c); Peters & Orejas-Miranda (1970); Orcés V. & Almendáriz (1989a); Rossman & Thomas (1979); Rossman & Kizirian (1993); Wallach (1995)	Known only from Huancabamba, Departamento Piura, Peru. Found on the Pacific slope of the Andes at about 1980 m.
S. ventrimaculatus* (Boulenger, 1885) Boulenger's Tree Snake	Peters (1960c, 1964b); Peters & Orejas-Miranda (1970); Amaral (1976); Cei (1993); Lema (1994b); Wallach (1995); Fernandes et al. (1998); Yanosky (1998); Giraudo (1999) *Remark: *Heterorhachis poecilolepis* Amaral, 1923 was placed in *Dipsas poecilolepis* by Peters (1960c). It was also listed as such in Peters & Orejas-Miranda (1970), where they showed it was known only from the type locality of Villa Bomfin in the state of Sao Paulo in Brazil, and listed its as *Dipsas poecilolepis*. Peters (1970), decided the *polylolepis* group, in which *D. poecilolepis* was included, was artificial and suggested that it could be a specimen of *Sibynomorphus ventrimaculatus* or *Sibynomorphus turgidus*. Amaral (1976) listed *Heterorhachis poecilolepis* as a valid taxon. Most recently, Fernandes et al. (1998), examined this snake and revalidated it as synonym of *Sibynomorphus ventrimaculatus*.	S Paraguay; Uruguay; and the state of Rio Grande do Sul in s Brazil into Misiones, Corrientes, and Formosa in ne Argentina.
S. williamsi Carrillo de Espinoza, 1974 Williams's Snail-eater	Carillo de Espinoza (1974); Orcés V. & Almendáriz (1989a); Rossman & Thomas (1979); Moreales et al. (1990); Henle & Ehrl (1991b)	Jicamarca in Lima, Peru

GENUS: *Simophis* W. Peters, 1860, is another genus of false coral snake that is found in Brazil and Paraguay. See Peters & Orejas-Miranda (1970); Amaral (1976); and Santos-Jordão & Fernandes-Bizerra (1996).

CONTENT: Two species. ★★

Simophis: Tri-colored False Coral Snake. Frank & Ramus (1995) called this the Snouted Snake.

Species-level Taxa	Citations and Remarks	Abbreviated Range
S. rhinostomus (Schlegel, 1837) São Paulo False Coral Snake	Peters & Orejas-Miranda (1970); Amaral (1976); Savage & Slowinski (1992); Santos-Jordão & Fernandes-Bizerra (1996); Argólo (1998); Freitas (1999) [as *Simophis rhynostoma*]	Brazil in Bahia, Minas Geraís, São Paulo, and Paraná; associated with the dryer regions.
S. rohdei (Boettger, 1885) Paraguay False Coral Snake	Boettger (1885); Peters & Orejas-Miranda (1970); Savage & Slowinski (1992); Yanosky (1998)	Paraguay.

GENUS: *Siphlophis* Fitzinger, 1843, is found from Costa Rica southward into South America to as far as Brazil and Bolivia. See Bailey *in* Peters & Orejas-Miranda (1970); Kofron (1980); Murphy (1997); Starace (1998); Kornacker (1999) and Köhler (2001).

Review assistance and important contributions for this section were provided by Thales de Lema.

CONTENT: Six species. ★★★★

Siphlophis: Liana Snake. Frank and Ramus (1995) call this the Spotted Night Snake.

Species-level Taxa	Citations and Remarks	Abbreviated Range
S. cervinus (Laurenti, 1768) Common Liana Snake	Underwood (1962); Hoge (1964); Amaral (1976); Duellman (1978); Dixon & Soini (1986); Nascimento et al. (1987); Villa et al. (1988); Almendariz (1991); Savage & Slowinski (1992); Murphy (1997); Starace (1998); Barrio et. al. (1998); Kornacker (1999); Boos (2001); Köhler (2001)	Panama into n South America to as far as c Bolivia, in Maranhão in Brazil; it is also present in Trinidad and the Guianas.
S. compressus* (Daudin, 1803) Red-eyed Tree Snake	Underwood (1962); Roze (1966); Bailey *in* Peters & Orejas-Miranda (1970); Amaral (1976) [as *Trypanurgos compressus*]; Duellman (1978); Lancini (1979); Dixon & Soini (1986); Savage & Villa (1986); Villa et al. (1988); Lancini & Kornacker (1989); Savage & Slowinski (1992); Murphy (1997); Starace (1998); Zaher & Prudente (1999); Kornacker (1999); Marques et al. (2001); Boos (2001) [as *Tripanurgos compressus*]; Köhler (2001) [as *Tripanurgos compressus*] *Remark: Until recently this snake was classified as *Tripanurgos compressus*. Zaher & Prudente (1999) reclassified it to *Siphlophis compressus*.	Discontinuous range from Costa Rica in Central America southward into nw South America to the mouth of the Amazon River into the state of Sergipe into the state of Rio de Janeiro, Brazil, and in c Bolivia and Trinidad.
S. leucocephalus (Günther, 1863) Günther's Liana Snake	Boulenger (1896); Bailey *in* Peters & Orejas-Miranda (1970)	EC Brazil in Bahía, Goiás and Minas Geraís.
S. longicaudatus (Andersson, 1901) Brazilian Liana Snake	Lema (1964); Hoge (1964); Bailey *in* Peters & Miranda (1970); Kofron (1980); Villa et al. (1988); Lema (1994b); Marques et al. (2001)	SE Brazil from Espirito Santo into Rio Grande do Sul.

Species-level Taxa	Citations and Remarks	Abbreviated Range
S. pulcher* (Raddi, 1820) Gunanbara Liana Snake	Bailey in Peters & Orejas-Miranda (1970); Amaral (1976) [as S. cervinus pulcher]; Lema (1994b); Freitas (1999) [as Syphlophis pulcher]; Marques et al. (2001) *Remark: See Bailey in Peters & Orejas-Miranda (1970) for comment about why this name has priority over S. geminatus Dumeril, Bibron & Dumeril, 1854.	SE Brazil in Gunanbara and s Minas Geraís into Rio Grande do Sul.
S. worontzowi (Prado, [1939] 1940) Rio Amana Liana Snake	Prado (1940); Bailey in Peters & Orejas-Miranda (1970); Freitas (1999) [as Tripanurgos compressus]	Known only from the type locality in Rio Amana in the state of Amazonas in Brazil.

GENUS: *Sonora* Baird & Girard, 1853, is from Mexico and the United States. See Stickel (1943); Smith & Taylor (1966); Conant (1975); Frost (1983b); Frost & Van Devender (1979); Savage & Slowinski (1992); Smith & Smith (1993); Degenhardt et al. (1996); and Dixon (2000a).

Review assistance and important contributions for this section were provided by Peter Holm, James R. Dixon and John E. Werler.

CONTENT: Three species. ★★★

Sonora: North American Groundsnake. Frank and Ramus (1995) call this the Desert Ground Snake.

Species-level Taxa	Citations and Remarks	Abbreviated Range
S. aemula (Cope, 1879) File-Tailed Groundsnake	Cope (1879); Nickerson & Heringhi (1966); Hardy & McDiarmid (1969); Savage & Slowinski (1992); Smith & Smith (1993)	NW Mexico in w and sw Chihuahua, e Sonora, and all of Sinaloa.
S. michoacanensis (Dugès in Cope, 1884) Michoacán Groundsnake	Stickel (1943); Smith & Taylor (1966); Echternacht (1973); Savage & Slowinski (1992); Smith & Smith (1993)	C and s Mexico.
S. m. michoacanensis (Dugès in Cope, 1884) Michoacán Groundsnake	Stickel (1943); Peters (1954); Smith & Taylor (1966); Smith & Smith (1993)	S Mexico, probably in the higher elevations of the lower Río Balsas Basin.
S. m. mutabilis Stickel, 1943 Mexican Groundsnake	Stickel (1943); Smith & Taylor (1966); Smith & Smith (1993)	C Mexico in the southern portion of the main Central Plateau from wc Jalisco through s Zacatecas and into the Districto Federal.
S. semiannulata* Baird & Girard, 1853 Sonoran Groundsnake or Variable Groundsnake	Conant (1975); Frost (1983b); Frost & Van Devender (1979); Tanner (1985); Stebbins (1985); Savage & Slowinski (1992); Degenhardt et al. (1996); Brown (1997); Tennant (1998); Conant & Collins (1998); McPeak (2000); Bartlett & Tennant (2000); Tennant & Bartlett (2000); Dixon (2000a)	Disjunct populations in sw United States southward into n Mexico from sw Missouri, s Kansas, Nevada, and California into nc Kansas, se Colorado, adjacent Oklahoma, Oregon, Idaho, Nevada, Utah, and Baja California.

Species-level Taxa	Citations and Remarks	Abbreviated Range
	*Remark: Smith & Smith (1993), Frost (1983b), Brown (1997), and Conant & Collins (1998) considered this to be a monotypic snake. Frost & Van Devander (1979) showed *S. semiannulata* to be a highly variable species throughout its range and with no subspecies. However, they did not take into consideration *taylori*. This was reviewed with James R. Dixon and John E. Werler (pers. comm. June 11, 1999), who agreed that *S. a. taylori* should remain as a valid subspecies. Based on this information, two subspecies are shown in this list.	
S. s. semiannulata Baird & Girard, 1853 Santa Rita Groundsnake	Smith & Taylor (1966); Smith & Smith (1993); Werler & Dixon (2000); Dixon (2000a)	Throughout the species range except where *S. s. taylori* is found.
*S. s. taylori** (Boulenger, 1894) South Texas Groundsnake	Smith & Taylor (1966); Conant (1975); Frost & Van Devander (1979); Smith & Smith (1993); Werler & Dixon (2000); Dixon (2000a) *Remark: Fide Frost & Van Devander (1979), subspecies are not valid; however, they did not review *S. e. taylori*. Based on Dixon (2000a), plus James R. Dixon and John E. Werler (pers. comm. February 2000), *taylori* is a valid subspecies. This subspecies was used by Werler & Dixon (2000).	S Texas into ne Mexico where it is primarily restricted to the Tamaulipian biotic province and prairies to the northeast in Mexico (see Dixon, 2000a).

GENUS: *Sordellina* Procter, 1923, is found in southern South America. See Peters & Orejas-Miranda (1970) and Amaral (1976).

CONTENT: One species. ★★★★

Sordellina: Brazilian Forest Snake. Frank and Ramus (1995) call this the Dotted Brownsnake.

Species-level Taxa	Citations and Remarks	Abbreviated Range
S. punctata (W. Peters, 1880) Brazilian Forest Snake	W. Peters (1880); Hoge (1958); Peters & Orejas-Miranda (1970); Amaral (1976); Marques et al. (2001)	SE Brazil in Rio de Janeiro, Santa Catarina, São Paulo and Paraná; associated with the Atlantic Forests of s Brazil; generally associated with low and tropical forests near water sources.

GENUS: *Spilotes* Wagler, 1830, is wide ranging from Mexico southward into South America as far as Brazil, Argentina, and Paraguay. See Amaral (1930d, 1976); Peters & Orejas-Miranda (1970); Savage & Villa (1986); Lancini & Kornacker (1989); Orcés & Almendáriz (1989b); Cei (1993); Smith & Smith (1993); Starace (1998); Kornacker (1999); Köhler (2001).

CONTENT: Often considered to be one species with about five subspecies. Two species are listed here ★★

Spilotes: Tiger Ratsnake

Species-level Taxa	Citations and Remarks	Abbreviated Range
S. megalolepis Günther, 1865 Ecuadorian Tiger Ratsnake	Günther (1865); Boulenger (1894); Amaral (1929); Peters & Orejas-Miranda (1970) [as a synonym of *S. p. pullatus*] Orcés & Almendáriz (1989b);	Lowlands of w Ecuador.
S. pullatus (Linnaeus, 1758) Tiger Ratsnake	Peters & Orejas-Miranda (1970); Dixon & Soini (1986); Wilson & Meyer (1995); Villa et al. (1988); Orcés & Almendáriz (1989b); Lee (1996, 2000); Campbell (1998a); Starace (1998); Renjifo & Lundberg (1999); Kornacker (1999); Marques et al. (2001); Köhler (2001)	S Mexico southward into much of northern South America to as far south as Argentina and Paraguay.
S. p. pullatus (Linnaeus, 1758) Common Tiger Ratsnake	Roze (1966); Peters & Orejas-Miranda (1970); Orcés & Almendáriz (1989b) Lancini & Kornacker (1989); Cei (1993); Lema (1994b); Murphy (1997); Starace (1998); Kornacker (1999); Freitas (1999); Mijares-Urrutia & Arends R. (2000)	Costa Rica in Central America southward into c South America in Paraguay and n Argentina in Misiones, Chaco, Formosa, and Salta, and the islands of Trinidad and Tobago. Barbour (1930) has it from se Nicaragua. Kornacker (1999) also shows it as found in Nicaragua. In Ecuador, it is isolated to the east.
S. p. anomalepis Bocourt, 1888 *in* Dumeril, Mocquard & Bocourt, 1870–1909 Southern Tiger Ratsnake	Barbour (1930); Hoge (1952a); Peters & Orejas-Miranda (1970); Abe & Fernandes (1977); Lema (1994b)	SE Brazil in Bahia and Rio Grande do Sul.
S. p. argusiformis Amaral, 1930 Northern Tiger Ratsnake	Amaral (1930); Peters & Orejas-Miranda (1970); Wilson & Meyer (1985)	Honduras in Central America into Nicaragua.
S. p. maculatus Amaral, 1930 Yellow Tiger Ratsnake	Amaral (1930, 1976); Peters & Orejas-Miranda (1970)	State of Sao Paulo, Brazil, in the Cubatão and Paranapiacaba mountains of the coast.
S. p. mexicanus (Laurenti, 1768) Mexican Tiger Ratsnake	Smith & Taylor (1966); Peters & Orejas-Miranda (1970); Alvarez del Toro (1982); Wilson & Meyer (1985); Pérez-Higareda & Smith (1991); Campbell (1998a); Meyer (2000); Lee (2000)	Tamaulipas and Oaxaca in Mexico southward into Guatemala, Belize, and Honduras.

GENUS: *Stenorrhina* Dumeril, 1853, is distributed from central Veracruz and central Guerrero in Mexico southward into northern South America to as far as Peru. See Smith & Taylor (1966); Peters & Orejas-Miranda (1970); Savage & Villa (1986); Lancini & Kornacker (1989); and Kornacker (1999); and Köhler (2001).

Important contributions for this section were provided by Peter Holm.

CONTENT: Two species. ★★

Stenorrhina: Middle American Scorpion-eating Snake, Latin American Spider-eating Snake, or Latin American Centipede-eating Snake. Frank and Ramus (1995) call this the Blood Snake.

Species-level Taxa	Citations and Remarks	Abbreviated Range
S. degenhardti* (Berthold, 1845) Degenhardt's Scorpion-eater	Savage (1980); Wilson & Meyer (1985); Savage & Villa (1986); Villa et al. (1988); Lancini & Kornacker (1989); Lee (1996, 2000); Stafford (1996); Campbell (1998a); Kornacker (1999); Köhler (2001) *Remark: Subspecies are not always recognized. Wilson & Meyer (1985) rejected all subspecies. Subspecies are noted here because they continue to be listed in various literature.	S Mexico southward through Central America into n South America in Venezuela and w Ecuador.
S. d. degenhardti (Berthold, 1845) Degenhardt's Scorpion-eater	Peters (1960b); Peters & Orejas-Miranda (1970); Freiberg (1982); Renjifo & Lundberg (1999)	Panama into w Colombia, Peru, and Ecuador.
S. d. mexicana* (Steindachner, 1867) Mexican Scorpion-eater	Smith (1943); Smith & Taylor (1966); Peters & Orejas-Miranda (1970); Wilson & Meyer (1985); Pérez-Higareda & Smith (1991); Campbell (1998a); Stafford & Meyer (2000); Lee (2000) *Remark: Higareda & Smith (1991), noted that "Wilson & Meyer (1985: 97–99, figs. 93–94) described the species and discuss the purported subspecies, rejecting all. It seems likely, however, that at least *S. d. mexicana*, the northern terminal population of this species, ranging northward from Venezuela and Pacific Ecuador, is valid."	S Mexico from c Veracruz into Guatemala and Belize.
S. d. ocellata Jan, 1876 *in* Jan & Sordelini, 1870–1881 Sangrita Scorpion-eater	Roze (1966); Peters & Orejas-Miranda (1970); Lancini (1979); Wilson & Meyer (1985); Lancini & Kornacker (1989); Kornacker (1999); Mijares-Urrutia & Arends R. (2000)	NC and nw Venezuela and possibly Colombia.
S. freminvilli* Dumeril, Bibron & Dumeril, 1854 Blood Snake	Stuart (1963); Peters & Orejas-Miranda (1970); Savage (1980); Alvarez del Toro (1982); Wilson & Meyer (1985); Savage & Villa (1986); Villa et al. (1988); Pérez-Higareda & Smith (1991); Savage & Slowinski (1992); Flores-Villela (1993); Campbell (1998a); Stafford & Meyer (2000); Lee (2000); Köhler (2001) *Remark: Subspecies are not normally considered valid. Peters & Orejas-Miranda (1970) did not use subspecies because Stuart (1963) pointed out the difficulties of defining subspecies and recommended use of the specific name only. Smith & Smith (1993) did not list subspecies.	Mexico from Guerrero and the Isthmus of Tehuantepec southward through Central America into Costa Rica.

GENUS: *Stilosoma* Brown, 1890, is endemic to Florida in the extreme southeastern United States. See Conant (1975); Highton (1976), Ashton & Ashton (1981); Dowling & Mason (1990) and Wright & Wright ([1957] 1994).

CONTENT: One species, sometimes considered to have up to three subspecies. ★★★

Stilosoma: Short-tailed Snake

Species-level Taxa	Citations and Remarks	Abbreviated Range
*S. extenuatum** Brown, 1890 Short-tailed Snake	Conant (1975); Highton (1976); Ashton & Ashton (1981); Tennant (1997b); Conant & Collins (1998); Tennant & Bartlett (2000)	The United States in Florida from Highlands county into Suwannee county.
	*Remark: Highton (1976) considered *Stilosoma* to be monotypic and synonymized the following subspecies that he originally recognized in Highton (1956) as: *S. e. extenuatum* Brown, 1890, from c peninsular Florida north to south, but not near the coast. Wright & Wright ([1957] 1994) showed it from the counties of Putnam, Marion, Lake, Seminole, Orange and Polk; *S. e. arenicola* Highton, 1956, from w peninsular Florida at the Gulf coast. Wright & Wright ([1957] 1994) listed it from the counties of Alachua and Levy; and, *S. e. multistictum* Highton, 1956, w peninsular Florida at the Gulf coast. Wright & Wright ([1957] 1994) listed them from the counties of Citrus, Sumter, Hernando, Pasco and Pinellas.	

GENUS: *Storeria* Baird & Girard, 1853, is found from southern Canada southward through the central and eastern United States and Mexico into Central America to as far as Guatemala and Honduras. See Trapido (1944); Conant (1975); Christman (1982); Dundee & Rossman (1989); Tennant (1998); Conant & Collins (1998); and Dixon (2000a).

Review assistance and important contributions for this section were provided by James A. Dixon.

CONTENT: Four species. ★★★★

Storeria: North American Brownsnake, Red-bellied Snake, and DeKay's Snake. Crother et al. (2000) call this the North American Brownsnake.

Species-level Taxa	Citations and Remarks	Abbreviated Range
S. dekayi (Holbrook, 1836 [1839?]) DeKay's Brownsnake	Trapido (1944); Conant (1975); Christman (1982); Villa et al. (1988); Dundee & Rossman (1998); Smith & Smith (1993); Campbell (1998a); Lee (2000); Dixon (2000a)	S Canada southward through the eastern United States into Central America; Lee (2000), noted there are apparently disjunct populations in e Chiapas, c Guatemala, and w Honduras.
S. d. dekayi (Holbrook, 1836 [1839?]) Northern Brownsnake	Conant (1975); Christman (1982); Conant & Collins (1998); Tennant & Bartlett (2000)	Quebec, Canada southward into the United states from s Maine into North Carolina.
S. d. anomala Dugès, 1888 Dugès' Brownsnake	Trapido (1944); Anderson (1961); Smith & Taylor (1966); Christman (1982); Wilson & Meyer (1985); Pérez-Higareda & Smith (1991); Smith & Smith (1993)	Foothills of c Veracruz, Mexico.

Species-level Taxa	Citations and Remarks	Abbreviated Range
S. d. limnetes Anderson, 1961 Marsh Brownsnake	Anderson (1961); Sabath & Sabath (1969); Conant (1975); Christman (1982); Dundee & Rossman (1989); Tennant (1997b, 1998); Conant & Collins (1998); Werler & Dixon (2000); Tennant & Bartlett (2000); Dixon (2000a)	The United States Gulf Coast of Texas, Louisiana, Mississippi, and Alabama to as far east as Mobile Bay.
S. d. temporalineata Trapido, 1944 Trapido's Brownsnake	Trapido (1944); Anderson (1961); Smith & Taylor (1966); Christman (1982); Pérez-Higareda & Smith (1991) Smith & Smith (1993)	Coastal plain of eastern Mexico from Tamaulipas southward into Guatemala.
S. d. texana Trapido, 1944 Texas Brownsnake	Trapido (1944); Smith & Taylor (1966); Conant (1975, 1978); Christman (1982); Dundee & Rossman (1989); Tennant (1998); Conant & Collins (1998); Dixon (2000a); Tennant & Bartlett (2000); Werler & Dixon (2000)	The United States in w Michigan and Minnesota to Texas and southward into se Mexico into the foothills of Hidalgo.
S. d. tropica Cope, 1885 Tropical Brownsnake	Trapido (1944); Anderson (1961); Peters & Orejas-Miranda (1970); Conant (1975); Christman (1982); Wilson & Meyer (1985); Campbell (1998a); Lee (2000)	Caribbean versant of n and c Guatemala to n Honduras; it is an isolated Central American subspecies.
*S. d. victa** Hay, 1892 Florida Brownsnake	Neill (1950); Conant (1975); Ashton & Ashton (1981); Christman (1982); Tennant (1997b); Conant & Collins (1998); Tennant & Bartlett (2000) *Remark: According to Crother et al. (2000), Christman (1980) presented evidence to suggest full specific status as *S. victa* for this snake; however, Christman (1982) continued to list *S. d. victa* as a subspecies in this later publication.	SE Georgia and Florida as far as the lower keys in the United States.
S. d. wrightorum Trapido, 1944 Midland Brownsnake	Conant (1975); Christman (1982); Dundee & Rossman (1989); Tennant (1997); Conant & Collins (1998); Tennant & Bartlett (2000)	N Mississippi Valley of the United States from e Wisconsin into the Carolinas and Gulf Coast into Louisiana.
S. hidalgoensis Taylor, 1942 Hidalgo or Mexican Yellow-bellied Brownsnake	Taylor (1942); Smith & Taylor (1966); Smith & Smith (1993)	C Mexican Plateau.
S. occipitomaculata (Storer, 1839) Red-bellied Snake	Conant (1975); Stebbins (1985); Dundee & Rossman (1989); Conant & Colllins (1998); Dixon (2000a)	Nova Scotia and e Saskatchewan in Canada southward through the United States in ne North Dakota, e South Dakota, Kansas, Oklahoma, and into Texas east of the Trinity River and into c Florida. There are disjunct populations in Wyoming and the Black Hills of South Dakota, c Nebraska, ne Missouri, and nw Ohio.
S. o. occipitomaculata (Storer, 1839) Northern Red-bellied Snake	Conant (1975); Ashton & Ashton (1981); Tennant (1998); Conant & Collins (1998); Tennant & Bartlett (2000)	Throughout the Red-Bellied Snake range except the Black Hills of South Dakota in the United States.

Species-level Taxa	Citations and Remarks	Abbreviated Range
S. o. obscura Trapido, 1944 Florida Red-bellied Snake	Conant (1975); Ashton & Ashton (1981); Dundee & Rossman (1989); Tennant (1997b, 1998); Conant & Collins (1998); Dixon (2000a); Tennant & Bartlett (2000); Werler & Dixon (2000)	E Texas coast eastward into c Florida in the United States.
S. o. pahasapae Smith, 1963 Black Hills Red-bellied Snake	Conant (1975); Stebbins (1985); Conant & Collins (1998); Bartlett & Tennant (2000); Tennant & Bartlett (2000)	Wooded areas of the Black Hills of extreme w South Dakota and Wyoming in the United States.
S. storerioides (Cope, 1865) Mexican Brownsnake	Davis & Smith (1953); Smith & Taylor (1966); Hardy & McDiarmid (1969); Tanner (1985); Smith & Smith (1993); Uribe-Peña et al. (1999)	C Mexican Plateau into the mountains of sw Chihuahua; it is reported from Chihuahua, Durango, Guerrero, Guanajuato, Jalisco, México, Michoacán, Morelos, Puebla, San Luis Potosí, and the Federal District.

GENUS: *Symphimus* Cope, (1869) 1870, is principally a Mexican snake that includes a few records from Belize. It may eventually be found in El Petén in Guatemala. See Smith & Taylor (1966); Rossman & Schaffer (1974); Williams & Wallach (1989); Lee (1996, 2000); and Köhler (2001).

CONTENT: Two species. ★★★★

Symphimus: White-lipped Snake

Species-level Taxa	Citations and Remarks	Abbreviated Range
S. leucostomus Cope, (1869) 1870 Isthmican White-lipped Snake	Smith & Taylor (1966); Rossman & Schaffer (1974); Villa et al. (1988); Williams & Wallach (1989); García & Ceballos (1994); Ramírez-Bautista (1994); Köhler (2001)	S Mexico in the Isthmus of Tehuantepec in Oaxaca and along the Pacific coast from Jalisco into Chiapas, from 200 to 1000 m.
S. mayae* (Gaige, 1936) Mayan Golden-backed Snake	Smith & Taylor (1966); Rossman & Schaffer (1974); Villa et al. (1988); Lee (1996, 2000); Campbell (1998a); Stafford & Meyer (2000); Köhler (2001) *Remarks: *Symphimus mayae* is sometimes found in older literature as *Opheodrys mayae* (see Smith & Taylor, 1966).	Yucatán peninsula in Mexico in the states of Yucatán and Quintana Roo, southward into Belize and Corozal districts of Belize; it probably also occurs in El Petén in Guatemala.

GENUS: *Sympholis* Cope, [1861] 1862, is restricted to western Mexico. See Smith & Taylor (1966); Hensley (1966); Hardy & McDiarmid (1969); and Williams & Wallach (1989).

Important contributions for this section came from Peter Holm.

CONTENT: One species. ★★

Sympholis: Mexican Short-tailed Snake

Species-level Taxa	Citations and Remarks	Abbreviated Range
*S. lippiens** Cope, 1861 (1862) Mexican Short-tailed Snake	Hensley (1966); Williams & Wallach (1989); Smith & Smith (1993) *Remark: Subspecies are questionable and are listed here for reference only.	W Mexico.
S. l. lippiens Cope, 1861 (1862) Mexican Banded Burrowing Snake	Hensley (1966); Hardy & McDiarmid (1969); Smith & Smith (1993)	W Mexico in w Jalisco and e Nayarit.
S. l. rectilimbus Hensley, 1966 Northern Mexican Short-tailed Snake	Hensley (1966); Nickerson & Heringhi (1966); Hardy & McDiarmid (1969); Tanner (1985); Smith & Smith (1993)	Sonora, Sinaloa and Chihuahua.

GENUS: *Synophis* Peracca, 1896, is found in the Upper Amazon Basin in Colombia and Ecuador. Peters & Orejas-Miranda (1970); Hillis (1990); Sheil (1998); and Dowling & Pinou (2003).

See also genus *Emmochliophis*.

CONTENT: Four species. ★★★★

Synophis: Upper Amazonian Basin Watersnake. Frank & Ramus (1995) called this the Fishing Snake.

Species-level Taxa	Citations and Remarks	Abbreviated Range
S. bicolor Peracca, 1896 Bicolored Watersnake	Nicéforo-María (1970); Peters & Orejas-Miranda (1970); Pérez-Santos & Moreno (1991) Williams & Wallach (1989); Hillis (1990); Sheil (1998); Sheil & Grant (2001)	Amazonian Ecuador from the Andean foothills.
S. calamitus Hillis, 1990 Hillis's Watersnake	Hillis (1990); Sheil (1998); Sheil & Grant (2001)	Amazonian Ecuador and Colombia. Associated with Pacific cloud forests.
S. lasallei (Niceforo-Maria, 1950) Lowland Watersnake	Nicéforo-María (1950) [as *Diaphorolepis*]; Peters & Orejas-Miranda (1970); Vanzolini (1986); Pérez-Santos & Moreno (1988 1991); Hillis (1990); Sheil (1998); Sheil & Grant (2001)	Amazonian lowland areas of Colombia and Ecuador but noted that the type is from 2200 m in Albán, Cundinamarca Province in Colombia.
S. plectovertebralis Sheil & Grant, 2001 Braided-back Watersnake	Sheil & Grant (2001)	Pacific versant of the Cordillera Occidental of w Colombia; in cloud forest at 1800 m.

GENUS: *Tachymenis* Wiegmann, 1835, [sometimes spelled *Tachymensis* or *Tachimenis*], is a snake found in higher elevations of western coastal South America in the Pacific versant of the Andes along the coast of Peru and Chile, into Amazonian Peru, Bolivia, and into nw Argentina. See Peters & Orejas-Miranda (1970), Vanzolini (1986), Cei (1986).

The Caenophidia

Review assistance and important contributions for this section were provided by Robert A. Thomas.

CONTENT: Five species. ★★

Tachymenis: Swift Andean Snake. Frank & Ramus (1995) called this the Slender Snake. However, anyone who has attempted to capture these snakes realize that the translation for the technical name, "swift snake," is also very appropriate.

Species-level Taxa	Citations and Remarks	Abbreviated Range
T. affinis Boulenger, 1896 Highland Swift Andean Snake	Boulenger (1896); Walker (1945); Peters & Orejas-Miranda (1970); Henle & Ehrl (1991b)	Highland valleys in Peru.
T. attenuata Walker, 1945 Walker's Scrub Snake	Walker (1945); Peters & Orejas-Miranda (1970); Henle & Ehrl (1991b)	E Andean slopes of Bolivia and s Peru.
T. a. attenuata Walker, 1945 Walker's Swift Snake	Walker (1945); Peters & Orejas-Miranda (1970)	Departamento Madre de Dios, Peru and Departamento Cochabamba, Bolivia.
T. a. boliviana Walker, 1945 Bolivian Slender Snake	Walker (1945); Peters & Orejas-Miranda (1970)	Edge of the Amazon basin in Bolivia.
T. chilensis (Schlegel, 1837) Chilean Swift Andean Snake	Donoso-Barros (1966); Ortiz (1973); Vanzolini (1986); Cei (1986, 1993)	C Chile with limited distribution into far w Argentina.
T. c. chilensis (Schlegel, 1837) Chilean Swift Andean Snake	Donoso-Barros (1966); Peters & Orejas-Miranda (1970); Ortiz (1973); Vanzolini (1986); Núñez (1992); Cei (1986, 1993)	C Chile from s of Santiago into Chiloé with limited range in Argentina to around San Carlos de Bariloche and the frontier of Neuquén.
T. c. coronellina (Werner, 1898) Central Chile Swift Andean Snake	Werner (1898); Peters & Orejas-Miranda (1970); Ortiz (1973); Vanzolini (1986); Cei (1986)	C Chile between Copiapó and Santiago.
T. c. melanura (Walker, 1945) Dark Swift Andean Snake	Donoso-Barros (1966); Silva et al. (1968); Peters & Orejas-Miranda (1970); Ortiz (1973); Cei (1986)	Coastal sc Chile from about 42°S to about 44°S including the islands of Chiloé and Calbuco.
*T. peruviana** Wiegmann, 1836 Peruvian Slender Snake	Boulenger (1896); Donoso-Barros (1966); Ortiz (1973); Cei (1986, 1993); Vanzolini (1986); Henle & Ehrl (1991b)	Coastal Peru and Chile, into Andean Peru and Bolivia to the altiplanicie of Junín into Oruro, Potosí, and Sucre to 3800 m, and in Chile in the Andean Tarapacá into Jujuy in Argentina.
	*Remark: Ortiz (1973) does not recognize subspecies. However, Vanzolini (1986) shows subspecies as valid. Cei (1993) lists subspecies in his literature history but does not discuss subspecies for this snake. Be-	

Species-level Taxa	Citations and Remarks	Abbreviated Range
	cause of the lack of current subspecies information, I have kept information on this snake to the species level. Subspecies if valid, seem to be limited to *T. p. peruviana* Wiegmann, 1834, from Coastal Peru into n Chile southward to as far as Antofagasta, Chile; and, *T. p. yutoensis* Miranda & Couturier, 1981, from Yuto in Jujuy, nw Argentina (see Miranda & Couturier)	
T. tarmensis Walker, 1945 Tarma Swift Snake	Walker (1945); Peters & Orejas-Miranda (1970); Henle & Ehrl (1991b)	Known only from the type locality in Tarma, Departamento Junín, Peru.

GENUS: *Taeniophallus* Cope, 1895, has been established around the *Rhadinaea "brevirostris"* group of snakes that had been classified in the genus *Echinanthera*. This group is found in South America from the northeast in Suriname southward into eastern Peru, Bolivia, Brazil, Argentina, and Paraguay. Di-Bernardo (1996) states that *T. nicagus* should be the only valid snake species in *Taeniophallus* and the rest belong to *Echinanthera*. See Di-Bernardo (1992, 1996); Myers & Cadle (1994); and Kornacker (1999). This work has followed Myers & Cadle (1994).

See also genus *Echinanthera*.

CONTENT: Seven species listed, but the valid species for *Taeniophallus* may be as few as one with the rest removed to *Echinanthera*. ★★★

Taeniophallus: South American Groundsnake or Savanna Racer

Species-level Taxa	Citations and Remarks	Abbreviated Range
T. affinis (Günther, 1858) Common Savanna Racer	Myers (1974); Amaral (1976); Di-Bernardo (1992); Myers & Cadle (1994); Lema (1994b); Argólo (1998); Freitas (1999) [as *Echynanthera affinis*]; Marques et al. (2001) [as *Echinanthera affinis*]	S and se Brazil in se Minas Geraís through Espirito Santo, Rio de Janiero, São Paulo, Paraná, Santa Catarina, into ne Rio Grande do Sul and Bahía.
T. bilineatus (Fischer, 1885) Two-lined Savanna Racer	Myers (1974); Di-Bernardo (1992); Myers & Cadle (1994); Lema (1994b); Marques et al. (2001) [as *Echinanthera bilineatus*]	S and se Brazil in s Minas Geraís through Espirito Santo, Rio de Janiero, São Paulo, Paraná, and Santa Catarina into ne Rio Grande do Sul.
T. brevirostris (W. Peters, 1863) Short-nosed Groundsnake	Myers (1974); Duellman (1978); Lancini & Kornacker (1989); Di-Bernardo (1992); Myers & Cadle (1994); Duellman & Mendelson III (1995); Starace (1998) [as *Rhadinaea brevirostris*]; Yuki (1999); Kornacker (1999)	Amazon Basin in Amazonas and Acre in Brazil, the Guianas, s Colombia, Bolivia, n Peru, and Zulia in Venezuela; southward along the eastern base of the Andes, and eastward through the Amazon Basin.
T. nicagus Cope, 1868 Guianan Savanna Racer	Myers (1974); Myers & Cadle (1994)	Suriname, French Guiana, and Brazil.
T. occipitalis (Jan, 1863) Spotted Savanna Racer	Hoge (1952b); Myers (1974); Amaral (1976); Cunha et al. (1985); Dixon & Soini (1986); Achaval (1987); Di-Bernardo (1992); Myers & Cadle (1994); Di-	NE Peru and Bolivia into s Brazil; and, into Paraguay and n Argentina in Córdoba, Santiago del Estero, Tucumán, Salta, Formosa, Chaco, and Corrientes; it

Species-level Taxa	Citations and Remarks	Abbreviated Range
	Bernardo (1996); Yuki & Santos (1998); Yanosky (1998); Leynaud & Bucher (1999); Freitas (1999) [as *Echynanthera occipitalis*]; Campos-Nogueira (2001)	seems to be absent from most of the Amazon Basin but does range north along the coast into ne Brazil, including Marajo and Mexicana Islands.
T. persimilis* (Cope, 1868 [1869]) Cope's Savanna Racer	Myers (1974); Di-Bernardo (1992); Myers & Cadle (1994); Marques et al. (2001) [as *Echinanthera persimilis*]	SE Brazil in São Paulo, Santa Catarina, Espirito Santo, Rio de Janiero, and Paraná.
	*Remark: *Rhadinaea beui* Prado, 1943 and *Rhadinaea insignissimus* (Amaral, 1926) are synonyms of *Rhadinaea persimilis*, which is a synonym of this snake (see Myers, 1974).	
T. poecilopogon (Cope, 1863) Pampa Savanna Racer	Myers (1974); Amaral (1976); Di-Bernardo (1992); Myers & Cadle (1994); Lema (1994b); Yanosky (1998)	Uruguay, se Paraguay into ne Argentina in Corrientes, Entre Ríos, Misiones and Buenos Aires as well as in s Brazil in São Paulo, Rio de Janiero, Minas Geraís, Paraná, Santa Catarina, and Rio Grande do Sul.

GENUS: *Tantalophis* Duellman, 1958, is a genus of a rare snake found in Oaxaca, Mexico and known only from a few specimens. See Duellman (1958) [as incertae sedis *Leptodeira* (?) *discolor* Günther]; Myers & Campbell (1981); Dowling & Jenner (1987), Williams & Wallach (1989); and Smith & Smith (1993).

Review assistance and important contributions for this section were provided by Jonathan A. Campbell.

Also see genera *Eridiphas*, *Hypsiglena*, *Leptodeira*, *Pseudoleptodeira*, and *Rhadinophanes*.

CONTENT: One species. ★★★★

Tantalophis: Oaxacan Cat-eyed Snake

Species-level Taxa	Citations and Remarks	Abbreviated Range
T. discolor (Günther, 1860) Oaxacan Cat-eyed Snake	Tanner (1944); Duellman (1958); Smith & Taylor (1966); Myers & Campbell (1981); Campbell (1981); Smith & Smith (1993)	Sierra Madre del Sur of Oaxaca, Mexico from 2400 to 2800 m in the Mexican highlands

GENUS: *Tantilla* Baird & Girard, 1853, contains the Black-headed Snake, Flathead Snake, Crowned Snake, or Centipede Snake. It ranges from the central United States southward through Mexico and Central America into South America as far as northern Argentina and Uruguay. See Smith & Taylor (1966); Peters & Orejas-Miranda (1970); Wilson & Meyer (1971); Lancini & Kornacker (1989); Wilson & Mena (1980); Wilson (1982a, 1982b, 1999); Stebbins (1985); Starace (1998); Kornacker (1999); Wilson & McCranie (1999); and Dixon (2000a).

Review assistance and important contributions for this section were provided by Larry D. Wilson.

CONTENT: A large, relatively complex genus with fifty-five species. ★★★★

Tantilla: Crowned Snake, Black-headed Snake, Centipede Snake; and Flat-headed Snake; Crother et al. (2000) called this group collectively the Black-headed Snakes for those in North America (The United States and Canada). Frank and Ramus (1995) call this the Centipede Snake.

Species-level Taxa	Citations and Remarks	Abbreviated Range
T. albiceps Barbour, 1925 Barbour's Centipede Snake	Barbour (1925); Peters & Orejas-Miranda (1970), Wilson et al. (1977); Wilson (1982a, 1982b, 1985b, 1999); Villa et al. (1988); Köhler (2001)	Known only from the holotype locality of Barro Colorado Island in Lago Gatún in the Canal Zone of Panama.
T. alticola (Boulenger, 1903) Brown Centipede Snake	Peters & Orejas-Miranda (1970); Savage (1980); Wilson (1982a, 1982b, 1986a, 1987a, 1999); Vanzolini (1986); Savage & Villa (1986); Villa et al. (1988); Köhler (2001)	Nicaragua, Costa Rica and Panama in Central America southward into nw Colombia in the Chocó region; from low to intermediate elevations of 91 to 2743 m.
T. andinista Wilson & Mena, 1980 Alausí Centipede Snake	Wilson & Mena (1980); Wilson (1982b, 1985c, 1987a, 1999); Vanzolini (1986)	Known only from the type locality near Alausí in Provencia Chimborazo, Ecuador; known only from the holotype that came from the Alausí Basin in Ecuador at 2600 to 2750 m.
T. atriceps (Günther, 1895) Mexican Black-headed Snake	Tanner (1966b); Conant (1975); Cole & Hardy (1981) (1983a); Wilson (1982b, 1999); Smith & Taylor (1966); Tennant (1998); Conant & Collins (1998); Dixon (2000a); Dixon et al. (2000); Werler & Dixon (2000); Tennant & Bartlett (2000)	Low, moderate, and intermediate elevations in Mexico from the states of se Chihuahua, s Coahuila, nw Nuevo León, n San Luis Potosí, n and e Durango, and ne Zacatecas, with disjunct populations in s Texas.
T. bairdi Stuart, 1941 Baird's Centipede Snake	Stuart (1941a, 1948, 1950); Peters & Orejas-Miranda (1970); Wilson (1982a, 1982b, 1985a, 1985e, 1999); Villa et al. (1988); Köhler (2001)	Mountains of the Guatemalan departments of Alta Verapaz and Baja Verapaz at moderate elevations of 1524 to 1550 m; the type locality is near Finca Chichén on the Chamelco Trail in Alta Verapaz, Guatemala. Wilson (1999) described the distribution as moderate and intermediate elevations (1524–1550 m) of the Caribbean versant of c Guatemala. Wilson (1985a) described a second specimen in Guatemala from 5 km south of La Union Barrios at Agua Zarcu at 1524 m in Dept. Baja Verapaz.
T. bocourti* (Günther, 1895) Bocourt's Black-headed Snake	Smith (1942c); Smith & Laufe (1945); Smith & Taylor (1945, 1966); Davis & Smith (1953); McDiarmid et al. (1976, 1992); Wilson (1982b, 1999); McDiarmid & Folke (1991); Pérez-Higareda & Smith (1991); García & Ceballos (1994); Ramírez-Bautista (1994) *Remark: Subspecies not normally listed but Smith & Smith (1993) gave subspecies as *T. b. bocourti* and *T. b. deviatrix*. See remark under *T. wilcoxi* on *deviatrix*. Ramírez-Bautista (1994), listed a subspecies of *T. b. bocourti*.	SC Mexican Plateau and the Sierra Madre del Sur; found from s Durango and the ne Sinaloa highlands to c Veracruz and the Guerrero coast; also reported in the Tres Marías Islands.
T. boipiranga Sawaya & Sazima, 2003 Highlands Centipede Snake	Sawaya & Sazima (2003)	The highlands (campos rupestres), of the Sierra do Cipó, Minas Geraís in se Brazil.

The Caenophidia

Species-level Taxa	Citations and Remarks	Abbreviated Range
T. brevicauda Mertens, 1952 Short-tailed Centipede Snake	Mertens (1952a, 1952b); Uzzell & Starrett (1958); Vanzolini (1986); Wilson (1970c, 1982a, 1982b, 1988b, 1999); Villa et al. (1988); Köhler (2001)	SC Guatemala and ne El Salvador in moderate to intermediate elevations, from 600 to 1750 m.
T. briggsi Savitzky & Smith, 1971 Briggs's Centipede Snake	Savitzky & Smith (1971); Wilson (1982a, 1982b, 1983a, 1985f, 1999); Pérez-Higareda & Smith (1991); Campbell (1998b)	Caribbean versant of the Isthmus of Tehuantepec in Oaxaca, Mexico; known only from the holotype.
*T. calamarina** Cope, 1867 Pacific Coast Centipede Snake	Smith (1942c); Peters (1954); Smith & Taylor (1945, 1966); McDiarmid, et al. (1976); Wilson & Meyer (1981); Wilson (1982b, 1988c, 1999); García & Ceballos (1994); Ramírez-Bautista (1994); Wilson & Campbell (2000) *Remark 1: *Geophis gertschi* Bogert & Porter, 1966, is a synonym (see Wilson & Meyer, 1981). *Remark 2: *T. martindelcampo* is a synonym (see Wilson & Meyer, 1981)	Edge of the Central Mexican Plateau from s Sinaloa, to Guerrero and Morelos; it is also found on the Tres Marías Islands; there is an unconfirmed report from Puebla.
T. capistrata Cope, 1876 Peruvian Centipede Snake	Schmidt & Walker (1943b); Peters & Orejas-Miranda (1970); Wilson & Mena (1980); Vanzolini (1986); Wilson (1990a, 1999); Pérez-Santos & Moreno (1991)	Low to intermediate elevations of nw coastal Peru and the valleys of the upper Río Marañon, Río Chincipe, and Río Chamaya, as well as the provinces of El Oroa and Loja in extreme s Ecuador.
T. cascadae Wilson & Meyer, 1981 Michoacán Centipede Snake	Wilson & Meyer (1981); Wilson (1982b, 1988e, 1999); Smith & Smith (1993)	Known only from the type locality in Michoacán, Mexico and only from the holotype.
T. coronadoi Hartweg, 1944 Guerreran Centipede Snake	Hartweg (1944); David & Dixon (1959); Smith & Taylor (1966); Wilson & Meyer (1981); Wilson (1982b, 1990b, 1999); Wilson & Campbell (2000)	Sierra Madre del Sur in c Guerrero, Mexico; it has been in only two localities in the Pacific versant.
T. coronata Baird & Girard, 1853 Southeastern Crowned Snake	Conant (1975); Ashton & Ashton (1981); Telford (1966) Wilson (1982b, 1999); Dundee & Rossman (1989); Wright & Wright ([1957] 1994); Tennant (1997b); Conant & Collins (1998); Tennant & Bartlett (2000)	E United States from sc Virginia and sw Kentucky into the Florida panhandle w into Louisiana. Isolated populations in s Indiana, adjacent Kentucky, and s Virginia.
*T. cucullata** Minton, 1956 Trans-Pecos Black-headed Snake	Conant (1975); Wilson (1982a, 1999); Smith & Smith (1993); Tennant (1998); Conant & Collins (1998); Dixon et al. (2000); Werler & Dixon (2000); Tennant & Bartlett (2000); Wilson et al. (2000a, b); Dixon (2000a) *Remark: Dixon et al. (2000) elevated this to full species status from *T. rubra cucullata* and reclassified *T. rubra diabola* Fouquette & Potter (1961), as a junior subjective synonym of this snake. Based on the Dixon et al. (2000) work, *T. cucullata* is a separate	The United States in the Trans-Pecos and Big Bend regions of sw Texas with populations in the Chisos and Davis Mountains, as well as Val Verde and Terrell counties.

Species-level Taxa	Citations and Remarks	Abbreviated Range
	from *T. rubra*, and, *T. diabola* is a synonym of *T. cucullata*. They demonstrated that *T. rubra* is limited to Mexico.	
T. cuniculator Smith, 1939 Mayan Black-headed Centipede-eater	Smith (1939, 1942c); Smith & Taylor (1966); Wilson et al. (1977); Wilson (1982a, 1982b, 1983a, 1985h, 1999); Villa et al. (1988); Lee (1996, 2000); Campbell (1998a, 1998b); Stafford & Meyer (2000); Köhler (2001)	Quintana Roo and Yucatán in Mexico and n Belize; found in low elevations.
T. deppei (Bocourt, 1883) Deppe's Centipede Snake	Smith (1942c); Hartweg (1944, 1966); Davis & Smith (1953); Wilson & Meyer (1981); Wilson et al. (1977); Wilson (1982b, 1988f, 1999); Wilson & Campbell (2000)	Intermediate elevations of the Pacific versant of Mexico in n Morelos and nw Oaxaca found from 1524 to 2438 m.
T. equatoriana Wilson & Mena, 1980 San Lorenzo Centipede Snake	Wilson & Mena (1980); Vanzolini (1986); Wilson (1982b; 1987a; 1988g, 1999)	Known only from the type locality of San Lorenzo in Provencia Esmeraldas on the coastal plain of extreme nw Ecuador.
T. flavilineata Smith & Burger, 1950 Yellow-lined Centipede Snake	Smith & Burger (1950); Wilson et al. (1977); Wilson (1982a, 1982b, 1983a 1985j, 1999); Campbell (1998b)	Mexico from intermediate elevations in c Oaxaca; from about 1890–2100 m.
T. gracilis* Baird & Girard, 1853 Flat-headed Snake	Conant (1975); Cole & Hardy (1981); Wilson (1982b, 1999); Dundee & Rossman (1989); Smith & Smith (1993); Tennant (1998); Conant & Collins (1998); Werler & Dixon (2000); Tennant & Bartlett (2000) *Remark: Although *T. gracilis* has been listed with subspecies in the past, most authorities do not consider them valid.	C United States from extreme s Illinois, Missouri, and e Kansas into s Texas and into Mexico in Coahuila; with isolated populations in nw Texas and extreme e Texas.
T. hobartsmithi* Taylor, 1936 [1937] Smith's Black-headed Snake	Smith & Taylor (1966); Wilson (1982b, 1999); Cole & Hardy (1981, 1983b); Stebbins (1985); Degenhardt et al. (1996); Brown (1997); Tennant (1998); Conant & Collins (1998); Wilson (1999); Werler & Dixon (2000); Bartlett & Tennant (2000); Tennant & Bartlett (2000); Dixon (2000a) *Remark: Smith & Smith (1993) list *T. utahensis* as a synonym. Bartlett & Tennant (2000), noted that this snake is called the Utah Black-headed Snake *Tantilla planiceps utahensis* and commented that taxonomic stability is not yet assured.	Very disjunct range in the United States from California through s Nevada and Utah, w Colorado, Arizona, s New Mexico, w Texas, and extreme nw Texas; in Mexico it is reported from w Sonora, e Chihuahua, and n Coahuila.
T. impensa Campbell, 1998 Guatemalan Centipede Snake	Campbell (1998b); Wilson & McCranie (1999); Wilson (1999); Köhler (2001)	Wilson (1999) described the range as near sea level to intermediate elevations of the Caribbean versant from e Chiapas, Mexico, through the mountains of c Guatemala to w Honduras.

The Caenophidia

Species-level Taxa	Citations and Remarks	Abbreviated Range
T. insulamontana Wilson & Mena, 1980 Río Jubones Valley Centipede Snake	Wilson & Mena (1980); Wilson (1982b, 1987a, 1990c, 1999); Vanzolini (1986)	Moderate to intermediate elevations of Río Jubones Valley in s Ecuador; found from 1250 to 2100 m in the Pacific versant.
T. jani* (Günther, 1895) Jan's Centipede Snake	Smith (1942c); Peters & Orejas-Miranda (1970); Wilson et al. (1977); Wilson & Meyer (1981); Alvarez del Toro (1982); Wilson (1982b, 1983a, 1984a, 1985i, 1999); Villa et al. (1988); Campbell (1998b); Wilson & McCranie (1999); Köhler (2001) [as *T. cuesta* and as *T. jani*]. Remark: According to Campbell (1998b), *T. cuesta* Wilson, 1982, known only from the type locality near San Rafael Pie de la Cuesta in the sw highlands of Guatemala, is a synonym.	Wilson (1999) stated that this snake is known only with certainty from the type locality of *Tantilla cuesta* at Finca San Julia, 1.5 km east of San Rafael de la Cuesta, elevation 1050 m, Depto. San Marcos, Guatemala.
T. johnsoni Wilson, Vaughan & Dixon, 1999 Johnson's Centipede Snake	Wilson et al. (1999); Wilson (1999)	Known only from the type locality of Musté, Municipo Motozintla, at about 450 m in Chiapas, Mexico.
T. lempira Wilson & Mena, 1980 Honduran or Mena's Centipede Snake	Wilson & Mena (1980); Wilson (1982a, 1982b, 1984a, 1990d, 1999); Wilson & Meyer (1982, 1985); Vanzolini (1986); Villa et al. (1988); Köhler (2001)	Moderate to intermediate elevations of the Pacific versant of sc Honduras; found from 1450–1730 m.
T. melanocephala* (Linnaeus, 1758) Common Black-headed Snake	Peters (1960); Roze (1966); Amaral (1976); Duellman (1978); Savage (1980); Wilson (1982b, 1987a, 1992a); Wilson & Mena (1980); Wilson & Meyer (1985); Dixon & Soini (1986); Savage & Villa (1986); Chippaux ([1986] 1987); Villa et al. (1988); Lancini & Kornacker (1989); Murphy (1997); Starace (1998); Monanelli & Alvarez (1998); Kornacker (1999); Freitas (1999); Mijares-Urrutia & Arends R. (2000); Boos (2001); Köhler (2001) *Remark 1: Wilson (1992a) refers to Wilson & Mena (1980), and confirms that no subspecies are recognized. However, he also points out that some workers have tacitly continued to utilize subspecific designation and referenced Boos (1984); Chippaux ([1986] 1987); Dixon & Soini (1986); and Pérez-Santos & Moreno (1988, 1992). Peters & Orejas-Miranda (1970) listed this with subspecies and showed *T. melancephala capistrata* Cope, 1876. *Remark 2: Wilson & Mena (1980), considered *T. fraseri* (Günther, 1895) of highland w slopes of the Andes and Quito Valley of n w Ecuador into Peru; *T. longifrontalis* (Boulenger, 1896), from the e slopes of the Andes in Colombia and Ecuador; *T. longifron-*	Guatemala in Central America southward into South America in s Peru, Bolivia, n Argentina, and Uruguay; it is also found on Trinidad and Tobago.

Species-level Taxa	Citations and Remarks	Abbreviated Range
	talis (Boulenger, 1896), from the e slopes of the Andes in Colombia and Ecuador; and, *T. mexicana* Günther, 1862), from the Pacific versant of Chiapas in s Mexico southward into Guatemala; and, *T. ruficeps* (Cope, 1894), to be a synonyms.	
T. miyatai Wilson & Knight *in* Wilson, 1987 Pichincha Centipede Snake or Miyata's Centipede Snake	Wilson (1987a, 1990c, 1999)	Only from the type locality in Pichincha at a moderate elevation of about 750 m in the Tropical Rainforest formation in the Andean foothills of nw Ecuador; known only from the holotype.
T. moesta (Günther, 1863) Black-bellied Centipede Snake	Smith & Taylor (1966); Peters & Orejas-Miranda (1970); Wilson (1982a, 1982b, 1988h, 1999); Villa et al. (1988); Lee (1996, 2000); Campbell (1998a); Köhler (2001)	Yucatán Peninsula, Mexico in the states of Quintana Roo and Yucatán, and into n El Petén, Guatemala; found in low and moderate elevations.
T. nigra (Boulenger, 1914) Chocó Centipede Snake	Amaral (1931b); Peters & Orejas-Miranda (1970); Wilson (1982b, 1984a, 1987a, 1992b, 1999)	Known only from the type locality near Peña Lisa, Condoto in Depto. Chocó w Colombia; known from the holotype.
*T. nigriceps** Kennicott, 1860 Plains Black-headed Snake	Conant (1975); Cole & Hardy (1981); Wilson (1982b, 1999); Tanner (1985); Stebbins (1985); Degenhardt et al. (1996); Tennant (1998); Conant & Collins (1998); Werler & Dixon (2000); Bartlett & Tennant (2000); Tennant & Bartlett (2000); Dixon (2000a) *Remark: Most authorities now treat this as a monotypic species.	Low, moderate, and intermediate elevations from sw Nebraska, e Colorado, and w Kansas southward to e and s New Mexico, se Arizona, and c and w Texas in the United States, southward to e Chihuahua, n Durango, and n Tamaulipas in Mexico.
T. oaxacae Wilson & Meyer, 1971 Oaxacan Centipede Snake	Wilson & Meyer (1971); Wilson (1982b, 1990f, 1999); Campbell (1998b)	Pacific versant of sc Oaxaca in Mexico in moderate to intermediate elevations.
T. oolitica Telford, 1966 Rim Rock Crowned Snake	Telford (1966, 1980a); Conant (1975); Wilson (1982b, 1999); Conant & Collins (1998); Tennant & Bartlett (2000)	The United States from Dade Co. southward into Key Largo in s Florida.
T. petersi Wilson, 1979 Peters's Centipede Snake	Wilson (1979, 1982b, 1987a, 1991b, 1999); Vanzolini (1986)	Intermediate elevations of the extreme n part of the Andean highlands of Ecuador; the type is from Provencia Imbabura.
T. planiceps (Blainville, 1835) Western Black-headed Snake	Smith & Taylor (1966); Cole & Hardy (1981, 1983c); Wilson (1982b, 1999); Stebbins (1985); Brown (1997); McPeak (2000); Bartlett & Tennant (2000) *Remark: Cole & Hardy (1983), Smith & Smith (1993), and Brown (1997), did not recognize subspecies. Wilson confirmed in a personal communication that subspecies should not be recognized.	SW United States from s California into s Baja California del Sur and Isla del Carmen in the Gulf of California; the range is very disjunct.

Species-level Taxa	Citations and Remarks	Abbreviated Range
T. relicta Telford, 1966 Florida Crowned Snake	Telford (1966, 1980b); Conant (1975); Ashton & Ashton (1981); Wilson (1982b, 1999); Tennant (1997b); Conant & Collins (1998); Tennant & Bartlett (2000)	Peninsular Florida from s Palm Beach, Highlands, and Charlotte counties north to Duval (?), Columbia, and Taylor counties.
T. r. relicta Telford, 1966 Peninsula Crowned Snake	Telford (1966, 1980b); Conant (1975); Ashton & Ashton (1981); Tennant (1997b); Conant & Collins (1998); Tennant & Bartlett (2000)	Central Ridge of peninsular Florida in disjunct populations from n into c Florida, and along the s Gulf Coast.
T. r. neilli Telford, 1966 Central Florida Crowned Snake	Telford (1966, 1980b); Conant (1975); Ashton & Ashton (1981); Tennant (1997b); Conant & Collins (1998); Tennant & Bartlett (2000)	NC Florida from Madison county southward into Hillsborough and Polk counties.
T. r. pamlica Telford, 1966 Coastal Dunes Crowned Snake	Telford (1966, 1980b); Conant (1975); Ashton & Ashton (1981); Tennant (1997b); Conant & Collins (1998); Tennant & Bartlett (2000)	Florida east coast from Cape Canaveral southward into s Palm Beach county.
*T. reticulata** Cope, 1860 Reticulated Centipede Snake	Wilson & Meyer (1971); Savage (1980); Wilson et al. (1977); Wilson (1982a, 1982b, 1983a, 1985k, 1987a, 1999); Savage & Villa (1986); Villa et al. (1988); Auth (1994); Campbell (1998b); Köhler (2001) *Remark: *T. virgata* (Günther, 1873), known only from Cartago, Costa Rica, is a synonym (Wilson, pers. comm. 1998).	Caribbean versant from se Nicaragua through Panama into nw Colombia in the Pacific and Caribbean versants; it is found in a disjunct distribution from sea level to about 1430 m.
*T. rubra** Cope, 1875 [1876] Red Black-headed Snake or La Rojilla	Smith (1942c); Smith & Taylor (1966); Conant (1975); Alvarez del Toro (1982); Wilson (1982b, 1999); Villa et al. (1988); Savage & Slowinski (1992); Roze (1996); Dixon et al. (2000); Köhler (2001) *Remark: *T. morgani* Hartweg, 1944, known only from the type locality of Necaxa in Pueblo, Mexico; and, *T. miniata* Cope, 1863, found in the foothills of c Veracruz, Mexico, are synonyms (see Dixon et al., 2000).	In the Atlantic versant of c Nuevo León, Mexico southward to w Guatemala; and, on the Pacific versant in Oaxaca, Mexico.
*T. schistosa** (Bocourt, 1883, *in* Dumeril, and *in* Mocquard & Bocourt, 1870–1909) Lesser or Red Earth Centipede-eater	Smith (1942c); Savage (1980); Wilson (1976, 1982a, 1982b, 1984a, 1987c, 1999); Wilson & Meyer (1985); Savage & Villa (1986); Lee (1996, 2000); Pérez-Higareda & Smith (1991); Campbell (1998b); Stafford & Meyer (2000); Köhler (2001) *Remark 1: Wilson (1982a), considered this species to be monotypic. *Remark 2: *T. phrenitica* and *T. phrenitica schistosa*, are synonyms (see Wilson, 1982a).	Veracruz and Oaxaca, Mexico on the Atlantic versant southward into Costa Rica and Panama.

Species-level Taxa	Citations and Remarks	Abbreviated Range
T. semicincta (Dumeril, Bibron & Dumeril, 1854) Banded False Coral Snake	Roze (1966); Peters & Orejas-Miranda (1970); Wilson (1976, 1982b, 1987a, 1990g, 1999); Villa et al. (1988); Lancini & Kornacker (1989); Esqueda & La Marca (1999); Kornacker (1999); Mijares-Urrutia & Arends R. (2000)	Colombia along the Caribbean Coast into Venezuela from Cartagena to Caracas; found at low elevations from sea level to 457 m.
T. sertula Wilson & Campbell, 2000 Wreathed Centipede Snake	Wilson & Campbell (2000)	Known only from the type locality, near the village of La Unión, Guerrero, Mexico on the Pacific coastal plain in forests.
T. shawi Taylor, 1949 Potosí Centipede Snake	Taylor (1949); Wilson (1976, 1982b, 1991a, 1999); Smith & Smith (1993); Campbell et al. (1995)	E slopes of the Sierra Madre Oriental in Mexico in extreme se San Luis Potosí and nw Veracruz.
T. slavensi Pérez-Higareda, Smith & Smith, 1985 Slaven's Centipede Snake	Pérez-Higareda et al. (1985); Pérez-Higareda & Smith (1991); Smith & Smith (1993); Campbell (1998b); Wilson (1999)	Low to moderate elevations in the Atlantic versant in the Los Tuxtlas region of se Veracruz, Mexico; found from 50 to 800 m.
T. striata Dunn, 1928 Striped Centipede Snake	Dunn (1928b); Smith (1942c); Smith & Taylor (1966); Wilson & Meyer (1971); Savitsky & Smith (1971); Wilson (1982b, 1983a, 1990h); Campbell (1998b)	Pacific slopes of the Isthmus of Tehuantepec in Oaxaca Mexico; found at low and moderate elevations.
T. supracincta* (W. Peters, 1863) Coastal Plain Centipede Snake	Peters & Orejas-Miranda (1970); Wilson (1982b, 1987, 1999); Savage & Slowinski (1992); Auth (1994); Köhler (2001) *Remark: Fide Wilson (1999), *T. annulata* Boettger, 1892 is a synonym (see Wilson, 1982a, 1982b, 1985c, 1999).	Caribbean versant of extreme se Nicaragua southward to c Panama, and on the Pacific versant in Costa Rica, and Ecuador on the coastal plain to Guayaquil.
T. taeniata (Bocourt, 1883, *in* Dumeril, & *in* Mocquard & Bocourt, 1870–1909) Central American Centipede Snake	Wilson (1974b, 1982a, 1982b, 1983b, 1999); Savage (1980); Wilson & Meyer (1971, 1985); Vanzolini (1986); Campbell (1998b); Smith et al. (1998); Wilson & McCranie (1999); Köhler (2001)	Both the Atlantic and Pacific versants in s Guatemala and w Honduras.
T. tayrae Wilson, 1983 Volcán Tacaná Centipede Snake	Wilson (1983a, 1984a, 1990i, 1999); Villa et al. (1988); Campbell (1998b); Köhler (2001)	SE Chiapas, Mexico; known from two localities at moderate elevations of 760 to 960 m on the slopes of Volcán Tacaná in the Pacific versant.
T. tecta Campbell & Smith, 1997 White-striped Centipede-eater	Campbell & Smith (1997); Campbell (1998a); Wilson (1999); Lee (2000); Köhler (2001)	Known only from the type locality reported as ne side of Laguna Yaxhá, El Petén, Guatemala.

Species-level Taxa	Citations and Remarks	Abbreviated Range
T. trilineata (W. Peters, 1880) Lost Centipede-eater	W. Peters (1880); Savitsky & Smith (1971); Peters & Orejas-Miranda (1970), Campbell (1998b); Wilson (1999)	Unknown; type locality was listed as "Brasilien" and was apparently in error, according to Smith & Williams (1966)
T. triseriata Smith & Smith, 1951 Three-lined Centipede Snake	Smith & Smith (1951); Villa et al. (1988); Smith et al. (1998); Wilson (1999); Köhler (2001)	Intermediate elevations of sc Oaxaca, Mexico.
T. tritaeniata* Smith & Williams, 1966 Guanaja Centipede Snake	Smith & Williams (1966a); Peters & Orejas-Miranda (1970); Wilson & McCranie (1999); Wilson (1999); Köhler (2001)	Isla Guanaja in the Bay Islands, Honduras.
T. vermiformis (Hallowell, 1861) Hallowell's Centipede Snake	Van Devender & Cole (1977); Wilson (1982a, 1982b, 1987d, 1999); Vanzolini (1986); Savage & Villa (1986); Villa et al. (1988); Wilson & Campbell (2000); Köhler (2001)	Pacific versant of El Salvador, nw Nicaragua into nw Costa Rica at low elevations.
T. vulcani Campbell, 1998 Volcano Centipede Snake	Campbell (1998b); Wilson & McCranie (1999); Wilson (1999); Köhler (2001)	Low to moderate elevations in the Pacific versant from e Oaxaca, Mexico into sc Guatemala.
T. wilcoxi* Stejneger, 1902 Chihuahuan Black-headed Snake	Smith & Taylor (1966); Cole & Hardy (1981); Wilson (1982b, 1999), Liner (1983); Tanner (1985); Stebbins (1985, 1992); McDiarmid (1992); Bartlett & Tennant (2000) *Remark: According to Bartlett & Tennant (2000), this snake is *T. wilcoxi wilcoxi,* and called the Huachuca Black-headed Snake. However, this is not tenable, according to McDiarmid (1992).	SW United States into Mexico from extreme s Arizona through sw Chihuahua, ne Sinaloa, c Durango, Zacatecas, se Coahuila, s Nuevo León, and w San Luis Potosí.
T. yaquia* Smith, 1942 Yaqui Black-headed Snake	Smith (1942c); Stebbins (1966, 1985); McDiarmid (1968, 1977); Cole & Hardy (1981); Wilson (1982b, 1999); Tanner (1966, 1985); Degenhardt et al. (1996); Bartlett & Tennant (2000) Remark: According to Cole & Hardy (1981), and Smith & Smith (1993), this species is monotypic. However, Degenhardt et al. (1996) shows subspecies so they are listed here.	SE Arizona in the United States southward into Nayarit, Mexico.
T. y. yaquia Smith, 1942 Yaqui Black-headed Snake	McDiarmid (1977); Tanner (1985); Cole & Hardy (1981)	Arizona in the United States southward into Nayarit, Mexico.
T. y. bogerti Hartweg, 1944 Bogert's Centipede Snake	Smith & Taylor (1966); Smith & Smith (1993); Degenhardt et al. (1996)	Known only from the type locality in Acaponeta, Nayarit in Mexico.

GENUS: *Tantillita* Smith, 1941, is confined to Mexico and Guatemala. This taxon was originally included as *Tantilla*. See Smith (1941a); Smith & Taylor (1966); Peters & Orejas-Miranda (1970); Wilson (1982a, 1988i); Smith & Smith (1993); Lee (2000); and Köhler (2001).

CONTENT: One to three species, depending on the authority being referenced. Three species are shown below. ★

Tantillita: Dwarf Short-tailed Snakes

Species-level Taxa	Citations and Remarks	Abbreviated Range
T. brevissima* (Taylor, [1936] 1937) Speckled Dwarf Short-tailed Snake	Smith (1941a); Alvarez del Toro (1982); Wilson (1982a, 1988i); Vanzolini (1986), Villa et al. (1988); Smith & Smith (1993); Köhler (2001) *Remark 1: Fide Wilson (1982b), this is *T. canula*. However, this it is now considered to be a valid species, based on a personal communication with Wilson. *Remark 2: Fide Wilson (1988j), *T. excubitor* Wilson, 1982, from Escuintla, Guatemala, is a synonym of *Tantillita brevissima*.	Restricted range from Oaxaca, Mexico into Escuintla, Guatemala; it is known only from three locations.
T. canula* (Cope, 1875) Yucatán Dwarf Short-tailed Snake	Smith & Taylor (1966); Peters & Orejas-Miranda (1970); Wilson (1982a, 1982b, 1988d); Smith et al. (1993); Smith & Smith (1993); Lee (1996, 2000); Campbell (1998a); Stafford & Meyer (2000); Köhler (2001) *Remark 1: Fide Wilson (1988d), there are no recognized subspecies. Historically, subspecies are only noted when listed as *Tantilla canula*. It is sometimes seen as *Tantilla canula* [reference Liner (1994) as most recent listing and as *Tantillita*]. Villa et al. (1988) consider *Tantillita canula* as valid. Remark 2: Peters and Orejas-Miranda (1970), quoted Neill & Allen (1961), that *brevis* Günther, probably should be considered a subspecies of *canula*, but received it too late to include in their work formally. *T. brevis* (Günther, 1895), from Belize is sometimes listed as *T. canula brevis* when *Tantilla canula* is listed instead of *Tantillita canula* or as *Tantilla canula brevis*. Fide Wilson (1986), this is *Tantillita*. Wilson (1988d, pers. comm. Wilson), noted that Neill & Allen (1961), recognized two subspecies, *canula* and *brevis*, but their analysis was flawed by lack of information on the holotype and only specimen of brevis and confusion of this species with *Tantillita lintoni*. Wilson (1982a), provided a synonymy, description, discussion of ecological and geographic distribution, and an unraveling of the taxonomy of the species.	Lowlands of the Yucatán peninsula, Mexico into Belize and n El Petén, Guatemala; in Mexico it is found in the states of Campeche, Quintana Roo, and Yucatán.

Species-level Taxa	Citations and Remarks	Abbreviated Range
T. lintoni (Smith, 1940) Linton's Dwarf Short-tailed Snake	Smith (1940); Neill & Allen (1961); Wilson & Meyer (1985); Wilson (1988i); Villa et al. (1988); Williams & Wallach (1989); Smith & Smith (1993); Lee (1996, 2000); Campbell (1998a); Köhler (1999)	Atlantic foothills of Chiapas, Mexico into El Petén in Guatemala and south into Belize, Honduras, and Nicaragua.
T. l. lintoni (Smith, 1940) Linton's Dwarf Short-tailed Snake	Smith (1940); Neill & Allen (1961); Wilson & Meyer (1985); Pérez-Higareda (1985); Wilson (1988i); Pérez-Higareda & Smith (1991); Smith & Smith (1993); Stafford & Meyer (2000); Lee (2000)	S Mexico southward into Guatemala and Central America.
T. l. rozellae Pérez-Higareda, 1985 Rozella's Dwarf Short-tailed Snake	Pérez-Higareda (1985); Wilson (1988i); Pérez-Higareda & Smith (1991); Smith & Smith (1993)	Mexican highlands.

GENUS: *Thalesius* Yuki, 1993, is a reclassification to a new genus for *Xenodon werneri* Eiselt, 1963 based on hemipenial morphology. The genus *X. werneri* would normally return to *Procteria* Werner, 1924, but the name is preoccupied by *Procteria* Davis, 1885, for a Coelenterata. The name *Thalesius* nom. nov. was proposed by Yuki to replace it, with the subfamily and tribe as *incertae sedis*. See Peters & Orejas-Miranda (1970); Hoogmoed (1985); Chippaux ([1986] 1987); Yuki (1993); and Starace (1998) [as *X. werneri*].

Review assistance for this section was provided by Thales de Lema.

See also genera *Xenodon* and *Waglerophis*.

CONTENT: One species. ★★★

Thalesius: Lema's Snake

Species-level Taxa	Citations and Remarks	Abbreviated Range
T. viridis* (Werner, 1924) Lema's Snake	Peters & Orejas-Miranda (1970) [as *Xenodon werneri*]; Hoogmoed (1985); Chippaux ([1986] 1987); Yuki (1994); Starace (1998) [maintained as *Xenodon werneri*]	Suriname and French Guiana.
	*Remark: Fide Yuki (1994), The hemipenian morphology of *Xenodon werneri* Eiselt, 1963, differs from the genus *Xenodon*, and as such, allows the species to be removed from the genus *Xenodon* Boie, 1827. Based on this, the species *X. werneri* would return to *Procteria viridis* Werner, 1924 if the genus *Procteria* Werner, 1924 was not preoccupied by *Procteria* Davis, 1885 (for a Coelenterata). The genus *Thalesius* nom. nov. was proposed in its place.	

GENUS: *Thamnodynastes* Wagler, 1830, is found from the Caribbean coast of northern South America, southward into northern Argentina. See Peters & Orejas-Miranda (1970); Vanzolini (1986); Lancini & Kornacker (1989); Myers & Donnelly (1996); Cei (1993); Lema (1994b); Gorzula & Ayarzagena (1995); Starace (1998); and Kornacker (1999). At the time of the Peters & Orejas-Miranda (1970) publication, they made a comment that Peters (1960b), had erroneously used *T. nattereri* (Mikan), for a specimen catalogued from Guayaquil, Ecuador, which might represent an undescribed species.

Review assistance and important contributions for this section were provided by Robert A. Thomas.

CONTENT: Twelve species. ★★★

Thamnodynastes: South American Large-eyed Snake; but sometimes referred to as the South American Mock Viper or American House Snake. Frank and Ramus (1995) call this the Coastal House Snake. I did not use their name because the snake is often encountered far from the coast.

Species-level Taxa	Citations and Remarks	Abbreviated Range
T. chaquensis Bergna & Alvarez, 1993 Chaco Large-eyed Snake	Giraudo (1996, 1999); Bergna & Alvarez (1993); Leynaud & Bucher (1999)	Ñeeubucu, Paraguay and into Argentina in n Santa Fe, Chaco, and e Formosa in Argentina.
T. chimanta Roze, 1958 Chimanta Large-eyed Snake	Roze (1958b, 1966); Peters & Orejas-Miranda (1970); Lancini & Kornacker (1989); Gorzula (1992); Pefaur (1992); Gorzula & Ayarzagena (1995); La Marca (1997); Gorzula & Señaris (1998); Kornacker (1999)	Chimantá Tepui in the state of Bolívar in se Venezuela; known only from the type locality.
T. corocoroensis Gorzula & Ayarzagena, 1995 Corocoro Large-eyed Snake	Gorzula & Ayarzagena (1995); La Marca (1997); Gorzula & Señaris (1998); Kornacker (1999)	Tepuí Corocoro, Bolívar, Venezuela.
T. duida Myers & Donnelly, 1996 Duida Large-eyed Snake	Myers & Donnelly (1996); La Marca (1997); Kornacker (1999)	Cerro Yaui, in nw Tepuis of Venezuela.
T. gambotensis Pérez-Santos & Moreno, 1989 Gambote Large-eyed Snake	Pérez-Santos & Moreno (1989); Thomas (in preparation)	Caribbean lowlands of n Colombia. The type is from Gambote.
T. hypoconia* (Cope, 1860) Argentine Large-eyed Snake	Cei, Bergna & Alvarez (1992); Cei (1993); McCranie (1996); Cacivio (1997); Yanosky (1998); Leynaud & Bucher (1999); Marques et al. (2001) *Remark: Shown as *Tachymensis hypoconia* in Peters & Orejas-Miranda (1970).	Argentina in Formosa, Chaco, Corrientes, Misiones, Santa Fe, Entre Ríos, Córdoba, and Buenos Aires.
T. marahuaquensis Gorzula & Ayarzagena, 1995 Marahuaca Large-eyed Snake	Gorzula & Ayarzagena (1995); La Marca (1997); Kornacker (1999)	Tepuí Marahuaca Norte, Estado Amazonas, Venezuela.

Species-level Taxa	Citations and Remarks	Abbreviated Range
T. pallidus Linnaeus, 1758 Common Large-eyed Snake	Marcuzzi (1951); Roze (1966); Peters & Orejas-Miranda (1970); Amaral (1976); Moonen et al. (1979); Dixon & Soini (1986); Lancini & Kornacker (1989); Gorzula & Ayarzagena (1995); La Marca (1997); Starace (1998); Yanosky (1998); Gorzula & Señaris (1998); Renjifo & Lundberg (1999); Kornacker (1999); Freitas (1999)	Peru, Venezuela, Brazil, and the Guianas; Marcuzzi (1951) described this snake in Venezuela as being reported from Puerto Ayacucho in the State of Amazonas; Amaral (1976) associated it with the Amazon Valley and ne sections of Brazil. Lancini & Kornacker (1989) extended its range to include Tepuí Guaiquinima in the state of Bolívar at 1370 m.
T. rutilus Prado, 1942 São Paulo Large-eyed Snake	Prado (1942c); Prado (1948); Peters & Orejas-Miranda (1970); Gorzula & Ayarzagena (1995)	State of São Paulo, Brazil.
T. strigatus (Günther, 1858) American Large-Eyed Snake	Roze (1966); Peters & Orejas-Miranda (1970); Amaral (1976); Achaval (1987); Cei (1993); Gorzula & Ayarzagena (1995); Lema (1994b); Yanosky (1998); Kornacker (1999); Marques et al. (2001)	SE and s Brazil, Paraguay, into Provencia Buenos Aires, Entre Rios, Chaco, Santa Fe, Formosa, Corrientes, and Misiones in n Argentina.
T. strigilis (Thunberg, 1787) Northern Large-eyed Snake	Roze (1966); Peters & Orejas-Miranda (1970); Amaral (1976); Lancini (1979); Rivero-Blanco & Dixon (1979); Vanzolini et al. (1980); Achaval (1987); Lancini & Kornacker (1989); Lema (1994b); La Marca (1997); Gorzula & Señaris (1998); Starace (1998); Yanosky (1998)	N South America in Colombia, Venezuela, the Guianas, Brazil, Paraguay, and Argentina.
T. yavi Myers & Donnelly, 1996 Cerro Yavi Large-eyed Snake	Myers & Donnelly (1996); La Marca (1997); Kornacker (1999)	Cerro Yaui, in the nw Tepuis of Venezuela.

GENUS: *Thamnophis* Fitzinger, 1843, is wide ranging from southern Canada southward through the United States, Mexico, and Central America to as far as Costa Rica and there is a large variation in individuals over a very large geographic range. The list uses Rossman et al. (1996) as the basis for the classification of *Thamnophis*. See Rossman *in* Peters & Orejas-Miranda (1970); Smith & Taylor (1966); Stebbins (1966); Conant (1961, 1975); Smith & Smith (1993); Wright & Wright ([1957] 1994); Rossman (1995); Rossman et al. (1996); and Dixon (2000a).

Review assistance and important contributions for this section were provided by Douglas A. Rossman.

Review Smith & Smith (1993:471) for a discussion about proposed "lumping" of *Thamnophis* and *Nerodia* by some authorities. *Thamnophis* are taxonomically very similar to *Nerodia*. Rossman (1995) identified two characteristics that distinguish *Thamnophis* from *Nerodia*. They are 1) the simple characteristic of a single cloacal scute (see Dundee, 1989); *Nerodia* has a divided anal plate; however, according to Rossman et al. (1996), the alternate character-state occur in fairly low frequencies and/or concentrated in localized populations; and, 2) Rossman(1995) identified a second characteristic to separate *Nerodia* from *Thamnophis*; in *Thamnophis*, there is typically a more enlarged first dorsal row (DSR-1).

At the time, Wright & Wright (1957) was the most complete work on *Thamnophis* but lacked information on the Mexican varieties. Rossman et al. (1996) is the first complete review listing all of the species and subspecies for *Thamnophis* in Canada, United States, Mexico, and Central America.

CONTENT: A large, complex group with thirty-one species. ★★★

Thamnophis: American Gartersnake and American Ribbonsnake

Species-level Taxa	Citations and Remarks	Abbreviated Range
T. atratus* (Kennicott *in* Cooper, 1860) Aquatic Gartersnake	Fitch (1936, 1984); Rossman (1979); Rossman & Stewart (1987); Rossman et al. (1996); Brown (1997); Boundy (1999) *Remark: Fide Rossman et al. (1996), *T. elegans aquaticus* Fox, 1951 sometimes listed as found from San Francisco Bay northward into nw California to the range of *T. a. hydrophilus* is actually an intergrade of *T. a. atratus* X *T. a. hydrophilus*.	Coastal California from Santa Barbara county northward along the coast to as far as the sw corner of Oregon; from sea level to 1920 m.
T. a. atratus (Kennicott, 1860) Santa Cruz Gartersnake	Fitch (1984); Rossman et al. (1996); Brown (1997); Boundy (1999); Bartlett & Tennant (2000)	Santa Cruz Mountains and s San Francisco Peninsula, from the San Andreas rift lakes to the San Lorenzo River watershed and Uvas Canyon; s Solano County and to s half of the c Santa Barbara County.
T. a. hydrophilus Fitch, 1936 Oregon Gartersnake	Fitch (1984); Rossman et al. (1996); Brown (1997); Boundy (1999); Bartlett & Tennant (2000)	Northern Coast Ranges of California and sw Oregon north of the Gualala River and Lake County, California.
T. a. zaxanthus Boundy, 1999 Diablo Range Gartersnake	Boundy (1999)	California in the Inner Coast Range from Napa and Solano to Santa Barbara counties, and the Santa Lucia Range; it is absent form the major valleys.
T. brachystoma (Cope, 1892) Short-headed Gartersnake	A. G. Smith (1945); Barton (1956); Conant (1958, 1975); Bothner (1963, 1976); Ernst & Barbour (1989); Wright & Wright ([1957] 1994); Rossman et al. (1996); Tennant & Bartlett (2000)	Upper Allegheny River drainage system of the Allegheny Plateau nw Pennsylvania into sw New York from 274 to 732 m, an introduced population has been established in Ohio.
T. butleri (Cope, 1889) Butler's Gartersnake	Conant (1938, 1975); A. G. Smith (1945); Barton (1956); Minton (1980); Wright & Wright ([1957] 1994); Rossman et al. (1996); Tennant & Bartlett (2000)	Rossman et al. (1996) described the range of this snake as "The continuous range . . . extends from c Ohio and c Indiana northward through e Michigan and the extreme s tip of Ontario at elevations from 152 to 457 m (based on Wright & Wright (1957). Geographically isolated populations occur in extreme se Wisconsin and the Luther Marsh area in cs Ontario."
T. chrysocephalus (Cope, 1885) Golden-headed Gartersnake	Smith (1942f); Smith & Taylor (1966); Pérez-Higareda & Smith (1991); Smith & Smith (1993); Rossman et al. (1996)	S Mexico in relatively high elevations of the s Sierra Madre Oriental in Puebla, Veracruz, the Mesa del Sur in Oaxaca, and Sierra Madre del Sur in Guerrero from 1219 to 3078 m.
T. couchi (Kennicott, 1859) Sierra Gartersnake	Fitch (1984); Stebbins (1985); Rossman & Stewart (1987); Savage & Slowinski (1992); Smith & Rossman et al. (1996); Brown (1997); Bartlett & Tennant (2000)	W United States from nc California southward along the Sierra Nevada into extreme w Nevada from the Pit and Sacramento rivers to the western end of the Tehachapi Mountains, there are populations in wc Nevada and Owens Valley, California; it is found from 91 to 2438 m.

The Caenophidia

Species-level Taxa	Citations and Remarks	Abbreviated Range
T. cyrtopsis (Kennicott, 1860) Black-necked Garter-snake	Milstead (1953); Fitch & Milstead (1961); Rossman *in* Peters & Orejas-Miranda (1970); Webb (1966, 1978, 1980, 19872b); Stebbins (1985); Villa et al. (1988); Smith & Smith (1993); Rossman et al. (1996); Köhler (2001)	SW United States from extreme Colorado and se Utah southward to New Mexico, Arizona and w and c Texas through tropical Mexico into Central America in w Honduras; from sea level to 2700 m.
T. c. cyrtopsis (Kennicott, 1860) Western Black-necked Gartersnake	Fitch & Milstead (1961); Conant (1975); Webb (1980); Smith & Smith (1993); Rossman et al. (1996); Conant & Collins (1998); Dixon (2000a); Werler & Dixon (2000); Bartlett & Tennant (2000); Tennant & Bartlett (2000)	SW United States south into c Mexico from se Utah and s Colorado; and w Arizona into Trans-Pecos Texas, and into Mexico in Sonora, Zacatecas, and San Luis Potosí; it seems to be absent from c Chihuahua; it is associated with the Sierra Madre Occidental and the Mexican Plateau to as far east as n Hidalgo.
T. c. collaris* (Jan, 1863) Tropical Black-necked Gartersnake	Fitch & Milstead (1961); Webb (1978, 1980); Pérez-Higareda & Smith (1991); Rossman et al. (1996); Rossman (1996); Köhler (2001) *Remark: *T. vicinus* Smith, 1942 from c Michoacán in Mexico, is a synonym of *T. cyrtopis collaris* (see Rossman et al., 1996 and Rossman, 1996).	In lowland and tropical areas in s Sonora southeastward through w and s Mexico into wc Guatemala.
T. c. ocellatus (Cope, 1880) Eastern Black-necked Gartersnake	Fitch & Milstead (1961); Conant (1975); Webb (1980); Smith & Smith (1993); Rossman et al. (1996); Conant & Collins (1998); Dixon (2000a); Werler & Dixon (2000); Tennant & Bartlett (2000)	Confined to the Edwards Plateau west into the Big Bend Country of Texas southward intergrading into Mexico.
T. elegans (Baird & Girard, 1853) Terrestrial Gartersnake	Fox (1951a); Fitch (1940, 1983); Rossman (1979); Smith & Smith (1993); Rossman et al. (1996); Brown (1997)	W Canada southward into w United States and continuing its range into c Mexico from sea level to 3990 m.
T. e. elegans (Baird & Girard, 1853) Mountain Gartersnake	Fitch (1940, 1983); Rossman et al. (1996); Brown (1997); Bartlett & Tennant (2000)	Sierras, Coast Ranges, and the Cascade Mountains in the w United States from c California (exclusive to the outer Coast Ranges) northward into sw Oregon with an isolated population in the San Bernardino Mountains of s California.
T. e. arizonae Tanner & Lowe, 1989 Arizona Wandering Gartersnake	Tanner & Lowe (1989); Rossman et al. (1996); Bartlett & Tennant (2000)	E Arizona in Apache and Navajo counties into w New Mexico in Catron and McKinley counties.
T. e. hueyi Van Denburgh & Slevin, 1923 San Pedro Martir Gartersnake	Van Denburgh & Slevin (1923); Fitch (1983); Smith & Smith (1993); Rossman et al. (1996)	Confined to Sierra San Pedro Mártir in Baja California from sea level to 3990 m.
T. e. terrestris Fox, 1951 Coast Gartersnake	Fox (1951a); Fitch (1983); Rossman et al. (1996); Brown (1997); Bartlett & Tennant (2000)	West coast of the United States in the outer Coast Ranges from Curry County, Oregon southward into Ventura county, California.
T. e. vagrans* (Baird & Girard, 1853) Intermountain Wandering Gartersnake	Fox (1951a); Fleharty (1967); Conant (1975); Fitch (1940, 1980, 1983); Smith & Smith (1993); Rossman et al. (1996); Brown (1997); Conant & Collins (1998); Bartlett & Tennant (2000); Tennant & Bartlett (2000)	Alberta, British Columbia and sw Saskatchewan, Canada southward into the United States into Nevada, New Mexico and c Arizona.

Species-level Taxa	Citations and Remarks	Abbreviated Range
	*Remark: Rossman et al. (1996) listed *T. e. nigrescens* Johnson, 1947 (found on Vancouver Island, and in the Peugeot Sound area of Washington and British Columbia), as a synonym of *T. e. vagrans*.	
T. e. vascotanneri Tanner & Lowe, 1989 Upper Basin Gartersnake	Tanner & Lowe (1989); Rossman et al. (2000); Bartlett & Tennant (2000)	Colorado and Green River drainages of e Utah.
T. eques (Reuss, 1834) Mexican Gartersnake	Conant (1963b); Tanner (1986); Smith & Smith (1993); Rossman et al. (1996)	SW United States in se Arizona and extreme sw New Mexico into Mexico; found from 53 to 2590 m.
T. e. eques (Reuss, 1834) Southern Mexican Gartersnake	Conant (1963b); Tanner (1986); Smith & Taylor (1966); Smith & Smith (1993); Rossman et al. (1996)	Chihuahua Desert into the Mexican Central Plateau from s Nayarit eastward along the Transverse Volcanic Axis to wc Veracruz; apparently with an isolated population in c Oaxaca.
T. e. megalops (Kennicott, 1860) Northern Mexican Gartersnake	Conant (1963b); Smith & Smith (1993); Rossman et al. (1996); Bartlett & Tennant (2000)	The United States in c Arizona and the lower Gila River in New Mexico southward into Mexico in the Sierra Madre Occidental into Guanajuato and east across the Mexican Plateau into Hidalgo with an isolated population in c Nuevo León.
T. e. virgatenuis Conant, 1963 Blue-striped Mexican Gartersnake	Conant (1963b, 1997); Tanner (1986); Smith & Smith (1993); Rossman et al. (1996)	3 isolated populations at high elevations in w then nw Chihuahua and sw Durango.
*T. errans** Smith, 1942 Mexican Wandering Gartersnake	Webb (1976); Fitch (1980); Smith & Smith (1993); Queiroz & Lawson (1994); Rossman et al. (1996)	NW Chihuahua into w Zacatecas in Mexico from 1860 to 2545 m.
T. exsul Rossman, 1969 Exiled Gartersnake	Rossman (1969, 1992); Liner (1992); Rossman et al. (1989); Rossman et al. (1996)	Known only from three locations in the Sierra Madre Oriental of se Coahuila, and se Nuevo León in Mexico from between 2650 to 2860 m.
T. fulvus (Bocourt, 1893 *in* Dumeril, Mocquard & Bocourt, 1870–1909) Meso-American or Highlands Gartersnake	Slevin (1939); Stuart (1948); Rossman (1965); Webb (1982b); Wilson & Meyer (1985); Villa et al. (1988); Rossman et al. (1996); Köhler (2001)	SE Mexico in c Chiapas southward through s Guatemala into sw Honduras and adjacent El Salvador from 1410 to 3353 m.
T. gigas Fitch, 1940 Giant Gartersnake	Fitch (1940, 1984); Stebbins (1985); Rossman & Stewart (1987); Rossman et al. (1996); Brown (1997); Bartlett & Tennant (2000)	C California throughout the Sacramento and San Joaquin valleys from sea level to 122 m.
T. godmani (Günther, 1894) Godman's Gartersnake	Smith (1942f); Pérez-Higareda & Smith (1991); Smith & Smith (1993); Rossman et al. (1996)	Mountains in s Mexico at relatively high elevations in s Sierra Madre Oriental in Puebla and Veracruz, the Mesa del Sur in Oaxaca, and Sierra Madre del Sur in Guerrero from 1768 to 3018 m.

The Caenophidia

Species-level Taxa	Citations and Remarks	Abbreviated Range
*T. hammondi** (Kennicott, 1860) Two-striped Gartersnake	Fitch (1940, 1984); Smith & Taylor (1966); Rossman (1979); Rossman & Stewart (1987); Smith & Smith (1993); McGuire & Grismer (1993); Rossman et al. (1996); Brown (1997); McPeak (2000); Bartlett & Tennant (2000) *Remark: McGuire & Grismer (1993), state that *T. digueti* is not taxonomically distinct from *T. hammondi*. However, *T. digueti* (Mocquard, 1899), from the Sierra de la Giganta in the southern half of Baja California, is a synonym of *T. hammondi* (see Rossman et al., 1996).	California from Monterey county southward into nw Baja California in Mexico with isolated populations in northern Baja California Sur and on Catalina Island off the California coast; from sea level to 2130 m.
T. marcianus (Baird & Girard, 1853) Checkered Gartersnake	Smith & Taylor (1966); Rossman *in* Peters & Orejas-Miranda (1970); Rossman (1971); Stebbins (1985); Savage & Villa (1986); Villa et al. (1988); Smith & Smith (1993); Rossman et al. (1996); Lee (1996, 2000); Brown (1997); Campbell (1998a); Dixon (2000a); Köhler (2001)	SW United States in Texas, Oklahoma, Kansas, Colorado, New Mexico, Arizona, and California southward into Baja California and through Mexico into lower Central America in n Costa Rica from sea level to 1640 m.
T. m. marcianus (Baird & Girard, 1853) Marcy's Checkered Gartersnake	Rossman (1971); Conant (1975); Pérez-Higareda & Smith (1991); Smith & Smith (1993); Rossman et al. (1996); Conant & Collins (1998); Werler & Dixon (2000); Bartlett & Tennant (2000); Tennant & Bartlett (2000)	C and sw United States, from sw Kansas westward into extreme se California and southward in n and e Mexico into n Veracruz; there is an isolated population in the Isthmus of Tehuantepec.
T. m. bovallii Dunn, 1940 Dunn's Checkered Gartersnake	Shreve & Gans (1958); Rossman *in* Peters & Orejas-Miranda (1970); Rossman (1971); Rossman et al. (1996); Köhler (2001)	Lower Central America around Lake Managua and Lake Nicaragua in Nicaragua southward into n Costa Rica.
T. m. praeocularis (Bocourt, 1892) Yucatecan Checkered Gartersnake	Dunn (1940b); Rossman *in* Peters & Orejas-Miranda (1970); Rossman (1971); Pérez-Higareda & Smith (1991); Rossman et al. (1996); Campbell (1998a); Lee (2000); Köhler (2001)	S Mexico in Quintana Roo in the Yucatán, Belize and ne Guatemala to Lake Yojoa in w Honduras.
*T. melanogaster** (W. Peters, 1864) Mexican Black-bellied Gartersnake	Tanner (1959, 1985); Conant (1963b); Chiasson & Lowe (1989); Smith & Smith (1993); Rossman et al. (1996) *Remark: Liner (1994) listed this taxon as *Nerodia melanogaster* but no details were given.	Wide ranging in Mexico from 1158 to 2545 m.
T. m. melanogaster (W. Peters, 1864) Mexican Black-bellied Gartersnake	Conant (1963b); Smith & Taylor (1966); Smith & Smith (1993); Rossman et al. (1996); Uribe-Peña et al. (1999)	Confined to the Valley of Mexico in c Mexico.
T. m. canescens Smith, 1942 Gray Black-bellied Gartersnake	Conant (1963b); Smith & Taylor (1966); Smith & Smith (1993); Rossman et al. (1996); Köhler (2001)	S and w Mexico from sw San Luis Potosí into Guanajuato in n Michoacán and westward through Jalisco and w Zacatecas into Durango.

Species-level Taxa	Citations and Remarks	Abbreviated Range
T. m. chihuahuaensis Tanner, 1959 Chihuahuan Black-bellied Gartersnake	Tanner (1959, 1986); Smith & Smith (1993); Rossman et al. (1996)	Bavispe and El Fuerte River Basins in w Chihuahua and possibly adjacent Sonora in Mexico.
T. m. linearis Smith, Nixon & Smith, 1950 Lined Black-bellied Gartersnake	Conant (1963b); Smith & Smith (1993); Rossman et al. (1996)	Confined to the valley of Toluca in Mexico.
T. mendax Walker, 1955 Tamaulipan Montane Gartersnake	Walker (1955); Kirk (1979); Rossman (1992); Smith & Smith (1993); Rossman et al. (1996)	Mexico, only from the Sierra de Guatemala Mountains of The Sierra Madre Oriental in sw Tamaulipas; it is found from 1050 to 2120 m; it frequents forested mountains in the humid pine-oak, lower humid oak-sweet gum, and upper and lower cloud forest zones.
T. nigronuchalis Thompson, 1957 Southern Durango Spotted Gartersnake	Thompson (1957); Tanner (1986); Rossman (1995); Rossman et al. (1996)	S Durango in the Río del Presidio drainage and one record from Arroyo in Los Mimbres in Durango in Río Del Tunal drainage from 2195 to 2743 m; Tanner (1986) suggested this snake is confined to high elevation basins.
T. ordinoides (Baird & Girard, 1852) Northwestern Gartersnake	Fitch (1940); Fox (1948); Smith & Smith (1993); Rossman (1995); Rossman et al. (1996); Brown (1997); Bartlett &; Tennant (2000)	SW Canada southward through the w United States into n Mexico. It is found on Vancouver Island and southward along the Pacific Coast from sw British Columbia to extreme nw California from sea level to 1370 m.
T. postremus Smith, 1942 Tepalcatepec Valley Gartersnake	Fitch & Milstead (1961); Webb (1978, 1980); Rossman (1992); Rossman et al. (1996)	Confined to the Tepalcatepec Valley of Michoacán in Mexico from 236 to 1067 m; restricted to the lowland arid tropical scrub forest.
T. proximus (Say *in* James, 1823) Western Ribbonsnake	Fitch (1948); Fox (1948); Rossman (1963, 1970a); Rossman *in* Peters & Orejas-Miranda (1970); Wilson & Meyer (1985); Savage & Villa (1986); Villa et al. (1988); Dundee & Rossman (1989); Rossman et al. (1996); Campbell (1998a); Conant & Collins (1998); Dixon (2000a); Lee (2000); Köhler (2001)	Central United States through e Mexico into c Costa Rica from sea level to 2438 m.
T. p. proximus (Say *in* James 1823) Orange-striped Ribbonsnake	Rossman (1963, 1970a); Smith & Taylor (1966); Conant (1975); Dundee & Rossman (1989); Rossman et al. (1996); Dixon (2000a); Werler & Dixon (2000); Tennant & Bartlett (2000)	C United States from Indiana and s Wisconsin westward into w Kansas and south into s Louisiana and e Texas.
T. p. alpinus Rossman, 1963 Chiapas Highland Ribbonsnake	Rossman (1963, 1970a); Alvarez del Toro (1982); Rossman et al. (1996); Köhler (2001)	S Mexico at high elevations in c Chiapas.
T. p. diabolicus Rossman, 1963 Arid Land Ribbonsnake	Rossman (1963, 1970a); Conant (1975); Rossman et al. (1996); Conant & Collins (1998); Dixon (2000a); Werler & Dixon (2000); Bartlett & Tennant (2000); Tennant & Bartlett (2000)	SW United States from se Colorado southward through the Pecos Valley in Texas into Mexico in Coahuila, Nuevo León, and wc Tamaulipas.

Species-level Taxa	Citations and Remarks	Abbreviated Range
T. p. orarius Rossman, 1963 Gulf Coast Ribbonsnake	Rossman (1963, 1970a); Conant (1975); Dundee & Rossman (1989); Rossman et al. (1996); Conant & Collins (1998); Dixon (2000a); Werler & Dixon (2000); Tennant & Bartlett (2000)	W Gulf Coast of the United States from se Louisiana through the Texas Gulf Coast into ne Tamaulipas in Mexico; this is a coastal form.
T. p. rubrilineatus Rossman, 1963 Red-striped Ribbonsnake	Rossman (1963, 1970a); Conant (1975); Rossman et al. (1996); Conant & Collins (1998); Dixon (2000a); Werler & Dixon (2000); Tennant & Bartlett (2000)	Confined to the Edwards Plateau of c Texas in the United States.
T. p. rutilorus (Cope, 1885) Mexican Coast Ribbonsnake	Rossman (1963, 1970a); Rossman *in* Peters & Orejas-Miranda (1970); Alvarez del Toro (1982); Wilson & Meyer (1985); Pérez-Higareda & Smith (1991); Smith & Smith (1993); Campbell (1998a); Lee (2000); Köhler (2001)	S Tamaulipas and coastal Guerrero in Mexico southward into Central American as far south as Costa Rica.
T. pulchrilatus (Cope, 1885) Yellow-throated Gartersnake	Fitch & Milstead (1961); Webb (1966, 1980); Rossman (1992); Smith & Smith (1993); Rossman et al. (1996)	Mexican highlands from Durango to Tamaulipas and Nuevo León and southward to as far south as Oaxaca in an arc that roughly outlines the Mexican Plateau; from 1372 to 2804 m; it is mainly an upland species.
T. radix (Baird & Girard, 1853) Plains Gartersnake	A. G. Smith (1949); P. W. Smith (1956); Conant (1975); Stebbins (1985); Rossman et al. (1996); Conant & Collins (1998); Bartlett & Tennant (2000) *Remark: According to Rossman et al. (1996), subspecies are not valid. However, subspecies can be found listed in Tennant & Bartlett (2000), Conant & Collins (1998); and Werler & Dixon (2000). This work follows Rossman et al. (1996).	S Canada in s Alberta southward into the United States in e Montana, Wyoming and Colorado to ne New Mexico and Oklahoma panhandle eastward through the Great Plains in s Wisconsin, n Illinois and nw Indiana with isolated populations in Ohio and Missouri and adjacent Illinois; from 120 to 2290 m.
T. rossmani* Conant, 2000 Rossman's Gartersnake	Conant (2000) *Remark: This new species is very similar to *T. eques*. According to Conant (2000:5), available evidence is fragmentary, but it indicates that survival of *T. rossmani* may be in grave danger, if it is not already extinct.	Known only from springs, seepage runs, and ditches near but not along the San Cayetano River, a small stream flowing northwestward to and beyond Tepic; near Tepic, Nayarit, Mexico.
T. rufipunctatus* (Cope *in* Yarrow, 1875) Narrow-headed Gartersnake	Thompson (1957); Smith & Taylor (1966); Fleharty (1967); Tanner (1985, 1986, 1990); Chiasson & Lowe (1989); Smith & Smith (1993); Rossman et al. (1996); Bartlett & Tanner (2000) *Remark: *T. angustirostris* (Kennicott, 1860), is listed as being found in se Arizona and sw New Mexico southward into Chihuahua, Coahuila, and Durango. According to Rossman et al. (1996), it should be treated as *nomen dubium*. Smith & Taylor (1966), noted that the type was the only known specimen and may be a hybrid between *rufipunctatus* and *melanogaster*. Smith & Smith (1993) list it as a synonym of *T. rufipunctatus*. See Rossman et al. (1996:244–245) for details.	Disjunct distribution from nc Arizona to sw New Mexico in the United States southward into nw Mexico into n Chihuahua and n and w Durango; from 701 to 2430 m.

Species-level Taxa	Citations and Remarks	Abbreviated Range
	*Remark 2: Tanner (1990), listed this species with three subspecies of: *T. r. rufipunctatus* (Cope *in* Yarrow, 1875) c Arizona e from Yavapai county into wc Mexico; and, *T. r. nigronuchalis* (Tanner, 1957) Durango in high altitudes basins that flow from near the Continental Divide westward in w Mexico; and, *T. r. unilabialis* (Thompson, 1985) from n Mexico in the Sierra Madre Occidental of c and w Chihuahua. Liner (1994) lists with subspecies but as *Nerodia rufipunctatus* with subspecies. This work follows Rossman et al. (1996) who stated that there are no subspecies. Also see Tanner ("1985", 1986) and Rossman (1995).	
T. sauritus (Linnaeus, 1766) Eastern Ribbonsnake	Rossman (1963, 1970b); Conant (1975); Dundee & Rossman (1989); Smith & Smith (1993); Wright & Wright ([1957] 1994); Rossman et al. (1996)	Ontario, Canada through all of the eastern third of the United States southward from sea level to 610 m west of the Mississippi River; it is closely associated with water. sources such as swamps, streams, roadside ditches, marshes, and bogs, generally in or near thickets or brushy vegetation.
T. s. sauritus (Linnaeus, 1766) Common Ribbonsnake	Klauber (1948); Rossman (1970b); Conant (1975); Dundee & Rossman (1989); Smith & Smith (1993); Rossman et al. (1996); Tennant (1997b); Conant & Collins (1998); Tennant & Bartlett (2000)	The United States from s New England west-southwestward through s Ohio and s Indiana to the Mississippi River, and southward through the rest of the e USA, excluding e Georgia and peninsular Florida.
T. s. nitae Rossman, 1963 Blue-striped Ribbonsnake	Rossman (1963, 1970b); Conant (1975); Rossman et al. (1996); Tennant (1997b); Conant & Collins (1998); Tennant & Bartlett (2000)	Restricted to the Gulf Coast of nw peninsular Florida in the United States.
T. s. sackeni (Kennicott, 1859) Peninsula Ribbonsnake	Klauber (1948); Rossman (1970b); Conant (1975); Rossman et al. (1996); Tennant (1997b); Conant & Collins (1998); Buckner & Franz (1998a); Powell & Henderson (1999); Tennant & Bartlett (2000)	SE United States in most of peninsular Florida and se Georgia to the southern tip of South Carolina. Recently reported from the Bahamas on New Providence Island.
T. s. septentrionalis Rossman, 1963 Northern Ribbonsnake	Rossman (1963, 1970b); Conant (1975); Rossman et al. (1996); Conant & Collins (1998); Tennant & Bartlett (2000)	Canada from Nova Scotia westward into Wisconsin in the United States.
T. scalaris Cope, [1860] 1861 Mexican Alpine Blotched Gartersnake	Smith & Taylor (1966); Pérez-Higareda & Smith (1991); Rossman et al. (1996); Rossman & Lara-Gongora (1991, 1997)	High elevations in mountains in Mexico at 2103 to 4273 m across the Transverse Volcanic Axis of c Mexico from Jalisco to Veracruz; found in high-elevation conifer forests of the central Mexican Plateau.
*T. scaliger** (Jan, 1863) Mesa Central Blotched Gartersnake	Smith & Taylor (1966); Rossman et al. (1996); Rossman & Lara-Gongora (1991, 1997); Quintero-Díaz (1999); Uribe-Peña et al. (1999) *Remark: Rossman et al. (1996) stated that this description is based on *Coluber sirtalis* Linnaeus, 1758, *sensu* Harlan, 1827.	Disjunct distribution in the c Mexican Mesa Central which lies immediately north of Transverse Volcanic Axis between 2288 to 2575 m.

Species-level Taxa	Citations and Remarks	Abbreviated Range
T. sirtalis* (Linnaeus, 1758) Common Gartersnake	Brown (1950); Rossman (1965c); Conant (1975); Fitch (1941, 1980); Stebbins (1985); Tanner (1988); Savage & Slowinski (1992); Rossman et al. (1996); Barry et al. (1996); Brown (1997); Conant & Collins (1998)	Canada southward through most of the United States into c Mexico. It is absent from Arizona; altitude ranges are from sea level to 2540 m.
T. s. sirtalis (Linnaeus, 1758) Eastern Gartersnake	Conant (1975); Rossman et al. (1996); Tennant (1997b); Buckner & Franz (1998b); Werler & Dixon (2000); Tennant & Bartlett (2000)	SC Canada and the United States in s New England southward into se Texas and Florida. Recently reported in the Bahamas on Abaco Island.
T. s. annectans Brown, 1950 Texas Gartersnake	Brown (1950); Conant (1975); Rossman et al. (1996); Conant & Collins (1998); Dixon (2000a); Werler & Dixon (2000); Tennant & Bartlett (2000)	WC Oklahoma to ec Texas to the e Texas panhandle and sw Kansas.
T. s. concinnus (Hallowell, 1852) Red-spotted Gartersnake	Ruthven (1908); Rossman et al. (1996); Bartlett & Tennant (2000)	NW United States from the Pacific coast of the outer Olympic Peninsula and the Colombia River Valley southward to the north side of San Francisco Bay and s San Francisco Peninsula to just north of San Diego.
T. s. dorsalis* (Baird & Girard, 1853) New Mexico Gartersnake	Webb (1966); Conant (1975); Smith & Smith (1993); Rossman et al. (1996); Conant & Collins (1998); Dixon (2000a); Bartlett & Tennant (2000) *Remark 1: T. s. ornata is a synonym, according to Smith & Smith (1993) with details as to why the name is used. *Remark 2: T. s. lowei Tanner, 1988 from Chihuahua is a synonym (see Rossman et al. 1996).	2 isolated populations, one in sw United States along the Grande River in s New Mexico and another in nw Chihuahua, Mexico.
T. s. fitchi Fox, 1951 Valley Gartersnake	Fitch (1951b); Rossman et al. (1996); Brown (1997); Bartlett & Tennant (2000)	W United States and Canada from se Alaska through s British Columbia into Idaho and the e 2/3 of Washington through nc Utah to c California to between Monterrey and Santa Barbara. It is excluded from Vancouver Island and the sw coast of British Columbia.
T. s. infernalis* (Blainville, 1835) San Francisco Gartersnake	Boundy & Rossman (1995); Rossman et al. (1996); Brown (1997); Bartlett & Tennant (2000); Crother et al. (2000) *Remark: T. s. tetrataenia Cope in Yarrow, 1875 is a synonym (see Rossman et al., 1996), based on Boundy & Rossman (1995). However, there has been a concerted effort in the herpetological community by various authorities to reestablish acceptance of T. s. tetrataenia as a valid subspecies. T. s. tetrataenia Cope in Yarrow, 1875, found in San Francisco peninsula, southward into Crystal Lake and on the coast to Point Año Nuevo, San Mateo county, California. Bartlett & Tennant (2000) listed T. s. tetrataenia in their work, but this was premature to the ICZN opinion being published. The same thing in Brown (1997). Crother et al. (2000) also preferred to use	Confined to coastal California on the San Francisco Peninsula in the United States.

Species-level Taxa	Citations and Remarks	Abbreviated Range
	T. s. infernalis until the ICZN made their decision. An opinion was formally given with the subspecific name being conserved by the designation of a neotype for *T. s. infernalis* as Opinion 1961 published 29 September 2000 in 57(3): 191–192 of the Bull. Zool. Nomenclature.	
T. s. pallidulus Allen, 1899 Maritime Gartersnake	Conant (1975); Smith & Smith (1993); Rossman et al. (1996); Conant & Collins (1998); Tennant & Bartlett (2000)	Canadian Maritime Provinces, in s Quebec and the adjacent n New England in the United States.
T. s. parietalis (Say *in* James, 1823) Red-sided Gartersnake	Ruthven (1908); Conant (1975); Rossman et al. (1996); Conant & Collins (1998); Dixon (2000a); Werler & Dixon (2000); Bartlett & Tennant (2000); Tennant & Bartlett (2000)	Throughout the Great Plains from w Canada at the lower edge of the North West Territory southward into the c United States into the Red River Valley between Texas and Oklahoma. There is a small population near Fort Smith in the Mackenzie District; fide Dixon (2000a), the status of this snake in Texas is unknown.
T. s. pickeringii (Baird & Girard, 1853) Peugeot Sound Gartersnake	Fitch (1941); Rossman et al. (1996); Bartlett & Tennant (2000)	SW British Columbia, Canada including Vancouver Island and in the western third of Washington in the United States.
T. s. semifasciatus (Cope, 1892) Chicago Gartersnake	P. W. Smith (1956); Conant (1975); Rossman et al. (1996); Conant & Collins (1998); Tennant & Bartlett (2000)	N Midwestern United States in ne Illinois and adjacent se Wisconsin into nw Indiana.
T. s. similis Rossman, 1965 Blue-striped Gartersnake	Rossman (1965c); Conant (1975); Rossman et al. (1996); Tennant (1997b); Conant & Collins (1998); Tennant & Bartlett (2000)	NW coastal peninsular Florida.
T. sumichrasti* (Cope, 1866) Sumichrast's Gartersnake	Pérez-Higareda & Smith (1991); Rossman (1992); Rossman et al. (1996) *Remark 1: Fide Rossman et al. (1996) subspecies are not valid. *Remark 2: *T. phenax* is simply a color morph of *sumichrasti* fide Rossman (1966).	E Mexico along the Sierra Madre Oriental from e Querrétaro and se San Luis Potosí to the vicinity of Cordóba in Veracruz; from between 1365 to 2305 m; Rossman (1992) found this species in montane forests in the humid immediate vicinity of small streams or narrow irrigation ditches with fast-moving water.
T. validus* (Kennicott, 1860) Mexican Pacific Lowlands Gartersnake	Conant (1946, 1969); Rossman & Eberle (1977); Stebbins (1985); McCranie & McAllister (1988); Chiasson & Lowe (1989); Smith & Smith (1993) García & Ceballos (1994); Ramírez-Bautista (1994); Queiroz & Rossman (1995); Rossman et al. (1996) *Remark: Fide Conant (1997:189), "The scientific name of this species (*Natrix* =*Nerodia*) *valida celaeno* is in dispute. Some authorities would place it with the garter snakes (*Thamnophis*), but recent research suggests that it may be a water snake after	Disjunct distribution in the Pacific drainage in the west coast of c Mexico from Durango into Guerrero, and another in the Cape region of Baja California; sea level to 1200 m.

Species-level Taxa	Citations and Remarks	Abbreviated Range
	all." Rossman et al. (1996) stated that the recent transfer of *Nerodia valida* to *Thamnophis* by Lawson (1987) on the biochemical grounds created the anomalous situation of a *Thamnophis* species characteristically possessing a divided anal plate. They pointed out that the jury is still out on the transfer, however, so it remains to be seen if *Thamnophis* will once again be distinguishable from *Nerodia* on the basis of the anal plate condition.	
T. v. validus (Kennicott, 1860) Pacific Gartersnake	Conant (1961, 1969); Fitch (1981); McCranie & McAllister (1988); Rossman et al. (1996)	Río Yaqui in s Sonora to about San Blas, Nayarit in Mexico.
T. v. celaeno (Cope, 1860) Cape Gartersnake	Conant (1961, 1969); Fitch (1981); McCranie & McAllister (1988); Rossman et al. (1996); McPeak (2000)	Confined to the Cape Region of Territorio Sur de Baja California in Mexico; found in the mountains of extreme s Baja California in intermittent streams that flow from the mountains to the sea in rocky arroyos with pools small streams.
T. v. isabelleae (Conant, 1953) Colima Gartersnake	Conant (1953, 1961, 1969); Peters (1954); Fitch (1981); McCranie & McAllister (1988); Rossman et al. (1996)	Coastal plain from sw Jalisco into the vicinity of Acapulco, Guerrero; it occurs in short stretches of narrow coastal plain in the Pacific versant, in floodplains at the mouths of rivers flowing into the Pacific, and attenuated lagoons paralleling the coast. It is found in both tropical deciduous forest and thorn forest.
T. v. thamnophisoides (Conant, 1961) Tepic Gartersnake	Conant (1961, 1969); Fitch (1981); McCranie & McAllister (1988); Rossman et al. (1996)	Altiplano of the Río San Cayetano near Tepic, Nayarit; associated with spring-fed meadows.

GENUS: *Toluca* Kennicott *in* Baird, 1859, is a Mexican snake found from the Mexican Plateau into the highlands. Goyenechea Mayer-Goyenechea (1995), in an unpublished master's thesis, synonymized *Toluca* with *Conopsis*. Although I agree with her assessment of the consolidation of both genera into *Conopsis*, because this is not considered formally published material, I retain *Toluca* as a valid genus. See Taylor & Smith (1942); Smith & Taylor (1966); Smith & Smith (1993); Goyenechea Mayer-Goyenechea (1995); Goyenechea & Flores-Villela (2000, 2002).

See also genera *Conopsis*, *Ficimia*, *Gyalopion*, and *Pseudoficimia*.

CONTENT: Four species depending on its status with *Conopsis*. If Goyenechea Mayer-Goyenechea (1995) formally publishes her thesis, then the four species of *Toluca* consolidate into three species of a five species *Conopsis* group. If *Toluca* remains valid it should have 3 stars, as a reliability rating. One star is given here because of the questions that Goyenechea Mayer-Goyenechea has raised. ★

Comment: Goyeneches & Flores-Villela (2002) defined why *Toluca* and *Conopsis* should be synonymized, and *Conopsis* with priority. Unfortunately the good work done by Goyenechea (1995) was not detailed to the species level so both have been considered valid species for the time being. See comment under *Conopsis*.

Toluca: Tolucan Groundsnake. Frank and Ramus (1995) also call this the Tolucan Ground Snake.

Species-level Taxa	Citations and Remarks	Abbreviated Range
T. amphisticha Smith & Laufe, 1945 Twin-spotted Tolucan Groundsnake	Smith & Laufe (1945); Smith & Smith (1993); Goyenechea Mayer-Goyenechea (1995)	Puebla in Mexico.
T. conica Taylor & Smith, 1942 Large-blotched Tolucan Groundsnake	Taylor & Smith (1942); Smith & Taylor (1966); Smith & Smith (1993); Goyenechea Mayer-Goyenechea (1995)	Highlands of c. Oaxaca and Sierra Madre del Sur in Guerrero.
T. lineata Kennicott *in* Baird, 1859 Lined Tolucan Groundsnake	Smith & Taylor (1966); Wilson & McCranie (1979); Smith & Smith (1993); Goyenechea Mayer-Goyenechea (1995)	Mexico c and s highlands.
T. l. lineata Kennicott *in* Baird, 1859 Lined Tolucan Groundsnake	Davis & Smith (1953; Smith & Taylor (1966); Pérez-Higareda & Smith (1991); Smith & Smith (1993); Goyenechea Mayer-Goyenechea (1995); Uribe-Peña et al. (1999)	C and s Mexico from s San Luis Potosí southward into Distrito Federal and westward into c Michoacán and eastward into Tlaxcala, not found in the deserts of c Puebla.
T. l. acuta (Cope *in* Ferrari-Pérez, 1886) Spotted Tolucan Groundsnake	Smith & Taylor (1966); Smith & Smith (1993); Goyenechea Mayer-Goyenechea (1995)	s Puebla and probably n Oaxaca in s Mexico.
T. l. varians (Jan, 1862) Lineless Tolucan Groundsnake	Smith & Taylor (1966); Pérez-Higareda & Smith (1991); Smith & Smith (1993); Goyenechea Mayer-Goyenechea (1995)	Mountains of c Veracruz and adjacent Puebla in se Mexico.
T. l. wetmorei Smith, 1943 Wetmore's Tolucan Groundsnake	Smith (1943); Smith & Taylor (1966); Pérez-Higareda & Smith (1991); Smith & Smith (1993); Goyenechea Mayer-Goyenechea (1995)	Extreme ne Puebla and adjacent Veracruz in Mexico.
T. megalodon Taylor & Smith, 1942 San Felipe Tolucan Groundsnake	Taylor & Smith (1942); Smith & Taylor (1966); Smith & Smith (1993); Goyenechea Mayer-Goyenechea (1995)	Known only from the type locality in Oaxaca, Mexico of Cerro San Felipe.

GENUS: *Tomodon* Duméril, 1853, can be found from central Brazil into Paraguay, Uruguay, and north central Argentina. See Peters & Orejas-Miranda (1970); Cei (1986, 1993); Lema (1994b); and Starace (1998). Van Wallach (pers. comm. April 11, 1999), reconfirmed that *Opisthoplus* (as *O. degener* W. Peters, 1882 from Quinta, Carazinho and the vicinity of Porto Alegre, Rio Grande do Sul, Brazil), is included in *Tomodon* as a synonym.

Review assistance and important contributions for this section were provided by Robert A. Thomas and Van Wallach.

See also the genus *Pseudotomodon*, especially the notes about its close association with *T. ocellatus*.

The Caenophidia

CONTENT: Two species. ★★★

Tomodon: South American Mock Pit Viper or South American Night Viper

Species-level Taxa	Citations and Remarks	Abbreviated Range
T. dorsatus Dumeril, Bibron & Dumeril, 1854 Brown-spotted Mock Viper	Peters & Orejas-Miranda (1970); Amaral (1976); Cei (1993); Lema (1994b); Cechin (1989); Starace (1998); Yanosky (1998); Marques et al. (2001)	C and se Brazil and Misiones in n Argentina as well as the Guianas in French Guiana.
T. ocellatus Dumeril, Bibron & Dumeril, 1854 Ocellated Night or Mock Viper	Peters & Orejas-Miranda (1970), Gallardo (1972); Amaral (1976); Cei (1993); Lema (1994b); Yanosky (1998); Lema (pers. comm. December 12, 1999)	Associated with the Argentine pampa and found in Buenos Aires, Córdoba, Santa Fe, Entre Ríos, Corrientes, and Misiones with dispersion into s Rio Grande do Sul in Brazil and n Uruguay; according to Cei (1993), it is found in Paraguay.

GENUS: *Tretanorhinus* Dumeril, Bibron and Dumeril, 1854, ranges from Mexico and the Caribbean, southward into South America as far south as Ecuador. See Peters & Orejas-Miranda (1970); Wilson & Hahn (1973); Savage & Villa (1986); Schwartz & Henderson (1985, 1991) and Köhler (2001).

Review assistance and important contributions for this section was provided by Brian I. Crother.

CONTENT: Four species. ★★

Tretanorhinus: Middle American Swampsnake, Cativo, and the Caribbean Watersnake. Frank and Ramus (1995) call this the Striped Swampsnake.

Species-level Taxa	Citations and Remarks	Abbreviated Range
T. mocquardi Bocourt, 1891 Panamá Swampsnake	Bocourt (1891); Peters & Orejas-Miranda (1970); Villa et al. (1988); Köhler (2001)	Panama.
T. nigroluteus* Cope, 1861 Orange-bellied Swampsnake	Dunn (1939); Conant (1975); Peters & Orejas-Miranda (1970); Wilson & Hahn (1973); Wilson & Meyer (1985); Villa et al. (1988); Pérez-Higareda & Smith (1991); Smith & Smith (1993); Bahena-Basave (1995); Campbell (1998a); Stafford & Meyer (2000); Lee (2000); Köhler (2001)	Quintana Roo and Tabasco in Mexico and through Central America into Colombia.
	*Remark: Wilson & Hahn (1973), and Smith & Smith (1993) did not recognize subspecies. When listed, the subspecies are shown as: *T. n. nigroleteus* Cope, 1861 from the Atlantic versant of Honduras, Nicaragua and Panama into Colombia; *T. n. lateralis* Bocourt, 1891 found in the Yucatán Peninsula in Mexico southward into Belize and Guatemala; and, *T. n. mertensi* Smith & Gillespie, 1965 from s Veracruz in Mexico south and east through El Petén in Guatemala	

Species-level Taxa	Citations and Remarks	Abbreviated Range
T. taeniatus Boulenger, 1903 Pacific Cativo	Boulenger (1903a); Peters (1960b); Peters & Orejas-Miranda (1970); Pérez-Santos & Moreno (1988, 1991)	Pacific lowlands of w Colombia and Ecuador.
T. variablis Dumeril, Bibron and Dumeril, 1854 Cuban Cativo	Schwartz & Henderson (1985, 1991); Estrada & Ruibal *in* Crother (1999)	Throughout Cuba and the Antilles including Isla de la Juventud and Grand Cayman.
T. v. variablis Dumeril, Bibron and Dumeril, 1854 Cuban Cativo	Schwartz & Henderson (1985, 1991); Estrada & Ruibal *in* Crother (1999)	Cuba and the Antilles near it, except in sw Oriente Province where it is replaced by *T. variablis binghami*; it is found in Cuba from La Habana Province in the west eastward into Guantánamo Province, except for the region about Sierra Maestra, Granma Province.
T. v. binghami (Schwartz & Ogren, 1956) Bingham's Cativo	Schwartz & Henderson (1985, 1991); Estrada & Ruibal *in* Crother (1999)	Cuba in sw Oriente Province from Manzanillo to Lago Ariguanabo, both in the lowlands and n foothills of the Sierra Mastra; Granma Province.
T. v. insulaepinorum Barbour, 1916 Isla de la Juventud Cativo	Schwartz & Henderson (1985, 1991); Estrada & Ruibal *in* Crother (1999)	Restricted to the Isla de la Juventud off the coast of Cuba.
T. v. lewisi (Grant, 1941) Cayman Watersnake	Grant (1941a); Schwartz & Henderson (1985, 1991)	Grand Cayman Island in the Caribbean.
T. v. wagleri (Jan, 1863) Wagler's Cativo	Schwartz & Henderson (1985, 1991); Estrada & Ruibal *in* Crother (1999)	Cuba in Pinar del Río Province.

GENUS: *Trimetopon* Cope, 1885, is found in Costa Rica and Panama. See Peters & Orejas-Miranda (1970); Savage (1980); Villa et al. (1988); Savage & Villa (1986); and Köhler (2001).

CONTENT: Six species. ★★

Trimetopon: Cope's Tropical Groundsnake. Frank and Ramus (1995) also call this the Tropical Groundsnake.

Species-level Taxa	Citations and Remarks	Abbreviated Range
T. barbouri Dunn, 1930 Barbour's Tropical Groundsnake	Dunn (1930a); Peters & Orejas-Miranda (1970); Villa et al. (1988); Köhler (2001)	Canal Zone in Panama.
T. gracile (Günther, 1872) Gracile Tropical Groundsnake	Taylor (1951); Peters & Orejas-Miranda (1970); Savage (1980); Savage & Villa (1986); Villa et al. (1988); Köhler (2001)	Costa Rica.

The Caenophidia

Species-level Taxa	Citations and Remarks	Abbreviated Range
T. pliolepis Cope, 1894 Cope's Tropical Groundsnake	Taylor (1951); Peters & Orejas-Miranda (1970); Savage (1980); Savage & Villa (1986); Villa et al. (1988); Köhler (2001)	Costa Rica.
T. simile Dunn, 1930 Reventazón Tropical Groundsnake	Dunn (1930a); Taylor (1951); Peters & Orejas-Miranda (1970); Savage (1980); Savage & Villa (1986); Villa et al. (1988); Köhler (2001)	Known only from the type locality of Reventazón in Costa Rica.
T. slevini Dunn, 1940 Slevin's Tropical Groundsnake	Dunn (1940); Peters & Orejas-Miranda (1970); Savage (1980); Mudde & Dijk (1985); Villa et al. (1988); Köhler (2001); Myers (2003)	Costa Rica and Panama. Myers (2003) does not believe this snake is found in Panama.
T. viquezi Dunn, 1937 Viquez's Tropical Groundsnake	Peters & Orejas-Miranda (1970); Savage (1980); Savage & Villa (1986); Villa et al. (1988); Köhler (2001)	Siquirres, Costa Rica.

GENUS: *Trimorphodon* Cope, 1861, is distributed from the southwestern United States from southern California into west Texas, and southward through the Pacific coast and Central Plateau of Mexico into Central America to as far as Costa Rica. See Smith & Taylor (1966); McDiarmid & Scott (1970); Scott & McDiarmid (1984a); Stebbins (1985); Wright & Wright ([1957] 1994); Degenhardt et al. (1996); Tennant (1998); Dixon (2000a); and Köhler (2001).

Review assistance and important contributions for this section were provided by James R. Dixon and Travis J. La Duc.

CONTENT: About three species. ★★

Trimorphodon: Lyresnake

Species-level Taxa	Citations and Remarks	Abbreviated Range
T. biscutatus (Dumeril, Bibron & Dumeril, 1854) Common Lyresnake	Taylor (1938 [1939]); Klauber (1940b); Smith (1941f); Smith & Taylor (1966); McDiarmid & Scott (1970); Gelbach (1971); Scott & McDiarmid (1984a, 1984b); Wilson & Meyer (1982, 1985); Stebbins (1985); Villa et al. (1988); Johnson (1989, 1990); García & Ceballos (1994); Degenhardt et al. (1996); Brown (1997); Dixon (2000a); Köhler (2001); Minton (1992)	SW United States in s California and sw Texas southward into Mexico in the Baja California peninsula, and from Sinaloa southward along the Pacific coast into Costa Rica; it is also found on the Atlantic side near Tuxtla Gutierrez.
T. b. biscutatus (Dumeril, Bibron & Dumeril, 1854) Western Lyresnake	Smith (1941f); Fouquette & Rossman (1963); Smith & Taylor (1966); Hardy & McDiarmid (1969); Peters & Orejas-Miranda (1970); Gelbach (1971); McDiarmid & Scott (1970); Alvarez del Toro (1982); Scott & McDiarmid (1984b); Pérez-Higareda & Smith (1986b, 1991); Ramírez-Bautista (1994) *Remark: *T. b. semirutus* is a synonym (Smith & Smith, 1993).	W Mexico from Sinaloa into Chiapas and Veracruz.

Species-level Taxa	Citations and Remarks	Abbreviated Range
T. b. lambda Cope, 1886 Sonoran Lyresnake	Klauber (1928b, 1940b); Taylor (1938 [1939]); Smith (1941f); Fouquette & Rossman (1963); Lowe (1964); Fowlie (1965); Smith & Taylor (1966); Tanner & Banta (1966); McDiarmid & Scott (1970); Gelbach (1971); Scott & McDiarmid (1984b); Wilson & Meyer (1982, 1985); Stebbins (1985); Tanner (1985); Lowe et al. (1986); Degenhardt et al. (1996); Brown (1997); Bartlett & Tennant (2000)	SW United States southward into Sonora Mexico.
T. b. lyrophanes (Cope, 1860) Baja California Lyresnake	Taylor (1938 [1939]); Smith (1941f); Smith & Taylor (1966); McDiarmid & Scott (1970); Gelbach (1971); Powers (1974); Scott & McDiarmid (1984b); Stebbins (1985, 1994); Wilson & Meyer (1982, 1985); Grismer et al. (1994); Wright & Wright ([1957] 1994)	S Baja California in Mexico.
T. b. paucimaculatus* Taylor, 1936 Taylor's Lyresnake	Smith & Taylor (1966); McDiarmid & Scott (1970); Scott & McDiarmid (1984b); Smith & Smith (1993) *Remark: The taxon *paucimaculatus* should probably be synonymized with *T. b. biscutatus* and work needs to be done on this (La Duc pers. comm. 8 June, 2000). It was shown in Scott & McDiarmid (1984b) as a synonym of *T. b. biscutatus*, and was not listed in Liner (1994).	C Sinaloa southward into Nayarit, Mexico.
T. b. quadruplex Smith, 1941 Central American Lyresnake	Smith (1941f); Smith & Taylor (1966); Peters & Orejas-Miranda (1970); Scott & McDiarmid (1984b); Wilson & Meyer (1985); Pérez-Higareda & Smith (1986)	Pacific versant of w Guatemala southward into Costa Rica.
T. b. vandenburghi Klauber, 1924 California Lyresnake	Klauber (1928b, 1940b); Taylor (1938 [1939]); Smith & Taylor (1966); McDiarmid & Scott (1970); Gelbach (1971); Scott & McDiarmid (1984b); Wilson & Meyer (1985); Stebbins (1985); Grismer et al. (1994); Wright & Wright ([1957] 1994); Brown (1997); Bartlett & Tennant (2000)	The United States in southern California south into Baja California, Mexico.
T. tau Cope, 1869 (1870) Mexican Lyresnake	Taylor (1938 [1939]); Smith (1941a); Smith & Taylor (1966); Hardy & McDiarmid (1969); McDiarmid & Scott (1970); Morafka (1977); Scott & McDiarmid (1984a, 1984c); Smith & Smith (1993); Vasquez-Díaz & Quintero-Díaz (1997)	NW, c and s Mexico along the slopes and foothills of Sierra Madre Occidental, Oriental, and Sur. They are also known from the Central Plateau and in foothills of the Balsas and Tepalcatepec basins.
T. t. tau* Cope, 1869 (1870) Mexican Lyresnake	Klauber (1928b); Taylor (1938 [1939]); Smith (1941f); P. W. Smith & Darling (1952); Fouquette & Rossman (1963); McDiarmid & Scott (1970); Scott & McDiarmid (1984c); Wright & Wright ([1957] 1994); Tanner (1985); Pérez-Higareda & Smith (1991); Dundee & Liner (1997) *Remark 1: *Trimorphodon collaris* Cope, 1876, from Veracruz, Mexico; *T. fasciolatus* Smith, 1941 from e Michoacán into Nayarit, Mexico-associated with the Upper Río Balsas Basin; and, *T. forbesi* Smith, 1941,	Chihuahua, Sinaloa, Sonora, and most of *T. tau* range and northeastward into Tamaulipas to the San Fernando area; associated with foothills and plateaus.

Species-level Taxa	Citations and Remarks	Abbreviated Range
	from s Puebla & n Oaxaca in Mexico are synonyms, (see Scott & McDiarmid, 1984c).	
	*Remark 2: *Trimorphodon tau upsilon* Cope, 1869, found from s Chihuahua southward into Michoacán and eastward into c Hidalgo in Mexico, is a synonym of *T. t. tau*, (see Scott & McDiarmid, 1984c). Smith & Smith (1993) separated them. According to Travis J. La Duc (pers. comm. 8 June, 2000), Scott & McDiarmid (1984c) are probably correct. Based on Scott & McDiarmid (1984c), and La Duc (2000 pers. comm.), *T. tau upsilon* is considered to be a synonym of *T. t. tau* in this work.	
T. t. latifascia W. Peters, 1870 Black-headed Lyre-snake	Taylor (1938 [1939]); Smith (1941f); Davis & Smith (1953); Peters (1954); Smith & Taylor (1966); McDiarmid & Scott (1970); Scott & McDiarmid (1984c); Smith & Smith (1993)	The southern part of the range in Morelos and se Puebla in Mexico; associated with the Balsas-Tepalcatepec Basin.
*T. vilkinsoni** Cope, 1886 Texas Lyresnake	Crimmins (1925); Klauber (1928b); Taylor (1938 [1939]); Klauber (1940b); Smith (1941f); Davis & Dixon (1957); Dixon et al. (1962); Jones & Findley (1963); Smith & Taylor (1966); McDiarmid & Scott (1970); Morafka (1977); Banicki & Webb (1982); Scott & McDiarmid (1984b); Stebbins (1985); Tanner (1985); Wright & Wright ([1957] 1994); Degenhardt et al. (1996); Tennant (1998); Conant & Collins (1998); Werler & Dixon (2000); Bartlett & Tennant (2000); Tennant & Bartlett (2000) [as *T. biscutatus vilkinsoni*]; Dixon (2000a) *Remark 1: Werler & Dixon (2000) listed this taxon as *T. lambda vilkinsoni*. Their classification follows Dixon et al. (1962). The junior author stated *lambda* and *vilkinsoni* from central and western Mexico are closely related and that *lambda* deserves its own species designation, with *vilkinsonii* as an associated subspecies. Bartlett & Tennant (2000), continued to follow the *T. biscutatus vilkinsonii* designation. *Remark 2: La Duc (pers. comm. 8 June, 2000), and J. Johnson agree that *T. b. vilkonsonii* should be elevated to *T. vilkinsonii*. Dixon (2000a), based on a pers. comm. with Travis J. La Duc, listed *T. vilkinsoni* as a distinct species. *Remark 3: Crother et al. (2000) did not list this based on Grismer et al. (1994) synonymizing it with *T. b. lyrophanes*.	SW United States in w and c New Mexico and w Texas, southward into ne Chihuahua, Mexico.

GENUS: *Tropidoclonion* Cope, 1860, is restricted to the United States. See Schmidt (1953); Conant (1975, 1978); Degenhardt et al. (1996); Tennant (1998); and Dixon (2000a).

Review assistance and important contributions for this section were provided by James R. Dixon.

CONTENT: One species with up to four subspecies. To quote Bartlett & Tennant (2000), "There is a tendency today to consider this diminutive garter snake relative simply as a variable, monotypic species." ★★

Tropidoclonion: Lined Snake

Species-level Taxa	Citations and Remarks	Abbreviated Range
T. lineatum* (Hallowell, 1856) Lined Snake	Conant (1975, 1978); Stebbins (1985); Degenhardt et al. (1996); Tennant (1988); Conant & Collins (1998); Werler & Dixon (2000); Bartlett & Tennant (2000); Tennant & Bartlett (2000); Dixon (2000a)	Disjunct populations in the Plains and central States of the United States from Colorado and Illinois and New Mexico and se South Dakota into sc Texas.
	*Remark 1: Subspecies are not recognized by many authorities today (see Tennant, 1998) based on recent field and DNA data. Although the consolidation back to a monotypic species may be valid, I have not seen the data, so prefer to retain this group with subspecies for now.	
	*Remark 2: Dixon (2000a) notes subspecies with a qualifying remark but, Werler & Dixon (2000) did not.	
	*Remark 3: Fide Tennant & Bartlett (2000), the subspecies have been consolidated back into synonymy with the species and they do not accept subspecies.	
T. l. lineatum (Hallowell, 1856) Northern Lined Snake	Smith (1965); Conant (1975, 1978); Stebbins (1985); Degenhardt et al. (1996); Dixon (2000a)	Plains states of the United States.
T. l. annectens Ramsey, 1953 Central Lined Snake	Smith (1965); Conant (1975, 1978); Dixon (2000a)	The United States from Texas in the ne woodlands.
T. l. mertensi* Smith, 1965 New Mexico Lined Snake	Smith (1965); Conant (1975); Degenhardt et al. (1996) Dixon (2000a)	Texas Panhandle into e New Mexico.
	*Remark: Smith & Chiszar (1994) removed this snake from the Texas fauna and synonymized it with *T. l. lineatum*. This action left some interesting gaps in the distribution which Smith & Chiszar (1994), state is due to soil types. This work follows the older line of thought with *mertensi* found in Texas.	
T. l. texanum Ramsey, 1953 Texas Lined Snake	Smith (1965); Conant (1975, (1978); Smith & Smith (1993); Dixon (2000a)	Restricted to the Texas coastal plains, central prairies and e Edwards Plateau in the United States.

GENUS: *Tropidodipsas* Günther, 1858, ranges from southern Mexico through Central America and into South America as far as into Panama. The genus *senu Tropidodipsas* Günther, 1858, is *Sibon* Fitzinger, 1826 according to

Kofron (1985). However, the genus *Tropidodipsas* was resurrected by Wallach (1995), and is listed as separate in this work to reflect the most recent published views of the classification of this group. See Peters (1960c); Smith & Taylor (1966); Peters & Orejas-Miranda (1970); Kofron (1980, 1985); Vanzolini (1986); Obst et al. (1988); Auth (1994); Wallach (1995); and Köhler (2001).

Review assistance and important contributions for this section were provided by Ronaldo Fernandes and Van Wallach.

Although I accept the *Tropidodipsas* resurrection, this action seems to be questioned by many others. Part of the reason for this disagreement in the profession, is that the tables Wallach provides do not make the differences between *Sibon* and *Tropidodipsas* obvious. For example, Campbell (1998a) still listed this same taxa as *Sibon*.

Wallach (1995) separates the *Tropidodipsas* from *Sibon* based on the following:
- Synapomorphic maxillary process (also found in *Sibon*)
- Dietary habits
- Absence of a tracheal lung (other dipsadini have well-developed tracheal lung as can be found in *Sibon*, *Dipsas*, and *Sibynomorphus*.
- The heart is positioned more caudal in other dipsadini than in *Tropidodipsas*.
- The heart-liver gap is shorter in other dipsadini than in *Tropidodipsas*.
- The posterior tip of the right lung is more caudal in other dipsadini than in *Tropidodipsas*.

See also genus *Sibon*; *Dipsas*, and *Sibynomorphus*.

CONTENT: About six species depending on if it is classified as a synonym of *Sibon* or not. ★★

Tropidodipsas: False Middle American Snail-eaters

Species-level Taxa	Citations and Remarks	Abbreviated Range
T. annulifera* (Boulenger, 1894) Western Snail-eating Snake	Shannon & Humpfrey (1959); Kofron (1980, 1988a); Savage & Slowinski (1992); Wallach (1995) *Remark: Van Wallach (pers. comm. April 11, 1999), *Exelencophis nelsoni* Slevin, 1926 from María Madre Island in the Tres Marías Islands in Mexico and *Geatractus tecpanecus* Dugès, 1896, named from a single specimen from the locality of Tecpan de Galeana in w Guerrero, Mexico, are synonyms. This is confirmed in the Smith & Smith (1993) checklist, therefore *Geatractus* has not been included as a synonym.	W Mexico in coastal lowlands and adjacent highlands in Sinaloa, Nayarit, Jalisco, Colima, Michoacán, and Guerrero.
T. fasciata (Günther, 1858) Banded Snail-eater	Peters & Orejas-Miranda (1970); Kofron (1985, 1987); Villa et al. (1988); Savage & Slowinski (1992); Wallach (1995); Lee (1996, 2000); Köhler (2001)	SE Mexico southward into Guatemala.
T. f. fasciata (Günther, 1858) Banded Snail-eater	Kofron (1980, 1981, 1987); Pérez-Higareda & Smith (1991); Smith & Smith (1993); Lee (2000)	Atlantic coastal plains and slopes of e Mexico in the Yucatán peninsula, the Isthmus of Tehuantepec, and the adjacent lowlands and highlands of c Guatemala.
T. f. guerreroensis (Taylor, 1939) Guerreran Snail-eater	Taylor (1939b); Smith & Taylor (1966); Kofron (1980, 1985, 1987); Smith & Smith (1993)	C Guerrero into w Oaxaca in Mexico.

Species-level Taxa	Citations and Remarks	Abbreviated Range
T. f. subannulata* (Müller, 1887) Southern Mexican Snail-sucker	Peters & Orejas-Miranda (1970); Alvarez del Toro (1982) *Remark: Not found in Smith & Smith (1993).	Upper elevations of the Atlantic versant from sw Chiapas and se Oaxaca in Mexico southward into Alta Verapaz in Guatemala.
T. fischeri* (Boulenger, 1894) Fischer's Snail-eater	Peters & Orejas-Miranda (1970); Kofron (1985); Wilson & Meyer (1985); Wilson et al. (1986); Villa et al. (1988); Savage & Slowinski (1992); Wallach (1995); Köhler (2001) *Remark: Kofron (1985) stated that the status of this snake is unclear.	Highlands of w Oaxaca and Chiapas se Mexico into Guatemala, w and n El Salvador, and probably w Honduras.
T. f. fischeri (Boulenger, 1894) Fischer's Snail-eater	Boulenger (1894); Kofron (1980); Smith & Smith (1993)	Most of the species range in se Mexico into c Guatemala.
T. f. kidderi (Stuart, 1942) Kidder's Snail-eater	Stuart (1942); Kofron (1980); Smith & Smith (1993)	Higher elevations of c Guatemala in Alta Verapaz and Baja Verapaz.
T. philippi (Jan, 1863) Philipp's Snail-eater	Kofron (1980); Savage & Slowinski (1992); García & Ceballos (1994); Ramírez-Bautista (1994); Wallach (1995)	Pacific lowlands and adjacent slopes of Sinaloa and Colima into Michoacán and Jalisco in Mexico.
T. sartorii (Cope, 1863) Terrestrial Snail-eating Snake	Smith (1943); Peters & Orejas-Miranda (1970); Kofron (1980, 1988a); Wilson & Meyer (1985); Villa et al. (1988); Savage & Slowinski (1992); Wallach (1995); Lee (1996, 2000); Campbell (1998a); Stafford & Meyer (2000); Köhler (2001)	San Luis Potosí and Chiapas in s Mexico into Guatemala, Belize, El Salvador, Ulua River Valley of Honduras, and Nicaragua.
T. s. sartorii (Cope, 1863) Terrestrial Snail-eating Snake or Sartorius' Snail-eater	P. W. Smith & Darling (1952); Smith & Taylor (1966); Peters & Orejas-Miranda (1970); Alvarez del Toro (1982); Pérez-Higareda & Smith (1991); Campbell (1998a); Lee (2000); Köhler (2001)	Mexico in the Atlantic versant from c Veracruz into Guatemala, including the Yucatán peninsula, it is found in the states of Campeche, San Luis Potosí, Tabasco, Veracruz, and Yucatán.
T. s. annulatus (W. Peters, 1870) Ringed Slug-eater	Smith (1943); Peters (1960); Smith & Taylor (1966); Peters & Orejas-Miranda (1970); Villa (1991b); Savage (1980); Alvarez del Toro (1982); Villa et al. (1988); Savage & McDiarmid (1992); Savage & Slowinski (1992) [as *Sibon annulatus*]; Campbell (1998a); Köhler (2001) [as *Sibon annulatus*]	Low and intermediate elevations of the Pacific versant of Chiapas, Mexico into Guatemala.
T. s. macdougalli (Smith, 1943) MacDougall's Snail-eater	Smith (1943); Smith & Taylor (1966); Kofron (1985); Alvarez del Toro (1982)	Known only from the type locality in Tehuantepec, Oaxaca in Mexico.
T. zweifeli (Liner & Wilson, 1970) Zweifel's Snail-eating Snake	Liner & Wilson (1970); Kofron (1980, 1988a); Savage & Slowinski (1992); Wallach (1995)	Mexico in Chilpancingo, Guerrero and Cuernavaca, Morelos.

GENUS: *Tropidodryas* Fitzinger, 1843, is found in Brazil. See Amaral (1976) [as *Philodryas*]; Thomas & Dixon (1977); Vanzolini (1986); and Lema (1994b).

Review assistance and important contributions for this section were provided by James R. Dixon and Robert A. Thomas.

CONTENT: Two species. ★★★★

Tropidodryas: Serra Snake. Frank and Ramus (1995) call this the Serra Snake.

Species-level Taxa	Citations and Remarks	Abbreviated Range
T. serra (Schlegel, 1837) Serra Snake	Amaral (1976); Thomas & Dixon (1977); Suzart-Argôlo (1999); Marques et al. (2001)	NW, c and sc Brazil.
T. striaticeps (Cope, 1870) Narrow-headed Serra Snake	Amaral (1976); Thomas & Dixon (1977); Lema (1994b); Suzart-Argôlo (1999); Freitas (1999); Marques et al. (2001)	SE Brazil.

GENUS: *Umbrivaga* Roze, 1964, is confined to South America. See Roze (1964b); Peters & Orejas-Miranda (1970); Markezich & Dixon (1979); Vanzolini (1986); Lancini & Kornacker (1989); and Kornacker (1999).

CONTENT: Three species. ★★★

Umbrivaga: Moss Snake.

Species-level Taxa	Citations and Remarks	Abbreviated Range
U. mertensi Roze, 1964 Merten's Moss Snake	Roze (1964b); Peters & Orejas-Miranda (1970); Lancini (1979); Markezich & Dixon (1979); Lancini & Kornacker (1989); La Marca (1997); Kornacker (1999)	Known only from the type locality in the Parque Nacional Henri Pittier (Rancho Grande) in the state of Aragua in Venezuela.
U. pyburni Markezich & Dixon, 1979 Meta Moss Snake	Markezich & Dixon (1979); Vanzolini (1986); Pérez-Santos & Moreno (1988)	Known only from the type locality in Loma Linda in Meta, Colombia.
U. pygmaea (Cope, 1868) Pygmy or Amazon Moss Snake	Markezich & Dixon (1979); Vanzolini (1986); Dixon & Soini (1986); Fernandes et al. (1999)	Colombia, Ecuador, Peru, into Amazonas in Brazil.

GENUS: *Uromacer* Dumeril, Bibron & Dumeril, 1854, in the Haitian part of Hispaniola. See Cochran (1941); Horn (1969); Maglio (1970); Henderson & Schwartz (1984); Schwartz & Henderson (1984a, 1991); and Dixon & Markezich (1992).

CONTENT: Three species. ★★★

Uromacer: Haitian Vinesnake. Frank & Ramus (1995) called this the Pointed Snake.

Species-level Taxa	Citations and Remarks	Abbreviated Range
U. catesbyi (Schlegel, 1837) Haitian Vinesnake	Cochran (1941); Maglio (1970); Henderson & Schwartz (1984); Schwartz & Henderson (1984a, 1984b, 1991); Dixon & Markezich (1992); Powell et al. *in* Crother (1999)	Hispaniola including its satellites from sea level to 1525 m.
U. c. catesbyi (Schlegel, 1837) Common Haitian Vinesnake	Henderson & Schwartz (1984); Schwartz & Henderson (1984b, 1991); Powell et al. *in* Crother (1999)	Haiti from Tiburon Peninsula eastward to about Momance.
U. c. cereolineatus Schwartz, 1970 Cayemite Vinesnake	Henderson & Schwartz (1984); Schwartz & Henderson (1984b, 1991); Powell et al. *in* Crother (1999)	Hispaniola on Île Grande Cayemite and Île Petite Cayemite?
U. c. frondicolor Schwartz, 1970 Gonave Vinesnake	Henderson & Schwartz (1984); Schwartz & Henderson (1984b, 1991); Powell et al. *in* Crother (1999)	Île de la Gonâve, Haiti.
U. c. hariolatus Schwartz, 1970 Northern Haiti Vinesnake	Henderson & Schwartz (1984); Schwartz & Henderson (1984b, 1991); Powell et al. *in* Crother (1999)	Haiti north of Plaine de Cul de Lac and west of Haiti into the Dominican Republic.
U. c. incháusteguii Schwartz, 1970 Saona Vinesnake	Henderson & Schwartz (1984); Schwartz & Henderson (1984b, 1991); Powell et al *in* Crother (1999)	Isla Saona, Dominican Republic.
U. c. insulaevaccarum Schwartz, 1970 Vache Vinesnake	Henderson & Schwartz (1984); Schwartz & Henderson (1984b, 1991); Powell et al. *in* Crother (1999)	Île-à-Vache, Haiti.
U. c. pampineus Schwartz, 1970 Dominican Republic Vinesnake	Henderson & Schwartz (1984); Schwartz & Henderson (1984b, 1991); Powell et al. *in* Crother (1999)	Throughout the Dominican Republic except the Valle de Neiba and Peninsula de Barahna and eastern Haiti.
U. frenatus (Günther, 1865) Slender Vinesnake	Cochran (1941); Maglio (1970); Henderson & Schwartz (1984); Schwartz & Henderson (1984a, 1984c, 1991); Dixon & Markezich (1992); Powell et al. *in* Crother (1999)	Hispaniola from sea level to 915 m.
U. f. frenatus (Günther, 1865) Slender Vinesnake	Schwartz (1979); Henderson & Schwartz (1984); Schwartz & Henderson (1984c, 1991); Powell et al. *in* Crother (1999)	Haiti in Tiburón Peninsula eastward into Jacmel in the south and Soliette in the north and onto the n slopes of the Sierra de Baoruco, and in Valle de Neiba, Îles Petite, Île-à-Vache and Grande Cayemite.
U. f. chlorauges Schwartz, 1976 Haitian Southern Border Vinesnake	Maglio (1970); Schwartz (1979); Henderson & Schwartz (1984); Schwartz & Henderson (1984c, 1991); Powell et al. *in* Crother (1999)	Haiti from Soliette into Barahona in the Dominican Republic.
U. f. dorsalis Dunn, 1920 Haitian Long-tailed Snake	Schwartz (1979); Henderson & Schwartz (1984); Schwartz & Henderson (1984c, 1991); Powell et al. *in* Crother (1999)	Île de la Gonâve; Haiti.

Species-level Taxa	Citations and Remarks	Abbreviated Range
U. f. wetmorei Cochran, 1931 Wetmore's Vinesnake	Schwartz (1979); Henderson & Schwartz (1984); Schwartz & Henderson (1984c, 1991); Powell et al. *in* Crother (1999)	Isla Beata and Peninsula de Barahona in the Dominican Republic.
U. oxyrhynchus (Dumeril, Bibron & Dumeril, 1854) Sharpnose Bush Snake	Maglio (1970); Schwartz & Thomas (1975); Henderson & Schwartz (1984); Schwartz & Henderson (1984d, 1991); Dixon & Markezich (1992); Powell et al. *in* Crother (1999)	Widespread in Hispaniola and satellite islands including Île de la Tortue, Isla Saone and Isla Catalina; found from sea level to 1220 m.

GENUS: *Uromacerina* Amaral, 1929, is found in Brazil. See Peters & Orejas-Miranda (1970); Amaral (1976); Vanzolini (1986); and Lema (1994b).

Important contributions for this section were provided by Robert A. Thomas.

CONTENT: One species. ★★★★

Uromacerina: Sharp Snake, Frank and Ramus (1995) call this the Sao Paulo Sharp Snake.

Species-level Taxa	Citations and Remarks	Abbreviated Range
U. ricardinii (Peracca, 1897) São Paulo Sharp Snake	Hoge (1959b); Peters & Orejas-Miranda (1970); Amaral (1976); Müller & Ritter (1978); Cunha & Nascimento (1982); Vanzolini (1986); Williams & Wallach (1989); Lema (1994b); Marques et al. (2001)	SE Brazil from São Paulo into Rio de Janiero, Paranà, and Rio Grande do Sul; there is an isolated population in Pará.

GENUS: *Urotheca* Bibron *in* de la Sagra, 1843 [1838–1857], is found from Tamaulipas and Oaxaca in Mexico southward through Central America into n South America in northern and western Colombia, northwestern Venezuela, and western Ecuador. The genus as shown here is a consolidation of the genus *Pliocercus* Cope, 1860 and select species of *Rhadinaea* extracted from the *Rhadinaea "laterstriga"* group. See Dunn (1957); Wilson & Myer (1985); Savage & Crother (1989); Savage & Lahanas (1989); Williams & Wallach (1989); Savage & Slowinski (1990, 1992); Roze (1996); Smith & Chiszar (1996, others in press); Smith et al. (1995); Wilson et al. (1996); Lee (1996, 2000); Wilson & McCranie (1997); and Köhler (2001).

Review assistance and important contributions for this section were provided by Brian I. Crother.

Special Note: The status of *Urotheca* is a very controversial subject. There are those authorities who argue for placing *Pliocercus* into *Urotheca*, and others who believe the two should remain separate. Peters & Orejas-Miranda recognized seven species in the genus *Pliocercus*, two tri-colored forms *(P. andrewsi* and *P. elapoides)* and five bicolored taxa *(P. annellatus, P. arubricus, P. bicolor, P. dimidiatus,* and *P. euryzonus)*. With a very organized reasoning, Savage & Crother (1989) reduced the number of species in *Pliocercus* and included it as a synonym of *Urotheca*. Peters & Orejas-Miranda (1970) stated that the *euryzonus* has a wide range in Amazonia but Savage & Crother (1989) could find no species or documentation that any member of this stock occurs east of the Andes in South America. Savage & Slowinski (1990, 1992) proposed a terminology for the basic pattern-types occurring in coral snakes and their mimics. Flores-Villela (1993) used *Pliocercus* in his checklist and did not mention *Urotheca*. Liner (1994) reflected *Urotheca* in his checklist but did not discuss why. Smith et al. (1995) separated the genera of *Pliocercus* and *Urotheca* based strictly on color. Smith & Chiszar (1996) separated them and proceeded to make what I perceive as a very unclear discussion for a group of species in *Pliocercus*. Lee (1996) acknowledged *Urotheca* and did not list *Pliocercus*. Wilson & McCranie (1997) defended Savage & Crother's conclusions that these snakes belong to *Urotheca*. Campbell (1998a) used *Pliocercus*.

In order to overcome some of the confusion, I reviewed the original descriptions of the two genera. In the description comparison it quickly became obvious that Cope (1860) gave a brief description of *Pliocercus* which falls into a *Urotheca* description. James R. Dixon (pers comm., December 15, 1999) agreed with the analysis and noted that in describing his new genus, Cope had compared it to *Rhadinaea* and apparently had not seen anything that looked like a *Urotheca*. The modern differentiation of the genera *Pliocercus* currently seems to be established strictly on color differences and not anatomical differences. Armed with this information, plus Cope's weak description of his type "*elapoides*," it is my opinion that *Pliocercus* should be considered as a synonym of *Urotheca*. Wilson & McCranie (1997) reviewed the Smith & Chiszar (1996) work, and recommended that Savage and Crother (1989) be followed for this genus which is what I have done. Also retained from Savage & Crother (1989), is their idea that many of the *Pliocercus* species are classified as synonyms under the polymorphic species *Urotheca elapoides*. The other implication for this consolidation, is that the two new species for *Pliocercus* listed in Smith and Chiszar, (1996), was reclassified to *Urotheca psycoides* Smith & Chiszar, 1996, and *Urotheca wilmarai* Smith, Perérez-Higareda & Chiszar, 1996. *Pliocercus* is noted in brackets where the name can be applied, although this is intended to note a genus alternative, it does not imply a subgenus although that is a possibility.

Smith & Chiszar (in press a, b, c, d, e, and f), in a series of CAAR papers to be published will update and reorganize their 1996 work. Dr. Chiszar (pers. comm. September 10, 2001) was kind enough to share their updated information with me. Although they continue to maintain ringed and/or red pattern mimicking for the *Pliocercus* various species versus cryptic, unicolor, or striped patterns in *Urotheca* as a primary diagnosis, Smith & Chiszar (in press a, b, c, d, e, f) define a total of 4 characteristics to differentiate *Pliocercus* from *Urotheca* in the genus description Smith & Chiszar (in press c). They also defined the various species in the remaining work (in press a, b, d, e, f). Since this work has not formally been published by the completion of this manuscript, I have not included detail of their work and continue to retain *Pliocercus* within *Urotheca*. I advise interested readers to obtain copies of the more current Smith & Chiszar (in press) work when it is published in CAAR and review it carefully to see their detailed assessments on the classifications and range maps. They have a complex of species and subspecies considerably different from those listed below and similar to their 1996 list. For the common names of the *Pliocercus*-group, I made the effort to retain the common names using the Smith & Chiszar (1996, in press a, b, c, d, e, f) names for the various species.

There are basically two groups within *Urotheca* under the classifications used below. They are the main *Urotheca* group of cryptic, unicolored, and striped snakes, and the *Pliocercus*-group coral snake mimics.

See also genus *Rhadinaea*.

CONTENT: Twelve species with no subspecies, depending on how the status of *Pliocercus* is accepted. ★

Urotheca: Common False Coral Snake and Neotropical Brownsnake. Greene (1997c) called these the Halloween Snakes and restricted them to two species.

Species-level Taxa	Citations and Remarks	Abbreviated Range
U. decipiens (Günther, 1893) Lower Central American Neotropical Brownsnake	Myers (1974); Savage & Villa (1986); Villa et al. (1988); Savage & Crother (1989); Savage & Lahanas (1989); Köhler (2001)	Lower Central America in Costa Rica and Panama into n Colombia in the Cordillera Central; between 15 to 1699 m.

The Caenophidia

Species-level Taxa	Citations and Remarks	Abbreviated Range
U. dumerili* (Bibron, 1855 [1840?]) Dumeril's Neotropical Brownsnake	Myers (1974); Savage & Crother (1989); Savage & Lahanas (1989) *Remark: see Myers (1974) for review of dates on Bibron publications.	The Río San Juan, in Departamento del Chocó in the Pacific versant of Colombia.
U. (*Pliocercus*) *elapoides** (Cope, 1860) Variegated False Coral Snake	Cope (1860); Bocourt (1891); Wilson & Meyer (1985); Savage & Crother (1989); Savage & Lahanas (1989); Smith et al. (1989); Campbell & Lamar (1989); Savage & Slowinski (1992); Smith & Smith (1993); Lee (1996, 2000); Smith & Chiszar (1996, 2000, in press a, b, e); Roze (1996); Stafford (1999); Stafford & Meyer (2000); Köhler (2001) [as *Pliocercus euryzonus*] *Remark 1: In this taxon I have included *U. aequalis* (Salvin, 1862) from Guatemala and Honduras; and, *U. bicolor* Smith, 1941 of northern Veracruz as synonyms (see Savage & Crother, 1989). *Remark 2: According to Savage & Crother (1980), the following names (all based on tricolor snakes) are regarded as junior synonyms of *U. elapoides*: *andrewsi, celatus, deppei, disastemus, hobartsmithi, laticollaris, occidentalis, pacificus, salvadorensis, salvinii, semicinctus,* and *tricinctus*.	C Mexico southward into n Central America into w Honduras and El Salvador.
U. (*Pliocercus*) *euryzona** (Cope, 1862) Cope's False Coral Snake	Peters (1960b); Peters & Orejas-Miranda (1970); Wilson & Meyer (1985); Savage & Villa (1986); Pérez-Higareda & Smith (1986a); Savage & Crother (1989); Savage & Lahanas (1989); Savage & Slowinski (1992); Smith & Smith (1993); Smith & Chiszar (1996, in press d); Renjifo & Lundberg (1999); Köhler (2001) [as *Pliocercus euryzonus*] *Remark: In this taxon I have included *Pliocercus annellatus* Taylor, 1951, *Pliocercus arubricus* Taylor, 1954, *Pliocercus dimidiatus* Cope, 1865, and *Pliocercus euryzonus* Cope, 1862 (see Savage & Crother, 1989).	N Nicaragua south into South America to as far as w Ecuador.
U. fulviceps (Cope, 1885 [1886]) Cope's Neotropical Brownsnake	Myers (1974); Savage & Villa (1986); Villa et al. (1988); Knight (1988); Savage & Crother (1989); Savage & Lahanas (1989); Kornacker (1999) [as *Rhadinaea fulviceps*]; Köhler (2001)	Lowlands of Costa Rica into nw Ecuador and the Magdalena drainage of c Colombia; from sea level to 500 m.
U. guentheri (Dunn, 1938) Guenther's Neotropical Brownsnake	Myers (1974); Savage & Villa (1986); Villa et al. (1988); Savage & Crother (1989); Köhler (2001); McCranie et al. (2001)	N Nicaragua into Costa Rica and c Panama. Recently reported from Honduras.
U. lateristriga (Berthold, 1859) Berthold's Neotropical Brownsnake	Andersson (1901); Dunn (1957); Myers (1974) Savage & Crother (1989); Savage & Lahanas (1989)	Andean cordilleras of Colombia and southward into Andean and western Ecuador.

Species-level Taxa	Citations and Remarks	Abbreviated Range
U. multilineata (W. Peters, 1863) Many-lined Neotropical Brownsnake	Myers (1974); Roze (1966); Savage & Crother (1989); Lancini & Kornacker (1989); Kornacker (1999) [as *Rhadinaea multilineata*]; Mijares-Urrutia & Arends R. (2000) [as *Rhadinaea multilineata*]	Caribbean versant of Colombia and Venezuela in montane forests in the Cordillera de la Costa.
U. myersi Savage & Lahanas, 1989 Myers's Neotropical Brownsnake	Savage & Lahanas (1989); Köhler (2001)	Costa Rica in the lower montane forests of the Cordillera de Talamanca; between 2000 to 2255 m.
U. pachyura* (Cope, 1875) Cope's Graceful Neotropical Brownsnake	Myers (1974); Savage & Villa (1986); Villa et al. (1988); Savage & Crother (1989); Savage & Lahanas (1989); Köhler (2001) *Remark: *Rhadinaea rubricollis* Taylor, 1954 from the type locality of Cinchona at Volcán Poás, Costa Rica at about 5500 feet is a synonym (see Savage & Crother, 1989).	Lower Central America into n South America from Costa Rica, Panama, Colombia, and Ecuador; from 5 to 1600 m.
U. (Pliocercus) psycoides* (Smith & Chiszar, 1996) Carib False Coral Snake	Smith & Chiszar (1996, in press e, g) *Remark: Smith & Chiszar (in press e, g) list this snake as a synonym of *Pliocercus elapoides aequalis* (Salvin, 1861).	S Venezuela possibly into Brazil fide Smith & Chiszar (1996). Smith & Chiszar (in press e) indicate that the type locality is unknown and given as Brasil or Venezuela, but apparently in error, probably c w Guatemala.
U. (Pliocercus) wilmarai* (Smith, Pérez-Higareda & Chiszar, 1996) Mara's False Coral Snake	Smith & Chiszar (1996, 2000, in press e) *Remark: Smith & Chiszar (2000, in press e) list this snake as *Pliocercus elapoides wilmarai* Smith, Pérez-Higareda & Chiszar.	S Veracruz in Mexico. The type locality is the Los Tuxtlas region of s Veracruz.

GENUS: *Virginia* Baird & Girard, 1853, is found in the United States from the Midwest and Texas Gulf Coast westward. *Haldea* is now included with this genus as a synonym. See Conant (1975); Rossman & Wallach (1991); Powell et al. (1992); Tennant (1998); Conant & Collins (1998); Tennant & Bartlett (2000); and Dixon (2000a).

CONTENT: Two species. ★★★★

Virginia: North American Earthsnake

Species-level Taxa	Citations and Remarks	Abbreviated Range
V. striatula (Linnaeus, 1766) Rough Earthsnake	Ashton & Ashton (1981); Rossman & Wallach (1991); Powell et al. (1994); Tennant (1997b, 1998); Conant & Collins (1998); Werler & Dixon (2000); Tennant & Bartlett (2000); Dixon (2000a)	Atlantic and Gulf coasts of the United States from c Texas into Missouri eastward into Florida and northward into North Carolina.
V. valeriae Baird & Girard, 1853 Smooth Earthsnake	Conant (1975); Dundee & Rossman (1989); Rossman & Wallach (1991); Powell et al. (1992); Conant & Collins (1998); Dixon (2000a)	E United States.

The Caenophidia

Species-level Taxa	Citations and Remarks	Abbreviated Range
V. v. valeriae Baird & Girard, 1853 Eastern Smooth Earthsnake	Conant (1975); Ashton & Ashton (1981); Powell et al. (1992); Tennant (1997b); Conant & Collins (1998); Tennant & Bartlett (2000)	The United States east of the Mississippi River.
V. v. elegans* Kennicott, 1859 Western Smooth Earthsnake	Conant (1975); Dundee & Rossman (1989); Powell et al. (1992); Tennant (1998); Conant & Collins (1998) Werler & Dixon (2000); Tennant & Bartlett (2000); Dixon (2000a) *Remark: Fide Dixon (2000a), the distributional records of *V. v. elegans* continues to defy his ability to recognize a geographic pattern.	Midwestern United States southward into the Gulf Coast from c Indiana and s Iowa southward into Texas and eastward into Florida.
V. v. pulchra (Richmond, 1954) Mountain Smooth Earthsnake	Conant (1975); Powell et al. (1992); Conant & Collins (1998); Tennant & Bartlett (2000)	CE United States.

GENUS: *Waglerophis* Romano & Hoge, 1973, is found in the southern part of South America. See Cei (1986, 1993); Lema (1994b); Ferreira (1997); Starace (1998), Dirksen & Duarte (1998); and Kornacker (1999).

Important contributions for this section were provided by Thales de Lema and Lutz Dirksen.

See also genera *Xenodon* and *Thalesius*.

CONTENT: One species with a high degree of polymorphism. ★★★★

Waglerophis: Wagler's False Pit Viper. Frank and Ramus (1995) call this the Wagler's Snake.

Species-level Taxa	Citations and Remarks	Abbreviated Range
W. merremi* (Wagler in Spix, 1824) Merrem's False Pit Viper	Romano & Hoge (1973); Amaral (1976); Vanzolini et al. (1980); Vanzolini (1986); Cei (1986, 1993); Lema (1994b); Vuoto (1996); Ferreira (1997); Starace (1998); Dirksen & Duarte (1998); Yanosky (1998); Leynaud & Bucher (1999); Kornacker (1999); Freitas (1999) *Remark: Older literature listed this snake as *Xenodon merremi* (see Amaral, 1976).	Wide ranging from the Guianas, Brazil, Paraguay, Bolivia, Peru, Ecuador, and c and n Argentina in Jujuy, Salta, Formosa, Chaco, Corrientes, Misiones, Catamarca, Tucumán, Santiago del Estero, La Rioja, San Juan, San Luis, Córdoba, Santa Fe, Entre Ríos, Buenos Aires, and Santa Fe.

GENUS: *Xenodon* Boie *in* Fitzinger, 1826, [genus originally *Ophis* Wagler, 1924 and preoccupied by *Ophis* Turton, 1807], ranges from central Mexico into South America to as far south as Argentina. See Smith & Taylor (1966); Peters & Orejas-Miranda (1970); Williams & Wallach (1989); Dixon (1983d); Vanzolini (1986); Savage & Villa (1986); Lancini & Kornacker (1989); Cei (1993); Lema (1994b); Yuki (1994); Ferreira (1997); Starace (1998); Kornacker (1999); and Köhler (2001).

Review assistance and important contributions for this section were provided by Robert A. Thomas, Thales de Lema and James R. Dixon.

Note: Also see genera *Thalesius* and *Waglerophis*.

CONTENT: Four species. ★★★

Xenodon: False Pit Viper or False Fer-de-Lance. Frank and Ramus (1995) call this the False Fer-de-Lance.

Species-level Taxa	Citations and Remarks	Abbreviated Range
X. guentheri Boulenger, 1894 Guenther's False Pit Viper	Peters & Orejas-Miranda (1970); Amaral (1976); Dixon (1983d); Lema (1994b); Ferreira (1997)	Santa Catarina in s Brazil.
X. neuwiedi Günther, 1863 Neuwied's False Pit Viper	Peters & Orejas-Miranda (1970); Amaral (1976); Cei (1993); Lema (1994b); Ferreira (1997); Yanosky (1998); Suzart-Argôlo (1999); Marques et al. (2001)	S and c Brazil, Paraguay and n Argentina in Misiones and possibly Corrientes.
X. rabdocephalus* (Wied, 1824) Common False Pit Viper	Peters (1960b); Peters & Orejas-Miranda (1970); Bocourt (1891); Dixon (1983d); Wilson & Meyer (1985); Villa et al. (1988); Lancini & Kornacker (1989); Lee (1996, 2000); Ferreira (1997); Campbell (1998a); Starace (1998); Yanosky (1998); Kornacker (1999); Freitas (1999); Köhler (2001) *Remark 1: *X. bertholdi* Jan, 1863 from Esparta, Costa Rica, is a synonym (see Villa et al., 1988). *Remark 2: Wilson & Meyer (1985) synonymized the two subspecies. Although I tend to agree with them, subspecies are listed here because they are still used by other authorities.	Guerrero and Veracruz in Mexico southward through Central America and South America as far as Bolivia.
X. r. rabdocephalus* (Wied, 1824) Common False Pit Viper	Peters & Orejas-Miranda (1970); Amaral (1976) [as *X. colubrinus*]; Dixon (1983d); Dixon & Soini (1986); Lancini & Kornacker (1989); Starace (1998); Kornacker (1999) *Remark: *Xenodon colubrinus* Günther, 1858 is a synonym (see Peters & Orejas-Miranda, 1970). *Xenodon suspectus* Cope, 1868 from e Peru is a synonym (see Dixon, 1983).	Nicaragua in Central America into South America in Venezuela, Colombia, Ecuador, Brazil, and Bolivia as well as the Guianas.
X. r. mexicanus Smith, 1940 Mexican False Pit Viper	Smith & Taylor (1966); Peters & Orejas-Miranda (1970); Henderson & Hoevers (1975); Alvarez del Toro (1982); Dixon (1983d); Wilson & Meyer (1985); Villa et al. (1988); Pérez-Higareda & Smith (1991); Stafford & Meyer (2000); Lee (2000)	Guerrero, Yucatán and Veracruz in s Mexico southward into the Atlantic versant of Belize and Guatemala; Smith & Taylor (1966) presumed that it would be found in El Salvador.

The Caenophidia

Species-level Taxa	Citations and Remarks	Abbreviated Range
X. severus (Linnaeus, 1758) Amazon False Pit Viper	Amaral, 1934 [1933–1934]; Peters (1960b); Roze (1966); Peters & Orejas-Miranda (1970); Amaral (1976); Duellman (1978); Fugler & Walls (1978); Dixon (1983d); Lancini & Kornacker (1989); Starace (1998); Mijares-Urrutia & Arenda R. (1999); Kornacker (1999)	Amazonian South America including s Venezuela, se Colombia, e Ecuador, Amazonas, Brazil, ne Bolivia, and the Guianas; recently reported from Falcón state in Venezuela.

GENUS: *Xenopholis* W. Peters, 1869, is a South American snake found in Brazil, French Guiana, Peru, Paraguay, Ecuador, Bolivia, and possibly Colombia. Hoge & Federsoni (1975), included *Paroxyrhopus* Schenkel, 1900 as a synonym. See Peters & Orejas-Miranda (1970); Vanzolini (1986); Cunha et al. (1985); Duellman & Mendelson III (1995); Starace (1998); Dowling & Pinou (2003).

CONTENT: Two species. ★★★★

Xenopholis: South American Flat-headed Snake. Frank and Ramus (1995) call this the Scaled Groundsnake.

Species-level Taxa	Citations and Remarks	Abbreviated Range
X. scalaris (Wucherer, 1861) Wucherer's Groundsnake	Peters (1960b); Peters & Orejas-Miranda (1970); Duellman (1978); Dixon & Soini (1986); Cunha et al. (1985); Savage & Slowinski (1992); Duellman & Mendelson III (1995); Starace (1995, 1998); Vidal et al. (1998)	Amazonian Peru, Bolivia, French Guiana, Ecuador, and Brazil.
*X. undulatus** (Jensen, [1900] 1901) Jensen's Groundsnake	Hoge & Federsoni (1975); Amaral (1976); Cunha et al. (1985); Chippaux ([1986] 1987); Yanosky (1998); Campos-Nogueira (2001) *Remark: *Paroxyrhopus reticulatus* Schenkel, 1900 is a synonym (see Hoge & Federsoni, 1975).	SC Brazil and Paraguay; generally associated with brushy areas.

GENUS: *Xenoxybelis* Rodrigues-Machado, 1993, until recently, this taxon was included in *Oxybelis*. It ranges from western South America in n French Guiana southward to c Bolivia, and westward to central southern Colombia and north of Ecuador southward through the Cordillera Oriental of the Andes into c Bolivia. See Rodrigues-Machado (1993) and Villa & McCranie (1995) for the most current work on the genus.

See also the genus *Oxybelis*.

CONTENT: Two species. ★★★★

Xenoxybelis: Sharp-nosed Snake

Species-level Taxa	Citations and Remarks	Abbreviated Range
*X. argenteus** (Daudin, 1803) Green-striped Vinesnake	Roze (1966); Peters & Orejas-Miranda (1970); Duellman (1978); Lancini (1979); Moonen et al. (1979); Dixon & Soini (1986); Nascimento et al. (1987); Lan-	Lowlands of n South America east of the Andes; the range includes Colombia, Venezuela, Ecuador, Brazil, Peru, Guyana; French Guiana, and Bolivia.

Species-level Taxa	Citations and Remarks	Abbreviated Range
	cini & Kornacker (1989); Coburn (1991); Rodrigues-Machado (1993); Villa & McCranie (1995); Greene (1997c); Starace (1998); Yanosky (1998); Kornacker (1999)	
	*Remark: This snake is found in older literature as *Oxybelis argenteus*. It continues to be listed as *Oxybelis argenteus* in such current work as by Starace (1998), and Kornacker (1999).	
X. boulengeri (Procter, 1923) Southern Sharp-nosed Snake	Procter (1923); Peters & Orejas-Miranda (1970) [as a synonym of *Oxybelis argenteus*]; Rodrigues-Machado (1993); Villa & McCranie (1995)	N and c South America; the type locality is reported to be Trinidad, Río Mamoré, Bolivia.

Family ELAPIDAE Boie, 1827

Subfamily MICRURINAE

These are commonly referred to as Coral Snakes.

This group of snakes is found in the Americas from the southern and southwestern United States southward into South America. Although there have been some changes since 1996, this section is based on Roze (1996).

The snakes found in the family Elapidae are very similar to Colubrids; however, all Elapids have permanently erect fangs positioned on the anterior ends of the principally immovable maxillae and fit into grooved slots in the buccal floor when the mouth is closed. The main difference between these snakes and the venomous Colubrids is that each fang contains an enclosed passage extending the length of the tooth from an opening at the base for entry of the venom to an opening near the tip, and venom production and storage are considerably more sophisticated.

Schmidt (1928a, 1933c, 1936a) appear to be the first works to organize Coral Snake information. Roze (1967) was the first work to comprehensively cover the New World Coral Snakes. He modernized this work in 1983, then updated it in 1994. Conant (1958, 1975) and Conant & Collins (1998), in conjunction with Campbell & Lamar (1989) expanded the knowledge of this complex group. Roze (1996) is the most recent revisor of the total complex of Coral Snakes.

CONTENT: 3 genera with about 71 species, restricted to the Americas. Some authorities consider *Leptomicrurus* to be a synonym of *Micrurus*. If this turns out to be the case, then there are two genera in the Americas.

GENUS: *Leptomicrurus* Schmidt, 1937 is found only in South America. This genus is placed in synonymy with *Micrurus* Wagler, 1824 by Romano (1972) and Campbell & Lamar (1989). Roze (1996) maintained the group separate from *Micrurus*. Campbell & Lamar (1989), and other authorities synonymized *Leptomicrurus* with *Micrurus*. I follow Roze (1996) as the most recent authority and revisor and he maintains them as separate. This has been further supported by Jorge da Silva & Sites (1999) statement that corals are arranged in three genera and noted *Leptomicrurus* as valid. See Peters and Orejas-Miranda (1970); Cunha & Nascimento (1982); Roze & Bernal-Carlo (1988); Campbell & Lamar (1989); Roze (1996); and Jorge da Silva & Sites (1999).

The Caenophidia

Review assistance and important contributions for this section were contributed by Janis A. Roze.

CONTENT: Two to three species depending on how *L. scutiventris* is classified. Three species are listed here. ★★

See also genus *Micrurus*.

Leptomicrurus: Black-backed Coral Snake

Species	Citations and Remarks	Abbreviated Range
L. collaris (Schlegel, 1837) Guianan Black-backed Coral Snake	Roze (1966, 1996); Peters & Orejas-Miranda (1970); Hoogmoed (1983); Chippaux ([1986] 1987); Roze & Bernal-Carlo (1988); Lancini & Kornacker (1989); Campbell & Lamar (1989) [as *Micrurus collaris*]; Grantsau (1991) [as *Micrurus collaris*]; Starace (1998); Kornacker (1999)	SE Venezuela, the Guianas, and probably into n Brazil in Pará.
L. c. collaris (Schlegel, 1837) Long Black-backed Coral Snake	Hoogmoed (1983); Chippaux ([1986] 1987); Roze & Bernal-Carlo (1988); Campbell & Lamar (1989) [as *Micrurus collaris*]; Roze (1996)	The Guianas and possibly Amazonian Pará in Brazil; it is found from sea level to about 700 m in lowland tropical rain forests and low montane wet forests.
L. c. breviventris Roze & Bernal-Carlo, 1988 Short Black-backed Coral Snake	Campbell & Lamar (1989) [as *Micrurus collaris*]; Roze (1983, 1996); Lancini (1979); Roze & Bernal-Carlo (1988); Kornacker (1999)	NW Guyana and se Venezuela; from 400 to 600 m in tropical low elevation rain forest.
L. narduccii (Jan, 1863) Andean Black-backed Coral Snake	Roze (1967, 1996); Peters & Orejas-Miranda (1970); Kempff-Mercado (1975); Duellman (1978); Roze & Bernal-Carlo (1988); Campbell & Lamar (1989) [as *Micrurus narduccii*]; Grantsau (1991) [as *Micrurus narduccii*]	Amazonian slopes of the Andes and upper Amazon in e Ecuador, Peru, n Bolivia, s Colombia. Probably present in Acre in Brazil.
L. n. narduccii (Jan, 1863) Bolivian Black-backed Coral Snake	Kempff-Mercado (1975); Roze & Bernal-Carlo (1988); Visinoni (1995); Roze (1996)	Bolivia; this snake is endemic to e Bolivia in the Santa Cruz region (see Visinoni, 1995); found between 300 to 600 m in subtropical low altitude forest altered by humans, and in intermediate vegetation.
L. n. melanotus (W. Peters, 1881) Slender Black-backed Coral Snake	W. Peters (1881); Duellman (1978); Roze & Bernal-Carlo (1988); Roze (1996)	E Ecuador, s Colombia, n Peru, and nw Brazil; between 150 to 1300 m in Amazonian rain forest and wet forests of the e Andean foothills.
L. scutiventris* (Cope, 1870) Little Black Coral Snake	Peters & Orejas-Miranda (1970) [as *L. schmidti*]; Romano (1972); Hoge & Romano-Hoge (1973); Dixon & Soini (1986); Roze & Bernal-Carlo (1988); Grantsau (1991) [as *Micrurus karlschmidti*]; Roze (1996); Campbell & Lamar (1989) [as *Micrurus karlschmidti*] *Remark 1: Both Lamar (pers. comm.), and Roze (1996) feel that this is the true *M. karlschmidti*. If transfered to *Micrurus* this becomes *M. karlschmidti* Romano, 1972 (*nomen novum*) (see Hoge & Ro-	SE Colombia, ne Ecuador, n Peru, east of the Andes, and nw Brazil; from 150 to 1200 m in Amazonian rain forest, wet low montane forest in Adnean valleys, and intermediate vegetation altered by humans.

Species-level Taxa	Citations and Remarks	Abbreviated Range
	mano, 1966; Romano, 1972; and Campbell & Lamar, 1989).	
	*Remark 2: *L. schmidti* Hoge & Romano, 1966, is from Amazonian Brazil (see Peters & Orejas-Miranda, 1970), and is only known from the type locality of Tapurucuara on Mount Uaupés in the state of Amazonas. Roze (1996) listed this as a synonym of *L. scutiventris*. Its classification as subspecies, *L. n. schmidti* (Hoge & Romano, 1966) does not seem accurate (see Peters & Orejas-Mirdanda, 1970). Harding & Welch (1980) called it *L. schmidti*. According to Roze (1996), *L. narducci schmidti* is a synonym.	

GENUS: *Micruroides* Schmidt, 1928 is distributed from the southwestern United States southward into northern Mexico, with a disjunct subspecies population in southern coastal Sinaloa. See Smith & Taylor (1966); Stebbins (1966); Roze (1974, 1996); Stebbins (1985); Campbell & Lamar (1989); Wright & Wright ([1957] 1994); and Degenhardt et al. (1996).

Review assistance and important contributions for this section were contributed by Janis A. Roze.

CONTENT: One species with three subspecies restricted to North America. ★★★★

Micruroides: Western Coral Snake

Species	Citations and Remarks	Abbreviated Range
M. euryxanthus (Kennicott, 1860) Sonoran Coral Snake	Zweifel & Norris (1955); Smith & Taylor (1966); Stebbins (1966, 1985); Roze (1974, 1996); Campbell & Lamar (1989); Degenhardt et al. (1996)	United States c and s Arizona to sw New Mexico, and w Mexico from n and w Chihuahua and Sonora, including Isla Tiburón to Mazatlán, Sinaloa.
M. e. euryxanthus (Kennicott, 1860) Arizona Coral Snake	Zweifel & Norris (1955); Roze (1974, 1996); Stebbins (1985); Campbell & Lamar (1989); Ernst (1992); Degenhardt et al. (1996); Bartlett & Tennant (2000)	SW United States into n Mexico into Sonora and western Chihuahua, including Tiburón Island. Roze (1996) noted a specimen with a questionable collecting record of El Paso, Texas; from sea level to about 2100 m in arid and semiarid plateau regions.
M. e. australis Zweifel & Norris, 1955 Sonoran Coral Snake	Zweifel & Norris (1955); Roze (1974, 1996); Campbell & Lamar (1989)	C and s Sonora, and sw Chihuahua, Mexico; from sea level to about 800 m in tropical dry and desert formations.
M. e. neglectus Roze, 1967 Sinaloan Coral Snake	Roze (1967, 1974, 1996); Hardy & McDiarmid (1969); Campbell & Lamar (1989)	North of Mazatlan in Sinaloa, Mexico; this is an allopatric group found from sea level to about 200 m in tropical semiarid and dry thorn forest.

GENUS: *Micrurus* Wagler *in* Spix, 1824, has a very wide distribution from the eastern and southern United States southward into South America to as far south as Bolivia, Argentina, Peru, and Uruguay. Vanzolini (1986), citing

Romer (1972), placed *Leptomicrurus* Schmidt, 1937 in the synonym of *Micrurus* Wagler, 1824. Campbell & Lamar (1989) agree with this reclassification. See Smith & Taylor (1966); Roze *in* Peters & Orejas-Miranda (1970); Cunha & Nascimento (1982); Campbell & Lamar (1989); Cei (1986, 1993); Smith & Smith (1978, 1993); Roze (1996); Starace (1998); and Kornacker (1999). This list is based on Roze (1996) as the most current work. Based on Roze's recommendations in a personal communication, his information has been updated with the *Micrurus frontalis* information from Jorge da Silva & Sites (1999) be used in conjunction with his work.

Review assistance and important contributions for this section were contributed by Janis A. Roze.

See also the genus *Leptomicrurus*.

CONTENT: Sixty-seven species listed below. ★★★

Micrurus: American Coral Snake

Species	Citations and Remarks	Abbreviated Range
*M. albicinctus** Amaral, 1925 White-banded Coral Snake	Roze (1967, 1970, 1996); Amaral (1976, 1978); Cunha & Nascimento (1982b, 1991); Vanzolini (1986) *Remark: According to Cunha & Nascimento (1982), this taxon is a synonym of *M. ornatissimus* (Jan, 1858). *M. langsdorffi ornatissimus* (see Peters & Orejas-Miranda, 1970 and Campbell & Lamar, 1989).	C Mato Grosso and Rondônia into the middle Amazon in Amazonas, Brazil; between 100 to 250 m in lowland forest.
*M. alleni** Schmidt, 1936 Arrow-headed Coral Snake	Roze (1967, 1970, 1983, 1996); Savage & Vial (1974); Villa (1984); Savage & Villa (1986); Villa et al. (1988); Campbell & Lamar (1989); McCranie (1993); Köhler (2001) *Remark: Roze (1996) does not recognize subspecies. Roze *in* Peters & Orejas-Miranda (1970) and Roze (1983) listed subspecies as *M. a. alleni* Schmidt, 1936, found in the Atlantic slopes of Nicaragua, Costa Rica, and nw Panama; and, *M. a. yatesi* Dunn, 1942 found on the Pacific slopes of Nicaragua, Costa Rica and sw Panama (see Campbell & Lamar, 1989).	Central America in Honduras, Nicaragua, Costa Rica, and w Panama (probably to Darién); from sea level to 1400 m in tropical lowland rain forest and humid low montane forest.
*M. altirostris** (Cope, 1859 [1860]) Uruguayan Coral Snake	Roze (1970, 1996); Campbell & Lamar (1989); Scrocchi (1990b); Cei (1993); Lema (1994a); Jorge da Silva & Sites (1999) *Remark: Jorge da Silva & Sites (1999) has elevated *M. frontalis altirostris* (Cope) full species status.	Brazilian state of Paraná southward to Uruguay, and the provinces of Misiones, Corrientes, and n Entre Ríos in Argentina and e Paraguay; from sea level to in pampas, savannas, and low deciduous forest and open fields which are frequently altered by humans.
M. ancoralis (Jan, 1872) Anchor Coral Snake	Roze (1967, 1970; 1983, 1996); Campbell & Lamar (1989); Köhler (2001)	Darién, Panama southward through Colombia into sw Ecuador; it is found along the Pacific lowlands and w slopes of the Andes.
M. a. ancoralis (Jan, 1872) Ecuadorian Anchor Coral Snake	Roze (1970, 1996); Campbell & Lamar (1989)	NW and w Ecuador southward to Bahía de Carquez; it is found from sea level to about 1000 m in tropical rain forest and low montane wet forest.

Species-level Taxa	Citations and Remarks	Abbreviated Range
M. a. jani Schmidt, 1936 Chocó Anchor Coral Snake	Schmidt (1936, 1955); Roze (1970, 1996); Campbell & Lamar (1989)	Panama and w Colombia; near sea level to about 1000 m or higher in tropical rain forest and humid low montane forest from Darién southward to the Pacific lowlands of the Chocó region and the w slopes of the Andes.
M. annellatus (W. Peters, 1871) Annulated Coral Snake	Roze (1967, 1970; 1983, 1996); Carillo de Espinosa (1983); Campbell & Lamar (1989)	Ecuador, Peru, and Bolivia; may eventually be found in w Brazil; found on the e slopes of the Andes and the upper Amazon.
M. a. annellatus* (Peters, 1871) Common Annulated Coral Snake	Schmidt (1954); Roze (1970, 1996); Carillo de Espinosa (1983); Campbell & Lamar (1989) *Remark: *M. annellatus montanus* is a synonym, (see Campbell & Lamar, 1989 and Roze, 1996).	S Ecuador to c Peru, and Bolivia; there is a questionable record from Leticia, Colombia; it is found in low montane rain forest of the upper Amazon and the e slopes of the Andes, from between 600 to 2000 m.
M. a. balzani (Boulenger, 1898) Yungas Coral Snake	Roze (1967, 1970, 1996); Campbell & Lamar (1989)	N and w Bolivia and probably adjacent Peru; between 500 to 2000 m in humid and dry regions in the e Andes and upper Amazon humid valleys.
M. a. bolivianus Roze, 1967 Bolivian Coral Snake	Roze (1967, 1970, 1996); Campbell & Lamar (1989)	E Andes and high Amazon Valleys in c Bolivia, between 1200 to 2200 m in mediam to high montane humid or dry forests.
M. averyi Schmidt, 1939 Black-headed Coral Snake	Schmidt (1939); Roze (1967, 1970, 1983, 1996); Campbell & Lamar (1989); Starace (1998)	S Guyana into Manaus, Amazonas in Brazil, and possibly Suriname and French Guiana; between 100 to 700 m in lowland and low montane, non-flooded rain forest of the c Amazon.
M. baliocoryphus* (Cope, 1859) Mesopotamian Coral Snake	Vanzolini (1986); Campbell & Lamar (1989); Scrocchi (1990); Roze (1996); Jorge da Silva & Sites (1999) *Remark 1: *Micrurus f. mesopotamicus* Barrio & Miranda, 1967, is a synonym, (see Cei, 1993 and Roze, 1996). *Remark 2: Jorge da Silva & Sites (1999) elevated *M. frontalis baliocoryphus* Cope to full species.	NE Argentina in the Argentine mesopotamic region through the provinces of Entre Ríos, Corrientes, and Misiones between Río Paraná and Río Uruguay; also present in sw Paraguay along the Río Paraguay from the provinces of Neembucu through Central and Paraguari to the Presidente Hayes, and the adjacent Argentine province of Formosa. There are no records from this taxon in Brazil or Uruguay. It is found in lowland pampas and deciduous formations.
M. bernadi (Cope, 1887) Blotched or Saddled Coral Snake	Smith & Taylor (1945, 1966); Schmidt (1958); Roze (1967, 1970, 1983, 1996); Campbell & Lamar (1989)	Mexico from the Atlantic versant of e Hidalgo, n Puebla and nw Veracruz, seems to prefer cloud forests from 50 to 2100 m.
M. bocourti (Jan, 1872) False Triad or Ecuadorian Coral Snake	Schmidt (1936c); Campbell & Lamar (1989); Roze (1967, 1970, 1983, 1989, 1996)	Pacific lowlands sw Ecuador into nw Peru; from sea level to about 1500 m in humid to dry lowland formations and dry w slopes of the Andes.
M. bogerti Roze, 1967 Coastal or Bogert's Coral Snake	Roze (1967, 1983, 1996); Villa et al. (1988); Campbell & Lamar (1989); Smith & Smith (1993); Köhler (2001)	Pacific coast of w Oaxaca, Mexico from about Puerto Angel into Tapanatepec; from sea level to 400 m in dry coastal thorn and scrub, and tropical deciduous forest.

Species-level Taxa	Citations and Remarks	Abbreviated Range
*M. brasiliensis** Roze, 1967 Brazilian Short-tailed Coral Snake	Roze (1970, 1996); Campbell & Lamar (1989); Jorge da Silva & Sites (1999); Freitas (1999) *Remark: *Micrurus frontalis brasiliensis* has been elevated to full species status by Jorge da Silva & Sites (1999).	CN Brazil, excluding Mato Grosso, according to Jorge da Silva & Sites (1999) there are no data available to support the geographic distribution in Mato Gross or Mato Grosso do Sul. Range is noted as a wide range east of the Río Araguaia between 10° and 14°S to 41°W and south to 16°S between 41° and 46°W.
M. browni Schmidt & Smith, 1943 Sierra Madre or Brown's Coral Snake	Smith & Taylor (1945, 1966); Roze (1967, 1970, 1983, 1996); Campbell & Lamar (1989); McCranie & Wilson (1991); Köhler (2001)	Mexican Plateau from Mexico City southward into the mountains of Guatemala; in Mexico the range includes the states of Guerrero, Oaxaca, and Chiapas.
M. b. browni Schmidt & Smith, 1943 Common Sierra Madre Coral Snake	Schmidt & Smith (1943); Roze (1967, 1970, 1983, 1996); Alvarez del Toro (1982); Campbell & Lamar (1989); McCranie & Wilson (1991); Köhler (2001)	Mexican Plateau from Mexico City, the state of Mexico and c Guerrero eastward on the Pacific side of the Sierra Madre del Sur to Oaxaca and Chiapas and southward into the western mountains of Guatemala; from 500 to over 2000 m in subhumid to dry high and intermediate montane areas.
M. b. importunus Roze, 1967 Antigua Coral Snake	Roze (1967, 1970, 1983, 1996); Campbell & Lamar (1989); Köhler (2001)	Antigua Basin in Guatemala from 1200 to 1800 m; known only from the Dueñas region.
M. b. taylori Schmidt & Smith, 1943 Acapulco Coral Snake	Schmidt & Smith (1943); Roze (1967, 1996); Campbell & Lamar (1989)	Guerrero, Mexico; sea level to 400 m in subhumid coastal tropical lowlands altered by humans around Acapulco.
M. catamayensis Roze, 1989 Catamayo Coral Snake	Roze (1989, 1996)	Catamayo Valley in Loja in s Ecuador from 1000 to 1800 m in high montane dry scrub and subhumid gallery forests.
*M. circinalis** (Dumeril, Bibron & Dumeril, 1854) Trinidad Northern Coral Snake	Schmidt (1936c); Roze (1966, 1967, 1970, 1983, 1989, 1996); Lancini (1979); Campbell & Lamar (1989); Murphy (1997); Kornacker (1999); Boos (2001) *Remark: According to Campbell & Lamar (1989), this taxon is *M. psyches circinalis*.	Trinidad (including Gasparee Island), ne Venezuela and n Guyana; from sea level to 400 m in lowland rain forest and low montane wet to intermediate forest.
M. clarki Schmidt, 1936 Clark's Coral Snake	Schmidt (1936a); Roze (1967, 1970, 1983, 1996); Savage & Vial (1974); Savage & Villa (1986); Villa et al. (1988); Campbell & Lamar (1989); Köhler (2001)	From se Costa Rica into Darién, Panama from sea level to about 900 m, there are unconfirmed records from w Colombia; found in intermediate and wet lowland forest and tropical rain forest.
M. corallinus (Merrem, 1820) Painted Coral Snake	Roze (1967, 1970, 1983, 1996); Amaral (1976); Freiberg (1984); Campbell & Lamar (1989); Scrocchi (1990b); Cei (1993); Lema (1994a); Freitas (1999); Marques et al. (2001)	From Minas Geraís and Espiritu Santo in e and c Brazil westward and southward to Rio Grande do Sul, and into e Paraguay, and ne Misiones, Argentina, and probably will also be found in Uruguay.
M. decoratus (Jan, 1858) Brazilian or Decorated Coral Snake	Schmidt (1936c); Roze (1967, 1970, 1983, 1996); Amaral (1976); Campbell & Lamar (1989); Lema (1994a); Marques et al. (2001)	E and se Brazil from Minas Geraís, São Paulo, and Rio de Janiero, southward into Rio Grande do Sul; from sea level to over 1000 m in humid lowland and low montane subtropical areas frequently altered by humans.

Species-level Taxa	Citations and Remarks	Abbreviated Range
*M. diana** Roze, 1983 Diana's Coral Snake	Roze (1983, 1994, 1996); Campbell & Lamar (1989); Jorge da Silva & Sites (1999) *Remark: According to Campbell & Lamar (1989), this taxon is *M. frontalis diana*. Silva & Sites (1999) elevated this taxon to full species.	Roze (1996) suggested this snake is restricted to the Serranía de Santiago of e Bolivia at about 700 m, Jorge da Silva & Sites (1999) have evidence that it ranges outside of this area. They have information on specimens from San Ramón, Bolivia, another from Serranía Huanchaca in the extreme ne province of Santa Cruz and another from an exactly unidentified location.
M. diastema (Dumeril, Bibron & Dumeril, 1854) Variable or Diastema Coral Snake	Schmidt & Andrews (1936); Smith & Taylor (1945); Roze (1967, 1970, 1996); Wilson & Meyer (1982, 1985); Alvarez del Toro (1982); Villa et al. (1988); Campbell & Lamar (1989); Lee (1996, 2000); Cruz (1997); Köhler (2001)	C Verzcruz, e Puebla, and n Oaxaca to the Isthmus of Tehuantepec and Yucatán and southward into c Central America in Belize and Guatemala, to as far south as Honduras.
M. d. distema (Dumeril, Bibron & Dumeril, 1854) Veracruz Coral Snake	Smith & Taylor (1966); Roze (1967, 1970, 1996); Campbell & Lamar (1989); Pérez-Higareda & Smith (1991)	C Verzcruz and e Puebla in s Mexico from sea level to about 1000 m; in wet lowland tropical forest and humid low montane formations, partially modified by humans.
M. d. affinis (Jan, 1858) Oaxacan Speckled Coral Snake	Roze (1967, 1996); Campbell & Lamar (1989)	N and c Oaxaca in Mexico from 500 to 1300 m in humid medium-altitude forests.
M. d. aglaeope (Cope, 1860) Splendid Coral Snake	Schmidt (1936a); Roze (1967, 1970, 1996); Wilson & Meyer (1985); Campbell & Lamar (1989); Köhler (2001)	Mountains of nw Honduras around Lago Yoja and the Ulua Valley in dry areas at moderate elevations.
M. d. alienus (Werner, 1903) Yucatán Coral Snake	Schmidt & Andrews (1936); Roze (1967, 1996); Campbell & Lamar (1989); Lee (2000); Köhler (2001)	Yucatán Penninsula including the Yucatán and n Quintana Roo; up to 300 m in dry and humid lowland forests modified by humans.
M. d. apiatus (Jan, 1858) Spotted-nose Coral Snake	Stuart (1948, 1950); Roze (1967, 1970, 1996); Alvarez del Toro (1982); Campbell & Lamar (1989); Köhler (2001)	N Guatemala in Alta Verapaz and Huehuetenango in humid tropical montane forests and grasslands; and possibly in the mountains of e Chiapas, Mexico; it may also eventaully be found in Tabasco and Veracruz; generally found between 600 to 1300 m.
M. d. macdougalli Roze, 1967 MacDougall's Coral Snake	Roze (1967, 1996); Powers (1974); Campbell & Lamar (1989); Köhler (2001)	Atlantic versant of e Oaxaca, Mexico in the Sierra Madre del Sur at about 600 m in subtropical broadleaf and scrub forest.
M. d. sapperi (Werner, 1903) Irregular Coral Snake	Roze (1967, 1970, 1996); Alvarez del Toro (1982); Campbell & Lamar (1989); Stafford & Meyer (2000); Lee (2000); Köhler (2001)	Yucatán Penninsula including Campeche, n Chiapas, e Tabasco, and sw Quintana Roo in Mexico southward into Belize and n and w Guatemala; in humid tropical lowland forest, tropical secondary forest, and grasslands.
M. dissoleucus (Cope, 1860) Pygmy Coral Snake	Schmidt (1936c); Roze (1967, 1970, 1983, 1989, 1996); Rivero-Blanco & Dixon (1979); Lancini (1979); Villa et al. (1988); Campbell & Lamar (1989); Kornacker (1999); Köhler (2001)	C Panama into n South America as far as e Venezuela and ne Colombia.

The Caenophidia

Species-level Taxa	Citations and Remarks	Abbreviated Range
M. d. dissoleucus (Cope, 1860) Venezuelan Pygmy Coral Snake	Schmidt (1936c); Roze (1966, 1996); Lancini (1979); Lancini & Kornacker (1989); Campbell & Lamar (1989); Kornacker (1999)	Norte de Santander, Colombia to Sucre and Delta Amacuro in e Venezuela; from sea level to 1000 m in llanos, savanna, and low montane dry and humid forest.
M. d. dunni Barbour, 1923 Panama Pygmy Coral Snake	Schmidt (1936c, 1955); Smith & Grant (1958a); Campbell & Lamar (1989); Roze (1996); Köhler (2001)	Panama in the lowland dry and humid forests of the Canal Zone and modified humid forest and savanna westward to Coclé and Herrera; from sea level to 250 m.
M. d. melanogenys (Cope, 1860) Santa Marta Pygmy Coral Snake	Schmidt (1936c, 1955); Campbell & Lamar (1989); Roze (1996)	Foothills of the Sierra Nevada of Santa Marta in Colombia; from sea level to about 200 m in low montane scrub and semidesert areas.
M. d. nigrirostris Schmidt, 1955 Barranquilla Pygmy Coral Snake	Schmidt (1955); Pérez-Santos & Moreno (1988); Campbell & Lamar (1989); Roze (1996)	Atlántico and the delta region of the lower Magdalena Valley in Colombia; near sea level in dry lowland forest and coastal scrub and humid forest.
M. distans (Kennicott, 1860) Clear-banded or West Mexican Coral Snake	Schmidt (1936a); Smith & Taylor (1945, 1966); Zweifel (1959); Roze (1967, 1983, 1996); Hardy & McDiarmid (1969); Tanner (1986); Campbell & Lamar (1989); García & Ceballos (1994)	W Mexico from sw Chihuahua and s Sonora through Sinaloa, Nayarit, Colima, and Jalisco and southward to the Río Balsas Basin in Michoacán and Guerrero
M. d. distans (Kennicott, 1860) Common Clear-banded Coral Snake	Campbell & Lamar (1989); Roze (1996)	W Mexico in s Sonora, s Chihuahua, and sw Sinaloa into n Nayarit; from sea level to 1500 m in tropical dry lowland and dry montane areas.
M. d. michoacanensis (Dugès, 1891) Michoacán Clear-banded Coral Snake	Schmidt & Smith (1943); Duellman (1961); Campbell & Lamar (1989); Roze (1996)	Balsas Basin in Michoacán and Guerrero, Mexico; between about 100 to 700 m in low dry montane tropical scrub forest.
M. d. oliveri Roze, 1967 Colima Clear-banded Coral Snake	Roze (1967, 1996); Casas-Andreu (1981); Campbell & Lamar (1989); Ramírez-Bautista (1994)	Colima and sw Jalisco, Mexico; from about sea level to 2000 m in dry lowland scrub forest.
M. d. zweifeli Roze, 1967 Zweifel's Coral Snake	Roze (1967, 1996); Campbell & Lamar (1989)	C and s Nayarit into adjacent Magdalena in Jalisco, Mexico; from sea level to 1200 m in dry to humid Moderate tropical areas.
M. dumerili (Jan, 1858) Dumeril's Coral Snake or Capuchin Coral Snake	Schmidt (1936c, 1955); Rendahl & Vestergren (1940); Roze (1955, 1966, 1967, 1970, 1983, 1996); Lancini (1979); Campbell & Lamar (1989); Kornacker (1999)	N Venezuela, n Colombia into nw Ecuador and probably into e Panama
M. d. dumerili (Jan, 1858) Common Capuchin Coral Snake	Schmidt (1936c, 1955); Rendahl & Vestergren (1940); Roze (1970, 1996); Campbell & Lamar (1989)	Lower Magdalena Valley in n Colombia; from sea level to about 2000 m in dry and intermediate tropical lowlands.

Species-level Taxa	Citations and Remarks	Abbreviated Range
M. d. antioquiensis Schmidt, 1936 Antioquian Coral Snake	Schmidt (1936c 1955); Roze (1970, 1996); Campbell & Lamar (1989)	Valle del Cauca in Medellín in c Colombia; 900 to 2600 m in humid to wet montane forest.
M. d. carinicauda Schmidt, 1936 Intermediate Capuchin Coral Snake	Schmidt (1936c); Roze (1955, 1966, 1970, 1996); Lancini (1979); Lancini & Kornacker (1989); Campbell & Lamar (1989); Kornacker (1999)	Norte de Santander and Santander in n Colombia into adjacent w Venezuela in Zulia, Táchira, and Apure; from 100 to 800 m in savanna and intermediate tropical formations.
M. d. colombianus (Griffin, 1916) Santa Marta Capuchin Coral Snake	Schmidt (1936c); Roze (1970, 1996); Campbell & Lamar (1989); Almendariz (1991)	Santa Marta Mountains of n Colombia; between 300 to 2200 m in tropical dry montane scrub areas and wet forest.
M. d. transandinus Schmidt, 1936 Transandean Capuchin Coral Snake	Schmidt (1936c); Roze (1970, 1996); Campbell & (1970); Campbell & Lamar (1989)	Pacific side of Andes in w Colombia into nw Ecuador, there is a record from Panama
M. d. venezuelensis Roze, 1989 Venezuelan Capuchin Coral Snake	Roze (1955, 1989, 1996); Lancini (1979 [photos under the name of *M. carinicauda* and *M. dumerilii carinicauda* respectively.] Kornacker (1999)	Cordillera de la Costa in nc Venezuela; sea level to about 1100 m in tropical lowland and humid montane forest altered by humans.
M. elegans (Jan, 1858) Elegant Coral Snake	Schmidt (1936a); Smith & Taylor (1945); Stuart (1948, 1950); Roze (1967, 1970, 1983, 1996); Alvarez del Toro (1982); Villa et al. (1988); Campbell & Lamar (1989); Köhler (2001)	C Veracruz and n Oaxaca to Tabasco and Chiapas in s Mexico into Guatemala.
M. e. elegans (Jan, 1858) Western Elegant Coral Snake	Smith & Taylor (1966); Alvarez del Toro (1982); Campbell & Lamar (1989); Pérez-Higareda & Smith (1990, 1991); Roze (1996)	C Veracruz and n Oaxaca in s Mexico; between 800 to 1700 m in wet and dry montane foest and cloud forest.
M. e. veraepacis Schmidt, 1933 Verapaz Elegant Coral Snake	Schmidt (1933); Roze (1970; 1996); Alvarez del Toro (1982); Campbell & Lamar (1989); Köhler (2001)	S Tabasco and Chiapas in Mexico southward into Alta Verapaz and Huehuetenango in Guatemala; between 600 to 1700 m in humid and dry montane areas.
M. ephippifer (Cope, 1886) Double-black Coral Snake	Smith & Taylor (1945, 1966) Roze (1967, 1983, 1989, 1996); Campbell & Lamar (1989)	S Mexico from the Isthmus of Tehuantepec into e Oaxaca
M. e. ephippifer (Cope, 1886) Tehuantepec Coral Snake	Schmidt (1958); Roze (1989, 1996)	S Mexico in Isthmus of Tehuantepec into Quiengola, Oaxaca; sea level to about 1500 m in dry lowlands and dry scrub montane areas.
M. e. zapotecus Roze, 1989 Zapotec Coral Snake	Roze (1989, 1996)	Sierra Madre del Sur west of Oaxaca, Mexico; between 1700 to 2400 m in humid and high mountain woodlands, pine-oak, and oak-madroño areas.

Species-level Taxa	Citations and Remarks	Abbreviated Range
*M. filiformis** (Günther, 1859) Slender or Thread Coral Snake	Schmidt (1955); Roze (1967, 1970, 1983, 1996); Amaral (1976); Cunha & Nascimento (1982); Carillo de Espinosa (1983); Dixon & Soini (1986); Chippaux ([1986] 1987); Campbell & Lamar (1989); Starace (1998) *Remark: According to Cunha & Nascimento (1982), subspecies are not valid, and subspecies are not shown in Roze (1996).	Amazon Basin of s Colombia, ne Peru, and into Pará and Maramhão in n Brazil, possibly will be found in the southern tip of Venezuela, Suriname, and French Guiana; from sea level to about 500 m in tropical lowland rain forest, secondary forest, low montane forest, and open fields.
M. frontalis (Dumeril, Bibron & Dumeril, 1854) Short-tailed Coral Snake	Schmidt (1936c); Roze (1967, 1970, 1983, 1994, 1996); Kempff-Mercado (1975); Campbell & Lamar (1989); Lema (1994a); Jorge da Silva & Sites (1999)	South of the Amazon Basin in c and s Brazil, Bolivia, and Paraguay.
M. f. frontalis (Dumeril, Bibron & Dumeril, 1854) Brazilian Giant Coral Snake	Schmidt (1936c); Roze (1970, 1996); Amaral (1976); Campbell & Lamar (1989); Lema (1994a); Silva & Sites (1999)	S Brazil and s Paraguay; sea level to about 700 m in lowland and low montane humid forest and deciduous forest altered by humans.
*M. f. multicinctus** Amaral, 1944 Southern Many-banded Coral Snake	Vanzolini (1986); Lema (1994a); Amaral (1976); Roze (1983, 1996); Campbell & Lamar (1989) *Remark: Silva & Sites (1999), noted that Roze (1996), interpreted *M. f. multicinctus* as a subspecies of *M. frontalis*. In earlier literature, it had been shown as an intergrade between *M. f. frontalis* and *M. f. altirostris* (see Shreve, 1953, and Scrocchi, 1990b).	E Brazil in e São Paulo, Paraná, Santa Catarina, and possibly into n Rio Grande do Sul.
*M. frontifasciatus** (Werner, 1927) Bolivian Triad Coral Snake	Roze (1967, 1970, 1983, 1996); Vanzolini (1986); Campbell & Lamar (1989) *Remark: Sometimes listed as the subspecies *M. lemniscatus frontifasciatus*.	N Bolivia and se Peru on e Andes slopes; between 600 to 2500 m in low and high mountain dry forest, cloud forest, and lowland gallery forest on the e slopes of the Andes.
M. fulvius (Linnaeus, 1766) Harlequin Coral Snake or North American Coral Snake	Smith & Taylor (1945); Roze (1967, 1970, 1983, 1996); Conant (1975); Roze & Tilger (1983); Campbell & Lamar (1989); Dundee & Rossman (1989); Wright & Wright ([1957] 1994); Tennant (1997b, 1998); Dixon (2000a) *Remark: Dixon (2000a) notes that Roze (1996) continues to recognize *M. f. tenere* as a race of *fulvius*, even though he stated that there is more to be learned through future analysis of the venom and the population genome. Liner (1994), Collins (1997), and Crother et al. (2000), recognized it as an allopatric, diagnostic species. According to these three authorities, those found east of the Mississippi River are *M. fulvius* and those west of the Mississippi	E and s United States from North Carolina to the southern tip Florida and westward into s Arkansas and e and s Texas and southward into ne and c Mexico in Coahuila, Nuevo León, and Tamaulipas, and to Aguas Calientes, Querétaro, Guanajuato, and Morelos in a variety of habitats.

Species-level Taxa	Citations and Remarks	Abbreviated Range
	River are *M. tener*. I continue to follow Roze (1996) as the most recent revisor until more data becomes available.	
M. f. fulvius (Linnaeus, 1766) Eastern Coral Snake	Conant (1975); Roze & Tilger (1983); Campbell & Lamar (1989); Dundee & Rossman (1989); Ernst (1992); Roze (1996); Conant & Collins (1998); Tennant & Bartlett (2000)	SE United States in se North Carolina southward into s Florida and westward through extreme eastern Louisiana; in a variety of dry and humid lowland habitats.
M. f. fitzingeri (Jan, 1858) Guanajuato Coral Snake	Schmidt (1933c); Roze (1967, 1983, 1996); Roze & Tilger (1983); Campbell & Lamar (1989)	Guanajuato and Querétaro into Morelos in Mexico, and possibly Zacatecas, Aguas Calientes, and s Coahuila; between 1000 to 3000 m in the dry desert Mexican Plateau.
M. f. maculatus Roze, 1967 Tampico Coral Snake	Roze (1967, 1996); Roze & Tilger (1983); Campbell & Lamar (1989); Pérez-Higareda & Smith (1991)	Lowlands in Tamaulipas, Mexico around Tampico; near sea level in swampy and humid lowlands.
M. f. microgalbineus Brown & Smith, 1942 Spotted Coral Snake	Taylor (1950); Roze (1967, 1996); Brown & Smith (1942); Roze & Tilger (1983); Campbell & Lamar (1989)	SW and s Tamaulipas to Guanajuato, and, c and e San Luis Potosí in Mexico; between 80 to 500 m; in dry subtropical and topical deciduous and humid forest.
M. f. tener* (Baird & Girard, 1853) Texas Coral Snake	Schmidt (1933c); Smith & Taylor (1966); Conant (1975); Roze & Tilger (1983); Campbell & Lamar (1989); Dundee & Rossman (1989); Ernst (1992); Roze (1996); Tennent (1998); Conant & Collins (1998); Werler & Dixon (2000); Tennant & Bartlett (2000); Dixon (2000)	The United States in s Arkansas and Louisiana into wc Texas and n Mexico to s Tamaulipas and westward into c Coahuila and Nuevo León. There is an unverified record from Sonora, Mexico. From sea level to 300 m in dry to humid lowlands west of the Mississippi River.
	*Remark: See the Dixon (2000a) remark under *M. fulvius*.	
M. hemprichi (Jan, 1858) Hemprich's or Worm-eating Coral Snake	Schmidt (1953b); Roze (1967, 1970, 1983, 1996); Cunha & Nascimento (1972, 1978, 1982a); Campbell & Lamar (1989); Starace (1998); Kornacker (1999)	E Colombia to s Venezuela, the Guianas, e Ecuador, Peru, n Bolivia, and the Amazon region of Rondônia into Pará in Brazil
M. h. hemprichi (Jan, 1858) Eastern Hemprich's or Eastern Worm-eating Coral Snake	Schmidt (1953b); Roze (1970, 1984, 1996); Cunha & Nascimento (1972, 1978); Lancini (1979); Lancini & Kornacker (1989); Campbell & Lamar (1989); Starace (1998); Kornacker (1999)	SE Venezuela, the Guianas, and into Brazil in Manaus, Amazonas; Balém, Pará. It is found from sea level to 250 m in lowland and low altitude tropical rain forest.
M. h. ortoni Schmidt, 1953 Western Hemprich's or Western Worm-eating Coral Snake	Schmidt (1953b); Roze (1967, 1970, 1996); Dixon & Soini (1986); Campbell & Lamar (1989); Lancini & Kornacker (1989); Almendariz (1991); Silva (1994); Kornacker (1999)	NW Brazil, e Ecuador, Peru, and, Amazonian s Colombia, Venezuela, and n Bolivia; from 100 to 1200 m in lowland and low montane tropical rain forest or humid forests of the upper Amazon region.
M. h. rondonianus (Roze & Jorge da Silva, 1990) Rondonia Coral Snake	Roze (1990, 1994, 1996); Roze & Silva (1990)	Rondônia, Brazil in tropical rain forest; found in the area of the Hydroelectric Plant at 80 to 100 m above sea level.

The Caenophidia

Species-level Taxa	Citations and Remarks	Abbreviated Range
M. hippocrepis (W. Peters, 1861) Maya or Belize Coral Snake	Schmidt (1933c); Stuart (1948); Roze (1967, 1970, 1983, 1996); Henderson & Hoevers (1975); Villa et al. (1988); Lee (1996, 2000); Campbell & Lamar (1989); Stafford & Meyer (2000); Köhler (2001)	N Guatamala and e Belize in Caribbean versant; from sea level to about 600 m in lowland rain forest and low montane humid valleys and pine forest.
M. ibiboboca (Merrem, 1820) Caatinga Coral Snake and Ibiboboca	Schmidt (1936c); Roze (1966, 1967, 1970, 1983, 1996); Amaral (1976); Chippaux ([1986] 1987); Campbell & Lamar (1989); Starace (1998); Freitas (1999); Marques et al. (2001)	E Brazil and disjunct populations in Suriname and possibly French Guiana; from sea level to about 1200 m in a wide variety of habitats.
M. isozonus (Cope, 1860) Equal-banded or Venezuelan Coral Snake	Schmidt (1936c); Roze (1955, 1966, 1967, 1970, 1983, 1996); Rivero-Blanco & Dixon (1979); Lancini (1979); Campbell & Lamar (1989); Lancini & Kornacker (1989); Kornacker (1999)	E Colombia into n, c and se Venezuela and adjacent Brazil in the Rio Cotinga region in Roraima, it is also found on Margarita Island; from sea level to 1200 m in llanos and humid forest.
M. langsdorffi Wagler *in* Spix, 1824 Langsdorff's Coral Snake	Schmidt (1936c, 1955); Rendahl & Vestergren (1941); Roze (1955, 1996); Duellman (1978); Cunha & Nascimento (1982); Dixon & Soini (1986); Campbell & Lamar (1989); Silva (1994)	N South America in s Colombia, ne Ecuador, nw Brazil, and ne Peru; between 80 to 450 m in lowland tropical rain forest and altered secondary forest.
M. laticollaris (W. Peters, 1870) Balsan or Double Collar Coral Snake	Schmidt (1936a, 1936c); Smith & Taylor (1945); Davis & Smith (1953); Roze (1967, 1983, 1989, 1996); Campbell & Lamar (1989); Smith & Smith (1993)	Río de las Balsas and Río Tepalcatepec basins in Michoacán, Guerrero, Jalisco, Morelos, Puebla, Oaxaca, and possibly the state of México.
M. l. laticollaris (W. Peters, 1870) Eastern Double Collar Coral Snake	Dullman (1961); Roze (1967, 1996); Campbell & Lamar (1989); Smith & Smith (1993)	Mexico in the Telpalcatepec-Balsas Basin in Michoacán and in Guerrero, Morelos, Puebla, and w Oaxaca; between 400 to 1800 m in dry low montane and transition humid areas.
M. l. maculirostris Roze, 1967 Western Double Collar Coral Snake	Roze (1967, 1996); Dixon & Webb (1965); Campbell & Lamar (1989); Smith & Smith (1993)	Colima and s Jalisco, Mexico; between 300 to 800 m in dry montane and scrub areas.
M. latifasciatus Schmidt, 1933 Broad-ringed or Long-banded Coral Snake	Schmidt (1933c); Schmidt & Smith (1943); Smith & Taylor (1945, 1966); Roze (1967, 1970, 1983, 1996); Alvarez del Toro (1982); Villa et al. (1988); Campbell & Lamar (1989); Smith & Smith (1993); Köhler (2001)	Pacific versant of s Mexico in Oaxaca and Chiapas into e Guatemala; near sea level to 1350 m in dry tropical lowlands to moderate and high humid montane areas.
M. lemniscatus (Linnaeus, 1758) Ribbon or South American Coral Snake	Roze (1967, 1970, 1983, 1996); Cunha & Nascimento (1973, 1978, 1982a); Kempff-Mercado (1975); Campbell & Lamar (1989); Roze & Silva (1990); Silva (1994); Starace (1998); Silva & Sites (1999); Kornacker (1999)	Amazon region east of the Andes and Amazon and Orinoco watersheds. Its range includes Trinidad and the Guianas; it may possibly be found in Paraguay; recently reported in Misiones, Argentina.
M. l. lemniscatus (Linnaeus, 1758) Guiana's Ribbon Coral Snake	Roze (1970, 1996); Cunha & Nascimento (1973, 1978, 1982a); Campbell & Lamar (1989); Starace (1998)	N Guianas into ne Brazil in Pará and Maranhão; from sea level to about 600 m in lowland rain forest, open humid forest, and drier cultivated lands.
M. l. carvalhoi Roze, 1967 Brazilian Ribbon Coral Snake	Roze (1967, 1996); Campbell & Lamar (1989); Freitas (1999)	Brazil from Paraná, São Paulo, Minas Geraís, and s Mato Grosso and its range extends north along the coast to Bahia and Rio Grande do Norte; it may also range into Paraguay; in lowland dry forest and low

Species-level Taxa	Citations and Remarks	Abbreviated Range
		montane semideciduous or humid forest; and marginally into the cerrado.
M. l. diutius Burger, 1955 Trinidad Ribbon Coral Snake	Roze (1955, 1966, 1967, 1970, 1996); Cunha & Nascimento (1973, 1978); Lancini (1979); Campbell & Lamar (1989); Lancini & Kornacker (1989); Murphy (1997); Starace (1998); Kornacker (1999); Boos (2001)	Amazonian Region of Ecuador, Peru, and Bolivia into Colombia, n Brazil, and e and se Venezuela and Orinoco drainage, and in the Guianas and Trinidad in low humid and rain forests.
M. l. helleri Schmidt & Schmidt, 1925 Western Ribbon Coral Snake	Roze (1967, 1970, 1987, 1996); Duellman (1978); Carillo de Espinosa (1983); Dixon & Soini (1986); Lancini & Kornacker (1989); Campbell & Lamar (1989); Almendariz (1991); Silva (1994); Murphy (1997); Kornacker (1999)	E slopes of the Andes on both sides of the Amazon in s Venezuela, Colombia, Peru, Ecuador, Bolivia, and in w Brazil; between 80 to about 1500 m in humid and rain forests.
M. limbatus Fraser, 1964 Tuxtlan Coral Snake	Fraser (1964); Smith & Taylor (1966); Roze (1967, 1983, 1996); Campbell & Lamar (1989); Greene et al. (1998)	Los Tuxtlas region of s Veracruz in Mexico.
M. l. limbatus Fraser, 1964 Tuxtlan Banded Coral Snake	Fraser (1964); Campbell & Lamar (1989); Pérez-Higareda & Smith (1990, 1991); Smith & Smith (1993); Roze (1996); Greene et al. (1998)	NW Los Tuxtlas around Volcán San Martín in Veracruz in Mexico; from 10 to 1500 m in tropical rain forest and low montane wet forest, and patchy grassland.
M. l. spilosomus Pérez-Higareda & Smith, 1990 Tuxtlan Spotted Coral Snake	Campbell & Lamar (1989); Pérez-Higareda & Smith (1990, 1991); Smith & Smith (1993); Roze (1996); Greene et al. (1998)	Sierra de Santa Marta, in e Los Tuxtlas region in Verzcruz in Mexico; 900 to 1000 m in low montane wet forest.
M. margaritiferus Roze, 1967 Peruvian Speckled Coral Snake	Roze (1967, 1970, 1983, 1996); Carillo de Espinosa (1983); Campbell & Lamar (1989)	Marañón and Santiago Valley on the eastern slopes of Andes in Peru; between 200 to 400 m in tropical rain forest.
M. medemi Roze, 1967 Villavicencio Coral Snake	Roze (1967, 1983, 1994, 1996); Pérez-Santos & Moreno (1988); Campbell & Lamar (1989)	Around Villavicencio in Departamento Meta, Colombia; between 250 to 600 m.
M. meridensis Roze, 1989 Mérida Pygmy Coral Snake	Roze (1989, 1994, 1996); Kornacker (1999)	Around Lagunilla in the w slopes of the Andes of Mérida, Venezuela at about 900 m in humid montane vegetation altered by humans.
M. mertensi Schmidt, 1936 Merten's Coral Snake or Peruvian Desert Coral Snake	Schmidt (1936c); Roze (1967, 1970, 1983, 1996); Carillo de Espinosa (1983); Campbell & Lamar (1989); Almendariz (1991)	SW Ecuador into wc Peru; from sea level to 1700 m in lowland desert and dry shrub valleys.
M. mipartitus (Dumeril, Bibron & Dumeril, 1854) Red-tailed Coral Snake	Schmidt (1936c); Rendahl & Vestergren (1940); Roze (1967, 1970, 1983, 1996); Villa et al. (1988); Campbell & Lamar (1989); Köhler (2001)	Darién, Panama into n South America in Colombia, n and nw Venezuela, and Ecuador; and possibly Peru.

The Caenophidia

Species-level Taxa	Citations and Remarks	Abbreviated Range
M. m. mipartitus (Dumeril, Bibron & Dumeril, 1854) Pacific Red-tailed Coral Snake	Schmidt (1955); Roze (1955, 1966, 1970, 1996); Campbell & Lamar (1989)	Darién, Panama southward into the Pacific lowlands of w Colombia; from sea level to 850 m in humid and rain forests.
M. m. anomalus (Boulenger, 1896) Santa Marta Red-tailed Coral Snake	Roze (1967, 1996); Roze in Peters & Orejas-Miranda (1970); Campbell & Lamar (1989); Kornacker (1999)	Sierra Nevada de Santa Marta in e Colombia eastward into the Andes of Mérida and the Sierra Perijá of Venezuela; between 500 to 2000 m in humid forests and savanna.
M. m. decussatus (Dumeril, Bibron & Dumeril, 1854) Andean Red-tailed Coral Snake	Schmidt (1955); Roze (1967, 1970, 1996); Campbell & Lamar (1989); Almendariz (1991)	C Colombia, w Ecuador, and possibly Peru; it has been reported from Isla Gorgona off the Pacific coast of Colombia; sea level to 2700 m in low to high humid forests and valleys.
M. m. popayanensis Ayerbe, Tidwell & Tidwell, 1990 Popayán Red-tailed Coral Snake	Ayerbe et al. (1990); Roze (1996)	Upper Río Cauca Valley in Colombia between 1300 to 1500 in wet montane forests.
M. m. semipartitus* (Jan, 1858) Venezuelan Red-tailed Coral Snake	Roze (1955, 1970, 1996); Lancini (1979); Lancini & Kornacker (1989); Campbell & Lamar (1989); Kornacker (1999); Golay et al. (1999) *Remark: Golay et al. (1999) proposed that *Micrurus mipartitus rozei* subsp. nov. Be used for the taxon currently known as *M. m. semipartitus* (Jan, 1858), which they consider to be an unjustified emendation of *Elaps mipartitus* Duméril et al., 1854. Their reasoning appears sound, I have maintained the more historically accepted nomenclature because I did not have time to research this presentation before this work went to press.	N Venezuela in an around the Cordillera de la Costa to valleys around San Felipe, Yaracuy, about 800 to 2000 m in humid and cloud forests.
M. multifasciatus (Jan, 1858) Many-banded Coral Snake	Roze (1967, 1983, 1996); Savage & Vial (1974); Savage (1980); Villa et al. (1988); Campbell & Lamar (1989); Köhler (2001)	S Central America from Nicaragua southward into the Canal Zone of Panama.
M. m. multifasciatus* (Jan, 1858) Panama Many-banded Coral Snake	Roze (1970, 1983, 1996); Campbell & Lamar (1989); Köhler (2001)	Panama from the Canal Zone south into Darién; from sea level to 400 m in humid to wet tropical forest, and areas altered by humans.
M. m. hertwigi* (Werner, 1897) Costa Rican Many-banded Coral Snake	Roze (1967, 1970, 1996); Villa (1972); Köhler (2001)	Caribbean slopes of c Nicaragua, nw Panama, and nw Costa Rica

Species-level Taxa	Citations and Remarks	Abbreviated Range
M. multiscutatus Rendahl & Vestergren, 1940 Cauca Coral Snake	Roze (1970, 1996); Campbell & Lamar (1989)	Dept. Cauca, Colombia on the Pacific side of the Cordillera Occidental from 150 to 900 m
M. nebularis Roze, 1987 Cloud Forest Coral Snake	Smith & Smith (1993); Roze (1987, 1989, 1996)	Sierra de Juárez of the Sierra Madre del Sur in c Oaxaca, Mexico; between 2100 to 2300 m in high mountain pine-oak and humid forest woodlands.
M. nigrocinctus (Girard, 1854) Central American Coral Snake	Schmidt (1933c, 1936c); Roze (1967, 1970, 1983, 1996); Henderson & Hoevers (1975); Savage & Villa (1986); Villa et al. (1988); Campbell & Lamar (1989); Lee (1996, 2000); Stafford & Meyer (2000), Köhler (2001)	S Chiapas in s Mexico southward through Central America into n South America to the nw tip of Colombia; it is also found on several offshore islands including Isla del Maíz Grande, Nicaragua, and Coiba, Taboga, San José, and San Miguel, Panama.
M. n. nigrocinctus (Girard, 1854) Common Central American Coral Snake	Schmidt (1933c); Roze (1970, 1983, 1996); Villa (1984); Wilson & Meyer (1985); Vanzolini (1986); Villa et al. (1988); Campbell & Lamar (1989); Köhler (2001)	Nicaragua, Panama, and Costa Rica into nw Colombia; sea level to about 1400 m in a wide variety of habitats.
M. n. babaspul Roze, 1967 Babaspul or Great Corn Island Coral Snake	Roze (1967, 1970, 1983, 1996); Villa (1984); Campbell & Lamar (1989); Köhler (2001)	Great Corn Island in Nicaragua in island rain forest.
M. n. coibensis Schmidt, 1936 Coiba Coral Snake	Schmidt (1936c); Roze (1970, 1996); Campbell & Lamar (1989); Köhler (2001)	Coiba Island in Panama in island vegetation.
M. n. divaricatus (Hallowell, 1854) Honduras Coral Snake	Schmidt (1933c); Roze (1970, 1996); Wilson & Meyer (1982, 1985); Campbell & Lamar (1989); Lee (2000); Köhler (2001)	N and c Honduras with one report from c Belize; from sea level to 1300 m in lowlands and wet low montane areas including short-transsavanna.
M. n. mosquitensis Schmidt, 1933 Mosquito Coral Snake	Schmidt (1933c); Roze (1970, 1996); Villa et al. (1988); Campbell & Lamar (1989); Köhler (2001)	Nicaragua, Costa Rica, and Panama; from sea level to about 1200 m in humid to wet lowland to montane forests.
*M. n. zunilensis** Schmidt, 1932 Zunil Coral Snake	Schmidt (1932a, 1936c); Davis & Smith (1953); Smith & Taylor (1966); Roze (1970, 1996); Alvarez del Toro (1982); Campbell & Lamar (1989); Köhler (2001) *Remark: *M. nigrocinctus ovandoensis* Schmidt & Smith, 1943, and *M. nigrocinctus wagneri* Mertens, 1941, are synonyms (see Smith & Taylor 1966, Smith & Smith 1993), and Roze 1996).	S Mexico, Guatamala, El Salvador, and possibly into Honduras; from sea level to about 1500 m in humid to wet lowland to montane tropical forest.
M. oligoanellatus Ayerbe López, 2002 Few-Ringed Coral Snake	Ayerbe & López (2002)	W versant of the Cordillera Occidental between 1000 and 1500 m of Vereda El Cocal and Reserva Natural Tambito in Colombia. They seem to prefer zones modified by humans between pastures and humid tropical forest.

The Caenophidia

Species-level Taxa	Citations and Remarks	Abbreviated Range
M. ornatissimus (Jan, 1858) Ornate Coral Snake	Schmidt (1936c); Roze (1967, 1970, 1983, 1996); Duellman (1978); Cunha & Nascimento (1982b); Campbell & Lamar (1989)	Amazon basin in s Colombia, e Ecuador, and se Peru with a questionable record from Pará, Brazil; recently reported from Mato Grosso; between 500 to 1200 m in tropical lowland to montane rain forest.
M. pachecogili Campbell, 2000 Pueblan High Desert Coral Snake	Campbell (2000)	High desert in Puebla, Mexico.
*M. paraensis** Cunha & Nascimento, 1973 Pará Coral Snake	Cunha & Nascimento (1973, 1977, 1982); Vanzolini (1986); Campbell & Lamar (1989); Roze & Silva (1990); Roze (1994, 1996); Strussman & Carvalho (1998) *Remark 1: Campbell & Lamar (1989) recognized this taxon as *M. psyches paraensis*. *Remark 2: *M. donosoi* Hoge et al. 1976 is as a synonym according to Roze (1996); however, Campbell & Lamar (1989) recognized it as *M. psyches donosoi*.	S Suriname and Brazil in Pará and w Maranhão into Rondônia in Brazil; from sea level to 400 m in coastal secondary and remnant rain forest.
M. peruvianus Schmidt, 1936 Peruvian Coral Snake	Schmidt (1936c); Roze (1967, 1970, 1983, 1996); Carillo de Espinosa (1983); Campbell & Lamar (1989)	NE Peru in the Chinchipe-Marañón valleys of Cajamarca and in Amazonas; from about 450 to 1500 m in dry scrub areaas on the e slopes of the Andes.
M. petersi Roze, 1967 Mountain Coral Snake or Peters's Coral Snake	Roze (1967, 1983, 1996); Vanzolini (1986); Campbell & Lamar (1989); Almendariz (1991)	E slopes of the Andes in Ecuador in the Santiago-Morona Province; between 1000 to 2800 m in humid moderate and high mountain forest.
M. proximans Smith & Chrapliwy, 1958 Nayarit Coral Snake	Zweifel (1959); Smith & Taylor (1966); Roze (1967, 1983, 1996); Campbell & Lamar (1989)	Nayarit, Mexico; from sea level to 1000 m in dry lowland and coastal thorn areas.
*M. psyches** (Daudin, 1803) Carib Coral Snake	Schmidt (1936c); Roze (1955, 1966, 1967, 1970, 1983, 1989, 1996); Rivero-Blanco & Dixon (1979); Campbell & Lamar (1989); Lancini & Kornacker (1989); Starace (1998); Kornacker (1999) *Remark: Roze (1996) did not recognize subspecies. Campbell & Lamar (1989) recognized six races of subspecies which Roze elevated to full species. These include *M. p. psyches* (Daudin, 1803); *M. p. circinalis* (Dumeril, Bibron and Dumeril, 1854); *M. p. donosoi* Hoge, Cordeiro & Romano, 1976; *M. p. medemi* Roze, 1967; *M. p. paraensis* Cunha & Nascimento, 1973; and *M. p. remotus* Roze, 1987.	NE South America in se Venezuela, n Guyana, Suriname, and French Guiana from about 50 to 500 m; rain forest, low montane wet forest, and gallery forest on the fringes of savanna.
M. putumayensis Lancini, 1962 Putumayo Coral Snake	Lancini (1962a, 1962b, 1982); Roze (1967, 1983, 1996); Carillo de Espinosa (1983); Dixon & Soini (1986); Campbell & Lamar (1989)	Dept. Loreto, Peru into the state of Amazonas, Brazil and s tip of Colombia; between 100 to 300 m in tropical rain forest of the upper Amazon.

Species-level Taxa	Citations and Remarks	Abbreviated Range
M. pyrrhocryptus* (Cope, 1862) Southern or Argentine Coral Snake	Campbell & Lamar (1989); Cei (1993); Roze (1996); da Silva & Sites (1999) *Remark: Both Cei (1986) and Campbell & Lamar (1989) listed this snake as *M. frontalis pyrrhocryptus*; but, Cei (1993) listed it as a full species. According to Silva & Sites (1999), it is a full species and they elevated *tricolor* to full species as well.	S South America from Argentina distributed to the west of Río Paraná and southward to the province of Río Negro, and ranges north to the province of Formosa, and northwestward into Paraguay (Departamento Boquerón) and Bolivia Departamento Santa Cruz), possibly through out the provinces of Salta and Jujuy.
M. remotus* Roze, 1987 Remote Coral Snake	Roze (1987, 1994, 1996); Campbell & Lamar (1989); Kornacker (1999) *Remark: Campbell & Lamar (1989) classified this species as a subspecies *M. psyches*.	E Colombia, s Venezuela, and Amazonas, Brazil; from 90 to 1500 m in lowland and low montane rain forest; and probably also cloud forest.
M. ruatanus (Günther, 1895) Roatán Coral Snake	Schmidt (1933c); Roze (1967, 1970, 1983, 1996); Wilson & Hahn (1973); Wilson (1984b); Wilson & Meyer (1982, 1985); Villa et al. (1988); Campbell & Lamar (1989); Wilson et al. (1992); Cruz (1997); Köhler (2001)	Restricted to the Bay Islands of Honduras in the Caribbean in tropical island vegetation altered by humans.
M. sangilensis Nicéforo-María, 1942 San Gil or Santander Coral Snake	Nicéforo-María (1942); Schmidt (1955); Roze (1967, 1970, 1983, 1996); Vanzolini (1986); Campbell & Lamar (1989); Roze (1996)	E Andes in Colombia in the Cordillera in Santander, Boyacá and n Cundinamarca; between 1000 to 2000 m in the Cordillera of the e Andes in cloud forest and humid high montane forest.
M. spixi Wagler *in* Spix, 1824 Amazonian Coral Snake	Schmidt & Walker (1943a); Roze (1955, 1966, 1967, 1970, 1983, 1996); Cunha & Nascimento (1973, 1978, 1982a); Kempff-Mercado (1975); Amaral (1976); Duellman (1978); Lancini (1979); Carillo de Espinsoa (1983); Chippaux ([1986] 1987); Dixon & Soini (1986); Campbell & Lamar (1989); Roze & Silva (1990); Starace (1998); Kornacker (1999)	Amazon and Orinoco drainages of Colombia, Venezuela, Ecuador, Peru, Brazil, Bolivia, and n Paraguay and possibly French Guiana
M. s. spixi Wagler *in* Spix, 1824 Central Amazon Coral Snake	Roze (1967, 1970, 1983, 1996); Kempff-Mercado (1975); Amaral (1976); Campbell & Lamar (1989)	C Amazonian Region of Brazil in Amazonas and Rondônia; between 50 to 150 m in lowland tropical rain forest and humid forest.
M. s. martiusi Schmidt, 1953 Black-headed Amazonian Coral Snake	Schmidt (1953c); Roze (1970, 1996); Cunha & Nascimento (1973, 1978, 1982b); Campbell & Lamar (1989)	State of Pará into Goiás and Mato Grosso in Brazil with one unverified record from Paraguay; from sea level to 300 m in tropical rain forest and humid secondary forest altered by humans.
M. s. obscurus (Jan, 1872) Black-necked Amazonian Coral Snake	Schmidt (1955); Roze (1970, 1996); Duellman (1978, 1986); Dixon & Soini (1986); Lancini & Kornacker (1989); Campbell & Lamar (1989); Almendariz (1991); Silva (1994); Kornacker (1999)	Upper Amazon and Orinoco Basins of s Colombia and Venezuela, e Ecuador, and Peru, and probably into n Bolivia east of the Andes and nw Brazil
M. s. princeps (Boulenger, 1905) Bolivian Amazonian Coral Snake	Schmidt (1953c); Roze (1970, 1996); Kempff-Mercado (1975); Campbell & Lamar (1989)	Amazonian and valleys of e Andes of Bolivia; between 500 to 1200 m in intermediate altitude humid forest and forest altered by humans.

The Caenophidia

Species-level Taxa	Citations and Remarks	Abbreviated Range
M. spurelli (Boulenger, 1914) Colombian Coral Snake	Schmidt (1955); Roze (1967, 1983, 1996); Campbell & Lamar (1989)	Chocó region in w Colombia; from sea level in lowland and low altitude rain forest of the Pacific versant. There is a questionable report from the llanos of Villavicencio.
M. steindachneri (Werner, 1901) Steindachner's Coral Snake or Piedmont Coral Snake	Roze (1967, 1983, 1996); Duellman (1979); Campbell & Lamar (1989)	E Ecuadorian Andes and adjacent Peru.
M. s. steindachneri (Werner, 1901) Santiago Piedmont Coral Snake	Roze (1967, 1983, 1996); Duellman (1979); Campbell & Lamar (1989)	S Ecuador in the upper Río Santiago watershed of the eastern slopes of the Andes and adjacent Peru; between 650 to 1100 m in humid and wet low and median montane areas in the valleys of Río Upano and Río Ayambis.
M. s. orcesi Roze, 1967 Pastaza Piedmont Coral Snake	Roze (1967, 1996); Roze *in* Peters & Orejas-Miranda (1970); Campbell & Lamar (1989)	E Andes in Provencia Pastaza and Napo, Ecuador; from 1200 to 2000 m in humid medium-high montane forests on the e slopes of the Andes in the upper Río Pastaza Basin.
M. stewarti Barbour & Amaral, 1928 Panamanian Coral Snake	Schmidt (1933c); Roze (1967, 1970, 1983, 1996); Savage & Vial (1974); Villa et al. (1988); Campbell & Lamar (1989); Köhler (2001)	C Panama around the Canal Zone; between 500 to 1200 m in low to moderate elevation of tropical rain forest and humid forest.
M. stuarti Roze, 1967 Volcano Coral Snake	Roze (1967, 1983, 1996); Villa et al. (1988); Campbell & Lamar (1989); Köhler (2001)	SW Guatemala from San Marcos to at least Suchitepequez; from 800 to 1600 m in humid montane and cloud forest altered by humans.
M. surinamensis (Cuvier, 1817) Aquatic Coral Snake	Schmidt (1936c, 1952); Roze (1967, 1970, 1983, 1996); Cunha & Nascimento (1973, 1978, 1982a); Kempff-Mercado (1975); Duellman (1978); Dixon & Soini (1986); Campbell & Lamar (1989); Roze & Silva (1990); Silva (1994); Starace (1998); Kornacker (1999)	The Guianas, Orinoco and Amazon drainage in s Colombia, Venezuela, n Brazil, e Ecuador, Peru, and n Bolivia; it is found in a wide variety of habitats near rivers and other water sources.
M. s. surinamensis (Cuvier, 1817) Amazonian Aquatic Coral Snake	Schmidt (1952); Roze *in* Peters & Orejas-Miranda (1970); Duellman (1978); Campbell & Lamar (1989); Roze (1996); Starace (1998)	Orinoco and Amazon drainages in s Colombia, e Ecuador, Peru, n Bolivia, n Brazil and the Guianas; sea level to about 500 m.
M. s. nattereri Schmidt, 1952 Venezuelan Aquatic Coral Snake	Schmidt (1952); Roze (1955, 1966, 1970, 1996); Lancini (1979); Lancini & Kornacker (1989); Campbell & Lamar (1989); Starace (1998); Kornacker (1999)	Río Orinoco-Negro Region, Venezuela and into Brazil; 50–250 m.
M. tricolor* (Hoge, 1957) Pantanal Coral Snake	Campbell & Lamar (1989); Roze (1983, 1996); Silva & Sites (1999) *Remark: According to Campbell & Lamar (1989) this is *M. frontalis tricolor*. Roze (1996) listed it as subspecies *M. pyrrhocryptus tricolor*. Silva & Sites (1999) elevated this snake to full species.	Restricted to the s-sw of the state of Mato Grosso do Sul and extending into e Bolivia in Departamento Santa Cruz. It may also have a distribution in Paraguay through the Departamento de Alto Paraguay and possibly Concepción; from 100 to 500 m in a wide variety of savanna and forest habitats.

Species-level Taxa	Citations and Remarks	Abbreviated Range
M. tschudii (Jan, 1858) Desert Coral Snake	Schmidt & Schmidt (1925); Schmidt (1936c); Roze (1967, 1970, 1983, 1996); Carillo de Espinosa (1983); Campbell & Lamar (1989)	Pacific versant of s Ecuador into s Peru, also probably nw Bolivia.
M. t. tschudii (Jan, 1858) Southern Desert Coral Snake	Schmidt & Schmidt (1925); Schmidt (1936c); Roze (1967, 1970, 1983, 1996); Carillo de Espinosa (1983); Campbell & Lamar (1989)	W Peru and probably nw Bolivia; from sea level to 400 m in desert and dry coastal valleys in the Pacific versant.
M. t. olssoni Schmidt & Schmidt, 1925 Northern Desert Coral Snake	Schmidt & Schmidt (1925); Schmidt (1936c); Parker (1938); Roze (1970, 1996); Campbell & Lamar (1989); Almendariz (1991)	Pacific versant of s Ecuador and nw Peru; from sea level to 1500 m in desert and dry valleys in the Pacific versant.

Family HYDROPHIIDAE Boie, 1827

These are commonly referred to as Sea Snakes.

CONTENT: This group is limited to two genera with two species in the Americas.

Subfamily LATICAUDINAE Cope, 1879

GENUS: *Laticauda* Laurenti, 1768, ranges from the Bay of Bengal to southern Japan into the north coast of Australia and through Oceania to the west coast of Nicaragua. They are not reported from the Atlantic Ocean. Campbell & Lamar (1989) have not been able to physically verify the presence of this genus from specimens collected in the Americas. See Donoso-Barros (1966), Peters & Orejas-Miranda (1970); Pickwell (1972); Campbell & Lamar (1989); Smith & Smith (1993).

Review assistance and important contributions for this section contributed by: Janis A. Roze and W. W. Lamar.

See also genus *Pelamis*.

CONTENT: One genus with one species in the Americas. ★★★★

Laticauda: Sea Krait, Flat-tailed Sea Snake, or Banded Sea Snake.

Species	Citations and Remarks	Abbreviated Range
L. colubrina (Schneider, 1799) Yellow-lipped Sea Krait	Peters & Orejas-Miranda (1970); Villa (1972); Pickwell (1972); Alvarez del Toro (1982); Villa et al. (1988); Campbell & Lamar (1989)	Indo-Pacific Oceans and reported by Villa (1962) from Nicaragua in w Central America. Campbell & Lamar (1989) were unable to confirm the specimen. Villa et al. (1988) lists its distribution in Middle America as being se Mexico, El Salvador, and Nicaragua.

Subfamily HYDROPHIINAE Fitzinger, 1843

GENUS: *Pelamis* Daudin, 1803 is widespread throughout the oceans of the world except the Atlantic. In the Americas, it has been reported from the Pacific coast from southern California in the United States, southward to Peru; it is also reported from Hawaii. See Donoso-Barros (1966); Smith & Taylor (1966); Peters & Orejas-Miranda (1970); Pickwell (1972); Stebbins (1985); Savage & Villa (1986); Campbell & Lamar (1989).

Review assistance and important contributions for this section were contributed by Janis A. Roze and W. W. Lamar.

See also genus *Laticauda*.

CONTENT: One monotypic genus. ★★★★

Pelamis: Pelagic Sea Snake

Species	Citations and Remarks	Abbreviated Range
P. platurus (Linnaeus, 1766) Yellow-bellied Sea Snake or Pelagic Sea Snake	Stejneger (1907); Donoso-Barros (1966); Peters & Orejas-Miranda (1970); Kropach (1975); Pickwell & Culotta (1980); Alvarez del Toro (1982); Stebbins (1985); Wilson & Meyer (1985); Savage & Villa (1986); Villa et al. (1988); Campbell & Lamar (1989); Ernst (1992); García & Ceballos (1994); Ramírez-Bautista (1994); Brown (1997); McPeak (2000); Bartlett & Tennant (2000); Köhler (2001)	Widespread in the Pacific and Indian Oceans into the west coast of the Americas from s California in the United States into Mexico and southward to Chile. Campbell & Lamar (1989) noted that *Pelamis* has been reported in the Easter Islands of Chile. Donoso-Barros (1966) reported it from waters of Isla Pascua in Chile. Stebbins (1985) discussed that this snake has been reported off the San Diego area in s California and has also been reported off the coast of the USA to as far north as San Clemente, Orange Co., California. Bartlett & Tennant (2000) list it from Hawaiian Islands.

Family VIPERIDAE Oppel, 1811

Subfamily CROTALINAE Oppel, [1810] 1811

These are commonly referred to as Pit Vipers.

Among other characteristics, the Viperids are differentiated by the presence of highly movable, enlarged, anterior hollow solenoglyphic fangs. The *Atractaspis* from the Old World which are defined as Colubrids, and are sometimes isolated into their own group which have very similar, but distinguishable fangs from the Vipers. The Pit Vipers are the subfamily Crotalinae and identified by the presence of paired infrared-sensing facial pits from which their name is derrived. They are found throughout the Americas, eastern Europe, and Asia.

Most Vipers and Pit Vipers have broad, triangular heads with bodies that tend to be heavy and accompanied by short tails. They have long solenoglyphic, or movable fangs that operate via a hinge mechanism which allows the fangs to be folded and stored at the roof of the mouth. Pit Vipers are differentiated from the general Viper group normally by the presence of a heat sensing pit to each side of the head positioned just below a line between the nose and eye.

CONTENT: Twelve genera with 101 species in the Americas.

GENUS: *Agkistrodon* Palisot de Beauvois, 1799, is the genus commonly referred to as the Moccasins. The genus is found in North and Central America and Asia. Gloyd & Conant (1990) completely reviews this genus in detail. See Smith & Taylor (1966); Gloyd *in* Peters & Orejas-Miranda (1970); Hoge & Romano-Hoge (1978/79); Campbell & Lamar (1989); Gloyd & Conant (1990); Smith & Smith (1993); McDiarmid et al. (1999); Dixon (2000a); Parkinson et al. (2000) and Köhler (2001).

CONTENT: An American and Asian genus with four species in the the Americas. ★★★★

Agkistrodon: Moccasin, American Copperhead and Cantil.

Species	Citations and Remarks	Abbreviated Range
A. bilineatus Günther, 1863 Cantil or Mexican Moccasin	Campbell & Lamar (1989); Gloyd & Conant (1990); Lee (1996, 2000); Peters & Orejas-Miranda (1970); Villa et al. (1988); García & Ceballos (1994); Campbell (1998); McDiarmid et al. (1999); Parkinson et al. (2000); Köhler (2001)	Mexico along both coasts southward, along the Pacific coast from s Sonora and on the Atlantic coast from n Tamaulipas; and se Nuevo León into the Yucatán peninsula and n Belize, into Central America as far south as Costa Rica.
A. b. bilineatus Günther, 1863 Common Cantil or Mexican Moccasin	Hoge & Romano-Hoge (1978/79); Campbell & Lamar (1989); Gloyd & Conant (1990); Peters & Orejas-Miranda (1970); Alvarez del Toro (1982); Ramírez-Bautista (1994); McDiarmid et al. (1999); Parkinson et al. (2000); Köhler (2001)	Pacific versant of Mexico, Guatemala, El Salvador and in the Rio Grijalva Valley of Chiapas, Mexico; from sea level to 1500 m in Michoacán.
A. b. howardgloydi Conant, 1984 Castellana or Gloyd's Cantil	Wilson & Meyer (1985); Campbell & Lamar (1989); Gloyd & Conant (1990); Smith & Smith (1993); McDiarmid et al. (1999); Cruz (1997); Parkinson et al. (2000); Köhler (2001)	Pacific coasts of Honduras, Nicaragua and Costa Rica; from sea level to 600 m in dry Pacific lowlands.
A. b. russeolus Gloyd, 1972 Yucatecan Cantil	Campbell & Lamar (1989); Gloyd & Conant (1990); Campbell (1998); McDiarmid et al. (1999); Stafford & Meyer (2000); Lee (2000); Parkinson et al. (2000); Köhler (2001)	Yucatán Peninsula in the states of Campeche, Quintana Roo, and Yucatán in Mexico and into n Belize.
A. contortrix (Linnaeus, 1766) American Copperhead	Conant (1975); Dundee & Rossman (1989); Gloyd & Conant (1990); Smith & Smith (1993); McDiarmid et al. (1999); Dixon (2000a); Parkinson et al. (2000)	E United States into very extreme n Mexico.
A. c. contortrix (Linnaeus, 1766) Southern Copperhead	Conant (1975); Hoge & Romano-Hoge (1978/79); Ashton & Ashton (1981); Dundee & Rossman (1989); Gloyd & Conant (1990); Ernst (1992); Smith & Smith (1993); Tennant (1997b, 1998); Conant & Collins (1998); McDiarmid et al. (1999); Dixon (2000a); Werler & Dixon (2000); Tennant & Bartlett (2000)	The United States from se Virginia southward ranging barely into Florida westward into e Texas northward into Oklahoma; it is associated with the Gulf States, the lower Mississippi Valley, and the Coastal Plains of the South Atlantic states.
A. c. laticinctus Gloyd & Conant, 1934 Broad-banded Copperhead	Conant (1975); Hoge & Romano-Hoge (1978/79); Gloyd & Conant (1990); Ernst (1992); Tennant (1998); Conant & Collins (1998); McDiarmid et al. (1999); Dixon (2000a); Werler & Dixon (2000); Tennant & Bartlett (2000)	The United States from southern Kansas southward through the middle of Oklahoma into c and sc Texas.
A. c. mokasen* Palisot de Beauvois, 1799 Northern Copperhead	Conant (1975); Hoge & Romano-Hoge (1978/79); Gloyd & Conant (1990); Ernst (1992); Conant & Collins (1998); McDiarmid et al. (1999); Tennant & Bartlett (2000)	The United States from Massachusetts into e Kansas southward and eastward into Alabama (except in the Marias des Cygnes-Osage River drainage).

Species-level Taxa	Citations and Remarks	Abbreviated Range
	*Remark: In older literature the subspecies name can be found spelled as "*mokeson*" (see Conant, 1958, and Hoge & Romano-Hoge, 1978/79).	
A. c. phaeogaster Gloyd, 1969 Osage Copperhead	Conant (1975); Hoge & Romano-Hoge (1978/79); Gloyd & Conant (1990); Ernst (1992); Conant & Collins (1998); McDiarmid et al. (1999); Tennant & Bartlett (2000)	The United States in e Kansas, extreme se Nebraska, and c Missouri, in and around the Marias des Cygnes-Osage River drainage area.
A. c. pictigaster Gloyd & Conant, 1943 Trans-Pecos Copperhead	Smith & Taylor (1966); Conant (1975); Hoge & Romano-Hoge (1978/79); Campbell & Lamar (1989); Gloyd & Conant (1990); Ernst (1992); Smith & Smith (1993); Tennant (1998); Conant & Collins (1998); McDiarmid et al. (1999); Dixon (2000a); Werler & Dixon (2000); Tennant & Bartlett (2000); Parkinson et al. (2000)	The United States and Mexico, from the Davis Mountains and the Big Bend Region in Texas southward into n Mexico in n Chihuahua and n Coahuila.
A. piscivorus (Lacépède, 1789) Cottonmouth	Conant (1975); Dundee & Rossman (1989); Gloyd & Conant (1990); McDiarmid et al. (1999); Dixon (2000a); Parkinson et al. (2000)	C, se and s United States from s Virginia southward through peninsular Florida and west into the midwestern United States into Texas, and some records from along the Rio Grande River.
A. p. piscivorus (Lacépède, 1789) Eastern Cottonmouth	Conant (1975), Hoge & Romano-Hoge (1978/79); Gloyd & Conant (1990); Ernst (1992); Tennant (1997b); Conant & Collins (1998); McDiarmid et al. (1999); Tennant & Bartlett (2000)	The United States in se Virginia southward through the Coastal Plains and lower Piedmont of the Carolinas (including banks, islands, and narrow peninsulas along the Atlantic Coast), westward across Georgia, and intergrading with other subspecies in ne Mississippi and the Florida Panhandle.
A. p. conanti Gloyd, 1969 Florida Cottonmouth	Conant (1975); Hoge & Romano-Hoge (1978/79); Ashton & Ashton (1981); Gloyd & Conant (1990); Ernst (1992); Tennant (1997b); Conant & Collins (1998); McDiarmid et al. (1999); Tennant & Bartlett (2000)	The United States in se Alabama, s Georgia, and s South Carolina southward through Florida into the Keys.
A. p. leucostoma (Troost, 1836) Western Cottonmouth	Conant (1975); Hoge & Romano-Hoge (1978/79); Dundee & Rossman (1989); Gloyd & Conant (1990); Ernst (1992); Tennant (1998); Conant & Collins (1998); McDiarmid et al. (1999); Dixon (2000a); Werler & Dixon (2000); Tennant & Bartlett (2000)	The United States from extreme s Illinois southward to Alabama, and across s Missouri, southward and southwestward across e Oklahoma into Louisiana, Mississippi, and the e half of Texas excluding s Texas.
A. taylori Burger & Robertson, 1951 Metapil or Taylor's Cantil	P. W. Smith & Darling (1952); Hoge & Romano-Hoge (1978/79); Campbell & Lamar (1989); Gloyd & Conant (1990); McDiarmid et al. (1999); Parkinson et al. (2000)	Restricted to Mexico in the states of Tamaulipas, Nuevo León, and San Luis Potosí in ne Mexico; it probably occurs in e Coahuila and extreme e Zacatecas.
	*Remark: Until recently this snake was classified as the subspecies *A. bilineatus taylori*. It was elevated to full species status by Parkinson et al. (2000) based on a re-evaluation of the comprehensive morphological and distributional analysis provided by Gloyd & Conant (1990) as well as mtDNA sequence comparisons.	

GENUS: *Atropoides* Werman, 1992, is found in Middle America from the mountains of e Mexico southward thorugh Central America into Panama. See Hoge & Romano-Hoge (1978/79) [as *Bothrops*]; Campbell & Lamar (1989, 1992); Werner (1992); Smith & Smith (1993); McDiarmid et al. (1999) and Köhler (2001). Smith & Smith (1993) show *Atropoides* as a *nomen nudum*. This listing follows Campbell & Lamar (1992).

See also *Bothriechis*, *Bothriopsis*, *Bothrocophias*, *Bothrops*, *Cerrophidion*, *Ophryacus*, and *Porthidium*.

CONTENT: Three species. ★★★

Atropoides: Jumping Pit Viper or Mano de Metate

Species	Citations and Remarks	Abbreviated Range
A. nummifer (Rüppel, 1845) Jumping Pit Viper	Peters & Orejas-Miranda (1970); Hoge & Romano-Hoge (1978/79); Wilson & Meyer (1985); Savage & Villa (1986); Villa et al. (1988); Campbell & Lamar (1989, 1992); Lee (1996, 2000); Cruz (1997); Campbell (1998); McDiarmid et al. (1999); Stafford & Meyer (2000); Köhler (2001)	S Mexico southward through Central America into Panama; from about 400 to 1600 m.
A. n. nummifer (Rüppell, 1845) Common Jumping Pit Viper	Peters & Orejas-Miranda (1970); Hoge & Romano-Hoge (1978/79); Campbell & Lamar (1989, 1992); Pérez-Higareda & Smith (1991); McDiarmid et al. (1999)	S Mexico in the Atlantic versant from San Luis Potosí southward to c Oaxaca.
A. n. mexicanus (Dumeril, Bibron & Dumeril, 1854) Mexican Jumping Pit Viper	Peters & Orejas-Miranda (1970); Hoge & Romano-Hoge (1978/79); Alvarez del Toro (1982); Campbell & Lamar (1989, 1992); Lee (1996, 2000); McDiarmid et al. (1999)	In the Caribbean drainage of Chiapas in s Mexico southward into Panama.
A. n. occiduus (Hoge, 1966) Western Jumping Pit Viper	Peters & Orejas-Miranda (1970); Hoge & Romano-Hoge (1978/79); Campbell & Lamar (1989, 1992); McDiarmid et al. (1999)	Pacific versant of Chiapas, Mexico, Guatemala, and El Salvador; and the Pacific and Atlantic versant in Oaxaca.
A. olmec (Pérez-Higareda, Smith & Julía-Zertuche, 1985) Olmecan Pit Viper	Campbell & Lamar (1989, 1992); Pérez-Higareda et al. (1985); Pérez-Higareda & Smith (1991); McDiarmid et al. (1999)	S Veracruz, Mexico on the low to moderate slopes of the Sierra de Los Tuxtlas; usually found between 500 to 1100 m, but may be found to over 1500 m; associated with lower montane wet forest and cloud forest.
A. picadoi (Dunn, 1939) Picado's Pit Viper	Taylor (1951); Hoge & Romano-Hoge (1978/79); Martínez (1983); Savage & Villa (1986); Campbell & Lamar (1989, 1992); Villa et al. (1988); Werman (1992); McDiarmid et al. (1999); Köhler (2001)	Mountains of Costa Rica and Panama, including the Cordillera de Tilarán, Cordillera Central, and the Cordillera de Talamance from 50 to about 1500 m in subtropical moist and wet forest, tropical moist and wet forest, and montane wet forest.

GENUS: *Bothriechis* W. Peters, 1859, occurs from southern Mexico into South America to as far south as Peru. See Stuart (1963); Peters & Orejas-Miranda (1970); Hoge & Romano-Hoge (1978/79) [as *Bothrops*]; Wilson & Meyer (1982, 1985); Campbell & Lamar (1989, 1992); Crother et al. (1992); Smith & Smith (1993); Solórzano et al. (1998); McDiarmid et al. (1999); Kornacker (1999) and Köhler (2001).

See also *Atropoides*, *Bothrops*, *Bothriopsis*, *Bothrocophias*, *Cerrophidion*, *Ophryacus*, and *Porthidium*.

CONTENT: Eight species. ★★★

Bothriechis: Palm Pit Viper

Species	Citations and Remarks	Abbreviated Range
B. aurifer (Salvin, 1860) Yellow-blotched Palm Pit Viper	Alvarez del Toro (1982); Campbell & Lamar (1989); McDiarmid et al. (1999); Köhler (2001)	Disjunct distribution in the highlands of Chiapas, Mexico southward into n Guatemala in cloud forests at 1200 to 2300 m and has been reported lower; it seems to prefer cloud forest but enters subtropical wet forest (pine-oak forest) along mesic ravines.
B. bicolor (Bocourt, 1868) Guatemalan Palm Pit Viper	Hoge & Romano-Hoge (1978/79); Alvarez del Toro (1982); Wilson & Meyer (1985); Villa et al. (1988); Campbell & Lamar (1989); Cruz (1997); McDiarmid et al. (1999); Köhler (2001)	Mountains of s Chiapas, Mexico southward into c Guatemala and through the Pacific versant of the s Volcanic Cordillera into nw Honduras; associated with rain and cloud forests from 500 to 2000 m.
B. lateralis W. Peters, 1862 Side-striped Palm Pit Viper	Hoge & Romano-Hoge (1978/79); Savage & Villa (1986); Villa et al. (1988); Campbell & Lamar (1989); McDiarmid et al. (1999); Köhler (2001)	Mountains of Costa Rica and w Panama in the Cordillera de Tilarán, the Cordillera Central, and the Cordillera de Talamance, into the provinces of Chiriquí and Veraguas; from 850 to 1980 m in lower montane moist and wet forest, and rain forest.
B. marchi (Barbour & Loveridge, 1929) March's Palm Pit Viper	Barbour & Loveridge (1929); Hoge & Romano-Hoge (1978/79); Wilson & Meyer (1985); Campbell & Lamar (1989); Wilson & McCranie (1992); McDiarmid et al. (1999); Köhler (2001)	NC and nw Honduras and e Guatemala on the Atlantic slopes in forests typically between 500 to 1500 m; in rain forest, lower montane wet forest, and cloud forest.
B. nigroviridis W. Peters, 1859 Black-speckled Palm Pit Viper	Barbour & Loveridge (1929); Hoge & Romano-Hoge (1978/79); Alvarez del Toro (1982); Savage & Villa (1986); Villa et al. (1988); Campbell & Lamar (1989); McDiarmid et al. (1999); Köhler (2001)	Mountains of Costa Rica and Panama in the Cordillera Central and Cordillera de Talamance in cloud forests from 1150 to 2400 m; found in lower montane wet forest, cloud forest, and high montane forest.
B. rowleyi (Bogert, 1968) Rowley's Palm Pit Viper	Bogert (1968); Hoge & Romano-Hoge (1978/79); Campbell & Lamar (1989); McDiarmid et al. (1999); Köhler (2001)	Rare in montane se Oaxaca and nw Chiapas, Mexico in cloud and pine-oak forests between 1500 to 1830 m.
B. schlegeli (Berthold, [1845] 1846) Eyelash Palm Pit Viper	Taylor (1951); Peters & Orejas-Miranda (1970); Hoge & Romano-Hoge (1978/79); Wilson & Meyer (1985); Alvarez del Toro (1982); Wilson & Meyer (1985); Savage & Villa (1986); Villa et al. (1988); Lancini & Kornacker (1989); Campbell & Lamar (1989); Crother et al. (1992); Lee (1996, 2000); Campbell (1998); Solórzano et al. (1998); McDiarmid et al. (1999); Renjifo & Lundberg (1999); Stafford & Meyer (2000); Kornacker (1999); Köhler (2001)	N Chiapas in s Mexico southward through Central Ameica n South America in Colombia, w Venezuela, Ecuador, and n Peru in lowlands and mesic forests from about sea level to 2640 m; it is not supposed to overlap the range of *B. superciliaris* at higher elevations in sw Costa Rica, but they are found in the lower elevations; in tropical moist forest, wet subtropical forest (cloud forest), and montane wet forest.
B. supraciliaris Taylor, 1954 Costa Rican Palm Pit Viper	Taylor (1954); Stuart (1963); Peters & Orejas-Miranda (1970); Hoge & Romano-Hoge (1978/79); Wilson & Meyer (1982, 1985); Solórzano et al. (1998); Köhler (2001)	SW Costa Rica, thought to be confined to the mid and higher elevations (800–1700 m) in the Valle del General and the Coto Brus altiplano.

GENUS: *Bothriopsis* W. Peters, 1861, is from the American neotropics in e Panama southward into the Amazon Basin in South America. See Hoge & Romano-Hoge (1978/79) [as *Bothrops*]; Campbell & Lamar (1989, 1992); Smith & Smith (1993); McDiarmid et al. (1999) and Köhler (2001).

See also *Atropoides*, *Bothriechis*, *Bothrocophias*, *Bothrops*, *Cerrophidion*, *Ophryacus*, and *Porthidium*.

CONTENT: Seven species. ★★★

Bothriopsis: Forest Pit Viper.

Species	Citations and Remarks	Abbreviated Range
B. bilineata* (Wied-Neuwied, 1821) Amazonian or Two-striped Forest Pit Viper	Peters & Orejas-Miranda (1970); Moonen et al. (1979); Campbell & Lamar (1989); Smith & Smith (1993); Starace (1998); McDiarmid et al. (1999); Kornacker (1999) *Remark: Although listed in their synonymy, subspecies were not considered valid in the work of McDiarmid et al. (1999).	Amazonian forests and the Atlantic forests of Brazil. Distribution is generally thought to be the Guianas, Brazil, s and se Venezuela, s Colombia, e Ecuador, n Bolivia, and e and ne Peru; an isolated population has been reported from se Brazil. It is most abundant in lowland rain forest, often in close association with waterways to 1000 m.
B. b. bilineata (Wied-Neuwied, 1821) Eastern Striped Forest Pit Viper	Peters & Orejas-Miranda (1970); Hoge & Romano-Hoge (1978/79); Campbell & Lamar (1989); Starace (1998); McDiarmid et al. (1999); Kornacker (1999); Freitas (1999); Marques et al. (2001) [as *Bothrops bilineatus*]	Amazon region of Brazil, Venezuela, and the Guianas; it is associated with equatorial forests to the e coast.
B. b. smaragdinae Hoge, 1966 Western Striped Forest Pit Viper	Peters & Orejas-Miranda (1970); Duellman (1978); Hoge & Romano-Hoge (1978/79); Dixon & Soini (1986); Lancini & Kornacker (1989); Campbell & Lamar (1989); McDiarmid et al. (1999); Kornacker (1999)	Amazon regions of Venezuela, Colombia, Bolivia, Peru, Ecuador, and Brazil.
B. medusa (Sternfeld, 1920) Venezuelan Forest Pit Viper or Tigra Mariposa	Peters & Orejas-Miranda (1970); Hoge & Romano-Hoge (1978/79); Lancini & Kornacker (1989); Campbell & Lamar (1989); McDiarmid et al. (1999); Kornacker (1999) [as *Bothrops medusa*]	Venezuela in the states of Aragua, Carabobo, and Bolivar, and the Distrito Federal; it is endemic to the central range of the Cordillera de la Costa from Caracas to Valencia; 1400 to over 2000 m with an unverified report from 2800 m; lower montane wet forest and cloud forest including areas of considerable cold.
B. oligolepis (Werner, 1901) Inca Forest Pit Viper	Peters & Orejas-Miranda (1970); Hoge & Romano-Hoge (1978/79); Campbell & Lamar (1989); McDiarmid et al. (1999)	E slopes of the Andes in Peru and Bolivia; presumed altitudinal distribution is 1000 to over 2000 m in cloud forest.
B. peruviana (Boulenger, 1903) Peruvian Forest Pit Viper	Peters & Orejas-Miranda (1970); Hoge & Romano-Hoge (1978/79); Campbell & Lamar (1989, 1992); McDiarmid et al. (1999)	SE Peru east of the Andes in the Cordillera de Carabaya in n Puno; it very likely occurs in the Cerros de Bala region of Dept. de La Paz, Bolivia. This is a montane form whose vertical limits have not been established; presumably it lives in wet montane forest.
B. pulchra* (W. Peters, 1862) Dusky Forest Pit Viper	Peters & Orejas-Miranda (1970); Hoge & Romano-Hoge (1978/79); Campbell & Lamar (1989, 1992); McDiarmid et al. (1999); ICZN Opinion 1039 (1999); Gutberlet & Campbell (2001)	E slopes of the Andeas from sc Colombia into s Ecuador in equitorial forests in Amazonian lowlands.

The Caenophidia

Species-level Taxa	Citations and Remarks	Abbreviated Range
	*Remark: McDiarmid et al. (1999) listed *B. albocarinata* Shreve, 1934 found from sc Colombia southward into Ecuador; and, *B. alticola* (Parker, 1934) found in Ecuador, to be synonyms of *B. pulchra*.	
*B. punctata** (García, 1896) Chocoan Forest Pit Viper	Peters & Orejas-Miranda (1970); Hoge & Romano-Hoge (1978/79); Villa et al. (1988); Campbell & Lamar (1992); McDiarmid et al. (1999); Köhler (2001) *Remark: According to Campbell & Lamar (1992), both *B. mahnerti* Schätti & Kramer, 199; and, *B. osbornei* Freire-Lascano, 1991, are synonyms.	Darién in Panamá in Central America southward along the Pacific coast to near the Ecuador-Peru border and extreme n Peru; to 2300 m in a wide variety of moist and wet forest.
*B. taeniata** Wagler *in* Spix, 1824 Castelnaud's Forest Pit Viper	Peters & Orejas-Miranda (1970) [as *B. castlenaudi*]; Duellman (1978) [as *B. castlenaudi*]; Cunha & Nascimento (1978); Campbell & Lamar (1989); McDiarmid et al. (1999); Starace (1998); Kornacker (1999) [as *Bothrops taeniata*] *Remark: Although listed in their synonymy, McDiarmid et al. (1999) did not comment on subspecies as listed below.	N South America in equitorial forests of Colombia, Peru, Venezuela, Brazil, Ecuador, French Guiana, Bolivia, and Guyana; from sea level to 1900 m in lowland and foothill rain forest and moist tropical forest.
*B. t. taeniata** Wagler *in* Spix, 1824 Speckled Forest Pit Viper	Peters & Orejas-Miranda (1970) [as *B. c. castlenaudi*]; Amaral (1976); [as *Bothrops castelnaudi*]; Hoge & Romano-Hoge (1978/1979) [as *Bothrops castelnaudi castelnaudi*]; Lancini & Kornacker (1989); Campbell & Lamar (1989); Kornacker (1999) *Remark: For some undetermined reason, Williams & Francini (1991) placed "*Bothrops taeniatus* Wagler" in synonymy with *Bothrops jararaca*. There was no explanation, and this reclassification is not accurate.	Widespread in equatorial forests South America of Brazil, Ecuador, Peru, Bolivia, Colombia, and Venezuela as for above species distribution.
B. t. lichenosa (Roze, 1958) Lichen-like Forest Pit Viper	Peters & Orejas-Miranda (1970) [as *B. c. lichenosus*]; Hoge & Romano-Hoge (1978/1979) [as *Bothrops castelnaudi lichenosus*]; Lancini (1979); Campbell & Lamar (1989), Lancini & Kornacker (1989); Kornacker (1999)	Venezuela; known from a single specimen that differs from the nominate race by having a lower number of ventral and subcaudal scutes. The type locality is Chimantá Tepui, Bolivar State, Venezuela.

GENUS: *Bothrocophias* Gutberlet & Campbell, 2001, ranges in mainland northwestern South America from Colombia southward into Bolivia and Brazil. See Hoge & Romano-Hoge (1978/79); Campbell & Lamar (1992); McDiarmid et al. (1999); and Gutberlet & Campbell (2001).

Review assistance and important contributions for this section contributed by Jonathan A. Campbell.

See also *Atropoides*, *Bothriechis*, *Bothriopsis*, *Bothrops*, *Cerrophidion*, *Ophryacus*, and *Porthidium*.

CONTENT: A distinctive clad of four species of South American pit vipers. ★★★★

Bothrocophias: Toadheaded Pit Viper

Species	Citations and Remarks	Abbreviated Range
B. campbelli (Freire-Lascano, 1991) Ecuadorian Toadheaded Pit Viper	Freire-Lascano (1991); McDiarmid et al. (1999); ICZN Opinion 1039 (1999); Gutberlet & Campbell (2001)	Pacific highland slopes of the Andes in Ecuador; found between 1300 to 2000 m.
B. hyoprora (Amaral, 1935) Amazonian Toadheaded Pit Viper	Peters & Orejas-Miranda (1970); Amaral (1976); Hoge & Romano-Hoge (1978/79); Dixon & Soini (1986); Campbell & Lamar (1989); McDiarmid et al. (1999); Gutberlet & Campbell (2001)	NW and c South America in e Ecuador, s Colombia, nw Brazil, ne Peru, and ne Bolivia; near sea level to at least 1000 m in equatorial forest of the Amazon Basin.
B. microphthalmus (Cope, 1875 [1876]) Small-eyed Toadheaded Pit Viper	Peters & Orejas-Miranda (1970); Hoge & Romano-Hoge (1978/79); Campbell & Lamar (1992); McDiarmid et al. (1999); Gutberlet & Campbell (2001)	Eastern versant of the Andes in Colombia, Ecuador, Peru, and possibly n Bolivia; an upland species found 1000 to 2350 m.
B. myersi Gutberlet & Campbell, 2001 Chocoan Toadheaded Pit Viper	Pérez-Santos & Moreno (1988) [*B. Pulcher* {in part}]; Schätti & Kramer (1993) [as *Porthidium almawebi* {in part}]; Golay et al. (1993) [*Porthidium almawebi* {in part}]; Gutberlet & Campbell (2001)	Pacific lowlands of Cuaca and Valle del Cauca, Colombia; Gutberlet & Campbell (2001) noted that it is possible that *B. myersi* occurs (or used to occur) throughout the Chocó region, but they knew of no records for Ecuador.

GENUS: *Bothrops* Wagler, 1824, is a complex group of venomous (often deadly) snakes found from Mexico in the north ranging southward through Central America and much of South America. It is found in the Antilles to as far north as Martinique. The next few pages note the most currently accepted species in *Bothrops*. Older literature referers to this genus as *Trimeresurus*. Some very closely related snakes, which were originally classified under *Bothrops*, recently have been partitioned into other genera including *Atropoides*, *Bothriechis*, *Bothriopsis*, *Bothrocophias*, *Cerrophidion*, *Ophryacus*, and *Porthidium*. See Smith & Taylor (1966); Peters & Orejas-Miranda (1970); Hoge & Romano (1981); Campbell & Lamar (1989, 1992); Grantsau (1990); Cei (1986, 1993); Smith & Smith (1993); Starace (1998); McDiarmid et al. (1999); Kornacker (1999) and Köhler (2001). Also see the *Incertae Sedis* list on page 55 of Peters & Orejas-Miranda (1970).

The *Bothrops*, and closely associated genera, are presently going through review by several authorities and the reader should expect major changes in this and the other related group in the next few years.

See also *Atropoides*, *Bothriechis*, *Bothriopsis*, *Bothrocophias*, *Cerrophidion*, *Ophryacus*, and *Porthidium*. Peters & Orejas-Miranda (1970); is the first detailed review of this genus, and then by Hoge & Romano-Hoge (1978/79), and those genera, at the time also known as *Bothrops*, closely associated with it.

CONTENT: A complex group of snakes consisting of about thirty species. Twenty-nine species are listed below. ★★

Bothrops: American Lancehead Pit Viper or Neotropical Pit Viper

Species	Citations and Remarks	Abbreviated Range
B. alternatus Dumeril, Bibron & Dumeril, 1854	Peters & Orejas-Miranda (1970); Amaral (1976); Hoge & Romano-Hoge (1978/79); Achaval (1987); Campbell & Lamar (1989); Lema (1994a); Cei (1986,	SE Brazil, Paraguay, Uruguay, and n Argentina; excluding the Amazon rainforest; from near sea level to about 700 m in a variety of habitats.

Species-level Taxa	Citations and Remarks	Abbreviated Range
Urutú or Crossed Lancehead Pit Viper	1993); McDiarmid et al. (1999); Leynaud & Bucher (1999)	
B. ammodytoides Leybold, 1873 Patagonian Lancehead Pit Viper	Peters & Orejas-Miranda (1970); Hoge & Romano-Hoge (1978/79); Campbell & Lamar (1989); Cei (1986, 1993); Scrocchi (1997); McDiarmid et al. (1999)	Argentina from Tucamán to Chubut and Patagonia; it is thought to be the southernmost ranging snake in the world found to about 47°S latitude; from sea level to at least 2000 m in a wide variety of habitats.
B. andianus Amaral, 1923 Andean Lancehead Pit Viper	Amaral (1923); Peters & Orejas-Miranda (1970); Hoge & Romano-Hoge (1978/79); Campbell & Lamar (1989); McDiarmid et al. (1999)	Known only from the Departamentos Cuzco and Puno in s Peru; the type is from Manchu Picchu between 8000 and 10,000 ft; it is found between 1800 to about 3300 m in montane and lower montane wet forest and is also found in the Peruvian high jungle.
B. asper* (Garman, 1884) Fer-de-Lance or Barba Amarilla	Peters & Orejas-Miranda (1970); Hoge & Romano-Hoge (1978/79); Alvarez del Toro (1982); Wilson & Meyer (1985) [as *Bothrops atrox*]; Campbell & Lamar (1989, 1992); Schätti & Kramer (1993); Lee (1996, 2000); Cruz (1997); Murphy (1997); Campbell (1998b); McDiarmid et al. (1999); & Lundberg (1999); Stafford & Meyer (2000); Boos (2001) [as *B. atrox*]; Köhler (2001)	Mexico in Tamaulipas and extreme sc San Luis Potosí southward into n South America east of the Andes into e Colombia, w Ecuador, Trinidad, and possibly Venezuela; from sea level to 2640 m in a variety of habitats.
	*Remark 1: Sometimes noted as *B. atrox asper* or *B. andinus asper*. Listed in Smith & Taylor (1966), as *B. atrox asper* (Garman) at the suggestion of Dunn.	
	*Remark 2: *Trigonocephalus xanthogrammus* (Cope, 1868), (see Campbell & Lamar, 1992). McDiarmid et al. (1999), stated that a petition is being prepared for submisison to the International Commission on Zoological Nomenclature to conserve the name *asper* over *xanthogrammus* and they continued to use the name *B. asper* for this snake.	
	*Remark 3: According to Campbell & Lamar (1989), *B. isabelae* is either *B. atrox asper,* or more likely, *B. atrox*. However, Kornacker (1999), still lists *B. isabelae* as a valid name.	
	*Remark 4: Lee (2000) stated "Taxonomic status of the species and its close relative *B. atrox* of South America is uncertain and confused. Some authorities treat the two as subspecies of a single species, with snakes in Mexico, Central America, and portions of South America considered a subspecies, *Bothrops atrox asper*. Others treat the two forms as specifically distinct, with no subspecies of *B. asper* recognized."	
B. atrox* (Linnaeus, 1758) Common Lancehead Pit Viper	Peters & Orejas-Miranda (1970); Hoge & Romano-Hoge (1978/79); Amaral (1976); Moonen et al. (1979); Dixon & Soini (1986); Campbell & Lamar (1989); Smith & Smith (1993); Schätti & Kramer (1993);	Regarded as restricted to the Amazon lowlands of n and c South America east of the Andes; range is generally thought to be the Guianas, Venezuela, Brazil, Colombia, e Ecuador, ne and e Peru, and n Bolivia.

Species-level Taxa	Citations and Remarks	Abbreviated Range
	Markezich & Taphorn (1993); Starace (1998); McDiarmid et al. (1999); Kornacker (1999)	Many authors consider *B. atrox* to be found in all countries in South America exclusive of Paraguay, Argentina, Uruguay, and Chile; from sea level to over 1500 m, with a questionable record of 2500 m in Colombia. According to Murphy (1997) it is not found in Trinidad. Boos (2001) listed it for Trinidad but the photos and description seems to be *B. asper*. Probably not on Trinidad.
	*Remark 1: Some authorities believe *B. atrox* actually encompasses a number of other nominal species including *B. colombiensis*, *B. isabelae*, *B. lecurus*, *B. marajoensis*, *B. moojenis* and *B. pradoi* (see Campbell & Lamar, 1992).	
	*Remark 2: Johnson & Dixon (1984), and Campbell & Lamar (1989), considered *Bothrops colombiensis* from northern Venezuela, to be a synonym of *B. atrox*. While Lancini & Kornacker (1989) show it to be *B. atrox atrox*.	
	*Remark 3: Subspecies of *B. atrox* most likely are not valid. When subspecies are shown, they are usually noted as *B. a. atrox* (Linnaeus, 1758), found in most of *B. atrox* range, and, *B. a. aidae* Sandner Montilla, 1981, said to be found in Venezuela, Colombia and some Caribbean islands (see Lancini & Kornacker, 1989).	
	*Remark 4: See Lee (2000) quoation above in *B. asper*.	
	*Remark 5: Three subspecies have been listed and are those that McDiarmid et al. (1999), list. Golay et. al. (1993), recognized two subspecies as *B. a. atrox* and *B. a. xanthogrammus* (Cope, 1868). See Remark 2 in *B. asper*.	
B. a. atrox * (Linnaeus, 1758) Common Lancehead Pit Viper	Vanzolini (1986); Lancini & Kornacker (1989); McDiarmid et al. (1999)	Most of *B. atrox* range.
	*Remark 1: McDiarmid et al. (1999) stated this seems to be valid but need further confirmation pending examination of the type and other specimens from the same general vicinity.	
	*Remark 2: Fide McDiarmid et al. (1999), *B. isabelae* from the Venezuela is a confirmed synonym of *B. atrox* and was placed in this subspecies by Golay et al. (1993).	
B. a. aidae * Sandner-Montilla, 1981 Sandner-Montilla's Common Lancehead Pit Viper	Vanzolini (1986); Lancini & Kornacker (1989); McDiarmid et al. (1999)	Venezuela, Colombia and some Caribbean islands; the type locality is noted as Agua Blanca in the jungles of Guatopo, Miranda in Venezuela at 490 m.
	*Remark 1: Included by McDiarmid et al. (1999) pending examination of the type and other other specimens from the same general vicinity.	
	*Remark 2: This taxon is listed as *Bothrops lanceolatus aidae* by Sandner-Montilla (1981).	

The Caenophidia

Species-level Taxa	Citations and Remarks	Abbreviated Range
B. a. nacaritae* Sandner-Montilla, 1990 Carabobo Common Lancehead Pit Viper	McDiarmid et al. (1999) *Remark 1: McDiarmid et al. (1999) included this subspecies, pending examination of the type and other specimens from the general vicinity. *Remark 2: Listed as *B. lanceolatus nacaritae* by Sandner Montilla (1990).	Venezuela, the type locality is described as Caria-prima to the north of Valencia at 900 m in woodlands; in the state of Carabobo.
B. barnetti Parker, 1938 Barnett's Lancehead Pit Viper	Parker (1938); Peters & Orejas-Miranda (1970); Hoge & Romano-Hoge (1978/79); Campbell & Lamar (1989); McDiarmid et al. (1999)	Northern Peru in two localities on the Pacific coast between Lobitos and Talara in Piura and the Rio Moche in La Libertad; associated with arid desert scrub regions.
B. brazili Hoge, 1954 Brazil's Lancehead Pit Viper	Hoge (1954); Peters & Orejas-Miranda (1970); Hoge & Romano-Hoge (1978/79); Dixon & Soini (1986); Lancini & Kornacker (1989); Campbell & Lamar (1989); Starace (1998); McDiarmid et al. (1999); Kornacker (1999)	Amazon Basin in n and e South America including the Guianas, s and se Venezuela, s Colombia, e Ecuador, ne and e Peru, n Bolivia, and in Brazil in Pará, Amazonas, Rondonia, and n Mato Grosso; vertical distribution is unknown but it has been collected at 460 m in Suriname. It seems to be restricted to humid leaf litter habitats and has been recorded only from elevated primary forest in low equitorial South America in the Amazon Basin.
B. caribbaeus* (Garman, 1887) St. Lucia Lancehead Pit Viper	Lazell (1964); Hoge & Romano-Hoge (1978/79); Gosner (1987); Schwartz & Henderson (1985, 1991); Campbell & Lamar (1989); Powell & Wittenberg (1998); McDiarmid et al. (1999); Censky & Kaiser *in* Crother (1999) *Remark: See the comment in McDiarmid et al. (1999), about the nomenclature on this snake.	Island of St. Lucia (WI) in the Lesser Antilles and is restricted to the lowlands to a maximum altitude of about 2000 m.
B. colombianus* Rendahl & Vestergren, 1940 Colombian Lancehead Pit Viper	Hoge & Romano-Hoge (1978/79); Campbell & Lamar (1992); McDiarmid et al. (1999) *Remark: This snake was recognized by Campbell & Lamar (1989) as a subspecies *B. microphthalmus*, and later was elevated to full species by Campbell & Lamar (1992).	W Colombia along the Pacific versant.
B. cotiara (Gomes, 1913) Cotiara	Peters & Orejas-Miranda (1970); Amaral (1976); Hoge & Romano-Hoge (1978/79); Campbell & Lamar (1989); Cei (1993); Lema (1994a); McDiarmid et al. (1999)	Remnant Araucaria forests in n Argentina in Misiones and s Brazil in se São Paulo, Paraná, Santa Catarina, and Rio Grande do Sul; to at least 1800 m in humid, temperate *Araucaria* forest and associated savanna.
B. erythromelas Amaral, 1923 Caatinga Lancehead Pit Viper	Peters & Orejas-Miranda (1970); Amaral (1923, 1976); Hoge & Romano-Hoge (1978/79); Campbell & Lamar (1989); McDiarmid et al. (1999); Freitas (1999)	Arid zones in ne Brazil in Alagoas, Bahia, Ceara, e Maranhão, Minas Geraís, Parnambuco, Piauí, Rio Grande do Norte, Paraíba, and Sergipe; to about 2000 m in a wide variety of habitats.
B. fonsecai Hoge & Belluomini, 1959	Peters & Orejas-Miranda (1970); Hoge & Romano-Hoge (1978/79); Campbell & Lamar (1989); McDiarmid et al. (1999); Marques et al. (2001)	Remnant Aruacaria forests in se Brazil in ne São Paulo, s Rio de Janiero, and s Minas Geraís; to a maximum of about 1800 m in well-drained Atlantic

Species-level Taxa	Citations and Remarks	Abbreviated Range
Fonseca's Lancehead Pit Viper		forest in zones dominated by *Araucaria*, *Podocarpus*, and other trees.
B. iglesiasi Amaral, 1923 Inglesia's Lancehead Pit Viper	Peters & Orejas-Miranda (1970); Amaral (1976); Hoge & Romano-Hoge (1978/79); Campbell & Lamar (1989); McDiarmid et al. (1999)	Northern part of state of Piaui in ne Brazil; abundant in sub-xerophytic sections; found in inland regions to a maximum of about 300 m in dry to semi-arid rocky areas.
B. insularis (Amaral, 1922) Golden Lancehead Pit Viper	Peters & Orejas-Miranda (1970); Amaral (1976); Hoge & Romano-Hoge (1978/79); Campbell & Lamar (1989); Duarte et al. (1995); McDiarmid et al. (1999); Marques et al. (2001)	Endemic to Queimada Grande Island in the state of São Paulo, Brazil; to about 200 m in scrubby forest and shrubs.
B. itapetinigae (Boulenger, 1907) Cotiarinha Lancehead Pit Viper	Amaral (1976); Hoge & Romano-Hoge (1978/79); Campbell & Lamar (1989); Grantsau (1990); McDiarmid et al. (1999)	Wide ranging over cs and s Brazil including Minas Geraís, s Goiás, Mato Grosso, São Paulo, Paraná, Rio Grande do Sul, Santa Catarina, and the Paraná Plateau; to a maximum of 1500 m in open fields and bushy areas.
B. jararaca* (Wied-Neuwied, 1824) Brazilian Lancehead Pit Viper or Jararacá	Peters & Orejas-Miranda (1970); Amaral (1976); Hoge & Romano-Hoge (1978/79); Campbell & Lamar (1989); Cei (1993); Lema (1994a); McDiarmid et al. (1999); Freitas (1999); Marques et al. (2001) *Remark: See McDiarmid et al. (1999) for details on the nomenclature.	S and se South America including s Bahia, Espirito Santo, Rio de Janiero, Minas Geraís, São Paulo, Paraná, Santa Catarina, Rio Grande do Sul, and e Mato Grosso in Brazil; in ne Paraguay and Misiones in Argentina; usually found in low to intermediate elevations in a variety of habitats.
B. jararacussu Lacerda, 1884 Jararacusu or Jararacuçu	Peters & Orejas-Miranda (1970); Amaral (1976); Hoge & Romano-Hoge (1978/79); Campbell & Lamar (1989); Cei (1993); Lema (1994a); McDiarmid et al. (1999); Freitas (1999); Marques et al. (2001)	Mainly in protected rainforests in sc South America including Minas Geraís in Brazil southward into Misiones in ne Argentina in Misiones, se Bolivia, and Paraguay; from sea level to about 700 m in a wide variety of habitats.
B. jonathani Harvey, 1994 Jonathan's Lancehead Pit Viper	Harvey (1994); McDiarmid et al. (1999)	Mountainous areas around Cochabamba in Bolivia in dry rocky, grasslands between 2800 to 3200 m
B. lanceolatus* (Bonnaterre, 1790) Martinique Lancehead Pit Viper	Lazell (1964); Hoge & Romano-Hoge (1978/79); Gosner (1987); Campbell & Lamar (1989); Schwartz & Henderson (1985, 1991); McDiarmid et al. (1999); Censky & Kaiser *in* Crother (1999) *Remark: Sandner-Montilla (1990) divides this into two subspecies, *B. lanceolatus lanceolatus* and *B. lanceolatus nacaritae* Sander Montilla (1990). This appears to be an artificial division, and the subspecies from the mainland of South America are generally not recognized to be *lanceolatus*. See *B. atrox* for McDiarmid et al. (1999) comments.	Island of Martinique (WI) in the Lesser Antilles; disjunct distribution with two isolated populations, one in the north and one in the south; confined to upland humid areas to about 1300 m.
B. leucurus* Wagler *in* Spix, 1824	Hoge & Romano-Hoge (1978/79); Campbell & Lamar (1989); McDiarmid et al. (1999); Freitas (1999)	NE coast of Brazil in Espirito Santo, Bahia, Sergipe, Alagoas, and Pernambuco.

The Caenophidia

Species-level Taxa	Citations and Remarks	Abbreviated Range
White-tailed Lancehead Pit Viper	*Remark: See McDiarmid et al. (1999) for a review of the nomenclature problem for this snake.	
B. lojanus Parker, 1930 Lojan Lancehead Pit Viper	Parker (1930); Peters & Orejas-Miranda (1970); Hoge & Romano-Hoge (1978/79); Campbell & Lamar (1989); McDiarmid et al. (1999)	Known only from the type locality in Loja in Proviencia Loja and Zamora-Chinchipe in s Ecuador from 2100 to 2250 m in arid temperate regions, primarily in montane dry forest.
B. marajoensis Hoge, 1966 Marajó Lancehead Pit Viper	Hoge (1966); Peters & Orejas-Miranda (1970); Hoge & Romano-Hoge (1978/79); Campbell & Lamar (1989); McDiarmid et al. (1999)	Brazil in the coastal lowlands of the Amazon Delta, in the savanna of Ilha Marajó in the state of Pará, Ampaná, and possibly southward along the coast of mainland Pará and questionably into equitorial regions of the state of Maranhão. It is associated with lowland savannas to not much above sea level.
B. moojeni Hoge, 1966 Caissaca	Hoge (1966); Peters & Orejas-Miranda (1970); Hoge & Romano-Hoge (1978/79); Campbell & Lamar (1989); Cei (1993); McDiarmid et al. (1999); Freitas (1999)	Misiones in Argentina and e Paraguay into c and se Brazil in Mato Grosso, Piauí, Paraná, Mato Grosso do Sul, Minas Geraís, Goiás, and Marahão, and possibly e Bolivia; found in lowlands to at least 1500 m in semi-arid and seasonally dry tropical savannas.
B. neuwiedi Wagler *in* Spix, 1824 Neuwied's Lancehead Pit Viper	Hoge (1966); Peters & Orejas-Miranda (1970); Hoge & Romano-Hoge (1978/79); Campbell & Lamar (1989); Lema (1994a); McDiarmid et al. (1999)	S South America east of the Andes and south of Latitude 10°S in the subtropical regions south of the Amazon forests in a variety of habitats from sea level to over 600 m.
B. n. neuwiedi Wagler *in* Spix, 1824 Maximilian's Lancehead Pit Viper	Peters & Orejas-Miranda (1970); Amaral (1976); Hoge & Romano-Hoge (1978/79); Campbell & Lamar (1989); Grantsau (1990); McDiarmid et al. (1999); Freitas (1999)	S state of Bahia in Brazil in low humid areas of the Rio Paraguçu. Campbell & Lamar (1989) discussed a "perplexing" isolated population in w Amazonas, Brazil.
B. n. bolivianus Amaral, 1927 Bolivian Lancehead Pit Viper	Amaral (1927c); Peters & Orejas-Miranda (1970); Kempff-Mercado (1975), Hoge & Romano-Hoge (1978/79); Grantsau (1990); Campbell & Lamar (1989); Leynaud & Bucher (1999)	Departamento Santa Cruz, Cochabamba, and Tarija in Bolivia and the adjacent area of w Mato Grosso, Brazil.
B. n. diporus Cope, 1862 Argentine Painted Lancehead Pit Viper	Peters & Orejas-Miranda (1970); Hoge & Romano-Hoge (1978/79); Campbell & Lamar (1989); Grantsau (1990); Cei (1986, 1993); Lema (1994a); Leynaud & Bucher (1999)	E and s Paraguay, and the states of Paraná and Mato Grosso de Sul in Brazil, and southward throughout Argentina to as far south as La Pampa, San Luis, Mendoza, and Santa Fe.
B. n. goyazensis Amaral, 1925 Goiaz Lancehead Pit Viper	Peters & Orejas-Miranda (1970); Amaral (1976); Hoge & Romano-Hoge (1978/79); Campbell & Lamar (1989); Grantsau (1990); McDiarmid et al. (1999)	Known only from the type locality of Ipamery in the state of Goiás, Brazil; in elevated, dry, cool regions.
B. n. lutzi (Miranda-Ribeiro, 1915) Lutz's Lancehead Pit Viper	Peters & Orejas-Miranda (1970); Amaral (1976); Hoge & Romano-Hoge (1978/79); Campbell & Lamar (1989); Grantsau (1990); McDiarmid et al. (1999); Freitas (1999)	Dry areas of n Bahia, Brazil; in the western district of the São Francisco river Basin; in warm areas between 300 to 500 m.
B. n. mattogrossensis Amaral, 1925	Peters & Orejas-Miranda (1970); Amaral (1976); Hoge & Romano-Hoge (1978/79); Campbell &	S Mato Grosso, Brazil; in low-lying, humid regions.

Species-level Taxa	Citations and Remarks	Abbreviated Range
Mato Grosso Lancehead Pit Viper	Lamar (1989); Grantsau (1990); McDiarmid et al. (1999)	
B. n. meridionalis Müller, 1885 Southern Neuwied's Lancehead Pit Viper	Peters & Orejas-Miranda (1970); Hoge & Romano-Hoge (1978/79); Campbell & Lamar (1989); Grantsau (1990); McDiarmid et al. (1999)	States of Rio de Janeiro, Guanabara and Espirito Santo, Brazil.
B. n. paranaensis Amaral, 1925 Paraná Lancehead Pit Viper	Peters & Orejas-Miranda (1970); Amaral (1976); Hoge & Romano-Hoge (1978/79); Campbell & Lamar (1989); Grantsau (1990); Lema (1994a); McDiarmid et al. (1999)	State of Paraná, Brazil and possibly adjacent Misiones in Argentina.
B. n. pauloensis Amaral, 1925 São Paulo Lancehead Pit Viper	Peters & Orejas-Miranda (1970); Amaral (1976); Hoge & Romano-Hoge (1978/79); Campbell & Lamar (1989); Grantsau (1990); McDiarmid et al. (1999)	SW and c part of the state of Sao Paulo, Brazil; between 200 to 300 m in humid, cool regions.
B. n. piauhyensis Amaral, 1925 Northern Neuwied's Lancehead Pit Viper	Peters & Orejas-Miranda (1970); Amaral (1976); Hoge & Romano-Hoge (1978/79); Campbell & Lamar (1989); Grantsau (1990); McDiarmid et al. (1999)	NE Brazil at least in Piauí, Pernambuco, Ceará, and s Maranhão; in xerophytic to subxerophytic areas.
B. n. pubescens (Cope, [1869] 1870) Uruguayan Painted Lancehead Pit Viper	Peters & Orejas-Miranda (1970); Amaral (1976); Hoge & Romano-Hoge (1978/79); Achaval (1987); Campbell & Lamar (1989); Grantsau (1990); Lema (1994a); McDiarmid et al. (1999)	State of Rio Grande do Sul, Brazil and Uruguay; in Brazil it is said to prefer low, humid, and cold places, while in Uruguay it is supposed to inhabit rockly ridges and rockpiles.
B. n. urutu (Liais, 1872) Painted Lancehead Pit Viper	Peters & Orejas-Miranda (1970); Amaral (1976); Hoge & Romano-Hoge (1978/79); Campbell & Lamar (1989); Grantsau (1990); McDiarmid et al. (1999)	SC Minas Geraís and n São Paulo, Brazil; in elevated, humid, cool areas.
*B. pictus** (Tschudi, 1845) Desert Lancehead Pit Viper	Peters & Orejas-Miranda (1970); Hoge & Romano-Hoge (1978/79); Campbell & Lamar (1989, 1992); McDiarmid et al. (1999)

*Remark: *B. roadingeri* is a synonym of *B. pictus* (see Campbell & Lamar, 1992). | Peru along the Pacific coast from sea level to 1800 m; on the lower w slopes of the Andes and throughout the coastal plain from Departamento de La Libertad southward into Departamento de Arequipa; to at least 2300 m in arid to semi-arid foothills, river valleys, and lower Andean slopes and dry rocky regions dotted with scrubby leguminous trees. |
B. pirajai Amaral, 1923 Bahia or Piraja's Lancehead Pit Viper	Amaral (1923, 1976); Peters & Orejas-Miranda (1970); Hoge & Romano-Hoge (1978/79); Campbell & Lamar (1989); McDiarmid et al. (1999); Freitas (1999)	Cacao growing areas in the Atlantic forests of the of c and s Bahia, Brazil., the Ilhéus-Itabuna region, and said to occur in ne Minas Geraís; presumably associated with lowland wet forest and lower montane wet forest.
B. pradoi (Hoge, 1948) Prado's Lancehead Pit Viper	Hoge (1948); Peters & Orejas-Miranda (1970); Hoge & Romano-Hoge (1978/79); Campbell & Lamar (1989); Schätti & Kramer (1993); McDiarmid et al. (1999)	Coast of se Brazil from Espirito Santo into s Bahia; found in lowlands to about 300 m in a variety of habitats.
B. sanctaecrucis Hoge, 1966 Santa Cruz Lancehead Pit Viper	Hoge (1966); Peters & Orejas-Miranda (1970); Hoge & Romano-Hoge (1978/79); Campbell & Lamar (1989); McDiarmid et al. (1999)	Amazonian lowlands of Bolivia in El Beni and Santa Cruz departments; lowlands to at least 450 m in lower montane wet forest and possibly rain forest.

Species-level Taxa	Citations and Remarks	Abbreviated Range
B. venezuelensis* Sandner-Montilla, 1952 Venezuelan Lancehead Pit Viper	Peters & Orejas-Miranda (1970); Hoge & Romano-Hoge (1978/79); Campbell & Lamar (1989); Lancini & Kornacker (1989); McDiarmid et al. (1999); Kornacker (1999) *Remark: *Bothrops pifanoi* Sandner-Montilla & Römer, 1961, and *Bothrops venezuelae* Sandner-Montilla, 1961, are synonyms of *B. venezuelensis* (see Peters & Orejas-Miranda, 1970, and McDiarmid et al., 1999).	N and c Venezuela; in the central range of the Cordillera de la Costa including the states of Carabobo, Aragua, the Federal District, and Miranda; and, in the Venezuelan Andes in Mérida, Trujillo, Lara, Falcón, Yaracuy, and Sucre. Usually found between 1000 to 2200 m, but there are reports of it being taken at 2800 m. Generally associated with lower montane wet forest and cloud forest, including temperate areas where cold weather conditions occur.

GENUS: *Cerrophidion* Campbell & Lamar, 1992 occurs from southern Mexico southward through Central America into western Panama. See Hoge & Romano-Hoge (1978/79) [as *Bothrops*]; Campbell (1985, 1988), Campbell & Lamar (1992); McDiarmid et al. (1999) and Köhler (2001).

See also *Atropoides, Bothriechis, Bothriopsis, Bothrocophias, Bothrops, Ophryacus*, and *Porthidium*.

CONTENT: Four species. ★★★★

Cerrophidion: Montane Pit Viper

Species	Citations and Remarks	Abbreviated Range
C. barbouri (Dunn, 1919) Barbour's Montane Pit Viper	Hoge & Romano-Hoge (1978/79); Campbell (1988); Campbell & Lamar (1989, 1992); McDiarmid et al. (1999)	SW Mexico where it is restricted to the Sierra Madre del Sur in the highlands of the state of Guerrero; between 2390 to 3300 m in pine-oak and cloud forest.
C. godmani (Günther, 1863) Godman's Montane Pit Viper	Peters & Orejas-Miranda (1970); Hoge & Romano-Hoge (1978/79); Alvarez del Toro (1982); Wilson & Meyer (1985); Savage & Villa (1986); Campbell (1988); Villa et al (1988); Campbell & Lamar (1989, 1992); Cruz (1997); McDiarmid et al. (1999); Köhler (2001)	Disjunct distribution in s Mexico in se Oaxaca and Chiapas southward into Central America as far south as Costa Rica and w Panama in moderate and high elevations; between 1600 to 3200 m in a variety of forest and meadow habitats.
C. petlalcalensis López-Luna, Vogt & Torre-Laranca, 1999 Veracruz Montane Pit Viper	López-Luna et al. (1999)	The holotype is from Cerro de Petlalcala, Municipo San Andres Tenejapun, Veracruz, México at 2100 m.
C. tzolzilorum (Campbell, 1985) Tzotzil Montane Pit Viper	Campbell (1985, 1988); Villa et al. (1988); Campbell & Lamar (1989); McDiarmid et al. (1999); Köhler (2001)	SE Mexico in the Meseta Central in Chiapas in the highlands at intermediate elevations on relatively flat tablelands, between 2050 to 2500 m in humid pine-oak forest.

GENUS: *Crotalus* Linnaeus, 1758, is widespread throughout the Americas from southern Canada southward to as far south as Argentina. See Gloyd (1940); Smith & Taylor (1966); Klauber (1956, 1972, 1997); Harris & Simmons (1978); Hoge & Romano-Hoge (1978/79); Campbell and Lamar (1989); McCranie (1993); Smith & Smith (1993);

Degenhardt et al. (1996); Greene *in* Klauber (1997a); Rubio (1998); Starace (1998); McDiarmid et al. (1999); and Dixon (2000a).

Gloyd (1940) was the first modern revisor to organize the information of the rattlesnakes. Klauber (1952, 1972, [1972] 1997), Hoge & Romano-Hoge (1978/79), and most recently McDiarmid, et al. (1999) brought the information up to date.

See also *Sistrurus*.

CONTENT: About twenty-five to thirty species. Twenty-seven are listed in the following species list. ★★

Crotalus: Rattlesnake or Cascabel.

Species	Citations and Remarks	Abbreviated Range
C. adamanteus Palisot de Beauvois, 1799 Eastern Diamond-backed Rattlesnake	Klauber (1972); Conant (1975); Hoge & Romano-Hoge (1978/79); McCranie (1980a); Ashton & Ashton (1981); Dundee & Rossman (1989); Ernst (1992); Wright & Wright ([1957] 1994); Tennant (1997b); Rubio (1998); Conant & Collins (1998); McDiarmid et al. (1999); Tennant & Bartlett (2000)	SE United States from se North Carolina southward into the Florida Keys and westward along the Gulf Coast states into s Mississippi and extreme se Louisiana.
*C. aquilus** Klauber, 1952 Queretaran Dusky Rattlesnake	Klauber (1972); Harris & Simmons (1978); Hoge & Romano-Hoge (1978/79); Campbell & Lamar (1989); Dorcas (1992); Rubio (1998); Smith & Smith (1999); Greene *in* Klauber (1997a); McDiarmid et al. (1999) *Remark 1: Much like *C. oaxacus*, *C. aquilus* has bounced back and forth between genera and species with widely varying opinions from the authorities. This classification reflects the most recent published opinions but expect that this could change. This listing is based on McDiarmid, et al. (1999). *Remark 2: Campbell and Lamar (1989) listed this as *C. triseriatus aquilus*. *Crotalus aquilus* is listed as a separate valid species by Harris & Simmons (1978); Flores-Villela (1993); Liner (1994) and Rubio (1998). *Remark 3: Campbell & Lamar (1989) listed *C. t. quadrangularis* Harris & Simmons, 1978, from Hidalgo, Mexico, as a synonym.	C Mexico in the eastern portion of the Mesa Central and parts of the Sierra Madre Oriental in the states of Hidalgo, San Luis Potosí, Guanajuato, México, and Michoacán.
*C. atrox** Baird & Girard, 1853 Western Diamond-backed Rattlesnake	Smith & Taylor (1966); Klauber (1972); Conant (1975); Hoge & Romano-Hoge (1978/79); Campbell & Lamar (1989); Ernst (1992); Degenhardt et al. (1996); Brown (1997); Rubio (1998); Tennant (1998); Conant & Collins (1998); McDiarmid et al. (1999); Dixon (2000a); Werler & Dixon (2000); Bartlett & Tennant (2000); Tennant & Bartlett (2000) *Remark: Most authorities consider this snake to be monotypic with large degrees of variation through	SW United States and widespread in the northern half of Mexico including Baja Californaia, it is found on Tiburón Island in the Gulf of California; from a northern range limit that spreads from se California to c Arkansas and southward into n Sinaloa, Hidalgo, and n Veracruz, with isolated populations in s Veracruz and se Oaxaca; it was accidentally introduced into Wisconsin where a small population persisted for several years, status of this population at this time is unclear.

Species-level Taxa	Citations and Remarks	Abbreviated Range
	out its range; however, Lowe et. al. (1986), and a few others considered *C. atrox* as polytypic with *C. atrox atrox* as the mainland form and *C. atrox tortugensis* for the population found on Isla Tortuga in the Gulf of California.	
C. basiliscus* (Cope, 1864) Mexican West Coast Rattlesnake	Gloyd (1948); Klauber (1972); Hoge & Romano-Hoge (1978/79); McCranie (1981); Campbell & Lamar (1989); García & Ceballos (1994); Ramírez-Bautista (1994); Rubio (1998); McDiarmid et al. (1999) *Remark: McCranie (1981), noted that *C. basiliscus* is probably conspecific with *C. molossus*.	W Mexico from extreme s Sonora southward along the Pacific coast into Michoacán; found in the states of Sonora, Sinaloa, Nayarit, Jalisco, Colima and Michoacán; although the species occurs mostly below 600 m, it has been reported from sea level to 2400 m. It is usually associated with thorn forest, tropical deciduous forest, and pine-oak forest.
C. catalinensis Cliff, 1954 Santa Catalina Island Rattlesnake	Cliff (1954); Klauber (1972); Hoge & Romano-Hoge (1978/79); Campbell & Lamar (1989); Rubio (1998); McDiarmid et al. (1999); McPeak (2000)	Santa Catalina Island off the east coast of Baja California del Sur in Mexico; found in desert habitat.
C. cerastes Hallowell, 1854 [1854/55] Sidewinder	Klauber (1972); Campbell & Lamar (1989); Brown (1997); Rubio (1998); McDiarmid et al. (1999)	Southwestern United States into c Mexico; Degenhardt et al. (1996) do not believe it is found in New Mexico so the range is generally considered to be e California, s Nevada, extreme sw Utah and w Arizona southward into nw Mexico in w Sonora and e Baja California del Norte to the Llano de San Pedro. It also occurs on Tiburón Island; from sea level to nearly 1830 m but usually seen below 1200 m and it is a snake of the desert but can also be encountered in other habitats.
C. c. cerastes Hallowell, 1854 [1854/55] Mojave Desert Sidewinder	Klauber (1972); Hoge & Romano-Hoge (1978/79); Campbell & Lamar (1989); Ernst (1992); Brown (1997); Rubio (1998); McDiarmid et al. (1999); Bartlett & Tennant (2000)	W United States where it occurs in desert regions from extreme sw Utah to ec California and southward to wc Arizona and adjacent California.
C. c. cercobombus Savage & Cliff, 1953 Sonoran Sidewinder	Klauber (1972); Hoge & Romano-Hoge (1978/79); Campbell & Lamar (1989); Ernst (1992); Rubio (1998); McDiarmid et al. (1999); Bartlett & Tennant (2000)	Arizona in the United Staes into wc Sonora, Mexico.
C. c. laterorepens Klauber, 1944 Colorado Desert Sidewinder	Smith & Taylor (1966); Klauber (1972); Hoge & Romano-Hoge (1978/79); Campbell & Lamar (1989); Ernst (1992); Brown (1997); Rubio (1998); McDiarmid et al. (1999); Bartlett & Tennant (2000)	W United States southward into e Baja California del Norte in Mexico and the panhandle of extreme nw Sonora.
C. durissus Linnaeus, 1758 Neotropical Rattlesnake	Peters & Orejas-Miranda (1970); Klauber (1972, 1997); Villa et al. (1988); Campbell & Lamar (1989); McCranie (1993); Lee (1996, 2000); Rubio (1998); Starace (1998); McDiarmid et al. (1999); Kornacker (1999); Marques et al. (2001); Köhler (2001)	From Tamaulipas and Nuevo León in Mexico southward into Central America and South America in a discontinuous range; extending into northern Argentina; they have not been reported from Ecuador or Chile. It is present on the Venezuelan islands of Margarita, Morro de la Iguana, and Tamarindo. The snake has not been confirmed from Panama, although it is alleged to be in the dry tablelands of Veraguas and

Species-level Taxa	Citations and Remarks	Abbreviated Range
		Chiriquí. It is usually found below 700 m but has been report 2200 m; found in a variety of habitats in semi-arid regions.
C. d. durissus Linnaeus, 1758 Central American Rattlesnake	Hoge (1966); Smith & Taylor (1966); Peters & Orejas-Miranda (1970); Klauber (1972); Amaral (1976); Hoge & Romano-Hoge (1978/79); Alvarez del Toro (1982); Wilson & Meyer (1985); Campbell & Lamar (1989); McCranie (1993); Rubio (1998); McDiarmid et al. (1999); Köhler (2001)	SE Mexico into Central America as far south as Costa Rica; in the Atlantic versant of Mexico it is found from about s Veracruz to w Honduras, and in the Pacific versant from about the Isthmus of Tehuantepec to sc Costa Rica.
C. d. cascavella Wagler *in* Spix, 1824 Northeastern Brazilian Rattlesnake	Hoge (1966); Peters & Orejas-Miranda (1970); Amaral (1976); Hoge & Romano-Hoge (1978/79); Campbell & Lamar (1989); Grantsau (1990); McCranie (1993); Greene *in* Klauber (1997a); Rubio (1998); McDiarmid et al. (1999); Freitas (1999)	NE Brazil in dry Caatinga areas of s Maranhão, Piauí, Ceará, Rio Grande do Norte southward into Paraíba, Pernambuco, Alagoas, Sergipe, Bahia, and ne Minas Geraís.
C. d. collilineatus Amaral, 1926 Central Brazilian Rattle snake	Hoge (1966); Peters & Orejas-Miranda (1970); Amaral (1976); Hoge & Romano-Hoge (1978/79); Campbell & Lamar (1989); Grantsau (1990); McCranie (1993); Greene *in* Klauber (1997a); Rubio (1998); McDiarmid et al. (1999)	C and ne Brazil in Rondônia, Mato Grosso, Goiás, sw Bahia, w Minas Geraís, São Paulo, and probably into w Paraná. There is an unsubstantiated report from Bolivia.
C. d. culminatus Klauber, 1952 Northwestern Neotropical Rattlesnake	Davis & Smith (1953); Klauber (1972); Hoge & Romano-Hoge (1978/79); Campbell & Lamar (1989); McCranie (1993); Rubio (1998); McDiarmid et al. (1999)	SW Mexico in s Michoacán to about the Isthmus of Tehuantepec.
C. d. cumanensis* Humbolt *in* Humbolt and Bonpland, 1833 Venezuelan Rattlesnake	Hoge (1966); Peters & Orejas-Miranda (1970); Hoge & Romano-Hoge (1978/79); Campbell & Lamar (1989); Lancini & Kornacker (1989); McCranie (1993); Greene *in* Klauber (1997a); Rubio (1998); McDiarmid et al. (1999); Kornacker (1999)	It is distributed along the Caribbean coastal plain from Departamento de Córdoba, Colombia, eastward to extreme e Venezuela, and southward e and s of the Andes through the llanos of both countries; Campbell & Lamar (1989) noted there may be an isolated population on the high plains of Boyacá, Colombia, and possibly w Guyana.

*Remark 1: The validity of *C. pifanorum* is highly suspect. At that time of their work, Campbell & Lamar (1989) could not distinguish this snake and they tentatively considered it to be a synonym or intergrade of *C. d. cumanensis*.

*Remark 2: McDiarmid et al. (1999) noted that McCranie (1993), included *C. pifanorum* Sandner-Montilla, 1980, known only from the type locality in the Distrito Infante in the state of Guárico in Venezuela, as a synonym. However, Kornacker (1999) lists as valid.

*Remark 3: The snake *C. maricelae* García-Pérez (1995) is called the Andean Rattlesnake as listed in Rodríguez & Rojas-Suárez (1999) and reported to be found in the Bolsón Árido de Lagunillas in the Cordillera de Mérida in the Andes of Venezuela.
The original description reported the main features

Species-level Taxa	Citations and Remarks	Abbreviated Range
	that are used to separate this snake are five prefrontal scales, dorsal coloration pattern that consists of rhomb-like design as in *C. durissus sp.*, but with greenish-brown background, and its small size. Since this description falls within the *C. durissus* description in Campbell & Lamar (1989), I attempted to have the holotype sent to Texas A&M University so that I could review it. Unfortunately the type was missing from the museum listed in the original paper and its whereabouts were unknown so I have not had the opportunity to see the holotype. However, the photos and description from the technical paper by García-Pérez describe a *C. durissus*. I discussed this with Jonathan Campbell who is also familiar with the García-Pérez (1995) paper and he agrees that the description sounds like a *C. durissus*. I feel it is probably *C. durissus cumanensis*. As a secondary note, the description is shown electronically but I have also been unable to obtain a hard copy of the published description.	
C. d. dryinas Linnaeus, 1758 Guianian Rattlesnake	Hoge (1966); Peters & Orejas-Miranda (1970); Hoge & Romano-Hoge (1978/79); Campbell & Lamar (1989); Grantsau (1990); McCranie (1993); Greene *in* Klauber (1997a); Rubio (1998); McDiarmid et al. (1999); Starace (1998)	Guianas in the n coastal portions of Guyana, Suriname and French Guiana into the interior of Amapá in Brazil
C. d. marajoensis Hoge, 1966 Marajoan Rattlesnake	Hoge (1966); Peters & Orejas-Miranda (1970); Hoge & Romano-Hoge (1978/79); Campbell & Lamar (1989); Grantasu (1990); McCranie (1993); Greene *in* Klauber (1997a); Rubio (1998); McDiarmid et al. (1999)	State of Pará, Brazil in the savannas in Ilha de Marajó.
C. d. ruruima Hoge, 1966 Mt. Roraima Rattlesnake	Hoge (1966); Peters & Orejas-Miranda (1970); Hoge & Romano-Hoge (1978/79) Campbell & Lamar (1989); Grantsau (1990); Lancini & Kornacker (1989); McCranie (1993); Greene *in* Klauber (1997a); Rubio (1998); McDiarmid et al. (1999); Kornacker (1999)	The slopes of Mt.Roraima and Mt.Cariman-Perú in the state of Bolivar, Venezuela. It will probably eventually be found in adjacent Guyana. A single specimen has been recorded from the territory of Roraima in n Brazil.
C. d. terrificus (Laurenti, 1768) South American Rattlesnake	Hoge (1966); Peters & Orejas-Miranda (1970); Klauber (1972); Amaral (1976); Hoge & Romano-Hoge (1978/79); Cei (1986, 1993); Achaval (1987); Campbell & Lamar (1989); Grantsau (1990); McCranie (1993); Lema (1994a); Rubio (1998); McDiarmid et al. (1999); Leynaud & Bucher (1999)	From the Cordillera de Carabay in Puno-Madre de Dios of Peru southward through c Bolivia, most of Paraguay, n Uruguay, northward into n Argentina, and se Brazil from Rio Grande do Sul and Mato Grosso northward into Minas Geraís and southward into Misiones, Argentina. It will probably be found in Ecuador.
*C. d. totonacus** Gloyd & Kauffeld, 1940 Totonacan Rattlesnake	Smith & Taylor (1966); Klauber (1972); Hoge & Romano-Hoge (1978/79); Campbell & Lamar (1989); Pérez-Higareda & Smith (1991); McCranie (1993); Rubio (1998); McDiarmid et al. (1999)	NE into se Mexico from s Tamaulipas, se San Luis Potosí, and n Veracruz.

Species-level Taxa	Citations and Remarks	Abbreviated Range
	*Remark: *C. durissus neoleonensis* Julioa-Zertuche, 1978 is considered *nomen nudum*, fide Campbell and Lamar (1989) and it is most likely a synonym of this snake. Smith & Smith (1993) also consider it to be *nomen nudum* and a synonym. As well as did McDiarmid et al. (1999).	
*C. d. trigonicus** Harris & Simmons, 1978 Rupununi Savanna Rattlesnake	Harris & Simmons (1978); Hoge & Romano-Hoge (1978/79) Vanzolini (1986); Campbell & Lamar (1989); McCranie (1993); Greene *in* Klauber (1997a); Rubio (1998); McDiarmid et al. (1999)	Rupununi Savanna of sw Guyana with a single report from the savannas of Tiriós in Amapá, Brazil and probably adjacent Roraima also in Brazil.
	*Remark: Rubio (1998) placed as synonym of *C. d. ruruima*. Campbell & Lamar (1989) questioned the validity of *C. d. trigonicus*. It is recognzied as a valid species by McDiarmid et al. (1999).	
C. d. tzabcan Klauber, 1952 Yucatán Rattlesnake	Klauber (1972); Hoge & Romano-Hoge (1978/79); Campbell & Lamar (1989); McCranie (1993); Lee (1996, 2000); Peters & Orejas-Miranda (1970); Liner (1994); Rubio (1998); Smith & Smith (1993); McDiarmid et al. (1999); Stafford & Meyer (2000); Köhler (2001)	Yucatán peninsula in Mexico southward into Central America in n Belize and n Guatemala.
*C. d. unicolor** Van Lidth de Jeude, 1887 Aruba Island Rattlesnake	Gloyd (1936c); Peter & Orejas-Miranda (1970); Hoge & Romano-Hoge (1978/79); McCranie (1983b); Campbell & Lamar (1989); Greene *in* Klauber (1997a); Rubio (1998); McDiarmid et al. (1999)	Restricted to Aruba Island in the Netherland Antilles off the coast of Venezuela.
	*Remark: In older literature this snake is listed as *C. unicolor* (see Klauber, 1972).	
*C. d. vegrandis** Klauber, 1941 Urorocan Rattlesnake	Hoge (1966); Peters & Orejas-Miranda (1970); Klauber (1972); Hoge & Romano-Hoge (1978/79); Lancini (1979); McCranie (1984); Lancini & Kornacker (1989); Campbell & Lamar (1989); Greene *in* Klauber (1997a); Rubio (1998); McDiarmid et al. (1999)	E Venezuela in s Anzoátegui and s Monagus in the Maturín Savanna.
	*Remark: In older literature this snake is listed as *C. vegrandis* (see Klauber, 1972; Lancini, 1979; McCranie, 1984; Lancini & Kornacker, 1989 and Kornacker, 1999).	
C. enyo (Cope, 1861) Baja California Rattlesnake	Klauber (1972); Stebbins (1985); Campbell & Lamar (1989); Beaman & Grismer (1994); Rubio (1998); McDiarmid et al. (1999)	W Baja California, Mexico throughout most of the Baja peninsula from the vicinity of San Telmo in Baja California Norte southward to the Cape and on the islands of San Francisco, Carmen, Partida and Cerralvo in the Gulf of California and on Margarita Island on the Pacific side. The snake is usually found in deserts, but can be encountered in other environments.

Species-level Taxa	Citations and Remarks	Abbreviated Range
C. e. enyo (Cope, 1861) Lower California Rattlesnake	Lowe & Norris (1954); Smith & Taylor (1966); Klauber (1972); Hoge & Romano-Hoge (1978/79); Campbell & Lamar (1989); Beaman & Grismer (1994); Rubio (1998); McDiarmid et al. (1999)	S Baja California, Mexico; from about El Rosario southward through the Peninsula, including several offshore islands.
*C. e. cerralvensis** Cliff, 1954 Cerralvo Island Rattlesnake	Hoge & Romano (1971); Klauber (1972); Hoge & Romano-Hoge (1978/79) Campbell & Lamar (1989); Beaman & Grismer (1994); Rubio (1998); McDiarmid et al. (1999); McPeak (2000) *Remark: According to Grismer (1999), *C. e. cerralvensis* is a synonym of *C. enyo*.	Endemic to Cerralvo Island in Gulf of California in Mexico.
*C. e. furvus** Lowe & Norris, 1954 Rosario Rattlesnake	Lowe & Norris (1954); Hoge & Romano (1971); Klauber (1972); Hoge & Romano-Hoge (1978/79); Campbell & Lamar (1989); Beaman & Grismer (1994); Rubio (1998); McDiarmid et al. (1999) *Remark: McDiarmid et al. (1999) quoted Beaman & Grismer (1994), as recognizing three subpecies, but, indicated that *Crotalus enyo enyo* and *Crotalus enyo furvus* "should be considered as the binomial of *C. enyo*" and that *Crotalus enyo cerralvensis* would best be considered as a full species. Since this does not seem to be popular opinion and there was a lack of detail for this justification, I have left it as the subspecies.	Baja California and mainland w Mexico from Río San Telmo to about El Rosario.
*C. exsul** Garman, 1884 Cedros Island Diamond Rattlesnake	Smith & Taylor (1966); Klauber (1972); Hoge & Romano-Hoge (1978/79) Campbell & Lamar (1989); Rubio (1998); McDiarmid et al (1999); McPeak (2000) *Remark: *Crotalus ruber* was reassigned as a subspecies of *Crotalus exsul* and the subspecies reassigned to follow suit by Grismer et al. (1994). Brown (1997) also listed this snake as *C. exsul ruber* (Kallert, 1827). Rubio (1998) felt that this was premature, and chose to retain *C. ruber* until the matter is settled by systematic herpetologists. McDiarmid et al. (1999) maintained *ruber* separate in their list, and the reader is advised to see their comments. Grismer (1999) stated Grismer et al. (1994) had placed *C. ruber* in synonmy of *C. exsul* and that an ICZN name was petitioned in 1995 to give precedence of *C. ruber* over *C. exsul*. Grismer noted that according to Article 80 of the International Code of Zoological Nomenclature, the most commonly used classification should be employed until a decision has been made by the commission. Opinion 1960 published 29 September 2000 in Vol. 57, Part 3: 189–190 of the Bulletin of Zoological Nomenclature gave *Crotalus ruber* Cope, 1892 (Reptilia: Serpentes) precedence over that of *Crotalus exsul* Garman, 1884.	Cedros Islands off nw Pacific coast of Baja California del Norte, Mexico.

Species-level Taxa	Citations and Remarks	Abbreviated Range
C. horridus* Linnaeus, 1758 Timber Rattlesnake	Klauber (1972); Ashton & Ashton (1981); Conant (1975); Collins & Knight (1980); Dundee & Rossman (1989); Smith & Smith (1993); Tennant (1997b); Rubio (1998); Tennant (1998); Conant & Collins (1998); McDiarmid et al. (1999); Dixon (2000a)	Eastern United States from se Minnesota and s Maine southward into Texas and n Florida, it is also found in s Canada.
	*Remark 1: Klauber (1936) originally recognized two subspecies. Collins & Knight (1980) and others recognized no subspecies. Brown & Ernst (1986) supported recognitition of two subspecies. I feel there is enough of a difference between the two types of snakes to justify subspecies.	
	*Remark 2: The uncertainties regarding the proper application of the Linnean names to rattlesnakes have been discussed by Klauber (1941c, 1948, 1972). For some years during the nineteenth century, the neotropical (Central and South American) rattler was often referred to as *C. horridus*, and the timber rattler, of the eastern United States, as *C. durissus*. To avoid confusion, Kluaber (1972) ignored these allocations in his checklist, since in present usage the allocation of these names is reveresed.	
C. h. horridus Linnaeus, 1758 Timber Rattlesnake	Klauber (1972); Conant (1975); Hoge & Romano-Hoge (1978/79); Brown & Ernst (1986); Ernst (1992); Rubio (1998); McDiarmid et al. (1999); Tennant & Bartlett (2000)	E and s United States including s New Hampshire and the Lake Champlin regions southward into n Georgia and westward into Illinois, sw Wisconsin, to nw Arkansas.
C. h. atricaudatus (Latreille, 1801 [1802]) Canebreak Rattlesnake	Gloyd (1936b); Klauber (1972); Conant (1975); Hoge & Romano-Hoge (1978/79); Brown & Ernst (1986); Dundee & Rossman (1989); Ernst (1992); Rubio (1998); McDiarmid et al. (1999); Werler & Dixon (2000); Tennant & Bartlett (2000)	S United States lowland from found from se Virginia into n Florida and westard into c Texas, northward into the Mississippi Valley into s Illinois.
C. intermedius Troschel *in* Müller, 1865 Mexican Small-headed Rattlesnake	Klauber (1972); Hoge & Romano-Hoge (1978/79); Campbell & Lamar (1989); McCranie (1991); Flores-Villela (1993); Liner (1994); Rubio (1998); Smith & Smith (1993); McDiarmid et al. (1999)	CS Mexico in the Sierra Juárez, Cerro San Felipe and surrounding mountains, Sierra de Cuatro Venados, Sierra Madre del Sur, Sierra de Mihuatlán and Sierra Madre del Sur; populations are disjunct in se Hidalgo, Puebla, sw Tlaxcala, wc Veracruz, Oaxaca, and Guerrero into Michoacán; its range is very disjunct, even within subspecies; from about 2000 to 3000 m in a wide variety of habitats.
C. i. intermedius Troschel *in* Müller, 1865 Totalcan Small-headed Rattlesnake	Klauber (1972); Hoge & Romano-Hoge (1978/79); Campbell & Lamar (1989); McCranie (1991); Pérez-Higareda & Smith (1991); Rubio (1998); McDiarmid et al. (1999)	EC Mexico from se Hidalgo, ne Puebla, and wc Veracruz.
C. i. gloydi Taylor, 1942 Oaxacan Small-headed Rattlesnake	Smith & Taylor (1966); Klauber (1972); Hoge & Romano-Hoge (1978/79); Campbell & Lamar (1989); McCranie (1991); Rubio (1998); McDiarmid et al. (1999)	Oaxaca and Michoacán, Mexico; the type locality is on Cerro San Felipe at 10,000 feet near Oaxaca, Oaxaca; the subspecies is found in three disjunct populations in the highlands.

Species-level Taxa	Citations and Remarks	Abbreviated Range
C. i. omiltemanus Günther, 1895 Omilteman Small-headed Rattlesnake	Smith & Taylor (1966); Klauber (1972); Hoge & Romano-Hoge (1978/79); Campbell & Lamar (1989); Rubio (1998); McDiarmid et al. (1999)	C Mexico in the Sierra Madre del Sur in the highlands of c Guerrero.
C. lannomi Tanner, 1966 Autlán Rattlesnake	Tanner (1966); Klauber (1972); Harris & Simmons (1978); Hoge & Romano-Hoge (1978/79); Tanner (1986); Campbell & Lamar (1989); Rubio (1998); McDiarmid et al. (1999)	Only known from a single specimen from the type locality in Jalisco in w Mexico; collected at about 1400 m on a ridge connecting the peaks of Cerro de Varón and Cerro del Muñeco.
C. lepidus (Kennicott, 1861) Rock Rattlesnake	Gloyd (1936d); Klauber (1972); Hoge & Romano-Hoge (1978/79); Campbell & Lamar (1989); Rubio (1998); McDiarmid et al. (1999); Dixon (2000a)	SW United Staets in se Arizona, s New Mexico, and sw Texas southward into nc Mexico in most of the Sierra Madre Occidental and n portion of the Mexican Plateau; in Mexico it is found in the states of ne Sonora, Chihuahua, Durango, ec and se Sinaloa, Zacatecas, and probably e Nayarit, n Jalisco, Aguascalientes, w San Luis Potosí, w Nuevo León, Coahuila, and sw Tamaulipas; from about 300 to 3000 m in a wide variety of habitats.
*C. l. lepidus** (Kennicott, 1861) Mottled Rock Rattlesnake	Smith & Taylor (1966); Klauber (1972); Conant (1975); Hoge & Romano-Hoge (1978/79); Campbell & Lamar (1989); Ernst (1992); Rubio (1998); Tennant (1998); Conant & Collins (1998); McDiarmid et al. (1999); Dixon (2000a); Werler & Dixon (2000); Bartlett & Tennant (2000); Tennant & Bartlett (2000) *Remark: Campbell & Lamar (1989) stated that *C. l. castaneus* Julia-Zertuche is a synonym of *C. l. lepidus*. McDiarmid et al. (1999) did not comment on the status of *C. l. castaneus*.	SW United States in se New Mexico and sw Texas southward across much of the e Mexican Plateau to San Luis Potosí.
C. l. klauberi Gloyd, 1936 Banded Rock Rattlesnake	Smith & Taylor (1966); Klauber (1972); Conant (1975); Hoge & Romano-Hoge (1978/79); Campbell & Lamar (1989); Ernst (1992); Rubio (1998); Tennant (1998); Conant & Collins (1998); McDiarmid et al. (1999); Dixon (2000a); Werler & Dixon (2000); Bartlett & Tennant (2000); Tennant & Bartlett (2000)	SW United States from se Arizona, se New Mexico, and extreme sw Texas southward across much of the w portion of the Mexican Plateau to n Jalisco.
*C. l. maculosus** Tanner, Dixon & Harris, 1972 Durangan Rock Rattlesnake	Hoge & Romano-Hoge (1978/79); Campbell & Lamar (1989); Greene *in* Klauber (1997a); Rubio (1998); McDiarmid et al. (1999)	Western drainages of the Sierra Madre Occidental in Durango and Sinaloa in Mexico.
C. l. morulus Klauber, 1952 Tamaulipian Rock Rattlesnake	Klauber (1972); Hoge & Romano-Hoge (1978/79); Campbell & Lamar (1989); Rubio (1998); McDiarmid et al. (1999) *Remark: Campbell & Lamar (1989), and Rubio (1998) stated that *C. lepidus castaneus* Julia-Zertuche is most likely a synonym of this snake.	Sierra Madre Oriental in Mexico in sw Tamaulipas, c Nuevo León, and se Coahuila.

Species-level Taxa	Citations and Remarks	Abbreviated Range
C. mitchellii (Cope, 1861) Speckled Rattlesnake	Klauber (1972); McCrystal & McCoid (1986); Campbell & Lamar (1989); Brown (1997); Rubio (1998); McDiarmid et al. (1999)	SW United States in ec and s California, sw Nevada, extreme sw Utah and w Arizona southward through most of peninsular Baja California in both Baja California Norte, Baja California Sur and on many of the islands in the Gulf of California in Mexico; usually associated with desert but can also be found in other habitat from sea level to 2440 m.
C. m. mitchelli (Cope, 1861) San Lucan Speckled Rattlesnake	Smith & Taylor (1966); Klauber (1972); Hoge & Romano-Hoge (1978/79); McCrystal & McCoid (1986); Campbell & Lamar (1989); Rubio (1998); McDiarmid et al. (1999)	Baja California del Sur and western mainland Mexico.
C. m. angelensis Klauber, 1963 Angel de la Guardia Rattlesnake	Klauber (1972); Hoge & Romano-Hoge (1978/79); McCrystal & McCoid (1986); Campbell & Lamar (1989); Rubio (1998); McDiarmid et al. (1999)	Angel de la Guardia Island in the Gulf of California, Mexico off the coast of Baja California.
C. m. muertensis* Klauber, 1949 El Muerto Island Rattlesnake	Klauber (1972); Hoge & Romano-Hoge (1978/79); McCrystal & McCoid (1986); Campbell & Lamar (1989); Rubio (1998); McDiarmid et al. (1999); McPeak (2000) *Remark: According to Grismer (1999), this should be a full species but did not give data to justify the change.	El Muerto Island in the Gulf of California, Mexico off the coast of Baja California.
C. m. pyrrhus (Cope, 1867) Southwestern Speckled Rattlesnake	Smith & Taylor (1966); Klauber (1972); Hoge & Romano-Hoge (1978/79); McCrystal & McCoid (1986); Campbell & Lamar (1989); Ernst (1992); Brown (1997); Rubio (1998); McDiarmid et al. (1999); Bartlett & Tennant (2000)	SW United States and the northern part of the peninsula of Baja California del Norte and n Sonora, Mexico.
C. m. stephensi Klauber, 1930 Panamint Rattlesnake	Klauber (1972); Hoge & Romano-Hoge (1978/79); McCrystal & McCoid (1986); Ernst (1992); Brown (1997); Rubio (1998); McDiarmid et al. (1999); Bartlett & Tennant (2000)	WS Nevada and c California in the United States; mainly in rocky, arid foothill, escarpment and outcropping habitats, but can also be found in other environments.
C. molossus Baird & Girard, 1853 Black-tailed Rattlesnake	Gloyd (1936a); Klauber (1972); Hoge & Romano-Hoge (1978/79); Price (1980); Campbell & Lamar (1989); Degenhardt et al. (1996); Rubio (1998); McDiarmid et al. (1999); Dixon (2000a)	SW United States in Arizona, New Mexico, and, w and c Texas into the Edwards Plateau southward through Mexico from Sonora, Chihuahua, and Coahuila, southward to the s Mexican Plateau and Mesa del Sur inc Oaxaca, and on Tiburón Island and San Estéban Island in the Gulf of California; from sea level to about 2930 m in a wide variety of habitats.
C. m. molossus Baird & Girard, 1853 Northern Black-tailed Rattlesnake	Smith & Taylor (1966); Conant (1975); Hoge & Romano-Hoge (1978/79); Price (1980); Campbell & Lamar (1989); Ernst (1992); Degenhardt et al. (1996); Tennant (1998); Rubio (1998); Conant & Collins (1998); McDiarmid et al. (1999); Dixon (2000); Werler & Dixon (2000); Bartlett & Tennant (2000); Tennant & Bartlett (2000)	SW USA from s Arizona, s New Mexico and Texas southward into Mexico in Chihuahua, Coahuila, Nuevo León, and Sonora, it is reported from Tiburón Island.

The Caenophidia 329

Species-level Taxa	Citations and Remarks	Abbreviated Range
C. m. estebanensis* Klauber, 1949 San Esteban Island Rattlesnake	Klauber (1972); Hoge & Romano-Hoge (1978/79); Price (1980); Campbell & Lamar (1989); Liner (1994); Rubio (1998); McDiarmid et al. (1999); McPeak (2000) *Remark: Grismer (1999) stated that *C. m. estebanensis* should be elevated to a full species but does not provide data to justify the change.	San Esteban Island in the Gulf of California, Mexico.
C. m. nigrescens Gloyd, 1936 Mexican Black-tailed Rattlesnake	Smith & Taylor (1966); Klauber (1972); Hoge & Romano-Hoge (1978/79); Price (1980); Campbell & Lamar (1989); Pérez-Higareda & Smith (1991); Rubio (1998); McDiarmid et al. (1999); Uribe-Peña et al. (1999)	C Mexico throughout the s Mexican Plateau.
C. m. oaxacus* Gloyd, 1948 Oaxacan Black-tailed Rattlesnake	Hoge & Romano-Hoge (1978/79); McCranie (1981); Campbell & Lamar (1989); Greene *in* Klauber (1997a); Rubio (1998); McDiarmid et al. (1999) *Remarks: Often listed as *C. basiliscus oaxacus* (see McCranie 1981 and Rubio 1998). However, this listing follows Campbell & Lamar (1989) who recognized this snake as *C. m. oaxacus*.	Mexico in the states of Oaxaca (central part) and Puebla.
C. polystictus (Cope, 1865) Mexican Lance-headed Rattlesnake	Gloyd (1940); Smith & Taylor (1966); Klauber (1972); McCranie (1976); Hoge & Romano-Hoge (1978/79); Campbell & Lamar (1989); Pérez-Higareda & Smith (1991); Rubio (1998); McDiarmid et al. (1999)	Mexico in the tableland of the southern portion of the Mexican Plateau from wc Veracruz westward across c Puebla, México, District Federal, s Guanajuato, n Michoacán, sw Querétaro, c and e Jalisco, and s Zacateca; it may also occur in Colima, Tlaxcala, and Aguascalientes, and literature reports it from Hidalgo. Normally found from 1450 to 2600 m in broad valleys, high gently rolling plains, and grassy meadows.
C. pricei Van Denburgh, 1895 Twin-spotted Rattlesnake	Smith & Taylor (1966); Klauber (1972); McCranie (1980b); Campbell & Lamar (1989); Rubio (1998); McDiarmid et al. (1999)	SW United States into n Mexico; Degenhardt et al. (1996), do not believe it is found in New Mexico; it is generally associated wtih pine-oak and boreal forest.
C. p. pricei Van Denburgh, 1895 Western Twin-spotted Rattlesnake	Smith & Taylor (1966); Klauber (1972); Hoge & Romano-Hoge (1978/79); McCranie (1980b); Campbell & Lamar (1989); Ernst (1992); Rubio (1998); McDiarmid et al. (1999); Bartlett & Tennant (2000)	SE Arizona in the United States southward into adjacent Mexico from Arizona in the Chiricahua, Huachuca, and Santa Rita mountains southward into the high mountains of w Chihuahua, Durango, e Sonora, and probably into e Sinaloa; it is closely assoicated with the Sierra Madre Occidental.
C. p. miquihuanus Gloyd, 1940 Eastern Twin-spotted Rattlesnake	Smith & Taylor (1966); Klauber (1972); Hoge & Romano-Hoge (1978/79); McCranie (1980b); Campbell & Lamar (1989); Rubio (1998); McDiarmid et al. (1999)	NE Mexico in the northern part of the Sierra Madre Oriental, in w Tamaulipas, and s Nuevo León.
C. pusillus Klauber, 1952 Tancitaran Dusky Rattlesnake	Klauber (1972); Hoge & Romano-Hoge (1978/79); McCranie (1983a); Campbell & Lamar (1989); Rubio (1998); McDiarmid et al. (1999)	WC Mexico in the highlands of Sierra de Coalcomán of sw Michoacán and the Transverse Volcanic Cordillera of wc Michoacán and adjacent Jalisco, and probably into ne Colima; it is found in two disjunct ranges; from 1525 to 2380 m usually in pine-oak forest.

Species-level Taxa	Citations and Remarks	Abbreviated Range
*C. ruber** Cope, 1892 Red Diamond Rattlesnake	Klauber (1972); Campbell & Lamar (1989); McDiarmid et al. (1999); Bartlett & Tennant (2000) [as *C. exsul ruber*] *Remark: *Crotalus ruber* was reassigned as a subspecies of *Crotalus exsul* and the subspecies reassigned to follow suit by Grismer et al. (1994). Brown (1997) also listed this snake as *C. exsul ruber* (Kallert, 1827). Rubio (1998) felt that this was premature, and chose to retain *C. ruber* until the matter is settled by systematic herpetologists. McDiarmid et al. (1999) maintained *ruber* separate in their list, and the reader is advised to see their comments. Grismer (1999) stated Grismer et al. (1994) had placed *C. ruber* in synonymy of *C. exsul* and that an ICZN name was petitioned in 1995 to give precedence of *C. ruber* over *C. exsul*. Grismer noted that according to Article 80 of the International Code of Zoological Nomenclature, the most commonly used classification should be employed until a decision has been made by the commission. Opinion 1960 published 29 September 2000 in Vol. 57, Part 3: 189-190 of the Bulletin of Zoological Nomenclature gave *Crotalus ruber* Cope, 1892 (Reptilia: Serpentes) precedence over that of *Crotalus exsul* Garman, 1884.	SW United States into Mexico through the Baja California Peninsula and attendant offshore islands; from sea level to 1500 m in desert, chaparral, pine-oak woodland and tropical deciduous forest.
*C. r. ruber** Cope, 1892 Red Diamond Rattlesnake	Smith & Taylor (1966); Klauber (1972); Harris & Simmons (1978); Hoge & Romano-Hoge (1978/79); Campbell & Lamar (1989); Ernst (1992); Smith & Smith (1993); Greene *in* Klauber (1997a); Rubio (1998); McDiarmid et al. (1999) *Remark 1: Sometimes found as *C. exsul exsul*. *Remark 2: To quote Campbell & Lamar (1989), "The populations occuring on Isla Monserrate and Isla Angel de la Guarda [*C. r. elegans* (Schmidt, 1922)] were recognized as endemic subspecies by Harris & Simmons (1978b). However, the justification for doing so was so feeble that their recommenations have not been taken seriously by subsequent investigators." McDiarmid et al. (1999) quoted Goaly et al. (1993), that this was a subspecies of *C. r. ruber*. *Remark 3: Grismer et al. (1994) pointed out that *C. r. lucasensis* Van Denburgh, 1920, the San Lucan Red Diamond Rattlesnake found on S Baja California del Sur, Mexico from about 26°N southward to the Cape. is not a valid subspecies. This was quoted by Greene *in* Klauber (1997a).	California in the United States southward into n Baja California del Norte to about 26°N latitude; and associated islands.

The Caenophidia 331

Species-level Taxa	Citations and Remarks	Abbreviated Range
*C. r. lorenzoensis** Radcliffe & Maslin, 1975 San Lorenzo Island Rattlesnake	Hoge & Romano-Hoge (1978/79); Campbell & Lamar (1989); Rubio (1998); Greene *in* Klauber (1997a); McDiarmid et al. (1999); McPeak (2000) *Remark: Grismer (1999) suggested this taxon should be a distinct species but did not give data to justify this recommendation.	Isla San Lorenzo del Sur in the Gulf of California, Mexico.
C. scutulatus (Kennicott, 1861) Mojave Rattlesnake	Klauber (1972); Price (1982b); Campbell & Lamar (1989); Degenhardt et al. (1996); Brown (1997); Rubio (1998); McDiarmid et al. (1999)	SW United States in s California, s Nevada, extreme sw Utah, Arizona, s New Mexico, and w Texas southward into Mexico to s Puebla from near sea level to about 2500 m; generally in desert, mesquite-grassland, and in the southern portion of its range in pine-oak forest.
C. s. scutulatus (Kennicott, 1861) Mojave Green Rattlesnake	Smith & Taylor (1966); Klauber (1972); Conant (1975); Hoge & Romano-Hoge (1978/79); Price (1982b); Campbell & Lamar (1989); Ernst (1992); Brown (1997); Tennant (1998); Conant & Collins (1998); Rubio (1998); McDiarmid et al. (1999); Dixon (2000a); Werler & Dixon (2000); Bartlett & Tennant (2000); Tennant & Bartlett (2000)	SW United States in California eastward to w Texas and southward into Querrétaro, Mexico inluding the states of Chihuahua, Coahuila, Durango, San Luis Potosí, Sonora, Tamaulipas, and Zacatecas.
C. s. salvini Günther, 1895 Huamantlan Rattlesnake	Smith & Taylor (1966); Klauber (1972); Hoge & Romano-Hoge (1978/79); Price (1982b); Campbell & Lamar (1989); Pérez-Higareda & Smith (1991); Rubio (1998); McDiarmid et al. (1999)	C Mexico from Hidalgo through Tlaxcala and se Puebla into sw Veracruz.
C. stejnegeri Dunn, 1919 Long-tailed Rattlesnake	Smith & Taylor (1966); Hardy & McDiarmid (1969); Klauber (1972); Harris & Simmons (1978); Hoge & Romano-Hoge (1978/79); Collins (1982); Campbell & Lamar (1989); Rubio (1998); McDiarmid et al. (1999)	In the rugged canyons and foothills of se Sinaloa and w Durango in nw in the western portion of the Sierra Madre Occidnetal in Mexico; it probably occurs in n Nayarit; from about 500 to 1200 m in tropical deciduous and pine-oak forest.
C. tigris Kennicott *in* Baird, 1859 Tiger Rattlesnake	Smith & Taylor (1966); Klauber (1972); Hoge & Romano-Hoge (1978/79) Campbell & Lamar (1989); Ernst (1992); Rubio (1998); McDiarmid et al. (1999); McPeak (2000); Bartlett & Tennant (2000)	SW United States in rocky foothills of mountains and their adjacent desert slopes of sc Arizona southward into ne and c Sonora in nw Mexico; Degenhardt et al. (1996) think *C. tigris* may range into New Mexico; it is also found on Isla Tiburón in the Gulf of California; near sea level to about 1465 m in rugged, rocky country in the Sonoran desert and mesquite-grassland, barely entering tropical deciduous forest in s Sonora.
*C. tortugensis** Van Denburgh & Slevin, 1921 Tortuga Island Diamond Rattlesnake	Smith & Taylor (1966); Klauber (1972); Hoge & Romano-Hoge (1978/79); Campbell & Lamar (1989); Rubio (1998); McDiarmid et al. (1999); McPeak (2000) *Remark: Sometimes listed as *C. atrox tortugensis* (see Golay et al., 1993).	Endemic to Tortuga Island off nw coast of Baja California del Sur, Mexico in desert.
C. transversus Taylor, 1944 Cross-Banded Mountain Rattlesnake	Davis & Smith (1953); Smith & Taylor (1966); Klauber (1972); Hoge & Romano-Hoge (1978/79) Armstrong & Murphy (1979); Campbell & Lamar (1989); Rubio (1998); McDiarmid et al. (1999)	C Mexico probably in the states of México, nw Morelos, and Distrito Federal from the Sierra Ajusco and the Sierra de Monte Alto of the Transverse volcanic Cordillera in temperate boreal forests above 2900 m;

Species-level Taxa	Citations and Remarks	Abbreviated Range
		McDiarmid et al. (1999) noted that a *Crotalus* from Cerro Tancítaro in Michoacán may represent *C. transversus*, but must be confirmed.
C. triseriatus (Wagler, 1830) Dusky Rattlesnake	Smith & Taylor (1966); Klauber (1972); Campbell & Lamar (1989); Dorcas (1992); Rubio (1998); McDiarmid et al. (1999)	C Mexico in the highlands of the Transverse Volcanic Cordillera along the southern edge of the Mexican Plateau from wc Veracruz westward to c Jalisco, in the east extending to as far north on the Mesa Central as s San Luis Potosí, and the west to the Sierra de Nayarit in Nayarit from about 1500 to over 4300 m in pine-oak forest, boreal forest, and mesquite grassland.
C. t. triseriatus* (Wagler, 1830) Central Plateau Dusky Rattlesnake	Davis & Smith (1953); Smith & Taylor (1966); Klauber (1972); Hoge & Romano-Hoge (1978/79); Campbell & Lamar (1989); Pérez-Higareda & Smith (1991); Rubio (1998); McDiarmid et al. (1999) Uribe-Peña et al. (1999) *Remark: Campbell & Lamar (1989) considered *C. t. anahuacus* Gloyd, 1940 as a synonym. It is associated with the southeastern part of the Mexican Plateau in the Distrito Federal and states of Mexico, Morealos, Puebla, and Veracruz.	SC part of the Mexican Plateau in c Mexico with a disjunct distribution in the highlands of the Transverse Volcanic Cordillera from wc Veracruz into the Volcán and El Nevado de Colima in sw Jalisco.
C. t. armstrongi Campbell, 1978 Armstrong's Dusky Rattlesnake	Campbell & Lamar (1989); Greene *in* Klauber (1997a); Rubio (1998); McDiarmid et al. (1999)	Mexico in the western part of the Mesa Central and probably the souternmost extension of the Sierra Madre Occidental in Nayarit.
C. viridis* (Rafinesque, 1818) Western Rattlesnake	Klauber (1972); Campbell & Lamar (1989); Brown (1997); Rubio (1998); McDiarmid et al. (1999); Dixon (2000a); Ashton (2000) *Remark1 : Known in older literature as *C. confluentes* (see Smith & Smith, 1993). *Remark 2: Pook et al. (2000) reviewed the historical biogeography infered from mitochondrial DNA sequence information. Since they considered this "infered" information, the older classficiations have been maintained here until further information becomes available.	SW Canada in s British Columbia, Alberta, and Saskatchewan southward into the United States in North Dakota, South Dakota, Nebraska, Kansas, Oklahoma, Texas, and all states west of this north-south line; and, southward into Mexico in n Coahuila, Chihuahua, w Baja California del Norte, extreme n Baja California del Sur, and Isla Coronado del Sur in the Pacific Ocean.
C. v. viridis (Rafinesque, 1818) Prairie Rattlesnake	Smith & Taylor (1966); Klauber (1972); Conant (1975); Hoge & Romano-Hoge (1978/79); Campbell & Lamar (1989); Ernst (1992); Rubio (1998); Tennant (1998); Conant & Collins (1998); McDiarmid et al. (1999); Dixon (2000a); Werler & Dixon (2000); Bartlett & Tennant (2000); Tennant & Bartlett (2000)	Rocky Mountain and Great Plains regions from se Alberta and sw Saskatchewan in Canada southward through the w and c United States into n Chihuahua, c Coahuila, and ne Sonora in nc Mexico.
C. v. abyssus Klauber, 1930 Grand Canyon Rattlesnake	Klauber (1972); Hoge & Romano-Hoge (1978/79); Ernst (1992); Rubio (1998); McDiarmid et al. (1999); Bartlett & Tennant (2000)	SW United States in the Grand Canyon of nw Arizona.

Species-level Taxa	Citations and Remarks	Abbreviated Range
C. v. caliginis Klauber, 1949 Coronado Island Rattlesnake	Klauber (1972); Hoge & Romano-Hoge (1978/79); Campbell & Lamar (1989); Rubio (1998); McDiarmid et al. (1999); McPeak (2000)	Isla Coronado del Sur off Baja California, Mexico; restricted to the interior of the canyon in piñon-juniper woodlands into desert-scrublands and mesquite-scrublands.
C. v. cerberus (Coues, 1875) Arizona Black Rattlesnake	Klauber (1972); Hoge & Romano-Hoge (1978/79); Ernst (1992); Smith & Smith (1993); Rubio (1998); McDiarmid et al. (1999); Bartlett & Tennant (2000)	C Arizona and adjacent w New Mexico in the United States; this is a montane form.
C. v. concolor Woodbury, 1929 Midget Faded Rattlesnake	Klauber (1972); Hoge & Romano-Hoge (1978/79) Ernst (1992); Rubio (1998); McDiarmid et al. (1999); Bartlett & Tennant (2000)	C western United States in e Utah, adjacent sw Wyoming, and w Colorado in suitable rocky-desert and plains habitat.
C. v. helleri (Meek, 1905) Southern Pacific Rattlesnake	Klauber (1972); Hoge & Romano-Hoge (1978/79); Campbell & Lamar (1989); Brown (1997); Rubio (1998); McDiarmid et al. (1999); McPeak (2000); Bartlett & Tennant (2000)	S California in the United States along the Pacific coast in the n half of Baja California, Mexico in a diverse variety of habitats.
C. v. lutosus Klauber, 1930 Great Basin Rattlesnake	Klauber (1972); Hoge & Romano-Hoge (1978/79); Ernst (1992); Brown (1997); Rubio (1998); McDiarmid et al. (1999); Bartlett & Tennant (2000)	Great Basin region of w United States ranging widely over the w United States in nw California eastward and northward to w Utah, s Wyoming, and se Oregon in a variety of habitats.
C. v. nuntius Klauber, 1935 Hopi Rattlesnake	Klauber (1972); Hoge & Romano-Hoge (1978/79); Ernst (1992); Rubio (1998); McDiarmid et al. (1999); Bartlett & Tennant (2000)	The United States in nw Arizona and immediately adjacent Utah and New Mexico, and perhaps to the immediate sw corner of Colorado; associated with arid and semi-arid plains and grasslands, but can also be found in open coniferous woodlands.
C. v. oreganus Holbrook, 1840 Northern Pacific Rattlesnake	Smith & Taylor (1966); Klauber (1972); Hoge & Romano-Hoge (1978/79); Ernst (1992); Brown (1997); Rubio (1998); McDiarmid et al. (1999); Bartlett & Tennant (2000)	NW United States into sw Canada from British Columbia southward into cw Idaho and c California; it is associated with rocky escarpments and outcroppings, rocky hillsides, and other rocky areas in grasslands and similar habitats in sw New Mexico and se Arizona.
C. willardi Meek, 1905 Ridge-nosed Rattlesnake	Smith & Taylor (1966); Klauber (1972); Campbell & Lamar (1989); Rubio (1998); McDiarmid et al. (1999)	SW United States from high elevations in the Santa Rita and Huachuca Mountains in Arizona and the Sierra Tarahumari and Sierra Madre mountains of e Sonora, w Chihuahua, Durango and w Zacatecas; from about 1660 to 2750 m in various habitats.
C. w. willardi Meek, 1905 Arizona Ridge-nosed Rattlesnake	Klauber (1972); Hoge & Romano-Hoge (1978/79); Campbell & Lamar (1989); Rubio (1998); McDiarmid et al. (1999); Bartlett & Tennant (2000)	S Arizona in the United States into n Sonora, Mexico; found in the Huachuca, Patagonia, and Santa Rita mountains of se Arizona, southward into the Sierra de los Ajos, Sierra de Cananca, and Sierra Azul of n Sonora, Mexico.
C. w. amabilis Anderson, 1962 Del Nido Ridge-nosed Rattlesnake	Klauber (1972); Hoge & Romano-Hoge (1978/79); Campbell & Lamar (1989); Rubio (1998); McDiarmid et al. (1999)	NC Chihuahua, Mexico; only from several canyons in the Sierra del Nido Mountains.

Species-level Taxa	Citations and Remarks	Abbreviated Range
C. w. meridionalis Klauber, 1949 Southern Ridge-nosed Rattlesnake	Klauber (1972); Hoge & Romano-Hoge (1978/79); Campbell & Lamar (1989); Rubio (1998); McDiarmid et al. (1999)	Mexico in s Durango, nw Zacatecas, and probably northward into the Sierra Madre Occidental into nw Durango.
C. w. obscurus Harris & Simmons, 1974 New Mexico Ridge-nosed Rattlesnake	Hoge & Romano-Hoge (1978/79); Campbell & Lamar (1989); Greene *in* Klauber (1997a); Rubio (1998); McDiarmid et al. (1999); Bartlett & Tennant (2000)	S New Mexico in the United States into nw Chihuahua, Mexico.
C. w. silus Klauber, 1949 West Chihuahuan Ridge-nosed Rattlesnake	Klauber (1972); Hoge & Romano-Hoge (1978/79); Campbell & Lamar (1989); Rubio (1998); McDiarmid et al. (1999)	Animas and Peloncillo mountains in s New Mexico in the United States, and in Mexico in the Sierra de San Luis Mountains in extreme ne Sonora and nw Chihuahua.

GENUS: *Lachesis* Daudin, 1803, was considered to be monotypic species as *Lachesis mutus* Linnaeus, 1766, until re-evaluated by Zamudio & Greene (1997). The snake ranges from Central America in Nicaragua southward becoming widespread in South America, although it is not plentiful anywhere. See Vial & Jiménez-Porras (1967); Peters & Orejas-Miranda (1970); Hoge & Romano-Hoge (1978/79); Savage & Villa (1986); Campbell & Lamar (1989); Grantsau (1990); Ripa (1994a, 1994b, 1999); Zamudio & Greene (1997); Starace (1998); Kornacker (1999); McDiarmid et al. (1999) and Köhler (2001).

Review assistance and important contributions for this section contributed by: Harry W. Greene.

CONTENT: Three speices. ★★★★

Lachesis: Bushmaster

Species	Citations and Remarks	Abbreviated Range
L. melanocephala Solórzano & Cerdas, 1986 Black-headed Bushmaster	Solórzano & Cerdas (1986); Campbell & Lamar (1989); Ripa (1994a, 1994b); Zamudio & Greene (1997); McDiarmid et al. (1999); Köhler (2001)	Confined to the southern Pacific lowlands of Costa Rica from sea level to about 1500 m in se Puntarenas Province.
L. muta (Linnaeus, 1766) South American Bushmaster	Taylor (1951); Peters & Orejas-Miranda (1970); Amaral (1976); Moonen et al. (1979); Campbell & Lamar (1989); Grantsau (1990); Zamudio & Greene (1997); Starace (1998); McDiarmid et al. (1999); Kornacker (1999)	Found only in South America and Trinidad and Tobago, widespread but not common anywhere; it is closely associated with the equitorial forests east of the Andes in Amazonia and the southern Atlantic coastasl forests in Brazil.
L. m. muta (Linnaeus, 1766) Amazonian Bushmaster	Peters & Orejas-Miranda (1970); Hoge & Romano-Hoge (1978/79); Martínez & Bolaños (1983); Campbell & Lamar (1989); Grantsau (1990); Zamudio & Greene (1997); Lancini & Kornacker (1989); Murphy (1997); Starace (1998); McDiarmid et al. (1999); Kornacker (1999); Boos (2001)	Equitorial forests of c and n South America in n Brazil, Trinidad, Guyana, Suriname, French Guiana, e Ecuador, e and s Venezuela, e Colombia, Peru, and n Bolivia. It is closely associated with the Amazon rainforest.
L. m. rhombeata* Wied, 1824 Atlantic Forest Bushmaster	Hoge & Romano-Hoge (1978/79); Campbell & Lamar (1989); Zamudio & Greene (1997); McDiarmid et al. (1999); Borges-Nojosa & Lima-Verde (1999); Freitas (1999); Marques et al. (2001)	Coastal Atlantic forests of se Brazil from Alagoas and s Rio Grande do Norte to Rio de Janeiro and Ceará.

Species-level Taxa	Citations and Remarks	Abbreviated Range
	*Remark: Hoge & Romano (1978) concluded that *L. m. noctivaga* Hoge, 1966 is a synonym.	
*L. stenophrys** Cope, 1875 Central American Bushmaster	Taylor (1951); Peters & Orejas-Miranda (1970); Hoge & Romano-Hoge (1978/79); Campbell & Lamar (1989); Ripa (1994a, 1994b); Zamudio & Greene (1997); McDiarmid et al. (1999); Köhler (2001)	Forests of Central America in the Atlantic versant from s Nicaragua, Costa Rica, into Panama then expanding its range to the Pacific lowlands in c Panama and southward into the Chocó lowlands along both coasts in n and w Colombia and nw Ecuador.
	*Remark: Bolaños (1982) thought bushmasters from c Panama belonged to *L. m. muta* based on the high ventral count. However, Campbell & Lamar (1989) stated the Chocó population in Colombia are probabaly referable to the widespread Central American taxon. According to Zamudio & Greene (1997), the bushmasters found from c Panama into the Pacific lowlands of Colombia and Ecuador remain problematic but seem to fit best genetically into *L. stenophrys* group.	

GENUS: *Ophryacus* Cope, 1887, is restricted to southern Mexico. See Hoge & Romano-Hoge (1978/79); Campbell & Lamar (1989, 1992); McDiarmid et al. (1999).

See also *Atropoides, Bothriechis, Bothriopsis, Bothrocophias, Bothrops, Cerrophidion,* and *Porthidium*.

CONTENT: Two species. ★★★★

Ophryacus: Mexican Horned Pit Viper or Torito.

Species	Citations and Remarks	Abbreviated Range
O. melanurus (Müller, 1924) Black-tailed Horned Pit Viper	Hoge & Romano-Hoge (1978/79); Campbell & Lamar (1989); Lee (1996); Gutberlet (1997); McDiarmid et al. (1999)	Mountains of s Mexico in s Puebla and c Oaxaca from 1600 to 2400 m; in high arid tropical scrub and tropical deciduous fores in the north and pine-oak forest in the south.
O. undulatus (Jan, 1859) Mexican Horned Pit Viper or Torito	Hoge & Romano-Hoge (1978/79); Pérez-Higareda & Smith (1991); Campbell & Lamar (1992); Flores-Villela et al. (1992); McDiarmid et al. (1999)	SW Mexico at intermediate elevations in the mountains of s Sierra Madre Oriental, Mesa del Sur and Sierra Madre del Sur; in pine-oak and cloud forests from about 1800 to 2800 m in Guerrero, wc Veracruz and Oaxaca; it is found west of the Isthmus of Tehuantepec. It was recently reported in Hidalgo.

GENUS: *Porthidium* Cope, 1871, occurs from southern Mexico southward through Central America into northern South America to n Venezuela and Ecuador. See Peters & Orejas-Miranda (1970); Hoge & Romano-Hoge (1978/79); Campbell & Lamar (1989, 1992); Campbell & Lewis (1992); McDiarmid et al. (1999) and Köhler (2001). See McDiarmid et al. (1999) for comments on the nomenclature of this snake.

See also *Atropoides, Bothriechis, Bothriopsis, Bothrocophias, Bothrops, Cerrophidion,* and *Ophryacus*.

CONTENT: Seven species. ★★★

Porthidium: Hog-nosed Pit Viper

Species	Citations and Remarks	Abbreviated Range
P. dunni (Hartweg & Oliver, 1938) Dunn's Hog-nosed Pit Viper	Hartweg & Oliver (1938); Hoge & Romano-Hoge (1978/79); Villa et al. (1988); Campbell & Lamar (1989); McDiarmid et al. (1999); Köhler (2001)	Pacific foothills and coastal plains of s Mexico in sw Oaxaca into extreme w Chiapas; near sea level to just over 500 m in tropical deciduous forest.
P. hespere (Campbell, 1976) Western Hog-nosed Pit Viper	Campbell (1976); Hoge & Romano-Hoge (1978/79); Campbell & Lamar (1989); McDiarmid et al. (1999)	Only known from the type locality in Colima, Mexico; in deciduous forest, fide Campbell & Lamar (1989), it is possible that the snake may range into the arid Tepalcatepec Valley in Michoacán and se Jalisco.
P. lansbergii (Schlegel, 1841) Lansberg's Hog-nosed Pit Viper	Roze (1959); Peters & Orejas-Miranda (1970); Hoge & Romano-Hoge (1978/79); Villa et al. (1988); Campbell & Lamar (1989); McDiarmid et al. (1999); Kornacker (1999); Köhler (2001)	E Central America in Panama into n South America in Colombia, Venezuela and Ecuador; to at least 1270 m in a wide variety of habitats.
P. l. lansbergi (Schlegel, 1841) Lansberg's Hog-nosed Pit Viper	Peters & Orejas-Miranda (1970); Hoge & Romano-Hoge (1978/79); Campbell & Lamar (1989); McDiarmid et al. (1999); Kornacker (1999)	E coast of Colombia and Panama; listed by Kornacker as being found in Venzuela.
P. l. arcosae Schätti & Krammer, 1993 Manabí Hog-nosed Pit Viper	Schätti & Krammer (1993); McDiarmid et al. (1999)	Manabí, Ecuador.
P. l. houtmanni Sandner-Montilla, 1989 Houtmann's Hog-nosed Pit Viper	Sandner-Montilla (1989); McDiarmid et al. (1999); Kornacker (1999)	Margarita Island, Venezuela.
P. l. rozei (Peters, 1968) Roze's Hog-nosed Pit Viper	Peters & Orejas-Miranda (1970); Lancini & Kornacker (1989); Campbell & Lamar (1989); McDiarmid et al. (1999); Kornacker (1999)	N Venezuela.
P. nasutum (Bocourt, 1868) Rainforest Hog-nosed Pit Viper	Peters & Orejas-Miranda (1970); Hoge & Romano-Hoge (1978/79); Alvarez del Toro (1982); Wilson & McCranie (1984); Wilson & Meyer (1985); Savage & Villa (1986); Villa et al. (1988); Campbell & Lamar (1989); Lee (1996, 2000); Cruz (1997); McDiarmid et al. (1999); Renjifo & Lundberg (1999); Stafford & Meyer (2000); Köhler (2001)	Veracruz and the Yucatán in s Mexico southward into nw South America into Colombia and Ecuador; it seems to be absent from the Darién region between Panama, and Colombia; typically from near sea level to 900, exceptionally to 1880 m in lowland rain forest and lower montane wet forest.
P. ophryomegas (Bocourt, 1868) Slender Hog-nosed Pit Viper	Schmidt & Andrews (1936); Peters & Orejas-Miranda (1970); Hoge & Romano-Hoge (1978/79); Wilson & Meyer (1985); Savage & Villa (1986); Solórzano et al. (1988); Villa et al. (1988); Campbell & Lamar (1989); Cruz (1997); McDiarmid et al. (1999); Köhler (2001)	Central America in sc Guatemala, El Salvador, Honduras, Nicaragua, Costa Rica, and possibly sw Panama; from sea level to about 1000 m, usually associated with seasonally dry forest.

Species-level Taxa	Citations and Remarks	Abbreviated Range
P. volcanicum Solórzano, 1994 [1995] Chinilla or Volcano Hog-nosed Pit Viper	Solórzano (1994); McDiarmid et al. (1999); Köhler (2001)	SW Costa Rica in Valle del General, Puntareana Providence.
P. yucatanicum (Smith, 1941) Yucatán Hog-nosed Pit Viper	Hoge & Romano-Hoge (1978/79); Campbell & Lamar (1989); Flores-Villela (1993); Villa et al. (1988); Liner (1994); Lee (1996, 2000); McDiarmid et al. (1999); Köhler (2001)	Northern half of the Yucatán Peninsula in Mexico; from just above sea level to about 250 m in tropical deciduous forest and thorn forest in a region composed of porous limestone covered by scrubby xerophytic vegetation.

GENUS: *Sistrurus* Garman, 1883 is found from s Ontario in Canada southward through the eastern and midwestern United States southward and westward into se Arizona, New Mexico and Texas, and southward into Mexico to the southern tip of the Mexican Plateau. See Gloyd (1940); Smith & Taylor (1966); Klauber (1956, 1972, 1997); Conant (1975); Hoge & Romano-Hoge (1978/79); Campbell & Armstrong (1979); Campbell & Lamar (1989); Greene *in* Klauber (1997a); Rubio (1998); Conant & Collins (1998); McDiarmid et al. (1999); and Dixon (2000a).

Gloyd (1940) was the first modern review of the rattlesnakes. Klauber (1952, 1972), and most recently McDiarmid (1999) brought the information up to date.

See also *Crotalus*.

CONTENT: Three species. ★★★

Sistrurus: Massasauga, Pygmy Rattlesnake, and Mexican Pygmy Rattlesnake; Crother et al. (2000) classifies all of the Canada and United States species as Pygmy Rattlesnakes.

Species	Citations and Remarks	Abbreviated Range
S. catenatus (Rafinesque, 1818) Massasauga	Klauber (1972); Conant (1975); Minton (1983); Smith & Smith (1993); Rubio (1998); McDiarmid et al. (1999); Dixon (2000a)	S Canada in Ontario and the United States and extreme n Mexico.
S. c. catenatus (Rafinesque, 1818) Eastern Massasauga	Klauber (1972); Conant (1975); Hoge & Romano-Hoge (1978/79); Minton (1983); Ernst (1992); Rubio (1998); Conant & Collins (1998); McDiarmid et al. (1999); Tennant & Bartlett (2000)	S Canada in s Ontario into n and midwestern United States, in Michigan, Wisonsin, Iowa, Illinois, Indiana, Ohio, w Pennsylvania, and New York.
S. c. edwardsi (Baird & Girard, 1853) Desert Massasauga	Klauber (1972); Conant (1975); Hoge & Romano-Hoge (1978/79); Minton (1983); Campbell & Lamar (1989); Ernst (1992); Rubio (1998); Tennant (1998); Conant & Collins (1998); McDiarmid et al. (1999); Dixon (2000a); Werler & Dixon (2000); Bartlett & Tennant (2000); Tennant & Bartlett (2000)	SW United States in s and c Texas westward to the se corner of Arizona and southward into n Mexico in Cuatro Ciénegas Basin in Coahuila and near Aramberri in Nuevo León. Campbell and Lamar (1989) feel that it occurs in n Tamaulipas, n Chihuahua and ne Sonora. There is an unconfirmed report from Durango; it ranges from sea level to over 1500 m.
S. c. tergeminus (Say *in* James, 1823) Western Massasauga	Smith & Taylor (1966); Klauber (1972); Conant (1975); Hoge & Romano-Hoge (1978/79); Minton (1983); Ernst (1992); Degenhardt et al. (1996); Ten-	Midwestern United States southward into n Mexico as far as Sonora to Tamaulipas; this snake has spotty distribution in Texas; in the United States it is found

Species-level Taxa	Citations and Remarks	Abbreviated Range
	nant (1998); Rubio (1998); Conant & Collins (1998); McDiarmid et al. (1999); Dixon (2000a); Werler & Dixon (2000); Tennant & Bartlett (2000)	in the Texas Gulf Coast through Oklahoma and Kansas, into parts of Nebraska, Iowa, and Missouri.
S. miliarius (Linnaeus, 1766) Pygmy Rattlesnake	Gloyd (1935, 1940); Klauber (1972); Conant (1975); Palmer (1978); Dundee & Rossman (1989); Rubio (1998); McDiarmid et al. (1999)	SE United States from s North Carolina southward into Florida and west into Texas and Oklahoma.
S. m. miliarius (Linnaeus, 1766) Carolina Pygmy Rattlesnake	Gloyd (1935); Klauber (1972); Conant (1975); Palmer (1978); Hoge & Romano-Hoge (1978/79); Ernst (1992); Rubio (1998); Conant & Collins (1998); McDiarmid et al. (1999); Tennant & Bartlett (2000)	SE United States; east of the Appalachian Mountains from c North Carolina to ne Georgia; generally associated with woodlands.
S. m. barbouri Gloyd, 1935 Dusky Pygmy Rattlesnake	Gloyd (1935); Klauber (1972); Conant (1975); Palmer (1978); Hoge & Romano-Hoge (1978/79); Ashton & Ashton (1981); Ernst (1992); Rubio (1998); Tennant (1997b); Conant & Collins (1998); McDiarmid et al. (1999); Tennant & Bartlett (2000)	SE United States throughout Florida, s Georgia, and c Alabama in a wide variety of habitats.
S. m. streckeri Gloyd, 1935 Western Pygmy Rattlesnake	Gloyd (1935); Klauber (1972); Conant (1975); Palmer (1978); Hoge & Romano-Hoge (1978/79); Dundee & Rossman (1989); Ernst (1992); Tennant (1998); Rubio (1998); Conant & Collins (1998); McDiarmid et al. (1999); Dixon (2000a); Werler & Dixon (2000); Tennant & Bartlett (2000)	SE and s United States in e Texas and Oklahoma to s Missouri and Tennessee, to the lower Louisiana Delta Range; it is not found in most of the Mississippi River flood plain, associated with forests to dry upland areas.
*S. ravus** (Cope, 1865) Mexican Pygmy Rattlesnake	Davis & Smith (1953); Smith & Taylor (1966); Klauber (1972); Campbell & Armstrong (1979); Campbell & Lamar (1989); Rubio (1998); McDiarmid et al. (1999); Uribe-Peña et al. (1999) *Remark: In the subspecies list, a fourth subspecies, *S. r. sinaloensis* Julia-Zertuche, 1983 had been shown in Campbell & Lamar (1989). They acknowledged that the subspecies had only been proposed at the time of their publication. They also noted that the description was described as "woefully incomplete and the locality data are likely suspect". Smith & Smith (1993) regarded this as *nomen nullum* and the subspecies was not listed in Rubio (1998) nor in McDiarmid et al. (1999).	Mexico in temperate montane to high elevation forests across the Mexican Plateau from the highlands of wc Veracruz southward through Sierra de Acatepec in Puebla and extreme n Oaxaca, Mesa del Sur in Oaxaca including Sierra Juárez and mountains extending from Cerro San Felipe, Sierra Mixe, and Sierra Madre del Sur in Guerrero from 1490 to about 3000 m.
*S. r. ravus** (Cope, 1865) Central Plateau Pygmy Rattlesnake	Hoge & Romano-Hoge (1978/79); Campbell & Armstrong (1979); Campbell & Lamar (1989); Pérez-Higareda & Smith (1991); Rubio (1998); McDiarmid et al. (1999) *Remark: Campbell & Armstrong (1979) consider *S. ravus lutescens*, as found in Hoge & Romano-Hoge (1978/79), to be *S. r. ravus*. Rubio (1998) did not list *S. r. lutescens*.	Mexican Plateau in altiplanicie meridional in the states of wc Puebla, Veracruz, Morelos, Tlaxcala, c Guerrero, c Oaxaca, and Hidalgo.
S. r. brunneus Harris & Simmons, 1978	Hoge & Romano-Hoge (1978/79); Campbell & Armstrong (1979); Liner (1994); Campbell & Lamar	Oaxaca in highlands of Mexico.

The Caenophidia

Species-level Taxa	Citations and Remarks	Abbreviated Range
Oaxacan Pygmy Rattlesnake	(1989); Smith & Smith (1993); Greene *in* Klauber (1997a); Rubio (1998); McDiarmid et al. (1999)	
S. r. exiguus Campbell & Armstrong, 1979 Guerreran Pygmy Rattlesnake	Campbell & Armstrong (1979); Campbell & Lamar (1979); Smith & Smith (1993); Greene *in* Klauber (1997a); Rubio (1998); McDiarmid et al. (1999)	C Mexico from Sierra Madre del Sur in Guerrero.

Incertae Sedis and Other Questions

Incertae Sedis and Other Questions

Species Name in Question	Reference and Comment	Abbreviated Range
*Atractus trihedrurus** Amaral, 1926	Savage (1960a); Peters & Orejas-Miranda (1970); Amaral (1976); Fernandes & Puorto (1994); Ferandes (1995a, 1996), Fernandes et al. (2000)	Known only from the type specimen and type locality listed simply as forest of southern Brazil. The type locality is reported as São Bento, Estado do Santa Catarina.
	*Remark: As noted above, Peters & Orejas-Miranda (1970) listed it as *incertae sedis* and commented that "Savage, Misc. Publ. Mus. Zool. Univ. Mich., 112, 1969, 31, indicated that this was probably not an *Atractus*, but that its generic status must be regarded as uncertain until the holotype can be re-examined." Vanzolini (1986) did not comment on its status. Recently, Fernandes et al. (2000) noted *A. serranus* and *A. trihedrurus* are probably sister taxa, sharing the punative synapomorphy of a deeply bilobate hemipene and other characteristics of unknown polarizition, such as unusually high number of teeth (9–10) and similar coloration pattern.	
	Fide Ronaldo Fernandes (pers. comm. 15 March, 2001), this definitely is an *Atractus*. Julio Cesar Moura Leite and his collegues at the Museu de História Natural Capão da Imbúia in Curitiba brought this to Ronaldo Fernandes' attention several years ago and they have been having a continuing exchange about the status of the animal since that time. Ronaldo Fernandes feels that it may need to be lumped with *A. serranus*, but until recently the group in Curitiba have disagreed. Work still needs to be completed on this species to identify it correct status.	
*Atractus vittatus** Boulenger, 1894 Venezuelan Striped Tellurian Snake	Roze (1966); Peters & Orejas-Miranda (1970), Lancini (1979); Lancini & Kornacker (1989); Kornacker (1999)	Venezuela, only from the type locality of Caracas
	*Remark: Most authorities do not consider this name valid. It was listed as *incertae sedis* in Peters	

Species Name in Question	Reference and Comment	Abbreviated Range
	& Orejas-Miranda (1970). Vanzolini (1986) only made a note to see Peters & Orejas-Miranda (1970) and did not comment. Lancini & Kornacker (1989) listed it as valid without data. Fide Kornacker (1999), this is a valid species; however, it does not seem to be listed anywhere else. It appears to be different from *A. crassicaudatus* (Dumeril, Bibron & Dumeril, 1854), (see Peters & Orejas-Miranda, 1970, and Auth, 1994), although in his synonymy, Kornacker lists *Rhabdosoma crassicaudatum* Günther, 1858.	
Calamaria favae Filippi, 1840	Discussed as *incertae sedis* in Peters & Orejas-Miranda (1970): "The taxon was called *Atractus favae* by Boulenger (1894), and its distribution was given by Boulenger as "Brazil?" It is not mentioned by Inger and Marx (1965) in their revision of *Calamaria,* and it has not been included in any recent works on Javan species. Savage (1960) regarded it as a *nomen dubium* of the genus *Atractus*. According to Peters & Orejas-Miranda (1970) recent authors on Brazilian species at that time had not mentioned it." In researching this taxon, I still find that it has not been mentioned in current works.	Type locality and distribution unknown.
Chionactis saxatilis Funk, 1967 Shovel-nosed Snake	Collins (July, 1997) shows *C. saxatilis* Funk, 1967 and calls it a "Mountain Shovelnose Snake" in the website for The Center for the North American Amphibians and Reptiles (CNAAR). Stebbins (1985) discusses *C. saxatilis* as being differentiated from other *Chionactis* by decreasing tail-to-total tail ratio with maturity, and in males having more ventrals than females. I feel that the information supplied in the original paper is insufficient data. I have been unable to find any more reference or primary information on this species that reviews the status of this genus compared to other *Chionactis* to validate the species. Therefore, I have listed the species here because I feel that this is an invalid designation and the reader should consider these snakes to be *C. occipitalis annulata* (Baird) based on Smith & Smith (1993) until the nomenclature is otherwise validated. Smith & Smith (1993) is a checklist reference but reflects a reasonable assessment of the snake. Bartlett & Tennant (2000) also chose to leave it out of their line-up but did mention "In 1967, *Chionactis saxatilis* (the Mountain Shovel-Nosed Snake) was described from the Gila Mountains of southwestern Arizona. Taxonomists have long questioned the validity of this taxon, and it has been virtually unmentioned in the ensuing years."	In the original literature of Funk (1967), the snake is described as being a mountain population found in extreme sw Arizona and known only from three unnamed peaks in the Gila Mountains. The snake had been collected between 1800 and 2400 feet. Funk (1967) described the type as being taken 2.5 air miles northeast of the Fortune Mine at an elevation of 2300 feet.

Incertae Sedis and Other Questions

Species Name in Question	Reference and Comment	Abbreviated Range
Cochliophagus isolepis Müller, 1924	Peters & Orejas-Miranda (1970) stated that this clearly is not a Dipsadine snake, although the genus *Cochliophagus* is a synonym of *Sibynomorphus*. They did not know where it belonged.	Type locality is shown as Middle or South America.
Coluber cinereus Linnaeus, 1758	Dixon (1983f) reviewed the *Liophis cobella* group of neotropical colubrid snakes and determined that *C. cinereus* Linnaeus cannot be allocated to any currently recognized colubrid genus. See Dixon (1983f) for a history of the name and Andersson's (1989) comments. Dixon (1983f) said that the status of *C. cinereus* may never be resolved and that the name should be regarded as a nomen dubium without allocation to species, until further information is obtained.	Andersson (1898) noted that the catalogues of 1802 and 1808 contain records of two speciems *C. cinereus* in the Royal Museum of Stockholm (Museum Drottningham). One is missing and it could be the type of *C. cinereus*.
Coluber minervae Linnaeus, 1754	Michaud & Dixon (1987) commented on this snake. They noted that it has been listed as a synonym of *Liophis lineatus* and Linnaeus lists *C. minervae* in the tenth edition of *Systema Naturae* (1758) and refers this 1754 publication for its description. Although originally described in 1754, the appearance of *C. minervae* in the tenth edition of *Systema Naturae* (1758) established its nomenclatural status. See the Michaud & Dixon (1987) paper for the history of this snake and why it is incertae sedis.	Type believed to be lost.
Crotalus tesselatus Herrmann, 1804	McDiarmid et al. (1999) reviewed Klauber (1936) who noted that *tesselatus* cannot be assigned to a current species even though it resembles *C. durissus* and *C. adamantus* in size and pattern and *C. triseriatus* in scale counts. Klauber 1972:104 and Klauber 1997:104 have "*tesselatus* Herrmann, 1804, *Nomen indeterminatum;* description to brief for recognition."	Unknown; type species not given.
Dendrophidion boshelli Dunn, 1944	Fide Lieb (1988), this is *incertae sedis*. It is similar to *D. percarinatum* but Lieb did not show them as synonyms. Known from only two specimens. It is found in citations of Dunn (1944c); Peters & Orejas-Miranda (1970); and Lieb (1988).	Known only from the type locality of Volcanes, Município de Caparrapi in Departamento Cundinamarca in Colombia.
Dipsas brevifacies (Cope, 1866) Short-faced Snail-sucker	Peters (1960c, 1964b); Smith & Taylor (1966); Peters & Orejas-Miranda (1970); Kofron (1982); Villa et al. (1988); Savage & Slowinski (1992); Flores-Villela (1993); Smith & Smith (1993); Liner (1994); Fernandes (1995b); Campbell (1998b); Stafford & Meyer (2000); Lee (2000); Köhler (2001)	N and e Yucatán peninsula in Mexico into c Belize.
Dipsas infrenalis Rosén, 1905 Rosen's Snail-eater	This snake was listed as *incertae sedis* by Peters (1960c). Peters & Orejas-Miranda (1970) noted it as *incertae sedis*.	Type locality and distribution unknown.

Species Name in Question	Reference and Comment	Abbreviated Range
Dipsas subaequalis Fischer, 1880	Boulenger (1896) suggested that this might be an *Imantodes*. Peters & Orejas-Miranda (1970) noted Boulenger's remark and said that the coloration as described is so different from any other species in that genus that this seems a most improbable generic assignment.	Type locality and distribution unknown.
Dromicus capensis Jan, 1863	Dixon (1980) questioned its validity.	Unknown.
Enicagnethus melanoauchen Jan, 1863	Myers (1974); Peters & Orejas-Miranda (1970); Dixon (1980); Vanzolini (1986) Remark: Dixon (1980) relegated it to *incertae sedis melanoauchen,* and it was not referenced in Dixon (1989).	Bahia, Brazil.
Erythrolamprus mentalis Werner, 1909	Fide Peters & Orejas-Miranda (1970), this taxon has been placed in *Coniophanes* by various authors and in *Rhadinaea* by Bailey, Pap. Mich. Acad. Sci, Arts, Letters, 24, 1938 (1939), 5, but all such actions have been tentative, awaiting re-examination of the type. This specimen was destroyed during World War II, and the taxon may never be properly allocated.	Type locality is Guatemala. Distribution is unknown. Stuart (1963) does not mention this snake in his checklist of Guatemala.
Geophis, species inquirenda Myers, 2003 [*G. brachycephalus, auctorum*]	Myers (2003)	Panama. Specimen from Bocas del Toro in La Loma.
Helicops leprieuri moesta Jan, 1865	Fide Peters & Orejas-Miranda (1970), Boulenger (1893) tentatively synonymized this variety with *Helicops modestus*, but this cannot be determined without question from Jan's description according to the type, which has not yet been located, must be re-examined for proper allocation of the taxon.	Type Locality is none given.
*Helicops septemvittatus** (Fischer, 1864)	*Species inquirendae* fide Smith & Taylor (1966). *Remark: Fide Smith & Taylor (1966), there were 5 cotypes in the Hamburg Museum but there is much doubt that the original series, comprising the only known specimens of the species, actually came from Mexico, therefore they listed this as *species inquirendae*. See Smith & Taylor (1966) notes on occurrence of *Helicops* in Mexico on their comment about *Helicops schistosus* (Daudin) which they seem to think is Asian and also treat as *species inquirendae*.	Mexico?
Herpetodryas annectens Werner, 1924	Fide Peters & Oreajs-Miranda (1970), Amaral (1929e) stated that this was not neotropical, and wrong in the generic assignment. Amaral did not attempt to assign it elsewhere.	Type locality presumed to be Brazil by Werner, since it was in a bottle with two other snakes for the country. Distribution unknown.

Species Name in Question	Reference and Comment	Abbreviated Range
Imantodes gemmistratus reticulatus (Müller, 1882)	Peters & Orejas-Miranda (1970) and Ponce-Campos & Huerta-Ortega (1998) stated that the range and type locality cannot be properly identified so they listed as *species iquirenda*. However, this subspecies is listed by Liner (1994) as present in Mexico. Fide Smith & Smith (1993), *I. g. oliveri* Smith, 1942 is a synonym. however, listed *I. g. oliveri* only as valid. It has been included here based on Liner (1994). It probably is not valid. It is also shown in the checklist section for reference because of outstanding questions.	Based on *I. g. oliveri* the range should be Tehuantepec area of Oaxaca in s Mexico into w Guatemala.
Leptognathus andrei Sauvage, 1884	Peters & Orejas-Miranda (1970) stated that this snake was placed in *incertae sedis* in the Dipsadinae by Peters (1960c).	Type locality is listed as New Grenada and distribution is known only from the type locality give on the holotype.
Leptognatus brevis Dumeril, Bibron & Dumeril, 1854	Peters & Orejas-Miranda (1970) stated this taxon may be synonymous with *Sibon nebulata nebulata* according to Peters, Beitr. Neotrop. Fauna, 4, 1964:49. Smith & Smith (1993) listed this as a valid species so it has been included in the main checklist as *Sibon brevis*. However, there is no way to confirm the validity of the nomenclature against a type or other reference specimens.	Type locality is Mexico and distribution is unknown except for the vague type locality given by Dumeril, Bibron & Dumeril, 1854.
Leptotyphlops gadowi Duellman, 1956	Although this species is listed in Smith & Smith (1993) as valid, most authorities do not seem to refer to it. Because of this, the validity is noted here as being in question.	S Mexico in Jalisco and Michoacán.
Leptotyphlops undecimstriatus (Schlegel, 1839)	Gray (1845); Boulenger (1893); Orejas-Miranda *in* Peters & Orejas-Miranda (1970); Hahn (1980a); McDiarmid et al. (1999).	Known only from the type locality listed as Santa Cruz de la Sierra in Bolivia. It appears that after Schlegel, 1839 described the species a complete description was given in Gray (1845:140) as *Epicta undecimstriata*. The next, and what appears to be the last time a complete description was given, was Boulenger (1893) who synomyzed the species *Epicta undecimstriata* from Gray (1845) (as now might be called *Letptotyphlops undecimestriatus*) with *Glauconia albifrons* (now *Leptotyphlops albifrons*). Orejas-Miranda *in* Peters & Orejas-Miranda (1970) resurrected the species but did not give a justification for this. In addition, the species was not included in the key so there is nothing to indicate why he did this. He listed only Schlegel, 1839 in his literature and did not list Gray (1845) or Boulenger (1983) and may not have realized that it had been synomyzed. It seems that later literature picked up *L. undecimstriata* without identifying Orejas-Miranda's error. I have synomyzed *L. undecimstratus* as found in Orejas-Miranda *in* Peters & Orejas-Miranda (1970); Hahn (1980a) and McDiarmid et al. (1999) with *L. albifrons* based on Boulenger (1893).

Species Name in Question	Reference and Comment	Abbreviated Range
*Liophis (Rhadinaea) amarali** Wettstein, 1930 Amaral's Legion Snake	Peters & Oreajs-Miranda (1970); Dixon (1980); Vanzolini (1986); Marques et al. (2001) *Remark: Fide Dixon (1980), this is *incertae sedis amarali*; however, Dixon did note "*Liophis amarali* is a *Rhadinaea* (its species relationship has not been determined at this time)." Marques et al. (2001) noted that this snake probably belongs in a distinct species.	Bahia and Minas Gerais to Santa Catarina in Brazil.
*Liophis californica** Jan, 1863	Dixon (1980) questions its validity. *Remark: Peters & Orejas-Miranda (1970) listed this snake as *Leimadophis poecilogyrus californica incertae sedis* (Jan, 1863). Amaral did not mention this snake.	Unknown, type shown as California.
Liophis reginae maculatus (Steindachner, 1867)	Dixon (1983)	*Species Inquirende*. Specimens lost, type series consists of at least two species and possibly two genera. Type locality is listed as "Ypenema" Brazil.
*Natrix cinnamomea** Wagler, 1824	Wiest (1978); Lancini (1979); Hoogmoed & Gruber (1983); Lancini & Kornacker (1989); Dixon et al. ([1993] 1995) *Note: It is discussed here because of its common appearance in literature and uncertain allocation. *Chironius cinnamomeus* is listed here as *Natrix cinnamomea incertae sedis*. Hoogmoed & Grüber (1983), examined all the extant Wagler specimens in the Munich Museum, and believed the type specimen of *Natrix cimmomeus* is lost. They concluded that it probably is not a *Chironius* based on Wiest (1978) and considering the elevated number of scale rows at midbody. Since this will probabaly never be resolved, they considered the name a *nomen dubium*. Dixon et al. ([1993] 1995: 194–195, 237) also concluded that this is not a *Chironius*. They supposed this snake is a *Pseustes*. It was originally classified by Wagler as *Natrix cinnamomea* Wagler, 1824 but to them it seems obvious from the original description that Wagler's *Natrix cinnamomea*. . . . is not a *Chironius*. Wager's description is very basic and the ventral and subcaudal counts must be consiered accurate in the absence of a type speciman. In this case, the ventral and subcaudal counts fall well below the range of those reported for *Pseustes* in descriptions that I could find in literature. Wiest (1978); Hoogmoed & Gruber (1983); and Dixon et al. ([1993]	Lower Amazon region of Brazil, Peru and the Guianas based on "*C. cinnamomeus*" (Wagler, 1824) as noted in Lancini (1979) and Lancini & Kornacker (1989).

Species Name in Question	Reference and Comment	Abbreviated Range
	1995) all agree that the combination *Chironius cinnamomeus* has been used incorrectly by several authors for reddish-brown specimens of *C. scurrulus* (Wagler). The scientific name seems to be commonly used for red-colored *Chironius*-like snakes but since the type has been lost this question will never be resolved.	
Rhadinaea steinbachi* Boulenger 1905	Peters & Orejas-Miranda (1970); Myers (1974); Dixon (1980, 1983); Vanzolini (1986) *Remark: Myers (1974:22) noted that *steinbachi* Boulenger was under investigation and can be conveniently (but also noted, improperly) be included in the genus *Liophis*, where they also have been placed by various authors. Based on Dixon (1983), this is *incerta sedis*. Originally classified as *Rhadinaea steinbachi* Boulenger, 1905, James R. Dixon (pers. comm. 30 June 2000) stated that this snake definitely is not a *Liophis*.	Vanzolini (1986) noted that the range should be se Bolivia.
Sibon annulata Boulenger, 1894	This snake is treated as *species inquirendae* by Smith & Taylor (1966) and they noted it may be possible that it is Mexican. Also reference Peters (1960c). However, *Sibon annulata* is noted in the Savage & Villa (1986) key.	It is only known from the type specimen.
Thanatophis patoquilla Posada Arango, 1889	Fide Peters & Orejas-Miranda (1970), this taxon is not mentioned in Boulenger (1896), or in Hoge, Mem. Inst. Butantan, 32, 1965 type locality; type specimen lost (1966), in their reviews of viperid snakes. Peters & Orejas-Miranda (1970) presumed it belongs to *Bothrops*, but were unsure of this. They noted that Blanchard, Bull. Sci. Zool. France, 1889, 347, suggested it may be a synonym of *Bothrops nigroviridis* (Peters). McDiarmid et al. (1999) noted that Niceforo-Maria (1938) suggested it might be a synonym of *Bothrops lansbergii* [=*Porthidium lansbergii* (Schelegel, 1841)]. They were unable to resolve the identification.	Type locality is the province of Medellin, Colombia; distribution only known from the type specimen which is lost.
Typhlops cinereus (Schneider, 1801)	McDiarmid et al. (1999)	Unknown.
Typhlops longissimus (Dumeril, Bibron & Dumeril, 1844)	McDiarmid et al. (1999)	Unknown.

Bibliography

Abalos, J. W., 1977: ¿Qué sabe usted de víboras? Editorial Losada, S.A., Buenos Aires, 15 de deciembre de 1977.

Abalos, J. W., E. C. Báez and R. Nadar, 1964: Serpientes de Santiago del Estero, Republica de Argentina. Acta Zool. Lillo., 20: 211-283.

Abbott, C. C., [1868] 1869: Catalogue of vertebrate animals of New Jersey. *In* Cook, George H., Geology of New Jersey, 1868: 751-830. [Reptiles, including amphibians: 799-805.]

Abe, A. S., and W. Fernandes, 1977: Polymorphism in *Spilotes pullatus anomalepis* Bocourt (Reptilia, Serpentes: Colubridae). J. Herpetol., 11 (1): 98-100.

Abuys, A., 1984: The snakes of Surinam, part IX: Subfamily Xenodontinae (Genera *Hydrops, Imantodes* and *Leimadophis*.) Litteratura Serpentium, 4 (2): 63-74.

Abuys, A., 1987: A new coral snake (genus *Micrurus*) from Surinam. Litt. Serp., 7: 215-220.

Acerrano, A., 1990: Poisonous Snakes: How dangerous are they? Sports Afield. June 1990: 63-66, 100-101.

Achával, F., 1973: El género *Clelia* en Uruguay. Trab. V Cong. Latinoamer. Zool., Montevideo, Uruguay, 1: 17-29.

Achával, F., 1980: El yacaré. Bol. Mus. Nac. Hist. Nat. de Montevideo, 2 (29): 5-6. Acosta Ortegon, J., 1938: El idioma Chibcha o Aborigen de Cundinamarca. Imprenta del Departamento, Bogota.

Achával, F., 1987: Lista de las especies de vertebrados del Uruguay—Reptiles. Facultad de Humanidades y Ciencias, Depto de Zoologia—Vertebrados. Dirección General de Extension Universitaria, Montevideo, Uruguay.

Achával, F., 1989: Lista de especies, de vertebrados del Uruguay. Parte 2. Anfibios, reptiles, aves y mamíferos. I–II + 1-41.

Achával, F., A. R. Melgarejo and M. Meneghel, 1978: Ofidios del area de influencia de Salto Grande (aspectos biologicos y referencias sobre ofidismo). V. Reunion sobre aspectos de desarrollo ambiental. Salto (R.O.U.)—Concordia (R.A.). 6-10 de noviembre, 1978, Comisión Técnica Mixta de Salto Grande. Gerencia Salud, Ecología y Desarrollo Regional, República Argentina-Republica Oriental del Uruguay, 5a RDA/78/26:1-31.

Agassiz, L., 1846: Nomenclatoris zoologici. Index universalis. Soloduri, fasc. 8: 1-393.

Ahrenfeldt, R. H., 1953: Two British anatomical studies on American reptiles (1650-1750)-I.-Hans Sloane: Comparative anatomy of the American crocodile. Herpetologica, 1953: 9 (2): 79-86.

Albert, R., and D. L. Shaul, 1985: A concise Hopi and English lexicon. John Benjamin Publication Company.

Aldridge, R. D., and D. Duvall, 2002: Evolution of the mating season in the pitvipers of North America. Herpetological Monographs, (16): 1-25.

Alemán, G. C., 1953a: Apuntes sobre reptiles y anfibios de la región Baruta-El Hatillo. Mem. Soc. Cien. Nat. La Salle (Caracas), Venezuela, 12 (31): 11-30.

Alemán G. C., 1953b: Contribución al estudio de los reptiles y batracios de la Sierra de Perijá. Mem. Soc. Cien. Nat. La Salle (Caracas), Venezuela 13 (3): 205-225.

Almendáriz, A., 1991: Anfibios y reptiles. *In* Ramiro Barriga; Anna Almendáriz and Luis Albuja: Lista de vertebrados del Ecuador: Peces de agua dulce, anfibios y reptiles, mamiferos. Politecnica (Biologia) 16 (3): 1-205 [Revista de Información Tecnico-Cientifico, Quito, Ecuador] XVI (3):140-162.

Alvarez, B.B, L. Rey and J. M. Cei, 1992: A new subspecies of the *reticulatus* group, genus *Atractus*, from southeastern South America (Serpientes, Colubridae), Bull. Mus. Reg. Sci. Nat. Torino, 10 (2): 249-256.

Álvarez, J. R. (Director), 1987a: Enciclopédico de Mexico. Tomo I. Secretaría de Educación Pública, Ciudad de México.

Álvarez, J. R. (Director), 1987b: Enciclopédico de Mexico. Tomo II. Secretaría de Educación Pública, Ciudad de México.

Álvarez, J. R. (Director), 1987c: Enciclopédico de Mexico. Tomo III. Secretaría de Educación Pública, Ciudad de México.

Álvarez, J. R. (Director), 1987d: Enciclopédico de Mexico. Tomo IV. Secretaría de Educación Pública, Ciudad de México.

Álvarez, J. R. (Director), 1987e: Enciclopédico de Mexico. Tomo V. Secretaría de Educación Pública, Ciudad de México.

Álvarez, J. R. (Director), 1987f: Enciclopédico de Mexico. Tomo VI. Secretaría de Educación Pública, Ciudad de México.

Álvarez, J. R. (Director), 1987g: Enciclopédico de Mexico. Tomo VII. Secretaría de Educación Pública, Ciudad de México.

Álvarez, J. R. (Director), 1988a: Enciclopédico de Mexico. Tomo X. Secretaría de Educación Pública, Ciudad de México.

Álvarez, J. R. (Director), 1988b: Enciclopédico de Mexico. Tomo XI. Secretaría de Educación Pública, Ciudad de México.

Álvarez, J. R. (Director), 1988c: Enciclopédico de Mexico. Tomo XII. Secretaría de Educación Pública, Ciudad de México.

Álvarez, J. R. (Director), 1988d: Enciclopédico de Mexico. Tomo XIII. Secretaría de Educación Pública, Ciudad de México.

Álvarez, J. R. (Director), 1988e: Enciclopédico de Mexico. Tomo XIV. Secretaría de Educación Pública, Ciudad de México.

Álvarez, J. R. (Director), 1999: Diccionario Enciclopédico de Baja California. Compañia Editora de Enciclopedias de México, S.A. de C.V., Instituto de Cultura de Baja California, Ciudad de México.

Alvarez del Toro, M., 1960: Reptiles of Chiapas. Chiapas, Mexico, Instituto Zoologico del Estado, First Edition, Tuxtla Gutérrez, Chiapas, Mexico.

Alvarez del Toro, M., 1972 [1973]: Los reptiles de Chiapas. 2nd edition, Gobierno del Estado de Chiapas, Tuxtla Gutierrez.

Alvarez del Toro, M., [1982] 1983: Los reptiles de Chiapas. 3rd edition, Colección Libros de Chiapas, Serie Especial, Mexico, Tuxtla Gutíerrez.

Amaral, A. do, 1921: Contribucão para o conhecimento dos ofidios do Brasil. A.-Parte 1. Quatro novas espécies de serpentes brasileiras. Anex. Mem. Inst. Butantan, Seção Ofiologia, 1 (1): 1-38.

Amaral, A. do, 1922a: Contribucão para o conhecimento dos ofidios do Brasil. A.-Parte II. Biologica da nova espécie, *Lachesis insularis*. Anex. Mem. Inst. Butantan, 1: 39-44.

Amaral, A. do, 1922b: Contribution toward the knowledge of snakes in Brazil. A. Part I. Four new species of Brazilian snakes. Anex. Mem. Inst. Butantan, 1: 49-81.

Amaral, A. do, 1923: New genera and species of snakes. Proc. New England Zool. Club, 8: 85-105.

Amaral, A. do, 1924a: *Helminthophis*. Proc. New England Zool. Club, 9: 25-30.

Amaral, A. do, 1924b: New genus and species of South American snakes contained in the United States National Museum. J. Washington Acad. Sci., 14: 200-202.

Amaral, A. do, 1925a: Ophidios de Mato Grosso (Contribução II para o conhecimento dos ophidios do Brasil). Comm. Linhas Telegr. Estrat. Mato Grosso ao Amazônas, Publ. 84, Annex 5, Hist. Nat. Zool.: 1-29. Comp. Melhoramentos de São Paulo.

Amaral, A. do, 1925b: South American snakes in the Collection of the United States National Museum. Proc. U.S. Natl. Mus. 67 (art. 24): 1-30.

Amaral, A. do, 1926a: Albinismo em "dorme-dorme", *Sibynomorphus turgidus* (Cope, 1868). Rev. Mus. Paulista, Tomo 15: 61-62.

Amaral, A. do, 1926b: 1. Nota de nomenclatura ophidiológica. Sobre o emprego de nome generico *Micrurus* em vez de *Elaps*. Rev. Mus. Paulista, 14: 3-6.

Amaral, A. do, 1926c: Novos generos e especies de ofidios brasileiros (contribução III para o conhecimento dos ophidios do Brasil). Arch. Mus. Nac. (Rio de Janeiro), Brazil, 26: 95-121.

Amaral, A. do, 1926d: Sobre os nomes genericos de ophidios, *Liophis* Wagler, 1830 e *Leimadophis* Fitzinger, 1843, Collectanea ophiligies. 9. pp. 77-78. In Amaral, A. do, Collectanea ophiologica. São Paulo, Rev. Mus. Paulista, 15: 1-110.

Amaral, A. do, 1926e: Studies of neotropical ophidia. II. On *Micrurus mipartitus* and allied forms. Proc. New England Zool. Club, 9: 61-66.

Amaral, A. do, 1926f: Studies of neotropical ophidia. III. On *Helminthophis flavoterminatus* (Peters, 1857). Proc. Biol. Soc. Washington, 39: 123-126, December 27, 1926.

Amaral, A. do, 1926 [1927]: Nombres vulgares de ophidios no Brasil. Bol. Mus. Nac., Rio de Janiero, II (2): 1-11.

Amaral, A. do, 1927a: Collectanea ophiologies. 2: Tres subespecies novas de *Micrurus corallinus*: *M. corallinus riesei*, *M. corallinus corallinus* y *M. corallinus dumerilii*. Collectanea ophiologies. 2. Rev. Mus. Paulista, 15: 13-25.

Amaral, A. do, 1927b: Collectanea ophiologies.3: Da invalidez da especie de Colubrideo Elapineo *Micrurus ibiboboca* (Merrem) e rediscripção de *M. lemniscatus* (L.), Rev. Mus. Paulista, 15: 29-40.

Amaral, A. do, 1927c: 2.Studies of neotropical ophidia. IV. A new form of Crotalidae from Bolivia. Bull. Antivenin Inst. Amer., 1: 5-6.

Amaral, A. do, 1927d: Studies of neotropical ophidia. V. Note on *Bothrops lansbergii* and *B. brachystoma*. Bull. Antivenin Inst. Amer., 1 (1): 22.

Amaral, A. do, 1927e: 8. The snake-bite problem in the United States and in Central America. Bull. Antivenin Inst. Amer., 1 (2): 31-35.

Amaral, A. do, 1927f: 13. Studies of neotropical ophidia. VII. An interesting collection of snakes from west Colombia. Bull. Antivenin Inst. Amer., 1 (2): 44-47.

Amaral, A. do, 1927g: 20. Studies of neotropic ophidia. VIII. *Trachyboa* Peters, 1860. Bull. Antivenin Inst. Amer., 1 (3): 86-87.

Amaral, A. do., 1927h: 21. Studies of neotropic ophidian. IX. *Anomalepis* Jan, 1861. Bull. Antivenin Inst. Amer., 1 (3): 86-87.

Amaral, A. do, 1928: 30. Studies of neotropical ophidia. XI. Snakes from the Santa Marta region, Colombia. Bull. Antivenin Inst. Amer., 7-8.

Amaral, A. do, 1929a: Da classificação e conceito de especie em ophiologia. Secretaria De Agricultura, Industria E Commercio Do Estado De São Paulo, São Paulo, Brazil, 1929: 1-2.

Amaral, A. do, 1929b: Estudos sobre ophidios neotropicos. XIX. Revasão do género *Spilotes* Wagler, 1830. Mem. Inst. Butantan, 4: 275-298.

Amaral, A. do, 1929c: Notas de ophiologica. Oficinas de Diario Oficial.

Amaral, A. do, 1929d: Studies of neotropical ophidia. XII. On the *Bothrops lansbergii* group. Bull. Antivenin Inst. Amer., 3: 19-27.

Amaral, A. do, 1929e: 71. Studies of neotropical ophidian. XIII. A new colubrinae snake in the collection of the Vienna Museum. Bull. Antivenin Inst. Amer., 3 (2): 40.

Amaral, A. do, [1929] 1930a: Estudos sôbre ophidios neotrópicos. XIX. Revisão de genero *Spilotes* Wagler, 1830. Mem. Inst. Butantan, 4: 273-298.

Amaral, A. do, [1929] 1930b: Estudos sôbre ophidios neotrópicos. XX. Revisão de genero *Phrynonax*, Cope, 1862. Mem. Inst. Butantan, 4: 299-320.

Amaral, A. do, [1929] 1930c: Estudos sôbre ophidios neotrópicos. XXI. Revisão de genero *Drymarchon* Fitzinger, 1843. Mem. Inst. Butantan, 4: 323-330.

Amaral, A. do, 1930a: Estudos sôbre ophidios neotrópicos. XVII. Valor systematico de varias formas de ophidios neotropicos. Mem. Inst. Butantan, (1929) 4: 3-68.

Amaral, A. do, 1930b: Estudos sôbre ophidios neotropicos. XVIII. Lista remissiva dos ophidios da regiao neotropica. Mem. Inst. Butantan, (1929) 4: 129-271.

Amaral, A. do, 1930c: Contribucão ao conhecimento dos ophidios do Brasil IV. Lista remissiva dos ophidios do Brasil. Mem. Inst. Butantan, 4: 69-125.

Amaral, A. do, 1930d: 85. Notes on *Spilotes pullatus*. Bull. Antivenin Inst Amer., 3: 96-99.

Amaral, A. do, 1930e: Serpientes venenosas Sudamericanas. Arch. Soc. Biol. Montevideo, 1930: 93-107

Amaral, A. do., 1930f: 91. Studies of neotropical ophidia. XIV. Notes on two colubrinae snakes. Bull. Antivenin Inst. Amer., 4 (1): 12-13.

Amaral. A. do., 1930g: 92. Studies of neotropical ophidia. XV. A rare Brazilian snake. Bull. Antivenin Inst. Amer., 4 (1): 13-16.

Amaral. A. do., 1930h: 94. Studies of neotropical ophidia. XVI. Two new snakes from Central Colombia. Bull. Antivenin Inst. Amer., 4 (2): 27-28.

Amaral. A. do., 1930i: 102. Studies of neotropical ophidia. XXIV. A new Brazilian snake. Bull. Antivenin Inst. Amer., 4 (3): 65.

Amaral, A. do, 1930j: Studies of neotropical ophidia. XXV. A new race of *Bothrops neuwiedii*. Bull. Antivenin Inst. Amer., 4 (3): 65-67.

Amaral, A. do, 1931a: 108. Studies of neotropical ophidia. XXIII. Additional notes on Colombian snakes. Bull. Antivenin Inst. Amer., 4 (4): 85-89.

Amaral, A. do, 1931b: 109. Studies of neotropical ophidia. XXVI. Ophidia of Colombia. Bull. Antivenin Inst. Amer., 4 (4): 89-94.

Amaral, A. do, 1932: 128. Studies in neotropical ophidia. XXVII. On two small collections of snakes from central Colombia. Bull. Antivenin Inst. Amer, 5 (3): 66-68.

Amaral, A. do, [1932] 1933a: Estudos sôbre ophidios neotrópicos. XXIX. Novas notas sôbre espécies da Colombia. Mem. Inst. Butantan, 7:103-123.

Amaral, A. do, [1932] 1933b: Contribucão ao conhecimento dos ophidios do Brasil. 5. Uma nova raça de *Bothrops neuwiedii*. Mem. Inst. Butantan, 7: 97-98.

Amaral, A. do, 1934: Sobre la especie *Bothrops alternata* D. & B. 1854 (Crotalidae). Variaçoes. Redescripção. Mem. Inst. Butantan, 1933-1934, 8: 161-182.

Amaral, A. do, [1933-1934] 1934: Estudos sôbre ophidios neotrópicos. XXX. Novo gênero e espécie de colubrideo na fauna da Colombia. Mem. Inst. Butantan, 8: 157-159.

Amaral, A. do, [1934] 1935: Collecta herpetologica no nordeste do Brasil. Mem. Inst. Butantan, 8: 185-192.

Amaral, A. do, 1935a: Collecta herpetologica no centro do Brasil. Mem. Inst. Butantan, 9: 234-246.

Amaral, A. do, 1935b: Collecta herpetologica no nordeste do Brasil (Contribuição II). Mem. Inst. Butantan, 9: 227-247.

Amaral, A. do, 1935c: Contribuição ao conhecimento dos ophidios do Brasil. VII. Novos generos e especies de Colubridae opisthoglyphos. Mem. Inst. Butantan, 9: 203-205.

Amaral, A. do, 1935d: Estudos sôbre ophidios neotrópicos. XXXII. Apontamentos sobre a fauna da Colombia. Mem. Inst. Butantan, 9: 208-217.

Amaral, A. do, 1935e: Estudos sôbre ophidios neotrópicos. XXXIII. Novas espécies de ophídios da Colombia. Mem. Inst. Butantan, 9: 219-223.

Amaral, A. do, 1936: Contribuição ao conhecimento dos ophidios de Brasil. VIII. Lista remissiva dos ophidios de Brasil (2. Edic.). Mem. Inst. Butantan, 1935-1936: 87-161

Amaral, A. do, 1937a: Check-list of the "ophidia" of Brazil. XIIe Congres Internatl. Zool. Lisbonne 1935: 1744-1761.

Amaral, A. do, 1937b: Contribuicão ao conhecimento dos ophidios do Brasil. VIII. Lista remissiva dos ophidios do Brasil. Mem. Inst. Butantan, 10: 87-162.

Amaral, A. do, 1937c: New species of ophidians from Colombia. Comptes Rendus. XII Congr. Int. Zool. (Lisbonne, 1935), 3: 1762-1767.

Amaral, A. do, 1944: Notas sobre a ofidiologia neotropica e Brasílica. XI. Subespecies de *Micrurus lemniscatus* (L.) e suas afinidades com *M. frontalis* (Dm. & Bibron). Pap. Avuls., Dept. Zool. São Paulo. 5 (11): 83-94.

Amaral, A. do, 1946: Cobras das serpentários de Instituto Butantan. São Paulo, Brazil, Inst. Butantan.

Amaral, A. do, 1948: Ofídos de Mato Grosso. Contribuição II para o conhecimento dos ofidios de Brasil. Pub. No. 84, Anexos No. 5, Historia Natural Zoologica. Comissão de Linhas Telegraficas Estratégicas de Mato Grosso as Amazonas, Impresa Nacional, Rio de Janiero, Brazil, 1948 (84): 1-43.

Amaral, A. do, 1954: Contribuicão do conhecimiento dos ofidios do Brasil. 12. Notas a respeito de *Helminthophis ternetzii* Boulenger, 1896. Mem. Inst. Butantan, 26: 191-195.

Amaral, A. do, 1955 [1954]: Contribução dos ophidios neotrópicos XXXVII. Subespécies de *Epicrates cenchria* (Lineu., 1758), Mem. Inst. Butantan, (1954) 26: 227-247.

Amaral, A. do, 1963: *Calamodontophis*: herpetological note. Copeia, 1963 (3): 580.

Amaral, A. do, 1964: Comment on the proposal to substitute the generic name *Dryadophis* Stuart, 1939, for *Mastigodryas* Amaral, 1934, Z.N. (S.) 1533. Bull. Zool. Nomencl., 21 (1): 13.

Amaral, A. do., 1973: Ofionimia amerindia na ofiodogia Brasiliense. Mem. Inst. Butantan, 37: 1-15.

Amaral, A. do, 1976: Serpentes do Brasil—iconografia colorido—Brazilian snakes: A color icongraphy. Edições Melhoramentos, Editora da Universidade de São Paulo.

Anderson, J. D., 1962: A new subspecies of the ridged-nosed rattlesnake, *Crotalus willardi*, from Chihuahua, Mexico. Copeia, 1962: 160-163.

Anderson, P., 1965: The reptiles of Missouri. Univ. Missouri Press, Columbia, Missouri.

Anderson, P. K., 1961: Variation in populations of brown snakes, genus *Storeria*, bordering the Gulf of Mexico. The Amer. Midl. Nat., University of Notre Dame Press, Notre Dame, Indiana, 66 (1): 235-249.

Andersson, L. G., 1898: List of reptiles and batrachians collected by the Swedish Expedition to Tierra del Fuego. Öef. Kongl. Vet. Akad. Torr., 7: 457-462.

Andersson, L. G., 1899: Catalogue of Linnaean type-specimens of snakes in the Royal Museum in Stockholm. Bih. Till Handl. K. Sv. Vetensk. Akad., 24: 1-35.

Andersson, L. G., 1901: Some species of snakes from Cameroon and South America, belonging to the collections of the Royal Museum in Stockholm. Bih. Till Handl. K. Sv. Vetensk. Akad., 27 (5): 1–26.

Andersson, L. G., 1916: Notes on the reptiles and batrachians in the Zoological Museum at Gothenburg. With an account of some new species. Götenburgs Kungl. Vent. Medd. Göteborgs Muse Zoologiska Afdelnig. 9, Stockholm, (9): 1–41.

Andersson, L. G., 1945: Batrachians from east Ecuador collected 1937, by Wm. Clarke-MacIntyre and Rolf Blomberg. Arkiv für Zoologi., 37 (2): 1–88.

Angel, Fernand, 1949: Petit atlas des amphibiens et reptiles. II. Sauriens-ophidiens. Éditions N. Boubée & Cie, Paris.

AEVRE-Assoc. Ecológica de Volta Redonda, SMSPMA, Prefeitura Municipal de Volta Redonda "Brasil: 70 espécies sao peçonhentas." Suplemento Agricola. EPD from Instituto Butantan, Divisao de Extensao Cultural, Sao Paulo, Brazil.

AEVRE-Assoc. Ecológica de Volta Redonda, SMSPMA, Prefeitura Municipal de Volta Redoda. "Evite acidentes-sempre é melhor prevenir do que remediar. Conhecondo os Hábitos das cobras, você pode veitar picadas," Suplemento Agricola. EPD from Instituto Butantan, Divisao de Exensao Cultura, Sao Paulo, Brazil.

Aoki, H., 1994: Nez Perce dictionary. Univ. California Press, Berkley, Los Angeles and London, Vol. 22.

Arana de Swadesh, E. B. Pérez Góznzalez, R. Escalante H., C. Robles U., R. D. Bruce S. and Y. L. Astra de Suárez, 1975: Las lenguas de Mexico, I. Instituto Nacional de Anthropología e Historia, Mexico, D.F.

Araujo, P., 1976/1977: Uma nova espécie do género *Neyra plectana* (Nematoda: Subuluroidea: Cosmocercidae) encontrada em ofídios. Mem. Inst. Butantan, 40/41: 259–264.

Arauz, P., 1960: El Pipil de la region de Los Itzalcos. Ministerio de Cultura-Depto. Editorial, San Salvador, El Salvador.

Argôlo, A. J. S., 1992: Considerações sôbre a ofidiofauna dos cacauais do sudeste da Bahia, Brasil. Monografia. Universidade Estadual de Santa Cruz, Ilhéus, Bahia.

Argôlo, A. J. S., 1998: *Simophis rhinostoma*. Herpertol. Rev., 29 (3): 178–179.

Argôlo, A. J. S., 1999: Geographic distribution—*Tropidodryas serra*. Herpetol. Rev., 30 (1): 55–56.

Argôlo, A. J. S., 1999: Geographic distribution—*Tropidodryas striaticeps*. Herpetol. Rev., 30 (1): 56.

Argôlo, A. J. S., 1999: Geographic distribution—*Xenodon neuwiedii*. Herpetol. Rev., 30 (1): 56.

Armstrong, B. L., and J. B. Murphy, 1979: The natural history of Mexican rattlesnakes. Univ. Kansas Publ., Mus. Nat. Hist., Lawrence, Kansas, Special Publication No.5, 14 December, 1979.

Arnold, R., 1973: What to do about bites & stings of venomous animals. New York, Collier Books.

Arriola, J. L., 1973: El libro de las geonimias de Guatemala—Diccionario etinologico. Editorial "José de Pineda Ibarra," Ministerio de Educación, Guatemala.

Arrlington, P. J., Jr., 1927: 17. Auto-hemorrhage in *Tropidophis semicinctus*. Bull. Antivenin Inst. Amer., 1 (1): 59.

Arroyo, V. M., 1966: Lenguas indígenas costaricenses. Editorial Costa Rica.

Arroyo, V. M., 1972: Lenguas indigenas costaricenses. Editorial Universitaria Centroamerica (EDUCA), 2nd edition, San José, Costa Rica.

Arzamendia, V., 1999: Geographic distribution—*Philodryas baroni* (Baron's Racer). Herpetol. Rev., 30 (1): 55.

Arzamendia, V., and A. R. Giraudo, 1999: Geographic distribution—*Liophis sagittifer modestus*. Herpetol. Rev. 30 (10): 54.

Ascêncio Machado, R., P. S. Bernarde and S. A. A. Morato, 1998: *Liophis miliaris* (common water snake)—prey. Herpetol. Rev., 29 (1): 45.

Ashton, K. G., 2000: Notes on the island population of the Western Rattlesnake, *Crotalus viridis*. Herpetological Review, 3(4): 214–217.

Ashton, K. G., and A. de Quiroz, 2001: Molecular systematics of the western rattlesnake, *Crotalus viridis* (Viperidae), with comments on the utility of the D-Loop in phylogenetic studies of snakes. Molecular Phylogenetics and Evolution, 21(2): 176–189.

Jr., Ashton, R. E., and P. Sawyer Ashton, 1981: Handbook of reptiles & amphibians of Florida, Part one, the snakes. Miami, Florida, Windward Publishing, Inc.

Atsatt, S. R., 1913: The reptiles of the San Jacinto area of southern California. Univ. California Publ. Zoöl., 12 (3): 31–50.

Auffenberg, W., 1949: *Coluber c. stejnegerianus* (Cope) in Texas. Herpetologica, 5 (4): 53–58.

Auffenberg, W., 1955: A reconsideration of the racer, *Coluber constrictor*, in eastern United States. Tulane Studies in Zoology, 2 (6): 89–155.

Auth, D. L., 1994: Checklist & bibliography of the amphibians and reptiles of Panama. Smithsonian Herpetol. Inform. Serv. No.98.

Auth, D. L., 1998: *Masticophis mentovarius mentovarius* (Neotropical Whipsnake). Herpetol. Rev., 29 (1): 54.

Auth, D. L., J. Mariaux, J. Clary, D. Chiszar, F. van Breukelen, and H. M. Smith, 1998: The report of the snake genus *Conophis* in South America is erroneous. Bull. Maryland Herpetol. Soc., 30 34 (4): 107–112.

Auth, D. L., B. C. Brown, H. M. Smith, and D. Chiszar, 1999: Geographic distribution: *Rhadinaea laureata*. Herpetol. Rev., 30 (4): 236.

Avila, L. J., 1996a: Geographic distribution—*Philodryas aestivus subcarinatus* (Brazilian Green Racer). Herpetol. Rev., 27 (3): 154.

Avila, L. J., 1996b: Geographic distribution—*Rhinobothryum lentiginosum (Coral Falsa)*. Herpetol. Rev., (3): 155.

Alveoli, L. J., 1997: Geographic distribution—*Pseudotomodon trigonatus* (False Tomodon Snake). Herpetol. Rev., 28 (2): 88.

Axtell, R. W., 1959: Amphibians and reptiles of the Black Gap Wildlife Management Area, Brewster County, Texas. Southwest Naturalist, 4 (2): 88–109.

Ayerbe, S., and F. J. López, 2002: Descripción de una nueva especie de serpiente coral (Elapidae: *Micrurus*). Mem. II Simposio de Investigación en Ciencias Biológicas, Departamento de Bilogica, Universidad del Cauca, Asociación colombiana de Ciencias Biológicas, Capitulo Popayán, Ponencia No.12, 2 pp., 1 fig., Popayán, February 12 and 13, 2002.

Ayerbe, S., M. A. Tidwell and M. Tidwell,

Bibliography

1990: Observaciones sobre la biología y comportamiento de las serpiente coral "Rabo de Aji": (*Micrurus mipartitus*). Descripción de una subespecie nueva. Noved. Colombianas, Popayán, 2: 30-41.

Badano, R., J. Luis and J. L. Badano Carbajal, 1965: Ofidismo en el Uruguay. Tercera ed. Aumentada y Corregida, Montevideo, Imprenta Morato, 1965.

Badger, D. (with photos by John Netherton): Snakes. Voyageur Press, 1999.

Bahena-Basave, H., 1995a: Geogrpahic distribution—*Micrurus browni*. Herpetol. Rev., 26 (1): 46.

Bahena-Basave, H., 1995b: Geographic distribution—*Tretanorhinus nigroluteus* (Orangebelly Swamp Snake). Herpetol. Rev., 26 (1): 47.

Bailey, J. R., 1937a: A new species of *Rhadinaea* from San Luis Potosi. Copeia, 1937 (2): 118-119.

Bailey, J. R., 1937b: A review of some recent *Tropidophis* material. Proc. New England Zool. Club, 16: 41-52

Bailey, J. R., 1937c: New forms of *Coniophanes* Hallowell, and the status of *Dromicus clavatus* Peters. Occ. Pap. Mus. Zool., Univ. Michigan, (362): 1-6.

Bailey, J. R., 1938 [1939]: A systematic revision of the snakes of the genus *Coniophanes*. Reprint from Pap. Michigan Acad. Sci., Arts Lett., 24 (Part II): 1-48.

Bailey, J. R., 1940: The Mexican snakes of the genus *Rhadinaea*. Occ. Pap. Mus. Zool., Univ. Michigan, (412): 1-19.

Bailey, J. R., 1955: The snakes of the genus *Chironius* in southeastern South America. Occ. Pap. Mus. Zool. Univ. Michigan, (571): 1-21.

Bailey, J. R., 1966: A redescription of the snake *Calamodontophis paucidens*. Copeia, 4 (1966): 885-886.

Bailey, J. R., 1967: The synthetic approach to colubrid classification. Herpetologica, 23 (2): 155-161.

Bailey, J. R., 1970: *Clelia*. *In* Peters and Orejas-Miranda (eds.), Catalogue of the Neotropical Squamata: Part I. Snakes, pp. 62-64, Bull. U.S. Natl. Mus.

Bailey, J. R., 1986: The *Oxyrhopus petola-petolarius* question continued. Bull. Maryland Herpetol. Soc., 22: 144-145.

Bailey, J. R. and A. L. de Carvalho, 1946: A new *Leptotyphlops* from Mato Grosso, with notes on *Leptotyphlops tenella* Klauber. Bol. Mus. Nac. Rio de Janeiro, Zoologia, 52: 1-7.

Bailey, V., 1905: Biological survey of Texas. North American Fauna, 25: 1-222.

Baird, S. F., 1852a: Characteristics of some new reptiles in the museum of the Smithsonian Institution. Proc. Acad. Nat. Sci. Philadelphia, 6: 68-70.

Baird, S. F., 1852b: Characteristics of some new reptiles in the museum of the Smithsonian Institution. Second Part. Proc. Acad. Nat. Sci. Philadelphia, 6: 125-129.

Baird, S. F., 1852c: Characteristics of some new reptiles in the museum of the Smithsonian Institution. Third Part. Proc. Acad. Nat. Sci. Philadelphia, 6: 173.

Baird, S. F., 1853: Catalogue of North American reptiles in the museum of the Smithsonian Institute. Part I. Serpents. Smithsonian Misc. Coll. 2, art. 5.

Baird, S. F., 1854: On the serpents of New-York. 7th Ann. Rept. Reagents Univ. State New York, appendix + G.: 95-124.

Baird, S. F., 1859a: Report on reptiles collected on the survey. *In* Explorations for a railroad route, from the Sacramento Valley to the Columbia River, part 4 (4): 9-13.

Baird, S. F., 1859b: Reptiles of the boundary, with notes by the naturalists on the survey. United States and Mexican Boundary Survey under the order of Lieut. Col. W. H. Emory, 3: 1-15.

Baird, S. F., and C. Girard, 1851 [1852]: Reptiles, p. 336-353. *In* Howard Stansbury. Exploration and survey of the Valley of the Great Salt Lake of Utah, including a reconnaissance of a new route through the Rocky Mountains. U.S. 32 Congress, Special Sen., Sen. Exec. Doc. 3, Lippincott, Grambo, and Co., Philadelphia.

Baird, S. F., and C. Girard, 1852a: Characteristics of some new reptiles in the Museum of the Smithsonian Institute. Parts I—III. Proc. Acad. Nat Sci, Philadelphia, 6: 68-70, 125-129, 173.

Baird, S. F., and C. Girard, 1852b: Description of new species of reptiles collected by the U.S. Exploring Expedition under the command of Capt. Charles Wilkes, USN, Part I. Proc. Acad. Nat. Sci., Philadelphia, 6: 174-177, 420-424.

Baird, S. F., and C. Girard, 1853: Catalogue of North American reptiles in the museum of the Smithsonian Institution. Part 1. Serpents. Smithsonian Misc. Colls., Smithsonian Institute, Washington, D.C., 2: xvi.

Baird, S. F., and C. Girard, 1853: Reptiles: *In* R. B. Marcey and G. B. McClellan (eds.): Exploration of the Red River of Louisiana in the year 1852. 32rd Congress, 2nd Session, Senate Exec. Doc. 8 (54): 217-244.

Baird, S. F., and C. Girard, 1854: Reptiles: *In* R. B. Marcey and G. B. McClellan (eds.): Exploration of the Red River of Louisiana in the year 1852. 33rd Congress, 1st Session, House of Representatives Executive Document: 188-215.

Baird, S. F., and C. Girard, 1859: Reptiles of the Boundary. *In* W. H. Emory. Report of the United States and Mexican Survey made under the direction of the Secretary of the Interior. 34th Congress, 1st Session, Sen. Ex. Doc., Vol.30.

Baker, R. H., R. G. Webb and E. Stern, 1971: Amphibians, reptiles, and mammals from north-central Chiapas. An. Inst. Biol. Univ. Nac. Auton., México, 42, Ser. Zool., 1: 77-86.

Baker, R. H., G. A. Mengden, and J. J. Bull, 1972: Karyotypic studies of thirty-eight species of North American snakes. Copeia, 1972 (2): 257-265.

Balée, W., 1999: Footprints of the forest-Ka'apor ethnobotany-the historical ecology of plant utilization by the Amazonian people. Colombia Univ. Press, New York.

Balouet, Jean-Christophe, 1990: Extinct species of the world, New York, Barron's.

Banicki, L. H. and R. G. Webb, 1982: Morphological variation of the Texas Lyre snake (*Trimorphodon biscutatus vilkinsoni*) from the Franklin Mountains, west Texas. Southwest. Naturalist, 27 (3): 321-324.

Banta, B. H., 1953: Some herpetological notes from southern Nevada. Herpetologica, 9 (2): 75-76.

Barbour, R. W, 1971: Amphibians and reptiles of Kentucky. Kentucky Nature Series, Univ. Press Kentucky, (2): 1-334.

Barbour, T., 1906a: Additional notes on Bahama snakes. Amer. Naturalist, 40: 229-232.

Barbour, T., 1906b: Vertebrata from the savannah of Panamá, IV. Reptilia and Amphibia. Bull. Mus. Comp. Zool., 46: 224-229.

Barbour, T., 1910: Notes on the herpetology of Jamaica. Bull. Mus. Comp. Zool., 52 (15): 272–301.

Barbour, T, 1914: A contribution to the zoogeography of the West Indies, with a special reference to amphibians and reptiles. Mem. Mus. Comp. Zool., 44 (2): 205–395.

Barbour, T., 1915: A new snake from southern Peru. Proc. Biol. Soc. Washington, 28: 149–150.

Barbour, T., 1916a: Amphibians and reptiles from Tobago. Proc. Biol. Soc. Washington, 29: 221–224.

Barbour, T., 1916b: Amphibians and reptiles of the West Indies. Zool. Jahr., 11 (4): 437–442.

Barbour, T., 1916c: The reptiles and amphibians of the Isle of Pines. Ann. Carnegie Mus. Nat. Hist., 10: 297–308.

Barbour, T., 1920: An addition to the American checklist. Copeia, 1920 (84): 68–69.

Barbour, T., 1921: On a small collection of reptiles from Argentina. Proc. Biol. Soc. Washington, 34: 139–142.

Barbour, T., 1925: A new frog and a new snake from Panama. Occ. Pap. Boston Soc. Nat. Hist., 5: 155–156.

Barbour, T., 1926: Reptiles and amphibians —their habits and adaptions. Houghton Mifflin Co., Boston and New York.

Barbour, T., 1930: Some faunistic changes in the Lesser Antilles. Bull. Antivenin Inst. Amer., 3 (4): 91–93.

Barbour, T., 1941: A new boa from the Bahamas. Proc. New England Zool. Club, 18: 61–65.

Barbour, T., 1943: A new water snake from North Carolina. Proc. New England Zool. Club, 22: 1–2.

Barbour, T., and A. do Amaral, 1924: Notes on some Central American snakes. Occ. Pap. Boston Soc. Nat. Hist., 5: 129–132.

Barbour, T., and A. do Amaral, 1928: A new elapid from western Panama. Bull. Antivenin Inst. Amer., 1: 100.

Barbour, T., and A. Loveridge, 1929a: Amphibians and reptiles [of the Corn Islands, Nicaragua]. Bull. Mus. Comp. Zool., 69: 138–146.

Barbour, T., and A. Loveridge, 1929b: 63. On some Honduran and Guatemalan snakes with the description of a new arboreal pit viper of the genus *Bothrops*. Bull. Antivenin Inst. Amer., 3 (1): 1–3.

Barbour, T., and G. K. Noble, 1920: Amphibians and reptiles from southern Peru collected by the Peruvian expedition of 1914–1915 under the auspices of Yale University and the National Geographic Society. Proc. U.S. Natl. Mus., 58: 609–620.

Barbour, T., and C. T. Ramsden, 1919: The herpetology of Cuba. Mem. Mus. Comp. Zool., 47: 71–213.

Barbour, T., and B. Shreve, 1936: New races of *Tropidophis* and of *Ameiva* from the Bahamas. Proc. New England Zool. Club., 40: 347–365.

Barigozzi, E., ed., 1982: Mechanisms of speciation. Alan R. Liss, New York.

Barker, D., and Barker, T., 1994: Boas in the spotlight. The Vivarium, 6 (2): 38–41, September/October 1994.

Barker, D. G., 1992: Variation, infraspecific relationships and biogeography of the ridge nose rattlesnake, *Crotalus willardi*. pp. 89–105 *in* J. A. Campbell and E. D. Brodie, Jr.: Biology of the pitvipers. Selva, Tyler, Texas.

Barker, W., 1964: Familiar reptiles and amphibians of America. Harper & Row Publs., New York.

Barrera Vásquez, A. (director); J. R. Bastarrachea Mauzano and W. Brito Sangores (Redactores); R. Vermont Salas, D. Azul Góngora and D. Azul Poot (Colaboradores), 1980: Diccionario Maya Codex—Maya-Español, Español-Maya. Ediciones Cordemes, Mérida, Yucatán, Mexico, 1 volume.

Barrio, A., R. F. Laurent and R. A. Thomas, 1976: The status of *Philodryas subcarinatus* Boulenger (Serpentes: Colubridae). J. Herpetol., 11 (2): 230–231.

Barrio, C. L., C. Brewer-Carías and O. Fuentes, 1999: Geographic Distribution—*Atractus torquatus* (Culebra Terrera Roja, Red Burrowing Snake). Herpetol. Rev. 30 (1): 53.

Barrio, C. L., L. F. Navarrete, O. Fuentes and R. Mattei, 1998: *Siphlophis cervinus* (Serpentes: Colubridae) en Venezuela. Acta Biol. Venezuela, 18 (1): 49–53.

Barry, S. J., M. R. Jennings and H. M. Smith, 1996: Current subspecific names for western *Thamnophis sirtalis*. Herpetol. Rev., 27 (4): 172–173.

Bartholomew, D. A. and L. C. Schoenhals, 1983: Bilingual dictionaries for indigenous languages. Summer Institute of Linguistics. Instituto Lingüístico de Verano, México, D.F., México, 370 pp.

Bartlett, R. D., 1987: On no! Rosies again! (The final (?) Word). Notes from the Northern Ohio Association of Herpetologists (NOAH), 14 (8): 4–6.

Bartlett, R. D., 1994: Kingsnakes of Florida. Tropical Fish Hobbyist, XLII (10) (460):116–126, June, 1994.

Bartlett, R., 1997a: Notes from the field-Miami and the Keys: A Troubled Trip. Reptiles, October, 1997, 5 (10): 18–22.

Bartlett, R., 1997b: Notes from the field-the Lake Wales Ridge and Lake Okeechobee. Reptiles, August 1997, 5 (8): 18–20.

Bartlett, R. D., 1997c: Reminiscing about Paraguayan herps. Reptiles, October 1997, 5 (10): 12–17.

Bartlett, R., 1998: The unsung beauty of Garter Snakes. Reptiles, January, 1998, 6 (1): 48–63.

Bartlett, R. D. and A. Tennant, 2000: Snakes of North America-western region. Gulf Publishing Company, Houston, Texas.

Barton, A. J., 1956: A statistical study of *Thamnophis brachystoma* (Cope) with comments on the kinship of *T. butleri* (Cope). Proc. Biol. Soc. Washington, 69: 71–82.

Barton, B. S., 1805a: [Notes on rattlesnake bite.] Philadelphia Med. Phys. J., 1 (section 3): 167–169.

Barton, B. S., 1805b: Ophiology [notes by the editor]. Philadelphia Med. Phys. J., 2 (part 1, section 3): 165–171.

Bartram, W., 1791: Travels through North and South Carolina, Georgia, east and west Florida. Philadelphia.

Basey, H. E., 1976: Discovering Sierra reptiles and amphibians. National Park Service, U.S. Department of Interior.

Battersby, J. C., 1938: Some snakes of the genus *Tropidophis*. Ann. Mag. Nat. Hist. Ser. 11, 1: 557–560.

Bauchot, Roland (editor), 1994: Snakes-a natural history. New York, New York, Sterling Publishing Company, Inc.

Bautista, A. R., and H. M. Smith, 1992: A new chromospecies of snake (*Pseudoleptodeira*) from Mexico. Bull. Maryland Herpetol. Soc., 28 (3): 83–98.

Baxter, G. T., and M. D. Stone, 1980: Amphibians and reptiles of Wyoming. Wyoming Game and Fish Dept., (16): 1–137.

Beaudet, Fr. G., 1995: Cree-English, English-Cree Dictionary. Wuerz Publishing Ltd, Winnipeg, Canada.

Bibliography

Beçak, W., S. M. Carneiro, and M. L. Beçak, 1971: Cariologia comparada em seis espécies de colubrídeos (Serpentes). (abstract). Ci. Cult. (São Paulo), 23: 126.

Beebe, W., 1919: The higher vertebrates of British Guiana with special reference to the fauna of Bartica District: No.7. List of amphibia, reptilia and mammalia. No. 8 Birds of Bartica District, No. 9. Lizards of the genus *Ameiva*. Zoologica, II (7): 205-227.

Beebe, W., 1924: Galápagos: world's end. Dover Publs., Inc., New York.

Beebe, W., 1946: 4. Field notes on the snakes of Karatabo, British Guiana, and Caripito, Venezuela, *in* Snakes of British Guiana and Venezuela. Preprint from Zoologica. 31 (Part 1): 11-52.

Behler, J. L., and F. W. King, 1979: The Audubon Society Field guide to North American reptiles & amphibians. New York, Alfred A. Knopf.

Beletsky, L., 1998: Costa Rica-The ecotravelers' wildlife guide. Academic Press, San Diego, California.

Beletsky, L., 1999a: Belize & Northern Guatemala-The ecotravelers' wildlife guide. Academic Press, San Diego, California.

Beletsky, L., 1999b: Tropical Mexico-The ecotravelers' wildlife guide-The Cancún region, Yucatán Peninsula, Oaxaca, Chiapas, and Tabasco. Academic Press, San Diego, California.

Bell, T., 1843: Reptiles. *In* Darwin, 1843: The zoology of the Voyage of H.M.S. "Beagle"... during the years of 1832 to 1836, under the command Captain Fitzroy. Part V. Reptiles. Smith Elder Pub., London, Part IV 5: I-iv, 1-51, pl.i-xix.

Bell, L. Neil, 1952: A new subspecies of the racer *Coluber constrictor*. Herpetologica, 8 (2): 21.

Bellagamba, P. J. and L. E. Vega, 1996: Geographic distribution—*Thamnodynastes hypoconia* (Cateye Coluber). Herpetol. Rev., 27 (1): 36.

Bellairs, Angus, 1970: The life of reptiles. Vol 1 and 2, New York, Universe Books, 590 pp.

Belluomini, H. E. and A. R. Hoge, 1959: Operação cesarinane realizada en *Eunectes murinus* (Linnaeus, 1758) (Serpentes). Mem. Inst. Butantan, 28: 187-194.

Belluomini, H. E., T. Veinert, F. Dissmann, A. R. Hoge, and A. M. Penha, 1976/1977: Notas biológicas a respeito do gênero *Eunectes* Wagler, 1930, "sucuris" (Serpentes: Boinae). Mem. Inst. Butantan, 40-41: 79-115.

Beltz, E., 1995: Citations for the original descriptions of North American amphibians and reptiles. SSAR Herp. Circ. (24): i-vi + 1-44.

Benavides Ruiz, R. Y., 1987: Herpetofauna en el centro-sur del municipio de Santiago, Nuevo León, México. Thesis, abril de 1987.

Benítez Gálvez, J. E., 1997: Los ofidios de Puebla. La ciudad de Puebla, Puebla, México, 121 pp.

Berg, C., 1884: Viajes al Tan dil y á la tinta del Eduardo L. Holmberg. Reptitiles, Amfibios. Acoas de la Académia Nacional de Ciencias en Córdoba, Tomo V, Buenos Aires, (Entrega Segunda) 1884: 93-97.

Berg, C., 1898: Contribuciones al conocimiento de la fauna erpetológica argentina y de los países limítrofes. An. Mus. Nac. Buenos Aires, 6: 1-35.

Berger, E., 2001: Original Texans? Austin area find adds to debate over early man. Houston Chronicle, Sunday, November 25, 2001, 101 (43): 1, 12A.

Berger, W. L., 1950: A preliminary study of the subspecies of the Jumping Viper, *Bothrops nummifer*. Bull. Chicago Acad. Sci., 9 (3): 59-67.

Berger, W. L., 1972: Genera of pitvipers (Serpentes: Crotalidae). Dissertation Abstacts International, 32 (10): 1-2.

Berger, W. L., and J. E. Werler, 1954: The subspecies of the Ring-necked Coffee Snake, *Ninia diademata*, and a short biological and taxonomic account of the genus. Univ. Kansas Sci. Bull., 36-Pt.II (10): 643-672.

Bernal-Carlo, A., 1994: A new subspecies of the colubrid snake *Liophis epinephelus* from Sierra Nevada de Santa Marta, Colombia. Bull. Maryland Herpetol. Soc., 30 (4): 186-189.

Bernal-Carlos, A., and J. A. Roze, 1994: A new *Leptophis* (Serpentes: Colubridae) from Sierra Nevada de Santa Marta, Colombia. Bull. Maryland Herpetol. Soc., 30 (1): 46-49.

Bernal-Carlo, A., and J. A. Roze, 1997: Snake genus *Atractus* (Colubridae) from Sierra Nevada de Santa Marta, Colombia, with a description of a new species. Bull. Maryland Herpetol. Soc., 33 (4): 165-170.

Bernarde, P. S., and J. C. de Moura-Leite, 1999: Geographic distribution—*Hydrodynastes gigas* (Sururucu do Pantanal). Herpetol. Rev., 30 (1): 54.

Berthold, A. W., 1846: Ueber verschiedene neue oder seltene Reptilien aus New Granada und Crustacien aus China. Abh. Ges. Wiss. Göttingen, 3: 3-32.

Beverly, R., 1722: The history of Virginia in four parts. 2nd Ed., London, pp [vi] + 284 + [xxii]. [reprint, 1855, Richmond, Virginia: pp xx + 264.]

Bevington, G., 1995: Maya For travelers and students. Univ. Texas Press, Austin, 239 pp.

Beyer, G. E., 1898: Observations on the life histories of certain snakes. Amer. Naturalist, 32: 17-24.

Beyer, G. E., 1900: Louisiana Herpetology. Proc. Louisiana Soc. Natur., Appendix 1, 1900: 25-46.

Bianchi, N. O., W. Beçak, M. A. S. Bianchi, M. L. Beçak, and M. N. Rabello, 1969: Chromosome replication in four species of snakes. Chromosoma (Berlin), 26: 188-200.

Bibron, F., 1870-1899: Mission scientifique au Mexique et dans l'Amérique Centrale, Troisième Partie, 1re Section (Reptiles): 1-xiv, 1-1012.

Bibron, G., 1840: *In* J. T. Cocteau and G. Bibron 1838-1843, Reptiles (5): 1-143 *In* R. De la Sagra: Historia física, politica y natural de isla de Cuba. Arthus Bertrand, Paris.

Bibron, G., 1843: Reptiles. *In* Ramón de la Sagra. Historie physique politique et naturelle de l'île de Cuba. Arthus Bertrand, Paris, 1843: 1-239.

Bills, G.D; B.Vallejo C. and R. C. Troike, 1969: An introduction to spoken Bolivian Quechua. Institute of Latin American Studies, Univ. Texas Press, Austin.

Biocca, E., 1996: Yanoáma-the story of Helena Valero, a girl kidnaped by Amazonian Indians. Kodasha International, New York.

Blainville, Henri Marie Duerotay de, 1835: Description espèces de reptiles de la California précédée de l'analyse d un système général d'herpétologie et d'amphibiologie. Nouv. Ann. Mus. Hist. Nat. Paris, 4: 232-296.

Blair, R. W., J. S. Robertson, L. Richman, G. Sansom, J. Salazar, J. Yool and A. Choc,

1981: Diccionario Español-Cakchiquel-Inglés. Garland Publishing, Inc., New York.

Blanchard, F. N., 1920a: A synopsis of the kingsnakes, genus *Lampropeltis* Fitzinger. Occ. Pap. Mus. Zool., Univ. Michigan, 87: 17.

Blanchard, F. N., 1920b: Three new snakes of the genus *Lampropeltis*. Occ. Pap. Mus. Zool., Univ. Michigan, (81): 1–10.

Blanchard, F. N., 1921: A Revision of the kingsnakes: genus *Lampropeltis*. Bull. U.S. Natl. Mus., 114: 1–260.

Blanchard, F. N., 1923: A new North American snake of the genus *Natrix*. Occ. Pap. Mus. Zool., Univ. Michigan, 140: 1–7.

Blanchard, F. N., 1924a: A new snake of the genus *Arizona*. Occ. Pap. Mus. Zool., Univ. Michigan, 150: 1–3.

Blanchard, F. N., 1924b: The snakes of the genus *Virginia*. Pap. Michigan Acad. Arts, Sci., and Lett., 3 (3): 343–365.

Blanchard, F. N., 1925: A key to the snakes of the United States, Canada and Lower California. Pap. Michigan Acad. Sci. Arts Letters, 4, part. 2.

Blanchard, F. N., 1938: Snakes of the genus *Tantilla* in the United States. Field Mus. Nat. His. (Zoology), 20 (28): 369–376.

Blanchard, F. N., 1942: The Ring-neck Snakes, Genus *Diadophis*. Bull. Chicago Acad. Sci., (1): 1–144.

Blaney, R. M., 1973: *Lampropeltis*. Cat. Amer. Amph. Rept., (150): 1–2.

Blaney, R. M., 1977: Systematics of the common kingsnake, *Lampropeltis getulus* (Linnaeus). Tulane Studies Zool. Bot., 19 (3–4): 47–103.

Blaney, R. M., 1979: *Lampropeltis calligaster*. Cat. Amer. Amph. Rept., (229): 1–2.

Blaney, R. M., and P. K. Blaney, 1978: Additional specimens of *Amastridium veliferum* Cope (Serpentes: Colubridae) from Chiapas, México. Southwest. Naturalist, 23: 692.

Blaney, R. M., and P. K. Blaney, 1979a: The *Nerodia sipedon* complex of water snakes in Mississippi and southeastern Louisiana. Herpetologica, 35: 350–359.

Blaney, R. M., and P. K. Blaney, 1979b: Variation in the coral snake, *Micrurus diastema*, in Quintana Roo, Mexico. Herpetologica, 35: 276–278.

Bleakney, J. S., 1958: A zoogeographical study of the amphibians and reptiles of eastern Canada. National Museum of Canada, Department of Northern Affairs and National Resources, Biological Series 54, Cat. No. R-93-155, (155): 1–119.

Blem, C. R., 1981: *Heterodon platyrhinos*. Cat. Amer. Amph. Rept., (282): 1–2.

Blumenbach, J. F., 1788: Beytrag zur Naturgeschichte der Schlangen. Magazin f.d. Neuste Aug. D. Physi u. Naturg. 5: 1–13.

Böckeler, W., 1988: Víboras del Chaco Paraguayo-Kleiner Führer über Schlangen des paraguayischen Chaco-A small guide to snakes of the Paraguayan Chaco. Zoologischen Institut del Universität D 2300 kiel, Bundesrepublik Deutschland (Germany).

Bocourt, M. F., 1868: Descriptions de quelques crotaliens vouveaux appartenant au genre *Bothrops*, recueillis dans la Guatémala. Ann. Sci. Nat. Paris, Ser. 5, 10: 201–202.

Bocourt, M. F., 1873–1897: Etudes sur les reptiles, mission scientifique au Mexique et dans l'Amérique Centrale. Recherches Zoologiques. Livr. 2–15: 33–860.

Bocourt, M. F., 1876: Note sur quelques reptiles de L'isthme de Tehuantepec (Mexique) donnés par M. Sumichrast au muséum. J. Zool. Paris, 5: 386–411.

Bocourt, M. F., 1878: Etudes sur les reptiles. Mission scientifique au Mexique et dans l'Amérique Centrale-Recherches Zooloques. Paris, Imprimerie Impériale, Livr. 5: 281–360.

Bocourt, F., 1879: Recherches zoologiques pour servir à l'histoire de la fauna de L'Amérique Centrale et du Mexique; Etudes sur les reptiles et les batraciens. Miss. Sci. Mexique et L'Amérique Centrale. Paris, Partie Troisiéme, Livraison IV (1879): 361–440 + pls. 21–22.

Bocourt, M. F., 1881: Recherches zoologiques pour servir à l'histoire de la fauna de L'Amérique Centrale et du Mexique; Etudes sur les reptiles et les batraciens. Miss. Sci. Mexique et L'Amérique Centrale. Paris., Partie Troisiéme, Livraison VII 1881): 441–448 + pls. 22 (D-J).

Bocourt, M. F., 1883: Mission scientifique au Mexique et dans l'Amerique Centrale. Études sur les reptiles. Paris. Livr. 9.

Bocourt, M. F., 1891: Note sur quelques ophidiens de l'Amérique intertropicale appartenant au genre *Tretanorhinus*. Naturaliste, (2) 5: 121–122, 208.

Bocourt, M. F., 1907–1909: Etudes sur les reptiles, mission scientifique au Mexique et dans l'Amérique Centrale. *In* Dumeril, Bocourt and Mocquard, Part 3, Section 1, I–XIV: 1–1012.

Boettger, O., 1891: Reptilien und Batrachier aus Bolivia. Zool. Anz., 14 (375): 343–347.

Boettger, O., 1893a: Katalog der Reptilien-Sammlung im Museum der Senckenbergischen Naturforschenden Gesellschaft. (Rhynchocephalen, Schildkroten, Krocodile, Eidechsen, Chamaleons). Druck von Gebruder Knauer, Frankfurt.

Boettger, O., 1893b: Reptilien und Batrachier aus Venezuela. Ber. Senck. Naturf. Ges., 1893: 35–42.

Boettger, O., 1894: Geschenke und Erwerbungen. A. Geschenke für die Reptilien und Batrachiersammlung. Ber. Senck. Naturf. Ges., 1894: 28–34.

Boettger, O., 1895: Sektionsberichte. Herpetologische Sektion. Ber. Senck. Naturf. Ges., 1895: 76–79.

Boettger, O., 1898: Katalog der Reptilien-Sammlung im Mussuem Senckenbergischen Naturforschenden Gesellschaft. II. Teil. Schlangen. A.M., Frankfurt, Druk von Gebruder Knauer, I-IX: 1–160.

Bogert, C. M., 1939a: Notes on snakes of the genus *Salvadora* with a redescription of a neglected Mexican species. Copeia, 1939: 140–147.

Bogert, C. M., 1939b: A study of the genus *Salvadora*, the Patch-nosed Snakes. Publ. Univ. California Los Angeles, Biol. Ser., 1 (10): 177–236.

Bogert, C. M., 1945: Two additional races of the Patchnose Snake, *Salvadora hexalepis*. Amer. Mus. Novitates, (1285): 1–14.

Bogert, C. M., 1947: The status of the genus *Leptodrymus* Amaral, with comments on modifications of Colubrid premaxillae. Amer. Mus. Novitates, (1382): 1–14.

Bogert, C. M., 1964: Snakes of the genera *Diaphorolepis* and *Synophis* and the colubrid subfamily Xenoderminae (Reptilia: Colubridae). Senck. Biol, Frankfurt am Main, 1,12.1964, 45: 509–531.

Bogert, C. M., 1968a: A new genus and species of dwarf boa from southern Mexico. Amer. Mus. Novitates, (2354): 1–38.

Bogert, C. M., 1968b: A new arboreal pit viper of the genus *Bothrops* from the

Bibliography

Isthmus of Tehuantepec, México. Amer. Mus. Novitates, (2341): 1-14.

Bogert, C. M., 1968c: The variations and affinities of the dwarf boas of the genus *Ungaliophis*. Amer. Mus. Novitates, 2340: 1-26.

Bogert, C. M., 1969: Boas-A paradoxical family. Animal Kingdom, 72 (4): 18-25.

Bogert, C. M., and W. G. Degenhardt, 1961: An addition to the fauna of the United States, the Chihuahua Ridgenosed Rattlesnake in New Mexico. Amer. Mus. Novitates, (2064): 1-15.

Bogert, C. M., and W. E. Duellman, 1963: A new genus and species of colubrid snake from the Mexican state of Oaxaca. Amer. Mus. Novitates, (2162): 1-15.

Bogert, C. M., and J. A. Oliver, 1945: A preliminary analysis of the herpetofauna of Sonora. Bull. Amer. Mus. Nat. Hist., New York, 83 (Article 6): 1-425 + 8 pl.

Bogert, C. M., and A. P. Porter, 1966a: A new species of *Geophis* (Serpentes, Colubridae) from the state of Colima, Mexico. Amer. Mus. Novitates, (2260): 1-10.

Bogert, C. M., and A. P. Porter, 1966b: The differential characteristics of the Mexican snakes related to *Geophis dubius* (Peters). Amer. Mus. Novitates, (2277): 1-19.

Boie, F., 1827: Bemerkungen über Merrem's Versuch eines Systems der Amphibien. Ite. Lieferung. Ophidier. Isis von Oken, 20 (6): 508-566.

Bolaños, R., 1984: Serpientes venoenos y ofidismo en Centroamerica. Editorial Universidad de Costa Rica, San Jose, Costa Rica.

Bolaños, R., and J. R. Montero, 1971: *Agkistrodon bilineatus* Günther from Costa Rica. Rev. Biol. Trop., 16: 277-279.

Boos, H. E. A., 1975a: Checklist of Trinidad snakes. J. Trinidad and Tobago Field Naturalist's Club, 1975: 22-28.

Boos, H. E. A., 1975b: The rediscovery of the yellow-tailed cribo, *Drymarchon corais*, in Trinidad. J. Trinidad and Tobago Field Naturalist's Club, 1975: 84-85.

Boos, H. E. A., 1983-1984: A new snake for Trinidad. Living World, J. Trinidad and Tobago Field Naturalist's Club, 1983-1984:12.

Boos, H. E. A., 1984: A consideration of the terrestrial reptile fauna on some offshore islands north west of Trinidad. Living World 1983-1984: 19-26.

Boos, H. E. A., 2001: The snakes of Trinidad & Tobago. Texas A&M Univ. Press, College Station, Texas.

Booth, E.S, 1959.: Amphibians and reptiles collected in Mexico and Central America from 1952 to 1958. Walla Walla Coll. Publ., Dept. Biol. Sci. (24): 1-9.

Borg, J., 1990: Watch out for brown tree snakes. Honolulu Advertiser. [Incomplete citation].

Borges-Nojosa, D. M., and J. S. Lima-Verde, 1999: Geographic distribution: *Lachesis muta rhombeata*. Herpetol. Rev., 30 (4): 235.

Borman, M. B., 1977: Cofan paragraph structure and function. pp. 289-338 *In* Longacre, R. E., ed., 1976: Discourse grammar—Studies in indigenous languages of Colombia, Panama, and Ecuador Part III.

Bothner, R. C., 1963: A hibernaculum of the short-headed garter snake. *Thamnophis brachystoma* (Cope). Copeia, 1963 (3): 572-573.

Bothner, R. C., 1976: *Thamnophis brachystoma* (Cope)—Short-headed garter snake. Cat. Amer. Amph. Rept., (190): 1-2.

Boulenger, G. A., 1882: Account of the reptiles and batrachians collected by Mr. Edward Whymper in Ecuador in 1879-1880. Ann. Mag. Nat. Hist., ser. 5, 9: 457-467.

Boulenger, G. A., 1891: Reptilia & batrachia, *In* Whymper, E. Supplementary appendix to travels amongst the great Andes of the Equator. London, John Murray, 1891: 128-136.

Boulenger, G. A., [1892] 1893: Reptiles and batrachians. Zool. Rec., 29: 1-41.

Boulenger, G. A., 1893: Catalogue of the snakes in the British Museum (Natural History)-Volume I, continuing the families Typhlopidæ, Glauconiidæ, Boidæ, Ilhysiidæ, Uropeltidæ, Xenopeltidæ, and Colubridæ aglyphæ, part. Trustees of the British Museum, London.

Boulenger, G. A., 1894: Catalogue of the snakes in the British Museum (Natural History)-Volume II. Trustees of the British Museum, London.

Boulenger, G. A., 1896: Catalogue of the snakes in the British Museum (Natural History)-Volume III. Trustees of the British Museum, London.

Boulenger, G. A., 1898a: A list of the reptiles and batrachians collected by the late Professor L. Balzan in Bolivia. Annali dell Museo Civico di Storia Naturale di Genova, 19 (2): 128-133.

Boulenger, G. A., 1898b: An account of the reptiles and batrachians collected by Mr. W. F. H. Rosenberg in western Ecuador. Proc. Zool. Soc. London, 1898: 107-127.

Boulenger, G. A., 1901: Further descriptions of new reptiles collected by Mr. P. O. Simmons in Peru and Bolivia. Ann. Mag. Nat. Hist., 7 (7): 394-403.

Boulenger, G. A., 1902a: Description of new batrachians and reptiles from north-western Ecuador. Ann. Mag. Nat. Hist., 7 (9): 51-57.

Boulenger, G. A., 1902b: Descriptions of new batrachians and reptiles from the Andes of Peru and Bolivia. Ann. Mag. Nat. Hist., 7 (10): 394-402.

Boulenger, G. A., 1902c: List of fishes, batrachians, and reptiles collected by the late Mr. P. O. Simons in the provinces of Mendoza and Cordova, Argentina. Ann. Mag. Nat. Hist., 7 (9): 336-339.

Boulenger, G. A., 1903a: Descriptions of new snakes in the collection of the British Museum. Ann. Mag. Nat. Hist. Ser. 7, 12: 350-354.

Boulenger, G. A., 1903b: On some batrachians and reptiles from Venezuela. Ann. Mag. Nat. His., 7 (II): 481-484.

Boulenger, G. A., 1904: Descriptions of three new snakes. Ann. Mag. Nat. Hist., 7 (13): 450-452.

Boulenger, G. A., 1905a: Description of a new snake from Venezuela. Ann. Mag. Nat. Hist. 7 (15): 561.

Boulenger, G. A., 1905b: Description of new snakes in the collection of British Museum. Ann. Mag. Nat. Hist., 7 (15): 453-456.

Boulenger, G. A., 1907: Description of a new pit-viper from Brazil. Ann. Mag. Nat. Hist. Ser. 7, 20: 338.

Boulenger, G. A., 1908: Description of new South-American Reptiles. Ann. Mag. Nat. Hist., 8 (1): 111-115.

Boulenger, G. A., 1912: Descriptions of new reptiles from the Andes of South America preserved in the British Museum. Ann. Mag. Nat. Hist. Ser. 8, 10: 420-424.

Boulenger, G. A., 1913: On a collection of batrachians and reptiles made by Dr. H. G. F. Spurrell, F.Z.S., in the Chocó,

Colombia. Proc. Zool. Soc. London, 1913: 1019-1038.

Boulenger, G. A., 1914: On a second collection of batrachians and reptiles made by Dr. H. G. F. Spurrell, F.Z.S. in the Chocó, Colombia. Proc. Zool. Soc. London, 1914: 813-817.

Boulenger, G. A., 1920: Descriptions of four new snakes in the collection of the British Museum. Ann. Mag. Nat. Hist., 9 (6): 108-111.

Boundy, J., 1999: Systematics of the garter snake *Thamnophis atratus* at the southern end of its range. Proc. California Acad. Sci., 51 (6): 311-336.

Boundy, J., and F. Burbrink, 1998: Snakes of Santa Rosa County, Florida: inadequate sampling and serendipity. Herpetol. Rev., 29 (1): 55-56.

Boundy, J., and D. A. Rossman, 1995: Allocation and status of the garter snake names *Coluber infernalis* Blainville, *Eutaenia sirtalis tetrataenia* Cope and *Eutaenia imperialis* Coues and Yarrow. Copeia, 1995 (1): 236-240.

Brasil, V, 1914: La défense contre L'ophidisme. 2nd Ed., Pocai Weiss & Co., São Paulo, Brazil, 319 pp.

Braswell, A. L., and W. M. Palmer, 1984: *Cemophora coccinea copei*. Herpetol. Rev., 15 (2): 49.

Brattstrom, B. H., 1952: Notes on a population of Leaf-nosed Snakes *Phyllorhynchus decurtatus perkinsi*. Herpetologica, 9 (2): 57-64.

Brattstrom, B. H., 1953: A ecological restriction of the type locality of the Western Worm Snake, *Leptotyphlops h. humilis*. Herpetologica, 8 (4): 180-181.

Brattstrom, B. H., 1955: Notes on the herpetology of the Revillagigedo Islands, Mexico. Amer. Midl. Naturalist, 54 (1): 219-229.

Brattstrom, B. H., and N. B. Adis, 1952: Notes on a collection of reptiles and amphibians from Oaxaca, Mexico. Herpetologica, 8 (3): 59-60.

Brattstrom, B. H., and T. R. Howell, 1954: Notes on some collections of reptiles and amphibians from Nicaragua. Herpetologica, 10: 114-123.

Brattstrom, B. H., and J. W. Warren, 1953: A new subspecies of racer, *Masticophis flagellum*, from the San Joaquin Valley of California. Herpetologica, 9 (4): 177-179.

Bravo, D. A., 1967: Diccionario Quichua Santiagueño Castellano. Instituto Amigos Del Libro Argentino, Buenos Aires.

Brazaitis, P., and M. E. Watanabe, 1992: Snakes of the world. New York, Crescent Books, 176 pp.

Breckenridge, W. J., 1944: Reptiles & amphibians of Minnesota. Minneapolis, Minnesota, Univ. Minnesota Press.

Breeder, C. M., Jr., 1946: Amphibians and reptiles of the Rio Chucunaque drainage, Darién, Panamá, with notes on their life histories and habitats. Bull. Amer. Mus. Nat Hist., 86: 375-436.

Breen, J. F., 1974: Encyclopedia of amphibians and reptiles. T.F.H. Pub., Neptune City, New Jersey.

Breul, M., 1999: Nouvelle espece du genre *Typhlops*, (serpentes, typhlopidae) l'ile de Saint-Barthelemy comparaison avec especes des Petites antilles. Bull. Mensuel de la Soc. Linn. de Lyon, 68 (20): 30-40.

Brimley, C. S., 1903: Notes on the reproduction of certain reptiles. Amer. Naturalist, 37: 261-264, 266.

Brimley, C. S., 1904: Further notes on the reproduction of reptiles. J. Elisha Mitchell Sci. Soc., 20: 139.

Brimley, C. S., 1905: Notes on the scutellation of the red king snake, *Ophibolus doliatus coccineus*. J. Elisha Mitchell Sci. Soc., 21: 145-147.

Brimley, C. S., 1907a: Artificial key to the species of snakes and lizards which are found in North Carolina. J. Elisha Mitchell Sci. Soc., 23: 141-149.

Brimley, C. S., 1926: Revised key and list of the amphibians and reptiles of North Carolina. J. Elisha Mitcheel Sci. Soc., 42: 75-93.

Briton, D., 1887: On the so-called Alagüilac language of Guatemala. PAPA, 24: 366-377.

Brongersma, L. D., 1933: A new species of *Leptotyphlops* From Surinam. Zoologische Mededeelingen, Leiden, XV: 175-174.

Brongersma, L. D., 1937: Herpetological Notes XIV-XVI. Zoologische Mededeelingen, Leiden, XX: 1-10 + 1 pl.

Brongersma, L. D., 1940: Snakes from the Leeward group Venezuela and eastern Colombia. Studies on the Fauna of Curacao, Aruba, Bonaire, and the Venezuelan islands. Utrecht, N. V. Drukkerii P. Den Boer. 2: 116-137.

Brongersma, L. D., 1954: On some snakes from the Republic of El Salvador. Proc. Koninkl. Nederland Akademie von Wetenschappen, Series C, 57 (2): 159-164, Amsterdam.

Brongersma, L. D., 1956a: On some reptiles and amphibians from Trinidad and Tobago, B.W.I. Proc. Koninklijke Nederland Akademie van Wetenschappen, Series C, 59 (2): 165-176.

Brongersma, L. D., 1956b: On some reptiles and amphibians from Trinidad and Tobago, B.W.I. II. Proc. Koninklijke Nederland Akademie van Wetenschappen, Series C, 59 (2): 176-188.

Brongersma, L. D., 1958: Some features of the dipsadinae and pareiane (Serpentes: Colubridae). Proc. Kon. Nederland Akademie Wet. C 61: 7-12.

Brongersma, L. D., 1959: Some snakes from the Lesser Antilles. Studies on the fauna of Curaçao and other Caribbean Islands. IX (37): 50-60 + plates I-Vb.

Brongersma, L. D., 1966: 9. Poisonous snakes of Surinam. Mem. Inst. Butantan Simp. Internac., 33 (1): 73-79.

Brons, H. A., 1882: Notes on the habits of some western snakes. Amer. Naturalist, 16: 564-567.

Brown, A. E., 1893: Notes on some snakes from Tropical America lately living in the collection of the zoological society of Philadelphia. Proc. Acad. Nat. Sci. Philadelphia, 45: 429-435.

Brown, A. E., 1901a: A new species of *Coluber* from western Texas. Proc. Acad. Nat. Sci. Philadelphia, 53: 492-495.

Brown, A. E., 1901b: A New Species of *Ophibolus* from Western Texas. Proc. Acad. Nat. Sci. Philadelphia, 53: 612-613 + pl. 34.

Brown, A. E., 1901c: A review of the genera and species of American snakes north of Mexico. Proc. Acad. Nat. Sci. Philadelphia, 53: 10-110.

Brown, A. E., 1903a: Texas reptiles and their faunal relationships. Proc. Acad. Nat. Sci. Philadelphia, 55: 543-558.

Brown, A. E., 1903b: Note on *Crotalus scutulatus* Kenn. Proc. Acad. Nat. Sci. Philadelphia, 55: 625.

Brown, A. E., 1908: Generic types of neararctic reptilia and amphibia. Proc. Acad. Nat. Sci. Philadelphia. 60: 112-127.

Brown, B. C., 1950: An annotated check list of the reptiles and amphibians of Texas. Waco: Baylor Univ. Studies.

Brown, B. C., and L. M. Brown, 1967: Notable records of Tamaulipan snakes. Texas J. Sci., 19: 323-326.

Brown, B. C., and H. M. Smith, 1942: A new subspecies of Mexican coral snake. Proc. Biol. Soc. Washington, 55: 63-66.

Brown, P. R., 1997: A field guide to snakes of California. Gulf's Fieldguide Series, Gulf Publishing Company, Houston, Texas, 215 pp.

Brown, R. W., 1956: Composition of scientific words—a manual of methods and a lexicon of materials for the practice of logotechnics. Revised edition 1956, released by Reese Press, Baltimore.

Brown, T. W., 1972: A new record of the Desert Striped Whipsnake (*Masticophis taeniatus taeniatus*) from the Providence Mountains regions of San Bernardino County, California. Herpetol. Rev., 4: 172.

Brown, T. W., and H. B. Lillywhite, 1992: Autecology of the Mojave desert sidewinder, *Crotalus cerastes cerastes*, at Kelso Dunes, Mojave Desert, California, USA. pp. 279-308 *in* J. A. Campbell and E. D. Brodie, Jr.: Biology of the pitvipers. Selva, Tyler, Texas.

Brown, W. S., 1992: Emergence, ingress, and seasonal captures at dens of northern timber rattlesnakes, *Crotalus horridus*. pp. 251-258 *in* J. A. Campell and E. D. Brodie, Jr.: Biology of the pitvipers. Selva, Tyler, Texas.

Brule, B., van den, [undated]: "Las serpientes de Guatemala," Homenaje al Cientifico Francés. Vol.V, Colección "Alianza Francesa," Quetzaltenango, Guatemala City, Guatemala (Central America.

Brule, B., van den, 1982: Ofidios venonosos de Guatemala. Guatemala, Centroamérica, Facultad de CC.QQ. y Farmacia Centro de Estudios Conservacionistas (CECON), Occ. Document Series (2): 1-73.

Buckner, S. D., and R. Franz, 1998a: *Thamnophis sauritius sackenii*. Geographic Distribution. Herp. Review, 29 (1): 55.

Buckner, S. D., and R. Franz, 1998b: *Thamnophis sirtalis sirtalis*—Geographic Distribution. Herp. Review, 29 (1): 55.

Buden, D. W., 1966: An evaluation of Jamaican *Dromicus* (Serpentes: Colubridae) with the description of a new species. Breviora, (238): 1-10.

Buden, D. W., 1975: Notes on *Epicrates chrysogaster* (Serpentes: Boidae) of the southern Bahamas, with description of a new subspecies. Herpetologica, 31 (2): 166-177.

Buckley, E. E., and N. Porges, 1958: Venoms. American Association for the Advancement of Science, 2nd printing, Washington, DC.

Buckner, S. D., and R. Franz, 1998: Geographic distribution: *Thamnophis sirtalis sirtalis*. Herpetol. Rev., 29 (1): 55.

Burbrink, F. T., R. Lawson, and J. B. Slowinski, 2000: Mitochondrial DNA phylogeography of the polytypic North American rat snake (*Elaphe obsoleta*): A critique of the subspecies concept. Evolution, 54 (6): 2107-2118.

Burger, M. R., 1996: Observations in courtship behavior in *Ungaliophis* Mueller. Bull. Chicago. Herpetol. Soc., 31 (4): 57-59.

Burger, W. L., 1950: A preliminary study of the subspecies of the Jumping Viper *Bothrops nummifer*, Bull. Chicago Acad. Sci., 9: 59-67.

Burger, W. L., 1952: A southern-most coast range locality for *Charina bottae*. Herpetologica, 8 (1): 1-2.

Burger, W. L., 1955: A new subspecies of coral snake, *Micrurus lemniscatus* from Venezuela, British Guiana and Trinidad; and a key for the identification of associated species of coral snakes. I-II. Bol. Soc. Venezolana Cien. Nac., 1: 35-50.

Burger, W. L., 1971: Genera of the pitvipers (Serpentes: Crotalidae). Ph.D. Dissertation Abstr. B32: 6119, Univ. Kansas, Lawrence, Kansas, April 1971.

Burger, W. L. *in* G. Pérez-Higareda, H. M. Smith and J. Juliá-Zertuche, 1985: A new jumping viper, *Porthidium olmec*, from southern Veracruz, Mexico (Serpentes: Viperidae). pp.103-104, Bull. Maryland Herpetol. Soc., 21: 97-106.

Burger, W. L., and M. M. Hensley, 1949: Notes on a collection of reptiles and amphibians from northwestern Sonora. Nat. Hist. Misc. Chicago, (35): 1-6.

Burger, W. L., and W. B. Robertson, 1951: A new subspecies of the Mexican moccasin, *Agkistrodon bilineatus*. Univ. Kansas Sci. Bull., 24: 213-218.

Burger, W. L., P. W. Smith and H. M. Smith, 1949: Notable records of reptiles and amphibians in Oklahoma, Arkansas, and Texas. J. Tennessee Acad. Sci., 24: 130-134.

Burger, W. L., and J. E. Werler, 1954: The subspecies of Ring-necked Coffee Snake, *Ninia diademata*, and a short biological and taxonomic account of the genus. Univ. Kansas Sci. Bull., 36 (2): 643-672.

Burtch, S., 1983: Diccionario Huitoto Murui. Tomo I, Ministerio de Educacion, Instituto Lingüístico de Verano, Pucallpa, Peru, 1983: 1-262.

Burtch, S., 1983: Diccionario Huitoto Murui. Tomo II, Ministerio de Educacion, Instituto Lingüístico de Verano, Pucallpa, Peru, (20): 1-166.

Bury, R. B., 1983: Geographic distribution: *Elaphe guttata emoryi*. Herpetol. Rev., 14 (4): 123.

Byington, Cyrus, 1915: A dictionary of the Choctaw language. Smithsonian Institute Bureau of American Ethnology Bulletin 46, (46): 1-611.

Cacivio, P. M., 1997: Geographic distribution—*Thamnodynastes hypoconia* (Housesnake). Herpetol. Rev., 28 (3): 160.

Cacivio, P. M., 1999: Geographic distribution-*Clelia bicolor*. Herpetol. Rev., 30 (2): 174.

Cadle, J. E., 1984a: Molecular systematics of Neotropical Xenodontine snakes: I. South American Xenodontines. Herpetologica, 40 (1): 8-20.

Cadle, J. E., 1984b: Molecular systematics of Neotropical Xenodontine snakes: II. Central American Xenodontines. Herpetologica, 40 (1): 21-20.

Cadle, J. E., 1988: Phylogenetic relationships among advanced snakes. Univ. California Pub. Zool., 119: I-x, 1-77.

Cadle, J. E., 1989: A new species of *Coniophanes* (Serpentes: Colubridae) from nw Peru. Herpetologica, 45 (4): 411-424.

Cadle, J. E., 1992: Phylogenetic relationships among vipers: Inmunological evidence. pp. 41-48 *In* Campbell and Brodie (eds.). Biology of the pitvipers. Selva, Tyler, Texas.

Cadle, J. E., 1998: The identity of *Leptophis varius* Fischer, 1884, and placement of *Liopholidophis pinguis* Parker, 1925, in its synonymy. J. Herpetol., 32 (3): 434-437.

Cagle, F. R., 1952: A key to the amphibians and reptiles of Louisiana. Tulane Univ., New Orleans, 1952: pp. v + 42.

Cain, A. J., 1993: Animal species and their evolution, with a new afterward by the

author. Princeton Science Library, Princeton Univ. Press, Princeton, New Jersey.
Camp, C. L., 1916: Notes on the local distribution and habits of the amphibian and reptiles of southeastern California in the vicinity of the Turtle Mountains. Univ. California Pub. Zoöl., Univ. California Press, Berkley, 12 (17): 503–544.
Campbell, G. R., and W. H. Stickel, 1939: Notes on the yellow-lipped snake. Copeia, 1939: 105.
Campbell, H., 1952: A note on the geographical incidence of poisonous snakes. Herpetologica, 8 (3): 56.
Campbell, H. W., 1978a: Endangered: Short-tailed snake, *Stilosoma extenuatum* (Brown). pp. 28–30. In R. W. McDiarmid, ed.: Rare and endangered biota of Florida. Vol. Three: Amphibians and reptiles. Univ. Presses Florida, Gainsville.
Campbell, H. W., 1978b: Threatened. Miami Black-headed Snake, *Tantilla oolitica* (Telford). pp. 45–46. In R. W. McDiarmid, ed.: Rare and endangered biota of Florida. Vol. Three: Amphibians and reptiles. Univ. Presses Florida, Gainsville.
Campbell, H. W., and S. P. Christman, 1982: The systematic status of *Phyllorhynchus decuratus porelli* Powers and Banta. J. Herpetol., 16: 182–183.
Campbell, H. W., and T. R. Howell, 1965: Herpetological records from Nicaragua. Herpetologica, 21: 130–140.
Campbell, H. W., and Simmons, R. S., 1962: Notes on some reptiles and amphibians from western Mexico. Bull. Southern California Acad. Sci., 61 (4): 193–203.
Campbell, J. A., 1976: A new terrestrial pit viper of the genus *Bothrops* (Reptilia, Serpentes, Crotalidae) from western Mexico. J. Herpetol., 10:151–160.
Campbell, J. A., 1979: A new rattlesnake (Reptilia, Serpentes, Viperidae) from Jalisco, Mexico. Trans. Kansas Acad. Sci., 81: 365–369.
Campbell, J. A., 1984: A new species of *Abronia* (Sauria: Anguidae) with comments on the herpetology of the highlands of southern Mexico. Herpetologica, 40 (4): 373–381.
Campbell, J. A., 1985: A new species of highland pit viper of the genus *Bothrops* from southern México. J. Herpetol., 1948: 54.
Campbell, J. A., 1988a: *Crotalus transversus.* Cat. Am. Amph. Rept., (450): 1–3.
Campbell, J. A., 1988b: The distribution, variation, natural history, and relationships of *Porthidium barbouri* (Viperidae). Acta Zoologica Mexicana, Nueva Serie, Instituto de Ecología, México, D.F. (NS), 26: 1–32.
Campbell, J. A., 1989: A new species of colubrid snake of the genus *Coniophanes* from the Highlands of Chiapas, Mexico. Proc. Biol. Soc. Washington, 102 (4): 1036–1044.
Campbell, J. A., 1998a: Amphibians and Reptiles of northern Guatemala, the Yucatán, and Belize. Univ. Oklahoma Press, Norman, Oklahoma.
Campbell, J. A., 1998b: Comments on the identifies of certain *Tantilla* (Squamata: Colubridae) from Guatemala, with descriptions of two new species. Scientific Papers, Nat. Hist. Mus. Univ. Kansas, (7): 1–14.
Campbell, J. A., 2000: A new species of venomous coral snake (Serpentes: Elapidae) from high desert in Pubela, Mexico. Proc. Biol. Soc. Washington, 113 (1): 291–297.
Campbell, J. A., and B. L. Armstrong, 1979: Geographic variation in the Mexican Pygmy Rattlesnake, *Sistrurus ravus*, with the description of a new species. Herpetologica, 35: 304–317.
Campbell, J. A., and E. D. Brodie, 1988: A new colubrid snake of the genus *Adelphicos* from Guatemala. Herpetologica, 44 (4): 416–422.
Campbell, J. A., and E. D. Brodie (eds.), 1992: Biology of the pitvipers. Tyler, Texas: Selva.
Campbell, J. A., and J. L. Camarillo R., 1992: The Oaxacan dwarf boa, *Exiliboa placata* (Serpentes: Tropidophiidae): Descriptive notes and life history. Caribbean J. Sci., 28 (1–2): 17–20.
Campbell, J. A., J. L. Camarillo R. and P. C. Ustach, 1995: Redescription and rediagnosis of *Tantilla shawi* (Serpentes: Colubridae) from the Sierra Madre Oriental of Mexico. Southwest. Naturalist, 40 (1): 120–123.
Campbell, J. A., and L. S. Ford 1982: Phylogentic relationships of the colubrid snakes of the genus *Adelphicos* in the highlands of Middle America. Occ. Pap. Mus. Univ. Kansas, Univ. Kansas, (100): 1–22.
Campbell, J.A, L. S. Ford and J. P. Karges, 1983: Resurrection of *Geophis anocularis* Dunn with comments on its relationships and natural history. Trans. Kansas Acad. Sci., 86 (1): 38–47.
Campbell, J. A., and W. W. Lamar, 1989: The venomous reptiles of Latin America. Comstock Publishing Associates, Ithaca, New York.
Campbell, J. A., and W. W. Lamar, 1992: Taxonomic status of miscellaneous Neotropical viperids, with the description of a new genus. Occ. Pap. Museum Texas Tech. Univ., (153): 1–31.
Campbell, J. A., and J. E. Murphy, 1977: A new species of *Geophis* (Reptilia, Serpentes, Colubridae) from the Sierra de Coalcomán, Michoacán, Mexico. J. Herpetol., 11: 397–403.
Campbell, J. A., and J. E. Murphy, 1984: Reproduction in five species of Paraguayan colubrids. Trans. Kansas Acad. Sci., 87 (1–2): 63–65.
Campbell, J. A., and E. N. Smith, 1998: A new genus and species of colubrid snake from the Sierra de Las Minas of Guatemala. Herpetologica, 54 (2): 207–220.
Campbell, J. A., and E. N. Smith, 2000: A new species of arboreal pitviper from the Atlantic versant of northern Central America. Revista de Biologia Tropical, 48: 1001–1003.
Campbell, J. A., and A. Solorzano, 1992: The distribution, variation, and natural history of the Middle American montane pitviper, *Porthidium godmani.* pp. 223–250 in J. A. Campbell and E. D. Brodie, Jr.: Biology of the pitvipers. Selva, Tyler, Texas.
Campbell, J. A., and J. P. Vannini, 1989: Distribution of amphibians and reptiles in Guatemala and Belize. Proc. Western Foundation of Vertebrate Zool., 4 (1): 1–21, July 1989.
Campbell, L., 1985: The Pipil language of El Salvador. Mouton Publishers, New York,
Campbell, T., and K. R. Campbell, 1999: *Arrhyton exiguum exiguum* (Virgin Islands Racer). Herpetol. Rev., 30 (2): 112.
Camper, J. D., 1990: Systematics of the striped whipsnake, *Masticophis bilinea-*

tus (Hallowell). Ph.D. Dissertation, Texas A&M Univ., College Station, Texas.

Camper, J. D., 1996a: *Masticophis bilineatus* Jan-Sonoran Whipsnake—Reptilia: Squamata: Serpentes: Colubridae. Cat. Amer. Amph. Rept., 633): 1-4.

Camper, J. D., 1996b: *Masticophis schotti* (Baird and Girard)-Schott's Whipsnake—Reptilia: Squamata: Serpentes: Colubridae. Cat. Amer. Amph. Rept., (638):1-4.

Camper, J. D., 1996c: *Masticophis taeniatus* (Hallowell)-Striped Whipsnake—Reptilia: Squamata: Serpentes: Colubridae. Cat. Amer. Amph. Rept., (639): 1-6.

Camper, J. D., and J. R. Dixon, 1990: High incidence of melanism in *Masticophis taeniatus girardi* (Reptilia: colubridae) from the Cuatro Cienegas Basin of Coahuila, Mexico. Texas J. Sci., 42 (2):202-204.

Camper, J. D., and J. R. Dixon, 1994: Geographic variations and systematics of the striped whipsnakes (*Masticophis taeniatus* complex; Reptilia: Serpentes: Colubridae). Ann. Carnegie Mus. 63 (1): 1-48.

Campos, M. de, 1936: Notas medicas e ethnographicas. Rio de Janeiro, Brazil, 1936: 182-189.

Campos-Mogueira, C. de, 2001: New records of Squamate reptiles in central Brazil. Cerrado II: Brasília Region. Herp. Review, 32 (4): 285-287.

Canseco-Marquez, L., and O. Flores-Villela, 1995: Geographic distribution—*Chersodromus liebmanni* (Liebmann's Earth Racer). Herpetol. Rev. 26 (2): 109.

Canseco-Marquez, L., G. Gutierrez-Mayen and J. Salazar-Arenas, 2000: New records and range extensions for amphibians and reptiles from Puebla, Mexico. Herpetological Review, 31 (4): 259-263.

Capula, M. (Edited by John L. Behler), 1989: Simon & Schuster's guide to reptiles & amphibians of the world. New York, Simon & Schuster/Fireside.

Caras, R., 1974: Venomous animals of the world. Englewood Cliffs, New Jersey, Printice-Hall, Inc.

Carmichael, P., and W. Williams, 1991: Florida's fabulous reptiles and amphibians. World Publications, Tampa, Florida.

Carpenter, C. C., and J. J. Krupa, 1989: Oklahoma herpetology, an annotated bibliography. Univ. Oklahoma Press, Norman, Oklahoma.

Carr, A. R., 1940: A contribution to the herpetology of Florida. Univ. Florida Pub. Biol. Sci. Services, 111 (1).

Carr, A. F., 1963: The reptiles. New York, Time Inc.

Carr, A. F., 1969: A naturalist at large. Natural History, J. Am. Mus. Nat. Hist., 78 (3): 18-24, 68-70.

Carrillo de Espinoza, N., 1966: Contribucion al conocimiento de los boideos peruanos (Boidea, Ophidia, Reptilia). Publ. Mus. Hist. Nat. "Javier Prado." Universidad Nacional Mayor de San Marcos, Lima, Peru, (21): 1-51.

Carrillo de Espinoza, N., 1970: Contribucion al conocimiento de los reptiles del Peru (Squamata, Crocodylia, Testudinata: Reptila), Publ. Mus. Hist. Nat. "Javier Prado," Universidad Nacional Mayor de San Marcos, Lima, Perú, (22): 1-64 June 1970.

Carrillo de Espinoza, N., 1974: *Sibynomorphus williamsi* Nov. Sp. (Serpentes: Colubridae). Publ. Mus. Hist. Nat. "Javier Prado." Universidad Nacional Mayor de San Marcos, Lima, Perú, (24): 1-11 F.

Carrillo de Espinoza, N., 1983: Contribucción al conocimiento de las serpientes venenosas del Peru de las familias viperidae, elapidae y hydrophiidae (Ophidia: Reptilia). Serie A-Zoologica, Publ. Mus. Hist. Nat. "Javier Prado," Univ. Mayor de San Marcos, Lima, Perú, (30): 1-55.

Carrillo de Espinoza, N., 1989: List of Peruvian colubridae. Given in London, UK 11-19 Sept., 1989. Dept. de Herpetología, Museo de Historia Natural, Universidad Nacional Mayor de San Marcos, Lima, Perú, 1989.

Carrillo de Espinoza, N., 1990: Nombres populares de los reptiles del Peru—Informes Y Notas. Boletin de Lima, Lima, Perú, (70): 23-28, Julio 1990.

Casas-Andreu, G., 1981: Lista preliminar de los anfibios y reptiles de la costa de Jalisco. Inst. Biol. Univ. Nac. Auton. México, 5 pp. (memo).

Casas-Andreu, G., 1982: Anfibios y reptiles de la costa suroeste del estado de Jalisco, con aspectos sobre su ecologia y biogeografia. Doctorial Dissertation.- Univ. Nac. Autón. Mexico, México, D.F.

Casas-Andreu, G., and W. López-Forment C., 1978: Notas sobre *Micrurus browni taylori* Schmidt and Smith, en Guerrero, México. An. Inst. Biol. Univ. Nac. Autón. Mexico, México, 49: 291-294.

Castro Astor, I. N., J. M. S. Correia and M. A. S. Alves, 1998: *Corallus hortulanus* (Deer Snake). Herpetol. Rev., 29 (1): 44.

Catesby, M., 1743: The natural history of Carolina, Florida and the Bahama Islands: containing the figures of birds, beasts, fishes, serpents, insects and plants, etc. Revised by Mr. Edwards, Vol. 2, London, W. Innys. Iv.

Cechin, S. T. Z., 1989: *Tomodon dorsatus*, Duméril, Bibron & Duméril 1854, um sinônimo sênior de *Opisthoplus degres* Peters 1883 (Serpentes: Colubridae: Tachymeninae). Comun. Mus. Cien. PUCRSC (Zool.), (11): 203-211.

Cei, J. M., 1979: The Patagonian herpetofauna. pp. 309-339. in W. E. Duellman, ed., 1979: The South American herpetofauna: Its origin, evolution, and dispersal. Mus. Nat. Hist. Univ. Kansas, Monogr. No.7.

Cei, J. M., 1987 [dated 1986]: Monografie IV-reptiles del centro, centro-oeste y sur de la Argentina-herpetofauna de las zonas áridas y semiáridas. Museo Regionale Di Scienze Naturali, Torino.

Cei, J. M., 1994 [dated 1993]: Monografie XIV-reptiles del noroeste, nordeste y este de la Argentina-herpetofauna de las selvas subtropicales, Puna y Pampas. Museo Regionale Di Scienze Naturali, Torino.

Censky, E. J., and H. Kaiser, 1999: The Lesser Antillean fauna. pp. 181-221. *In* B. I. Crother: Caribbean amphibians and reptiles, Academic Press, San Diego.

Censky, E. J., and C. J. McCoy, 1988: Female reproductive cycles of five species of snakes (Reptilia: Colubridae) from the Yucatan Peninsula. Biotropica, 20: 326-333.

Chagnon, N. A., 1992: Yanomamö-The last days of Eden. Harcourt Brace Jovanovich, Publishers, San Diego, New York, London.

Chandler, C. R., and P. J. Tolson, 1990: Habitat use by a boid snake, *Epicrates monensis*, and its anoline prey, *Anolis cristatellus*. J. Herpetol., 24 (2): 151-157.

Charmant, T., 1995: Creole-English; English-

Creole dictionary. Hippocrene Concise Dictionary, Hippocrene Books, New York.

Chaves Mora, F., J. Alvarado Alvarado, R. Aymerich Blen and A. Solórzano Lopez, 1989: Aspectos baricos sobre los serpientes de Costa Rica. Oficina de Publicaciones de la Universidad de Costa Rica.

Chernela, J. M., 1993: The Wanano Indians of the Brazilian Amazon: A sense of space. Univ. Texas Press, Austin.

Chiasson, R. B., and C. H. Lowe, 1989: Ultrastructural scale patterns in *Nerodia* and *Thamnophis*. J. Herpetol., 23 (2): 109–118.

Chippaux, J. P., 1987 [1986]: Les serpentes de la Guyane Française. Fauna Tropicale XXVII, Paris.

Christie, M. I., 1998: *Clelia rustica* (colubridae) en el sur del Neuquen, Argentina. Cuadernos de Herpetologica, December 1998, 12 (2): 38.

Christman, S. P., 1982: *Storeria dekayi* (Holbrook). Cat. Amer. Amph. Rept., (306): 1–4.

Chiszar, D., J. D. Drew and H. M. Smith, 1992: A range extension of the Great Plains Rat Snake in eastern Colorado, Bull. Chicago Herpetol. Soc., 27 (8): 165.

Chiszar, D., R. K. K. Lee, C. W. Radcliffe and H. M. Smith, 1992: Searching behaviors by rattlesnakes following predatory strikes. pp. 369–382 in J. A. Campbell and E. D. Brodie, Jr.: Biology of the pitvipers. Selva, Tyler, Texas.

Clark, D. R., 1974: The Western Ribbon Snake *(Thamnophis proximus)*: Ecology of a Texas population. Herpetologica, 30: 372–379.

Clark, D. R., Jr., and R. R. Flee, 1976: The Rough Earth Snake (*Virginia striatula*): ecology of a Texas population. Southwest. Naturalist, 20 (4): 467–478.

Clark, H. L., 1903: The water snakes of southern Michigan. Amer. Naturalist, 37: 1–23.

Clark, Lawrence E., 1960: Vocabularío Popoluca de Sayula. Instituto Lingüístico de Verano.

Clay, W. M., 1938a: A new water snake of the genus *Natrix* from Mexico. Ann. Carnegie Mus., 27: 251–255.

Clay, W. M., 1938b: A synopsis of the North American water snakes of the genus *Natrix*. Copeia, (4): 173–182.

Clemente Perroud, P., and H. M. Chouvenc, 1967: Diccionario Castellano Kechwa; Kechwa Castellano. Seminario San Alfonso: Padres Redentoristas, Peru.

Cliburn, J. W., 1956: The taxonomic relations of the water snakes *Natrix taxispilota* and *rhombifera*. Herpetologica, 12: 198–200.

Cliburn, J. W., 1975: Geographic distribution: *Coluber constrictor latrunculus*. Herpetol. Rev., 14 (4): 123, 1.

Cliff, F. S., 1954: Snakes of the islands in the Gulf of California. Mexico. Trans. San Diego Soc. Nat. Hist., 12: 67–98.

Coborn, J., 1985: Howell beginner's guide to snakes. New York, Howell Book House Inc.

Coborn, J., 1991: The atlas of snakes of the world. Neptune Plaza City, New Jersey, T.F.H. Pub.

Coborn, J., 1994: The mini-atlas of snakes of the world. Neptune Plaza City, New Jersey, T.F.H. Pub.

Cochran, D. M., 1932: A new snake *Ialtris parishi*, from the Republic of Haiti. Proc. Biol. Soc. Washington, 45: 189–190.

Cochran, D. M., 1935: New reptiles and amphibians collected in Haiti by P. J. Darlington. Proc. Boston Soc. Nat. Hist., 40: 367–375.

Cochran, D. M., 1938: Reptiles and amphibians from the Lesser Antilles collected by Dr. S. T. Danforth. Proc. Biol. Soc. Washington, 51: 147–156.

Cochran, D. M., 1941: The herpetology of Hispaniola. Bull. U.S. Natl. Mus. 177: 1–398.

Cochran, D. M., 1946: Notes on the herpetology of the Pearl Islands, Panama. Smithsonian Misc. Collect., 106: 1–8.

Cochran, D. M., 1961: Type specimens of reptiles and amphibians in the U.S. Natl. Museum. Bull. U.S. Natl. Mus., 220, Washington, D.C.

Cochran, D. M., and C. J. Goin, 1970: The new field book of reptiles & amphibians. New York, G. P. Putnam's Sons.

Coe, Michael D., 1993: The Maya. Thames and Hudson, London, 5th edition.

Cogger, H. G., 1988: Reptiles and amphibians of Australia. Revised and expanded, Reed Books PTY, Ltd., New South Wales, Australia.

Cogger, H. G., and R. G. Zweifel (eds.), 1992: Reptiles and amphibians. Smithmark Pub., New York.

Cole, C., Jr., and L. M. Hardy, 1981: Systematics of the North American colubrid snakes related to *Tantilla planiceps* (Blainville). Bull. Amer. Mus. Nat. Hist., 171: 199–248.

Cole, C., Jr., and L. M. Hardy, 1983a: *Tantilla atriceps*. Cat. Amer. Amph. Rept., (317): 1–2.

Cole, C. J., and L. M. Hardy, 1983b: *Tantilla hobartsmithi*. Cat. Amer. Amph. Rept., (318): 1–2.

Cole, C. J., and L. M. Hardy, 1983c: *Tantilla planiceps*. Cat. Amer. Amph. Rept., (319): 1–2.

Colli, G. R., and A. K. Péres, Jr., 1997: Geographic distribution—*Chironius flavolineatus*. Herpetol. Rev., 28 (3): 158.

Colli, G. R., and A. K. Péres, Jr., 1998: *Chironius flavolineatus* (cobra-cipó, Boettger's Sipo). Herpetol. Rev., 29 (1): 53.

Collins, J. T., 1992: Reply to Grobman on variation in *Opheodrys aestivus*. Herpetol. Rev. 23 (1):15–16.

Collins, J. T., 1990: Standard common and current scientific names for North American amphibians and reptiles-Third Edition. SSAR, Herpetological Circular (19): 1–39.

Collins, J. T., 1993: Amphibians and reptiles of Kansas. Univ. Kansas Mus. Nat. Hist., Publ. Education Series, (13): xx–397.

Collins, J. T., J. E. Huheey, J. L. Knight and Hobart M. Smith, 1978: Standard common & scientific names for North American amphibians and reptiles. SSAR, Miscellaneous Pub., Herpetol. Circ. (7): 1–36.

Collins, J. T., R. Conant, J. E. Huheey, J. L. Knight, E. M. Rundquist and H. M. Smith, 1982: Standard common and current scientific names for North American amphibians and reptiles-Second Edition. SSAR, Herpetol. Circ. (12): 1–28.

Collins, J. T. (Ed.), 1990: Standard common and current scientific names for North American amphibians and reptiles-third edition. SSAR Herpetol. Cir. (19): 1–41.

Coloma, L. A., and S. R. Ron, 2001: Ecuador megadiverso: anfibios, reptiles, aves, mamíferos—megadiverse Ecuador: amphibians, reptiles, birds, and mammals. Serie de Divulgación 1, Mus. Zool., Centro de Biodiversidad y Ambiente, Departamento de Ciencias Biológicas, Pontifica Universidad Católica del Ecuador.

Bibliography

Conant, R. C., 1930: Field notes of a collecting trip. Bull. Antivenin Inst. Amer., 4: 60-64.

Conant, R. C., 1936: Semiaquatic snakes of the genus *Thamnophis* from the isolated drainage system of the Río Nazas and adjacent areas in Mexico. Copeia, 1963: 473-499.

Conant, R. C., 1937: *Alsophis* from new islands with the description of a new subspecies. Proc. New England Zoöl. Club, 16: 81-83.

Conant, R. C., 1938: The reptiles of Ohio. Amer. Midl. Naturalist, 20 (1): 1-200.

Conant, R. C., 1940: A new subspecies of the fox snake, *Elaphe vulpina* (Baird and Girard). Herpetologica, 2 (1): 1-14.

Conant, R. C., 1943: Studies on North American water snakes-I: *Natrix kirtlandii* (Kennicott). Amer. Midl. Naturalist, 29 (2): 313-341.

Conant, R. C., 1946: Studies on North American water snakes—II: The subspecies of *Natrix valida*. Amer. Midl. Naturalist, 35: 250-275.

Conant, R. C., 1949: Two new races of *Natrix erythrogaster*. Copeia, 1949: 1-15.

Conant, R. C., 1953: Three new water snakes of the genus *Natrix* from Mexico. Nat. Hist. Misc., Chicago Acad. Sci., (126): 1-9.

Conant, R. C., 1956: A review of two rare pine snakes from the Gulf Coastal Plain. Amer. Mus. Novitates, (1781): 1-31.

Conant, R. C., 1958: A field guide to reptiles & amphibians of the United States and Canada East of the 100th meridian. Houghton Mifflin Company, Boston, The Riverside Press, Cambridge-The Peterson Field Guide Series.

Conant, R. C., 1961: A new water snake from Mexico, with notes on anal plates and apical pits in *Natrix* and *Thamnophis*. Amer. Mus. Novitates, (2060): 1-22.

Conant, R. C., 1963a: Another new water snake of the genus *Natrix* from the Mexican Plateau. Proc. Biol. Soc. Washington, 76: 169-172.

Conant, R. C., 1963b: Evidence for the specific status of the water snake *Natrix fasciata*. Amer. Mus. Novitates, (2122): 1-38.

Conant, R. C., 1965: Notes & comments on toads, lizards, and snakes from Mexico. Amer. Mus. Novitates, (2205): 1-38.

Conant, R. C., 1966: A second record for *Ungaliophis continentalis* from Mexico. Herpetologica, 226: 157-160.

Conant, R. C., 1969: A review of the water snakes of the genus *Natrix* in Mexico. Bull. Amer. Mus. Nat. Hist., 142: Art.1: 1-140, New York.

Conant, R. C., 1975: A field guide to reptiles & amphibians, eastern & central North America. 2nd Edition, Boston, Houghton Mifflin Co., Boston.

Conant, R. C., 1982: The origin of the name "Cantil" for *Agkistrodon bilineatus*, Herpetol. Rev., 13: 118.

Conant, R. C., 1984: A new subspecies of the pit viper *Agkistrodon bilineatus* (Reptilia: Viperidae) from Central America. Proc. Biol. Soc. Washington, 97: 1-135.

Conant, R. C., 2000: A new species of garter snake from western Mexico. Occ. Pap. Mus. Nat. Sci., Louisiana State Univ., Baton Rough, Louisiana, May 2000, (76): 1-7.

Conant, R. C., and W. Bridges, 1939: What snake is that? A field guide to the snakes of the U.S. east of the Rocky Mountains. D. Appleton-Century, Co., New York.

Conant, R. C., F. R. Cagle, C. J. Goin, C. H. Lowe, Jr., W. T. Neill, M. G. Netting, K. P. Schmidt, C. E. Shaw, R. C. Stebbins and C. M. Bogert, 1956: Common names for North American amphibians and reptiles plus an index to common names. Copeia, 1956 (3): 172-185.

Conant, R. C., and W. M. Clay, 1937: A new subspecies of water snake from islands in Lake Erie. Occ. Pap. Mus. Zool., Univ. Michigan, (346): 1-9.

Conant, R. C., and J. T. Collins, 1991: A field guide to reptiles and amphibians: eastern and central North America. Boston, Houghton Mifflin.

Conant, R. C., and J. T. Collins, 1998: A field guide to reptiles and amphibians-eastern and central North America. Third Edition, Expanded, The Peterson Field Guide Series, Houghton Mifflin Company, Boston-New York.

Contreras Arquieta, A., 1989: *Adelphicos quadrivirgatus newmanorum* (Serpentes: Colubridae), nuevo registro genérico para Nuevo Léon. Publ. Biol., Facultad de Ciéncias Biológicas, Universidad Autónoma de Nuevo Léon, México, 3: 35-36.

Contreras Balderas, S., F. Gonzalez Saldivar, D. L. Villarreal, and A. Contreras Arquieta, 1995: Listado preliminar de la fauna silvestre del estado de Nuevo Leon, México. Consejo Consultivo Estatal para la Preservación y Fomento de la Flora y Fauna Silvestre de Nuevo Leon. Comisión Consultiva Tecnica Subcomision de Fauna Silvestre. Impresora Monterrey, S.A., 152 pp.

Cook. D. G., and F. J. Aldridge, 1984: *Coluber constrictor priapus*. Herpetol. Rev., 15 (2): 49.

Cook, F. A., 1954: Snakes of Mississippi. State Game and Fish Commission, Publ. Relations Department, Jackson, Mississippi.

Cook, F. R., 1984: Introduction to Canadian amphibians and Reptiles. Mus. Nat. Science, Mus. Canada, Canada, 200 pp.

Cooper, J. G., 1860: Report upon the reptiles collected on the Survey. No.4. *In* 36th Congress, 1st Session, Senate, EX. Doc.-Reports of explorations and surveys to ascertain the most practicable and economical routes for a railroad from the Mississippi River to the Pacific Ocean.-Made under the direction of the Secretary of War, in 1853-1855, according to Acts of Congress of March 3, 1853, May 31, 1854, and August 5, 1854. Volume XII. Book II. Thomas H. Ford, Printer, Washington, D.C.

Cope, E. D., 1859: Catalogue of the venomous serpents in the Museum of the Academy of Natural Sciences of Philadelphia, with notes on the families, genera and species. Proc. Acad. Nat. Sci. Philadelphia, 11: 332-347.

Cope, E. D., 1859: Notes and descriptions of foreign reptiles. Proc. Acad. Nat. Sci. Philadelphia, 11: 294-297.

Cope, E. D., 1860a: An enumeration of the genera and species of Rattlesnakes, with synonymy and references. Smithsonian Contrib. Knowl., 12: 119-126.

Cope, E. D., 1860b: Catalogue of colubridae in the Museum of the Academy of Natural Sciences in Philadelphia with notes and descriptions of new species. Part 2. Proc. Acad. Nat. Sci. Philadelphia, 12: 241-266.

Cope, E. D., 1860c: Catalogue of the colubridae in the Museum of the Academy of Natural Sciences of Philadelphia. Part 3. Proc. Acad. Nat. Sci. Philadelphia, 12: 553-556.

Cope, E. D., 1860d: Descriptions of reptiles from tropical America and Asia. Proc. Acad. Nat. Sci. Philadelphia, 12: 368–374.

Cope, E. D., 1860e: Notes and descriptions of a new and little known species of American reptiles. Proc. Acad. Nat. Sci. Philadelphia, 12: 339–345.

Cope, E. D., 1860f: Supplement to "A catalogue of the venomous serpents in the Museum of the Academy of Natural Sciences of Philadelphia, with notes on the families, genera, and species." Proc. Acad. Nat. Sci. Philadelphia, 1860: 72–74.

Cope, E. D., 1860 [1861a]: Catalogue of the colubridae in the Museum of the Academy of Natural Sciences of Philadelphia. Proc. Acad. Nat. Sci. Philadelphia, 12: 74–79, 241–266, 553–566.

Cope, E. D., 1861b: Contributions to the ophiology of lower California, Mexico, and Central America. Proc. Acad. Nat. Sci. Philadelphia, 13: 292–306.

Cope, E. D., 1861c: Some remarks defining the following species of Reptilia Squamata. Proc. Acad. Nat. Sci. Philadelphia, 1861: 75–76

Cope, E. D., 1862a: Catalogue of the reptiles obtained during the explorations of Parana, Paraguay, Bermejo and Uruguay rivers, by Capt. Thos. J. Page, U.S.N.; and of those procured by Lieut. N. Michler, U.S. Top. Eng., Commander of the expedition conducting the survey of the Atrato River. Proc. Acad. Nat. Sci Philadelphia, 14: 346–359.

Cope, E. D., 1862b: Synopsis of the species *Holcosus* and *Ameiva*, with diagnosis of new West Indian and South American colubridae. Proc. Acad. Nat. Sci. Philadelphia, 14: 60–82.

Cope, E. D., 1863: Descriptions of new American squamata in the museum of the Smithsonian Institution, Washington. Proc. Acad. Nat. Sci. Philadelphia, 15: 100–106.

Cope, E. D., 1864: Contributions to the herpetology of tropical America. Proc. Acad. Nat. Sci. Philadelphia, 16: 166–181.

Cope, E. D., 1866a: Fourth contribution to the herpetology of tropical America I. The collection made by direction of the governor of Yucatan, Jose Salazar Starregui, by Arthur Scott, naturalist of the commission and sent to the Smithsonian Institution. Proc. Acad. Nat. Sci. Philadelphia, 18: 123–132.

Cope, E. D., 1866b: Fifth contribution to the herpetology of tropical America. Proc. Acad. Nat. Sci. Philadelphia, 18: 317–323.

Cope, E. D., 1868a: An examination of the reptilia and batrachia obtained by the Orton expedition to Ecuador and the upper Amazon, with notes on other species. Proc. Acad. Nat. Sci. Philadelphia, 20: 96–140.

Cope, E. D., 1868b: Additional descriptions of Neotropical reptilia and batrachian not previously known. Proc. Acad. Nat. Sci. Philadelphia, 1868: 119–138.

Cope, E. D., [dated 1869] 1870: Seventh contribution to the herpetology of tropical America. Proc. Amer. Philos. Soc., 11: 147–169.

Cope, E. D., 1870: Third contribution to the herpetology of tropical America. Proc. Amer. Philos. Soc., 11: 147–169.

Cope, E. D., 1871a: Eighth contribution to the herpetology of tropical America. Proc. Amer. Philos. Soc., 11: 553–559.

Cope, E. D., 1871b: Ninth contribution to the herpetology of tropical America. Proc. Acad. Nat. Sci. Philadelphia, 23: 200–224.

Cope, E.D, 1874: Descriptions of some species of reptiles obtained by Dr. John F. Bransford, assistant surgeon, United States Navy, while attached to the Nicaraguan surveying expedition in 1873. Proc. Acad. Nat. Sci. Philadelphia, 26: 64–72.

Cope, E. D., 1875a: Check-list of North American batracia and reptilia; with a systematic list of the higher groups, and an essay on geographic distribution based on specimens contained in the U.S. National Museum. Washington, D.C., Bull. U.S. Natl. Mus. GPO-1–104.

Cope, E. D., 1875b [1876 preprint]: On batrachia and reptilia of Costa Rica with notes on the herpetology and icthyology of Nicaragua and Peru. Extracts J. Acad. Nat. Sci., Philadelphia.

Cope, E. D., 1876a: On the batrachia and reptilia of Costa Rica with notes on the herpetology and ichthyology of Nicaragua and Peru. J. Acad. Nat. Sci. Philadelphia, Ser.2, 8 (part 2): 23–28.

Cope, E. D., 1876b: Report on the reptiles brought by Professor James Orton from the middle and upper Amazon, and western Peru. J. Acad. Nat. Sci. Philadelphia, (2) 8: 159–188.

Cope, E. D., 1877: Synopsis of the cold-blooded vertebrata procured by Professor James Orton during his exploration in Peru in 1876–77. Proc. Amer. Philos. Soc., 17: 33–49.

Cope, E. D., 1879: Eleventh contribution to the herpetology of tropical America. Proc. Acad. Nat. Sci. Philadelphia, 18 (4): 261–277.

Cope, E. D., 1883: Notes on the geographic distribution of batrachia and reptilia in western North America. Proc. Acad. Nat. Sci. Philadelphia, 35: 10–35.

Cope, E. D., 1884: [Review of] Garman's North American reptiles and batrachians. Amer. Naturalist, 1884: 513–515.

Cope, E. D., 1885 [dated 1884]: Twelfth contribution to the herpetology of tropical America. Proc. Amer. Philos. Soc., 22: 167–194.

Cope, E. D., 1885: A contribution to the herpetology of Mexico. Proc. Amer. Philos. Soc., 22: 379–404.

Cope, E. D., 1886 [dated 1885]: Thirteenth contribution to the herpetology of tropical America. Proc. Amer. Philos. Soc., 23: 271–287.

Cope, E. D., 1887: Catalogue of batrachians and reptiles of Central America and México. Bull. U.S. Natl. Mus., 32: 1–98.

Cope, E. D., 1888: Catalogue of batrachia and reptilia brought by William Taylor from San Diego, Texas. Proc. U.S. Natl. Mus., 11: 395–398.

Cope, E. D., 1889: Scientific results of explorations by the U.S. Fish Commission steamer Albatross. No. 111. Report on the batrachians and reptiles collected in 1887–'88. Proc. U.S. Natl. Mus., 12: 141–147.

Cope, E. D., 1892a: A critical review of the characters and variations of snakes of North America. Proc. U.S. Natl. Mus., 14: 589–694.

Cope, E. D., 1892b: The batrachia and reptilia of northwestern Texas. Proc. Acad. Nat. Sci. Philadelphia, 44: 331–337

Cope, E. D., 1893: Prodromus of a new system of the non-venomous snakes. Amer. Naturalist, 27: 477–483

Cope, E. D., 1894a: The Batrachia and

Reptilia of the University of Pennsylvania West Indian Expedition of 1890 and 1891. Proc. Acad. Nat. Sci., Philadelphia, 1884: 429-442.

Cope, E. D., 1894b: The classification of snakes. Amer. Naturalist, 28 (334): 831-844.

Cope, E. D., 1894c: Third addition to a knowledge of the batrachia and reptilia of Costa Rica. Proc. Acad. Nat. Sci., Philadelphia, 1894: 194-206.

Cope, E. D., 1895 [dated 1894]: Classification of the ophidia. Trans. Amer. Philos. Soc., 18: 186-219.

Cope, E. D., 1895a: On some new North American snakes. Amer. Naturalist, 29: 676-680.

Cope, E. D., 1895b: The classification of the ophidia. Trans. Amer. Philos. Soc. 18 (3): 186-219.

Cope, E. D., 1896: The geographical distribution of batrachia and reptilia in North America. Amer. Naturalist, 30: 886-902, 1003-1026.

Cope, E. D., 1898 [1900]: The crocodilians, lizards, and snakes of North America. Ann. Rept. U.S. Natl. Mus.

Cope, E. D., 1899: Contributions to the Herpetology of New Granada and Argentina. The Philadelphia Museums Scientific Bull. No. 1, (1): 3-22, plate I-II.

Cordero, Luis, 1967: Diccionario Quichua-Español, Español-Quichua. Tomo XXIII, octubre-diciembre No.4, Analis de la Universidad de Cuenca, Cuenca, Ecuador, (4): i-xli; 1-257.

Corn, M. J., [1974] 1975: Report on the first certain collection of *Ungaliophis panamensis* from Costa Rica. Caribbean J. Sci., 14: 167-175.

Corrington, J. D., 1929: Herpetology of the Columbia, South Carolina, region. Copeia, 172 (1929): 58-83.

Costa Prudente, A. L., da and R. A. Brandão, 1998: *Gomesophis brasiliensis* (Brazilian Burrowing Snake). Herpetol. Rev., 29 (2): 112.

Coues, E., 1875: Synopsis of the reptiles and batrachians of Arizona; with critical and field notes, and on extensive synonymy. Rep. Geog. Explor. Surv. West 100th Meridian by Geo. M. Wheeler, 5: 585-633.

Coues, E., and H. C. Yarrow, 1878: Note on the herpetology of Dakota and Montana. Bull. U.S. Geol. And Geog. Survey. 4 (1) (art.11): 259-291.

Cox, D. C., and W. W. Tanner, 1995: Snakes of Utah. Brigham Young Univ., Provo, Utah, 92 pp.

Cragin, F. W., 1881: A preliminary catalog of Kansas reptiles and batrachia. Trans. Kansas Acad. Sci., 7: 114-123.

Cranston, T., 1994: Natural history of the Sierra Mountain Kingsnake (*Lampropeltis zonata multicincta*), The Vivarium, 6 (3): 38-43, 47.

Crawford, J. M., 1975: Studies in southeastern Indian languages. Univ. Georgia Press, Athens.

Crema, M. (Coordinador), 1968: Ecologia vegetal fauna-estudio de Caracas-Volume I. Univerisdad Central de Venezuela. Edicion de la Biblioteca, Caracas.

Crimmins, M. L., 1925: An addition to the herpetological fauna of the United States. Copeia, 1925: 7.

Crombie, R. I., 1999: Jamaica. pp. 63-92. *In* B. I. Crother ed., 1999 Caribbean amphibians and reptiles. Academic Press, New York.

Crother, B. I., 1999 ed.: Caribbean amphibians and reptiles. Academic Press, New York, 495 pp.

Crother, B. I., J. A. Campbell and D. M. Hillis, 1992: Phylogeny and historical biogeography of the palm-pitvipers, genus *Bothriechis*: biochemical and morphological evidence. pp. 1-20. *In* Campbell & Brodie, Jr. (Eds.) Biology of the Pitvipers. Selva, Tyler, Texas.

Cruz, F. B., J. B. Schulte and P. Bellagamba, 1999: New distributional records and natural history for reptiles from southern Argentina. Herpetol. Rev., 30 (30): 182-183.

Cruz, G. A., 1997: Serpientes venenosas de Honduras. Universidad Nacional Autónoma de Honduras, Editorial Universitaria, Tegucigalpa, M. D. C., Honduras, C. A., Agosto, 1997, 160 pp.

Cruz, G. A., L. D. Wilson and J. Espinosa, 1979: Two additions to the reptile fauna of Honduras, *Eumeces managuae* Dunn and *Agkistrodon bilineatus* (Günther) with comments on *Pelamis platurus* (Linnaeus). Herpetol. Rev., 10: 26-27.

Cruz D., G. A. and L. D. Wilson, 1986: *Oxybelis brevirostris* (Cope): An addition to the snake fauna of Honduras. Herpetol. Rev., 17: 7-8.

Cunha, O. R., da and F. P. Nascimento, 1970: Ofídios de Amazonia. II. *Liophis miliaris* (Linneu, 1758) na Amazônia norte oriental (Território Federal do Amapá) (Ophidia, Colubridae). Bol. Mus. Paraense Emilio Goeldi (Zool.), (70): 1-6.

Cunha, O. R., da and F. P. Nascimento, 1972: Ofídios da Amazônia. 3. Sôbre a ocorrência de *Bothrops lichenosus* Roze, 1958 no Brasil (Ophidia, Crotalidae). Rev. Bras. Biol., 32: 27-32.

Cunha, O. R., da and F. P. Nascimento, 1973: Ofídios da Amazônia. IV. As cobras corais (gênero *Micrurus*) da região leste do Pará. (Ophidia, Elapidae). Nota preliminar. Pub. Avuls. Mus. Emilio Goeldi, Sesq., 20: 273-286.

Cunha, O. R., da and F. P. Nascimento, 1975: Ofídios da Amazônia, VIII. A ocorrência de *Rhinobothryum lentiginosum* (Scopoli, 1785) nas proximidades de Belém, Pará (Ophidia, Colubridae). Bol. Mus. Paraense Emilio Goeldi, Nova Serie (Zool.), 84: 1-6, pl.

Cunha, O. R., da and F. P. Nascimento, 1976: Ofídios de Amazonia. IX. O gênero *Liophis* Wagler, 1830, na região leste do Pará (Ophidia: Colubridae). Bol. Mus. Paraense Emilio Goeldi, Nova Serie (Zool.), (85): 1-34.

Cunha, O. R., da and F. P. Nascimento, 1978: Ofídios da Amazônia. X. As cobras da região leste do Pará, Belém. Bol. Mus. Paraense Emílio Goeldi, Nova Serie (Zool.), Avuls., (31): 1-218.

Cunha, O. R., da and F. P. Nascimento, 1982a: Ofidios da Amazônia. XIV.-As espécies de *Micrurus, Bothrops, Lachesis* e *Crotalus* do sul do Pará e oeste do Maranhão, incluindo áreas de cerrado deste estado. (Ophidia: Elapidae e Viperidae).Bol. Mus. Paraense Emilio Goeldi, Nova Serie (Zool.), (112): 1-58.

Cunha, O. R., da and F. P. Nascimento, 1982b: Ofidios da Amazônia. XVI.—A espécie *Uromacerina ricardini* (Peracca, 1897) na Amazônia oriental (leste do Pará) (Ophidia: Colubridae). Bol. Mus. Paraense Emilio Goeldi, Nova Serie (Zool.), (113): 1-9.

Cunha, O. R., da and F. P. Nascimento, 1983 [1982b]: Ofidios da Amazônia. XV.-As espécies de *Chironius* da Amazônia Oriental (Pará, Amapá e

Maranhão) (Ophidia: Colubridae). Mem. Inst. Butantan, 46: 139–172.

Cunha, O. R., da and F. P. Nascimento, 1982c: Ofidios da Amazônia. XVII. Revalidação e redescrição de *Micrurus ornatissimus* (Jan, 1858) diferenciada de *M. langsdorffi* (Wagler, 1824) e distribuição geográfica das duas espécies (Ophidia: Elapidae) Meise, do Amazonas (Ophidia: Elapidae). Bol. Mus. Paraense Emílio Goeldi, Nova Serie (Zool.), (116): 1–17.

Cunha, O. R., da and F. P. Nascimento, 1983: Ofidios da Amazonia XX-As expecies de *Atractus* Wagler, 1828, na Amazonia oriental and Maranhão (Ophidia, Colubridae). Bol. Mus. Paraense Emílio Goeldi, Nova Serie (Zool.), (123): 1–40.

Cunha, O. R., da, F. P. Nascimento, and TCS de Avila-Oire, 1985: Os repteis da área de Carajás, Pará, Brasil (Testudinese e Squamata) I. Contribuições do Museu Paraense Emilio Goeldi ao Projeto Carajás. Mus. Paraense Emilio Goeldi, (40): 1–92.

Cunha, O. R., da, and F. P. Nascimento, 1991: Ofidios da Amazônia. XXII. Revalidação e redescrição de *Micrurus albicinctus* Amaral, de Rondônia, e sôbre a validade de *Micrurus waehnerorum* Meise, do Amazonas (Ophidia: Elapidae). Bol. Mus. Paraense Emílio Goeldi, Nova Serie (Zool.), 7 (1): 43–52.

Cunha, O. R., da, F. P. Nascimento and T. C. S. Ávila-Pires, 1985: Os réptiles da área de Carajás, Pará, Brasil (Testudinese e Squamata). Contribuições do Museu Paraense "Emílio Goeldi" ao Projeto Carajás, Bol. Mus. Paraense Emílio Goeldi, Nova Serie (Zool.), 40: 1–92.

Cunha, O. R., da, F. P. Nascimento and A. R. Hoge, 1980: Ofidios da Amazônia. XII. Uma subespécie do *Sibynomorphus mikani* do noroeste do Maranháo (Ophidia: Colubridae: Dipsadinae). Bol. Mus. Paraense Emílio Goeldi, Nova Serie (Zool.), 103: 1–15.

Cunningham, J. D., 1966: Observations on the taxonomy and natural history of the rubber boa, *Charina Bottae*. Southwest. Naturalist, 1: 298–299.

Curran, C. H., and Carl Kaulfield, 1937: Snakes and their ways. New York, Harper & Brothers Publishers.

Curtis, L., and D. W. Tinkle, 1951: The striped water snake, *Natrix rigida* (Say), in Texas. Field and Lab., 19 (2): 72–74.

Curtiss, A., 1947: Prevalence of snakes in Haiti. Herpetologica, 3 (4): 224.

Cuvier, G. L. C. F. D., [1816] 1817: Le Règne animal distribué d'après son organisation, pour servir de base à l'hostoire naturelle des animaux et d'introduction à l'anatomie comparée. Paris, 8: i–xviii, 1–532.

Cuvier, G., 1829: Le règne animal distribué d'après son organisation. Tome II, contenant les reptiles, les poissons, les molusquera et les annelides. Nouvelle edition, Paris, Déterville, crochard, xv.

Cuvier, G., 1831: The class reptilia arranged by the Baron Cuvier—order ophidia: The third order of reptiles. *In* E. Griffith & E. Pidgeon, Whittaker, Treachor, and Co., Ave-Maria-Lane, London, 1831: 241–387.

Dalgle, J. O., 1984: A Dictionary of the Cajun language. Edwards Brothers, Inc., Ann Arbor, Michigan.

D'Agostini, F. M., 1998: Variação da fidoe de *Philodryas aestivus* (Duméril, Bibron et Duméril, 1854) e a invalidação das subespecies (Serpentes, Colubridae, Xenodontinae, Philodryadini). Biociencias, Porto Alegre, 6: 169–182.

Daniel, H., 1949: Las serpientes en Colombia. Rev. Fac. Nacion. Agropn. Medellin, 10: 301–333.

Dart, R. C., J. T. McNally, D. W. Spaite and R. Gustafson, 1992: The sequelae of pitviper poisoning in the United States. Pp. 395–404 *in* J. A. Campbell and E. D. Brodie, Jr.: Biology of the pitvipers. Selva, Tyler, Texas.

Daudin, F. M., 1801–1803: Histoire naturelle, générale et particuliére des reptiles. 8 volumes, F. Dufort, Paris.

Davenport, J. W., 1943: A field book of the snakes of Bexar County, Texas and vicinity -A simplified key and notes on their behavior. Witte Mem. Mus., Brackenridge Park, San Antonio, Texas.

David, D. D., 1932: Occurrence of *Thamnophis butleri* in Wisconsin. Copeia, 1932: 113–118.

David, P., and A. Dubois, 2001: A herpetological analysis of Shaw and Nodder's *Vivarium Naturae or The Naturalist's Miscellany* (1789-1813), a 24-volume series on natural history. Int. Soc. History Bibliog. Herpetol., 2 (2): 5–39.

David, P., and I. Ineich, 1999: Les serpents venimeux du monde: Systémamatuque et répartition. Dumerilia, Association des Amis du Laboratoire des Reptiles et Amphibians du Muséum (AALRAM), 3: 1–500.

Davis, N. S., Jr., and F. L. Rice, 1883: Descriptive catalogue of North American batrachia and reptilia east of the Mississippi River. Bull. Illinois State Lab. Nat. Hist., (5): 27–44.

Davis, W. B., and J. R. Dixon, 1957: Notes on Mexican snakes (Ophidia). Southwest. Naturalist, 2: 19–27.

Davis, W. B., and J. R. Dixon, 1959: Snakes of the Chilpancingo region, Mexico. Proc. Biol. Soc. Washington, 72: 79–92.

Davis, W. B., and H. M. Smith, 1953: Snakes of the Mexican state of Morelos. Herpetologica, 8: 133–143.

Day, D., 1989: Vanished species. New York, Gallery Books, 1989 revision of an imprint.

Degenhardt, W. G., C. W. Painter and A. H. Price, 1996: Amphibians & reptiles of New Mexico. Univ. New Mexico Press, Albuquerque.

Degerbol, M., 1923: Description of a new snake genus *Glauconia* from Mendosa. Vidensk. Meddr. dansk Naturhist. Foren. Kjöbenhaun, 76: 213–214.

DeGraaf and D. D. Rudis with drawings by A. Rorer, 1983: Amphibians and reptiles of New England-habitats and natural history. Univ. Massachusetts Press, Amherst.

Deiques, C. H., and S. T. Z. Cechin, [1990] 1991: O status de *Helicops carinicaudus* (Wied 1825) (Serpentes: Colubridae). Acta Biol. Leopoldensia, 12 (2): 313–326.

DeKay, J. E., 1842: Zoology of New York, or the New York fauna. Reptiles and amphibia. Pt. 3: 59–72; pt. 4: plates 19–22. *In* Natural History of New York, Albany.

De Lisle, H. F., 1977: The Mexican kingsnake. A scientific mystery story. Herpetology, Sept., 1977, 9 (3): 3–7.

De Lisle, H. F., 1978: Key to the snakes of Baja California. Herpetology, 9 (4): 11–20.

De Lisle, H. F., P. R. Brown, B. Kaufman, and B. M. McGurty (eds.), 1988: Proceedings of the Conference on Califor-

Bibliography

nia Herpetology-1987. Southwest. Herpetol. Soc., Special Publ., November 1, 1988, (4): 1–143.

Densmore III, L. D., F. L. Rose and S. J. Kain, 1992: Mitochrondrial DNA evolution and speciation in water snakes (genus *Nerodia*) with special references to *Nerodia harteri*. Herpetologica, 48 (1): 60–68.

DePoe, C., J. B. Funderberg, Jr., and T. L. Quay, 1936.: The reptiles and amphibians of North Carolina. J. Elisha Mitchell Scientific Soc., 77 (2): 125–136.

Deptula, W., 1997: Ghost of the Great Plains. Reptiles, October, 5 (10): 32–46.

Despax, R., 1910: Mission geodesique de l'Equateur. Collections recueillies par M. Le Dr. Rivet. Liste des ophidiens et description des especies nouvelles. Bull. Mus. Natl. d'Hist. Nat., 16: 368–376.

Despax, R., 1911: Reptiles et batraciens de l'Equateur recueillis par M. le Dr. Rivet. pp. 17–44. *In* Ministère de l'Instruction Publique, Mission du Service Geógraphique de l'Armée pour la mesure d'un Arc de Méridien Equatorial en Amérique du Sud (1899–1906). Vol.9, Paris, Gauthier-Villars.

Diaz Olivera, A., 1995: Instituto Nacional de Salud, Lima—Peru. General discussions

Diaz, J. A., and D. del C. Huacuz Elias, 1996: Guia ilustrada de los anfibios y reptiles mas comunes de la Reserva Colonia-Maruata en la costa de Michoacán, México. Universidad Michoacana de San Nicolás de Hidealgo, 1st Ed., Morelia, Michoacán, México.

Di-Bernardo, M., 1992: Revalidation of the genus *Echinanthera* Cope, 1894, and its conceptual amplification (Serpentes, Colubridae). Comunicaçoes do Museu de Ciencias da PUCRS. Serie Zoologica, 5 (13): 225–256.

Di-Bernardo, M., 1994: Uma nova espécie de *Echinanthera* Cope, 1894 (Serpentes, Colubridae) do nordeste do Brasil, Biociências. Porto Alegre, 2 (2): 75–81.

Di-Bernardo, M., 1996: A new species of the Neotropical snake genus *Echinanthera* Cope, 1864 from southeastern Brazil (Serpentes, Colubridae). The Snake, 27: 120–126.

Di-Bernardo, M., and T. de Lema, 1986: O gênero *Rhadinaea* Cope, 1863, no Brasil meridional. II. *Rhadinaea persimilis* (Cope, 1869) (Serpentes, Coloubridae). Acta Biol. Leopoldensia, VIII, (1): 101–122.

Di-Bernardo, M., and T. de Lema, 1987: O gênero *Rhadinaea* Cope, 1863, no Brasil meridional. I. *Rhadinaea poecilopogyrus* Cope, 1863 (Serpentes, Colubridae). Acta Biol. Leopoldensia, 9 (2): 203–226.

Di-Bernardo, M., and T. de Lema, 1988: O gênero *Rhadinaea* Cope, 1863, no Brasil meridional. III. *Rhadinaea affinis* (G. 1858) (Serpentes, Colubridae). Acta Biol. Leopoldensia, 10 (2): 223–252.

Di-Bernardo, M., and T. de Lema, [1990] 1991: O gênero *Rhadinaea* Cope, 1863 no Brasil meridional. IV. *Rhadinaea bilineata* (Fischer, 1885) (Serpentes, Colubridae). Acta Biol. Leopoldensia, 12 (2): 359–392.

Di-Bernardo, S., and M. Di-Bernardo, 1996: Considerações sistemáticas sobre as espécies dos gêneros *Echinanthera* Cope, 1894 e *Taeniophallus* Cope, 1895 (Serpentes, Colubridae). Libro de Resumes, IV Congreso Latinoamericana de Herpetologia 14–19 de Octubre de 1996, Santiago, Chile, 125.

Diller, L. V., and R. L. Wallace 2002: Growth, reproduction, and survival in a population of *Crotalus viridis oreganus* in north central Idaho. Herpetological Monographs, (16): 26–45.

Dirksen, L., 2001: Rekordexemplare und Menschenfresser. Draco, Berlin, Germany, 2 (5): 26–31.

Dirksen, L., and M. Auliya, 2001: Zur Systematik und Biologie der Riesenschlangen (Boidae). Draco, Berlin, Germany, 2 (5): 4–10.

Dirksen, L., and W. Böhme, 1998a: Studien an Anakondas 1: Indizien für natürliche Bastardierung zwischen de Großen Anakonda (*Eunectes murinus*) und der Paraguay-Anakonda (*Eunectes notaeus*) in Bolivien, mit Anmerkungen zur Taxonomie der Gattung *Eunectes* (Reptilia: Squamta: Serpentes: Boidae). Zool. Abh. Mus. Teirkd. Dresden, 50 (4): 45–58.

Dirksen, L., and W. Böhme, 1998b: Studien an Anakondas 2: Zum taxonomischen Status von *Eunectes murinus gigas* (Latreille, 1801) (Serpentes: Boidae), mit neuen Ergebnissen zur Gattung *Eunectes* Wagler, 1830.—Studies on anacondas 2: On the taxonomic status of *Eunectes murinus gigas* (Latreille, 1801) (Serpentes: Boidae), with new findings on the genus *Eunectes* Wagler, 1830. Salamandra, Rheinbach, 34 (4): 359–374

Dirksen, L., E. Buongermini, Strüssman, C., and T. Waller, 1998c: Protective balling-posture behavior in the genus *Eunectes* Wagler, 1830 (Serpentes: Boidae). Herpetol. Nat. Hist., 6 (2): 151–155.

Dirksen, L., and M. R. Duarte, 1998: Polymorphismus in *Waglerophis merremii* (Wagler, 1824). Kurzmitteilung, Herpetofauna, 20 (115): 20.

Ditmars, R. L., [1902] 1903: Observations on the development of reptiles, with notes on feeding reptiles in captivity. New York Zool. Soc., 7th Ann. Rept., (1902): 145–153.

Ditmars, R. L., 1905: A new species of rattlesnake. Ninth Ann. Report, New York Zool. Soc., 1904: 197–200.

Ditmars, R. L., 1907: The reptile book. Doubleday, Page and Co., New York; 1907: I–xxxii, 1–472.

Ditmars, R. L., 1910a: Our poisonous snakes. Independent, 68: 1379–1385.

Ditmars, R. L., 1910b: Reptiles of the world: tortoises and turtles, crocodilians, lizards and snakes of the eastern and western hemisphere. New York, Sturgis and Walton, Co.

Ditmars, R. L., 1922: Reptiles of the world —tortoises and turtles, crocodilians, lizards and snakes of the eastern and western hemispheres. The Macmillian Co., New York.

Ditmars, R. L., 1929: Serpents of the eastern U.S. New York Zool. Soc. Bull., 32: 83–120.

Ditmars, R. L., 1936: The reptiles of North America. Doubleday, Doran and Co., New York.

Ditmars, R. L., 1939: The reptiles of North America, Doubleday, Doran and Co., New York.

Ditmars, R. L., 1946: The reptiles of North America. Doubleday, New York.

Ditmars, R. L., 1957: Snakes of the world. 9th Printing, New York, The MacMillan Co.

Dixon, J. R., 1960: A new name for the snake *Arizona elegans arizonae*. Southwest. Naturalist, 5 (4): 226.

Dixon, J. R., 1965: A taxonomic reevaluation of the Night Snakes *Hypsiglena ochrorhyncha* and relatives. Southwest. Naturalist, 10 (2): 125-131.

Dixon, J. R., 1967: Amphibians & reptiles of Los Angeles County California, Science Series 23, Los Angeles County Mus., Zoology No.10, May, 1967, (10): 1-64.

Dixon, J. R., 1979: Origin and distribution of reptiles in lowland tropical rainforests of South America. *In* W. E. Duellman (ed.), The South American herpetofauna: Its origin, evolution, and dispersal, Monog. Mus. Nat. Hist., Univ. Kansas, (7): 217-240.

Dixon, J. R., 1980: The Neotropical colubrid snake genus *Liophis*. The generic concept. Contrib. Biol. Geol., Milwaukee Publ. Mus., (31): 1-40.

Dixon, J. R., 1981a: Notes on Mexican *Rhadinaea* (Serpentes: Colubridae.) Southwest. Naturalist, 26 (4): 436-437.

Dixon, J. R., 1981b: The Neotropical colubrid snake genus *Liophis*: The Eastern Caribbean complex. Copeia, 1981 (2): 296-304.

Dixon, J. R., 1983a: Systematics of *Liophis reginae* and *L. williamsi* (Serpentes, Colubridae), with a description of a new species. Ann. Carnegie Mus., 52 (6): 113-138.

Dixon, J. R., 1983b: Systematics of the Latin American snake *Liophis epinephalus* (Serpentes: Colubridae). pp: 132-149. *In* A. G. J. Rhodin and K. Miyamata, eds, Advances in Herpetology and Evolutionary Biology. Mus. Comp. Zoo., Howard, Massachusetts.

Dixon, J. R., 1983c: Taxonomic revision of the common water snakes of South America. American Philos. Soc. Grant, (143): 34.

Dixon, J.R, 1983d: Taxonomic status of the Brazilian colubrid snake, *Xenodon suspectus* Cope. Texas J. Sci., 35 (3): 257-260.

Dixon, J. R., 1983e: Taxonomic status of South American snakes *Liophis miliaris*, *L.amazonicus*, *L. chrysostomus*, *L. mossorensis* and *L. purpurans* (Colubridae: Serpentes). Copeia, 1983 (3): 791-802.

Dixon, J. R., 1983f: The *Liophis cobella* group of the Neotropical colubrid snake genus *Liophis*. J. Herpetol., 17 (2): 149-165.

Dixon, J. R., 1985a: A new species of the colubrid snake genus *Liophis* from Brazil. Proc. Biol. Soc. Washington, 98 (2): 295-302.

Dixon, J. R., 1985b: A review of *Liophis anomalus* and *Liophis elegantissimus*, and the description of a new species (Serpentes: Colubridae). Copeia, 1985 (3): 565-573.

Dixon, J. R., 1987a: Amphibians & reptiles of Texas with keys, taxonomic synopses, bibliography & distribution Maps. 1st Edition, College Station, Texas, Texas A&M Univ. Press.

Dixon, J. R., 1987b: Taxonomy and geographic variations of *Liophis typhlus* and related "green" species of South America (Serpentes: Colubridae). Ann. Carnegie Mus., 52 (6): 113-138.

Dixon, J. R., 1989: A key and checklist to the Neotropical snake genus *Liophis* with country lists and maps. Smithsonian Herpetol. Info. Serv., (79): 1-28.

Dixon, J. R., 1991: Geographic variation and taxonomy of *Liophis almadensis* (Wagler) (Serpentes: Colubridae), and description of a new species of *Liophis* from Argentina and Bolivia. Texas J. Sci., 43 (3): 225-236.

Dixon, J. R., 1993: Supplement to the literature for the "amphibians and reptiles of Texas", 1987. Smithsonian Herpetol. Info. Serv. (94): 1-43.

Dixon, J. R., 2000a: Amphibians and reptiles of Texas. 2nd ed., with keys, taxonomic synopses, bibliography, and distribution maps. Texas A&M Press, College Station, Texas.

Dixon, J. R., 2000b: Ecuadorian, Peruvian, and Bolivian snakes of the *Liophis taeniurus* complex with descriptions of two new species. Copeia, 2000 (2): 482-490.

Dixon, J. R., and R. H. Dean, 1986: Status of the southern populations of the night snake (*Hypsiglena*: Colubridae) exclusive of California and Baja California. Southwest. Naturalist, 31 (3): 307-318.

Dixon, J. R., and R. R. Fleet, 1976: *Arizona elegans* Reptilia; Squamata: Serpentes: Colubridae. Cat Amer. Amph. Rept., (179): 1-4.

Dixon, J. R., and C. M. Fugler, 1959: Systematic status of two Mexican species of the genus *Gyalopion* Cope. Herpetologica, 15: 163-164.

Dixon, J. R., and F. S. Hendricks, 1979: The wormsnakes (family Typhlopidae) of the neotropics, exclusive of the Antilles. Zoologische Verhandelingen, (173):1-39.

Dixon, J. R., and C. A. Ketchersid, 1969: The status of the Mexican genus *Schmidtophis* Taylor (Colubridae). J. Herpetol., 3 (3-4): 163-165.

Dixon, J. R., C. A. Ketchersid and C. S. Lieb, 1972: The herpetofauna of Queretaro, Mexico, with remarks on taxonomic problems. Southwest. Naturalist, 16 (3&4): 225-237.

Dixon, J. R., and C. P. Kofron, 1983: The Central and South American anomalepid snakes of the genus *Liotyphlops*. Amphibia-Reptilia 4: 241-264.

Dixon, J. R., and C. S. Lieb, 1972: A new nightsnake from Mexico (Serpentes: Colubridae). Contributions in Science, Nat. Hist. Mus. Los Angeles County, (22): 1-7.

Dixon, J. R., and E. J. Michaud, 1992: Shaw's Blackbacked Snake (*Liophis melanotus*) (Serpentes: Colubridae) of northern South America. J. Herpetol., 26 (3): 250-259.

Dixon, J. R., and A. L. Markezich, 1979: Rediscovery of *Liophis taeniurus* Tschudi (Reptilia, Serpentes, Colubridae) and its relationship to other Andean colubrid snakes. J. Herpetol., 13 (3): 317-320.

Dixon, J. R., and A. L. Markezich, 1992: Taxonomy and geographic variation of *Liophis poecilogyrus* (Wied) from South America (Serpentes: Colubridae). Texas J. Sci., 44 (2): 131-166.

Dixon, J. R., and P. A. Medica, 1965: Noteworthy records of reptiles from New Mexico. Herpetologica, 21: 72-75.

Dixon, J. R., M. Sabbath, and R. Worthington, 1962: Comments on snakes from central and western Mexico. Herpetologica, 18: 91-100.

Dixon, J. R., and P. Soini, 1977: The reptiles of the upper Amazon Basin, Iquitos Region, Peru. II. Crocodilians, turtles and snakes. Milwaukee Publ. Mus., 12: 1-91.

Dixon, J. R., and P. Soini, 1986: The reptiles of the upper Amazon Basin, Iquitos Region, Peru. I and II. 2nd ed. Crocodilians, turtles and snakes. Milwaukee Publ. Mus.

Dixon, J. R., and R. A. Thomas, 1974: A dichromatic population of snake *Geophis latifrontalis*, with comments on the status of *Geophis semiannulatus*, J. Herpetol., 8 (3): 271-273.

Dixon, J. R., and R. A., Thomas, 1982: The

Bibliography

status of the Argentina colubrid snakes *Liophis sagittifer* and *L. trifascalis*. Herpetologica, 38 (3): 389-395.

Dixon J. R., and R. A. Thomas, 1985: A new species of South American water snake (genus *Liophis*) from southeastern Brazil. Herpetologica, 41 (3): 259-262.

Dixon, J. R., R. A. Thomas and H. W. Greene, 1976: Status of the Neotropical snake *Rhabdosoma poeppigi* Jan, with notes on variation in *Atractus elaps* (Günther). Herpetologica 32: 221-227.

Dixon, J. R., and B. L. Tipton, 2003: *Liophis miliaris intermedius* (Henle and Ehrl, 1991) is actually *Liophis reginae* (Serpentes: Colubridae). J. Herpetol., 37 (1): 191.

Dixon, J. R. and B. L. Tipton, (in press): *Dryadophis* versus *Mastigadryas* (Ophidia: Colubridae): A proposed resolution Herp. Review.

Dixon, J. R., R. K. Vaughan, and L. D. Wilson, 2000: The taxonomy of *Tantilla rubra* and allied taxa (serpentes: Colubridae). Southwest. Naturalist, 45 (2): 141-153.

Dixon, J. R., and R. G. Webb, 1965: *Micrurus latiicollaris* Peters from Jalisco, Mexico. Southwest. Naturalist, 10: 77.

Dixon, J. R., J. A. Wiest Jr. and J. M. Cei, 1993: Revision of the neotropical snake genus *Chironius* Fitzinger (Serpentes, Colubridae). Mus. Reg. Sci. Nat., Torino, 1995: 279 pp.

Dixon, J. R., A. A. Yanosky, and C. Mercolli, 1993: *Typhlops brongersmianus* Vanzolini and *Liophis almadensis* (Wagler): two new records for the snake fauna of the province of Formosa (Argentina). Herpetol. J., England, 3 (2): 72.

Doan, T. M., and W. Arizabal Arriaga, 1999: Natural History Notes—*Imantodes cenchoa* (Chunk-Headed Toad Snake)—Aggregation. Herpetol. Rev., 30 (2): 102.

Dobie, J. F., 1965: Rattlesnakes. Boston, Mass., Little, Brown & Co.

Dodge, N. N., 1981: Poisonous dwellers of the desert. Popular Series, Southwest Parks and Monuments Assn., Globe, Arizona, 17th. Ed., (3): 1-40.

Donelly, M. A., and C. W. Myers, 1991: Herpetological results of the 1990 Venezuelan Expedition to the summit of Cerro Guaiquinima, with new tepui species. Amer. Mus. Novitates (3017): 1-54.

Donoso-Barros, R., 1962: Los ofidios chilenos. Not. Mens. Mus. Hist. Nat., 6 (66): 3-8.

Donoso-Barros, R., 1965a: Las serpientes. Mus. Nac. Hist. Nat., Santiago de Chile. Pub. Ocas., 2: 1-24.

Donoso-Barros, R., 1965b: Nuevos reptiles y anfibios de Venezuela. Not. Mens. Mus. Hist. Nat., Santiago-Chile, 9 (102): 1-3.

Donoso-Barros, R., 1966: Reptiles de Chile. Ediciones De La Universidad de Chile, Santiago de Chile.

Donoso-Barros, R., 1968: Contribución al conocimiento del género *Chironius* (Serpentes: Colubridae). Bol. Soc. Biol. Concepción, XLII: 189-197.

Dorcas, M. E., 1992: Relationships among montane populations of *Crotalus lepidus* and *Crotalus triseriatus*. pp. 71-87 *In* J. A. Campbell and E. D. Brodie, Jr.: Biology of the Pitvipers, Selva, Tyler, Texas.

Dorcas, M. E., J. W. Gibbons, and H. G. Dowling, 1998: *Seminatrix* Cope—Black Swamp Snake. Cat. Amer. Amph. Rept., (679): 1-5.

Dorsey and Swanton, eds, 1912: Bureau of American Ethnology: A dictionary of the Biloxi and Ofo Languages. Smithsonian Institute. Bureau of American Ethnology, Bulletin no. 47, U.S. Government Printing Office, Washington, D.C.

Douglas, M. E., M. R. Douglas, G. W. Schuett, L. W. Porras and A. T. Holycross, 2002: Phylogeography of the western rattlesnake (*Crotalus viridis*) complex, with emphasis on the Colorado Plateau. pp. 15-50 in G. W. Schuett, H. Höggren, M. E. Douglas and H. W. Greene (eds.). Biology of the vipers. Eagle Mountain Publishing, Eagle Mountain, Utah.

Dowling, H. G., 1950: Studies of the Black Swamp Snake *Seminatrix pygea* (Cope), with description of two new species. Misc. Pub., Mus. Zool., Univ. Michigan, Ann Arbor, (76): 1-38.

Dowling, H. G., 1951a: A practical solution of the nomenclature of the Common Garter Snake and the Ribbonsnake. Copeia, 1951 (4): 309-310.

Dowling, H. G., 1951b: A taxonomic study of the ratsnakes, genus *Elaphe* Fitzinger. I. The status of the name *Senticolis laetus* Baird and Girard (1853). Copeia, 1951 (1): 39-44.

Dowling, H. G., 1952a: A taxonomic study of the ratsnakes, genus *Elaphe* Fitzinger. II. The subspecies of *Elaphe flavirufa* (Cope). Occ. Pap. Mus. Zool., Univ. Michigan, (540): 1-14.

Dowling, H. G., 1952b: A taxonomic study of the ratsnakes, genus *Elaphe* Fitzinger. IV. A check list of the American forms. Occ. Pap. Mus. Zool., Univ. Michigan, (541): 1-12.

Dowling, H. G., 1953: A taxonomic study of the ratsnakes, genus *Elaphe* Fitzinger. V. The Rosalie section. Occ. Pap. Mus. Zool., Univ. Michigan, (583): 1-22.

Dowling, H. G., 1957: A taxonomic study of the ratsnakes, genus *Elaphe* Fitzinger. Occ. Pap. Mus. Zool., Univ. Michigan, (583): 1-22.

Dowling, H. G., 1958: A taxonomic study of the ratsnakes. VI. Validation of the genera *Gonyosoma* Wagler and *Elaphe* Fitzinger. Copeia, 1958 (1): 29-40.

Dowling, H. G., 1959a: Classification of the serpents: A critical review. Copeia, 1959 (1): 38-52.

Dowling, H. G., 1959b: The Spotted Racer *Coluber constrictor anthicus* Cope, in Arkansas. Southwest. Naturalist. 4 (1): 40-43.

Dowling, H. G., 1960a: A Bushmaster in the zoo again. Animal Kingdom, 63 (3): 109-111, May-June, 1960.

Dowling, H. G., 1960b: A taxonomic study of the ratsnakes, genus *Elaphe* Fitzinger. VII. The *Triaspis* section. Zoologica, 45 (Part 2): 53-80.

Dowling, H. G., 1961: Serpents of the night. Herpetol. Digest, 1 (1): 3-5, 1961 Summer Issue.

Dowling, H. G., 1964: Zoologist tour of Martinique. Animal Kingdom, December 1964: 180-185.

Dowling, H.G, 1965a: Our air-conditioned Bushmaster, Animal Kingdom, 64 (5): 155-156.

Dowling, H. G., 1965b: The puzzle of the *Bothrops*; or, a tangle of serpents. Animal Kingdom, February 1965, 63 (1): 18-21.

Dowling, H. G., 1990: Genetic and taxonomic relations of the Short-tailed Snakes, genus *Stilosoma*. J. Zool., London, 221: 77-85.

Dowling, H. G., and W. E. Duellman, 1978: Systematic herpetology: A synopsis of families and higher categories. HISS Pub., Pub. Herpetol., No.7, New York.

Dowling, H. G., and I. Fries, 1987: A taxonomic study of the ratsnakes. VIII. A proposed new genus for *Elaphe triaspis* (Cope). Herpetologica, 43 (2): 200–207.

Dowling, H. G., and F. W. Gibson, 1970: Relationship of the Neotropical snakes *Hydrodynastes bicinctus* and *Cyclagras gigas*. Herpetol. Rev. 2 (2): 37–38.

Dowling, H. G., R. Highton; G. C. Maha and L. R. Maxson, 1983: Biochemical evaluation of colubrid snake phylogeny. J. Zool., London, 201: 309–329.

Dowling, H. G., and J. V. Jenner, 1987: Taxonomy of American xenodontine snakes. II. The status and relationships of *Pseudoleptodeira*. Herpetologica, 43 (2): 199–200.

Dowling, H. G., S. A. Minton and F. E. Russell, 1965: Poisonous snakes of the world, a manual for use by U.S. Amphibious Forces. Washington, D.C., U.S. Government Printing Office.

Dowling, H. G., and T. Pinou, 2003: Xenodermatid snakes in America. Herpetological Review, 34(1): 20–23.

Dowling, H. G., and R. M. Price, 1988: A proposed new genus for *Elaphe subocularis* and *Elaphe rosaliae*. The Snake New York, 20:52–63.

Downs, F. L., 1961: Generic reallocation of *Tropidodipsas leucomelas* Werner. Copeia, 4 (1961): 113, 383–387.

Downs, F. L., 1967: Intrageneric relationalysis among colubrid snakes of the genus *Geophis* Wagler. Misc. Publ. Mus. Zool., Univ. Michigan, No. 131: 1–193.

Duarte, M. R., G. Puorto and F. L. Franco, 1995: A biological survey of the pit viper *Bothrops insularis* Amaral (Serpentes, Viperidae). An endemic and threatened offshore island snake of southeastern Brazil. Studies on Neotropical Fauna and Environment, 30 (1): 1–13.

Duellman, W. E., 1956: A new snake of the genus *Leptotyphlops* from Michoacán, México. Copeia, 1956 (2): 93–94.

Duellman, W. E., 1957: Notes on snakes from the Mexican state of Sinaloa. Herpetologica, 13: 237–240.

Duellman, W. E., 1958a: A monographic study of the Colubrid snake genus *Leptodeira*. Bull. Amer. Mus. Nat. His., 114 (1): 1–152.

Duellman, W. E., 1958b: A preliminary analysis of the herpetofauna of Colima, Mexico. Occ. Pap. Mus. Zool., Univ. Michigan, (589): 1–22.

Duellman, W. E., 1958c: Systematic status of the Colubrid snake, *Leptodeira discolor* Günther. Univ. Kansas Publ. Mus. Nat. Hist., 11 (1): 1–9.

Duellman, W. E., 1959: Two new snakes, genus *Geophis*, from Michoacán, Mexico. Occ. Pap. Mus. Zool., Univ. Michigan, (605): 1–9.

Duellman, W. E., 1960a: A record size for *Drymarchon corias melanurus*. Copeia, 1960: 367–368.

Duellman, W. E., 1960b: A taxonomic study of the middle American snake *Pituophis deppei*. Univ. Kansas Publ. Mus. Nat. Hist., 10: 599–610.

Duellman, W. E., 1961: The amphibians and reptiles of Michoacán, Mexico. Univ. Kansas Publ. Mus. Nat. Hist., 15: 1–148.

Duellman, W. E., 1963: Amphibians and reptiles of the rainforests of southern El Petén, Guatemala. Univ. Kansas Publ. Mus. Nat. Hist., 15 (5): 205–249.

Duellman, W. E., 1965: Amphibians and reptiles from the Yucatán Peninsula, México. Univ. Kansas Publ. Mus. Nat. Hist., 15: 577–614.

Duellman, W. E., 1966: Remarks on the systematic status of certain Mexican snakes of the genus *Leptodeira*. Herpetologica, 22 (2): 97–106.

Duellman, W. E., 1978: The biology of an equatorial herpetofauna in Amazonian Ecuador. Lawrence, Kansas, Univ. Kansas, (65): 1–352.

Duellman, W. E., 1979: The South American herpetofauna: Its origin evolution and dispersal. Mus. Nat. Hist. Univ. Kansas, Lawrence, Kansas, (7): 1–485.

Duellman, W. E., 1982: A revision of the colubrid snakes of the genus *Tantilla* of Central America. Milwaukee Publ. Mus. Contr. Biol. Geol., (52): 1–77.

Duellman, W. E., and J. R. Mendelsen, III, 1995: Amphibians and reptiles from northern Departamento Loreto, Peru: Taxonomy and biogeography. Univ. Kansas Sci. Bull., 55 (10): 329–376.

Duellman, W. E., and A. Schwartz, 1958: Amphibians and reptiles of southern Florida. Bull. Florida State Mus., 3 (5): 181–324.

Duellman, W. E., and J. E. Werler, 1955: Variation and relationships of the colubrid snake *Leptodeira frenata*. Occ. Pap. Mus. Zool., Univ. Michigan, (570): 1–9.

Dugès, A., 1869: Catálogo de animales vertebrados observados en la república Mexicana. Naturaleza, 1: 137–145, erratas p. 414.

Dugès, A., 1891: *Elaps diastema*, var. *Michoacanensis*. Naturaleza, 1 (2): 87.

Dugès, A., 1876–1877: Erpetología del Valle De México. Naturaleza, Ser. 2, 4: 1–29.

Dugès, A., 1885: Nota sobre las coralillas (*Elaps* Schneider). Naturaleza, 7: 200–203.

Dugès, A., 1888: Sur deux espéces nouvelles des ophidiens de Mexique. Proc. Amer. Philos. Soc., 25: 181–183.

Dugès, A., 1890: Dos nuevas especies de ofidios Mexicanos. Naturaleza, ser. 2, 1: 402–403.

Dugès, A., 1896: Reptiles y batracios de los Estados Unidos Mexicanos. Naturaleza, ser. 2, 2: 479–485.

Duméril, A. M. C., and G. Bibron, 1834–1840: Erpetologie générale, ou historie naturelle complète des reptiles. Ed. Roret. Paris, 6°–1844, I–XII.

Duméril, A. M. C., and G. Bibron, 1836: Erpétologie générale ou histoire naturelle compléte des reptiles. Volume 3. Librairie Encyclopéedique de Roret.

Duméril, A. M. C., and G. Bibron, 1837: Erpétologie générale ou histoire naturelle compléte des reptiles. Volume 4. Librairie Encyclopéedique de Roret.

Duméril, A. M. C., G. Bibron, and A. Duméril, 1854: Erpétologie générale ou histoire naturelle compléte des reptiles. Volume 7, Part II. Librairie Encyclopéedique de Roret.

Duméril, A. M. C., and M. F. Bocourt, 1886: *In* A. Duméril, M. R. Bocourt, and F. Mocquard, Études sur les reptiles et les batraciens. *In* M. Milne Édwards (dir.), Recherches zoologiques pour servir a l'historie de la faune de l'Amérique Centrale et du Mexique. Mission scientifique au Mexique et dans l'Amérique Centrale, recherches zool., Traisiéne Partie 3, sect. 1. Imprimerie Nat., Paris.

Duméril, A. M. C., M. F. Bocourt, and F. Mocquard, 1870–1909. Mission scientifique au Mexique et dans L'Amerique Centrale... Étude sur les reptiles. 2 vols. 1978 reprint Arno Press, New York. *In* M. Milne Édwards (dir.), Recherches zoologiques pour servir a l'historie de la faune de l'Amérique Centrale et du Mexique. Mission scientifique au Mexique et dans l'Amérique Centrale, recherches zool., Traisiéne Partie 3, sect. 1. Imprimerie Nat., Paris.

Bibliography

Duméril, A. M. C., and M. A. Duméril, 1851: Catalogue méthodique dela collection de reptiles de Muséum d'Histoire Naturelle de Paris. 2 parts, Paris.

Dundee, H. A., 1989: Inconsistencies, inaccuracies, and inadequacies in herpetological methodology and terminology, with suggestions for conformity. Herpetol. Rev., 20: 62-65.

Dundee, H. A., and E. A. Liner, 1997: Geographic distribution: *Trimorphodon tau tau* (Mexican Lyre Snake). Herpetol. Rev., 28 (4): 211.

Dundee, H. A., and D. A. Rossman, 1989: The amphibians and reptiles of Louisiana. Louisiana State Univ. Press, Baton Rouge.

Dundee, H. A., D. A. White and V. Rico-Gray, 1986: Observations on the distribution and biology of some Yucatán Peninsula amphibians and reptiles. Bull. Maryland Herpetol. Soc., 22: 37-150.

Dunn, E. R., 1915: List of reptiles and amphibians from Clark County, Va. Copeia (25) 1915: 62-63.

Dunn, E. R., 1919: Two new crotaline snakes from western Mexico. Proc. Biol. Soc. Washington, 32: 213-216.

Dunn, E. R., 1920: A new *Geophis* from Mexico. Proc. Biol. Soc. Washington, 33: 27.

Dunn, E. R., 1922a: New Central American snakes in the American Museum of Natural History. Amer. Mus. Novitates, (314): 1-4.

Dunn. E. R., 1922b: Two new South American snakes. Proc. Biol. Soc. Washington, 35: 219-220.

Dunn, E. R., 1923: Some snakes from northwestern Peru. Proc. Biol. Soc. Washington, 36: 185-188.

Dunn, E. R., 1924: *Amastridium*, a neglected genus of snakes. Proc. U.S. Natl. Mus., Washington Government Printing Office, 65 (11): 1-3.

Dunn, E. R., 1928a: A tentative key and arrangement of the American genera of colubridae. Bull. Antivenin Inst. Amer. 2 (1): 18-23.

Dunn, E. R., 1928b: New Central American snakes in the American Museum of Natural History, Amer. Mus. Novitates, (314): 1-4.

Dunn, E. R., 1930a: New snakes from Costa Rica and Panama. Occ. Pap. Boston Soc. Nat. Hist., 5: 329-332.

Dunn, E. R., 1930b: Notes on *Bothrops lansbergii* and *Bothrops ophryomegas*. Bull. Antivenin Inst. Amer., 2: 28-30.

Dunn, E. R., 1932a: The status of the snake genus *Rhadinaea* Cope. Occ. Pap. Mus. Zool. Univ. Michigan, (251): 1-2.

Dunn, E. R., 1932b: The status of *Tropidoclonion lineatum*. Proc. Biol Soc. Washington, 45: 195-198.

Dunn, E. R., 1933a: Amphibians and reptiles from El Valle de Antón, Panamá. Occ. Pap. Boston Soc. Nat. Hist., 8: 65-79.

Dunn, E. R., 1933b: Notes on *Coluber oaxaca*, and *Masticophis mentovarius*. Copeia, 1933 (4): 214.

Dunn, E. R., 1935: The snakes of the genus *Ninia*. Proc. Natl. Acad. Sci., 21 (1): 9-12.

Dunn, E. R., 1936a: Notes on North American *Leptodeira*. Proc. Natl. Acad. Sci., 22 (12): 689-698.

Dunn, E. R., 1936b: The amphibians and reptiles of the Mexican Expedition of 1934. Acad. Nat. Sci. Philadelphia, 88: 471-477.

Dunn, E. R., 1937: Notes on tropical *Lampropeltis*. Occ. Paps. Mus. Zool. Univ. Michigan, (353): 1-11.

Dunn, E. R., 1938: The snake genus *Enulius* Cope. Proc. Acad. Nat. Sci. Philadelphia, 89: 415-418.

Dunn, E.R, 1939a: A new pit viper from Costa Rica. Proc. Biol. Soc. Washington, 52: 165-166.

Dunn, E. R., 1939b: Mainland forms for the snake genus *Tretanorhinus* Cope. Copeia, 1939: 212-217.

Dunn, E. R., 1940a: New and noteworthy herpetological material from Panama. Proc. Acad. Nat. Sci. Philadelphia, 92: 105-122.

Dunn, E. R., 1940b: Notes on some American lizards and snakes in the Museum at Goteborg. Herpetologica, 1: 189-194.

Dunn, E. R., 1941: A review of the snake genus *Adelphicos*. Proc. Rochester Acad. Sci., 8: 175-186.

Dunn, E. R., 1942: New or noteworthy snakes from Panama. Notulae Naturae, 108: 1-8.

Dunn, E. R., 1943: Notes in Colombian herpetology, I: A new snake of the genus *Rhadinaea*. Caldasia, (8): 307-308.

Dunn, E. R., 1944a: A new snake of the genus *Hydrops* from Colombia. Caldasia, III (11): 71-72.

Dunn, E. R., 1944b: A review of the Colombia snakes of the families Typhlopidae and Leptotyphlopidae. Caldaisa, 3 (11): 47-55.

Dunn, E. R., 1944c: A revision of the Colombian snakes of the genera *Leimadophis, Lygophis, Liophis, Rhadinaea* and *Pliocercus*, with a note on Colombian *Coniophanes*. Caldasia, 2 (10): 479-495.

Dunn, E. R., 1944d: *Dugandia*, a new snake genus for *Coluber bicinctus* Heerman. Caldasia, 3 (11): 69-70.

Dunn, E. R., 1944e: Los géneros de anfibios y reptiles de Colombia. II. Caldasia, 3: 73-110.

Dunn, E. R., 1944f: Los géneros de anfibios y reptiles de Colombia, III Tercera parte. Reptiles. Orden de las Serpientes. Caldasia, 3: 155-224.

Dunn, E. R., 1944g: Notes in Colombian herpetology, III. The Snake genus *Dendrophidion* in Colombia. Caldasia, 2 (10): 474-477.

Dunn, E. R., 1946: *Atractus sanctaemartae*, a new species of snakes from the Sierra Nevada de Santa Marta, Colombia. Occ. Pap. Mus. Zool. Univ. Michigan, (493): 1-6.

Dunn, E. R., 1947: Snakes of the Lérida Farm (Chirqui Volcano, western Panamá). Copeia, 1947: 153-157.

Dunn, E. R., 1951: The status of the snake genera *Dipsas* and *Sibon*, A Problem for "Quantum Evolution." Evolution, V (4): 355-358.

Dunn, E. R., 1954: The coral snake "mimic" problem in Panamá. Evolution, VIII (28): 97-102.

Dunn, E. R., 1957: Contributions to the herpetology of Colombia—1943-1946. Private Printing-Reprinted from the Revista de la Academia Colombiana de Ciencias Exactas, Físicas y Naturales, and from Caldasia-Boletín del Instituto de Ciencias Naturales de la Universiadad Nacional de Colombia-Bogota.

Dunn, E. R., and J. R. Bailey, 1939: Snakes from the uplands of the Canal Zone and of Darien. Bull. Mus. Comp. Zool., 86: 1-22.

Dunn, E. R., and R. Conant, 1936: Notes on anacondas, with descriptions of two new species. Acad. Nat. Sci. Philadelphia, 88: 503-506.

Dunn, E. R., and H. G. Dowling, 1957: A Neotropical snake genus *Nothopsis* Cope, Copeia, 1957 (4): 255-261.

Dunn, E. R., and J. T. Emlen, Jr., 1932: Reptiles and amphibians from Honduras. Acad. Nat. Sci. Philadelphia, 84: 21-32.

Dunn, E. R., and L. H. Saxe, 1950: Results of the Catherwood-Chaplin West Indies Expedition, 1948. Part V. Amphibians and reptiles of San Andres and Providencia. Proc. Acad. Nat. Sci. Philadelphia, 52: 141-165.

Dunn, E. R., and L. C. Stuart, 1951: Comments on some recent restrictions of type localities of certain South and Central American amphibians and reptiles. Copeia, 1951 (1): 55-61.

Dunson, W., 1975: The biology of sea snakes. Baltimore, Univ. Park Press.

Durán A., H., 1967: Pequeño diccionario Castellano-Quechua, Quechua-Castellano. Lima, Peru.

Duval, D., S. J. Arnold and G. W. Schuett, 1992: Pitviper mating systems: ecological potential, sexual selection, and microevolution. pp. 321-326 in J. A. Campbell and E. D. Brodie, Jr.: Biology of the pitvipers. Selva, Tyler, Texas.

Echternacht, A. C., 1973: The color pattern of *Sonora michoacanensis* (Dugès) (Serpentes: Colubridae) and its bearing on the origin of the species. Breviora, 410: 1-18.

Eckerman, C. M., 1996: Variation, systematics, and interspecific position of *Heterodon nasicus* (Serpentes: Xenodontidae). Masters Thesis, University of Texas, El Paso.

Edgren, R. A., 1952: A synopsis of the snakes of the genus *Heterodon*, with the diagnosis of a new race of *Heterodon nasicus* Baird and Girard. Chicago Acad. Sci. Nat. Hist. Misc., 112: 1-4.

Edgren, R. A., 1955: The natural history of the hognosed snakes, genus *Heterodon*: A review. Herpetologica, 11: 105-117.

El Comercio. 1993: Naturaleza Impresa. Lima, Peru, 30 de Juinio de 1993, Section C.

Ellis, M. D., 1975: Dangerous plants, snakes, arthropods and marine life of Texas. U.S. Department of Health, Education, and Welfare.

Emsley, M. G., 1963a: A consideration of the snakes recorded from Trinidad. Copeia, 1963 (3): 576-577.

Emsley, M. G., 1963b: The re-discovery of Cope's *Liophis perfuscus* in Barbados. Copeia, 1963: 577-579.

Emsley, M. G., 1966a: The mimetic significance of the snake *Erythrolamprus ocellatus* Peters from Trinidad. Evolution, 20: 663-664.

Emsley, M. G., 1966b: The status of the snake *Erythrolamprus ocellatus* Peters. Copeia, 1966: 128-129.

Emsley, M. G., 1977: Snakes, and Trinidad and Tobago. Bull. Maryland Herpetol. Soc., 13 (4): 201-304.

Englemann, W.-E. and F. J. Obst, 1981: Snake biology, behavior and relationship to man. Exeter Books, New York.

Erize, E., 1960: Diccionario comentado Mapuche-Español. Cuadernos Del Sur, Instituto de Humanidades, Universidad Nacional del Sur, Buenos Aires, Argentina.

Ernst, C. H., 1992: Venomous reptiles of North America. Smithsonian Institution Press, Washington, D.C.

Ernst, C. H., and R. W. Barbour, 1989: Snakes of eastern North America. George Mason Univ. Press, Fairfax, VA.

Ernst, C. H., and G. R. Zug, 1996: Snakes in question-The Smithsonian answer book. Smithsonian Institution Press, Washington, D.C. and London.

ESEA Bilingual Education program. 1976: The language research department of the Northern Cheyenne. Title VII, Lame Deer, Montana.

Esqueda, L. F., and E. La Marca, 1999: New reptilian species records from the Cordillera de Mérida, Andes of Venezuela. Herpetol. Rev., 30 (4): 238-240.

Estrada, A. R., and R. Ruibal, 1999: A review of Cuban herpetology. pp. 31-62. In B. I. Crother (ed.), Caribbean amphibians and reptiles. Academic Press, New York.

Etheridge, R., 1961: Additions to the herpetological fauna of Isla Cerralvo in the Gulf of California, Mexico. Herpetologica, 17: 57-60.

Fabian, M. E., 1971: Contribution to the knowledge of the cranial osteology of *Liophis* Wagler (Serpentes, Colubridae). Museu Rio-Grandense de Ciencias Naturais, Porto Alegre, Brazil.

Fantham, H. B., and A. Porter, 1954: The endoparasites of some North American snakes and their effects on the ophidia. Proc. Zool. Soc. London, 123(4): 867-898.

Fawcett, P. H., Col., 1953: Lost trails, lost cities—from his manuscripts, letters, and other records, selected and arranged by Brian Fawcett. Funk & Wagnals Company, New York.

Feio, R. N., P. S. Santos, R. Fernandes and T. S. de Freitas, 1999: *Chironius flavolineatus* (NCN). Courtship. Herpetol. Rev., 30 (2): 99.

Ferguson, D. E., 1952: The distribution of amphibians and reptiles of Wallowa County, Oregon. Herpetologica, 8 (3): 66-70.

Ferguson, D. E., 1954: An annotated list of the amphibians and reptiles of Union County, Oregon. Herpetologîca, 10 (3): 149-152.

Fernandes, D. S., F. L. Franco and V. J. Germano, 1999: Geographic distribution- *Umbrivaga pygmaea*. Herpetol. Rev., 30 (2): 175.

Fernandes, R., 1995a: A new species of snake in the genus *Atractus* (Colubridae: Xenodontinae) from northeastern Brazil. J. Herpetol., 29 (3): 416-419.

Fernandes, R., 1995b: Phylogeny of the Dipsadine snakes. Ph.D. dissertation, Univ. Texas Arlington, December 1995, i-ix + 1-116 pp.

Fernandes, R., 1995c: Variation and taxonomy of the *Atractus reticulatus* complex (Serpentes: Colubridae). Comun. Mus. Cienc. Tec. PUCRS, Ser. Zool., (8): 37-53.

Fernandes, R., 2000: Geographic variation of the Brazilian Atlantic Forest *Atractus maculatus* (Günther, 1858) with revalidation of *Rhabdosoma zebrinum* Jan 1862 (Serpentes: Colubridae). Bol. Mus. Nac., 419: 1-8.

Fernandes, R., and A. J. S. Argôlo, 1999: Rediscovery of *Atractus guentheri* (Wucherer, 1861) (Serpentes, Colubridae) in southeastern Bahia, Brazil, Bol. Mus. Nac., Nova Série, Rio de Janeiro-Brazil, Zoologia, (397): 1-5.

Fernandes, R., E. M. X. Freire, and G. Puorto, 2000: Geographic variation of the Brazilian Atlantic Rain Forest snake *Atractus maculatus* (Günther, 1858) with the revalidation of *Rhabdosoma zebrinum* Jan, 1862 (Serpentes: Colubridae). Bol Mus. Nac., Nova Série, Zool., Rio de Janeiro, (419): 1-8.

Fernandes, R., and G. Puorto, 1993: A new species of *Atractus* from Brazil and the status of *A. guntheri* (Serpentes: Colubridae). Mem. Inst. Butantan, 55: 7-14, Supplement 1, 1993.

Bibliography

Fernandes, R., M. Porto and U. Caramaschi, 1997 [1998]: The Taxonomic Status of *Heterorhachis poecilolepis* Amaral, 1923. J. Herpetol., (1): 139–141.

Ferrarezzi, H., 1993a: Nota sobre o gênero *Phalotris* com revisão do grupo *nasutus* e descrição de três espécies (Serpentes, Colubridae), Xenodontinae). Mem. Inst. Butantan, 55 (1): 21–38.

Ferrarezzi, H., 1993b: Sistemática filogenética de *Elapomorphus*, *Phalotris* e *Apostolepis* (Serpentes, Colubridae, Xenodontinae). São Paulo, Dissertation Maestrado, Pós-grad. Ciênc. Biológicas-Instituto Biociências, Univ. São Paulo.

Ferreira, V. L., 1997: Aspectos zoogeográficos do *Xenodon* Boié, 1826 e *Waglerophis* Romano & Hoge, 1973 (Serpentes, Colubridae, Xenodontinae). Biociências, Porto Alegre, 5(2): 109–139.

Ferreira, V. L., and A. Ballestrin Outeiral, 1998: Natural History Notes—*Mastigodryas bifossatus* (Jararacu do Banhado)—tail breakage. Herpetol. Rev., 29 (4): 242.

Ferrerezzi, H., and E. M. X. Freire, 2001: New Species of *Bothrops* Wagler, 1824 from the Atlantic forest of northeastern Brazil (Serpentes, Viperidae, Crotalinae). Bol Mus. Nac., Nova Serie, Zoologia, 440:1–10.

Firestone, H., 1965: Description and classification of Sirionó. Mouton & Co., The Hauge, Series Practica XVI.

Fischer, J. G., 1882: Herpetologische Bemerkungen vorzugsweise uber Stücke der Sammlung des Naturhistorischen Museums in Bremen. Abh. Naturwiss. Ver. Bremen, 7: 225–238.

Fischer, J. G., 1880: Neue Amphibien und Reptilien. Arch. Naturgesch., 46: 215–227.

Fischer, J. G., [1884] 1885: Ichthyologische und herpetologische Bemerkungen V. Herpetologische Bemerkungen, Jb. Hamb. Wiss. Anst., [1884]: 82–119.

Fischer, J. G., 1885: Ichthyologische und herpetologisches. Bemerkungen, Jahrb. Hamburgischen Wiss. Anst., II Jahrgang, 1885: 47–121 + pls. 1–4.

Fischer, J. G., 1886: Herpetologische Notizen. Abhandl. Gebiete Naturwiss, 9: 51–67.

Fisher, C. B., 1973: Status of the Flat-headed Snake, *Tantilla gracilis* Baird and Girard, in Louisiana. J. Herpetol., 7: 136–137.

Fitch, C. B., 1940: A biogeographical study of the *ordinoides* artenkreis of garter snakes (genus *Thamnophis*). Univ. California Pub. Zool., 44: 1–150.

Fitch, H. S., 1936: Amphibians and reptiles of the rogue River Basin, Oregan. Amer. Midl. Naturalist, 17 (3): 634–652.

Fitch, H. S., 1941: Geographic variation in garter snakes of the species *Thamnophis sirtalis* in the Pacific Coast region of North America. Amer. Midl. Naturalist, 26 (3): 570–592.

Fitch, H. S., 1960: Autecology of the copperhead. Univ. Kansas Publ. Mus. Nat. Hist., 3 (4): 85–288.

Fitch, H. S., 1980a: Remarks concerning certain western garter snakes of the *Thamnophis elegans* complex. Trans. Kansas Acad. Sci., 83: 106–113.

Fitch, H. S., 1980b: *Thamnophis sirtalis* (Linnaeus)—Common garter snake. Cat. Amer. Amph. Rept., (270): 1–4.

Fitch, H. S., 1981: Sexual size differences in reptiles. Univ. Kansas Mus. Nat. Hist., Misc. Pub., (70): 1–72.

Fitch, H. S., 1983: *Thamnophis elegans* (Baird and Girard)—Western Terrestrial Garter Snake. Cat. Amer. Amph. Rept., (320): 1–4.

Fitch, H. S., 1984: *Thamnophis couchii* (Kennicott)—Western Aquatic Garter Snake. Cat. Amer. Amph. Rept., (351): 1–3.

Fitch, H. S., and W. W. Milstead, 1961: An older name for *Thamnophis cyrtopsis* (Kennicott). Copeia, 1961 (1): 112.

Fitzinger, L. J. F. J., 1826: Neue classification der Reptilien nach ihern naturlichen Verwantschaften nebst einer Verwandtschafts-Tafel und einem Verzeichnisse der Reptilien-Sammlung des K.K. zoologischen Museums zu Wien. Vienna, J. G. Huebner, 66 pp. + plates.

Fitzinger, L. J. F. J., 1843: System reptilium Fausciculus primus Ambylglossae. Braumuller et Seidel, Vienna.

Flanagan, W. D. III, and R. A. Odum, 2000: *Leptodeira bakeri* (Aruban Cat-Eyed Snake)—reproduction. Herpetol. Rev., 31 (1): 46.

Fleharty, E. D., 1967: Comparative ecology of *Thamnophis elegans*, *T. cyrtopsis*, and *T. rufipunctatus* in New Mexico. Southwest. Naturalist, 12: 207–230.

Flores-Villela, O., 1993a: Herpetofauna mexicana- lista anotada de las especies de anfibios y reptiles de México, cambios taxonómicos recientes, y nuevas especies-annotated list of the species of amphibians and reptiles of Mexico, recent taxonomic changes, and new species. Spec. Pub. No.17, Carnegie Mus. Nat. Hist., 17: 1–73.

Flores-Villela, O., 1993b: Herpetofauna of Mexico-distribution and endemism. pp. 253–280. *In* T. P. Ramamoorthy, R. Bye, A. Lott and J. Fa (eds.), Oxford Press, New York and Oxford.

Flores-Villela, O., and Y. P. Gerez, 1988: Conservación en México: Sintesis sobre vertebrados terrestres, vegetación, y use del suelo. Institutio Nacional de Investigaciones sobre Recursos Bioticos Conservación Internacional.

Flores-Villela, O., and Y. P. Gerez, 1994: Biodiversidad y conservación en Mexíco: Vertebrados, vegetación e uso del suelo. 2nd edition, Univ. Nac. Auto. Mexico, Mexico City.

Flores-Villela, O., A. Loeza-Gorichi and A. Pérez-Nuñez, 1995: Geographic distribution - *Geophis nigrocinctus* (Sierra Coaleoman Earth Snake). Herpetol. Rev., 26 (2): 109.

Flores-Villela, O.; F. Mendoza; E. Hernandez; M. Mancilla; E. Godinez and I. Goyencochea Mayer, 1992: *Ophryacus undulatus* in the Mexican state of Hidalgo. Texas J. Sci., 44 (2): 249–250.

Flores-Villela, O. A., F. Mendoza Quijano and G. Gonzalez Porter, 1995: Recopilacion de claves para la determinación de anfibios y reptiles de Mexico. Publicaciones Esp. Mus. Zool. Univ. Nac. Auton. Mexico, 10, 1995: i–xv, 1–285.

Flury, A., 1950: A new kingsnake from Trans-Pecos, Texas. Copeia, 1950 (3): 215–217.

Forcart, L., 1951: Nomenclature remarks on some generic names of the snake family boidae. Herpetologica, 7: 197–200.

Forster, K., 1977: The narrative folklore discourse in Border Cuna. pp. 1–23. *In* Longacre, R. E., ed., 1977: Discourse grammar—Studies in indigenous languages of Colombia, Panama, and Ecuador Part II.

Forster, J. R., 1771: Travels through the part of America formerly called Louisiana, 1:364. *In* J. B. Bossou: A catalogue of animals in North America.

Fouquette, M. J. Jr., and J. Delahoussaye, 1966: Noteworthy herpetological records from Louisiana. Southwest. Naturalist, 11: 137–139.

Fouquette, M. J. Jr., and F. E. Porter, Jr., 1961: A new black-headed snake from southwestern Texas. Copeia, 1961 (2): 92-96.

Fouquette, M. J. Jr., and D. A. Rossman, 1963: Noteworthy records of Mexican amphibians and reptiles in the Florida State Museum and the Texas Natural History Collection. Herpetologica, 19 (3): 185-201.

Fowler, H. W., [1906] 1907: Part II. The amphibian and reptiles of New Jersey. 1907: 23-409 + 69 plates. *In* The annual report of the New Jersey State Museum including a list of the species received during the year-Financial report with a report of the amphibians and reptiles of New Jersey and a supplement to the "Fishes of New Jersey." MacCrellish & Quigley, Trenton, New Jersey.

Fowlie, J. A., 1965: Snakes of Arizona. McGraw Hill, Inc., distributed by Azul Press, 164 pp.

Fox. W., 1951a: Relationships among the garter snakes of the *Thamnophis elegans* rassenkreis. Univ. California Publ. Zool., 50: 485-530.

Fox, W., 1951b: The status of the garter snake *Thamnophis sirtalis tetrataenia*. Copeia, 1951: 257-267.

Franck, H. A., 1921: Working north from Patagonia. Garden City Pub. Co., Inc. Garden City, New York.

Franco, F. L., O. A. V. Marques and G. Puorto, 1997: Two new species of colubrid snakes of the genus *Clelia* from Brazil. J. Herpetol., 31 (4): 483-490.

Franco, F. L., E. L. Salomão, M. Borges-Martins, M. Di-Bernardo, M. D. Meneghel and S. Carreira, 2001 [2000]: New Record of *Calamodontophis paucidens* (Serpentes, Colubridae, Xenodontinae) from Brazil and Uruguay. Cuad. herpetol., 14 (2): 155-159.

Frank, N., 1994: The American Milksnake *Lampropeltis triangulum*. Reptile & Amphibian Mag. Sept./Oct., 1994: 20-33.

Frank, N., and E. Ramus, 1995: A complete guide to scientific and common names of reptiles and amphibians of the world. N G Publishing, Inc., Pottsville, Pennsylvania.

Frank, W., 1978: Boas & other non-venomous snakes. Neptune, New Jersey, T.F.H. Pub., Inc.

Franklin, C. J., and J. Franklin, 1999: Geographic distribution—*Pseustes poecilonotus* (Bird-eating Tree Snake). Herpetol. Rev., 30 (2): 114.

Fraser, D. F., 1964: *Micrurus limbatus*, a new coral snake from Veracruz, Mexico. Copeia, 1964 (3): 570-573.

Frantz, D. G., and N. J. Russell, 1989: Blackfoot dictionary of stems, roots and affixes. Univ. Toronto Press, Toronto.

Frazer, D. F., 1973: Variation in the coral snake *Micrurus diastema*. Copeia, 1973: 1-17.

Freed, P., 1996: Paraguay "sleepy" herp country. Reptiles & Amphibians Mag., May/June 1996: 66-75.

Freiberg, M., 1939: Enumeración sistemática de los reptiles de Entre Rios y lista de los ejemplares que los representan en el Museo de Entre Rios. Mem. Mus. Entre Rios, Zool., (11): 1-28.

Freiberg, M., 1951: Nuevo hallazgo de *Leptotyphlops borrichiana* Degergol en Argentina. Physis, 20 (58): 259-262.

Freiberg, M., 1954: Vida de batracios y reptiles Sudamericanos. Editores: Cesarini Hnos., Buenos Aires, 12 marzo 1954: 1-192 pp + plates I-XLIV.

Freiberg, M., 1982: Snakes of South America. Neptune, New Jersey, T.F.H. Pub., Inc., 189 pp.

Freiberg, M., 1984: El mundo de los ofidios. Buenos Aires, República Argentina, Editorial Albatros, Hipólito Yrigoyen 3920-T.E.981-1161.

Freiberg, M. A., and B. R. Orejas-Miranda, 1968: Un nuevo Leptotyphlopidae de la República Argentina. Physis. Buenos Aires, 28 (76): 145-147

Freiberg, M., and J. G. Walls, 1984: The world of venomous animals. T.F.H. Pub., Inc., Neptune City, New Jersey.

Freire, E. M. X., 1998: *Sibon nebulata*. Herpetol. Rev., 29 (3): 178

Freire, E. M. X., 1999a: Geographic distribution—*Dendrophidion dendrophis*. Herpetol. Rev., 30 (1): 53.

Freire, E. M. X., 1999b: Geographic distribution—*Oxyrhopus guibei* (False Coral Snake). Herpetol. Rev., 30 (1): 55.

Freire-Lascano, A., 1991: Dos nuevas especies de *Bothrops* en Ecuador. Pub. Trab. Cient. Ecuador, Univ. Téc. Macala.

Freitas, M. A., 1999: Serpentes da Bahia e do Brasil. Co-Edição do Autor/Editora DALL, Feira de Santana, Bahia.

Fritts, T. H., 1994: Does the brown tree snake pose a threat to Florida and its tourist industry?: pp 71-72. *In* D. C. Schmitz and T. C. Brown eds., 1994: An assessment of invasive non-indigenous species in Florida's public lands. Bureau of Aquatic Plant Management, Technical Reports TSS-94-100.

Fritts, T. H., 1987: The brown tree snake in Guam as a threat to other Pacific Islands. U.S. Fish & Wildlife Services Research Information Bulletin 69-87.

Froom, B., 1972: The snakes of Canada. McClelland & Stewart Ltd., Toronto, 128 pp.

Frost, D. R., 1983a: The relationships of the ground snakes (genus *Sonora*) in Baja California, Mexico. Trans. Kansas Acad. Sci., 86: 31-37.

Frost, D. R., 1983b: *Sonora semiannulata*. Cat. Amer. Amph. Rept., (333): 1-4.

Frost, D. R., and T. R. Van Devender, 1979: The relationships of the groundsnakes *Sonora semiannulata* and *S. episcopa* (Serpentes: Colubridae). Occ. Pap. Mus. Zool., Louisiana State Univ., (52): 1-9.

Fuentes, O., and C. L. Barrio, 1999a: Geographic distribution—*Erythrolamprus pseudocorallus* (Falsa Coral de Maracaibo, Maracaibo's False Coral Snake). Herpetol. Rev., 30 (10): 53.

Fuentes, O., and C. L. Barrio, 1999b: Geographic distribution—*Liophis poecilogyrus*. Herpetol. Rev., 30 (10); 54.

Fuentes, O., F. Navarrete, and A. Rodriguez-Acosta, 1999: El genero *Liophis* Wagler, 1930 (Serpentes: Colubridae) en Venezuela. Serpentarium del Instituto de Medicina Tropical, Universidad Central de Venezuela. Coleccién de Herpetología. Mus. Cienc. Natur. Caracas, 5th Congreso Latinoamericano de Herpetología. Montevideo, Uruguay, 12 to 17 December, 1999, Publicacion Extra—Mus. Nac. Hist. Nat., Montevideo, Uruguay, (50) 1999: 63.

Fulger, C. M., 1951: The distribution of the amphibians and reptiles of the Mexican state of San Luis Potosí. M.S. Thesis, Louisiana State Univ., Department of Zoology, Baton Rouge, Louisiana.

Fugler, C. M., 1960: New herpetological records for British Honduras. Texas J. Sci., 12: 8-13.

Fugler, C. M., and J. R. Dixon, 1961: Notes on the herpetofauna of the El Dorado area of Sinaloa, Mexico. Publ. Mus., Michigan State Univ., East Lansing, Michigan, 2 (1): 1–23.

Fugler, C. M., and A. B. Walls, 1978: Snakes of the Upano Valley of Amazonia Ecuador. J. Tennessee Acad. Sci., 53 (3): 81–87.

Funk, R. S., 1967: A new colubrid snake of the genus *Chionactis* from Arizona. Southwest. Naturalist, 12 (2): 180–188.

Furbee-Losee, L., 1976: The correct language Tojolabal-a grammar with ethnographic notes. Garland Publishers, Inc., New York.

Gadow, H., 1901: Amphibia and reptiles. MacMillian and Co., Ltd, New York.

Gaige, H. T., 1936: Some reptiles and amphibians from Yucatan and Campeche, Mexico. Pub. Carnegie Inst. (457): 289–304.

Gallardo, J. M., 1972: Observaciones, biolóogicas sobre una falsa yarará, *Tomodon ocellatus* Duméril, Bibron et Duméril. Neotropica, 18 (56): 57–65.

Gallardo, J. M., 1979: Composición, distribución y origin de la hérpetofauna Chaqueña. In W. E. Duellman, ed.: The South American herpetofauna: Its origin, evolution and dispersal. Monogr. Mus. Nat. Hist., Univ. Kansas, (7): 299–307.

Gans, C., 1964: A redescription of, and geographic variation in, *Liophis miliaris* Linné, the common water snake of southeastern South America. Amer. Mus. Novitates, (2178): 1–58.

Garbíras, A., 1883: Ligeras observaciones y apuntamientos sobre las serpientes. Imprenta Bolivar, Caracas, Venezuela.

García, A., and M. Valtierra-Azotla, 1996a: Geographic distribution—*Ficimia publia* (Blotched Hognose Snake). Herpetol. Rev., 27 (2): 88.

García, A., and M. Valtierra-Azotla, 1996b: Geographic distribution—*Sibon annuliferus* (Western Snail-eating Snake). Herpetol. Rev., 27 (2): 89.

García, E., 1896: Los ofidios venenosos del Cauca. Métodos empíricos y racionales empleados contra los accidentes producidos por la mordedura de esos reptiles. Librería Colombiana, Cali.

García, A., and G. Ceballos, 1994: Guia de campo de los reptiles y anfibios de la costa de Jalisco, Mexico-Field guide to the reptiles and amphibians of the Jalisco Coast, Mexico. Mexico City, Fundacion Ecologica de Cuixmala, A.C., Insitituto de Biologia, Univ. Nac. Autonomia Mexico.

García-Pérez, J. E., 1995: Una nueva especie de cascabel (Serpentes: Crotalidae) para el Bolsón Arido de Lagunillas, Cordillera de Mérida, Venezuela—A new species of rattlesnake (Serpentes: Crotalidae) from the: Bolsón Arido de Lagunillas" Cordillera de Mérida, Venezuela. Revista Ecología Latino Americana, 3 (1–3) Art. 2: 7–12.

Garel, T., and S. Matola, 1995: A field guide to the snakes of Belize. The Belize Zoo and Tropical Education Center, 1995.

Garman, S., 1877: Reptiles and batrachians collected by Allen Lesley, Esq., on the Isthmus of Panama. Proc. Boston Soc. Nat. Hist., 1876. 18: 402–413.

Garman, S., 1881: New and little-known reptiles and fishes in the museum collections. Bull. Mus. Comp. Zool., 8: 85–93.

Garman, S., 1883: The reptiles and batrachians of North America. Part 1, Ophidia. Mem. Mus. Comp. Zool. 8 (3).

Garman, S., 1884: The North American reptiles and batrachians. A list of the species occurring north of the Isthmus of Tehuantepec, with references. Bull. Essex Inst., 16: 1–46.

Garman, S. [dated 1883] 1884: The reptiles and batrachians of North America. Mem. Mus. Comp. Zool., 8 (2).

Garman, S., 1887a: On West Indian reptiles and batrachians in the Museum of Comparative Zoology. Bull. Essex Inst., 19: 1–24.

Garman, S., 1887b: On West Indian reptiles in the Museum of Comparative Zoology, at Cambridge. Mass. Proc. Amer. Philos. Soc., 24: 278–286.

Garman, S., 1887c: Reptiles and batrachians from Texas and Mexico. Bull. Essex. Inst., 19: 119–138.

Garman, S., 1892a: On reptiles collected by Dr. Geo. Baur near Guayaquil, Ecuador. Bull. Essex Inst., 24: 88–95.

Garman, S., 1892b: On Texas reptiles. Bull. Essex Inst., 24: 1–12.

Garman, S., 1892c: The reptiles of the Galapagos Islands from the collections of Dr. Geo. Baur. Bull. Essex Inst., 24: 73–87.

Garman, S., 1992c: The reptilian rattle. Science, 20 (492): 16–17.

Garman, S., 1908: The reptiles of Easter Island. Bull. Mus. Comp. Zool., 52: 1–13.

Garrigues, R., 1994a: Give the poor snake a break, Costa Rica Today. Thursday, June 9, 1994: 30.

Garrigues, R., 1994b: An image problem for snakes, Costa Rica Today. Thursday, June 9, 1994: 14.

Garstka, W. R., 1982: Systematics of the *mexicana* species group of the colubrid genus *Lampropeltis* with an hypothesis mimicry. Breviora, Mus. Comp. Zool., (466): 1–35.

Gasc, J. P., and M. T. Rodrigues, 1979: Une nouvelle espèce du genre *Atractus* (Colubridae, Serpentes) de la Guyane Française. Bull. Mus. Nat. Hist. Paris, 1 (2): 547–557.

Gasc, J. P., and M. T. Rodrigues, [1979] 1980: Liste préliminaire des serpentes de la Guyane Française. Bull. Mus. Natl. Hist. Nat. Paris, 4 (2) A (2): 559–598.

Gasc, J. P., and M. T. Rodrigues, 1980: Sur la presence du genere *Geophis* (Colubridae, Serpentes) dans la region guyanaise. Description d'une nouvelle espece de Guyane francaise. Bull. Mus. Natl. D'hist. Nat., Section A Zoologie Biologie et Ecologie, Animales, 1 (4): 1121–1130.

Gaywood, M., and I. Spellerberg, 1999: Snakes. WorldLife Library. Voyageur Press, 1999, 72 pp.

Gehlbach, F. R., 1956: Annotated records of southwestern amphibians and reptiles. Trans. Kansas Acad. Sci., 59 (3): 364–372.

Gehlbach, F. R., 1965: Herpetology of the Zuni Mountain region, northwestern New Mexico. Proc. U.S. Natl. Mus., 116 (3305): 243–332.

Gehlbach, F. R., 1967: *Lampropeltis mexicana*. Cat. Amer. Amph. Rep., (55): 1–2.

Gelbach, F. R., 1971: Lyre snakes in the *Trimorphodon biscutatus* complex: a taxonomic resume. Herpetologica, 27: 200–211.

Gelbach, F. R., 1974: Evolutionary relations of southwestern ringneck snakes (*Diadophis punctatus*). Herpetologica, 30 (2): 140–148.

Gelbach, F. R., and J. Baker, 1962: Kingsnakes allied with *Lampropeltis mexicanus*: Taxonomy and natural history. Copeia, 1962: 291–300.

Gehlbach, F. R., and B. B. Collette, 1957: A contribution to the herpetofauna of the highlands of Oaxaca and Puebla, Mexico. Herpetologica, 13: 227-231.

Gervais, 1843: D'Orbigny's Dict. Univ. Hist. Nat., 3: 191.

Gibbons, W., 1983: Their blood runs cold. Tuscaloosa, Alabama, Univ. Alabama Press, 164 pp.

Gibbons, J. W., and R. D. Semlitsch, 1991: Guide to the reptiles and amphibians of the Savannah River Site. Univ. Georgia Press, Athens, Georgia.

Girard, C., 1854: Abstract of a report to Lieut. James M. Gillis, USN., upon the reptiles collected during the USN astronomical expedition to Chile. Proc. Acad. Nat. Sci. Philadelphia, 7: 226-227.

Girard, C., 1855: Reptiles and fishes: pp. 207-253. *In* U.S. Naval Astronomical Expedition to the southern hemisphere during the years 1849-1850-1851-. 1852. Vol. 2. AOP Nicholson, Printer, Washington, D.C.

Girard, C., [1857] 1858: Herpetology: Reptiles. *In* United States exploring expedition during the years 1838, 1840, 1841, 1842, under the command of Charles Wilkes, U.S.N., J. B. Lippincott and Co., Philadelphia, reprint.

Girard, C., 1859: Herpetological notices. Proc. Acad. Nat. Sci. Philadelphia, 11: 169-170.

Giraudo, A. R., 1996: Geographic distribution—*Thamnodynastes chaquensis* (Chaco Coastal House Snake). Herpetol. Rev., 27 (4): 215.

Giraudo, A. R., 1997a: Geographic distribution—*Clelia rustica* (Mussurana). Herpetol. Rev., 28 (3): 158-159.

Giraudo, A. R., 1997b: Geographic distribution—*Lystrophis matogrossensis* (South American Hognose Snake). Herpetol. Rev., 28 (3): 159.

Giraudo, A. R., 1999: New records of snakes from Argentina. Herpetol. Rev., 30: 179-181.

Giraudo, A. R.; G. A. Couturier and M. Di Bernardo, 1996: *Echinanthera cyanopleura* (Cope, 1895), A new record for the ophiodauna of Argentina (Serpentes: Colubridae). Cuadernos de Herpetologica, 10 (1-2): 72.

Giraudo, A. R., and G. S. Scrocchi, 1998: A new species of *Apostolepis* (Serpentes: Colubridae) and comments on the genus in Argentina. Herpetologica, 54 (4): 470-476.

Githens, T. S., 1935: Snake-bite in the U.S. Sci. Monthly., 41: 163-170.

Glaucia, M.,F. Pontes, M. Di-Bernardo, 1988: Registros sobre aspectos reproductivos de serpentes oviparas neotropicais (Serpentes: Colubridae e Elapidae). Comun. Mus. Ciénc. PUCRS, Sér. Zool. Porto Alegre, 1 (5): 123-149.

Glidewell, J., 1974: Records of the snake *Coluber constrictor* (Reptilia: Colubridae) from New Mexico and the Chihuahua Desert of Texas. Southwest. Naturalist, 19 (2): 215-217.

Gliesch, R., 1925: As cobras do estado do Rio Grande do Sul. Almanak Agri. Brasileiro, 1925: 97-118.

Gloyd, H. K., 1935: The subspecies of *Sistrurus miliarius*. Occ. Pap. Mus. Zool., Univ. Michigan, 322: 1-7.

Gloyd, H. K., 1936a: A Mexican subspecies of *Crotalus molossus* Baird and Girard. Occ. Pap. Mus. Zool. Univ. Michigan, (325): 1-5.

Gloyd, H. K., 1936b: The canebreak rattlesnake. Copeia, 1935 (4): 175-178.

Gloyd, H. K., 1936c: The status of *Crotalus unicolor* Van Lidth De Jeude and *Crotalus pulvis* Ditmars. Herpetologica. 1: 65-68.

Gloyd, H. K., 1936d: The subspecies of *Crotalus lepidus*. Occ. Pap. Mus. Zool. Univ. Michigan, (337): 1-5.

Gloyd, H. K., 1938: The subspecies of the copperhead *Agkistrodon mokasen* Beauvois. Bull. Chicago Acad. Sci., 6: 163-166.

Gloyd, H. K., 1940: The rattlesnakes, genera *Sistrurus* and *Crotalus*. A study in zoogeography and evolution. Spec. Pub. Chicago Acad. Sci. (4): 1-270 + 11 pls.

Gloyd, H. K., 1948: Description of a neglected subspecies of rattlesnake from Mexico. Nat. Hist. Misc., (17): 1-4.

Gloyd, H. K., 1955: A review of the massasaugas, *Sistrurus catenatus*, of the southwestern United States (Serpentes: Crotalidae). Bull. Chicago Acad. Sci., 10 (6): 83-98.

Gloyd, H. K., 1956: Distribution of the Mexican Moccasin: A correction. Copeia, 1956: 259.

Gloyd, H. K., 1969: Two additional subspecies of North American crotalid snakes. Bull. Chicago Acad. Sci., 10 (12): 185-195

Gloyd, H. K., 1972: A subspecies of *Agkistrodon bilineatus* (Serpentes: Crotalidae) on the Yucatán Peninsula, México. Proc. Biol. Soc. Washington, 84: 327-334.

Gloyd, H. K., and R. C. Conant, 1934a: The Broad-banded Copperhead; a new subspecies of *Agkistrodon mokasen*. Occ. Pap. Mus. Zool., Univ. Michigan, (283): 1-6.

Gloyd, H. K., and R. C. Conant, 1934b: The taxonomic status, range, and natural history of Schott's Racer. Occ. Pap. Mus. Zool., Univ. Michigan, (287): 1-17.

Gloyd, H. K., and R. C. Conant, 1943: A synopsis of the American forms of *Agkistrodon* (copperheads and moccasins). Bull. Chicago Acad. Sci., 7: 147-170.

Gloyd, H. K., and R. C. Conant, 1990: Snakes of the *Agkistrodon* Complex, a monographic Review. I, SSAR Contrib. Herpetol. 6: vi-614.

Gloyd, H. K., and C. F. Kauffeld, 1940: A new rattlesnake from Mexico. Bull. Chicago Acad. Sci., 6: 11-14.

Gmelin, J. F. *in* Linnaeus, 1788: Systema Naturae.13. Ed. Laurentii Salvi Homiae I: 3: 1033-1125.

Gmelin, J. F., 1789: Caroli a Linné Systemus Naturae. Vol. I (III): 1035-1514.

Godman, F. D., 1915: Biologia Centrali-Americana. Zoology, botany, and archaeology. Introductory Volume. London, Taylor and Francis.

Goff, C. C., 1936: Distribution and variation of a new subspecies of water snake, *Natrix cyclopion floridana*, with a discussion of its relationships. Occ. Pap. Mus. Zool. Univ. Michigan, 327:109.

Golay, P., D. Chiszar, H. M. Smith and F. Van Brukelen, 1999: The proper name for the Venezuelan Red-tailed Coral Snake—El nombre correcto de la Coral Cola Roja Venezolana. Acta Biol. Venez., 19 (4): 73-75.

Golay, P., H. M. Smith, D. G. Broadly, J. R. Dixon, C. McGarthy, J. C. Rage, M. Toriba Schátti, 1993: Endoglyphs and other major venomous snakes of the world. A checklist. Published by Azemiop, S.A., Herpetological Data Center, Geneva, Switzerland.

Goldberg, S. R., 1995: Reproduction in the Western Patchnose Snake, *Salvadora hexalepis* and the Mountain Patchnose Snake, *Salvadora grahamiae* (Colubri-

Bibliography

dae) from Arizona. Southwest. Naturalist, 40 (1): 119-120.

Goldstein, R. C., 1941: Notes on the mud snake in Florida. Copeia, 1941: 49-50.

Gomes, J. F., 1913: Uma nova cobra venenosa do Brasil. Ann. Paulistas Med. Cir. (São Paulo), 1: 65-67.

Gomes, J. F., 1915: Contribuição para o conhecimento dos ophídios do Brasil-Descrição de quatro especies novas e um novo genero de opistoglifos-Contribution to the knowledge of Brazilian snakes-description of four new species and a new genus of opisthoglyphs. Ann. Paulistas Med. Circ. São Paulo, Inst. Seroterapico de Butantan, Junho 1915, IV (6): 121-129.

Gomes, J. F. 1918a: Contribução para o conhecimento dos ophídios do Brasil. II. Ophídios do Museu Rocha (Ceará). Rev. Mus. Paulista, 10: 505-527.

Gomes, J. F., 1918b: Contribução para o conhecimento dos ophídios do Brasil. III (1). Ophídios do Museu Paraense. Mem. Inst. Butantan, 1: 57-77.

González Sponga, M. A., 1971: *Atractus emigdioi* (Serpentes: Colubridae) nueva especies para los Andes de Venezuela. Departamento de Tecnología Audiovisual, Instituto Pedagógico, monografas Cientificos "Augusto Pi Suñer," Caracas, noviembre de 1971," (3): 1-11.

Goodman, J. D., 1953: Further evidence of the venomous nature of the saliva of *Hypsiglena ochrorhyncha*. Herpetologica, 9 (4): 174-176.

Gorham, S. W., 1970: The amphibians and reptiles of New Brunswick. The New Brunswick Mus., Saint John, N. B., (6): 1-30.

Gorzula, S., 1982: Life History: *Leptodeira annulata ashmeadii*. Envenomation. Herpetol. Rev., 13 (2): 47.

Gorzula, S., 1987: Una revisión de los orígenes de la fauna de vertebrados del Pantepui. Pantpui, 3: 4-10.

Gorzula, S., 1992: La herpetofauna del macizo de Chimantá. pp 267-280. *In* El Macizo de Chimantá, Escudo de Guayana, Venezuela. O. Huber (ed.), C. Todtman Editores, Caracas.

Gorzula, S. and J. Ayarzagena, 1995: Dos nuevas especies del genero *Thamnodynastes* (Serpentes; Colubridae) de los Tepuyes de la Guayana Venezolana. Publicaciones de la Associación de Amigos de Doñana, Sevilla-España, June, 1995, (6): 1-17.

Gorzula, S. and J. C. Señaris, 1998: Contribution to the herpetofauna of the Venezuelan Guayana I. A data base. Scientia Guaianae, 8: 1-268 + 129 plates.

Gosner, K. L., 1987: Observations on Lesser Antillean pit vipers. J. Herpetol., 21 (1): 78-80.

Gotch, A. F., 1986: Reptiles-their Latin names explained-A guide to the animal classification. Blanford Press, New York.

Gotch, A. F., 1995: Latin names explained-A guide to the scientific classification of reptiles, birds and mammals. Facts On File.

Goulard, Jean-Pierre *In* Fernando Santos y Frederica Barclay (Eds.), 1994: Guiá Etnográfica de la Alta Amazonia. Vol I, Mai huna, Yagua, Ticuna. "Los Ticunas." FLACSO-Sede Ecuador, 454 pp.

Goyenechea Mayer-Goyenechea, Irene. 1995: Revisión taxanomica de los generos *Conopsis* Gunther y *Toluca* Kennicott, (Reptilia:Colubridae). Universidad Nacional Autonoma de Mexíco, Masters of Science Thesis (Biologica Animal), Mexico, D.F.

Goyenechea, L., and O. Flores-Villela, 2000: Designation of a neotype for *Conopsis nasus* (Serpentes: Colubridae). Copeia, 2000 (1): 285-287.

Goyenechea, L., and O. Flores-Villela, 2002: Taxonomic status of the snake genera *Conopsis* and *Toluca* (Colubridae). J. Herpetol. 36 (1): 92-95.

Granberry, J., 1991: Amazonian origins and affiliations of the Timucua language. pp. 195-242. *In* Kay, Mary Ritchie (Ed.): Language changes in South American Indian Languages. Univ. Pennsylvania Press, Philadelphia.

Grant, C., 1932: Notes on the boas of Puerto Rico and Mona. J. Dept. Agr. Puerto Rico, 16: 327-329.

Grant, C., 1933: Notes on *Epicrates inornatus* (Reinhardt). Copeia, 1933 (4): 224-225.

Grant, C., 1937: Herpetological notes with new species from the American and British Virgin Islands. J. Dept. Agr. Rico, 16 (3): 339-346.

Grant, C., 1940: Notes on the reptiles and amphibians of Jamaica, with diagnosis of new species and subspecies. Jamaica To-day, London and Aylesbury, Chapter 15: 151-157.

Grant, C., 1941a: The herpetology of the Cayman Islands [including the results of the Oxford University Cayman Islands Biological Expedition 1938] with an appendage on The Cayman Islands and marine turtles by C. Bernard Lewis, Bull. Inst. Jamaica, Sci. Ser., (2): 1-65.

Grant, C., 1941b: The herpetology of Jamaica, II. The reptiles. Bull. Inst. Jamaica Sci. Ser., (1): 61-148, 3 plates.

Grant, C., 1946: Notes on *Tretanorhinus* of Cuba and the Isle of Pines. Herpetologica, 30: 102-117.

Grant, C.; H. M. Smith and D. Alayo Pastor, 1959: The status of snakes of the genus *Arrhyton* in Cuba. Herpetologica, 15: 129-133.

Grantsau, R., 1991: As cobras venenosas do Brasil. São Bernardo do Campo, São Paulo, Direito da 1° Ediçao en português-1° Tiragem-1000 Exemplares, Bandeirante S.A. Gráfica e Editora, Mercedes-Benz.

Gray, L., 1831: A synopsis of the species of the class reptilia. *In* The animal kingdom arranged in conformity with its organization by G. Cuvier, with additional descriptions of all the species hitherto named, and of many not before noticed. ed. E. Griffith and E. Pidgeon, Appendix to E. London, Whittaker, Treacher, and Co.

Gray, J. E., 1829: Synopsis generum reptilium et amphibiorum. Isis von Oken, 22: 187-206.

Gray, J. E., 1831-1845: Zoological Miscellany. London.

Gray, J. E., 1831 [1996]: A synopsis of the species of the class reptilis. HerpPrint International, Bredell, South Africa.

Gray, J. E., 1840: Catalogue of the species of reptiles collected in Cuba by W. S. MacLeory, Esq.-with some notes on their habits extracted from his M.S. Ann. Mag. Nat. Hist., 5: 108-115.

Gray, J. E., 1842a: Synopsis of the species of prehensile-tailed snakes, of the family Boidae. Zool. Misc., (2): 41-46.

Gray, J. E., 1842b: Synopsis of the species of rattle-snakes, or family of Crotalidae. Zool. Misc. Parts 2/3: 47-51.

Gray, J. E., 1845: Catalogue of the specimens of lizards in the collection of the

British Museum. Printed by order of the Trustees, London, 1845: 1–289.

Gray, J. E., 1849: Catalogue of the specimens of snakes in the collection of the British Museum. Trustees of the British Museum, London.

Greene, H. W., 1976: Coral snake mimicry: does it occur: Herp. Review, 7 2: 85.

Greene, H. W., 1992: The ecological and behavioral context for pitviper evolution. pp. 107–117 in J. A. Campbell and E. D. Brodie, Jr.: Biology of the pitvipers. Selva, Tyler, Texas.

Greene, H. W., 1997a: A new forward. In Klauber, L. M., 1997: Rattlesnakes, their habits, life histories, & influence on mankind. Volumes 1 & 2, 2nd Edition, Berkeley and Los Angeles, Univ. California Press. Berkeley

Greene, H. W., 1997b: Snakes and others: past, present, and future. Reptile and Amphibian Mag., July/August 1997, (49): 13–23.

Greene, H. W., 1997c: Snakes-the evolution of mystery in nature. Univ. California Press. Berkeley.

Greene, H. W., and J. A. Campbell, 1992: The future of pitvipers. pp. 421–428 in J. A. Campbell and E. D. Brodie, Jr.: Biology of the pitvipers. Selva, Tyler, Texas.

Greene, H. W., R. F. Wilkinson, Jr., and R. Powell, 1998: *Micrurus limbatus* Fraser—Tuxtlan coral snake. Cat. Amer. Amph. Rept., (678): 1–3.

Greer, A. E., 1965: A new subspecies of *Clelia clelia* (Serpentes: Colubridae) from the island of Grenada. Breviora, Mus. Comp. Zool., (223): 1–6.

Grenard, S., 1994: Snakebite: are "non-poisonous" colubrids really harmless? Reptile & Amphibian Mag. Sept./Oct., 1994: 51–63.

Griehl, K., [date unknown]: Schlangen, Riesenschlangen und Nattern im Terrarium, Alles über Anschaffung, Pflege, Ernährung und Krankheiten. Munich, Gräfe und Unzer GmbH.

Griffin, L. E., 1916: A catalogue of the ophidia from South America at present (June, 1916) contained in the Carnegie Museum, with description of some new species. Mem. Carnegie Mus., for 1915, 7 (3):163–228.

Grinnel, J., and C. L. Camp, 1917: A distributional list of the amphibians and reptiles of California. Univ. California Publ. Zoöl., 17: 127–208.

Griswold, B., 1994: The "Trans-Pecos Rat Snake". Reptile & Amphibians Mag., May-June, 1994: 64–69.

Grismer, L. L., 1990a: *Masticophis aurigulus* (Cope) Baja California Striped Whipsnake. Cat. Amer. Amph. Rept., (499): 1–2.

Grismer, L. L., 1990b: Relationships, taxonomy and biogeography of the *Masticophis lateralis* complex in Baja California, Mexico. Herpetologica, 46 (1): 66–77.

Grismer, L. L., 1994: The origin and evolution of the peninsular herpetofauna of Baja California, México. Herpetol. Nat. Hist., 2: 51–106.

Grismer, L. L., 1996: Geographic distribution—*Eridiphas slevini slevini* (Slevin's Night Snake). Herpetol. Rev., 27 (1): 33.

Grismer, L. L., 1997: Peninsula paradise: the natural wonders of Baja. Fauna, 1 (1): 54–67.

Grismer, L. L., 1999: An evolutionary classification of reptiles on islands in the Gulf of California; México Herpetologica, 55 (4): 446–449.

Grismer, L. L., 2001: Comments on the taxonomy of gopher snakes from Baja California, México: a reply to Rodríguez-Robles and de Jesús-Escobar. Herpetol. Rev., 32 (2): 81–83.

Grismer, L. L., J. A., McGuire, and B. D. Hollingsworth, 1994: A report on the herpetofauna of the Vizcaino Peninsula, Baja California, Mexico, with a discussion of its biogeographic and taxonomic implications. Bull. Southern California. Acad. Sci., 93: 45–80.

Grobman, A. B., 1941: A contribution to the knowledge of variation in *Opheodrys vernalis* (Harlan), with the description of a new subspecies. Misc. Pub., Mus. Zool., Univ. Michigan, (50): 1–38.

Grobman, A. B., 1984: Scutellation variation in *Opeodrys aestivus*. Bull. Florida State Mus., 29: 153–170.

Grobman, A., 1992a: On races, clines, and common names in *Opheodrys*. Herpetol. Rev., 23 (1): 14–15.

Grobman, A. B., 1992b: Metamerism in the snake *Opheodrys vernalis*, with a description of a new subspecies. J. Herpetol., 26 (2): 175–186.

Grocott, R. G., and G. G. Sandler, 1958: The poisonous snakes of Panama. Panama Canal Printing Plant, Mount Hope, Canal Zone, October, 1958.

Grogen, W. L. Jr., and W. W. Tanner, 1974: Range extension of the Long-nosed Snake *Rhinocheilus l. lecontei*, into east-central Utah. Great Basin Naturalist, 34 (3): 238–240.

Groves, J. D., 1995: Reproduction and feeding behavior of *Oxybelis wilsoni*, a new species of a vine snake (Serpentes: Colubridae). Rev. Biol. Trop., 43 (1–3): 307–309.

Grzimek, B., (ed.), 1974: Le monde anima. Tome VI. Reptiles. Stauffacher S.A., Zurich, 585 pp.

Grzimek, B. (Editor), 1984: Grzimek's animal life encyclopedia, volume 6, Reptiles. New York, Van Nostrand Reinhold Co.

Guasco Milanes, F., 1942: Las serpientes del paraiso mexicano. Gráfico Panamericana, S.R.L., Pánuco, Mexico, D.F.

Guido, A. R., G. A. Couturier and M. DiBernardo, 1996: *Echinanthera cyanopleura* (Copeia, 1885), a new record for the ophidiofauna of Argentina (Serpentes; Colubridae). Cuadernos de Herpetologia, 10 (1–2): 72.

Gunter, R, 1971: The Big Thicket—a challenge for conservation. Jenkins Publishing Co., Austin.

Günther, A. C. L. G., 1858: Catalogue of colubrine snakes in the collection of the British Museum. London, Taylor and Francis.

Günther, A. C. L. G., 1859: On the genus of *Elaps* of Wagler. Proc. Zool. Soc., London, 1859: 79–89.

Günther, A. C. L. G., 1860: Description of *Leptodeira torquata*, a new snake from Central America. Ann. Mag. Nat. Hist., (ser. 3): 169–171 + 1 plate.

Günther, A. C. L. G., 1861a: Account of the reptiles sent by Dr. Wucherer from Bahia. Proc. Zool. Soc., London, 1861: 12–18.

Günther, A. C. L. G., 1861b: On a new subspecies of the family Boidae. Proc. Zool. Soc., London, p. 142.

Günther, A. C. L. G., 1863: Third account of the snakes in the collection of the British Museum. Ann. Mag. Nat. Hist. Ser. 3, 12: 348–365.

Günther, A. C. L. G., 1865: Fourth account of new snakes in the collection of the British Museum. Ann. Mag. Hist., 15 (3): 89–98.

Bibliography

Günther, A. C. L. G., 1868: Sixth account of new species of snakes in the collection of the British Museum. Ann. Mag. Nat. Hist., 4 (1): 413-429.

Günther, A. C. L. G., 1872: Seventh account of new species of snakes in the collection of the British Museum. Ann. Mag. Nat. Hist., Ser. 4, 9: 13-37

Günther, A. C. L. G., 1885-1902: Biologia centrali-americana reptilia and batrachia. SSAR Facsimile Reprints in Herpetology, 1987: 326 pp +76 plates.

Guppy, Nicholas, 1958: Wai-Wai—through the forests north of the Amazon. E. P. Dutton & Co., Inc., New York, 373 pp.

Gutberlet, R. L. Jr., 1995: A new locality for Rowley's Palm Pitviper (Serpentes: Viperidae), a Mexican relict. Southwest. Naturalist, 40 (1): 124-125.

Gutberlet, R. L. Jr., 1998: The phylogenetic position of the Mexican Black-Tailed Pitviper (Squamata: Viperidae: Crotalinae). Herpetologica, 54 (2): 184-206.

Gutberlet, R. L., and J. A. Campbell, 2001: Generic recognition for a neglected lineage of South American pitvipers (Squamata, Viperidae, Crotalinae), with the description of a new species from the Colombian Choco. Amer. Mus. Novitates, (3316): 1-15.

Guthrie, J. E., 1920: The snakes of Iowa. Agric. Sta., Iowa State College, (239): 145-192.

Gutierrez, José María; Richard T. Taylor, Bolaños, 1979: Cariotipos de diez especies de serpientes costarricenses de la familia viperidae. Rev. Biol. Trop., 27 (2): 309-319.

Gutierrez Lopez, B., 1902: Contribucion al estado de las serpientes pozoñosas de Venezuela. Tesis para El Doctorado en Medicina y Cirugia, Facultad de Medicinas de Caracas, Tipografia Guttenberg.

Guyer, C., and M. S. Laska, 1996: Natural history notes—*Coluber* (=*Masticophis*) *mentovarious* (Tropical Racer)—predation. Herpetol. Rev., 27 (4): 203.

Haas, G. H., 1964: Anatomical observations of the head of *Liotyphlops albirostris* (Typhlopidae, Ophidia), Acta. Zool., Stockholm, 45: 1-62.

Hagmann, G., 1909: Die Reptilien der Insel Mexicana, Amazonestrom. Zool. Jahrb., Abt. Syst., 28: 472-504.

Hahn, D. E., 1971: Noteworthy herpetological records from Honduras. Herpetol. Rev., 3: 111-112.

Hahn, D. E., 1974: *Leptotyphlops dulcis*. Cat. Amer. Amph. Rept., (231): 1-2

Hahn, D. E., 1979a: Leptotyphlopidae Stejneger, *Leptotyphlops* Fitzinger—Slender Blind Snake. Cat. Amer. Amph. Rept., (230): 1-4.

Hahn, D. E., 1979b: *Leptotyphlops dulcis*. Cat. Amer. Amph. Rept., (231): 1-2.

Hahn, D. E., 1979b: *Leptotyphlops humilis*. Cat. Amer. Amph. Rept., (232): 1-4.

Hahn, D. E., 1980a: Anomalepidae, Leptotyphlopidae, Typhlopidae. Das Tierreich, Lieferung 104, Liste der rezenten Amphibien und Reptilien, Walter de Gruyter, Berlin-New York, 93 pp.

Hahn, D. E., 1980b: *Leptotyphlops maximus*. Cat. Amer. Amph. Rept., (244): 1.

Hall, C. W., 1951: Notes on a small herpetological collection from Guerrero. Univ. Kansas Sci. Bull., 34—Part 1 (4): 201-212.

Hall, R. A., 1966: Pidgin and Creole languages. Cornell Univ. Press, Ithaca, New York, 188 pp.

Hall, R. J., 1969: Ecological observations on Graham's Watersnake (*Regina grahami*) Baird and Girard). Amer. Midl. Naturalist, 81: 156-163.

Halliday, T., and K. Adler (Editors), 1988: The encyclopedia of reptiles & amphibians. New York, Facts On File.

Hallowell, E., 1845: Description of reptiles from South America, supposed to be new. Proc. Acad. Nat. Sci. Philadelphia, 2: 241-247.

Hallowell, E., 1852: Descriptions of new species of reptiles inhabiting North America. Proc. Acad. Nat. Sci. Philadelphia, 6: 177-182.

Hallowell, E., 1854a: Descriptions of new reptiles from California. Proc. Acad. Nat. Sci. Philadelphia, 7: 91-97.

Hallowell, E., 1854b: Notices of new reptiles from Texas. Proc. Acad. Nat. Sci Philadelphia, 7: 192-193.

Hallowell, E., 1854c: Remarks on the geographical distribution of reptiles, with descriptions of several species supposed to be new, and corrections of former papers. Proc. Acad. Nat. Sci. Philadelphia, 7: 98-105.

Hallowell, E., 1855: Contributions to South American herpetology. J. Acad. Nat. Sci. Philadelphia, 2 (3): 33-36.

Hallowell, E., 1856a: Note on the collection of reptiles from the neighborhood of San Antonio, Texas, recently presented to the Academy of Natural Sciences by Dr. A. Heerman. Proc. Acad. Nat. Sci. Philadelphia, 8: 306-310.

Hallowell, E., 1856b: Notice of a collection of reptiles from Kansas and Nebraska, presented to the Academy of Natural Sciences, by Dr. Hammond, USA. Proc. Acad. Nat. Sci. Philadelphia, 8: 238-253.

Hallowell, E., 1859: Report on reptiles collected on the survey. *In* Report of Lieutenant J. G. Parke—"Expl. Surv. R. R. Mississippi R. Pacific Ocean." Rept. 10 (4): 1-27.

Hallowell, E., 1860 [1861]: Report upon reptilia of the North Pacific Exploring Expedition under command of Capt. John Rogers, USA. Proc. Acad. Nat. Sci. Philadelphia, 1860: 480-509.

Hallowell, E. *in* Cope, 1860: Proc. Acad. Nat. Sci. Philadelphia, 112: 248.

Halstead, B. W., 1957: Dangerous marine animals. Cambridge, Maryland, Cornell Maritime Press.

Halstead, B. W., 1970: Poisonous and venomous marine animals of the world-Volume 3-Vertebrates (Continued). United States Government Printing Office-Washington, D.C., 1001 pp.

Hammerson, G. A., 1982: Amphibians and reptiles in Colorado. Colorado Division of Wildlife, Denver, Colorado, June.

Harding, J. H., 1997: Amphibians and reptiles of the Great Lakes Region. Univ. Michigan Press, Ann Arbor.

Harding, K. A., 1995: A new species of tree snake of the genus *Leptophis* Bell 1825 from Mount Aripo, Trinidad. Trop. Zool., 8: 221-226.

Harding, K. A., and K. R. G. Welch, 1980: Venomous snakes of the world, A Checklist. Pergamon Press, New York.

Hardy, D. L., 1982a: Envenomation by the Northern Blacktail Rattlesnake (*Crotalus molossus molossus*): Report of two cases and the *in vitro* effects of the venom on fibrinolysis and platelet aggregation. Toxicon, 20 (2): 487-493.

Hardy, D. L., 1982b: Snakebite update: *Crotalus* envenomation in Tucson, 1973-1980 and comments on the new Australian method of first aid for elapid snakebites. Presented September 22, 1982, Annual Conference, American Association of Zoological Parks and

Aquariums, Scottsdale, Arizona, AAZPA 1982 Ann., 1982: 430-435.

Hardy, D. L., 1983: Envenomation by the Mojave Rattlesnake (*Crotalus scutulatus scutulatus*) in southern Arizona, U.S.A. Toxicon, 21 (1): 111-118.

Hardy, D. L., 1985a: Overview of rattlesnake bite treatment. Second Annual Southwest Poison Symposium, Scottsdale, Arizona, October 5-6, 1982, Revised September, 1985: 1-7.

Hardy, D. L., 1985b: Rattlesnake envenomation in Tucson, Arizona: 1973-1980. Toxicon, 23 (4): 573.

Hardy, D. L., 1985c: Venomous snakebite in North America: Some current views on management. Abstracts and papers-1985, Ann. Meet., Amer. Assoc. Zoo Veterinarians, 1985: 102-104.

Hardy, D. L., 1986: Fatal rattlesnake envenomation in Arizona: 1969-1984. Clinical Toxicology, 24 (1): 1-10.

Hardy, D. L. Sr., 1992: A review of first aid measures for pitviper bite in North America with an appraisal of Extractor™ suction and stun gun electroshock. pp. 405-414 *in* J. A. Campbell and E. D. Brodie, Jr.: Biology of the pitvipers. Selva, Tyler, Texas.

Hardy, G. A., 1954: The natural history of the Forbidden Plateau Area Vancouver Islands, British Colombia. Provincial Museum of Natural History and Anthropology, Report for the year 1954, Prov. British Colombia, Dept. Education, 1954: 24-63.

Hardy, J., 1952: The crowned snake, *Tantilla coronata coronata*, in North Carolina, Copeia, 1952 (3): 188.

Hardy, J. D., 1982: Biogeography of Tobago, West Indies, with special reference to amphibians and reptiles: A review. Bull. Maryland Herpetol. Soc., 18: 37-142.

Hardy, J. D. Jr., and H. A. E. Boos, 1995: Snakes of the genus *Erythrolamprus* (Serpentes: Colubridae) from Trinidad and Tobago, West Indies. Bull. Maryland Herpetol. Soc., 31 (3): 158-190.

Hardy, K. A., and K. R. G. Welch, 1980: Venomous snakes of the world-a checklist. Pergamon Press, New York.

Hardy, L. M., 1963: Description of a new species of snake (Genus *Dryadophis*) from México. Copeia, 1963 (4): 669-672.

Hardy, L. M., 1964: A replacement name for *Dryadophis fasciatus* Hardy. Copeia, (1964) (4): 714.

Hardy, L. M., 1968: Western amphibians and reptiles. The Biologist, Dept. Biology, Univ. New Mexico, Albuquerque, New Mexico, 1 (1-2), January 1968.

Hardy, L. M., 1970: Systematic revisions of the genera *Pseudoficimia, Gyalopion, and Ficimia* (Serpentes, Colubridae). Dissertation Abstracts International, 30 (12).

Hardy, L. M., 1972: A systematic revision of the genus *Pseudoficimia* (Serpentes: Colubridae). J. Herpetol. 6 (1): 53-69.

Hardy, L. M., 1973: *Pseudoficimia, P. frontalis*. Cat. Amer. Amph. Rept., (146): 1-2.

Hardy, L. M., 1975a: A systematic revision of the colubrid snake genus *Ficimia*. J. Herpetol., 9 (2): 133-168.

Hardy, L. M., 1975b: A systematic revision of the colubrid snake genus *Gyalopion*. J. Herpetol., 9 (1): 107-132.

Hardy, L. M., 1976a: *Ficimia streckeri* Taylor—Mexican Hook-nose Snake. Cat. Amer. Amph. Rep., (181): 1-2.

Hardy, L. M., 1976b: *Gyalopion, G. cranum, G. quadrangularis*. Cat. Amer. Amph. Rept., (182): 1-4.

Hardy, L. M., 1978: *Ficimia olivacea* Gray—Brown Hook-nosed Snake. Cat. Amer. Amph. Rept., (219): 1-2.

Hardy, L. M., 1979a: *Ficimia ramirezi*. Cat. Amer. Amph. Rept., (228): 1.

Hardy, L. M., 1979b: Checklist of the amphibians and reptiles of Caddo and Bossier Parishes, Louisiana. Bull. Mus. Life Science, LSU in Shreveport, (2): 1-11.

Hardy, L. M., 1980a: *Ficimia publia* Cope—Blotched Hooknose Snake. Cat. Amer. Amph. Rept., (254): 1-2.

Hardy, L. M., 1980b: *Ficimia ruspator* Smith and Taylor—Guerrero Hook-nosed Snake. Cat. Amer. Amph. Rept., (243): 1.

Hardy, L. M., 1980c: *Ficimia variegata* (Günther) Tehuantepec Hook-nosed Snake. Cat. Amer. Amph. Rept., (296): 1-2.

Hardy, L. M., 1990: *Ficimia* Gray—Southern Hook-nosed Snake. Cat. Amer. Amph. Rept., (471): 1-5.

Hardy, L. M., 1995: Checklist of the amphibians and reptiles of the Caddo Lake Watershed in Texas and Louisiana. Bull. Mus. Sci., Louisiana State Univ. Shreveport, 10: 1-31.

Hardy, L. M., and C. J. Cole, 1967: The colubrid snake *Tantilla armillata* Cope in Nicaragua. J. Arizona Acad. Sci., 4 (3):194-196.

Hardy, L. M., and C. J. Cole, 1968: Morphology variation in a population of the snake, *Tantilla gracilis* Baird and Girard. Univ. Kansas Pub. Mus. Nat. His., Lawrence, Kansas, 17 (15): 613-629.

Hardy, L. M., and R. W. McDiarmid, 1969: The amphibians and reptiles of Sinaloa, México. Univ. Kansas Pub. Mus. Nat. Hist., 18 (3): 39-252.

Hardy, R., 1938: An annotated list of reptiles and amphibians of Carbon County, Utah. Utah Acad. Sci., Arts, Literatures, 15: 99-102.

Harlan, R., 1826-1827: Genera of North American reptiles, and a synopsis of the species. J. Acad. Nat. Sci. Philadelphia, (1) 5 :317-372 (1826), (1) 6 (1): 7-38 (1827).

Harlan, R., 1827a: American herpetology. Philadelphia.

Harlan, R., 1827b: Genera of North American reptilia and a synopsis of the species. J. Acad. Nat. Sci. Philadelphia, (1) part 2, 5: 345-372.

Harrington, J. P., 1928: Vocabulary of the Kiowa language. Smithsonian Inst., Bureau of American Ethnology. (84): 1-255.

Harris, H. S. Jr., and R. S. Simmons, 1972a: A checklist of the rattlesnakes (*Crotalus durissus* group) of South America. Bull. Maryland Herpetol. Soc., 8: 27-32.

Harris, H. S. Jr., and R. S. Simmons, 1972b: A checklist of the Neotropical species and subspecies of the *Crotalus durissus* group. Bull. Maryland Herpetol. Soc., 8: 33-40.

Harris, H. S. Jr., and R. S. Simmons, 1972c: Keys to the Neotropical species and subspecies of the *Crotalus Durissus* Group. Bulletin Maryland Herpetol. Soc., 8 (2): 33-40.

Harris, H. S. Jr., and R. S. Simmons, 1978a (dated 1976/1977): A new subspecies of *Crotalus durissus* (Serpentes: Crotalidae) from the Rupununi savanna of Southwestern Guyana. Mem. Inst. Butantan, 40/41: 305-311.

Harris, H. S. Jr., and R. S. Simmons, 1978b: A preliminary account of the rattlesnakes with the description of four new subspecies. Bull. Maryland Herpetol. Soc., 14 (3): 1-214.

Bibliography

Harrison, H. H., 1971: The world of the snake. Philadelphia and New York, J. B. Lippincot & Co.

Harvey, M. B., 1999: Revision of Bolivian *Apostolepis* (Squamata: Colubridae). Copeia, (2): 376-381.

Hartweg, N., 1940: Description of *Salvadora intermedia*, new species, with remarks on the *grahamiae* Group. Copeia, 1940 (4): 256-259.

Hartweg, N., 1944: Remarks on some Mexican snakes of the genus *Tantilla*. Occ. Pap. Mus. Zool., Univ. Michigan, (486): 1-9.

Hartweg, N., 1959: A new colubrid snake of the genus *Geophis* from Michoacán. Occ. Pap. Mus. Zool.-Univ. Michigan, (601): 1-5.

Hartweg, N., and J. A. Oliver, 1938: A contribution to the herpetology of the Isthmus of Tehuantepec-III. Three new snakes from the Pacific slope. Occ. Pap. Mus. Zool., Univ. Michigan, (390): 1-8.

Hartweg, N., and J. A. Oliver, 1940: A contribution to the herpetology of the Isthmus of Tehuantepec-IV. Misc. Pub. Mus. Zool., Univ. Michigan, (47): 1-31.

Harvey, M. B., 1994: A new species of Montane Pitviper (Serpentes: Viperidae: *Bothrops*) from Cochabamba, Bolivia. Proc. Biol. Soc. Washington, 107 (1): 60-66.

Harvey, M. B., 1999: Revision of Bolivian *Apostolepis* (Squamata: Colubridae). Copeia, 1999 (2): 388-409.

Harvey, M. B., L. Gonzales A., and G. J. Scrocchi, 2001: New species of *Apostolepis* (Squamata: Colubridae) from the Gran Chaco in southern Bolivia. Copeia, 2001 (2): 501-507.

Hasler, J. A., 1961: Tetradialectologia Nauha. A William Cameron Townsend, Mexico, Instituto Lingüístico de Verano, 1961: 455-464.

Hay, O. P., 1887: The massasauga and its habits. Amer. Naturalist, 1st Ser., 7: 200-203.

Hay, O. P., 1892: On the ejection of blood from the eyes of horned toads. Proc. U.S. Natl. Mus., 15: 375-378.

Hay, O. P., 1893: On the breeding habits, eggs, and young of certain snakes. Proc. U.S. Natl. Mus., 15: 385-397.

Hayes, W. K., I. I. Kaiser and D. Duvall, 1992: The mass of venom expanded by prairie rattlesnakes when feeding on rodent preys. pp. 383-388 *in* J. A. Campbell and E. D. Brodie, Jr.: Biology of the pitvipers. Selva, Tyler, Texas.

Hedges, S. B., A. R. Estrada and L. M. Diaz, 1999: New snake (*Tropidophis*) from western Cuba. Copeia, 1999 (2): 376-381.

Hedges, S. B., and O. H. Garrido, 1992: Cuban snakes of the genus *Arrhyton*: two new species and a reconsideration *A. redimitum* Cope. Herpetologica, 48: 168-177.

Hedges, S. B., and O. H. Garrido, 1999: A new snake of the genus *Tropidophis* (Tropidophidae) from central Cuba. J. Herpetol., 33 (3): 436-441.

Hedges, S. Blair and O. H. Garrido, 2002: A new snake of the genus *Tropidophis* (Tropidophiidae) from eastern Cuba. J. Herpetol. 36 (2): 157-161.

Hedges, S. Blair, O. H. Garrido and L. M. Díaz, 2001: A new banded snake of the genus *Tropidophis* (Tropidophiidae) from north-central Cuba. J. Herpetol. 35 (4): 615-617.

Hedges, S. B., and R. Thomas, 1989: Supplement to West Indian amphibians & reptiles: A Checklist. Milwaukee Publ. Mus., Contrib. Biol. Geol., (77): 1-11.

Hellmich, W., 1953a: Contribuciones al conocimiento de los ofidios de Venezuela—2.Sobre la subespecie venezolana de *Coluber (Masticophis) mentovarius* (Dumeril et. Bibron). Acta Biologica Venezuelica. Univ. Central Venezuela, 16 junio 1953, Caracas, Venezuela, 1 (Art.8): 146-154.

Hellmich, W., 1953b: Contribuciones al conocimiento de los ofidios de Venezuela-1. *Spilotes pullatus pullatus* (Linne). Acta Biologica Venezuelica. Univ. Central Venezuela, 16 junio 1953, Caracas, Venezuela., 1 (Art.8): 141-145.

Helm, T., 1965: A world of snakes. New York, Dodd, Mead & Co.

Henderson, J., and J. P. Harrington, 1914: Ethnology of the Tewa Indians. Smithsonian Inst. Bureau of American Ethnology, Bulletin 56, (56): 1-76.

Henderson, R. W., 1974: Aspects of the ecology of the Neotropical vine snake, *Oxybelis aeneus* (Wagler). Herpetologica, 30 (1): 19-24.

Henderson, R. W., 1976: A new insular subspecies of the colubrid snake *Leptophis mexicanus* (Reptilia, Serpentes, Colubridae) from Belize, J. Herpetol., 10: 329-331.

Henderson, R. W., 1978: Notes on *Agkistrodon bilineatus* (Reptilia, Serpentes: Viperidae) in Belize. J. Herpetol., 12: 412-413.

Henderson, R. W., 1979: Variation in the snake *Tretanorhinus nigroluteus lateralis* in Belize with notes on breeding tubercles. Herpetologica, 35: 245-248.

Henderson, R. W., 1984: *Scaphiodontophis* (Serpentes: Colubridae); natural history and test of a mimicry-related hypothesis. pp. 185-194. *In* R. R. Seigel, *et. al.*, eds, Vertebrate ecology and systematics, Mus. Nat. Hist., Univ. Kansas Spec. Pub., 10: 1-278.

Henderson, R. W., 1991: Distribution and preliminary interpretations of geographic variation in the Neotropical tree boa, *Corallus enydris*: a progress report. Bull. Chicago Herpetol. Soc., 26 (5): 105-110.

Henderson, R. W., 1993a: *Corallus* Daudin. Cat. Amer. Amph. Rept., (572): 1-2.

Henderson, R. W., 1993b: *Corallus annulatus*—Reptilia: Squamata: Serpentes: Boidae. Cat. Amer. Amph. Rept. (573): 1-4.

Henderson, R. W., 1993c: *Corallus caninus* (Linnaeus). Cat. Amer. Amph. Rept., (574): 1-4.

Henderson, R. W., 1993d: *Corallus cropanii* (Hoge). Cat. Amer. Amph. Rept., (575): 1.

Henderson, R. W., 1993e: *Corallus enydris* (Linnaeus). Cat. Amer. Amph. Rept., (576): 1-6.

Henderson, R. W., 1997a: An irascible aerialist -The common tree boas of the America tropics. Fauna, 1 (1): 16-25.

Henderson, R. W., 1997b: A taxonomic review of the *Corallus hortulanus* complex of Neotropical tree boas. Caribbean J. Sci., 3 (3-4): 198-221.

Henderson, R. W., and M. H. Binder, 1980: The ecology and behavior of vine snakes (*Ahaetulla, Oxybelis, Thelotornis, Uromacer*): a review. Milwaukee Publ. Mus. Contrib. Biol. Geol., 37: 1-38.

Henderson, R. W., and H. E. A. Boos, 1993-1994: The tree boa (*Corallus enydris*) on Trinidad and Tobago. Living World, J. Trinidad and Tobago Field Naturalist's Club, 1993-1994: 3-5.

Henderson, R. W., J. R. Dixon and P. Soini, 1979: Resource partitioning in Amazonian snake communities. Milwaukee Publ. Mus. Contrib. Biol. Geol., 22: 1-11.

Henderson, R. W., and G. T. Haas, 1993:

Status of the West Indian snake *Chironius vincenti*. Oryx, 27 (3): 181-184.

Henderson, R. W., and L. G. Hoevers, 1975: A checklist and key to the amphibians and reptiles of Belize, Central America. Milwaukee Publ. Mus. Contrib. Biol. Geol., (5): 1-63.

Henderson, R. W., and L. G. Hoevers, 1979: Variation in the snake *Tretanorhinus nigroluteus lateralis* in Belize with notes on breeding tubercles. Herpetologica, 35 (3): 245-248.

Henderson, R. W., L. G. Hoevers and L. D. Wilson, 1977: A new species of *Sibon* (Reptilia, Serpentes, Colubridae) from Belize, Central America. J. Herpetol., 11: 77-79.

Henderson, R. W., and R. Powell, 1990a: *Alsophis ater* (Gosse)—Jamaican Racer. Cat. Amer. Amph. Rept., (633): 1-2.

Henderson, R. W., and R. Powell, 1990b: *Alsophis sanctaecrucis* Cope-St. Croix racer. Cat. Amer. Amph. Rept., (634): 1-2.

Henderson, R. W., and R. Powell, 1996: Reptilia: Squamata: Serpente: Colubridae: *Alsophis ater*. Cat. Amer. Amph. Rept., (633): 1-2.

Henderson, R. W., R. Powell, J. C. Daltry and M. L. Day, 1996: Reptilia: Squamata: Serpentes: Colubridae. *Alsophis antiguae* Parker—Antiguan Racer. Cat. Amer. Amph. Rept., (632): 1-3.

Henderson, R. W., and G. Puorto, 1993: *Corallus cropanii*. Cat. Amer. Amph. Rept., (572): 1-2.

Henderson, R. W., and R. A. Sajdek, 1986: West Indian racers: disappearing act or a second chance? Lore (Milwaukee Publ. Mus.), 36 (3): 13-18.

Henderson, R. W., R. A. Sajdek and R. M. Henderson, 1988: The rediscovery of the West Indian snake *Chironius vincenti*. Amphibia-Reptilis, 9: 415-422.

Henderson, R. W., and A. Schwartz, 1984: A guide to the identification of the amphibians and reptiles of Hispaniola. Milwaukee Publ. Mus., Spec. Pub. (4) :1-70.

Henderson, R. W., A. Schwartz, and S.J Icháustegui, 1984: Guia para la identificación de los anfibios y reptilés de la hispaniola. Mus. Nac. Hist. Nat., Serie Monog. No.1, Santo Domingo, República Dominicana.

Henderson, R. W., T. Waller, P. M. Icucci, G. Puorto and R. W. Bourgeois, 1995: Ecological correlates and pattern in the distribution of Neotropical boines (serpentes: boidae): a preliminary assessment. Herpetol. Nat. Hist., 3 (1): 15-21.

Henle, K., and A. Ehrl, 1991: Zur Reptilien fauna Perus nebst Beschreibung eines neuen *Anolis* (Iguanidae) und zweier Schlangen (Colubridae). Bonner Zoolo. Beitr. 42 (2): 143-180.

Henriksen, L. A., and S. H. Levinsohn, 1977: Progression and prominence in Cuaiquer discourse. pp 43-67. *In* Longacre, R. E., ed., 1977: Discourse grammar—Studies in indigenous languages of Colombia, Panama, and Ecuador Part II.

Hensley, M. M., 1950: Results of a herpetological reconnaissance in extreme southwestern Arizona and adjacent Sonora, with a description of a new subspecies of the Sonoran Whipsnake, *Masticophis bilineatus*. Trans. Kansas Acad. Sci., 53: 270-288.

Henzl, M. J., 1991: Reptiliengesellschaften eines amazonischen Inselgebirges (Serranía de Sira, Peru): Höhenverbreitung, Habitatnutzung und biogeographische Beziehungen. Ph.D. thesis for the University of Wien, Germany.

Hernández Garcia, E., and F. Mendoza Quijano, 1994: *Rhadinaea marcellae*. Herpetol. Rev., 25 (1): 34.

Hernández-Ibarra, X., A. Ramírez-Bautista and R. Torres-Cervantes, 1999: Geographic distribution: *Ficimia hardyi*. Herpetol. Rev., 30 (4): 235.

Hibbitts, T. D., 1994: Geographic variation and evolutionary relationships of the ringneck snakes (Genus *Diadophis*). Masters Thesis, University of Texas at Arlington, August 1994, 121 pp.

Hidalgo, H. N., 1979: Range extension of the snake *Sibon anthracops* Cope in El Salvador. Herpetol. Rev., 10: 103.

Hidalgo, H. N., 1980: Occurrence of *Pelamis platurus* (Linnaeus) in El Salvador. Herpetol. Rev., 11: 117.

Hidalgo, H. N., 1981a: Additions to the snake fauna of El Salvador. Trans. Kansas Acad. Sci. 84: 55-58.

Hidalgo, H. N., 1981b: Additions to the snake fauna of El Salvador. Herpetol. Rev., 12: 67-68.

Highton, R., 1976: *Stilosoma, S. extenuatum*. Cat. Amer. Amph. Rept., (183): 1-2.

Hilken, G., and R. Schlepper, 1998: Der *Lampropeltis mexicana*-Komplex (Serpentes: Colubridae): Naturgeschichte und Terrarienhaltung. Salamandra, 34 (2): 97-124.

Hillis, D. M., 1990: A new species of xenodontine colubrid snake of the genus *Synophis* from Ecuador and the phylogeny of the genera *Synophis* and *Emmochiliophis*. Occ. Pap. Mus. Nat. Hist. Univ. Kansas, (134): 1-23.

Hillis, D. M., and S. L. Campbell, 1982: New localities for *Tantilla rubra cucullata* (Colubridae) and the distribution of its two morphotypes. Southwest. Naturalist, 27 (2): 220-221.

Hoevers, L. G., 1967: Herpetological collections from the Atkinson-Maduni-Laluni area, east Demerara, Guyana. Timehri, 43: 34-50.

Hoevers, L. G., and R. W. Henderson, 1974: Additions to the herpetofauna of Belize (British Honduras). Milwaukee Publ. Mus. Contrib. Biol. Geol., Milwaukee Publ. Mus., (2): 1-6.

Hoge, A. R., 1948 [dated 1947]: Notas erpetológicas. 3. Uma nova espécie de *Trimeresurus*. Mem. Inst. Butantan, 20: 193-202.

Hoge, A. R., 1952a: Notas erpetológicas, 1a. Contribuição ao conhecimento dos ofidios do Brasil central. Mem. Inst. Butantan. 24 (2): 179-214.

Hoge, A. R., 1952b: Notas erpetológicas, 2a. Contribuição ao conhecimento dos ofidios do Brasil central. Mem. Inst. Butantan. 24 (2): 215-227.

Hoge, A. R., 1952c: Notas erpetológicas revalidação de *Thamnodynastes strigatus* (Günther, 1858). Mem. Inst. Butantan, 24 (2): 146-154.

Hoge, A. R., 1952d: Snakes of the Uaupés Region. Mem. Inst. Butantan, 24 (2): 225-230.

Hoge, A. R., 1953a: A new *Bothrops* from Brasil. *Bothrops brazili*, sp. Nov. Mem. Inst. Butantan. 25 (1): 15-22.

Hoge, A. R., 1953b: Notes on *Lygophis* Fitzinger with revalidation of two subspecies. Mem. Inst. Butantan, 24: 245-268.

Hoge, A. R., 1953c: Notas erpetológicas. Revalidação de *Thamnodynastes strigatus* (Günther, 1858). Mem. Inst. Butantan, 24 (2): 157-172.

Hoge, A. R., 1954: Notas erpetologicas. Una nova subespecies de *Leimadophis*

Bibliography

reginae. Mem. Inst. Butantan, 24: 241-244.

Hoge, A. R., 1956 [dated 1955]: Uma nova espécie de *Micrurus* (Serp. Elap.) do Brasil. Mem. Inst. Butantan, 27: 67-72.

Hoge, A. R., 1958: Três notas sôbre serpentes brasileiras. I. Sôbre a posição genérica de *Coluber bicinctus* Hermann, 1804 e *Xenodon gigas* Duméril, 1853 (Colubridae). II. Sôbre a posição sistemática de *Enicognathus joberti* Sauvage, 1884 (Colubridae). III. Dimorfismo sexual em *Micrurus s. surinamensis* (Cuvier, 1817) (Elapidae). Pap. Avuls. Zool., São Paulo, 13: 221-224.

Hoge, A. R., 1959a: Êtude sur *Apostolepis coronata* (Sauvage 1877) et *Apostolepis quinquelineata* Boulenger 1896. Mem. Inst. Butantan, 28 (1957-1958): 73-76 + pl.

Hoge, A. R., 1959b: Êtude sur *Uromacerina ricardinii* (Peracca) (Serpentes). Mem. Inst. Butantan,. 28 (1957-1958): 83-84.

Hoge, A. R., 1964 [1960-1962]: Sur la position systemitique de quelques serpents du genre *Siphlophis* Fitzinger 1843. Mem. Inst. Butantan, 30: 35-50.

Hoge, A. R., 1964: Sur la position systematique de quelques serpents du generes *Siphlophis* Fitzinger 1843 (Serpentes). Mem. Inst. Butantan, 30: 35-50.

Hoge, A. R., 1965: Preliminary account on Neotropical crotalinae [Serpentes, Viperidae]. Mem. Inst. Butantan, 32: 109-184 + XX plates.

Hoge, A. R., 1966: 199. Notes on *Hydrodynastes* (Serpents-Colubridae). Ciência e Cultura, São Paulo, 18: 143.

Hoge, A. R., [1969] 1970: Notes on the holotype of *Dipsas indica cisticeps* (Boettger) (Serpentes-Dipsadinae). Mem. Inst. Butantan, 34 (1969): 87-88.

Hoge, A.R and H. E. Belluomini., [dated 1957/58] 1959: Uma nova espécie de *Bothrops* do Brasil (Serpentes). Mem. Inst. Butantan, 28: 195-206.

Hoge, A. R., Belluomini, H. and G. Schreiber, 1954: Intersexuality in a high isolated population of snakes. Atti Del IX Congresso Internazionale di Genetica, 1954: 964-965.

Hoge, A. R., C. L. Cordeiro, and S. A. R. W. D. L. Romano, 1976: A new species of *Micrurus* from Brazil (Serpentes Elapidae). Soc. Brasil Prog. Ci., 28: 417-418.

Hoge, A. R., and P. A. Federsoni, Jr., 1975: Notes on *Xenopholis* Peters and *Paraxoyrhopus* Schenkel (Serpentes: Colubridae). Mem. Inst. Butantan, 38 (1974): 137-146.

Hoge, A. R., and A. Garcia, 1949a: Notas erpetológicas 4. Sobre caracteres sexuais secundarios nas serpentes. Mem. Inst. Butantan, 21 (1948): 55-66.

Hoge, A. R., and A. Garcia, 1949b: Notas erpetológicas 5. Notas sobre *Elapomorphus nasutus* Gomes, 1915. Mem. Inst. Butantan, 21: 67-76.

Hoge, A. R., and A. R. Lancini, 1960: Notas sobre la ubicación de la tierra typical de varias especies de "serpentes" colectadas por M. Beauperthuis en la "Cote Ferme" y en la "Province de Venezuela." Bol. Soc. Venezolana Ci. Nat. Caracas, 6-7: 58-62.

Hoge, A. R., and A. R. Lancini, 1962: Sinopsis de las serpientes venenosas de Venezuela. Editora Grafa, C. A., Pub. Ocas. Mus. Ci. Nat., Zoología (1): 1-24, Caracas, 28 de Febrero de 1962.

Hoge, A. R., I. L. Laporta and S. A. Romano Hoge, [1978/1979] 1981: Notes on *Sibynomorphis mikanii* Schlegel, 1837. Mem. Inst. Butantan, 42/43: 175-178.

Hoge, A. R., and J. S. Lima-Verde, 1972: *Liophis mossoroensis* nov. sp. do Brasil (Serpentes: Colubridae). Mem. Inst. Butantan, 36 (1972): 215-220.

Hoge, A. R., and S. A. Romano, [dated 1965] 1966: *Leptomicrurus* in Brasil (Serpentes-Elapidae). Mem. Inst. Butantan, 32: 1-8.

Hoge, A. R., and S. A. Romano, [1969] 1970: A new species of *Chironius* (Serpentes-Colubridae). Mem. Inst. Butantan, 34: 93-96.

Hoge, A. R., and S. A. R. W. D. I. Romano, 1971: Neotropical pit vipers, sea snakes and coral snakes. pp. 211-293. *In* W. Bucherl and E. Buckley (eds.), Venomous animals and their venoms. Vol. 2: Venomous vertebrates. Academic Press, New York.

Hoge, A. R., and S. Romano-Hoge, [1972] 1973: Sinopse das serpentes peçonhentas do Brasil. Serpentes, Elapoidae e Viperidae. Mem. Inst. Butantan, 36 [1972]: 109-207.

Hoge, A. R., N. Pereira Santos, C. Heitor, L. A. Lopes and I. Menezos de Souza, [1972] 1973: Serpentes coletada pelo projeto Rondon VII em Iauareté, Brasil. Mem. Inst. Butantan, 36: 221-232.

Hoge, A. R., and S. A. L. Romano, [1975] 1976: A new subspecies of *Dipsas indica* from Brasil [Serpentes, Colubridae, Dipsidinae]. Mem. Inst. Butantan, 39: 51-60.

Hoge, A. R., and S. A. R. W. L. Romano-Hoge, [dated 1978-1979] 1981: Poisonous snakes of the world. Part I-Checklist of the pit vipers, Viperoidea, Viperidae, Crotalinae., Mem. Inst. Butantan, 42/43: 179-310.

Hoge, A. R., A. R. W. D. L. Romano, and C. L. Codeiro, [1976/1977] 1978: Contribuição ao conhecimento das serpentes do Maranhao, Brasil (Serpentes, Boidae, Colubridae e Viperidae). Mem. Inst. Butantan, 40/41: 37-52.

Hoge, A. R., C. R. Russo, M. C. Santos and M. F. D. Furtadio, [1978/1979] 1981: Snakes collected by "Porjecto Rondon XXII"—Piauí, Brazil. Mem. Inst. Butantan, 42/4: 87-94.

Holbrook, J. E., 1832-1836: North American herpetology. Dobson, Philadelphia, 3 volumes.

Holbrook, J., 1842: North American herpetology; or, a description of the reptiles inhabiting the United States. Volume I, II, III, IV and V. J. Dobson, Philadelphia; SSAR reprint 1976. Misc. Pub. SSAR Facsimile Reprints in Herpetology, 1976, five volumes in one.

Holland, R. L., 1977: Geographic distribution. *Masticophis taeniatus taeniatus*. Herpetol. Rev. 8: 13.

Holm, J. A., 1988: Pidgins and Creoles-Volume I-Theory and structure. Cambridge Univ. Press, Cambridge.

Holm, J. A., 1989: Pidgins and Creoles-Volume II-Reference survey. Cambridge Univ. Press, Cambridge, pp. 259-704.

Holm, P. A., and G. A. Cruz D., 1994: A new species of *Rhadinaea* (Colubridae) from a cloud forest in northern Honduras. Herpetologica, 50 (1): 15-23.

Holman, J. A., 2000: Fossil snakes of North America—origin, evolution, distribution, paleoecology. Indiana Univ. Press, Bloomington and Indianapolis.

Holman, J. A., and H. P. Arai, 1962: Illinois range extension of *Lygosoma laterale* (Say) and *Natrix kirtlandi* (Kennicott). Herpetologica, 18 (3): 210.

Hoogmoed, M. S., 1973: Notes on the herpetofauna of Surinam IV. Biogeographiea 4; Junk, The Hague.

Hoogmoed, M. S., 1977: On a new species of *Leptotyphlops* from Surinam, with notes on the other Surinam species of the genus (Leptotyphlopidae, Serpentes). Notes on the herpetofauna of Surinam V., Zoologische Mededelingen, Rijksmuseum Von Natuurlijke, Leiden, E. J. Brill, 51 (7): 99–123.

Hoogmoed, M. S., 1980: Revision of the genus *Atractus* in Surinam, with the resurrection of two species (Colubridae, Reptilia). Notes on the Herpetofauna of Surinam VII. (175): 1–47, Leiden, E. J. Brill, 14 maart 1980.

Hoogmoed, M. S., [dated 1982] 1983: Snakes of the Guiana region. Mem. Inst. Butantan, 46: 219–254.

Hoogmoed, M. S., 1997: Rediscovery of a forgotten snake in an unexpected place and remarks on a small herpetological collection from southeastern Brazil. Zoologische Mededelingen (Leiden), July, 1997, 71 (1–18): 63–81.

Hoogmoed, M. S., and U. Gruber, 1983: Spix and Wagler type specimens of reptiles and amphibians in the natural history museum in Munich (Germany) and Leiden (The Nederlands). Spixiana, Suppl. 9: 319–415.

Horn, H. S., 1969: Polymorphism and evolution of the Hispaniolan snake genus *Uromacer* (Colubridae). Breviora, (324): 1–23.

Howes, P. G., 1970: Photographer in the rain-forests. P. G. Howes and Assoc., Noroton, Conn., xx + 218 pp.

Hoyt, D. L., 1964: The rediscovery of the snake *Leptophis modestus*. Copeia, (1964): 214–215.

Hueck, K., and P. Seibert, 1972: Mapa de la vegetación de America del Sur. Gustav Fischer Verlag, Stuttgart, Germany.

Huheey, J. E., 1959: Distribution and variation in the glossy water snake, *Natrix rigida* (Say). Copeia, 1959: 303–311.

Humbolt, A. von, 1811: Sur deux nouvelles sepecies de crotales. Rec. D'obs. Zool. Anat. Comp., II: 1–8.

Hummelinck, P. W., 1940: Studies of the fauna of Curaçao, Aruba, Bonaire and the Venezuelan Islands. J. The Hague, (3), 1: 1–130.

Hunziker, R., 2000: Dry moccasins-Cantils. Reptile & Amphibian Hobbyist, April, 2000, 5 (8): 29–35.

Hurter, J., 1893: Catalogue of reptiles and batrachians found in the vicinity of St. Louis, Mo. Trans. Acad. Sci. St. Louis, 6 (11): 251–261.

Hurtado, J. L., and D. Blanco, 1994 [1995]: Nuevo registro de ofidios del bosque nublado del Valle de Q'osnipata, Cusco. Bol. Lima, 16 (91–96): 49–52.

Ihering, R. von, [dated 1910] 1911: As cobras do Brazil. Priveira parte. Rev. Mus. Paulista, São Paulo, 8: 273–379.

Inger, R. F., and P. J. Clark, 1943: Partition of the genus *Coluber*. Copeia, 1943: 141–145. International Code of Zoological Nomenclature, 1985: Univ. California Press, Berkeley, California.

International Commission on Zoological Nomenclature, 1999: Opinion 1939: *Trigonocephalus pulcher* Peters, 1862 (currently *Bothrops pulcher*, *Bothriechis pulcher* or *Bothrops pulcher*; Reptilia, Serpentes): defined by the holotype, and not a neotype; *Bothrops campbelli* Freire Lascano, 1991: specific name placed on the official list. Bull. Zool. Nomenclature, 56: 218–220.

International Commission on Zoological Nomenclature, 2000: Opinion 1960: *Crotalus ruber* Cope, 1892 (Reptilia, Serpentes): specific name given precedence over that of *Crotalus exsul* Garman, 1884. Bull. Zool. Nomenclature, 57 (3): 189–190.

International Commission on Zoological Nomenclature, 2000: Opinion 1961: *Coluber infernalis* Blainville, 1835 and *Eutaenia sirtalis tetrataenia* Cope in Yarrow, 1875 (currently *Thamnophis sirtalis infernalis* and *T. s. tetrataenia*; Reptilia, Serpentes): subspecific names conserved by the designation of a neotype for *T. s. infernalis*. Bull. Zool. Nomenclature, 57 (3): 191–192.

Jacobson, E. R., and J. M. Gaskin, 1992: Paramyxovirus infection of viperid snakes, pp. 415–419 in J. A. Campbell and E. D. Brodie, Jr.: Biology of the pitvipers. Selva, Tyler, Texas.

Jaffe, M., 1996: Brown Tree Snake stirs hiss-teria. Houston Chronicle, Section B: 6, Monday, April, 23, 1996.

James, E., 1822–1823: Account of an expedition from Pittsburgh to the Rocky Mountains. 2 volumes. (Expedition under Major Stephen H. Long). H. C. Cavey and I. Lee, Philadelphia

Jameson, D. L., and A. G. Flury, 1949: The reptiles and amphibians of the Sierra Vieja Range of Southwestern Texas. Texas J. Sci., 1 (2): 54–77.

Jan, G., 1858a: Plan d'une iconographie descriptive des ophidiens. Rev. Mag. Zool., 11: 148–157.

Jan, G., 1858b: Plan d'une iconographie descriptive des ophidiens et description sommaire de nouvelles espèces de serpents. Rev. Mag. Zool., ser. 2, 10: 438–449, 514–527.

Jan, G., 1859a: Additions et rectifications aux Plan et Prodrome de l'iconograpie descriptive des ophidiens. Rev. Mag. Zool. Ser. 2: 505–512.

Jan, G., 1859b: Plan d'une iconographie descriptive des ophidiens et description sommaire de nouvelles espèces de serpents. Rev. Mag. Zool., Ser. 2, 10: 122–130, 148–157.

Jan, G., 1859c: Spix's Serpentes brasilienses Beurtheilt nach Autopsie der original Exemplare und auf die Nomenclatur von Duméril und Bibron zurückgeführt. Arch. Naturg., 25: 272–275.

Jan, G., 1862a: Enumerazione sistematica delle specie d'ofidi del gruppo Calamaridae. Arch. Zool., l'Anat. Fis., 2: 1–76.

Jan, G., 1862b: Prodromus de Iconographie générale. II. Parte. Calamaridae. Arch. Zool. Anat. Phys. 2 (1): 1–76

Jan, G., 1863a: Elenco sistematico degli ofidi descritti e disegnati per l'iconographie générale. Milano, Lombardi I-VIII.

Jan, G., 1863b: Enumerazione sistematico degli ofidi appartenenti al gruppo Coronellidae. Arch. Zool., l'Anatomia e la Fisiologia, 2 (2): 211–330.

Jan, G., 1863c: Prodromus de iconographie générale. II. Parte. Coronellidae. Arch. Zool. Anat. Phys. 2 (2): 1–120.

Jan, G., 1864a: Iconographie genérale des ophidiens. Milan, Vol. 1, livr. 12.

Jan, G., 1864b: Iconographie genérale des ophidiens. Première famille. Les Typhlopiens. Beillièr, Paris.

Jan, G., 1865a: Enumerazione sistematica degli ofidi appartenenti ai gruppo potamohilidae. Arch. Zool. Anat. Fis., 3 (2): 201–265.

Jan, G., 1865b: Prodromus de iconographie générale. II. Parte. Potamophilidae. Arch. Zool. Anat. Phys. 3 (2): 1–50.

Jan, G., 1872: Iconographie générale des

ophidiens. *In* Jan and F. Sordelli, 1860–1861. 3 vols. 50 livr., 300 pls. Milan:chez les auteurs. Londen: Bailliere Tindal and Cox. Paris.

Jan, G., and F. Sordelli, 1860–1866: Iconographie générale des ophidiens. 3 vols., Milan. covig.

Jan, G., And F. Sordelli, 1867: Iconographie générale des ophidiens. Livr. 24, Paris, Bailliere, 6 pls.

Jan, G., and F. Sordelli, 1860–1882: Iconographie générale des ophidiens. 3 vols., Livraisons 1–50, J. B. Bailliáre et Fils, Paris: *In* John. Cramer-Weinhaim reprint, 1961, Wheldon & Wesley, Ltd. and Hafner Publishing Co., New York.

Jenner, J. V., and H. G. Dowling, 1985: Taxonomy of American xenodontine snakes: the tribe Pseudoboini. Herpetologica, 41: 161–172.

Jennings, M. A., 1983: *Masticophis lateralis*. Cat. Amer. Amph. Rept., (343): 1–2.

Jiménez, A., and J. M. Savage, 1962: A new blind snake (genus *Typhlops*) from Costa Rica. Rev. Biol. Trop. 10 (2): 199–203.

Johnson, J. D., [1973] 1974: New records of reptiles and amphibians from Chiapas, México. Trans. Kansas Acad. Sci., 76: 223–225.

Johnson, J. D., 1977: The taxonomy and distribution of the Neotropical whipsnake *Masticophis mentovarius* (Reptilia, Serpentes, Colubridae), J. Herpetol., 11 (3): 287–309.

Johnson, J. D., 1978: First record of *Manolepis putnami* (Serpentes: Colubridae) from Chiapas, México. Southwest. Naturalist, 23: 538.

Johnson, J. D., 1979: Taxonomic status and distribution of *Geophis laticinctus* (Colubridae) in southern México. Southwest. Naturalist, 24: 698–701.

Johnson, J. D., 1982: *Masticophis mentovarius*. Cat. Amer. Amph. Rept. (295): 1–4.

Johnson, J. D., 1989: A biogeographic analysis of the herpetofauna of northwestern nuclear Central America. Milwaukee Publ. Mus. Contrib. Geol. Biol., 76: 1–66.

Johnson, J. D., 1990: Biogeographic aspects of the herpetofauna of the Central Depression of Chiapas, Mexico, with comments on surrounding areas. Southwest. Naturalist, 35 (3): 268–278.

Johnson, J. D., and J. R. Dixon, 1984: Taxonomic status of the Venezuelan Macagua, *Bothrops colombiensis*. J. Herpetol., 18: 329–332.

Jones, C., and J. S. Findley, 1963: Second record of the lyre snake *Trimorphodon vilkinsoni* in New Mexico. Southwest. Naturalist, 8: 175–177.

Jones, J., 1856: Investigations chemical and physiological relative to certain American vertebrates. Smithsonian Contr., 8: 1–137.

Jordan, D. S., 1876: Manual of the vertebrates of the northern U.S. Chicago. 1876: 172–184. Later eds.: 1878: 172–184; 1894: 188–200

Jordan, D. S., and B. N. Van Vleck, 1874: A popular key to the birds, reptiles, batrachians, and fishes of the northern United States east of the Mississippi River. Appleton, Wisconsin, 1874: 35–42.

Jones, S., and A. Ruiz Salvador, 1974: Spanish one. D. Van Nostrand Company, New York.

Jorge da Silva, N. Jr., and Jack W. Sites, Jr.: Revision of the *Micrurus frontalis* complex (Serpentes: Elapidae). Herpetol. Monog., 1999, (13): 142–194.

Jover Peralta, A., and T. Osuna, 1952: Diccionario, Guaraní-Español; Español-Guaraní. Editorial Tupã, Buenos Aires, Argentina.

Juliá-Zertuche, J., 1983: Una nueva subespecie de víbora de cascabel pigmea de la Sierra de Sinaloa. Resumenes del Sexto Cong. Nac. Zool., Mazatlán, Sinaloa, México: (1 page without pagnation).

Juliá-Zertuche, J., and C. H. Treviño, 1978: Una nueva subespecie de *Crotalus lepidus* encontrada en Nuevo León, Resumenes del Segundo Cong. Nac. Zool., Monterrey, Nuevo León, México. 1978: 60.

Juliá-Zertuche, J., and M. Varela-J., 1978: Una *Bothrops* de México, nueva para la ciéncia. Mem. Primer Congreso Nacional Zool., Escuela Nac. Agric. (Univ. Autón. Chapingo), Chapingo, Mexico 1977: 209–210.

Jusayú, M. A., 1977: Diccionario de la lengua Guajira: Guajiro-Castellano. Universidad Catolica Andres Bello, Centro de Lenguas Indigenas, Caracas.

Karkabi, B., 2000: Gullah gems. Houston Chronicle, Lifestyle & Entertainment, Section D, Thursday, June 29, 2000, 1D and 4D.

Karol, J. S. (Ed.), and S. L. Rozman (Asst. Ed.), 1974: Everyday Lakota, an English-Sioux dictionary for beginners. Rosebud Educational Society, April, 1974 revision.

Kauffeld, C. F., 1953: Snakes have personality. Animaland, Staten Island Zool. Soc., 20 (5): 2–4.

Kauffeld, C. F., 1958: The reptile department in 1958. Animaland, Staten Island Zool. Soc., 24 (2): 5–6.

Keifer, M. C., 1998: Geographic distribution: *Pseudablabes agassizii*. Herpetol. Rev., 29 (1): 54.

Keiser, E. D., 1974: A systematic study of the Neotropical vine snake, *Oxybelis aeneus* (Wagler). Texas Memorial Mus. Bull. (22): 1–51.

Keiser, E. D., 1982: *Oxybelis aeneus* (Wagler)-Neotropical Vine Snake. Cat. Amer. Amph. Rept., 305: 1–4.

Keiser, E. D. Jr., and L. D. Wilson, 1979: Checklist & key to the amphibians and reptiles of Louisiana. 2nd Ed., Bull. I, Lafayette Nat. Hist. Mus., Lafayette, Louisiana.

Kempff-Mercado, N., 1975: Ofidios de Bolivia. Acad. Nac. Cien. Bolivia, La Paz-Bolivia. [Special note: the copy used in this work was given to the author by the Kempff-Mercado family. Some of the notations are made from the handwritten corrections that were intended for future publication to update information from the 1975 publication.]

Kennicott, R., 1859: Notes on *Coluber calligaster* of Say, and a description of new species of serpents in the collection of the North Western University of Evanston, Ill., Proc. Acad. Nat. Sci, Philadelphia, 77 (7–8): 98–100.

Kennicott, R., 1860a: Descriptions of new species of North American serpents in the Museum of the Smithsonian Institution, Washington. Proc. Acad. Nat. Sci. Philadelphia, 12: 328–338.

Kennicott, R., 1860b: Report upon the reptiles collected on the survey: pp. 292–306 + 10 plates. *In* Cooper, Reports of explorations and surveys, to ascertain the most practicable and economical route for a railroad from the Mississippi River to the Pacific Ocean. Vol. 12, Book 2, Pt. 3, (4): 292–306.

Kennicott, R., 1861: On three new forms of rattlesnakes. Proc. Acad. Nat. Sci. Philadelphia, 12: 206–208.

Kerr, R., 1802: The natural history of oviparous quadrupeds and serpents. 4 vols., Weimar, 1802.

Kiemele Muro, M., 1975: Vocabulario Mazahua-Español y Español-Mazahua. Biblioteca Enciclopédia del Estado de México, México.

Kirn, A. J., W. L. Burger and H. M. Smith, 1949: The subspecies of *Tantilla gracilis*. Amer. Midl. Naturalist, 42: 238–251.

Kirn, A. J., W. L. Burger and H. M. Smith, 1949: The subspecies of *Tantilla gracilis*. Amer. Midl. Naturalist, 42 (1): 238–251.

Kimball, G. D., 1994: Koasati dictionary. Univ. Nebraska, Lincoln and London, 406 pp.

Kimele Muro, M. 1975: Vocabulario Mazahua-Español y Español-Mazahua. Biblioteca Enciclopédica del Estado de México., México.

Kirk, J. J., 1979: *Thamnophis ordinoides* (Baird and Girard)—Northwestern Garter Snake. Cat. Amer. Amph. Rept., (233): 1–2.

Kirtland, J. P., 1838: Report on the zoölogy of Ohio. 2nd Ann. Report. Geol. Survey Ohio. 1838: 157–200.

Klauber, L. M., 1928a: The collection of snake venom. Bull. Antivenin Inst. Amer., 2: 11–18.

Klauber, L. M., 1928b: The *Trimorphodon* (Lyre Snake) of California, with notes on the species of the adjacent areas. Trans. San Diego Soc. Nat. Hist., 5 (11): 183–194.

Klauber, L. M., 1930a: Differential characteristics of southwestern rattlesnakes allied to *Crotalus atrox*. Bull. Zoöl. Soc. San Diego, (6): 1–70.

Klauber, L. M., 1930b: New and renamed subspecies of *Crotalus confluentus*, Say, with remarks on related species. Trans. San Diego Soc. Nat. Hist., 6 (3): 95–144.

Klauber, L. M., 1931a: A new subspecies of the California Boa, with notes on the genus *Lichanura*. Trans. San Diego Soc. Nat. Hist., 6 (20): 305–318.

Klauber, L. M., 1931b: Notes on the worm snakes of the southwest, with descriptions of two new subspecies. Trans. San Diego Soc. Nat. Hist., 6 (23): 333–352.

Klauber, L. M., 1932: A herpetological review of the Hopi dance. Bull. Zool. Soc. San Diego, (9): 1–92.

Klauber, L. M., 1934: Annotated list of the amphibians and reptiles of the southern border of California. Bull. Zoöl. Soc. San Diego, (11): 1–28.

Klauber, L. M., 1935a: Notes on herpetological field collecting. San Diego Soc. Nat. Hist. Coll. Leaflet. 3rd Rev., (1): 1–10.

Klauber, L. M., 1935b: *Phyllorhynchus* the Leaf-nosed Snake. Bull. Zool. Soc. San Diego, (12): 1–31.

Klauber, L. M., 1936a: A key to the rattlesnakes. Trans. San Diego Soc. Nat. Hist., 8: 185–276.

Klauber, L. M., 1936b: *Crotalus mitchelli* the Speckled Rattlesnake. Trans. San Diego Soc. Nat. Hist., 8: 149–184.

Klauber, L. M., 1938: Notes from a herpetological diary, I. Copeia, (1938): 191–197.

Klauber, L. M., 1939a: A new subspecies of the western worm snake. Trans. San Diego Soc. Nat. Hist., 9: 67–68.

Klauber, L. M., 1939b: Three new worm snakes of the genus *Leptotyphlops*. Trans. San Diego Soc. Nat. Hist., 9 (14): 59–66.

Klauber, L. M., 1940a: Notes from an herpetological diary, II. Copeia, (1940): 15–18.

Klauber, L. M., 1940b: The lyre snakes (genus *Trimorphodon*) of the United States. Trans. San Diego Soc. Nat. Hist., 9: 163–194.

Klauber, L. M., 1940c: The worm snakes of the genus *Leptotyphlops* in the United States and Northern Mexico. Trans. San Diego Soc. Nat. Hist., IX (8): 87–162.

Klauber, L. M., 1941a: A new species of rattlesnake from Venezuela. Trans. San Diego Soc. Nat. Hist., 9 (30): 333–336.

Klauber, L. M., 1941b: The long-nosed snakes of the genus *Rhinocheilus*. Trans. San Diego Soc. Nat. Hist., 9: 289–332.

Klauber, L. M., 1941c: The rattlesnakes listed by Linnaeus in 1758. Bull. Zoöl. Soc. San Diego, (7): 1941: 81–95.

Klauber, L. M., 1942a: A new snake of the genus *Sonora* from lower California, Mexico. Trans. San Diego Nat. Hist., 10 (4): 69–70.

Klauber, L. M., 1942b: The state of the black whip snake. Copeia, 1942 (2): 88–97.

Klauber, L. M., 1943a: The coral king snakes of the Pacific Coast. Trans. San Diego Soc. Nat. Hist., 10 (6): 75–82.

Klauber, L. M., 1943b: The subspecies of the rubber snake, *Charina*. Trans. San Diego Soc. Nat. Hist., 10 (7): 83–90.

Klauber, L. M., 1946a: A new gopher snake (*Pituophis*) from Santa Cruz Island, California. Trans. San Diego Soc. Nat. Hist., 11 (2): 41–46.

Klauber, L. M., 1946b: The gopher snakes of Baja California, with description of a new species of *Pituophis catenifer*. Trans. San Diego Soc. Nat. Hist., 11 (1): 1–40.

Klauber, L. M., 1946c: The Glossy Snake, *Arizona*, with descriptions of new subspecies. Trans. San Diego Soc. Nat. Hist., 10: 311–398.

Klauber, L. M., 1947: Classification and ranges of the gopher snakes of the genus *Pituophis* in the western United States. Bull. San Diego Zool. Soc., 22:1–81.

Klauber, L. M., 1948: Some misapplications of the Linnaean names applied to American snakes. Copeia, 1948: 1–14.

Klauber, L. M., 1949: Some new and revived subspecies of rattlesnakes. Trans. San Diego Soc. Nat. Hist., 11: 61–116.

Klauber, L. M., 1951: The Shovel-nosed Snake, *Chionactis* with descriptions of two new species. Trans. San Diego Soc. Nat. Hist., 11 (9): 141–204.

Klauber, L. M., 1956: Rattlesnakes, their habits, life histories, & influence on mankind. Volumes 1 & 2, Univ. California Press, xxx + 1–708; xviii + 709–1476.

Klauber, L. M., 1963: A new insular subspecies of the speckled rattlesnake. Trans. San Diego Soc. Nat. Hist., 13: 73–80.

Klauber, L. M., 1972: Rattlesnakes, their habits, life histories, & influence on mankind. Volumes 1 & 2, 2nd Edition, Berkeley and Los Angeles, Univ. California Press, 1–xxx; 1–1533 pp.

Klauber, L. M., 1997: Rattlesnakes, their habits, life histories, & influence on mankind. Volumes 1 & 2, 2nd Edition with a new forward by H. W. Greene, Berkeley, Los Angeles, and London, Univ. California Press, 1–xlvi; 1–1533 pp.

Kliment, P., 1993: Ein Beitrag zur Biologie der Kubanischen Natter *Alsophis cantherigerus* (Bibron, 1840). Sauria, 15 (4): 27–31.

Kluge, A. G., 1984: Type-specimens of reptiles in the University of Michigan Museum of Zoology. Misc. Pub. Mus. Zool., Univ. Michigan, (167): i–ii, 1–85.

Bibliography

Kluge, A. G., 1991: Boine Snake phylogeny and research cycles. Misc. Pub. Mus. Zool. Univ. Michigan, 178: 1–58

Kluge, A. G., 1993: *Calabarina* and the phylogeny of erycine snakes. 3001, Zool. J. Linnaean Soc., 107: 293–351.

Knight, A., L. D. Densmore III and E. D. Rael, 1992: Molecular systematics of the *Agkistrodon* complex. pp. 49–69 *in* J. A. Campbell and E. D. Brodie, Jr.: Biology of the pitvipers. Selva, Tyler, Texas.

Knight, J. L., 1988: Notes on *Rhadinaea fulviceps* Cope (Serpentes: Colubridae) from Ecuador. J. Herpetol., 22 (3): 344–345.

Kock, E. D., and C. R. Peterson, 1995: Amphibians and reptiles of Yellowstone and Grand Teton National Parks. Univ. Utah Press, Salt Lake City, Utah.

Kofron, C. P., 1980: A revision of the Central America slug-eating snakes. *In* The *Tropidodipsas* complex of the genus *Sibon* (Serpentes, Colubrinae). Dissertation submitted to the Graduation College of Texas A & M University, August 1980.

Kofron, C. P., 1982: A review of the Mexican snail-eating snakes, *Dipsas brevifacies* and *Dipsas gaigae*. J. Herpetol., 16: 270–286.

Kofron, C. P., 1985a: A new snake of the genus *Tropidodipsas* from Mexico. J. Washington Acad. Sci., 33: 371–373.

Kofron, C. P., 1985b: Review of Central American colubrid snakes, *Sibon fischeri* & *S. carri*. Copeia, 1985 (1):164–174.

Kofron, C. P., 1985c: Systematics of Neotropical gastropod-eating snake genera *Tropidodipsas* and *Sibon*. J. Herpetol., 19: 84–92.

Kofron, C. P., 1987: Systematics of Neotropical gastropod-eating snakes: The *fasciata* group of the genus *Sibon*. J. Herpetol., 21 (3): 210–225.

Kofron, C. P., 1988a: Systematics of Neotropical gastropod-eating Snakes: the *sartorii* group of the genus *Sibon*. Amphibia-Reptilia, 9: 145–168.

Kofron, C. P., 1988b: The Central and South American blindsnakes of the genus *Anomalepis*. Amphibia-Reptilia, 9 (1): 7–13.

Kofron, C. P., 1989: Systematics of Neotropical gastropod-eating snakes the *dimidata* group of the genus *Sibon*, with comments on the *nebulata* group. Amphibia-Reptilia, 11: 207–223.

Köhler, G, 1997: Geographic distribution —*Ungaliophis continentalis* (Isthmian Dwarf Boa). Herpetol. Rev., 28 (4): 211.

Köhler, G., 1999: Geographic distribution —*Tantillita lintoni* (Linton's Dwarf Short-tail Snake). Herpetol. Rev., 30 (1): 55.

Köhler, G., 2001: Reptilien und Amphibien Mittelamerikas. Band 2: Schlangen-Doppelschleichen. Offenbach.

Köhler, G., and J. R. McCranie, 1999: A new species of colubrid snake of the *Rhadinaea godmani* group from Cerro Saslaya, Nicaragua (Reptilia, Serpentes, Colubridae). Senckenbergiana biol., 79 (2).

Köhler, G., and J. R. McCranie, 1999: Taxonomic status of *Rhadinaea pinicola* Mertens (serpentes: colubridae). Copeia, 1999 (2): 529–530.

Kornacker, P.M, 1999: Checklist and key to the snakes of Venezuela—Lista sistemática y clave para las serpientes de Venezuela. 1. Edition—Rheinbach, Germany: Pako—Verlag, 270 pp.

Koslowsky, J., 1895: Batracios y reptiles de Rioja y Catamarca (Republica Argentina) recogidos durante los meses de Febrero a Mayo de 1895 (expedición del Director del Museo). Rev. Mus. La Plata, 6: 357–370.

Koslowsky, J., 1896: Sobre algunos reptiles de Patagonia y otras regiones argentinas. Rev. Mus. La Plata, 7: 445–454.

Koslowsky, J., 1898a: Enumeración sistemática y distribución geográfica de los reptiles argentinos. Rev. Mus. La Plata, 8: 161–200.

Koslowsky, J., 1898b: Ofidios de Mato-Grosso (Brasil). Rev. Mus. La Plata, 8: 3–32.

Kretzschmar, S., 1996: New data on *Leptotyphlops unguirostris* (Boulenger, 1902) (Serpentes: Leptotyphlopidae) [in Spanish]. Acta Zool. Lillana, 43 (2): 275–279.

Kretzschmar, S., 1998: *Liotyphlops ternetzii*. Herpetol. Rev., 29 (2): 114.

Kropach, C., 1975: The yellow-bellied sea snake, *Pelamis*, in the eastern Pacific. pp185–213. *In* Dunson, W., ed.: The Biology of Sea Snakes. Univ. Park Press, Baltimore.

Krywicki, Jarad, 2001: Sunlight in the rain forest. Reptile Magazine, March, 2001, 9 (3): 10–12, 14, 16, 18, 22, 24–27.

Kundert, F., 1984: Das Neue Schlangenbuch in Farbe. 2nd Edition, Rüschlikon-Zurich, Albert Müller Verlag AG, 39 pp.

La Flesche, F., 1932: A dictionary of the Osage language. Smithsonian Institute Bureau of American Ethnolgy, Bulletin 109, Republished 1976 by Scholarly Press, Inc. St. Clair Shores, Michigan, (109); 1–406.

Lacépède, B. G. E., 1788: Histoire naturelle des quadrupèdes ovipares et des serpens. Paris, 2 vol. (Vol.1: 1–651).

Lacépède, B. G. E., 1789: Historie naturelle des serpens. Vol. 2. Paris.

LaDuc, T. J., 1995: The nomenclatural status and gender of *Adelphicos*. J. Herpetol., 29 (1): 141–142.

LaDuc, T. J., 1996: A taxonomic revision of the *Adelphicos quadrivirgatum* species group (Serpentes: Colubridae). Masters of Science Thesis, University of Texas at El Paso, May 1996.

LaGrone, G. C., 1974: Basic conversational Spanish. Holt, Rinehart and Winston, New York.

Lamar, W. W., 1997: Rustlers. An except from the field journals of William W. Lamar. Fauna, 1 (1): 34–39.

Lamar, W. W., with photographs by Pete Carmichael and Gail Shumway, 1997a: The world's most spectacular reptiles & amphibians. World Publications, Tampa, Florida.

Lamar, W. W., 1997b: Notes and comments-checklist and common names of the reptiles of the Peruvian Lower Amazon. Herpetol. Nat. His., 5 (12): 73–76.

La Marca, E., 1997: Lista actualizada de los reptiles de Venezuela. *In* La Marca, E. (Ed.), 1997: Vertebrados de Venezuela. Mus. Cien. Tecno. Mérida, Venezuela, 123–142.

La Marca, E., and P. J. Soriano, 1995: *Epicrates cenchria cenchria*. Herpetol. Rev., 26 (2): 109.

La Marca, E., P. J. Soriano and R. Casado B., 1995: Geographic distribution—*Drymarchon corais melanurus* (Palancacoate). Herpetol. Rev., 26 (2): 109.

Lancini, A. R., 1962a: Un cambio de nombre para una serpiente coral (Elapidae: *Micrurus*) del Peru. Pub. Ocas. Mus. Cien. Nat., Caracas, Zool., 3:1.

Lancini, A. R., 1962b: Una nueva especie de serpiente coral (Serpentes: Elapidea) del Peru. Pub. Ocas. Mus. Cien. Nat., Caracas, Zool., 2: 1-3.

Lancini, R., 1964: Contribución al conocimiento de la distribución en Venezuela de *Helicops scalaris* y descripción de *Helicops hogei* (Serpentes: Colubridae), nueva especie de serpiente semiacuática para la Ciencia. Pub. Ocas. Mus. Cien. Nat. Caracas, Zool. 7: 1-3.

Lancini, R., 1969: *Atractus mariselae*, una nueva especie de serpiente minadora de los Andes de Venezuela. Pub. Ocas. Mus. Cien. Nat. Caracas Zool., 15: 1-6.

Lancini, R., A. R., 1979: Serpientes de Venezuela. Caracas, Ernesto Armitano (Editor), 262 pp.

Lancini V., A. R. and P. M. Kornacker, 1989: Die Schlangen Von Venezuela. Caracas, Venezuela, Verlag Armitano Editores C. A.

Lando, R. V., and E. E. Williams, 1969: Notes on the herpetology of the U.S. Naval Base at Guantanamo Bay, Cuba. Stud. Fauna Curacao and Caribbean Isl., 31 (116): 159-201.

Landy, M. J., D. A. Langebartel, E. O. Moll and H. M. Smith, 1966: A collection of snakes from Volcán Tacaná, Chiápas, Mexico. J. Ohio Herpetol. Soc., 5 (3): 93-101.

Langebartel, D. A., and H. M. Smith, 1954: Summary of the Norris Collection of reptiles and amphibians from Sonora, Mexico. Herpetologica, 10 (2): 125-136.

Langhammer, J. K., 1983: A new subspecies of boa constrictor, *Boa constrictor melanogaster*, from Ecuador. Tropical Fish Hobbyist, 32 (4): 70-79.

Latreille, P. A., 1802: L'Amérique septentrionale. Histoire Naturelle des Reptiles. Paris.

Laurent, R., 1949: Notes sur quelques reptiles appartenant a la collection de l'Institut Royal des Sciences Naturelles de Belgique. Bull. Inst. Roy. Sci. Nat. Belgique, 25 (9): 1-20.

Laurent, R. F., 1973: A parallel survey of equatorial amphibians and reptiles in Africa and South America, B. J. Meggers, E. S. Ayensu & W. P. Duckworth (eds.), Tropical Forest Ecosystems in Africa and South America: A Comparative Review, Smithsonian Institute Press, Washington, 1973: 259-266.

Laurent, R. F., 1984: El genero *Leptotyphlops* en la colección de la Funcación Migule Lillo. Acta Zool. Lilloana, 38 (1): 29-34.

Laurent, R. F., and E. M. Teran, 1981: Lista de anfibios y reptiles de la Provincia de Tucumán. Fund. Miguel Lilloana Misc., 71: 1-15.

Laurenti, J. N., 1768: Specimen medicum exhibens synopsin reptillum emendatum cum experimentis circa venena et antidota reptillium austriacorum, quod authorite et consensu. Vienna, Joan. Thomae.

Lawler, H. E., and T. R. Van Defender, 1996: Natural History Notes—*Sympholis lippiens* (Banded Burrowing Snake). Herpetol. Rev., Herpetological Review, December 1996, 27 (4): 205.

Lawson, R., 1987: Molecular studies of Thamnophiine snakes: 1. The phylogeny of the genus *Nerodia*. J. Herpetol., 21: 140-157.

Lawson, R., and C. S. Lieb, 1990: Variation and hydridization in *Elaphe bairdi* (Serpentes: Colubridae). J. Herpetol., 24 (3): 280-292.

Laycock, G., 1966: The alien animals. The American Museum of Natural History, The Natural History Press, Garden City, New York.

Lazcano Villarreal, D. and R. D. Jacobo Galván, 1999: Capítulo V. Fauna. Sección anfibios y reptiles. pp 47-51. In L. J. Galán Wong, J. A. García Salas, M. de J. Flores Hinojosa, U. de la Garza Hinojosa, and H. A. Luna Olivera (eds.), 1999, Santa Catarina hacia el Siglo XXI-pasado, presente y futuro. Univ. Autón. Nuevo León, Monterrey, México.

Lazell, J. D. Jr., 1964: The Lesser Antillean representatives of *Bothrops* and *Constrictor*. Bull. Mus. Comp. Zool., 132 (3): 247-273.

Lazell, J. D. Jr., 1967: Wiederentdeckung von zwei angeblich ausgestorbenen Schlangenarten der westindischen Inseln. Salamandra, 3: 91-97.

Lazell, J. D. Jr., 1976: This broken archipelago-Cape Cod and the Islands, amphibians and reptiles. Demeter Press, New York.

Lazell, J., 1992: Taxonomic tyranny and the exoteric. Herpetol. Rev., 23 (1): 14.

Lee, J. C., 1980: An ecogeographic analysis of the herpetofauna of the Yucatán Peninsula. Univ. Kansas Mus. Nat. Hist., Misc. Pub. (67): 1-75, Feb. 29, 1980.

Lee, J. C., 1990: Creatures of the Maya, Natural History. New York, 1/90 January 1990: 45-50.

Lee, J. C., 1996: The amphibians and reptiles of the Yucatán Peninsula. Comstock Publishing Associates, Ithaca and London.

Lee, J. C., 2000: A field guide to the amphibians and reptiles of the Maya world, the lowlands of Mexico, northern Guatemala and Belize. Comstock Publishing Association-a Division of Cornell Univ. Press, Ithaca.

Legler, J. M., 1959: A new snake of the genus *Geophis* from Chihuahua, Mexico. Univ. Kansas Pub. Mus. Nat. Hist., 11: 327-334.

Lehmann, H. D., 2001: Quo vadis, Boidenhaltung in Menschenhan. Draco, 2 (5): 20-25.

Leher, E., and C. Aguilar, 2000: *Rhinobothryum lentiginosum* (Amazon Banded Snake). Herpetol. Rev., 31 (1): 57.

Leher, E., and R. Fernandez, 2000: *Ninia hudsoni* (Amazon Coffee Snake). Herpetol. Rev., 31 (1): 56.

Lehman, H. D., 1970: Beobactungen bei der Haltung und Aufzucht von *Trachyboa boulengeri* (Serpentes, Boidae). Salamandra, 6 (1-2): 32-42.

Lema, T. de, 1958: Notas sôbre os répteis do estado Rio Grande do Sul-Brasil. Notas 1a IV. Iheringia, Zool., (10): 1-31.

Lema, T. de, 1961: Notas sôbre os répteis do estado Rio Grande do Sul-Brasil. Notas 1X a XI. Iheringia, Zool., (17): 1-19.

Lema, T. de, 1962: Ocorrência de *Philodryas arnaldoi* (Amaral, 1932) no estado do Rio Grande do Sul, Brasil (Serpents, Colubridae). Iheringia, Zool., (22): 1-4.

Lema, T. de, 1964: Uma nova espécie de serpente do gênero *Siphlophis* Fitzinger, 1843 do Brasil meriodional (Colubridae, Xenodontinae). Rev. Brasil Biol., Rio de Janeiro, 24 (2): 221-228.

Lema, T. de, 1967: Nôvo gênero e espécie de serpente, opistoglifo-donte do Brasil meridional ("Colubridae," "Colubrinae"). Iheringia, Zool., (35): 61-74.

Lema, T. de, 1970: Sôbre o status de *Elapomophus bilineatus* Duméril, Bibron & Duméril, 1854 curiosa serpente subterrânea. Iheringia, Zool., (38): 89-118.

Lema, T. de, 1971: Serpentes peçonhentas

do Rio Grande do Sul. Iherengia, Zool., 4 (1): 25-32.

Lema, T. de, 1973: Ocorrência de *Uromacerina ricardinii* (Peracca, 1897) no Rio Grande do Sul e contribuição ao conhecimento dessa rara serpente (Ophidia, colubridae). Iheringia, Zool., 8 (44): 64-73.

Lema, T. de, 1978a: Cobras não venenosas que matam. Natureza em Revista, junho, 1978, (4): 38-46.

Lema, T. de, 1978b: Invalidação de *Elapomorphus bollei* Mertens 1954 e o status de *Elapomorphus spegazzinii* Boulenger 1913 (Ophidia: Colubridae). Comun. Mus. Ci. PUCRGS, vol. 16/17: 1-16.

Lema, T. de, 1978c: Novas especies de opistoglifontes do genero *Apostolepis* Cope 1861 do Paraguai (Ophidia: Colubridae: Colubrinae). Comun. Mus. Ci. PUCRGS, (18/19): 27-49.

Lema., T. de, 1978d: O status de *Elapomorphus suspectus* Amaral 1924 (Ophidia: colubridae). Comun. Mus. Ci. PUCRGS, (16/17): 1-16.

Lema, T. de, 1978e: Ocorrêmcia de *Tantilla melanocephala* (L., 1758) no Rio Grande do Sul, Brasil, e o "status" de *Tantilla capistrata* Cope, 1876 (Ophidia: Colubridae). Comun. Mus. Ci. PUCRGS, (18/19): 1-25.

Lema, T. de, 1978f: Sobre o status de *Elapomorphus bilineatus* Duméril, Bibron & Duméril, 1854, curiosa serpente subterrânea. Iheringia. Sér. Zool., 38: 89-118.

Lema, T. de, 1979a: *Elapomorphus punctata*, nova espécie de colubridae para a Argentina (Ophidia). Rev. Brasil Biol., 39 (4): 835-853.

Lema, T. de, 1979b: Sobre a validade das nomes *Elapomorphus bilineatus* Duméril, Bibron & Duméril, 1854 e *E. lemniscatus* Duméril, Bibron & Duméril, 1854 (Ophidia: Colubridae). Iheringia, Zool., (54): 77-81.

Lema, T. de, 1982: Fauna de serpentes da província Pampeana e inter-relações com as províncias limítrofes. Mem. Inst. Butantan, 46: 173-182.

Lema, T. de, 1984a: Concentração de herpetofauna na regiao de Paraná médio. Resumes XI. Congr. Bras. Zool., (1984): 308-309.

Lema, T. de, 1984b: Relaçoes herpetofaunísticas de Rio Grande do Sul com os países vizinhos. Veritas. Porto Alegre, setiembre 1984, 29 (115): 421-429.

Lema, T. de., 1984c: Sobre o gênero *Elapomorphus* Wiegmann, 1843 (Serpentes, Colubridae, Elapomorphinae). Iheringia, Zool., (64): 53-86.

Lema, T. de, 1985: Aspectos biológicos de *Elapomorphus* (*Phalotris*) *lemniscata* Duméril, Bibron & Duméril, 1854 (Serpentes, Colubridae, Elapomorphinae). Iheringi, Zool., (65): 57-64.

Lema, T. de, 1987: Lista preliminar das serpentes registradas para o estado do Rio Grande do Sul (Brasil meridional) (Reptilia, Lepidosauria, Squamata). Acta Biol. Lepoldendisa, 9 (2): 225-240.

Lema, T. de, 1989a: A nomenclatura vulgar das espécies de serpentes ocorrentes no estado Rio Grande do Sul, Brasil, e a proposição de sua unificação (Reptilia, Serpentes). Acta Biol. Leopoldensia, 11 (1): 25-46.

Lema, T. de, 1989b: Notas sobre a biologia de duas espécies de *Elapomorphus* Wiegmann, 1843 (Serpentes, Colubridae, Elapomorphinae). Iheringi, Zool., (69): 61-69.

Lema, T. de, 1989c: Serpentes do complexo *Liophis lineatus* (Linnaeus, 1758) no Brasil nordeste (Serpentes, Colubridae, Colubrinae). Acta Biol. Leopoldensia, 11 (2): 251-271.

Lema, T. de, 1990: Considerações sobre a herpetofauna das terras baixas de clim a temperado do estado do RS. Veritas, Pôrto Alegre, março 1990, 35 (137): 99-107.

Lema, T. de, 1990 [1991]: Anomalias na escamção de duas espécies de *Elapomorphus* da América do sul meridional (Serpentes: Colubridae: Elaphmorphini). Acta Biol. Leopoldensia, ano 12 (2): 339-358.

Lema, T. de, 1992: Presença de *Elapomorphus quinquelineata* (Raddi) no extremo-sul do Brasil e a ocorrência de rara anomalia. (Serpentes, Colubridae, Xenodontinae, Elaphmorphini). Comm. Mus. Ci. PUCRS, Zool., 5 (1): 1-7.

Lema, T. de, 1993: Polimofismo em *Apostolepis dimidiata* (Jan, 1862) com a invalidaçao de *Apostolepis villaricae* Lema, 1978 e *Apostolepis barrioi* Lema, 1978 (Serpentes: Colubridae: Xenodontinae: Elapomorphini). Acta Biol. Leopoldensia, 15 (1): 35-52.

Lema, T. de, 1994a: Descrição de um exemplar cem bifurcação axial de *Echinanthera cyanopleura* (Cope, 1885) (Serpentes, Colubridae, Xenodontinae). Acta Biol. Lepoldensia, 16 (2): 113-117.

Lema, T. de, 1994b: Lista comentada dos répteis ocorrentes no Rio Grande do Sul, Brasil. Comun. Mus. Ci. PUCRS, Zool., 7: 41-150.

Lema, T. de, 1994c: Lista comentada dos répteis occorentes no Rio Grande do Sul, Brasil, Comun, Mus. Ci. PUCRS, Zool., 92 (3153): 349-395.

Lema, T. de, 1997a [1998]: A redescription of the tropical Brazilian serpent *Apostolepis nigrolineata* (Peters, 1869), (Colubridae: Elapomorphinae), synonymous with *A. pymi* Boulenger, 1903. Studies on Neotropical Fauna and Environment, Swets & Zeitlinger Publishers, The Nederlands, 32:193-199.

Lema, T. de, 1997b: Review of the snake genus *Apostolepis*. Abstracts of the 3rd World Congress on Herpetology, Prague, 1997: 51.

Lema, T. de, 1999: *Phalotris nasutus*. Herpetol. Rev., 30 (2): 175.

Lema, T. de, and M. E. Fabián-Beurmann, 1977: Levantamento preliminar dos répteis da região da fronteira Brasil-Uruguai. Iheringia, Zool., (50): 61-92.

Lema, T. de, M. E. Fabián-Beurmann, M. Leitão de Araujo, M. L. Machado Alves and M. Ibarra Vieira, 1980: Lista de répteis encontrados na região da Grande Porto Alegre, estado do Rio Grande do Sul, Brasil, Iheringia, Zool., (55): 27-36.

Lema, T. de, and R. Fernandes, 1997: The status of *Apostolepis sanctaeritae* Werner, 1924, and its revalidation (Serpentes: Colubridae: Elapomorphini). O status de *Apostolepis santaeritae* Werner, 1924 e sua revalidação (Serpentes: Colubridae: Elapomorphini). Acta Biol. Leopoldensia, 19 (1): 51-59.

Lema, T. de, and C. Hofstadler Deiques, 1992: Contrabuição ao conhecimento da "cobra espada d'agua", *Ptychophis flavovirgatus*, Gomes, (Serpentes, Colubridae, Xenodontinae, Tachymenini). Comun. Mus. Ci. PUCRS, Zool., 5 (6): 55-83.

Lema, T. de, and C. Hofstadler Deiques, 1995: Estudo revisivo de *Elapmorphus lepidus* Reinhardt com a invalidação de *E. wuchereri* Gunther, *E. accadens* Jan e *E. coronatus* Sauvage mediante análise tipológica e a ostelogia craniana (Serpentes, Colubridae, Xenodontinae, Elapamorphinae). Bio. Ciênc., Porto Alegre, jun. 1995, 3 (1): 91–143.-Errata of pages 136 and 137 with help from the Fundação de Amparo à Pesquisa do Estado do Rio Grande do Sul-FAPERGS, Porto Alegre (help from the author Lema, T., de-Processo no. 93.1140.7).

Lema, T. de, M. Ibarra Vieria and M. Leitão de Araújo, 1984: Fauna reptilana do norte da Grande Porto Alegre, Rio Grande do Sul, Brasil. Revta. Bras. Zool., São Paulo, 2 (4): 203–227, 31–xii.

Lema, T. de, and L. A. Matschulat Ely, 1979: Considerações sobre *Pseudoboa haasi* (Boettger, 1905) no extremo sul do Brasil (Ophidia: Colubridae). Iheringia, Sér. Zool., Porto Alegre, (54): 53–56.

Lema, T. de, and M. F. Renner, 1998: O status de *Apostolepis quinquelineata* Boulenger, 1896, *A. pymi* Boulenger, 1903, *E. rondoni* Amaral, 1925 (Serpentes, Colubridae, Elapomorphini). Biociências, Porto Alegre, 6 (1): 99–121.

Lemos Barbosa, A., 1967: Pequeño vocabulário Tupi-Português. Livraria São José, 204 pp.

Lemos-Espinal, J. A., H. M. Smith, R. E. Smith, R. Geoffrey, D. Chiszar, 1992: A herpetological collection from northern Chihuahua, Mexico. Bull. Chicago Herpetol. Soc. 32 (9): 198–201.

Leviton, A. E., 1953: Catalogue of the amphibian and reptile types in Natural History Museum of Stanford University. Herpetologica, 8 (4): 121–132.

Leviton, A. E., 1970: Reptiles and amphibians of North America. Animal Life Series of North America Series. Doubleday & Co., Inc., New York.

Leviton, A. E., and B. H. Banta, 1956: Catalogue of the amphibian and reptile types in the Natural History Museum of Stanford University, suppl.1, Herpetologica, 12 (3): 213–219.

Leviton, A. E., and B. H. Banta, 1964: Midwinter reconnaissance of the herpetofauna of the Cape region of Baja California, Mexico. Proc. California Acad. Sci, (Ser.4), 30: 127–156.

Leynaud, G. C., and E. H. Bucher, 1999: La fauna de serpientes del Chaco sudamericano: Diversidad, distribucion geografica y estado de conservacion. Acad. Nac. Cien., Córdoba, Argentina, Misc. (98):1–46.

Lichtenstein, H., 1823: Verseichnifs der doubletten des zoologischen museums der Konigl. Universitat zu Berlin nebst Beschriebubgvieler bisher unbekannten Arten von Saugethieren, Bogein, Amphibien und Fischen. T. Trautwein, Berlin.

Lichtenstein, M. H. C., 1856: Nomenclature reptilium et amphibiorum Museu Zoolgici Berolinensis. Amenverzeichniss der in der Zoologischen Sammmlung der Königlichen Universität zu Berlin aufgestellten Arten von Reptilien und Amphibien nach ihren Ordnungen, Familien und Gattungen. Berline, Köningl. Akad. Wiss.

Lieb, C. S., 1988: Systematic status of the Neotropical snake *Dendrophidion dendrophis* and *D. nuchalis* (Colubridae). Herpetologica, 44 (2): 162–174.

Lieb, C. S., 1991a: *Dendrophidion nuchale*. Cat. Amer. Amph. Rept., (520): 1–2.

Lieb, C. S., 1991b: *Dendrophidion paucicarinatum*. Cat. Amer. Amph. Rept., (521): 1–2.

Lieb, C. S., 1991c: *Dendrophidion vinitor*. Cat. Amer. Amph. Rept., (522): 1–2.

Lieb, C. S., 1996: Reptilia: Squamata: Serpentes: Colubridae: *Dendrophidion percarinatum*. Cat. Amer. Amph. Rept., (636): 1–2.

Lillywhite, H. B., and A. W. Smits, 1992: The cardiovascular adaptations of viperid snakes. pp. 143–153 in J. A. Campell and E. D. Brodie, Jr.: Biology of the pitvipers. Selva, Tyler, Texas.

Liner, E. A., 1960: A new subspecies of false coral snake (*Pliocercus elapoides*) from San Luis Potosí, Mexico. The Southwest. Naturalist, 5 (4): 217–220.

Liner, E. A., 1964: Notes on four small herpetological collections from Mexico. I. Introduction, turtles and snakes. The Southwest. Naturalist, 8 (4): 221–227.

Liner, E. A., 1983: *Tantilla wilcoxi*. Cat. Amer. Amph. Rept., (345): 1–2.

Liner, E. A., 1992: *Thamnophis exsul* Rossman—Montane Garter Snake, culebra de agua nomada de montaña. Cat. Amer. Amph. Rept., (549): 1–2.

Liner, E. A., 1994: Scientific & common names for the amphibians and reptiles of Mexico in English & Spanish. Nombres científicos y comunes en ingles y español de los anfibios y los reptiles de México. SSAR Herpetol. Circ. (23): 1–113, June 1994.

Liner, E. A., 1996a: Addenda to checklist of scientific and common names of Mexican amphibians and reptiles. Herpetol. Rev., 27 (3): 128–129.

Liner, E. A., 1996b: Herpetological type material for Nuevo León, México. Bull. Chicago Herpetol. Soc., 31 (9): 168–171.

Liner, E. A., 1996c: *Rhadinaea montana* Smith—Nuevo León Graceful Brown Snake-Hojarasquera de Nuevo León. Cat. Amer. Amph. Rept., (640): 1–2.

Liner, E. A., 1997: The herpetofauna of Terrebonne Parish, La. Bull. Chicago Herpetol. Soc., 32 (8): 169–172.

Liner, E. A., A. H. Chaney and R. W. Johnson, 1982: Geographic distribution: *Elaphe subocularis*. Herpetol. Rev., 13: 52–53.

Liner, E. A., and A. H. Chaney, 1987: Life history notes: *Rhadinaea montana* (Nuevo León Yellow-lipped Snake)—habitat. Herpetol. Rev., 18: 37.

Liner, E. A., and L. D. Wilson, 1970: Changes in the name and generic status of the Mexican snake *Chersodromus annulatus* Zweifel (Colubridae). Copeia, 1970: 786–788.

Linnaeus, C., 1758: Systema Naturae per regna tria naturae, secundum classes, ordines, genera, species cum characteribus, differentiis, synonymis, locis. Tenth ed. Vol.1, L. Salvivus, Stockholm.

Linnaeus, C., 1766: Systema naturae per regna tria naturae, secundum classes, ordines, genera, species cum characteribus, differentiis, synonymis, locis. Twelfth ed. Stockholm, L. Salvivus.

Lions, M. L., and B. B. Alvarez, 1996a: Geographic distribution—*Sibynomorphus lavillai* (La Villa's Treesnake). Herpetol. Rev., 27 (4): 214.

Lions, M. L., and B. B. Alvarez, 1996b: Geographic distribution—*Sibynomorphus ventrimaculatus* (Boulenger's Tree Snake). Herpetol. Rev., 27 (4): 214.

Lions, M. L., and B. B. Alvarez, 1997: Geographic Distribution—*Psomophis genimaculatus* (Spirit Ground Snake). Herpetol. Rev., 28 (1): 52.

Lips, K., and J. M. Savage, 1994: A new fossorial snake of the genus *Geophis* (Reptilia: Serpentes: Colubridae) from the Cordillera De Talamanca of Costa Rica. Proc. Biol. Soc. Washington, 107 (2): 410–416.

Bibliography

Livezey, R. L., and Peckham, R. S., 1953: Some snakes from San Marcos, Guatemala. Herpetologica, 8 (4): 175-177.

Lizot, J., 1985: Yanomani—daily life in the Venezuelan forest. Cambridge Universtiy Press, Paris.

Löding, H. P., 1922: A preliminary catalogue of Alabama amphibians and reptiles. Geol. Survey Alabama Mus. Paper, 5: 7-59.

Loennberg, E., 1894: Notes on reptiles and batrachians collected in Florida in 1892 and 1893. Proc. U.S. Natl. Mus., 17 (1003): 317-339.

Logier, E. B. S., and G. C. Toner, 1961: Check list of the amphibians and reptiles of Canada and Alaska. Royal Ontario Mus., (53): 1-92.

Longacre, R. E., 1976: Discourse grammar —Studies in indigenous languages of Colombia, Panama, and Ecuador. Part I: Summer Institute of Linguistics Publications in Linguistics and Related Fields. 4 parts. 1-445 pp.

Longacre, R. E., 1977a: Discourse grammar —Studies in indigenous languages of Colombia, Panama, and Ecuador. Part II: Summer Institute of Linguistics Publications in Linguistics and Related Fields. i-x; 1-299.

Longacre, R. E. (Ed.), 1977b: Discourse grammar—Studies in indigenous languages of Colombia, Panama, and Ecuador. Part III: Summer Institute of Linguistics Publications in Linguistics and Related Fields. i-vii; 1-377.

López-Luna, M. A., R. C. Vogt and M. A. de la Torre-Loranca, 1999: A new species of montane pit viper from Veracruz, México. Herpetologica, 55 (3): 382-389.

Loukotka, Č., 1968: Classification of South American Indian languages. Reference Services Vol.7, Latin American Center, University of California-UCLA-Los Angeles, 453 pp.

Love, K. with photos by B. Love, 1998: Hypomelanism in colubrids. The Vivarium, 9 (1): 32-36.

Lowe, C. H. Jr., and K. S. Norris, 1954: Analysis of the herpetofauna of Baja California, Mexico. III. New and revived reptilian subspecies of Isla de San Esteban, Gulf of California, Sonora, Mexico, with notes on other satellite islands of Isla Tiburón. Herpetologica, 1: 89-96.

Lowe, C. H. Jr., and K. S. Norris, 1955: Analysis of the herpetofauna of Baja California, Mexico, III. New and revived reptilian subspecies of Isla de San Esteban, Gulf of California, Sonora, Mexico, with notes on other satellite islands of Isla Tiburon. Herpetologica, 11 (2): 89-96.

Lowe, C. H. Jr., C. R. Schwalbe and T. B. Johnson, 1986: The venomous reptiles of Arizona. Arizona Game and Fish Department, Phoenix.

Lowe, C. H. Jr., and W. H. Woodin, III, 1954: A new racer (genus *Masticophis*) from Arizona and Sonora, Mexico. Proc. Biol. Soc. Washington, 67: 247-250.

Lönnberg, E., 1902: On a collection of snakes from north-western Argentine and Bolivia, containing new species. Ann. Mag. Nat. Hist., 10: 457-462.

Lutz, A., and O. Mellol, 1923: Duas novas espécies de colubrideos brasileiros. Folia Med. (Rio de Janeiro), 4: 2-3.

Lynch, J. D., and Smith, H. M., 1965: New or unusual amphibians and reptiles from Oaxaca, Mexico. Herpetologica, 21: 168-177.

Lynch, J. D., and Smith, H. M., 1966: New or unusual amphibians and reptiles from Oaxaca, Mexico. II. Trans. Kansas Acad. Sci., 69 (1): 58-75.

Lynn, W. G., and C. Grant, 1940: The herpetology of Jamaica. Bull Inst. Jamaica —Science Series. (1): 1-148.

Macazanga Ordoño, C., [date unidentified]: Diccionario de zoologia Nahuatl— Introducción, iconografía, diccionario y vocabulario español-náhuatl, del autor. Editorial Innovación, S.A., Mexico D. F.

Machado, O., 1945a: Estudios comparativos sobre ophidios de Brasil. Typhlopidae, Leptotyphlopidae, Boidae, Aniliidae. Bol. Inst. Vital Brazil, 5 (2): 17-36.

Machado, O., 1945b: Estudo comparativo das Elapideas do Brasil. Bol. Inst. Vital Brasil, 5 (2): 37-46.

Machado, O., 1945c: Estudo comparativo das Crotalideas do Brasil. Bol. Inst. Vital Brasil, 5 (2): 47-66.

Machado, O., 1945d: Variedade rara de *Erythrolamprus aesculapii* encontrada no Estado do Rio. Bol. Inst. Vital Brazil, 5 (2): 77-78.

Machado, R. A., P. S. Bernarde, and Morato, S. A. A., 1998: Natural History notes: *Liophis miliaris* (Common Water Snake). Prey. Herpetol. Rev., 29: 45.

Maglio, V. J., 1970: West Indian xenodontine colubrid snakes: Their probable origin, phylogeny, and zoogeography. Bull. Mus. Comp. Zool., 141 (1): 1-53.

Mahrdt, C. R., 1969: First record of the snake *Coniophanes schmidti* from the state of Chiapas, México. Herpetologica, 25:125.

Malnate, E. V., 1960: Systematic division and evolution of the colubrid snake genus *Natrix*, with comments on the subfamily Natricinae. Proc. Acad. Nat. Sci. Philadelphia, 112: 41-71.

Malnate, E. V., 1971: A catalog of primary types in the herpetological collections of the Academy of Natural Sciences of Philadelphia, Proc. Acad. Nat. Sci. Philadelphia, 123 (9): 345-375.

Mijares-Urrutia, A., and A. Arenda R., 1999: Additional new regional and local records of amphibians and reptiles from the state of Falcón, Venezuela. Herpetol. Rev., 30 (2): 115.

Mancilla-Moreno, M., and A. Rendon Rojoas, 1995: Geographic distribution— *Nina sebae sebae* (Common Redback Coffee Snake). Herpetol. Rev., 26 (3): 157.

Mancilla-Moreno, M., and W. Schmidt Ballardo, 1995: Geographic distribution— *Leptophis mexicanus septemtrionalis* (Tamaulipian Parrot Snake). Herpetol., 26 (3): 156.

Mancilla-Moreno, M., and J. L. Camarillo Rangel, 1998: Geographic distribution. *Chersodromus liebmanni*. Herpetol. Rev., 29 (1): 52-53.

Mansen, R., 1883 (?): Fonemas del Guajiro. B. B. Bekar.

Mántica, C., 1973: El habla nicaragüense— Estudio morfológico y semántico. First Edition. Editorial Universitaria Centroamerica (EDUCA).

Manzanilla, J., A. Mijares-Urrutia and R. Rivero, 1998: *Rhadinaea fulviceps*. Herpetol. Rev., 29 (2): 115.

Mara, W. P., 1993: Venomous snakes of the world. T.F.H. Pub., Neptune City, New Jersey.

Mara, W. P., 1994a: Garter & ribbon snakes. T.F.H. Pubs., Inc, Neptune City, New Jersey.

Mara, W. P., 1994b: Pine snakes-a complete guide. T.F.H. Pubs., Inc, Neptune City, New Jersey.

Mara, W. P., 1995: Racers, whipsnakes, and indigos. T.F.H. Pubs., Inc., Neptune City, New Jersey.

Mara, W. P., 1996: Greensnakes. T.F.H. Pub., Inc. Neptune City, New Jersey.

Marais, J., 1997: Snakes. Barnes & Noble, Inc., New York.

Marcuzzi, G., 1951: Ofidios existentes en las colecciones de los museos de Caracas. Noved. Cien. La Salle, Ser. Zool., (3): 1–20.

Marcy, R. B., 1854: Exploration of the Red River of Louisiana, in the year 1852. A. O. P. Nicholson, Public Printer.

Markel, R. G., 1980: The kingsnakes: an annotated checklist. Bull. Chicago Herpetol. Soc., 14 (4): 101–116.

Markel, R. G., 1990: Kingsnakes & milksnakes. Neptune City, New Jersey, T.F.H. Pub.

Markel, R. G., 1994: Selected breeding of California kingsnakes. Reptile & Amphibian Mag. July/August 1994, (29): 8–16.

Markel, R. G., and R. D. Bartlett, 1995: Kingsnakes and milksnakes-everything about purchase, care, nutrition, breeding, behavior, and training. Barron's Educational Series, Inc. Hauppauge, N.Y.

Markezich, A. J., 1976: A reassessment of variation in, and a redescription of, the South American snake, *Leimadophis poecilogyrus* (Wied). MS Thesis, Univ. Illinois, Chicago, 129 pp.

Markezich, A. J., and J. R. Dixon, 1979: A new South American species of snake and comments on the genus *Umbrivaga*. Copeia, 1979 (4): 698–701.

Markezich, A. J., and P. Parrillo, 1999: Natural History Notes—*Leptodeira annulata* (False Mapanare, Banded Cat-eyed Snake)—Predation. Herpetol. Rev., 30 (1): 46–47.

Markezich, A. J., and D. C. Taphorn, 1993: A variational analysis of populations of *Bothrops* (Serpentes: Viperidae) from western Venezuela. J. Herpetol., 27 (3): 248–254.

Marques, L. B., and T. de Lema, 1992: Estudio comparativo da osteologia craniana de *Clelia occipitolutea* (Dumeril, Bibron et Dumeril, 1854) e *C. rustica* (Cope, 1878). (Serpentes: Colubridae: Xenodontinae: Pseudoboini). Acta Biol. Leopoldensia, 14 (1): 27–54.

Marques, O. A. V., 1996a: Biologia reprodutiva de cobra-coral *Erythrolamprus aesculapii*, no sudeste do Brasil. Rev. Brasileira Zool., 13: 747–753.

Marques, O. A. V., 1996b: Natural History notes: *Sordellina punctata* diet. Herpetol. Rev., 27 (3): 147.

Marques, O. A. V., 2000: Natural History notes: *Uromacerina ricardinii* (Vine Snake). Predation and prey. Herpetol. Rev., 31: 180–181.

Marques, O. A. V., and M. E. Calleffo, 1997: Geographic distribution—*Pseustes sulphureus*. Herpetol. Rev., 28 (3): 160.

Marques, O. A. V., and F. L. Franco, 1998: *Philodryas viridissimus* (Common Green Racer). Herpetol. Rev., 29 (1): 54.

Marques, O. A. V., M. Martins and I. Sazima, 2002: A new insular species of pitviper from Brazil, with comments on evolutionary biology and conservation of the *Bothrops jararaca* group (Serpentes, Viperidae). Herpetologica, 58: 303–312.

Marques, O. A. V., and Puorto, G, 1996: Geographic distribution—*Chironius laevicollis* (Brazilian Cipó). Herpetol. Rev., 27 (4): 212.

Marques, O. A. V., André Eterovic, and Ivan Sazima, 2001: Serpentes da mata Atlântica—guia ilustrado para a Serra do Mar. Holos, Editoria Ltda-ME, São Paulo, Brazil.

Martí de Tortajada, J., 1942: Los batracios y los reptiles. Espasa-Calpe, S.A., Madrid, Libros de la Naturaleza.

Martin, C. L., 1997: When racers were ringnecks. Reptiles, October, 1997, 5 (10): 60–67.

Martin, P. S., 1955: Herpetological records from the Gómez Farías region of southwestern Tamaulipas, México. Copeia, 1955: 173–180.

Martin, W. H., 1992: Phenology of the timber rattlesnakes (*Crotalus horridus*) in an unglaciated section of the Appalachian mountains. pp. 259–277 *in* J. A. Campbell and E. D. Brodie, Jr.: Biology of the pitvipers. Selva, Tyler, Texas.

Martín del Campo, R., 1984: Herpetología mexicana antigua. II. Nomenclatura y taxonomía de las serpientes. An. Inst. Biol. Univ. Natl. Autón. México, Ser. Zool., 54 (1): 177–198.

Martínez, C. V., 1983: Panamá: nuevo ámbito de distribución para la serpiente venenosa *Bothrops picadoi* (Dunn). Conciencia (Universidad de Panamá), 10: 26–27.

Martínez, C. V., and R. Bolaños, [1982] 1983: The bushmaster, *Lachesis muta muta* (Linnaeus) (Ophidia: Viperidae) in Panamá. Rev. Biol. Trop., 30 [1982]: 100–101.

Martinez, F. A., J. C. Troiano, D. Selles, E. Prada, N. Fescina and D. Jara, 1994: *Tratrema stenocotyle* (Cohn, 1902), Goodman, 1956 (Trematode-Plagiorchidae en *Bothrops neuwiedi diporus* (Ophidia-Crotalidae). X. Reunion de Comunicaciones Herpetológicas, Mar de Plata-Pcia. de Buenos Aires, Argentina.

Martínez, S., and L. Cedas, 1986: Captive reproduction of the Mussurana, *Clelia clelia* (Daudin) from Costa Rica. Herpetol. Rev., 17: 12–13.

Martins, M., and M. E. Oliveria, 1993: The snakes of the genus *Atractus* Wagler (Reptilia: Serpentes: Colubridae) from the Manaus region of Amazonia, Brazil. Zool. Mededelingen (Leiden), 67 (1-2): 21–40.

Martof, B. S., 1956: Amphibians and reptiles of Georgia. Univ. Georgia Press, 74 pp.

Mattison, C., 1989: Notes on shovel-nosed snakes and sand snakes, *Chionactis* and *Chilomeniscus*. British Herpetol. Soc. Bull., (28): 25–30.

Marx, H., 1958: Catalogue of type specimens of reptiles and amphibians in the Chicago Nat. Hist. Mus. Fieldiana, Zool., 36 (4): 409–496.

Marx, H., 1960: A new colubrid snake of the genus *Atractus*. Fieldiana, Zool., 39 (38): 411–413.

Maslin, T. P., 1953: The status of the whipsnake *Masticophis flagellum* (Shaw) in Colorado. Herpetologica, 9 (4): 193–200.

Matola, S., and T. Garel, 1995: A field guide to the snakes of Belize. World Wildlife Fund-US, Corporación Gráfica, Costa Rica.

Mattei, R., and C. L. Barrio, 1999: Geographic distribution—*Oxyrhopus formosus*. Herpetol. Rev., 30 (1): 55.

Mattison, C., 1990: A-Z of snake keeping. New York, Sterling Publishing Co,.Inc.

Mattison, C., 1987: The care of reptiles & amphibians in captivity. New York, Sterling Publishing Co., Inc.

Mattison, C., 1995: The encyclopedia of

Bibliography

snakes. Facts on File, Inc., New York, New York.

Mayer, G. C., and J. D. Lazell, Jr., 1988: Distributional records for reptiles and amphibians from the Puerto Rican Bank. Herpetol. Rev., 19 (1): 22–23.

Mayr, E., 1942: Systematics and the origin of species. Columbia Univ. Press, New York.

Mayr, E., 1949: Speciation and selection. Proc. Amer. Philos. Soc., 93: 514–519.

Mayr, E., 1982: Processes of speciation in animals. pp 1–19. *In* Barigozzi, 1982, Alan R. Liss, New York.

Mayr, E., 1992: A local flora and the biological species concept. Amer. J. Botany, 79: 222–238.

McAllister, C. T., 1985: *Nerodia rhombifera*. Cat. Amer. Amph. Rept., (376): 1–4.

McCleary, R. J. R., and R. W. McDiarmid, 1993: *Phyllorhynchus decuratus*. Cat. Amer. Amph. Rept., (580): 1–7.

McCoid, M. J., T. H. Fritts and E. W. I. Campbell, 1994: A Brown Tree Snake (Colubridae: *Boiga irregularis*) sighting in Texas. Texas J. Sci., 46 (4): 365–368.

McCoid, M. J., J. W. Sites Jr. and J. R. Dixon, 1980: An additional specimen of *Chersodromus rubriventris* (Colubridae). Southwest. Naturalist, 25 (3): 429.

McCoy, C. J., 1969: Snakes of the genus *Coniophanes* (Colubridae) from the Yucatan Peninsula, Mexico. Copeia, 1969 (4): 847–849.

McCoy, C. J., 1971: Comments on Bolivian *Atractus* (Serpentes: Colubridae). Herpetologica, 27 (3): 314–316.

McCoy, C. J., 1986: Results of the Carnegie Museum of Natural History Expeditions to Belize. I. Systemcatic status and geographic distribution of *Sibon neilli* (Reptilia, Serpentes). Ann. Carnegie. Mus., 55: 117–123.

McCoy, C. J. Jr., B. A. Branson and M. E. Sisk, 1960: *Natrix valida* in Sonora, Mexico. Herpetologica, 16: 130.

McCoy, C. J., and E. J. Censky, 1992: Biology of the Yucatan hognosed pitviper, *Porthidium yucatanicum*. pp. 217–222. *In* J. A. Campbell and E. D. Brodie, Jr.: Biology of the pitvipers. Selva, Tyler, Texas.

McCoy, C. J., E. J. Censky and R. R. Van Devender, 1986: Distribution records for amphibians and reptiles in Belize, Central America. Herpetol. Rev., 17: 28–29.

McCranie, J. R., 1980a: *Crotalus adamanteus* Beauvois-Eastern Diamondback Rattlesnake. Cat. Amer. Amph. Rept., (252): 1–2.

McCranie, J. R., 1980b: *Crotalus pricei* Van Denburgh-Twin-spotted Rattlesnake. Cat. Amer. Amph. Rept., (266): 1–2.

McCranie, J. R., 1980c: *Drymarchon, D. corais*. Cat. Amer. Amph. Rept., (267): 1–4.

McCranie, J. R., 1983a: *Crotalus pusillus* Klauber—Southwestern Mexican Dusky Rattlesnake. Cat. Amer. Amph. Rept., (313): 1–2.

McCranie, J. R., 1983b: *Crotalus unicolor* Van Lidth de Jeude—Aruba Island Rattlesnake. Cat. Amer. Amph. Rept., (389): 1–2.

McCranie, J. R., 1983c: *Nerodia taxispilota*. Cat. Amer. Amph. Rept., (331): 1–2.

McCranie, J. R., 1984: *Crotalus vegrandis* Klauber—Uracoan Rattlesnake. Cat. Amer. Amph. Rept., (350): 1–2.

McCranie, J. R., 1990: *Nerodia erythrogaster*. Cat. Amer. Amph. Rept., (500): 1–8.

McCranie, J. R., 1993a: Additions to the herpetofauna of Honduras. Caribbean J. Sci., 29: 254–255.

McCranie, J. R., 1993b: *Crotalus durissus*. Cat. Amer. Amph. Rept., (577): 1–11.

McCranie, J. R., 1996: Geographic distribution—*Ungaliophis continentalis* (Isthmian Dwarf Boa: Boilla). Herpetol. Rev., 27 (1): 36.

McCranie, J. R., 1999: Notes on the type series of *Oxybelis wilsoni* Villa and McCranie, Herpetol. Rev., 30 (1): 11.

McCranie, J. R., and C. T. McAllister, 1988: *Nerodia valida* (Kennicott)—Mexican West Coast Water Snake. Amer. Amph. Rept., (431): 1–3.

McCranie, J. R., and G. Köhler, 1999: Two new species of colubrid snakes of the genus *Enulius* from Islas de la Bahía, Honduras. Carib. J. Sci., 35 (1–2): 14–22.

McCranie, J. R., and J. Villa, 1993: A new genus for the snake *Enulius sclateri* (Colubridae: Xenodontinae). Amphibia-Reptilia, 14 (3): 261–267.

McCranie, J. R., and L. D. Wilson, 1990: Annotated bibliography to the herpetofauna of the pine-oak woodlands of the Sierra Madre Occidental, Mexico. Smithsonian Herpetol. Survey, (84): 1–16.

McCranie, J. R., and L. D. Wilson, 1991: *Geophis fulvoguttatus* Mertens and *Micrurus browni* Schmidt and Schmidt: additions to the snake fauna of Honduras. Amphibia-Reptilia, 12: 112–114.

McCranie, J. R., and L. D. Wilson, 1991a: *Rhadinaea kinkelini*. Cat. Amer. Amph. Rept., (523): 1–2.

McCranie, J. R., and L. D. Wilson, 1991b: *Rhadinaea montecristi*. Cat. Amer. Amph. Rept., (524): 1–2.

McCranie, J. R., and L. D. Wilson, 1992: *Rhadinaea godmani*. Cat. Amer. Amph. Rept., (546): 1–3.

McCranie, J. R., and L. D. Wilson, 1993: *Leptophis modestus*. Cat. Amer. Amph. Rept., (578): 1–2.

McCranie, J. R., and LD. Wilson, 1995: Two new species of colubrid snakes of the genus *Ninia* from Central America. J. Herpetol., 29 (2): 224–232.

McCranie, J. R., and L. D. Wilson, 2001: Taxonomic status of *Typhlops stadelmani* Schmidt (Serpentes: Typhlopidae). Copeia 2001 (3): 820–822.

McCranie, J. R., L. D. Wilson and S. W. Gotte, 2001: Three new country records for Honduran snakes.

McCrystal, H. K., and M. J. McCoid, 1986: *Crotalus mitchellii* (Cope)-Speckled Rattlesnake. Cat. Amer. Amph. Rept., (388): 1–4.

McDaniel, R. V., and J. P. Karges, 1983: *Farancia abacura*. Cat. Amer. Amph. Rept., (314): 1–2.

McDiarmid, R. W., 1968: Variation, distribution and systematic status of the black-headed snake *Tantilla yaquia* Smith. Bull. Southern California Acad. Sci., 67 (3): 159–177.

McDiarmid, R. W., 1977: *Tantilla yaquia*. Cat. Amer. Amph. Rept., (198): 1–2.

McDiarmid, R. W., 1992: Systematic status of the San Luis Potosi Black-Headed Snake *Tantilla deviatrix* Barbour Colubridae. Southwest. Naturalist, 37 (3): 303–307.

McDiarmid, R. W., and R. L. Bezy, 1971: The colubrid snake *Enulius oligostichus* in Western Mexico. Copeia, 1971 (2): 350–351.

McDiarmid, R. W., J. A. Campbell and T. A. Touré, 1999: Snake species of the world- a taxonomic and geographic reference. Vol. I. The Herpetologists' League, Washington, D.C., 2 July 1999, 511 pp.

McDiarmid, R. W., J. F. Copp, and D. E. Breedlove, 1976: Notes on the herpetofauna of western México: New records from Sinaloa and the Tres Marías Islands. Contrib. Sci., Mus. Nat. Hist., Los Angeles Co., (275): 1–17.

McDiarmid, R. W., and S. H. Folke, 1991: *Tantilla bocourti*. Cat. Amer. Amph. Rept., (526): 1–3.

McDiarmid, R. W., and R. J. R. McCleary, 1993: *Phyllorhynchus*. Cat. Amer. Amph. Rept., (579): 1–5.

McDiarmid, R. W., and N.J. Scott Jr, 1970: Geographic variation and systematic status of Mexican Lyre snakes of the *Trimorphodon tau* group (Colubridae). Los Angles County Mus. Contrib. Sci., (179): 1–42.

McDiarmid, R. W., T. S. Touré, and J. M. Savage, 1966: The proper name of the Neotropical tree boa often referred to as *Corallus enydris* (Serpentes: Boidae). J. Herpetol., 30 (3): 320–326.

McDowell, S. B., 1961:[Review of] Systematic division and evolution of the colubrid snake genus *Natrix,* with comments on the subfamily Natricinae, by Edmond V. Malnate. Copeia, 1961 (4): 502–506.

McGuire, J. A., and L. L. Grismer, 1993: The taxonomy and biogeorgaphy of *Thamnophis hammondii* and *Thamnophis digueti* (Reptilia: Squamata: Colubridae) in Baja California, Mexico. Herpetologica, 49 (3): 354–465.

McKeown, S., 1978: Hawaiian reptiles & amphibians. Honolulu, Oriental Publishing Co.

McPeak, R., 2000: Amphibians and reptiles of Baja California. Sea Challengers, Monterey, California.

Meade, G. O., 1941: The natural history of the mud snake. Sci. Monthly, 63 (1): 21–29.

Mecham, J. S., 1983: *Nerodia harteri*. Cat. Amer. Amph. Rept., (330): 1–2.

Medica, P. A., 1975: *Rhinocheilus*. Cat. Amer. Amph. Rept., (175): 1–4.

Medica, P. A., and B. G. Maza, 1974: Geographic distribution. *Masticophis bilineatus bilineatus*. Herpetol. Rev., (5): 70.

Medina, F. L., 1973: Nota sobre *Bothrops pictus* (Tschudi, 1895). Rev. Biol., 2 (2): 191–201.

Meek, S. E., 1905: An annotated list of a collection of reptiles from southern California and northern Lower California. Zoöl. Ser. Field Mus. Nat. Hist., 7 (1): 1–19, pls. 1–3.

Meek, S.E, 1910: Notes on batrachians and reptiles from the islands north of Venezuela. Field Mus. Nat. Hist. Zool. Ser., 7 (12): 415–418.

Mehrtens, J. M., 1987: Living snakes of the world in color. New York, Sterling Publishing Co., Inc.

Meise, W., 1938: Eine neue Korallenschlange aus dem Amazonasgebiet. Zool. Anz., 123 (): 20–22.

Mejenes-López, S. de M. A., F. Mendoza-Quijano, C. Madero-Vega and A. K. Molleda-Contreras, 1999: Geographic distribution: *Coniophanes imperialis imperialis*. Herpetol. Rev., 30 (4): 235.

Mendelson J. R. III, 1990: Notes on a collection of amphibians and reptiles from Pueblo Viejo, Alta Verapaz, Guatemala. Fundación Interamericana Publi. Investig. Trop. Ocas. (3): 1–18.

Mendelson J. R. III, and D. A. Kirziran, 1995: Geographic variation in *Rhadinaea hempsteadae* (Serpentes: Colubridae) with the description of a new species from Chiapas, Mexico. Herpetologica, 51 (3): 301–313.

Menden, F., 1969: El desarrollo de la herpetologia en Colombia. Revi. Acad. Colombiana Cienc. Exactas, Fisicas y Nat., XIII (50): 149–199.

Mendoza Quijano, F. and H. M. Smith, 1993: A new species of Hooknose Snake, *Ficimia* (Reptilia, Serpentes). J. Herpetol., 27 (4): 406–410.

Meneghel, M. D., and F. Achaval, 1997: Geographic distribution—*Leptophis ahaetulla marginatus* (Parrot Snake). Herpetol. Rev., 28 (2): 98.

Merker, G., and C. Merker, 1998: A rosy pet. Reptiles USA, Annual edition, 1998: 118–126.

Merrem, B., 1790: Beytrage zur Naturgeschichte. [Also published under the title "Beytrage zur Naturgeschichte der Amphibien.] Fasc. 1-3, Duisburg, Lemgo & Essen.

Merrem, B., 1820: Versuch eines Systems der Amphibien Tentamen systematis amphibiorum. Marburg.

Merriam, C. H., 1892: The geographical distribution of life in North America. Proc. Biol. Soc. Washington, 4: 199–243.

Mertens, R., 1926: Herpetologische Mitteilungen VIII-XV. Senckenb., 8 (3–4): 137–155.

Mertens, R., 1930: Zoologische Ergebnisse eine Reise von Otto Conde. 2. Amphibien und Reptilien. Folia Zool. Hydrobiol. Univ. Lettlends, Riga, 1 (2): 161–166.

Mertens, R., 1942. Amphibian und Reptilian I. Austeute der Hamburger Sudperu-Expedition. pp. 277–287 *In* E.Titscheck, ed., Beitrage zur Fauna Perus. Hamburg, Vol.2.

Mertens, R., 1952a: Die amphibien und reptilien von El Salvador. Abhand. Senckenb. Naturfors. Gesell., (487): 1–120.

Mertens, R., 1952b: Weitere neue Reptilien aus El Salvador. Zool. Anz., 149: 133–138.

Mertens, R., 1956: Das Problem der Mimikry bei Korallenschlangen. Zool. Jahrb. (Syst.), 84 (6): 541–576.

Mertens, R., 1960: Über die Schlangen der Galapagos. Senckenb. Biol., 41: 133–141.

Mertens, R., 1969: Herpetologische Beobachtungen auf der Insel Tobago. Salamandra, 5 (1–2): 63–70.

Mertens, R., 1970: Herpetologische neues von der insel Tobago. Salamandra, 6 (1–2): 42–44.

Mertens, R., 1973a: Bemerkenswerte schlanknattern der neotropicshen Gattung *Leptophis*. Stud. Neotrop. Fauna, 8: 141–154.

Mertens, R., 1973b: Uber falsche Korallennater auf Trinidad and Tobago. Salamandra, 9 (3–4): 161–163.

Mertens, R., and H. G. Dowling, 1952: The Identity of the Snake *Pityophis intermedius*. Senckenb., Band 33, 15 Nov. 1952, (416): 197–201.

Meyer, J. R., and L. D. Wilson, 1971: A distributional checklist of the amphibians of Honduras. Los Angeles County Mus. Nat. Hist., Contrib. Sci. (218): 1–47.

Meyer, J. R., and L. D. Wilson, 1972: Taxonomic studies and notes on some Honduran amphibians and reptiles. Bull. Southern California Acad. Sci., 70: 106–114.

Meylan, P. A., 1985: *Heterodon simus*.Cat. Amer. Amph. Rept., (375): 1–2.

Michaud, E. J. III, 1984: A taxonomic revision of the *Liophis lineatus* complex (Reptilia: Colubridae) of Central and South America. M.S. Thesis, May 1984.

Michaud, E. J., and J. R. Dixon, 1989: Prey items of 20 species of the Neotropical colubrid snake genus *Liophis*. Herpetol. Rev., 20 (2): 39-41.

Michaud, E. J., and J. R. Dixon, 1987: Taxonomic Revision of the *Liophis lineatus* Complex (Reptilia: Colubridae) of Central and South America. Contrib. Biol. Geol., Milwaukee Publ. Museum, (71): 1-26.

Michelon, O. (Ed.), 1976: Diccionario de San Francisco. Akademische Druck—U. Verlagsanstalt, Graz-Austria, 1, 2 volumes.

Mijares-Urrutia, A., A. L.Markezich and R. A. Arends, 1995: Discovery of *Leptodeira bakeri* Ruthven (Serpentes: Colubridae) in the Paraguana Peninsula, northeastern Venezuela. With description and biological comments [in Spanish]. Caribbean J. Sci., 31 (1-2): 77-82.

Mijares-Urrutia, A., and A. Arends R., 2000: Herpetofauna of Estado Falcón, Northwestern Venezuela: A checklist with geographical and ecological data. Smithsonian Herpetological Infor. Serv., (123): 1-29.

Miller, M., and K. Taube, 1993: The gods and symbols of ancient Mexico and the Maya-An illustrated dictionary of Mesoamerican religion. Thames and Hudson, London.

Milstead, W. W., 1953: Geographic variation in the garter snakes, *Thamnophis cyrtopsis*. Texas J. Sci., 5 (3): 348-379.

Milstead, W. W., J. S. Mecham and H. McClintock, 1950: The amphibians and reptiles of the Stockton Plateau in northern Terrell County, Texas. Texas J. Sci., 2 (4): 543-562.

Minton, S. A., 1992: Serologic relationships among pitvipers: evidence from plasma albumins and immunodiffusion. pp. 155-161 in J. A. Campbell and E. D. Brodie, Jr.: Biology of the pitvipers. Selva, Tyler, Texas.

Minton, S. A. Jr., 1956: A new snake of the genus *Tantilla* from west Texas. Fieldiana Zool., 34: 449-452.

Minton, S. A. Jr., 1958 [1959]: Observations on amphibians and reptiles of the Big Bend region of Texas. Southwest. Naturalist, 3: 28-54.

Minton, S. A. Jr., 1980: *Thamnophis butleri* (Cope)—Butler's Garter Snake. Cat. Amer. Amph. Rept., (258): 1-2.

Minton, S. A. Jr., 1983: *Sistrurus catenatus* (Rafinesque)—Massasauga. Cat. Amer. Amph. Rept., (332): 1-2.

Minton, S. A. Jr., J. M. Cisneros and B. M. De Cervantes, 1997: Ophiophagy by an arthropod-eating snake. Bull. Chicago Herpetol. Soc., 32 (12): 253.

Minton, S. A. Jr., and M. R. Minton, 1969: Venomous reptiles. New York, Charles Scribner's Sons.

Minton, S. A. Jr., and M. R. Minton, 1991: Life history note: *Masticophis mentovarius* (Neotropical Whipsnake)—Reproduction. Herpetol. Rev., 22 (3): 100-101.

Minton, S. A., and H. M. Smith, 1960: A new subspecies of *Coniophanes fissidens* and notes on Central American amphibians and reptiles. Herpetologica, 16: 103-111.

Miranda, M. E., and C. A. Couturier, 1981: Una nueva subespecie de *Tachymenis peruviana* Wiegmann 1835 (Ophisia: Boigidae) para la Argentina. Comun. Mus. Argent. Cie. Nat. B. Rivadavia, Zool., 4 (10): 79-83.

Miranda, M. E., G. A. Couturier and J. D. Williams, 1982: Guía de los ofidios Bonaerenses. Asociación Cooperadora Jardín Zoológico de La Plata, La Plata, Argentina, 71 pp + Indice Sistematico.

Mitchell, J. C., 1980: Notes on *Lampropeltis triangulum* (Colubridae) from northern Jalisco, Mexico. Southwest. Naturalist, 25 (2): 269.

Mtichell, J. C., 1982a: *Farancia*. Cat. Amer. Amph. Rept., (292): 1-2.

Mitchell, J. C., 1982b: *Farancia erythrogramma*. Cat. Amer. Amph. Rept., (293): 1-2.

Mitchell, J. C., 1994: The reptiles of Virginia. Smithsonian Inst. Press., Washington and London.

Mitchell, J. D., 1903: The poisonous snakes of Texas, with notes on their habits. Trans. Texas Acad. Sci., 5 [part 1] (2): 21-48.

Mittleman, M. B., 1952: Another interpretation of *Coluber doliatus* Linnaeus. Herpetologica, 8 (2): 22-25.

Mocquard, F., 1887: Sur une nouvellie espèce d'*Elaps, E. heterochilus*. Bull. Soc. Philom., Paris, 7 (11): 39-41.

Mocquard, F., 1899: Reptiles et batraciens recueillis au Mexique par M. Léon Diguet en 1896 et 1897. Bull. Soc. Philo—Math, Paris, (9) 1: 154-169, pl. 1.

Mocquard, F., 1909: Etudes sur les septiles. Mission scientifique au Mexique et dans l'Amérique Centrale. Recherches zoologiques. Part 3. Paris, Imprimerie Impériale. [Livrasion 16, 1908. pp.861-932; Livraison 17: 933-1012.], 1908-1909.

Mole, R. R., 1892: *Eunectes murinus*. J. Trinidad Field Naturalist's Club, 1 (3): 56-58.

Mole, R. R., 1924: Trinidad snakes. Proc. Zool. Soc. London, (1): 235-278.

Mole, R. R., 1894: Biological notes upon some ophidia of Trinidad, B. W. I., with a preliminary list of the species recorded from the island. Proc. Zool. Soc. London, 1894: 499-518.

Mole, R. R., and R. W. Urich, 1894: Biological notes upon some of the ophidia of Trinidad, B. W. I., with a preliminary list of the species recorded from the island. Proc. Zool. Soc. London, 1894: 499-518.

Molina R., C. R. and G. A. Rivas F., 1996: *Rhinobothryum lentiginosum* (coral falsa). Herpetol. Rev., 27 (3): 155.

Monroy-Ibarra, A. C., S. de M. Mejenes-Lopéz, F. Mendoza-Quijenes, 1996: Geographic distribution—*Dryadophis melanolomus veracrucis*. Herpetol. Rev., 27 (4): 212.

Montanelli, S., and B. B. Alvarez, 1998: Geographic distribution—*Tantilla melanocephala*. Herpetol. Rev., 29 (3): 179.

Moonen, J., W. Eriks and K. van Duersen, 1979: Surinaamse Slangeninkleur. C. Kerstan & Co.N. V., Paramaibo, Surinam, July 1979.

Moore, B. J., 1977: Some discourse features of Hupda Macu. pp 25-42. In Longacre, R. E., ed., 1977: Discourse grammar—Studies in indigenous languages of Colombia, Panama, and Ecuador Part II.

Moore, J. E., 1953: The hog-nosed snake in Alberta. Herpetologica, 9 (4): 173.

Morafka, D. J., 1977: A biographical analysis of the Chihuahuan Desert through its herpetofauna. The Hague, Dr. W. Junk B. V., Publishers.

Morales, V. R., N. Carrillo and H. Ortega, 1990: "El material tipo de peces, anfibios y reptiles en el museo de historia natural de la Universidad Nacional Mayor de San Marcos," Serie A Zoologica, Lima, Peru, Publi. Univ. Nac. Mayor San Marcos, 15 set. (33): 1-7.

Morato, S. A. A., J. C.-de Moura-Leite,

A. L. C. Prudente and R. S. Bérnils, 1995: A new species of *Pseudoboa* Schneider, 1801 from southeastern Brazil, (Serpentes: Colubridae: Xenodontinae: Pseudoboini). Biociêcias, 3 (2): 253-264.

Morato de Carvalho, C., 2002: Descrição de uma nova espécie de *Micrurus* do Estado de Roraima, Brasil (Serpentes, Elapidae). Pap. Avul. Zool., Mus. Zool, Univ. São Paulo, 42 (8): 183-192.

More, B. R., 1966: Diccionario Castellano-Colorado, Colorado-Castellano. Instituto Lingüístico de Verano, Quito, Ecuador, pp. 95-221.

Moreno Mora, M., [1955] 1956: Diccionario etimologico y comparado del Kichua del Ecuador. Tomo I. Editorial de la Casa de la Cultura Ecuatoriana, Núcleo del Azuay. 15 May 1956, 375 pp.

Moreno Mora, M., 1967: Diccionario etimologico y comparado del Kichua del Ecuador. Tomo II. Editorial de la Casa de la Cultura Ecuatoriana, Núcleo del Azuay. 13 July 1967, 284 pp.

Morris, P. A., 1944: They hop and crawl. The Jacques Cattell Press, Lancaster, PA.

Morris, P. A., 1974: An introduction to the reptiles and amphibians of the United States. Dover Publications, Inc., New York.

Mortof, B. S., W. H. Palmer, J. R. Bailey and J. R. Harrison III, 1980: Amphibians and reptiles of the Carolinas and Virginia. Univ. North Carolina Press, Chapel Hill.

Mosonyi, E. E. and J. C. Mosonyi, 1999: Manual de Lenguas Ingígenos de Venezuela. Vols. I and II, Fundación Bigott, Serie Orígens, Vol. I: 1-336, Vol. II: 339-664.

Mount, R. H., 1975: The reptiles and amphibians of Alabama. Auburn Univ., Auburn, Alabama, September, 1975.

Moura-Leite, J. C. de, S. A. A. Morato and R. S. Bérnils, 1996: New records of reptiles from the state of Paraná, Brazil. Herpetol. Rev., 27 (4): 216-217.

Moura-Leite, J. C. de and P. S. Bernarde, 1999: Geographic distribution—*Waglerophis merremi* (Boipeve). Herpetol. Rev., 30 (1): 56.

Mudde, P., and M. Van Kijk, 1985: Herpetologische waarneminger in Costa Rica (13). Slangen (Serpentes). Lacerta 43: 176-180.

Mugica, P. C., 1969: Aprenda El Guajiro-Gramática y Vocabularios. Vicario Apostólica de Ríohacha, Guajira, Colombia, noviembre 30, 1969.

Müller, F., 1878: Katalog der im Museum und Universitätskabinet zu Basel aufgestellten Amphibien und Reptilien. Verh. Nat. Ges. Basel, 6: 561-709.

Müller, F., 1880: Erster Nachtrag zum Katalog der herpetologischen Sammlung des Basler Museums. Verh. Nat. Ges. Basel, 7: 120-165.

Müller, F., [1881] 1882: Erster Nachtrag zum Katalog der herpetologischen Sammlung des Basler Museums. Verh. Natur. Ges. Basel, 7: 120-165

Müller, F., 1882: Zweiter Nachtrag zum Katalog der herpetologischen Sammulung des Basler Museums. Verh. Naurt. Ges. Basel., 7: 165-175.

Müller, F., 1883: Dritter Nachtrag zum Katalog der herpetologischen Sammlung des Basler Museums. Verh. Natur. Ges. Basel, 7: 274-297.

Müller, F., 1885: Vierter Nachtrag zum Katalog der herpetologischen Sammlung des Basler Museum. Verh. Natur. Ges. Basal, 7: 668-717.

Müller, F., *in* Schenkel, 1901:Achter Nachtrag zum Katalog der herpetologischen Sammulung des Basler Museums. Verh. Natur. Ges. Basel, 13 (1): 142-199.

Müller, L., 1926: Neue Reptilien und Batrachier der zoologischen Sammulung des bayerischen Staates. Zool. Anz., 7/8: 192-200.

Müller, L., 1927: Amphibien und Reptilien der Ausbeute Prof. Breslau's in Brasilien 1913-1914. Abh. Senckenb. Natur. Ges., 40: 259-304.

Müller, P., 1974: *Clelia clelia plumbea* von der Insel Florianópolis (Santa Catarina, Brasilien) (Serpentes, Colubridae). Salamandra, 10 (1): 43.

Müller, P., and C. Ritter, 1978: Erstnachweis von *Uromancerina ricardinii* (Peracca 1897) für den Staat von Santa Catarina (Brasilien). Salamandra, 14: 44.

Muñoz, M. C., T. Escalona, W. F. Holmstrom, Jr., and R. W. Henderson, 1997: Notes on the reproduction of a Venezuelan whipsnake, *Masticophis mentovarius suborbitalis*. Bull. Chicago Herpetol. Soc., 32 (7): 146.

Munro, P., and C. Willmend, 1994: Chickasaw-An analytical dictionary. Univ. Oklahoma Press, Norman.

Murphy, J. C., 1980: The lyre snakes. Bull. Chicago Herpetol. Soc., 15 (1): 24-28.

Murphy, J. C., 1997: Amphibians and reptiles of Trinidad and Tobago. Krieger Publishing Co., Malabar, Florida.

Murphy, R. W., 1975: Two new blind snakes (Serpentes: Leptotyphlopidae) from Baja California, Mexico with a contribution to the biogeography of peninsular and insular herpetofauna. Proc. California Acad. Sci., Ser. 4, 40 (5): 93-107.

Murphy, R. W., 1983a: A distributional checklist of the reptiles and amphibians on the islands in the Sea of Cortez. pp 429-437. *In* T. J. Case and M. L. Cody, eds.: Island biogeography in the Sea of Cortez, Univ. California Press, Berkeley.

Murphy, R. W., 1983b: Paleobiogeography and genetic differentiation of the Baja California herpetofauna. Occ. Pap. California Acad. Sci., 137: 1-48.

Murphy, R. W., 1983c: The reptiles: Origins and evolution. pp 130-158. *In* T. J. Case and M. L. Cody, eds.: Island biogeography in the Sea of Cortez, Univ. California Press, Berkeley.

Murphy, R. W., and J. R. Ottley, 1980: A genetic evaluation of the leafnose snake *Phyllorhynchus arenicolus*. J. Herpetol., 14 (3): 263-268.

Murphy, R. W., and J. R. Ottley, 1984: Distribution of amphibians and reptiles on islands in the Gulf of California. Ann. Carnegie Mus., 53 (8): 207-230.

Myers, C. W., 1965: Biology of the Ringneck Snake, *Diadophis punctatus*, in Florida. Bull. Florida State Mus., Biol. Sci., 10 (2): 43-90.

Myers, C. W., 1966a: A new species of colubrid snake, Genus *Coniophanes*, from Darién, Panama. Copeia, 1966 (4): 665-668.

Myers, C. W., 1966b: *Lygophis boursieri* (Jan 1862), A snake new to the fauna of Colombia. Copeia, 1966 (4): 886-888.

Myers, C. W., 1967: The Pine Woods Snake, *Rhadinaea flavilata* (Cope): Bull. Florida State Mus., Biol. Sci., 11 (2).

Myers, C. W., 1969a: South American snakes related to *Lygophis boursieri*: A reappraisal of *Rhadinaea antioquiensis*, *Rhadinaea tristriata*, *Coronella whymperi*, and *Liophis atahuallpae*. Amer. Mus. Novitates, (2385): 1-27.

Myers, C. W., 1969b: Snakes of the genus

Myers, C. W., 1973: A new genus for Andean snakes related to *Lygophis boursieri* and a New Species (Colubrid). Amer. Mus. Novitates, (2522): 1-37.

Myers, C. W., 1974: The systematics of *Rhadinaea* (Colubridae), a genus of New World snakes. Bull. Amer. Mus. Nat. Hist., 153 (1): 1-262.

Myers, C. W., 1982: Blunt-headed vine snakes (*Imantodes*) in Panama, including a new species and other revisionary notes. Amer. Mus. Novitates, (2738): 1-50.

Myers, C. W., 1986: An enigmatic new snake from the Peruvian Andes, with notes on the Xenodontini (Colubridae: Xenodontinae). Amer. Mus. Novitates, (2853): 1-12.

Myers, C. W., 2003: Rare Snakes—Five new species from eastern Panama: Reviews of northern *Atractus* and southern *Geophis* (Colubridae: Dipsadinae). Am. Mus. Novitates, (3391): 1-47.

Myers, C. W., and J. E. Caddle, 1994: A new genus for South American Snakes related to *Rhadinaea obtusa* Cope (Colubridae) and resurrection of *Taeniophallus* Cope for the "*Rhadinaea*" *brevirostris* group. Amer. Mus. Novitates, (3102): 1-33.

Myers, C. W., and J. A. Campbell, 1981: A new genus and species of colubrid snake from the Sierra Madre del Sur of Guerrero, Mexico. Amer. Mus. Novitates, (2708): 1-20.

Myers, C. W., and M. A. Donnelly, 1996: A new herpetofauna from Cerro Yavi, Venezuela: first results of the Robert G. Goelet American Museum-TERRAMAR Expedition to the northeastern Tepuis. Amer. Mus. Novitates, (3172): 1-56.

Myers, C. W., and M. S. Hoogmoed, 1974: Zoogeographic and taxonomic status of the South American snake *Tachymenis surinamensis* (Colubridae). Zool. Meded. Leiden, 48 (17): 187-194.

Myers, C. W., and A. S. Rand, 1969: Checklist of amphibians and reptiles of Barro Colorado Island, Panama, with comments on faunal change and sampling. Smithsonian Contrib. Zool., Washington, D.C., (10): 1-11.

Myers, G. S., 1926: A synopsis for the identification of the amphibians and reptiles of Indiana. Proc. Indiana Acad. Sci., 34: 277-294.

Nakamura, E. L. and H. Smith, 1966: A comparative study of selected characters in certain American species of watersnakes. Trans. Kansas Acad. Sci., 63 (2): 102-113.

Nascimento, F. P., T. C. S. Avila-Pires and O. R. Rodrigues da Cunha, 1987: Os réptiles da área de Carajás, Pará, Brasil (Squamata). II. Bol. Mus. Paraense Emilio Goeldi, Ser. Zool., 3: 33-65.

Nascimento, F. P. do, T. C. S. Avila-Pires and O. R. da Cunha, 1988: Réptiles squamata de Rondônia e Mato Grosso coletados atraves do programa polonoroeste. Bol. Mus. Paraense Emilio Goeldi, Ser. Zool., 4 (1): 21-66.

Neill, W. T., 1950: The status of the Florida Brown Snake, *Storeria victa*. Copeia, 1950 (2): 155-156.

Neill, W. T., 1954: Ranges and taxonomic allocations of amphibians and reptiles in the southeastern United States. Pub. Research Station Ross Allen's Rept. Inst., I (1954): 75-96.

Neill, W. T., 1964a: Taxonomy, natural history and zoogeography of the rainbow snake, *Farancia erytrogramma* (Palisot de Beauvois). Amer. Midl. Naturalist, 71 (2): 257-295.

Neill, W. T., 1964b: The phylogenetic position of *Dugandia bicincta* (Serpentes: Colubridae). Herpetologica, 20 (3): 194-197 + 3 figs.

Neill, W. T., 1965: New and noteworthy amphibians and reptiles from British Honduras. Bull. Florida State Mus., 9 (3): 77-130.

Neill, W. T., and R. Allen, 1959: Studies on the amphibians & reptiles of British Honduras. Volume 2, No.1, Silver Springs, Florida, Pub. Research Division Ross Allen's Rept. Inst., Inc., November 10, 1959, 2 (1): 1-76.

Neill, W. T., and R. Allen, 1959: Additions to the British Honduras herpetofaunal list. Herpetologica, 15: 235-240.

Neill, W. T., and R. Allen, 1961: Studies on the herpetology of British Honduras. Herpetologica, 17 (1): 37-52.

Neill, W. T., and R. Allen, 1962: Reptiles of the Cambridge Expedition to British Honduras, 1959-1960. Herpetologica, 18: 79-91.

Nellis, D. W., R. L. Norton and W. P. MacLean, 1983: On the biogeography of the Virgin Islands Boa, *Epicrates monensis granti*. J. Herpetol., 17 (4): 413-417.

Nelson, C. E., 1966: Systematics and distribution of snakes of the Central American genus *Hydromorphus* (Colubridae). Texas J. Sci., 18: 365-371.

Nelson, C. E., and J. R. Meyer, 1967: Variation and distribution of the Middle American snake genus *Loxocemus* Cope (Boidae?). Southwest. Naturalist, 12: 439-453.

Nelson, E. W., 1922: Lower California and its natural resources. Mem. Nat. Acad. Sci, Washington, 6 (1): 1-194, pls. 1-35, maps (1921).

Netting, M. G., and C. J. Goin, 1944: Another new boa of the genus *Epicrates* from the Bahamas. Ann. Carnegie Mus. 30 (6): 71-76.

Nicéforo-María, S. C., 1942: Los ofidios de Colombia. Rev. Acad. Colombiana Cien. Exactas, Fis. Nat., 5: 84-101.

Nicéforo-María, S. C., 1950: *Diaphorolepis lasallei*—Contribución al conocimiento de los ofidios de Colombia. Rev. Acad. Colombiana Cien., 7 (28): 517-518.

Nicéforo-María, S. C., 1956: 1. Mastozoologia —Sección Cientifica. Bol. Inst. La Salle, (196): 1-8.

Nicéforo-María, S. C., 1958: Dos casos de albinismo en los ofidios Colombia. Sección Herpetologica-Reptilia Serpentes. Bol. Inst. La Salle, XLV (198): 1-16.

Nicéforo-María, S. C., 1964: Herpetologicá. Bol. Inst. La Salle (Bogotá), 204: 129-135.

Nicéforo-María, S. C., 1970: Contribución al conocimiento de los ofidios de Colombia-Herpetologia. Bol. Inst. La Salle, (210): 1-6.

Nicéforo-María, S. C., 1975: Contribución al estudio de las serpientes de Colombia II. Bol. Inst. La Salle (Bogotá), 215: 1-4

Nickerson, M. A., and H. L. Heringhi, 1966: Three noteworthy colubrids from southern Sonora, Mexico. Great Basin Naturalist, 26 (3-4): 136-140.

Nieto-Montes de Oca, A., and J. R. Mendelson, III, 1997: Variation in *Rhadinaea marcellae* (Squamata: Colubridae), a poorly known species from the Sierra Madre Oriental of Mexico. J. Herpetol., 31 (1): 124-127.

Noble, G. K., and G. C. Klingel, 1932: The reptiles of Great Inagua Island, British

West Indies. Amer. Mus. Novitates, 549: 1–25.

Nogales, Graciela Q., de, Rosa E. de Rosquellas and Germán Montecinos R., 1992: Nueva geografía de Bolivia. "PROINSA" Empresa Editora—La Paz, Bolivia.

Norman, D. R., with photos by Lawrence Naylor, 1994: Amfibios y reptiles del Cháco Paraguayo-Tomo I-amphibians and reptiles of the Paraguayan Chaco. Volume I. 1st ed., San Jose, Costa Rica.

Norton, R. L., 1993: *Alsophis portoricensis richardi*. Herpetol. Rev., 24 (1): 34.

Núñez, H., 1992: Geographical data of Chilean lizards and snakes in the Museo Nacional de Historia Natural Santiago, Chile. Smithsonian Herpetological Infor. Serv., (91): 1–29.

Nussbaum, R. A., E. D. Brodie Jr., and R. M. Storm, 1983: Amphibians and reptiles of the Pacific Northwest. A Northwest Naturalist Book, Univ. Press Idaho, Moscow, Idaho.

Oca-M., A. N. de, and J. R. Mendelson, 1997: Variation in *Rhadinaea marcellae* (Squamata: Colubridae), a poorly known species from the Sierra Madre Oriental of Mexico. J. Herpetol., 31 (1): 124–127.

Ogawa, B., 1991: Troubled by increased discoveries of brown tree snakes in Hawaii, Alaska introduces tougher legislation. Senator Daniel K. Akaka, News 2 pp.

Oldham, J. C., and Hobart M. Smith, 1991: The generic status of the Smooth Green Snake, *Opheodrys vernalis*. Bull. Maryland Herpetol. Soc., 27 (4): 201–215.

Oliver, J. A., 1937: Notes on a collection of amphibians and reptiles from the state of Colima, Mexico. Occ. Pap. Mus. Zool. Univ. Michigan, (360): 1–28.

Oliver, J. A., 1942a: A checklist of the snakes of the genus *Leptophis*, with descriptions of new forms. Occ. Pap. Mus. Zool. Univ. Michigan, (462): 1–19.

Oliver, J. A., 1942B: The relationships and zoogeography of the genus *Thalerophis* Oliver. Bull. Amer. Mus. Nat. Hist., 92: 157–280.

Oliver, J. A., 1947: The seasonal incidence of snakes. Amer. Mus. Novitates, 1363: 1–14.

Oliver, J. A., 1948: The relationships and zoogeography of the genus *Thalerophis* Oliver. Bull. Amer. Mus. Nat. Hist., 92 (4): 157–280.

Oliver, J. A., 1967: The natural history of North American amphibians and reptiles. D. Van Nostrand Co., Inc., Princeton, New Jersey, November, 1967.

Olson, E. R., 1977: Evidence for the species status of Baird's Ratsnake. Texas J. Sci., 29 (1–2): 79–84.

Orcés V. G., 1948: Notas sobre los ofidios venenosos del Ecuador. Pub. Escu. Polit. Nac. (Quito), 3: 231–250.

Orcés V. G., and A. Almendariz, 1987: Sistematica Y distribucion de las serpientes Dipsadinae del grupo Oreas. Depart. Cien. Biol., Escu. Polit. Nac., Politecnica, 12 (4).

Orcés V. G., and A. Almendariz, 1989a: Acerca de la sistematica de *Spilotes megalolepis* Gunther (Serpentes-Colubridae). Politecnica, Departamento de Ciencias Biologicas, Escuela Politecnica Nacional, Quito, Ecaudor, XIV (2): 69–73.

Orcés V. G., and A. Almendariz, 1989b: Presencia en el Ecuador de los colubridos del genero *Sibynomorphus*. Depart. Cien. Biol., Escu. Polit. Nac., Politecnica, Quito, Ecuador, XIV (2): 57–67.

Orejas-Miranda, B. R., 1958: Dos espécies de ofidios nuevos para el Uruguay. Com. Zool. Mus. Montevideo, 4 (79): 1958: 1–6, 2 pls.

Orejas-Miranda, B. R., 1959a: The snake genus *Lystrophis* in Uruguay. Copeia, 1966 (2): 193–205.

Orejas-Miranda, B. R., 1959b: Una nueva subespecie del genero *Philodryas* del Uruguay. Com. Zool. Mus. Hist. Nat. Montevideo. 4 (82): 1–3.

Orejas-Miranda, B. R., 1961: Una nueva especie de ofidio de la familia Leptotyphlopidae. Act. Biol. Venezuelica, 3 (5): 83–97.

Orejas-Miranda, B. R., 1962: Descripción del hemipenis de *Leptotyphlops muñoai* Orejas-Miranda, 1961. Com. Zool. Mus. Hist. Nat. Montevideo, 7 (7): 1–5, 2 pls.

Orejas-Miranda, B. R., 1964: Dos nuevos *Leptotyphlopidae* de Sur America. Comun. Zool. Mus. Hist. Nat. Montevideo, Uruguay, 8 (103): 1–7 + 3 plates.

Orejas-Miranda, B. R., 1966a: Notas sobre la familia Leptotyphlopidae, I–II. I—Revalidación de *Leptotyphlops cupiensis* Bailey & Carvalho, 1946 II-Sinonimización de *L. ihlei* Brongersma, 1933 con *L. macrolepis* (Peters, 1881). Comun. Zool. Mus. Hist. Nat. Montevideo. 9 (108): 1–3.

Orejas-Miranda, B. R., 1966b: The snake genus *Lystrophis* in Uruguay. Copeia, 1966 (2): 193–205.

Orejas-Miranda, B. R., 1967: El genero *Leptotyphlops* en la region Amazonica. Mus. Nac. Hist. Nat., Montevideo, Uruguay, Atlas do Simposio sôbre a Biota Amazônica. Vol.5 (Zoologia): 421–442.

Orejas-Miranda, B. R., 1969: Tres nuevos *Leptotyphlops* (Reptilia: Serpentes). Comun. Zool. Mus. Hist. Nat. Montevideo. 10 (124): 1–11.

Orejas-Miranda, B. R., 1973: Observaciones sobre un caso de albinismo de *Leptotyphlops munoai*. Bol. Soc. Zool. Montevideo, 2: 36.

Orejas-Miranda, B. R., and D. Garcia, 1967: Observaciones sobre una puesta de *Philodryas patagoniensis* (Girard, 1857) = *P. schotti* (Schlegel, 1837). Neotropica, 13 (40): 41–46.

Orejas-Miranda, B. R., and G. Peters, 1970: Eine neue Schlankblindschlange (Serpentes: Leptotyphlopidae) aus Ecuador. Mitt. Zool. Mus. Berlin, Akadem'e-Verlag, 46 (2): 439–441.

Orejas-Miranda, B. R., R. Roux-Esteve and J. Guibe, 1970: Un nouveau genre de Leptotyphlopides (Ophidia) *Rhinoleptus koniagui* (Villiers). Comun. Zool. Mus. Hist. Nat. Montevideo, 10 (127): 1–4

Orejas-Miranda, B. R., and G. R. Zug, 1974: A new tricolor *Leptotyphlops* (Reptilia: Serpentes) from Peru. Proc. Biol. Soc. Washington, 87 (16): 167–174.

Ortenburger, A. I., 1923: A note on the genera *Coluber* and *Masticophis*, and a description of a new species of *Masticophis*. Occ. Pap. Mus. Zool. Univ. Michigan, 139: 1–14.

Ortenburger, A. I., 1926: A preliminary list of the snakes of Oklahoma. Proc. Oklahoma Acad. Sci., 5 (1925): 83–87.

Ortenburger, A. I., [1926] 1927a: A key to the snakes of Oklahoma. Proc. Oklahoma Acad. Sci. 6 (1926): 197–218.

Ortenburger, A. I., [1926] 1927b: A report on the amphibians and reptiles of Okla-

Bibliography

homa. Proc. Oklahoma Acad. Sci., 6 (1926): 89-100.

Ortenburger, A. I., 1928: The whip snakes and racers—genera *Masticophis* and *Coluber*. Univ. Michigan, Ann Arbor.

Ortenburger, A. I., 1930: A key to the lizards and snakes of Oklahoma: Vol. II—Biological survey No. 4. Univ. Oklahoma Press., Norman, II (4): 209-239.

Ortiz, J. C., 1973: Étude sur le statut taxinomique de *Tachymenis peruviana* Wiegmann et *Tachymenis chilensis* (Schlegel) (Serpentes: Colubridae). Bull. Mus. Nat. 7 d'Hist. Nat., Zool. 110, 3e série (146): 1021-1039, mai-junin 1973.

Osborn, H. F., 1931: Cope: master naturalist. The life and letters of Edward Drinker Cope with a bibliography of his writings classified by subject. Princeton U. Press, Princeton, NJ.

O'Shea, M., 1996: A guide to the snakes of Papua New Guinea. Independent Publishing, Port Moresby, Paupa New Guinea.

O'Shea, M. T., 1986: Geographic distribution—*Nothopsis rugosus*. Herpetol. Rev., 17: 27.

O'Shea, M., and T. Halliday, 2002: Reptiles and amphibians. Doring Kindersley Handbooks. 1st American Ed., Reprint with corrections. New York.

O'Shea, M. T., and A. F. Stimson, 1993: An abberrant specimen of *Drymobius rhombifer* (Colubridae: Colubrinae): a new generic record for Brazil. Herpetol. J., 3 (2): 70-71.

Ottley, J. R., and W. W. Tanner, 1978: New range and a new subspecies for the snake *Eridiphas slevini*. Great Basin Naturalist, 38 (4): 406-410.

Padilla Garcia, U., and F. Mendoza Quijano, 1996a: Geographic distribution—*Imantodes gemmistratus* (Central American Tree Snake). Herpetol. Rev., 27 (4): 213.

Padilla Garcia, U., and F. Mendoza Quijano, 1996b: *Leptophis mexicanus septentrionalis* (Tamaulipan parrot snake). Herpetol. Rev., 27 (4): 213-214.

Pagini, E., and T. de Lema, 1987: Reeniontro de *Calamodontophis paucidens* (Amaral, 1936) e controbuiçao ao conhecimento do gênero e da espécie (Serpentes, Colubridae, Tachymeninae). Com. Mus. Ciênc. PUCRS, Porto Alegre, (47): 195-208.

Painter, C. W., P. W. Hyder and G. Swinform, 1992: Three species new to the herpetofauna of New Mexico. Herpetol. Rev., 23 (2): 64.

Palau, M. y, B. Sáiz, 1989: Moxos-descripciones exactas e historia fiel de los indios, animales y plantas de la provincia de Moxos en el virreinato del Perú por Lázaro de Ribera 1786-1794. Ministro de Agricultura, Pesca y Alimentación, Ediciones el Viso.

Palmer, W. M., 1978: *Sistrurus miliarius* (Linnaeus)—Pygmy rattlesnake. Cat. Amer. Amph. Rept., (220): 1-2.

Palmer, W. M., and A. L. Braswell, 1995: Reptiles of North Carolina. Univ. North Carolina Press, Chapel Hill, North Carolina.

Paolillo, A. O., 1986: *Helicops hagmanni* (Mapanare de Agua). Herpetol. Rev., 17 (2): 49.

Parker, H. W., 1926: Description of a new snake from Trinidad. Ann. Mag. Nat. Hist., 9 (18): 205-207.

Parker, H. W., 1930: Two new reptiles from southern Ecuador. Ann. Mag. Nat. Hist., 10 (5): 568-571.

Parker, H. W., 1933: Some amphibians and reptiles from the Lesser Antilles. Ann. Mag. Nat. Hist., Ser. 10, 11: 151-158.

Parker, H. W., 1935: The frogs, lizards and snakes of British Guiana. Proc. Zool. Soc. London, 3: 505-530.

Parker, H. W., 1936: A collection of reptiles and amphibians from the Upper Orinoco. Mus. Roy. Hist. Nat. Belgique, 12 (26): 1-4.

Parker, H. W., 1938: The vertical distribution of some reptiles and amphibians in southern Ecuador. Ann. Mag. Nat. Hist., 11 (2): 438-450.

Parker, H. W., 1940: Undescribed anatomical structures and new species of reptiles and amphibians. Ann. Mag. Nat. Hist., 11 (5): 257-274.

Parker, H. W., 1963: Snakes of the world, their ways & means of living. New York, Dover Publications, Inc.

Parker, W. S., 1982: *Masticophis taeniatus*. Cat. Amer. Amph. Rept., (304): 1-4.

Parker, W. S., and W. S. Brown, 1980: Comparative ecology of two colubrid snakes, *Masticophis t. taeniatus* and *Pituophis melanoleucus deserticola*, in northern Utah. Milwaukee Mus. Publ. Biol. Geol. (7): vii + 104 pp.

Parker, W. S., and W. S. Brown, 1984: Growth, reproduction and demography of the racer, *Coluber constrictor mormon*, in northern Utah. *In* R. A., Seigel, et al. (Eds), Vertebrate ecology and systematics. A tribute to Henry S. Fitch. Spec. Pubs Kansas Mus. Nat. Hist., No. 10.

Parker, W. S., and S. Sweet, 1990: *Pituophis melanoleucus*. Cat. Amer. Amph. Rept., 474: 1-8.

Parkinson, C. L., K. R. Zamudio and H. W. Greene, 2000: Phylogeography of the pitviper clade *Agkistrodon*: historical ecology, species, status, and conservation of cantils. Molecular Ecology, 9: 411-420.

Patrick, D., and I. Ineich, 1999: Les serpents venimieux du monde: systématique et réparition. Dumerilia, 3: 1-499.

Patzelt, E., 1989: Fauna del Ecuador. Banco Central del Ecuador, Quito, Ecuador.

Paynal, N., and C. Sosa, 1993: Mundos amazonicos—pueblos y culturas de la amazonia Ecuatoriana. Fundación Sinchi Sacha, Quito, Ecuador.

Payne, D. L., 1981: The pholonogy and morphology of Axininca Campa. Summer Institute of Linguistics, University of Texas at Arlington.

Pefaur, J. E., 1992: Checklist and bibliography (1960-1985) of the Venezuelan herpetofauna. Smithsonian Herpetological Information Service, 89: 1-54.

Peracca, M. G., 1896: Rettili et amphibi raccolti nel Darien et a Panamá dal Dott. E. Festa. Boll. Mus. Zool. Comp. Anat. Univ. Torino, 11 (235): 1-4.

Peracca, M. G., 1910: Descrizione de alcune nuove specie di ofidii del Museo Zoologico della R. Universitá di Napoli, Ann. Mus. Zool. Univ. Napoli, n.s., 3: 1-3.

Peracca, M. G., 1912: Reptiles et batrachiens de Colombie. *In* Fuhrmann, O. & E. Mayor, Voyage d'exploration scientifique en Colombie. Vol. V des Mémoires de la Société neuchàteloise des Sciences naturelles. Imprimerie Attinger Fréres, Neuchatel, 5: 96-111.

Peracca, M. G., 1914: Reptiles et batrachiens de Colombie. *In* Fuhrmann, O. & E. Mayor 1914, Voyage d'exploration scientifique en Colombie. Meu. Soc. Neuchatel Sci. Nat., 5: 1-1090.

Peraria de Souza, G. M.; J. Rodrigues de Lima and L. C. Aveline, 1991: Sistema de informação de recursos natuarais e meío ambiente. Vol.3, Sistematizaçõ de dades sôbre a fauna de nomes vulgares de anfíbios e réptileis brasileiros. Seríe Estudos E Pesquisas Em Geociências. Supserie Recursos Naturals e Meío Ambiente. I. Maio 1991, Rio de Janiero, No. 03/91.

Pérez-Bravo, G., [1976/1977] 1981: Segundo hallazgo de *Helicops hogei* Lancini, 1964 (Serpentes: Colubridae). Mem. Inst. Butantan, 40/41: 313–315.

Pérez-Higareda, G., M. A. López-Luna and H. M. Smith, 2002: A new snake related to *Sibon sanniola* (Serpentes: Dipsadidae) from Los Tuxtlas, Veracruz, Mexico. Bull. Maryland Herp. Soc., 38 (2): 62-65.

Pérez-Higareda, G., and D. Navarro L., 1980: The faunistic districts of the low plains of Verazcruz, Mexico, based on reptilian and mammalian data. Bull. Maryland Herpetol. Soc., 16 (2): 54-69.

Pérez-Higareda, G., and H. M. Smith, 1986a: The status of Los Tuxtlas (Mexico) false coral snakes (*Pliocercus*). Bull. Maryland Herpetol. Soc., 22: 125-130.

Pérez-Higareda, G., and H. M. Smith, 1986b: *Trimorphodon biscutatus* (Reptilia: Serpentes) on the Atlantic versant in southern Mexico, Bull. Maryland Herpetol. Soc., 22 (4): 179-180.

Pérez-Higareda, G., and H. M. Smith, 1988: Notes on two species of *Geophis* (Serpentes) of southern Mexico. Southwest. Naturalist, 33 (3): 388-390.

Pérez-Higareda, G., and H. M. Smith, 1989a: Termite nest incubation of the eggs of the Mexican colubrid snake *Adelphicos quadrivirgatus*. Herpetol. Rev., 20 (1): 5-6.

Pérez-Higareda, G., and H. M. Smith, 1989b: The distribution of *Leptophis diplotropis* (Serpentes: Colubridae). Bull. Maryland Herpetol. Soc., 25 (3): 73-76.

Pérez-Higareda, G., and H. M. Smith, 1990: The endemic coral snakes of the Los Tuxtlas region, southern Veracruz, Mexico. Bull. Maryland Herpetol. Soc., 26 (1): 5-13.

Pérez-Higareda, G., and H. M. Smith, 1991: Ofidiofauna de Veracruz análisis taxonómico y zoolgeográfica: Ofidiofauna of Veracruz-taxonomical and zoogeographical analysis. Univ. Nac. Autón. México, México, D. F., Publicaciones Especiales 7: 1-122.

Pérez-Higareda, G., H. M. Smith and M. A. López-Luna, 2001: A new *Geophis* (Reptilia: Serpentes) from southern Veracruz, Mexico. Bull. Maryland Herp. Soc., 37 (2): 42-48.

Pérez-Higareda, G.; H. M. Smith and J. Juliá-Zertuche, 1985: A new jumping viper, *Porthidium olmec*, from southern Veracruz, Mexico (Serpentes: Viperidae). Bull. Maryland Herpetol. Soc., 21: 97-106.

Pérez-Santos, C., 1986a: Las serpientes del Atlantico. Mus. Nac. Cien. Nat., Madrid, Spain.

Pérez-Santos, C., 1986b: Las serpientes del Tolima. Mus. Nac. Cien. Nat., Madrid, Spain.

Pérez-Santos, C. E., 1986c: Zoogeografia de los ophidia en Colombia. Tomo I and Tomo II. Doctorial Thesis No. 46/86. Printed in Editorial de la Universidad Complatense de Madrid, Spain, November 3, 1986, Volume 1: 1-487; Volume 2: 488-971.

Pérez-Santos, C., and A. G. Moreno, 1988: Ofidios de Colombia, monografie VI. Mus. Reg. Sci. Nat. Torino, Italy.

Pérez-Santos, C., and A. G. Moreno, 1989: Una nueva especie de *Thamnodynastes* (Serpentes: Colubridae) en el norte de Colombia. Mus. Reg. Sci. Nat. Boll. (Turin), 7 (1): 1-9.

Pérez-Santos, C., and A. G. Moreno, [1991] 1992: Serpentes de Ecuador, Monografica XI. Mus. Reg. Sci. Nat., Torino, Italy.

Perkins, C. B., 1949a: A key to the snakes of the United States. Bull. Zool. Soc. San Diego, (24): 1-79.

Perkins, C. B., 1949b: Longevity of snakes in captivity in the United States. Copeia, 1949: 223.

Perkins, C. B., 1949c: The snakes of San Diego County with descriptions and key. Bull. Zool. Soc. San Diego, (23): 1-77.

Pessoa, S. B., P. de Baisi, and G. Puorto, 1974: Nota sobre a frequencia de hemoparasitas em serpentes do Brasil. Mem. Inst. Butantan, 38: 69-118.

Peters, J. A., 1950: A new snake of the genus *Coniophanes* from Veracruz, Mexico, Copeia, 1950 (4): 279-280.

Peters, J. A., 1952: Catalogue of type specimens in the herpetological collections of the University of Michigan Museum of Zoology. Occ. Pap. Mus. Zool. Univ. Michigan, (539): 1-55.

Peters, J. A., 1954: The amphibians and reptiles of the coast and coastal sierra of Michoacán, Mexico. Occ. Paps. Mus. Zool. Univ. Michigan, (554): 1-37.

Peters, J. A., 1955: The proper citation for certain species described by Tschudi. Copeia, 1955 (1): 70.

Peters, J. A., 1956: An analysis of variation in a South American snake, Catesby's snail sucker (*Dipsas catesbyi* Sentzen). Amer. Mus. Novitates, (1783): 1-41.

Peters, J. A., 1957a: A new snake species of the snake genus *Sibon* from Ecuador. Copeia, 1957 (2): 109-111.

Peters, J. A., 1957b: Taxonomic notes on Ecuadorian snakes in the American Museum of Natural History. Amer. Mus. Novitates, (1851): 1-13.

Peters, J. A., 1960a: *Leptophis cupreus* Cope—A valid South American colubrid species. Beitr. Z. Neotrop. Fauna, II (2): 139-141.

Peters, J. A., 1960b: The snakes of Ecuador, a check list and key. Bull. Mus. Comp. Zool., 122 (9): 491-541.

Peters, J.A, 1960c: The snakes of the family Dipsadinae. Misc. Pub. Mus. Zool. Univ. Michigan, (114): 1-224, May 25, 1960.

Peters, J. A., 1964a: Dictionary of herpetology. Hafner Publishing Co., New York.

Peters, J. A., 1964b: Supplemental notes on snakes of the subfamily dipsadinae (Reptilia: Colubridae). Beiträge Zur Neotropischen Fauna, IV, Band, Heft 1, Gustav Fischer Verlag, Stuttgart, 1964: 45-50.

Peters, J. A., 1968: A replacement name for *Bothrops lansbergii venezuelensis* Roze, 1959 (Viperidae, Serpentes). Proc. Biol. Soc. Washington, D.C., 81: 319-322.

Peters, J. A., 1970a: A note on the generic names *Cyclagras* Cope and *Lejosophis* Jan (Reptilia: Serpentes). Proc. Biol. Soc. Washington, D.C., 83 (67): 847-850.

Peters, J. A., 1970b: Generic position of the South American snakes *Tropidodipsas perijanensis*. Copeia, 1970: 394-395.

Peters, J. A., and G. Orcés-V., 1960: *Leptophis cupreus* Cope-a valid South American colubrid species. Beitr. Z. Neotrop. Fauna, II (2): 139-141, January 15, 1960.

Bibliography

Peters, J. A., and B. Orejas-Miranda, 1970: Catalogue of the neotropical squamata—Part I. Snakes. U.S. Natl. Mus. Bull. 297.

Peters, W., [dated 1859] 1860a: Über die von Hm. Hoffmann in Costa Rica gesammelten und an das Königl. Zoologische Museum gesandten Schlangen. Monatsber. Preuss. Akad. Wiss., Berlin, 1860: 275-278.

Peters, W., 1860b: Eine neue Gattung von Riesenschlangen vor, welche von einem gebornen Preussen, Arn. Carl. Reiss, in Guyaquil nebst mehrereh anderen werthvollen Naturalien dem zoologischen Museum Zugesandt worden ist.Mber. Königl. Akad. Wiss. Berlin, 1860: 200-202.

Peters, W., 1861a: [Eine Mittheilung über neue Schlangen des Königl. zoologischen Museums: *Typhlops striolatus, Geophidium dubium, Streptophorus (Ninia) maculatus, Elaps hipporcrepis.*] Mber. Königl. Akad. Wiss. Berlin, 1861: 922-925.

Peters, W., 1861b: Über eine Sammlung von Schlangen aus Huanusco in Mexico welche ds Königl. Zoologische Museum kurzlich von Dr. Hille erworden hat.Mber. Königl. Akad. Wiss. Berlin, 1861: 460-462.

Peters, W., 1861c: Über zwei neue von Hrn. Dr. Gundlach auf Cuba entdeckte Schlangen, *Tropidonotus cubanus* und *Cryptodacus vittatus*. Mber. Königl. Akad. Wiss. Berlin, 1861: 1001-1004.

Peters, W., 1862: Über neue Schlangen des Königl. Zoologische Museums: *Typhlops striolatus. Geophidium dubium, Streotiogiryn (Ninia) maculatus, Elaps hippocrepis*. Mber. Königl. Akad. Wiss. Berlin, 1861: 922-925.

Peters, W., 1863a (dated 1862): Ueber die craniologischen Verschiedenheiten der Grubenottern (*Trigonocephali*) und über eine veuve Art der Gattung *Bothriechis*.Mber. Königl. Akad. Wiss. Berlin, 1863: 670-674.

Peters, W., 1863b: Über einige neue oder weniger bekannte Schlangenarten des zoologischen Museums zu Berlin. Mber. Königl. Akad. Wiss. Berlin, 1863: 272-289.

Peters, W., 1863c: Über einige neue oder wenig bekannte Schlangenarten der zoologischen Museums zu Berlin.Mber. Königl. Akad. Wiss. Berlin, 1864: 272-289.

Peters, W., [1864] 1865: Über einige neue Säugethiera (*Mormops, Macrotus, Vesperus, Molossus, Capromys*), Amphibien (*Plathydactylus, Otocryptis, Euprepes, Ungalia, Dromicus, Tropidonotus, Xenodon, Hylodes*), und Fische (*Sillago, Sebastes, Channa, Myctophum, Carassius, Barbus, Capoëta, Sauæcuchelys, Leptocephalus*).Mber. Königl. Akad. Wiss. Berlin, [1864]: 381-399.

Peters, W., 1867: Herpetologische Notizen.Mber. Königl. Akad. Wiss. Berlin, 1867: 13-37.

Peters, W., 1868: Über neue Säugetiere (*Colobus, Rhinalopus, Vesperus*) und neue oder weniger bekannte Amphibien (*Hemidactylus, Herpetodryas, Spilotes, Elaphis, Lamprophis, Erythrolamprus*). Mber. Königl. Akad. Wiss. Berlin, 1868 (December): 63 7-642.

Peters, W., 1869: Ueber neue Gattungen und neue oder weniger bekannte Arten von Amphibien (*Eremias, Dicrodon, Euprepes, Lygosoma, Typhlops, Eryx, Rhynchonyx, Elapomorphus, Achalinus, Coronella, Dromicus, Xenopholis, Anoplodipas, Spilotes, Tropidonotus*). Mber. Königl. Akad. Wiss. Berlin, 1869: 432-447.

Peters, W., 1871: Mittheilung über eine von Hrn. Dr. Robert Abendroth in dem Hochlande von Peru germachte Sammlung von Amphibien, welche derselbe dem Königl. Zoologischen Museum geschenkt hat. Monatsber. K. Preussichen Adad. Wiss. Berlin, 1871: 397-404.

Peters, W., 1877: Über die von Herr Prof. Dr. K. Möbius, 1874 auf den Markasenen und Seychellen, sowie über die von Herr Dr. Sachs im vorigen Jahr in Venezuela gesammellten Amphibien. Mber. Königl. Akad. Wiss. Berlin, 1877: 455-460.

Peters, W., 1880: Eine Mittheilung über neue oder weniger bekannte Amphibien des Berliner Zoologischen Museums (*Leposoma dispar, Monopeltis, Phractogonus jugularis, Typhlops depressus, Leptocalamus drilineatu, Xenodon punctatus, Elaphomorphus erythronotus, Hylomantis fallax)*.Mber. Königl. Akad. Wiss. Berlin, 1880: 217-224.

Peters, W., 1881a: Einige herpetologische Mittheilungen. Ber. Ges. Natur.uende Berlin, 1881 (4): 69-72.

Peters, W., 1881b: Über das Vorkommen schildförmiger Verbreitungen der Dornfortsatze bei Schlangen und über neue oder weniger bekannte Arten dieser Abtheilung der Reptilien, Ber. Ges. Naturf. Freunde, Berlin, 1881 (3): 49-52.

Peters, W., 1882: Über eine neue Gattung und Art der Vippernattern, *Dinodipsas angulifera*, aus Südamerika. Sber. Königl. Preuss. Akad. Wiss. Berlin, 40: 893-896.

Phelps, T, 1981: Poisonous Snakes. Blandford Press, Poole, Dorset, UK.

Philippi, R. A., 1873: Ueber die *Boa* der westlichen Provinzen des argentinischen Republik. Zeitsch. Gesammte Naturwiss., Halle, 41: 127-130.

Philippi, R. A., 1899: Sobre las serpientes de Chile. An. Univ. Chile, 104: 715-725.

Phillips, C. A., R. A. Branden, and E. O. Moll, 1999: Field guide to amphibians and reptiles of Illinois. Manuel 8. Illinois Nat. Hist. Survey. Champaign, Illinois, August.

Picado T., C., 1976: Serpientes venenosas de Costa Rica sus venenos seroterapia antiofidia. Editorial Universidad de Costa Rica, San Jose, Costa Rica.

Pickwell, G. V., 1972: The venomous sea snakes. Fauna, 4: 17-32.

Pickwell, G. V. and W. A. Culotta, 1980: *Pelamis, P. platurus*. Cat. Amer. Amph. Rept., (255): 1-4.

Pinou, T., C. A. Haas and L. R. Maxson, 1995: Geographic variation of serum albumin in the monotypical snake genus *Diadophis* (Colubridae: Xenodontinae). J. Herpetology, 29 (1): 105-110.

Pisani, G. R., J. T. Collins and S. R. Edwards, 1973: A re-evaluation of the subspecies of *Crotalus horridus*. Trans. Kansas Acad. Sci., 75 (3): 255-263.

Platt, D. R., 1969: Natural history of the hognose snakes *Heterodon platyrhinos* and *Heterodon nasicus*. Univ. Kansas Pub. Mus. Nat. Hist., 18 (4): 253-420.

Platt, D. R., 1983: *Heterodon*. Cat. Amer. Amph. Rept., (315): 1-2.

Platt, D. R., 1985: History and spelling of the name *Heterodon platirhinos*. J. Herpetol., 19 (2): 417-418.

Platt, S. G., and T. R. Rainwater, 1998: Distribution records and life history notes for amphibians and reptiles in Belize. Herpetol. Rev., 29 (4): 250-251.

Plenge, H., 1993: Peru Vida Silvestre/Wildlife. Lima, Peru, Geo-Foto.

Ponce-Campos, P., and S. M. Huerta-Ortega, 1998a: *Dryadophis cliftoni* (Clifton's Lizard Eater). Herpetol. Rev., 29 (3): 176.

Ponce-Campos, P., and S. M. Huerta-Ortega, 1998b: *Imantodes gracillimus* (Slender Blunthead Treesnake). Herpetol. Rev., 29 (3): 177.

Ponce-Campos, P., and H. M. Smith, 2001: A review of the Stripeless Snake (*Coniophaes lateritus*) complex of Mexico. Bull. Maryland Herp. Soc., 37 (1): 10-17.

Pook, K. E., W. Woster, and R. S. Thorpe, 2000: Historical biogeography of the western rattlesnake (Serpentes: Viperidae: *Crotalus viridis*), inferred from mitochondrial DNA sequence information. Pub. Molecular Philogenetics Evol., 15 (2): 269-282.

Pope, C. H., 1937: Snakes alive and how they live. Alfred A. Knoft, New York. reviews: by Gloyd, Prog. Chicago Acad. Sci., 8: 31-32; also, Science Newsletter, 32 (Sept. 11): 175; and Scientific American, 158 (1938): 58; Herpetologica, 1: 120.

Pope, C. H., 1955: The reptile world. A natural history of the snakes, lizards, turtles, and crocodilians. Alfred A. Knopf.

Pope, C. H., 1969: The giant snakes. 4th Edition, New York, Alfred A. Knopf.

Pope, C. H., 1971: The reptile world. 6th Printing, New York, Alfred A. Knopf.

Pope, J., 1792: A tour through the southern and western territories of the United States of North-America; the Spanish Dominions on the River Mississippi, and the Floridas. Richmond, Virginia, pp. 104 pp. + iv. [reprint, New York, 1888, 104 pp. + 4].

Porras, L., J. R. McCranie, and L. D. Wilson, 1981: The systematics and distribution of the hognose viper *Bothrops nasuta* Bocourt (Serpentes: Viperidae). Tulane Stud. Zool. Bot., 22 (2): 85-107.

Porter, K., 1972: Herpetology. W. B. Saunders Company, Philadelphia, Pennsylvania, 524 pp.

Porto, M., and U. Caramaschi, 1988: Notes on the taxonomic status, biology, and distribution of *Ptychophis flavovirgatus* Gomes 1915 (Ophidia, Colubridae). Ann. Acad. Bras. Cien., 60 (4): 471-475.

Porto, M., and R. Fernandes, 1996: Variations and natural history of the snail-eating snake *Dipsas neivai* (Colubridae: Xenodontinae). J. Herpetol., 30 (2): 269-271.

Pough, F. H., R. M. Andrews, J. E. Cadle, M. Crump, A. H. Savitzky, and D. Wells, 1998: Herpetology. Prentice-Hall, Upper Saddle River, N.J.

Powell, E., B. Bruun, D. Kleiman and B. Kypta, 1970: Animals of the world, North America. New York, The Hamlyn Publishing Group, Ltd.

Powell, R., 1990: *Elaphe vulpina*. Cat. Amer. Amph. Rept., (470): 1-3.

Powell, R., 1993: Comments on the taxonomic arrangement of some Hispaniolan amphibians and reptiles. Herpetol. Rev., 24: 135-137.

Powell, R., J. T. Collins and L. D. Fish, 1992: *Virginia valeriae* Baird and Girard—Smooth Earth Snake. Cat. Amer. Amph. Rept., (552): 1-6.

Powell, R., J. T. Collins and L. D. Fish, 1994: *Virginia striatula* (Linnaeus)—Rough Earth Snake. Cat. Amer. Amph. Rept., (599): 1-6.

Powell, R., J. T. Collins and E. D. Hooper, Jr., 1998: A key to the amphibians and reptiles of the continental United States and Canada. Univ. Kansas Press, Lawrence, Kansas.

Powell, R., and R. W. Henderson, 1988: *Alsophis melanichnus*. Cat. Amer. Amph. Rept., (660): 1-2.

Powell, R., and R. W. Henderson, 1994a: *Ialtris* Cope. Cat. Amer. Amph. Rept., (590): 1-2.

Powell, R., and R. W. Henderson, 1994b: *Ialtris aygrtus* Schwartz & Rossman. Cat. Amer. Amph. Rept., (591): 1-2.

Powell, R., and R. W. Henderson, 1994c: *Ialtris parishi* Cochran. Cat. Amer. Amph. Rept., (593): 1-2.

Powell, R., and R. W. Henderson, 1998: *Alsophis anomalus* (Peters)—Hispaniolan Racer. Cat. Amer. Amph. Rept., (659): 1-2.

Powell, R., and R. W. Henderson, 1999: Addenda to the checklist of West Indian amphibians and reptiles. Herp. Review, 30 (3): 137-139.

Powell, R., J. A. Ottenwalder and S. J. Incháustegui, 1999: The Hispaniolan herpetolofauna: diversity, endemism, and historical perspectives, with comments on Navassa Island. 1999: 93-168. *In* B. I. Crother (ed.), Caribbean amphibians and reptiles, Academic Press.

Powers, A., 1974: The systematic status of *Trimorphodon lyrophanes* Cope on Cerralvo Island, Gulf of California, Mexico. Bull. Maryland Herpetol. Soc., 10 (2): 53-55.

Powers, A. L. and B. H. Banta, 1974: Description of a *Phyllorhynchus* from Cerralvo Island, Gulf of California, Mexico. Great Basin Naturalist, 34: 241-244.

Prado, A., [1939] 1940: Notas ofidologicas 4—Cinco especies novas de serpentes colombianas de genero *Atractus* Wagler. Mem. Inst. Butantan, 13: 5, 15-23.

Prado, A., [1940] 1941a: Notas ofiológicas. 6. Uma nova espécie de colubrídeo áglifo da Colômbia. Mem. Inst. Butantan, 14 (1940): 13-15.

Prado, A., [1940] 1941b: Notas ofiológicas. 7. Sobre a determinação de *Elapomorphus trilineatus* Boulenger e afins. Mem. Inst. Butantan, 14 (1940): 17-23, pls.

Prado, A., [1940] 1941c: Notas ofiológicas. 8. Dois novos *Atractus* da Colombia. Mem. Inst. Butantan, 14 (1940): 25-28, pl.

Prado, A., 1941b: Sôbre una rasa do serpente neotropica. Mem. Inst. Butantan, 15: 373.

Prado, A., [1941] 1942: Notas ofiolólogicas 12. Considerações en torno de dois *Atractus* da Colômbia, con a descrição de uma nova espécie. Mem. Inst. Butantan, 15 [1941]: 377-380.

Prado, A., 1942a: Notas ofiológicas. 11. Sobre uma raça de serpente neotrópica. Mem. Inst. Butantan, 15 (1941): 373-375.

Prado, A., 1942b: Notas ofiológicas. 12. Consideração em torno de dois *Atractus* da Colombia com a descrição de uma nova espécie. Mem. Inst. Butantan, 15 (1941): 377-380, pl.

Prado, A., 1942c: Serpentes do genero *Dryophylax*, con a descriçao de uma nova especie. Ciencia (Mexico), 3 (7): 204-205.

Prado, A., 1943a: Notas ofiloógicas. 13. Redescrição de duas serpentes colombianas. Mem. Inst. Butantan, 16 (1942): 1-6 + pl.

Prado, A., 1943b: Notas ofiológicas. 14. Comentários acerca de algumas serpentes opistóglifas do gênero *Apostolepis*, com a descrição de uma nova espécie. Mem. Inst. Butantan, 16 (1942): 7-12 + pl.

Prado, A., 1944: Notas ofidiológicas 15. Serpentes do genero *Dryophylax*, com a redescriçao de uma nova especie. Mem. Inst. Butantan, 17 (1943): 1-5.

Prado, A., 1945a: Serpentes do Brasil, São Paulo, Edit. Sitios e Fazendas, Biblioteca Agropecuaria. 134 pp.

Prado, A., 1945b: Um novo *Atractus* da Colômbia. Ciencia (Mexico), 6 (2): 61.

Prado, A., [1945] 1946a: Notas ofidiologicas. 18. A. Posição do genero *Rhadinaea* em sistematica. Mem. Inst. Butantan, 18: 109–114.

Prado, A., [1945] 1946b: Notas ofidiologicas. 19. *Atractus* da Colombia, com a descrição de tres novas especies. Mem. Inst. Butantan, 18: 493–498.

Prado, A., 1948: Notas ofidiológicas 20. Descriçao do alotipo de *Dryophylax rutilus* Prado, 1942. Mem. Inst. Butantan, 20 (1947): 189–192.

Price, A. H., 1980: *Crotalus molossus* Baird and Girard-Black-tailed Rattlesnake. Cat. Amer. Amph. Rept., (242): 1–2.

Price, A. H., 1982a: *Crotalus molossus*. Cat. Amer. Amph. Rept., (242): 1–2.

Price, A. H., 1982b: *Crotalus scutulatus*. Cat. Amer. Amph. Rept., (291): 1–2.

Price, A. H., 1998: Poisonous snakes of Texas. Texas Parks and Wildlife Press, Austin, Texas.

Price, R., 1983: Microdermatoglyphica: The *Liodytes*—*Regina* problem. J. Herpetol., 17: 292–294.

Price, R., 1987: Disjunct occurrence of mole snakes in peninsular Florida, and the description of a new subspecies of *Lampropeltis calligaster*. Bull. Chicago Herpetol. Soc., 22 (9): 148.

Price, R. M., 1990a: *Bogertophis* Dowling and Price—Trans-Pecos and Santa Rosalia Rat Snake. Cat. Amer. Amph. Rept., (497): 1–2.

Price, R. M., 1990b: *Bogertophis rosalie* (Mocquard)—Santa Rosalia Rat Snake. Cat. Amer. Amph. Rept., (498): 1–3.

Price, R. M., 1991: *Senticolis, S. triaspis*. Cat. Amer. Amph. Rept., (525): 1–4.

Price, R. M., and P. Russo, 1991: Revisionary comments on the genus *Boa* with the description of a new subspecies of *Boa constrictor* from Peru. The Snake, 23: 29–35.

Procter, J. B., 1923: On new and rare reptiles from South America. Proc. Zool. Soc. London, 1923: 1061–1068.

Puorto, G., and H. Ferrarezzi, 1993: Uma nova espécie de *Phalotris* Cope, 1862, com comentários sobre o grupo *bilineatus* (Serpentes, Colubridae, Xenodontinae). Mem. Inst. Butantan, v. 55, Supl. 1: 39–46.

Puorto, G., and R. W. Henderson, 1994: Ecologically significant distribution records for the common tree snake (*Corallus enydris*) in Brazil. Herpetol. Nat. Hist., 2 (2): 89–91.

Puorto, G., M. G. Salomão, R. D. G. Theakston, R. S. Thorpe, D. A. Warrell and W. Wüster, 2001: Combining mitochondrial DNA sequence and morphological data to infer species boundaries: phylogeography of lancehead pitvipers in the Brazilian Atlantic fores, and the status of *Bothrops pradoi* (Squamata: Serpentes: Viperidae). J. Evolutionary Biol., 14 (4): 527–538.

Quaini, R. O., and V. Arzamendia, 1998: *Philodryas olfersii* (Lichtenstein's Green Racer). Herpetol. Rev., 29 (1): 54.

Queiroz Alves, F. and A. J. Suzat Argôlo, 1998: Geographic distribution—*Dipsas indica petersi*. Herpetol. Rev., 29 (3): 176.

Queiroz, K., de and R. Lawson, 1994: Phylogenetic relationships of the garter snakes based on DNA sequence and allozyme variation. Biol. J. Linnaean Soc., 53: 209–229.

Quinn, H. R., 1983: Two new subspecies of *Lampropeltis triangulum* from México. Trans. Kansas Acad. Sci., 86: 113–135.

Quintero-Díaz, G., 1999: Geographic distribution: *Thamnophis scaliger*. Herpetol. Rev., 30 (4): 237.

Rabb, G. B., and M. Hymen, 1973: Major ecological and geographic patterns in the evolution of colubrid snakes. Evolution, 27: 69–83, figs. 1–4.

Radcliffe, C. W., and T. P. Maslin, 1975: A new subspecies of the Red Rattlesnake, *Crotalus ruber*, from San Lorenzo Sur Island, Baja California Norte, Mexico. Copeia, 1975 (3): 490–493.

Raddi, G., 1820: Die alcune species nuove di rettili e piante Brasilien. Atti Soc., Ital. Sci. Modena, 18: 1–39.

Rael, E. D., J. D. Johnson, O. Molina and H. K. McCrystal, 1992: Distribution of a Mojave toxin-like protein in rock rattlesnakes (*Crotalus lepidus*) venom. pp. 163–168 in J. A. Campbell and E. D. Brodie, Jr.: Biology of the pitvipers. Selva, Tyler, Texas.

Rahak, I., 1995: Malo znamy hroznys *Boa constrictor orophias*. Akvarium Terarium, 38 (6): 39–42.

Ramírez-Bautista, A., 1978: La distribución ecológica de los anfibios y reptiles de la región de "Los Tuxtlas", Veracruz. Congreso Nac. Zool., México, 1: 22–24.

Ramírez-Bautista, A., 1994: Manual y claves ilustradas de los anfibios y reptiles de la región de Chamela, Jalisco, México. Mexico, D. F., Departamento de. Zoologia., Instituto de Biologia., Univ. Nac. Autón. México, México, D. F.

Ramírez-Bautista, A., and C. Balderas-Valdivia, 1998: *Conophis vittatus vittatus* (Road Guarder, Guardo Camino). Herpetol. Rev. 29 (2): 102.

Ramírez-Bautista, A., and H. M. Smith, 1992: A new chromospecies of snake (*Pseudoleptodeira*) from Mexico. Bull. Maryland Herpetol. Soc., 28 (3): 83–98.

Ramomorthy, T. P., R. Bye, A. Lot and J. Fa (eds.), 1993: Biological diversity of Mexico, origins and distribution. Oxford Press, New York and Oxford.

Raun, G. G., 1972: A guide to Texas snakes. Museum notes No.9, 2nd. Printing, Texas Memorial Museum.

Raymond, L. R., and L. M. Hardy, 1983: Taxonomic status of the corn snake, *Elaphe guttata* (Linnaeus) (Colubridae), in Louisiana and eastern Texas. Southwest. Naturalist, 28 (1): 105–107.

Reichling, S. B., 1995: The taxonomic status of the Louisiana pine snake (*Pituophis melanoleucus ruthveni*) and its relevance to the evolutionary species concept. J. Herpetol., 29 (2): 186–198.

Reinert, H. K., 1992: Radiotelemetric field studies of pitvipers: data acquisition and analysis. pp. 185–197 in J. A. Campbell and E. D. Brodie, Jr.: Biology of the pitvipers. Selva, Tyler, Texas.

Reinhardt, J. T., 1843: Beskrivelse af nogle nye Slangerter. Viden. Sel. Noturvid. Og Mathem. Afh., Kjobenhavn, 10: 233–279.

Reinhardt, J. T., and C. F. Lütken, 1862: Bidragtil det vestindiske Öeriges og navnligen tilde dansk-vesindiske Oers Herpetologie. Vidensk. Meddel. Naturhist. Foren. Kjobenhaven, 1018: 153–291.

Renjifo, J. M., and M. Lundberg, 1999: Guia de campo anfibios y reptiles de Urrá. Skanska, Editorial Colina, Medellin, Colombia.

Rendahl, H., and G. Vestergren, [1940] 1941: Notes on Colombian snakes. Ark. Zool., 33A: 1–16.

Rendahl, H., and G. Vestergren, 1941: On a small collection of snakes from Ecuador. Ark. für Zool., Stockholm, 33A (5): 1–16

Restrepo T., J. H. and J. W. Wright, 1987: A new species of colubrid snake genus

Geophis from Colombia. J. Herpetol., 21 (3): 191–196.
Reuss, A., 1834: Zoologische Miscellen. Reptilien, Ophidier. Mus. Senckenbergiana, Frankfurt, 1: 129–162.
Rey, L. R., and M. L. Lions, 1997a: Geographic distribution—*Atractus taeniatus* (Ground Snake). Herpetol. Rev., 28 (1): 51.
Rey, L. R., and M. L. Lions, 1997b: Geographic distribution—*Oxyrhopus rhombifera inaequifasciatus* (Falsa Coral, False Coral Snake). Herpetol. Rev., 28 (2): 160.
Reynolds, R. P., and M. S. Foster, 1992: Four new species of frogs and one new species of snake from the Chapare region of Bolivia, with notes on other species. Herpetol. Rev., (6): 83–104.
Rice, R. M., and P. Russo, 1991: Revisionary comments on the genus *Boa* with the description of a new subspecies of *Boa constrictor* from Peru. Snake, 23: 29–35.
Richmond, N. D., 1955: The blind snakes (*Typhlops*) of Bimini, Bahamas Isls., BWI, with descriptions of a new species. Amer. Mus. Novitates, (1734): 1–7.
Richmond, N. D., 1964: The blind snakes (*Typhlops*) of Haiti with descriptions of three new species. Breviora, (202): 1–12.
Richmond, N.D, 1965: A new species of blind snake *Typhlops* from Trinidad. Proc. Biol. Soc. Washington, 78: 121–124.
Ride, W. D. L., C. W. Sabrosky, G. Bernardi and R. V. Melville, 1985: International code of zoological nomenclature. Third Edition adopted by the XX General Assembly of the International Union of Biological Sciences, February 1985.
Riemer, W. J., 1954: A new subspecies of the snake *Masticophis lateralis* from California. Copeia, 1954 (1): 45–48.
Riley, J., 1985: Life history: *Tripanurgos compressus* eggs. Herpetol. Rev., 16 (1): 29.
Ripa, D., 1994a: Reproduction of the Central American bushmaster (*Lachesis muta melanocephala* and *Lachesis muta stenophrys*) for the first time in captivity. Bull. Chicago Herpetol. Soc., 29: 165–183.
Ripa, D., 1994b: The reproduction of the Central American Bushmasters (*Lachesis muta melanocephala*) and the Black-headed Bushmaster (*Lachesis muta stenophrys*) for the first time in captivity. Vivarium, 5 (5): 36–37.
Ripa, D., 1999: Keys to the understanding the bushmasters (genus *Lachesis* Daudin, 1803). Bull. Chicago Herpetol. Soc., 34: 45–92.
Rivas, J, 1999: Tracking the anaconda. National Geographic, Washington, D.C., January, 1999, 195 (1): 62–69.
Rivas-Fuenmaeyer, G., and C. R. Molina R, 1998: Geographic Distribution—*Pseustes shropshirei*. Herpetol. Rev., 29 (3): 178.
Rivero, J. A., 1978: Los anfibios y reptiles de Puerto Rico-The amphibians and reptiles of Puerto Rico. Departamento de Biología, Univ. Puerto Rico, Editorial Universitaria, Mayagüez, Puerto Rico.
Rivero, J. A., 1998: Los amfibios y reptiles de Puerto Rico—The amphibians and reptiles of Puerto Rico. 2nd. edition—revised, Editorial de la Univ. Puerto Rico.
Rivero-Blanco, C., and J. R. Dixon: Origin and distribution of the herpetolofauna of the dry lowland regions of northern South America. (17). pp 281–289. *In* Duellman, W. E., ed., 1979: The South American herpetofauna: Its origin, evolution and dispersal. Univ. Kansas Mus. Nat. Hist. Monograph 17.
Robb, J., and H. M. Smith, 1993: The systematic position of the group of snake genera allied to *Anomalepis*. Nat. Hist. Misc. Chicago Acad. Sci., (184): 1–8.
Robinson, M. D., 1993: Death and dancing on the sun-baked dunes of Namibia. Nat. Hist. Amer. Mus. Nat. Hist., 102 (8): 28–30.
Rodrigues, A., 1986: Linguas brasileiras-para o conhecimento das linguas indigenas. Edicoes Loyola, São Paulo, Brazil.
Rodrigues, M. T., [1992] 1993a: Herpetofauna das dunas interiores do Rio São Francisco: Bahia: Brasil.5. duas novas especies de *Apostolepis* (Ophidia; Colubridae). Mem. Inst. Butantan (São Paulo), 54 (2): 53–59.
Rodrigues, M. T., [1992] 1993b: Herpetofauna of palaeoquaternary sand dunes of the middle São Francisco River: Bahia: Brazil.6. Two new species of *Phimophis* (Serpentes: Colubridae) with notes on the origins of psammophilic adaptions. Pap. Avuls. Zool. (São Paulo), 38 (11): 187–198.
Rodrigues, M. T., 1997: Lizards, snakes & amphibians from the Quarternary sand dunes of the Middle Rio São Francisco, Bahia, Brazil. J. Herpetol., 30 (4): 513–523.
Rodrigues, M. T., and G. Puorto, 1994: On the second specimen of *Leptotyphlops brasilensis* Laurent, 1949 (Serpentes: Leptotyphlopidae). J. Herpetol., 28 (3): 393–394.
Rodrigues-Machado, S., 1993: A new genus of Amazonian vine snake (Xenodontinae: Alsophiini). Acta Biol. Leopoldensia, 15 (2): 99–108.
Rodrigues-Machado, S., and M. DiBernardo, 1996: Análise filogenetica das espé do gênero *Oxybelis* Wagler, 1830 (Serpentes, Colubridae). Resumes do XXI Congresso Brasileiro de Zoología 5 a 9 de fevereiro de 1996. Porto Alegre, RS, (953): 1.
Rodríguez-Robles, J. A., and J. M. De Jesús-Escobar, 2000: Molecular Systematics of New World gopher, bull, and pine-snakes (*Pituophis*: Colubridae), a transcontinental species complex. Molecular Phylogenetics and Evolution, 14 (1): 35–50.
Rodríguez, J. P., and F. Rojas-Suárez, 1999: Libro rojo de la fauna venezolana. 2nd Ed., PROVITA, Fundación Polar Wildlife Conservación Society, PROFAUNA-MARNR UINC.
Röhl, E., 1942: Fauna descriptiva de Venezuela (Vertebrados). Tipografía Americana, Caracas, Venezuela.
Rolker, A. W., 1903: The story of the snake. McClure's Mag., 21: 280–290.
Romano, S. A. R. W. D. L., [1971] 1972: Notes of *Leptomicrurus* Schmidt (Serpentes, Elapidae). Mem. Inst. Butantan, 1971: 111–115.
Romano, S. A. R. W. D. L., and A. R. Hoge, 1973: Nota sobre *Xenodon* e *Ophis* (Serpentes, Colubridae). Mem. Inst. Butantan, 36 (1972): 209–214.
Romero Castillo, M., E. Arana de Swadesh, C. Robles U., R. D. Bruce S. and J. M. Lope Blanch, 1975: Las lenguas de México, II. Instituto Nacional de Anthropología e Historia.
Romero Martinez, O. (Ed.), 1996: Lista de anfibias y reptiles del Muso de Historia

Bibliography

Natural del Paraguay (mayo, 1980-septiembre 1995) *in*-Colecciones de Flora y Fauna del Museo Nacional de HistoriA Natural del Paraguay. Museo Nacional de Historia Natural del Paraguay, San Lorenzo, Paraguay, Feb. 1996.

de Roodt, A., J. Dolab, E. Gould, J. C. Troiano, J. Gould, J. C. Vidal and L. Segre, 1996: Actividad procoagulate del veneno de *Bothrops alternatus* (Vibora de Cruz), *Bothrops neuwiedi* (Yarara Chica) y *Bothrops ammodytoides* (Yara ñata) el plasma del hombre y de algunos animales domesticos. IV Congreso Latinamericana de Herpelologíca, 14-19 October, 1999, Santiago de Chile, Chile.

Roquette-Pinto, E., 1975: Rondônia. 6th ed., Ed. Nacional/INL, São Paulo, Brazil, 1975: 274-277.

Rosanía González, J. V., 1960-1965: Notas sobre taxonomía de las especies de serpientes venenosas de la Serrianía de El Avila. Manuel del Excursionista Cientifico, Caracas, Tipografica El Pilar, entre 1960-1965.

Rosen, 1905: List of snakes in the Zoological Museum of Lund and Malmö, with description of new species and a new genus. Ann. Mag. Nat. Hist. London, 15 (7): 168-181.

Rossman, D. A., 1961: Nomenclatural status of the Neotropical subspecies of the colubrid snake, *Thamnophis sauritus*. Notulae Naturae, Acad. Nat. Sci. Philadelphia, (340): 1-2.

Rossman, D. A., 1962: *Thamnophis proximans* (Say), a valid species of Garter Snake. Copeia, 1962 (4): 744-748.

Rossman, D.A, 1963a: Relationships and taxonomic status of the North American natricine snake genera, *Liodytes*, *Regina*, and *Clonophis*. Occ. Pap. Mus. Zool., Louisiana State Univ., (29): 1-29.

Rossman, D. A., 1963b: The colubrid snake genus *Thamnophis*: a revision of the *Sauritus* group. Bull. Florida State Mus. Biol. Sci., Univ. Florida, Gainesville, 7 (3): 178.

Rossman, D. A., 1965a: Identity and relationships of the Mexican Garter Snake *Thamnophis sumichrasti* (Cope). Copeia, 1965 (2): 242-244.

Rossman, D. A., 1965b: Two new colubrid snakes of the genus *Rhadinaea* from southern Mexico. Occ. Pap. Mus. Zool., Louisiana State Univ., (32): 1-8.

Rossman, D. A., 1965c: A new subspecies of the common garter snake, *Thamnophis sirtalis*, from the Florida Gulf Coast. Proc. Louisiana Acad. Sci, 27: 67-73.

Rossman, D. A., 1966: Evidence for conspecificity of the Mexican Garter Snakes *Thamnophis phenax* (Cope) and *Thamnophis sumichrasti* (Cope). Herpetologica, 22 (4): 303-307.

Rossman, D. A., 1968: Identity of *Helicops wettsteini* Amaral (Serpentes: Colubridae). Herpetologica, 24 (3): 262-263.

Rossman, D. A., 1969: A new natracine snake of the genus *Thamnophis* from northern Mexico. Occ. Pap. Mus. Zool., Louisiana State Univ., (39): 1-4.

Rossman, D. A., 1970a: *Thamnophis proximus*. Cat. Amer. Amph. Rept., (98): 1-2.

Rossman, D. A., 1970b: *Thamnophis sauritus* (Linnaeus)—Eastern Ribbon Snake. Cat. Amer. Amph. Rept., (99): 1-2.

Rossman, D. A., 1971: Systematics of the Neotropical population of *Thamnophis marcianus* (Serpentes: Colubridae). Occ. Pap. Mus. Zool., Louisiana State Univ., Baton Rouge, (41): 1-13.

Rossman, D. A., 1973: Evidence for the conspecificity of *Carphophis anoenus* (Say) and *Carphophis vermis* (Kennicott). J. Herpetol., 7: 140-141.

Rossman, D. A., 1973 [1974]: Miscellaneous notes on the South American water snake genus *Helicops*. HISS—New J., 1 (6): 189-191.

Rossman, D. A., 1975: Redescription of the South American colubrid snake *Helicops hagmanni* Roux. Herpetologica, 31: 414-418.

Rossman, D. A., 1976: Revision of the South America colubrid snakes of the *Helicops pastaze* group. Occ. Pap. Mus. Zool. Louisiana State Univ., 50: 1-15, 15.

Rossman, D. A., 1979: Morphological evidence for taxonomic partitioning of the *Thamnophis elegans* complex (Serpentes, Colubridae). Occ. Pap. Mus. Zool. Louisiana State Univ., (55): 1-12.

Rossman, D. A., 1984: Life history. *Helicops angulatus*. Reproduction. Herpetol. Rev., 15 (2): 20.

Rossman, D. A., 1985: Geographic distribution: *Adelophis copei*. Herpetol. Rev., 16 (3): 84.

Rossman, D. A., 1992: Taxonomic status and relationships of the Tamaulipian Montane Garter Snake, *Thamnophis mendax* Walker, 1955. Proc. Louisiana Acad. Sc., 55: 1-14.

Rossman, D. A., 1995: A second external character for distinguishing Garter Snakes (*Thamnophis*) from Water Snakes (*Nerodia*). Herpetol. Rev., 26 (4): 182.

Rossman, D. A., 1996: Identity and taxonomic status of the Mexican Garter Snake, *Thamnophis vicinus*. Proc. Biol. Soc. Washington, 109 (1): 10-16.

Rossman, D. A., 2000: Identity of the snake *Calopisma septemvittatum* Fischer, 1879, and redescription of the lectotype of *Helicops infrataeniatus* Jan, 1865. Mitt. Hamb. Zool. Mus. Inst., 97: 123-129.

Rossman, D. A., 2002: Variation in the Xenodontid water snake *Itelicops scalaris* Jan, and the Status of *H. hogei* Lancini. Occ. Pap. Mus. Nat. Hist., Lousiana State University, (78): 76-19.

Rossman, D. A., and A. S. Abe, 1979: Comments on the taxonomic status of *Helicops yacu* (Serpentes: Colubridae). Proc. Louisiana Acad. Sci., 42: 7-9.

Rossman, D. A., and R. M. Blaney, 1968: A new natracine snake of the genus *Adelophis* from western Mexico. Occ. Pap. Mus. Zool., Louisiana State Univ., 35: 1-12.

Rossman, D. A., and J. R. Dixon, 1975: A new colubrid snake from genus *Helicops* from Peru. Herpetologica, 3 (4): 412-414.

Rossman, D. A., and G. W. Eberle, 1977: Partition of the genus *Natrix*, with preliminary observations on evolutionary trends in natricine snakes, Herpetologica. 33: 34-43.

Rossman, D. A., and R. L. Erwin, 1980: Geographic variation in the snake *Storeria occipitomaculata* (Serpentes: Colubridae) in southeastern United States. Brimleyana, 4: 95-102.

Rossman, D. A., N. B. Ford and R.A Seigel, 1996: The garter snakes, evolution and ecology. Univ. Oklahoma Press, Norman.

Rossman, D. A., and D. A. Kizirian, 1993: Variations in the Peruvian dipsidine snakes ..*Sibynomorphus oneilli* and *S. vagus*. J. Herpetol., 27 (1): 87-90.

Rossman, D. A., and G. Lara-Gongora, 1997: Variation in the Mexican garter snakes *Thamnophis scalaris* Cope and the taxonomic status of *T. scaliger*

(Jan). Occ. Pap. Mus. Nat. Sci., Louisiana State Univ., 74: 1–14.

Rossman, D. A., E. A. Liner, C. H. Trevino, and A. H. Chaney, 1989: Redescription of the garter snake *Thamnophis exsul* Rossman, 1969 (Serpentes: Colubridae). Proc. Biol. Soc. Washington, 102 (2): 507–514.

Rossman, D. A., and R. Powell, 1985: *Clonophis, C. kirtlandi*. Cat. Amer. Amph. Rept., (364): 1–2.

Rossman, D., and G. C. Schaeffer, 1974: Generic status of *Opheodrys mayae*, a colubrid snake endemic to the Yucatán Peninsula. Occ. Pap. Mus. Zool., Louisiana State Univ., (45): 1–12.

Rossman, D. A., and G. R. Stewart, 1987: Taxonomic reevaluation of *Thamnophis couchii* (Serpentes: Colubridae). Occ. Pap. Mus. Zool., Louisiana State Univ., (63): 1–25.

Rossman, D. A., and R. Thomas, 1979: A new dipsadine snake of the genus *Sibynomorphus* from Peru. Occ. Pap. Mus. Zool, Louisiana State Univ., (54): 1–6.

Rossman, D. A., and V. Wallach, 1987: *Adelphicos* Dugès—Mountain Meadow Snake, *A. copei, A. foxi*. Cat. Amer. Amph. Rept., (408): 1–2.

Rossman, D. A., and V. Wallach, 1991: *Virginia*. Cat. Amer. Amph. Rept., (529): 1–4.

Rossman, D. A., and L. D. Wilson, 1965: Comments on the revival of the colubrid snake subfamily Heterodontidae. Herpetologica, 20: 284–285.

Rossman, N. T., and D. A. Rossman, 1982: Comparative visceral topography of the New World snake tribe Thamnophiini (Colubridae, Natricinae). Tulane Stud. Zool. Bot., 23 (2): 123–164.

Rouby, A., and O. Riedmayer, 1983: Shurar Chicham-gamática Shuar. *In* Luis M. Caillet, Mundo Shuar.

Roux, J., 1907: Revisión de quelques espéces de reptiles et amphibiens du Perou. Rev. Suisse Zool., 15: 293–303.

Roux, J., 1926a: Contribución a l'erpetologie du Venezuela. Verh. Natur. Ges. Basel, 38: 259–260.

Roux, J., 1926b: Notes d'erpetologie sud'-americaine. 1. Sur une collection de Reptiles et d'Amphibiens de l'ile de la Trinite. Revue Suisse de Zoologie, 33: 291–299.

Roux, J., 1926c: Sur une nouvelle espéce de *Typhlops. T. lehneri* n. sp. du Vénézuéla. Revue Suisse Zool. 33 (4): 298–299.

Roux-Estève, R., 1983: Les spécimens-types du genere *Micrurus* (Elapidae) conservés au Muséum National d'Histoire naturelle de Paris. Mem. Inst. Butantan, 1982, 46: 79–94.

Roze, J. A., 1952a: Colección de reptiles del Professor Scorza, de Venezuela. Acta Biologica Venezuelica. Facultad de Ciencas Matematicas y Naturales Universidad de Central de Venezuela, 8 octubre 1952, Caracas, Venezuela, 1 (5): 93–116.

Roze, J. A., 1952b: Contribucción al conocimiento de los ofidios de las familias Typhlopidae y Leptotyphlopidae en Venezuela. Mem. Soc. Cien. Nat. La Salle, 32 (12): 143–158.

Roze, J. A., 1953a: Ofidios de Camurí, Chico, Macuto, D. F., Venezuela colectados por el Rvdo. Padre Cornelius Vogl. Bol. Soc. Venezuela Cien. Nat. Caracas, 14 (79): 200–211.

Roze, J. A., 1953b: The rassenkreis *Coluber (Masticophis) mentovarius* (Dumeril, Bibron et Dumeril), 1854. Herpetologica, 9 (3): 113–120.

Roze, J. A., 1954: Nota preliminar sobre los ophidios de la expedicion Franco-Venezuelana al Alto Orinoco. Arch. Venezol. Patol. Trop. Parasit. Med. II (2): 218–226.

Roze, J. A., 1955a: Ofidios colecionados por la expedicion Franco-Venezolana al Alto Orinoco: 1951 a 1952. Bol. Soc. Venezolana Cien. Nat., 1 (3–4): 179–195.

Roze, J. A., 1955b: Revisión de las corales del género *Micrurus* (Serpentes: Elapidae) de Venezuela. Acta Biol. Venezuelica, 1 (17): 453–500.

Roze, J. A., 1956a: La herpetofauna de las islas los Roques y la Orchila. Mem. Soc. Cien. Nat. La Salle, 1956: 79–86.

Roze, J. A., 1956b: Ofidios colecionados por la expedición Franco-Venezolana al alto Orinoco 1951–1952. Bol. Mus. Cien. Nat., I: 179–195.

Roze, J. A., 1957: Resumen de una revision del genero *Hydrops* (Wagler) 1830 (Serpentes: Colubridae). Acta Biol. Venezuelica, 2 (8): 1–95.

Roze, J. A., 1958a: A new species of the genus *Urotheca* (Serpentes, Colubridae) from Venezuela. Breviora, (88): 1–5.

Roze, J. A., 1958b: Los reptiles del Auyantepui, Venezuela, basándose en las colecciones de las expediciones de Phelps-Tate, del American Museum of Natural History, 1937–1938, y de la Universidad Central de Venezuela, 1956. Acta Biol. Venezuelica, 2 (22): 261–264.

Roze, J. A., 1958c: Los reptiles del Chimantá Tepui (Estado Bolívar, Venezuela) colectadas por la expedición botánica del Chicago Natural History Museum. Acta Biol. Venezuelica, 2 (25): 299–314.

Roze, J. A., 1958c: On Hallowell's type specimens of reptiles from Venezuela in the collection of the Academy of Natural Sciences of Philadelphia. Notulae Naturae, (309): 1–4.

Roze, J. A., 1958d: Resultados zoologicos de la expedición de la Universidad Central de Venezuela a la region de Auyan Tepui, Venezuela, basandose en las colecciones de las expediciones de Phelps-Tate, del American Museum of Natural History, 1937–1938, y de la Universidad Central de Venezuela, 1956. Acta Biol. Venezuelica, 2: 243–270.

Roze, J. A., 1959: Taxonomic notes on a collection of Venezuelan reptiles in the American Museum of Natural History. Amer. Mus. Novitates, (1934): 1–14.

Roze, J. A., 1958a: A new species of the genus *Urotheca* (Serpentes: Colubridae), from Venezuela. Breviora, (88): 1–5.

Roze, J. A., 1958b: Los reptiles del Chimanta Tepui (Estado Bolivar, Venezuela) colectados por la Expedicion Botanica del Chicago Natural History Museum. Acta Biologica Venezuelica, Universidad Central de Venezuela, Caracas, Venezuela, 2 (25): 299–314, 15 December, 1958.

Roze, J. A., 1958c: Resultados zoologicos de la expedicion de la Universidad de Venezuela a la region del Auyantepui en la Guyana Venezolana, abril de 1956.5. Los Reptiles del Auyantepui, Venezuela, basandose en las colecciones de las expediciones de Phelps-Tate, de American Museum of Natural History 1937–1938, y de la Universidad Central de Venezuela, 1956. Acta Biol. Venezuelica. 2 (22): 243–270.

Roze, J. A., 1959a: El género *Erythrolamprus* Wagler (Serpentes: Colubridae) en Venezuela. Acta Biol. Venezuelica, 2 (35): 523–534.

Roze, J. A., 1959b: Una nueva especie del género *Drymarchon* (Serpentes: Colubridae) de la isla Margarita, Venezuela. Noved. Cien. La Salle, Caracas, ser. Zool. 25: 1–4.

Roze, J. A., 1961: El género *Atractus* (Serpentes: Colubridae) en Venezuela. Acta Biol. Venzuelica, 3 (7): 103–119.

Roze, J. A., 1964a: La herpetologia de la Isle de Margarita, Venezuela. Mem. Soc. Cien. Nat. LaSalle, 69 (24): 209–241.

Roze, J. A., 1964b: The snakes of the *Leimadophis-Urotheca-Liophis* complex from Parque Nacional Henri Pittier (Rancho Grande), Venezuela, with a description of a new genus and species. Escuela de Biologia, Universidad Central de Venezuela, y The American Museum of Natural History, New York, 1964: 533–542.

Roze, J. A., 1966: La taxonomia y zoogeografia de los ofidios en Venezuela, Caracas, Universidad Central de Venezuela, Ediciones de la Biblioteca.

Roze, J. A., 1967: A check list of the new world venomous coral snakes (Elapidae), with descriptions of new forms. Amer. Mus. Novitates, (2287): 1–60.

Roze, J. A., 1969: Una nueva coral falsa del Género *Scaphiodontophis* (Serpentes: Colubridae) de Colombia. Caldasia (Bogota), 10 (48): 355–363.

Roze, J. A., 1970: Ciencia, fantasia sobre las serpientes de Venezuela. Caracas, Venezuela, Editorial Fondo de Cultura Cientifica:, S.R.L.

Roze, J. A., 1974: *Micruroides* Schmidt, *M. euryxanthus*—Western coral snake. Cat. Amer. Amph. Rept., (163): 1–4.

Roze, J. A., 1983: New world coral snakes (Elapidae): a taxonomic and biological summary. Mem. Inst. Butantan, 46: 305–338.

Roze, J. A., 1987: Summary of coral snakes (Elapidae) from Cerro de la Neblina, Venezuela, with description of a new subspecies. Revue Francaise D'Aquariologie Herpetologie, 14 (3): 109–112.

Roze, J. A., 1989: New species and subspecies of coral snake, genus *Micrurus* (Elapidae), with notes on type specimens of several species, Amer. Mus. Novitates, (2932): 1–15.

Roze, J. A., 1994: Notes on taxonomy of venomous coral snakes (Elapidae) of South America. Bull. Maryland Herpetol. Soc., 30: 177–185.

Roze, J. A., 1996: Coral snakes of the Americas—biology, identification, and venoms. Krieger Publishing Company, Malabar, Florida.

Roze, J. A., and A. Bernarl-Carlo, [1987] 1988: Las serpientes corales venonas del género *Leptomicrurus* (Serpentes, Elapidae) de Suramerica un descripcion de una nueva subspecies. Torino, Italy, Museo Regionale di Scienze Naturali, Estratto dal Bollettino Museo Regional di Scienze Naturali, 5 (2): 573–608.

Roze, J. A., and N. Jorge da Silva, 1990: Coral snakes (Serpentes, Elapidae) from hydroelectric power plant of Samuel, Rondonia, Brazil; with Description of a New Species. Bull. Maryland Herpetol. Soc., 26 (4): 169–176.

Roze, J. A., and G. M. Tilger, 1983: *Micrurus fulvius* (Linnaeus) North American coral snake. Cat. Amer. Amph. Rept., (316):1–2.

Roze, J. A., and C. P. Trebbau M., 1958: Un nuevo género de corales venenosas (*Leptomicrurus*) para Venezuela. Act. Cient. Venez., 9: 128–130.

Rubio, Manny, 1998: Rattlesnake—portrait of a predator. Smithsonian Institution Press, Washington, D.C.

Rufino, N., 1998: Geographic distribution—*Liophis poecilogyrus schotti*. Herpetol. Rev., 29 (3): 177.

Rufino, N., 1999: Geographic distribution—*Liophis almadensis* (Jararaquinha do Campo). Herpetol. Rev., 30 (2): 113.

Russell, F. E., 1983: Snake venom poisoning. Great Neck, New York, Scholium International.

Russell, M., 2001: Road cruising in the Amazon. Reptiles Magazine, March, 2001, 9 (3): 44–57.

Ruthven, A.G, 1907.: A collection of reptiles and amphibians from southern New Mexico and Arizona. Bull. Amer. Mus. Nat. Hist., 23: 483–603.

Ruthven, A. G., 1908: Variations and genetic relationships of the garter-snakes. Bull. U.S. Natl. Mus., (61): i–xii, 1–201.

Ruthven, A. G., 1910: Review of R. L. Ditmars, Reptiles of the world. New York. Science. 34: 54–55.

Ruthven, A. G., 1912: The amphibians and reptiles collected by the University of Michigan-Walker Expedition in southern Veracruz, Mexico, Zool. Jahrb., Abt. Syst., 32: 295–332.

Ruthven, A. G., 1922: The amphibians and reptiles of the Sierra Nevada de Santa Marta, Colombia. Misc. Pub. Mus. Zool. Univ. Michigan, (8): 1–69.

Ruthven, A. G., 1932: Notes on the amphibians and reptiles of Utah. Occ. Pap. Mus. Zool. Univ. Michigan, (243): 1–4.

Ruthven, A. G., 1936: *Leptodeira bakeri*, n. sp. Occ. Pap. Mus. Zool. Univ. Michigan, (330): 1–2.

Ruthven, A. G., and H. T. Gaige, 1915: The reptiles and amphibians collected in northeastern Nevada by the Walker-Newcomb Expedition of the University of Michigan. Occ. Pap. Mus. Zool. Univ. Michigan, (8): 1–33 + 5 pl.

Ruthven, A. G., C. Thompson and H. T. Gaige, 1928: The herpetology of Michigan. Michigan Handbook Series, No. 3, Published by the Univ. Michigan, (3): 1–229.

Sabath, M. D., and L. E. Sabath, 1969: Morphological intergradiation of gulf coastal brown snakes, *Storeria dekayi* and *Storeria tropica*. Univ. Notre Dame Press, Notre Dame, Indiana, 31 (1): 148–155, June 1969.

Sackett, J. T., 1940: Preliminary report on results of the West Indies—Guatemala Expedition of 1940 for the Academy of Natural Sciences of Philadelphia, Part I. A new blind snake of the genus *Typhlops*. Notulae Naturae, Amer. Midl. Naturalist, (48): 1–2.

Sajdak, R. A., Status of the Caribbean Racer (genus *Alsophis*) in the Lesser Antilles, Oryx, 25: 33–38.

Sajdak, R. A., and R. W. Henderson: Notes on the eggs and young of *Antillophis parvifrons stygius* (Reptilia, Serpentes, Colubridae), Florida Sci, 45: 200–204.

Salazar, F. A., 1967: The innocent assassins. E. P. Dutton & Co., Inc., New York, 256 pp. Salomêo, E. L. and M. Di-Bernardo, 1995: *Philodryas olfersii*: Um cobra comum que mata caso registrado na área da 8a. Delegacia Regional de Saúde. Arquivos da SBZ/Sorocaba-SP/No. 14/15/16/Maio de 1995.

Salvin, O., 1860: On the reptiles of Guatemala. Proc. Zool. Soc. London, 1860: 451–461.

Sandner Montilla, M. F., 1976: Una nueva

especie de genero *Bothrops* (Serpientes: *Crotalidae*, *Lachesinae*) de la Gran Sabana, Edo. Bolivar, Venezuela. Mem. Cien. Ofid.: 1-4.

Sandner Montilla, M. F., 1979: Una nueva especie de genero *Bothrops* (Serpientes: *Bothropinae*) de la region de Guanare, estado Portuguesa, Venezuela. Mem. Cien. Ofid. 4: 1-19.

Sandner Montilla, M. F., 1981: Una nueva especie de genero *Crotalus* (Serpientes: *Crotalidae*, *Crotalinae*) del sur de estado Guarico, Venezuela. Mem. Cien. Ofid. 6: 1-15.

Sandner Montilla, M. F., 1989: Una nueva subespecie de *Bothrops lansbergi* (Schlegel, 1841) de la familia Crotalidae: *Bothrops lansbergi hutmani* n ssp. Mem. Cien. Ofid., 9: 1-16.

Sanderson, W. E., 1993: Additional evidence for the specific status of *Nerodia cyclopion* and *Nerodia floridiana* (Reptilia: Colubridae). Brimleyana, 1993 (19): 83-94.

Santos, E., 1942: Anfíbois e répteis do Brasil (vida e costumes). F. Briguiet & Cia., Rio de Janiero.

Santos, E., 1943: As cobras venenosas, como conhece-las e evitá-las tratamento do ofidismo. Bibliotecan Agrícola Popular Brasileira, Edição Da Chácaras e Quintais, São Paulo, Brasil.

Santos, E., 1955: Anfíbos e répteis do Brasil. 2nd. Ed., F. Briguiet & CIA.-Editores, Rio de Janiero, Brazil.

Santos, E., 1981: Anfíbos e répties do Brasil (vida e costumbres). 3rd. Ed., Editora Litatiaia Limitad, Belo Horizonte, Brazil.

Santos-Jordão, R. dos, and A. Fernandes-Bizerra, 1996: Reproducão, dimorfismo sexual e atividade de *Simophis rhinostoma* (Serpentes, Colubridae). Rev. Bras. Biol., 56 (3): 507-512.

Sarmiento Rodriguez, F., 1987: Desde la selva..hasta el mar. Antologia Ceologica Del Ecuador. Ministro de Energia y Minas, casa de Cultura Ecuatoriana. Year 7, No.2, April 1987, Publicatciones de Museo de Ciencias Naturales Ecuatoriana, Quito, Ecuador.

Sasa, M., 1993: Distribution and reproduction of the gray earth snake *Geophis brachycephalus* (Serpentes, Colubridae) in Costa Rica. Rev. Biol. Trop., 41 (2): 295-297.

Savage, J. M., 1949: Notes on a Central American snake; *Conophis lineatus dunni* Smith, with a Record From Honduras. Trans. Kansas Acad. Sci., 52 (4): 483-486.

Savage, J. M., 1955: Descriptions of new colubrid snakes, genus *Atractus*, from Ecuador. Proc. Biol. Soc. Washington, 68: 11-20.

Savage, J. M., 1960a: A revision of the Ecuadorian snakes of the Colubrid genus *Atractus*. Misc. Pub. Mus. Zool., Univ. Michigan, 49:112: 1-86.

Savage, J. M., 1960b: Evolution of a peninsular herpetofauna. Syst. Zool., 93: 184-212.

Savage, J. M., 1973: A preliminary handlist of the herpetofauna of Costa Rica. Univ. Southern California, Los Angeles, California.

Savage, J. M., 1974: Type localities for species of amphibians and reptiles described from Costa Rica. Rev. Biol. Trop., 22 (1): 71-122.

Savage, J. M., 1980: A handlist with preliminary keys to the herpetofauna of Costa Rica. Allan Hancock Foundation, August 1980.

Savage, J.M, 1981.: A new species of the secretive Colubrid snake genus *Geophis* from Costa Rica. Copeia, 1981: 549-553.

Savage, J. M., 1996: Evolution of coloration, urotomy and coral snake mimicry in the snake genus *Scaphiodontophis* (Serpentes: Colubridae). Biol. J. Linnaean Soc., 57: 129-194.

Savage, J. M., and F. S. Cliff, 1953: A new subspecies of sidewinder; *Crotalus cerastes*, from Arizona. Nat. His. Misc. Chicago Acad. Sci., (119): 1-7.

Savage, J. M., and F. S. Cliff, 1954: A new snake, *Phyllorhynchus areincola*, From The Gulf of California, Mexico. Proc. Biol. Soc. Washington, 67: 69-76

Savage, J. M., and B. I. Crother, 1989: The status of *Pliocercus* and *Urotheca* (Serpentes: Colubrid), with a review of included species of coral snake mimics. Zool. J. Linnaean Soc., 95: 335-362 with 4 figs.

Savage, J. M., and M. A. Donnelly, 1988: Variation and systematics in the Colubrid snakes of the genus *Hydromorphus*. Amphibia-Reptilia, 9: 289-300,

Savage, J. M., and P. N. Lahanas, 1989: A new species of Colubrid snake (Genus *Urotheca*) from the Cordillera de Talamanca of Costa Rica. Copeia, 1989 (4): 892-896.

Savage, J. M., and P. N. Lahanas, 1991: On the species of the Colubrid snake genus *Ninia* in Costa Rica and western Panama. Herpetologica, 47 (1): 37-53.

Savage, J. M., and R. W. McDiarmid, 1992: Rediscovery of the Central American Colubrid snake, *Sibon argus*, with comments on related species in the region. Copeia, 1992 (2): 421-432.

Savage, J. M., and N.J. Scott, 1987: The *Imantodes* (Serpentes: Colubridae) of Costa Rica: Two or three species? Rev. Biol. Trop., 33: 07-132.

Savage, J. M., and J. B. Slowinski, 1990: Short note: A simple consistent terminology for the basic colour patterns of the venomous coral snakes and their mimics. Herpetol. J., 1: 530-532.

Savage, J. M., and J. B. Slowinski, 1992: The colouration of the venomous coral snakes (family Elapidae) and their mimics (family Aniliidae and Colubridae). Biol. J. Linnaean Soc., 45: 235-253.

Savage, J. M., and J. B. Slowinski, 1996: Evolution of coloration, urotomy and coral snake mimicry in the snake genus *Scaphiodontophis* (Serpentes: Colubridae). Biol. J. Linnaean Soc., 57: 129-194.

Savage, J. M., and J. L. Vial, 1974: The venomous coral snakes (Genus *Micrurus*) of Costa Rica, Rev. Biol. Trop., 21: 295-394.

Savage, J. M., and J. Villa R., 1986: Introduction to the herpetofauna of Costa Rica. SSAR, Contrib. Herpetol., (3): 1-207, Athens, Ohio.

Savitzky, A., 1992: Embryonic development of the maxillary and prefrontal bones of crotalinae snakes. pp. 119-141 in J. A. Campell and E. D. Brodie, Jr.: Biology of the pitvipers. Selva, Tyler, Texas.

Savitzky, B. A. C., 1992: Laboratory studies on piscivory in an opportunistic pitviper, the cottonmouth, *Agkistrodon piscivorous*. pp. 347-368 in J. A. Campell and E. D. Brodie, Jr.: Biology of the pitvipers. Selva, Tyler, Texas.

Savitzky, R. H., and H. M. Smith, 1971: A new snake for Mexico of the *taeniata* group of *Tantilla*. J. Herpetol., 5: 167-171.

Sawaya, R. J., and I. Sazima, 2003: A new

Bibliography

species of *Tantilla* (Serpentes: Colubridae) from southeastern Brazil. Herpetologica, 59 (1): 119-126.

Saxton, D., and L. Saxton, 1969: Papago and Pima to English: English to Papago and Pima. Univ. Arizona Press.

Say, T., 1823: Descriptions of amphibians and reptiles. From account of an expedition from Pittsburgh to the Rocky Mountains, commanded by Major S. H. Long. Compiled by Edwin James Philadelphia, H. C. Perry and I. Lea, 2 volumes: 1-442.

Say, T., 1825: Descriptions of three new species of *Coluber*, inhabiting the United States. J. Acad. Sci. Philadelphia, 1: 237-241.

Sazima, I., 1992: Natural history of the jararaca pitviper, *Bothrops jararaca*, in southeastern Brazil. Pp. 199-216 *in* J. A. Campbell and E. D. Brodie, Jr.: Biology of the pitvipers. Selva, Tyler, Texas.

Schankel, E., 1901: Achter Nachtrag zum Katalog der herpetologischen Sammlung des Basler Museums. Separatabdruck Varh. Naturf. Gesell. Basel, 8 (1): 142-199.

Schargel, W. E., and J. E. García-Pérez, 2002: A new species and a new record of *Atractus* (Serpentes: Colubridae) from the Andes of Venezuela. J. Herpetol. 36 (3): 398-402.

Schätti, B., E. Kramer and J. M. Touzet, 1990: Systematic remarks on a rare crotalid snake from Ecuador, *Bothriechis albocarinata* (Shreve), with some comments on the generic arrangement of arboreal Neotropical pitvipers. Rev. Suisse Zool., 97: 877-885.

Schätti, B. and E. Kramer, 1993: Ecuadorianische Grubenottern der Gattungen *Bothriechis*, *Bothrops* und *Porthidium* (Serpentes: Viperidae). Revue Suisse Zool., 100 (2): 235-278, juin 1993.

Schätti, B., and L. D. Wilson, 1986: *Coluber*. Cat. Amer. Amph. Rept., (399): 1-4.

Schenk, E. T., and J. H. McMasters, 1936: Procedure in Taxonomy-Including a reprint in translation of the Règles Internationales de la Nomenclature Zoölogique (International code of Zoölogical Nomenclautre) with titles of and notes on the opinions rendered to the present date (1907 to 1947). Completely indexed by A. M. Keen and S. W. Muller (eds.), 1948, Stanford Univ. Press, Stanford, California, 1948: 1-93.

Schlegel, H., 1837: Essai sur la Physionomie Serpens. II. Partie descriptive. Arnz & Comp, Leiden.

Schlegel, H., 1837-1844: Abbildungen veuer oder unvollständig bekannter Amphibien, nach der Natur oder dem Leben entworfen, herausgegeben und mit einem erläuternden Texte begleitet. Arnz & Comp., Düseldorf.

Schlegel, H., 1841: Description d'une nouvelle espèce du genere Trigonocéphale (*Trigonocephalus Lansbergii*), Mag. Zool. (Paris), 3: 1-3.

Schmidt, D., 2000: Kornnattern und Erdnattern—*Elaphe guttata* and *Elaphe obsoleta*. Terrarien Bibliothek, Berlin.

Schmidt, K. P., 1921: Notes on the herpetology of Santo Domingo. Bull. Amer. Mus. Nat. Hist., 44: 7-20.

Schmidt, K. P., 1924: The amphibians and reptiles of Mona Island, West Indies. Field Mus. Nat. Hist. Ser. Zool., 12: 149-163.

Schmidt, K. P., 1926: The herpetology of Mona Island, West Indies. Pub. Field Mus. Nat. Hist., Zool., 12: 217-224.

Schmidt, K. P., 1928a: 47. Notes on American coral snakes. Bull. Antivenin Inst. America, II (3): 63-64, October 1928.

Schmidt, K. P., 1928b: Scientific survey of Porto Rico and the Virgin Islands-Vol.X.-Survey Porto Rico and Virgin Islands.-Part 1-Amphibians and land reptiles of Porto Rico, with a list of those reported from the Virgin Islands. New York Acad. Sci., 10: 1-160.

Schmidt, K. P., 1932a: A new subspecies of coral snake from Guatemala. Proc. California Acad. Sci., 20 (7): 265-267.

Schmidt, K. P., 1932b: VII A new subspecies of coral snake from Guatemala. Proc. California Acad. Sci., 4th Ser. XX (7): 265-267.

Schmidt, K. P., 1932c: Reptiles and amphibians of the Mandel Venezuelan Expedition. Field Mus. Nat. Hist., Zool. Ser., 18 (7): 157-163.

Schmidt, K. P., 1932d: Stomach contents of some American coral snakes, with the description of a new species of *Geophis*. Copeia, 1932: 6-9.

Schmidt, K. P., 1933a: Amphibians and reptiles collected by the Smithsonian biological survey of the Panama Canal Zone. Smithsonian Misc. Coll., [3181] 89 (1): 1-20.

Schmidt, K. P., 1933b: New reptiles and amphibians from Honduras. Field Mus. Nat. Hist. Zool. Ser. Chicago, 20 (4): 15-22.

Schmidt, K. P., 1933c: Preliminary account of the coral snakes of Central America and Mexico. Zool. Ser. Field Mus. Nat. Hist. Chicago, 20: 29-40.

Schmidt, K. P., 1936a: Notes on Central American and Mexican coral snakes. Field Mus. Nat. Hist. Zool. Ser. 20, (20): 205-216.

Schmidt, K. P., 1936b: New amphibians and reptiles from Honduras in the Museum of Comparative Zoology. Proc. Biol. Soc. Washington, 49: 43-50.

Schmidt, K. P., 1936c: Preliminary account of coral snakes of South America. Field Mus. Nat. Hist. Zool. Ser.20, (20): 189-203.

Schmidt, K. P., 1937: The history of *Elaps collaris* Schlegel. Field Mus. Nat. Hist., Zool. Ser., 20 (26): 361-364.

Schmidt, K. P., 1939: A new coral snake from British Guiana. Field. Mus. Nat. Hist., Zool. Ser., 24 (6): 45-47.

Schmidt, K. P., 1940: Notes on Texan snakes of the genus *Salvadora*. Field Mus. Nat. Hist. Zool. Ser., 24 (12): 143-150.

Schmidt, K. P., 1941: The amphibians and reptiles of British Honduras. Field Mus. Nat. Hist. Zool. Ser, 22: 475-510.

Schmidt, K. P., 1952: The Surinam coral snake *Micrurus surinamensis*. Fieldiana, Zool. 34 (4): 25-34.

Schmidt, K. P., 1953a: A check list of North American amphibians and reptiles. 6th edition. Amer. Soc. Ichthyologists Herpetologists.

Schmidt, K. P., 1953b: Hemprich's coral snake, *Micrurus hemprichi*. Fieldiana, Zool., 34 (13): 165-170.

Schmidt, K. P., 1953c: The Amazonian coral snake *Micrurus spixi*. Fieldiana, Zool., 34 (13): 171-180.

Schmidt, K. P., 1955: Coral snakes of the genus *Micrurus* in Colombia. Fieldiana, Zool., 34 (34): 337-359.

Schmidt, K. P., 1957: The venomous coral snakes of Trinidad. Fieldiana, Zool., 39: 55-63.

Schmidt, K. P., 1958: Some rare or little known Mexican coral snakes. Fieldiana Zool, 39: 201-202.

Schmidt, K. P., and E. W. Andrews, 1936: Notes on snakes from Yucatán. Field Mus. Nat. Hist. Zool. Ser, 20: 167-187.

Schmidt, K. P., and D. D. Davis, 1941: Field book of snakes of the United States and Canada. G. P. Putnam's Sons, New York.

Schmidt, K. P., and R. F. Inger, 1957: Living reptiles of the world. Doubleday & Co., Garden City, New York.

Schmidt, K. P., and A. S. Rand, 1957: Geographic variation in the Central American colubrine snake, *Ninia sebae*. Fieldiana, Zool., 39 (10): 73-84, October 31, 1957.

Schmidt, K. P., and F. J. W. Schmidt, 1925: New coral snakes from Peru. Reports on results of the Captain Marshall Field Expeditions. Pub. Field. Mus. Nat. Hist., Zool. Ser., 12 (10): 129-134.

Schmidt, K. P., and H. M. Smith, 1943: Notes on coral snakes from México. Field Museum Nat. Hist. Zool. Ser., 29: 25-31.

Schmidt, K. P., and T. F. Smith, 1944: Amphibians and reptiles of the Big Bend region of Texas. Field Mus. Nat. Hist. Zool. Ser., 29: 75-96.

Schmidt, K. P., and W. Walker, 1943a: Peruvian snakes from the University of Arequipa. Field. Mus. Nat. Hist. Zool. Ser., 24 (26): 278-296.

Schmidt, K. P., and W. Walker, 1943b: Snakes of the Peruvian coastal region. Field. Mus. Nat. Hist. Zool. Ser., 27 (24): 297-324.

Schmidt, K. P., and W. Walker, 1943c: Three new snakes from the Peruvian Andes. Pub. Field Mus. Nat. Hist. (Zool. Ser.), 24 (28): 325-329.

Schneider, J.G, 1799: Historiae amphibiorum naturalis et literariae. Frriederici Frommann, Jena. 1: 1-266.

Schneider, J. G., 1801: Historiae amphibiorum naturalis et literariae. Fasciculus secundus continens Crocodilos, Scincos, Chamaesuros, Boas, Pseudoboas, Elaps, Angues, Amphisbaenas et Caecilias. Friederici Frommann, Jenae.

Schoenhals, L. C., 1988: A Spanish-English glossary of Mexican flora and fauna—Summer Institute of Linguistics. Mexico.

Schuett, G. W., 1992: Is long-term sperm storage an important component of the reproductive biology of temperate pitvipers? pp. 169-184 *in* J. A. Campbell and E. D. Brodie, Jr.: Biology of the pitvipers. Selva, Tyler, Texas.

Schuller, R. R., 1911: Tijuna (Tukuna) margem esquerda do Solimões entre a fronteira do Brasil e Peru, cituado por Nimuendaju. Revista Americana.

Schulte, R., 1999: Amphibians and reptile species of San Martin, North East Peru. Inibico-Instituto de Investigacion de la Biologia de las Cordilleras Orientales, Website, May 7, 1999.

Schulz, K.D, 1996: A monograph of the colubrid snakes of the genus *Elaphe* Fitzinger. Koeltz Scientific Books.

Schulz, K. D., 1996: Eine Monographie de Schlangengattung *Elaphe* Fitzinger. Busmaster, Berg (CH): 1-460.

Schumann G. O., 1973: La lengua Chol, de Tila (Chiapas). Centro de Estudios Mayas, Cuaderno 8, UNAM Coordinación de Humanidades, México, Centro de Estudios Mayas, Cuaderno 8, 1973.

Schwanor, T. D., and R. H. Mount, 1976: Systematic and ecological relationships of the water snakes *Natrix sipedon* and *N. fasciata* in Alabama and the Florida Panhandle. Occ. Pap. Mus. Nat. Hist, Univ. Kansas, (45): 1-44.

Schwartz, A., 1957: A new species of boa (genus *Tropidophis*) from western Cuba. Amer. Mus. Novitates, (1839): 1-8.

Schwartz, A., 1970: *Darlingtonia* (Serpentes, Colubridae): a new subspecies. Herpetol., 26: 324-331.

Schwartz, A., 1971: A systematic review of the Hispaniolan snake genus *Hypsirhynchus*. Stud. Fauna Curaçao Carib. Isl., 35: 63-94.

Schwartz, A., 1975: Variation in the Antillean boid snake *Tropidophis haetianus* Cope. J. Herpetol., 9 (3): 303-311.

Schwartz, A., 1979: The herpetofauna of Île à Cabrit, Haïti, with the description of two new subspecies. Herpetologica, 35 (3): 248-255.

Schwartz, A., 1986: *Darlingtonia, D. haetiana*. Cat. Amer. Amph. Rept., (390): 1-2.

Schwartz, A., and R. Etheridge, 1954: New and additional herpetological records from the North Carolina coastal plain. Herpetologica, 10 (3): 167-171.

Schwartz, A., and O. H. Garrido, 1975: A reconsideration of some Cuban *Tropidophis* (Serpentes: Boidae). Proc. Biol. Soc. Washington, 88 (9): 77-90.

Schwartz, A., and O. H. Garrido, 1981: A review of the Cuban members of the genus *Arrhyton* (Reptilia, Serpentes, Colubridae). Ann. Carnegie Mus., 50: 207-230.

Schwartz, A., and R. W. Henderson, 1984a: *Uromacer*. Cat. Amer. Amph. Rept., (355): 1.

Schwartz, A., and R. W. Henderson, 1984b: *Uromacer catesbyi*. Cat. Amer. Amph. Rept., (356): 1-2.

Schwartz, A., and R. W. Henderson, 1984c: *Uromacer frenatus*. Cat. Amer. Amph. Rept., (357): 1.

Schwartz, A., and R. W. Henderson, 1984d: *Uromacer oxyrhynchus*. Cat. Amer. Amph. Rept., (358): 1-2.

Schwartz, A., and R. W. Henderson, 1985: A guide to the identification of the amphibians & reptiles of the West Indies exclusive of Hispaniola. Milwaukee Publ. Mus.

Schwartz, A., and R. W. Henderson, 1988: West Indies amphibians & reptiles: A checklist. Milwaukee Publ. Mus. Contrib. Biol. Geol., (74): 1-264.

Schwartz, A., and R. W. Henderson, 1991: Amphibians and reptiles of the West Indies, description, distribution and natural history. Univ. Florida Press, Gainesville, Florida.

Schwartz, A., and R. J. Marsh, 1960: A review of the *pardalis-maculatus* complex of the boid genus *Tropidophis* of the West Indies. Bull. Mus. Comp. Zool., 137: 255-309.

Schwartz, A., and D. A. Rossman, 1976: A review of the Hispaniolan colubrid snake genus *Ialtris*. Stud. Fauna Curaçao Caribbean Isl., 50: 76-102.

Schwartz, A., and Richard Thomas, 1960: Four new snakes (*Tropidophis, Dromicus, Alsophis*) from the Isla de Pinos and Cuba. Herpetologica, 16 (2): 73-90.

Schwartz, A., and Richard Thomas, 1975: A checklist of West Indian amphibians & reptiles. Carnegie Mus. Nat. Hist., Pittsburgh, Pennsylvania.

Schwartz, A., Richard Thomas and L. D. Ober, 1978: First supplement to a checklist of West Indian amphibians and reptiles. Carnegie Mus. Nat. Hist., Spec. Pub. (5): 1-35.

Schwenk, K., 1995: The serpent's tongue. Natural History, 104 (4): 48-54, April, 1995. Scopoli, 1788: Deliciae florae et faunae insubricae, ticino:Monast. S. Salvatoris, 3: 1-87.

Scott, N.J., and R. W. McDiarmid, 1984a: *Trimorphodon*. Cat. Amer. Amph. Rept., (352): 1–2.

Scott, N.J., and R. W. McDiarmid, 1984b: *Trimorphodon biscutatus*. Cat. Amer. Amph. Rept., (353): 1–4.

Scott, N.J., and R. W. McDiarmid, 1984c: *Trimorphodon tau*. Cat. Amer. Amph. Rept., (354): 1–2.

Scrocchi, G. J., 1990a: Contribution to the study of the Leptotyphlopidae of Argentina II. New data on *Leptotyphlops australis* Peters and Orejas-Miranda 1968 [in Spanish]. Acta Zool. Lilloana, 39 (2): 113–114.

Scrocchi, G. J., 1990b: The genus *Micrurus* (Serpentes: Elapidae) in the Argentine Republic. Mus. Reg. Sci. Nat. Boll. (Torino), 8 (2): 343–368.

Scrocchi, G. J., 1997: Acerca de la localidad tipo de *Bothrops ammodytoides* Leybold (Serpentes, Viperidae) y *Pseudotomodon trigonatus* (Leybold) (Serpentes: Colubridae). Cuadernos de Herpetologica, 11 (1–2): 69–70.

Scrocchi, G. J., and J. M. Cei, 1991: A new species of the genus *Atractus* from nw Argentina (Serpentes, Colubridae). Mus. Geg. Sci. Nat. Boll. (Torino), 9 (1): 205–208.

Scrocchi, G. J., and F. B. Cruz, 1993: Description of a new species of the genus *Lystrophis* Cope and a revalidation of *Lystrophis pulcher* (Jan, 1863), (Serpente, colubridae). Pap. Avuls. Zool. (São Paulo), 38 (10): 171–185.

Scrocchi, G., and M. Vinas, 1990: El género *Clelia* (Serpentes: Colubridae) en la República Argentina revision y comentários. Boll. Mus. Reg. Sci. Nat. Terino, 8: 487–499.

Sebeok, T. A., (Ed.), 1976: Native languages of the Americas. Vol. I. Plenum Press, New York and London.

Sebeok, T. A., (Ed.), 1977: Native languages of the Americas, Vol II. Plenum Press, New York and London.

Secor, S. M., 1992: A preliminary analysis of the movement and home range size of the sidewinder, *Crotalus cerastes*. pp. 389–393 *in* J. A. Campbell and E. D. Brodie, Jr.: Biology of the pitvipers. Selva, Tyler, Texas.

Seidel, M. E., and L. D. Wilson, 1979: Geographic distribution: *Coluber constrictor*. Herpetol. Rev., 10 (2): 60.

Seigel, R. A., and Collins, J. T., 1993: Snakes ecology & behavior. New York, New York, McGraw-Hill, Inc.

Señaris, J. C., 1998: A new species of *Typhlophis* (Serpentes: Anomalepididae) from Bolívar State, Venezuela. Amphibia-Reptilia, 19 (3): 303–310.

Sentzen, U. I., 1796: Ophiologische fragmente. *In* Meyer's Zool. Arch., 2: 49–74.

Seoane, V. L., 1881: Neue Boiden-gattung und—Art von den Philippinen. Abh. Senckenb. Naturf. Ges., 12: 217–224.

Serié, P., 1914: Notes d'erpetologie. Description d'une variete de *Philodryas baroni* Berg. Ann. Mus. Nac. Buenos Aires, 26: 227–230.

Serié, P., 1915a: Notas sobre la erpetologica del Paraguay. I. Colección de ofidios del Dr. Fiebrig. Physis, 1: 573–581.

Serié, P., 1915b: Notas sobre la erpetologia del Paraguay. II. Addenda a los ofidios de la *Fauna Paraguaya* de W. Bertoni. Physis, 1: 581–582.

Serié, P., 1915c: Suplemento a la fauna erpetologica Argentina. An. Mus. Nac. Buenos Aires, 27: 93–109.

Serié, P., 1916 [1919]: Enumeración de los ofidios de Tucumán. Prim. Reun. Nac. Soc., Argentina Cien. Nat. Buenos Aires, 1916: 418–420.

Serié, P., 1919: Notas sobre las serpientes venenosas de la Argentina. 1a. Reun. Nac. Soc. Argentina Cien. Nat., 1919: 100–417.

Serié, P., 1921: Catálogo de los ofidios Argentinos. An. Soc. Cien. Argentina, Buenos Aires, 92: 145–192.

Serié, P., 1936: Nueva enumeración sistemática y distribución geográphica de los ofidios Argentinos. Inst. Mus. Univ. Nac. La Plata, Obra Cincenten, (50): 33–61.

Sexton, O. J., P. Jacobson and J. E. Bramble, 1992: Geographic variation in some activities associated with hibernation in nearartic pitvipers. pp. 337–346 *in* J. A. Campbell and E. D. Brodie, Jr.: Biology of the pitvipers. Selva, Tyler, Texas.

Shannon, F. A., and F. L. Humpfrey, 1959: A new subspecies of *Phyllorhynchus browni* from Sinaloa. Herpetologica, 15 (3): 145–148.

Shannon, F. A., and H. M. Smith, 1949: Herpetological results of the University of Illinois Field Expedition, spring 1949. I. Introduction, Testudines, Serpentes. Trans. Kansas Acad. Sci., 52: 494–509.

Shaw, C., and S. Campbell, 1974: Snakes of the American west. Alfred A. Knopf, New York.

Shaw, G., 1802: General zoology, or systematic natural history. Vol. III-Part II. Amphibia. G. Kearsley, London, 1802: vii + 446, 313–615 + pls. 87–140.

Sheil, C. A., 1998: *Emmochiliophis miops*: redescription of *Synophis miops* (Boulenger, 1989). J. Herpetol., 32 (4): 604–607.

Sheil, C. A., and T. Grant, 2001: A new species of Colubrid snake (*Synophis*) from western Colombia. J Herpetol., 35 (2): 204–209.

Sheplan, B. R., and A. Schwartz, 1974: Hispaniolan boas of the genus *Epicrates* (Serpentes, Boidae) and their Antillean relationships. Ann. Carnegie Mus., 45 (5): 57–143.

Shreve, B., 1934: Notes on Ecuadorian snakes. Occ. Pap. Boston Soc. Nat. Hist., 8: 125–131.

Shreve, B., 1947: On Venezuelan reptiles and amphibians collected by Dr. H. G. Kugler. Bull. Mus. Comp. Zool., 99 (5): 518–537.

Shreve, B., 1953: Notes on the races of *Micrurus frontalis* (Duméril, Duméril, and Bibron). Breviora, (16): 1–6.

Shreve, B., and C. Gans, 1958: *Thamnophis bovallii* Dunn rediscovered (Reptilia: Serpentes). Breviora, (63): 1–7.

Simpson, G. G., 1951: The species concept. Evolution, 5: 285–298.

Silva, F., A. Veloso, J. Solervicens, and J. C. Ortiz, 1968: Investigaciones zoológicas en el Parque Nacional Vicente Perez Rosales y zona de Pargua. Not. Mens. Mus. Hist. Nat. Santiago, 12 (148): 1–12.

Silva Haad, J. J., 1994: Los *Micrurus* de la Amazonia colombiana. Biología y toxicología experimental de sus venenos. Colombia Amazónica, 7 (1–2): 41–138.

Siméon, R., 1977: Diccionario de la lengua Nahuatl o mexicana, Siglo Vientiuno editores S.A. de C.V.

Simon, H., 1979: Easy identification guide to North American snakes. Dodd, Mead & Co., New York.

Slevin, J. R., 1934: A handbook of reptiles and amphibians of the Pacific states including certain eastern species. California Acad. Sci., Spec. Pub.

Slevin, J. R., 1936: A new Central American snake. Proc. California Acad. Sci., Ser. 4, 23 (4): 79–81.

Slevin, J. R., 1939: Notes on a collection of reptiles and amphibians from Guatemala. I. Snakes. Proc. California Acad. Sci., Ser. 4, 23 (26): 393–414.

Slevin, J. R., 1942: Notes on a collection of reptiles from Boquete, Panama, with the description of a new species of *Hydromorphus*. Proc. California Acad. Sci., 23: 463–480.

Slevin, J. R., and A. E. Leviton, 1956: Holotype specimens of reptiles and amphibians in the collection of the California Academy of Sciences. Proc. California Acad. Sci., (4), 15 (3): 195–207.

Smith, A. G., 1945: The status of *Thamnophis butleri* Cope and a redescription of *Thamnophis brachystoma* (Cope). Proc. Biol. Soc. Washington, 58: 147–154.

Smith, A. G., 1949: The subspecies of the plains garter snake, *Thamnophis radix*. Bull. Chicago Acad. Sci., 8: 285–300.

Smith, B. E., and J. A. Campbell, 1996: The systematic status of Guatemalan populations of snakes allied with *Ninia maculata* (Reptilia: Colubridae). Proc. Biol. Soc. Washington, 109 (4): 749–754.

Smith, E. N., 1995: *Geophis rhodogaster* (Colubridae), an addition to the snake fauna of Mexico. Southwest. Naturalist, 40 (1): 123–124.

Smith, E. N., and M. E. Avecedo, 1997: The northernmost distribution of *Corallus annulatus* (Boidae), with comments on its natural history. Southwest. Naturalist, 42 (3): 347–349.

Smith, E. N., and J. A. Campbell, 1994: A new species of *Rhadinaea* (Colubridae) from the Caribbean versant of Guatemala. Occ. Pap. Mus. Nat. Hist. Univ. Kansas, (167): 1–9.

Smith, E. N., and J. A. Campbell, 1996: The systematic status of Guatemalan populations of snakes allied with *Ninia maculata* (Reptilia: Colubridae). Proc. Biol. Soc. of Washington, 109 (4): 749–754.

Smith, H. M., 1938: A review of the snake genus *Farancia*. Copeia, 1938: 110–117.

Smith, H. M., 1939: Notes on Mexican reptiles and amphibians. Field Mus. Nat. Hist. Zool. Ser., 24 (4): 15–35.

Smith, H. M., 1940: Description of new lizards and snakes from Mexico and Guatemala. Proc. Biol. Soc. Washington, 53: 55–64.

Smith, H. M., 1941a: A new genus of Central American snakes related to *Tantilla*. J. Washington Acad. Sci., 31: 115–117.

Smith, H. M., 1941b: A new name for the Mexican snakes of the genus *Dendrophidion*. Proc. Biol. Soc. Washington, 54: 73–76.

Smith, H. M., 1941c: A new species of snake related to *Rhadinaea*. Copeia, 1941 (1): 7–10.

Smith, H. M., 1941d: A review of the subspecies of the Indigo Snake, *Drymarchon corias*. J. Washington Acad. Sci., 31 (11): 466–481.

Smith, H. M., 1941e: Notes on Mexican snakes of the genus *Geophis*. Smithsonian Misc. Coll., 99 (19): 1–6.

Smith, H. M., 1941f: Notes on the snake genus *Trimorphodon*. Proc. U.S. Natl. Mus., 91: 149–168.

Smith, H. M., 1941g: On the Mexican snakes of the genus *Pliocercus*. Proc. Biol. Soc. Washington, 54: 119–124.

Smith, H. M., 1941h: Revision of indigo snake *Drymarchon corais*. J. Washington Acad. Sci., 31: 466–481.

Smith, H. M., 1941i: The Mexican subspecies of the snake *Coniophanes fissidens*. Proc. U.S. Natl. Mus., 91 (3127): 107–111.

Smith, H. M., 1942a: Additional notes on Mexican snakes of the genera *Pliocercus*. Proc. Biol. Soc. Washington, 55: 159–164.

Smith, H. M., 1942b: A new race of *Ninia* from Mexico. Copeia, 1942 (3): 152–154.

Smith, H. M., 1942c: A resume of the Mexican snakes of the genus *Tantilla*. Zoologica, 27: 33–42.

Smith, H. M., 1942d: A review of the snake genus *Adelphicos* and remarks on the Mexican king snakes of the *Triangulum* group. Proc. Rochester Acad. Sc., 8: 175–207.

Smith, H.M, 1942e: Mexican herpetology miscellany. Proc. U.S. Natl. Mus., Washington, 92 (3153): 349–395.

Smith, H. M., 1942f: Notes on *Masticophis mentovarius*. Copeia, 1942 (2): 85–88.

Smith, H. M., 1942g: The synonymy of the garter snakes (*Thamnophis*), with notes on Mexican and Central American species. Zoologica, 27 (Parts 3 and 4): 97–123.

Smith, H. M., 1943: Summary of the collection of snakes and crocodilians made in Mexico under the Walter Rathbone Bacon traveling scholarship. Proc. U.S. Natl. Mus., 93 (3169): 393–504.

Smith, H. M., 1944: Snakes of the Hoogstraal Expedition to northern Mexico. Field Mus. Nat. Hist Zool. Ser., 29 (8): 135–153.

Smith, H. M., 1950a: Handbook of the amphibians and reptiles of Kansas. Univ. Kansas Mus. Nat. Hist. Misc. Pub., 2: 1–336.

Smith, H. M., 1950b: Type localities of Mexican reptiles and amphibians. Kansas Univ. Sci. Bull., 33 (8): 313–380.

Smith, H.M, 1952: Commentary on the identity of *Coluber doliatus*. Herpetologica, 8 (2): 26–27.

Smith, H. M., 1958: Handlist of the snakes of Panama. Herpetologica, 14: 222–224.

Smith, H. M., 1962: The subspecies of *Tantilla schistosa* of Middle America (Reptila: Serpentes). Herpetologica, 18 (1): 13–18.

Smith, H. M., 1965: Two new Colubrid snakes from the United States and Mexico. J. Ohio Herpetol. Soc., 5 (1): 1–4.

Smith, H. M., 1987: Current nomenclature for the names and material cited in Günther's Reptilia and Batrachia volume of the Biologia Centrali-Americana. *In* A. C. L. G. Günther, Biologia Centrali-Americana rept. & bio. SSAR Reprint. xxxiii–li.

Smith, H. M., and M. Alvarez del Toro, 1962: Notulae herpetologicae Chiapasiae III. Herpetologica, 18 (2): 101–107.

Smith, H. M., R. G. Arndt and W. C. Sherbrook, 1967: A new snake of the genus *Enulius* from Mexico. Nat. Hist. Misc., (186): 1–4.

Smith, H. M., and E. D. Brodie, Jr., 1982: A guide to the field identification—reptiles of North America. Golden Press, Western Publishing Co., New York, 240 pp.

Smith, H. M., and D. Chiszar, 2001: A new subspecies of cantil (*Agkistrodon bilineatus*) from central Veracruz, Mexico (Reptilia: Serpentes). Bull. Maryland Herpetolo. Soc., 37: 130–136.

Smith, H. M., F. van Breukelen, D. L. Auth, and D. Chiszar, 1998: A subspecies of the Texas Blind Snake (*Leptoptyhlops dulcis*) with supraoculars. Southwest. Naturalist, 43 (4): 437–440.

Smith, H. M., F. van Breukelen, and D. Chiszar, 1997: Geographic distribution—*Tantilla rubra* (Red Blackhead Snake). Herpetol. Rev., 28 (4): 210.

Smith, H. M., and H. L. Buechner, 1947: The influence of the Balcones Escarpment on the distribution of amphibians and reptiles in Texas. Bull. Chicago Acad. Sci., 8: 1-16.

Smith, H. M., and L. Burger, 1950: A new snake (*Tantilla*) from Mexico. Herpetologica, 6: 117-119.

Smith, H. M., and D. Chiszar, 1992a: A Mexican genus of tropidophinae snakes. Bull. of Maryland Herpetol. Soc., 28 (1): 19-28.

Smith, H. M., and D. Chiszar, 1992: A second locality for *Geophis sallei* (Reptilia: Serpentes). Bull. Maryland Herpetol. Soc., 28 (1): 16-18.

Smith, H.M, and D. Chiszar, 1993: Apparent intergradation in Texas between the subspecies of the Texas Blind Snake (*Leptotyphlops dulcis*). Bull. Maryland Herpetol. Soc., 29 (4): 143-155.

Smith, H. M., and D. Chiszar, 1994: Variation in the Lined Snake (*Tropidoclonion lineatum*) in northern Texas. Bull. Maryland Herpetol. Soc. 30 (1): 6-13.

Smith, H. M., and D. Chiszar, 1996: Species-group taxa of the false coral snake genus *Pliocercus*. Ramus Publishing, Inc., July 15, 1996, Pottsville, PA.

Smith, H. M., and D. Chiszar, 2000: The identity of the isolated Los Tuxtlas population of the false coral snake, *Pliocercus elapoides*, of southern Veracruz, Mexico. J. Colorado-Wyoming Acad. Sci., 32 (1): 18.

Smith, H. M., and D. Chiszar, (in press) a: *Pliocercus andrewsi* Smith—Andrews' False Coral Snake. Cat. Amer. Amph. Rept.

Smith, H. M., and D. Chiszar, (in press) b: *Pliocercus bicolor* Smith—Northern False Coral Snake. Cat. Amer. Amph. Rept.

Smith, H. M., and D. Chiszar, (in press) c: *Pliocercus* Cope—False Coral Snakes. Cat. Amer. Amph. Rept., 735: 1-9.

Smith, H. M., and D. Chiszar, (in press) d: *Pliocercus dimidiatus* Cope—Central American False Coral Snake. Cat. Amer. Amph. Rept.

Smith, H. M., and D. Chiszar, (in press) e: *Pliocercus elapoides* Cope—Variegated False Coral Snake. Cat. Amer. Amph. Rept.

Smith, H. M., and D. Chiszar, (in press) f: *Pliocercus euryzonus* Cope—Cope's False Coral Snake. Cat. Amer. Amph. Rept.

Smith, H. M., and D. Chiszar, (in press) g: Reassessment of a nominal species of false coral snakes, *Pliocercus psychoides* (Reptilia, Serpentes). Southwest. Naturalist, 46 (1): in press.

Smith, H. M., D. Chiszar and F. v. Breukelen, 1998: Resurrection of *Tantilla triseriata* (Reptilia: Serpentes) of Mexico. Southwest. Naturalist, 43 (3): 374-375.

Smith, H. M., D. Chiszar and M. Mancolla Moreno, 2001: Nomenclature of the Earth Snake (*Adelphicos*) of the *A. quadrivirgatus* complex. Bull. Maryland Herp. Soc., 37 (2): 39-41.

Smith, H. M., D. Chiszar and Pérez-Higareda, 1989: Variation in the northern subspecies of *Pliocercus elapoides* (Reptilia, Serpentes). Bull. Chicago Herpetol. Soc., 24: 112-116.

Smith, H. M., D. Chiszar, J. R. Staley, II and K. Tepedelen, 1994: Populational relationships in the corn snake *Elaphe guttata* (Reptilia: Serpentes). Texas J. Sci., 46 (3): 259-292.

Smith, H. M., and P. S. Chrapiwy, 1957: A new subspecies of the Central American false coral snake (*Pliocercus*). Herpetologica, 13: 233-235.

Smith, H. M., and P. David, 1999: Introduction to Shaw 1802 SSAR reprint. (See under Shaw.)

Smith, H. M., and J. R. Dixon, [1987] 1988: The amphibians and reptiles of Texas: a guide to records needed for Mexico. Bull. Maryland Herpetol. Soc., 23 (3): 154-157.

Smith, H. M., J. R. Dixon and V. Wallach, 1993: *Dromicus giganteus* Jan (Reptilia: Serpentes) is a nomen nudum. Bull. Maryland Herpetol. Soc., 29 (3): 77-79.

Smith, H. M., and K. Fitzgerald, G. Pérez-Higareda and D. Chiszar, 1986: A taxonomic rearrangement of the snakes of the genus *Scaphiodontophis*. Bull. Maryland Herpetol. Soc., 22: 159-166.

Smith, H. M., and O. Flores-Villela, 1993: Variation in two species (*G. bicolor, G. duellmani*) of Mexico earth snakes (*Geophis*). Bull. Maryland Herpetol. Soc. 29 (1): 20-23.

Smith, H. M., O. Flores Villela and D, Chiszar, 1993: New variational extremes for *Tantilla calamarina* and a locality record correction for *Conophis vittatus viduus* (Reptilia: Serpentes). Bull. Maryland Herpetol. Soc., 29 (1): 1-3.

Smith, H. M., O. Flores Villela and D. Chiszar, 1993: The generic allocation of *Tantilla canula* (Reptilea: Serpente). Bull. Maryland Herpetol. Soc., 29 (3): 126-129.

Smith, H. M., and J. Gillespie, 1965: Two new Colubrid snakes from the United States and Mexico. J. Ohio Herpetol. Soc., 5 (1): 1-4.

Smith, H. M., and C. Grant, 1958a: New and noteworthy snakes from Panamá. Herpetologica, 14: 207-215.

Smith, H. M., and C. Grant, 1958b: Noteworthy herpetiles from Jalisco, Mexico. Herpetologica, 14: 18-23.

Smith, H. M., and C. Grant, 1958c: The proper names for some Cuban snakes and analysis of dates of publication of Ramón de la Sagra's historia natural de Cuba and of Fitzinger's systema reptilium. Herpetoleuca., 14 (4): 215-222.

Smith, H. M., and R. L. Holland, 1969: Two new snakes of the genus *Geophis* from México. Trans. Kansas Acad. Sci., 72: 47-53.

Smith, H. M., and J. E. Huheey, 1960: The water snake genus *Regina*. Trans. Kansas Acad. Sci., 63 (3): 156-164.

Smith, H. M., and J. P. Kennedy, 1951: *Pituophis melanoleucus ruthveni* in Eastern Texas and its bearing on the status of *P. catenifer*. Herpetologica, 7 (part 3): 93-96.

Smith, H. M., and M. J. Landy, 1965: New and unusual snakes of the genus *Pliocercus* from Oaxaca, Mexico. Nat. Hist. Misc. Chicago Acad. Sci. (183): 1-4.

Smith, H. M., and D. A. Langebartel, 1949: Notes on a collection of reptiles and amphibians from the Isthmus of Tehuantepec, Oaxaca. J. Washington Acad. Sci., 39 (12): 409-416.

Smith, H. M., and D. A. Langebartel, 1951: A new geographic race of leaf-nosed snake from Sonora, Mexico. Herpetologica, 7 (part 4): 181-184.

Smith, H. M., and K. R. Larson, 1974a: The name of the Baja California Cape wormsnake. Great Basin Naturalist, 34 (2): 94-96.

Smith, H. M., and K. R. Larsen, 1974b: The nomincal snake genera *Mastigodryas* Amaral 1934 and *Dryadophis* Stuart 1939. Great Basin Naturalist, 33: 276.

Smith, H. M., and L. E. Laufe, 1945: Notes on a herpetological collection from Oaxaca. Herpetologica, 3 (1): 1-13.

Smith, H. M., J. D. Lynch and R. Altig,

1965: New and noteworthy herpetozoa from southern Mexico. Nat. Hist. Misc., (180): 1–4.

Smith, H. M., and E. O. Moll, 1969: A taxonomic rearrangement of the pit vipers of the *Bothrops nigroviridis* complex of southern México. J. Herpetol., 3: 151–155.

Smith, H. M., C. A. Pague and D. Chiszar, 1996: A brood of lined snakes (*Tropidoclonion lineatum*) from Southeastern Colorado. Bull. Maryland Herpetol. Soc., 32 (1): 24–27.

Smith, H. M., and G. Pérez-Higareda, 1986: The proper name for the southern Atlantic coast subspecies of the lagartijera, *Dryadophis* (Reptilia: Serpentes). Bull. Maryland Herpetol. Soc., 22 (2): 51–55.

Smith, H. M., and G. Pérez-Higareda, 1991: Clara evidencia de la coespecificidad de *Geophis dubius* y *Geophis rostrialis* (Reptilia: Serpentes). Bol. Sociedad Herpetologica Mexicana, 32 (2): 39–40.

Smith, H. M., A. Ramírez-Bautista and D. Chiszar, 1995: The nomenclatural status of Müller's two varietal names of *Dipsas cenchoa* (Reptilia: Serpentes). Bull. Maryland Herpetol. Soc., 31 (4): 198–203.

Smith, H. M., and P. W. A. Smith, 1951: A new snake (*Tantilla*) from the Isthmus of Tehuantepec, México. Proc. Biol. Soc. Washington, 64: 97–100.

Smith, H. M., and R. B. Smith, 1973: Synopsis of the herpetofauna of Mexico. Vol. II. Analysis of the literature exclusive of the Mexican axolotl. John Johnson, North Bennington, Vt., 1973: vii - xxxiii, 1–367.

Smith, H. M., and R. B. Smith, 1976a: Synopsis of the herpetofauna of Mexico. Vol. III. Source analysis and index for Mexican reptiles. John Johnson, North Benington, Vermont., 1976: 1–23, AM-1 through COR-4.

Smith, H. M., and R. B. Smith, 1976b: Synopsis of the herpetofauna of Mexico, Volume III. Source analysis and index for Mexican Reptiles. North Bennington, Vermont, John Johnson, 1976: S-A-1 to S-G-7.

Smith, H. M., and R. B. Smith, 1993: Synopsis of the herpetofauna of Mexico. Vol. VII, bibliographic addendum IV and index, bibliographic addenda II-IV, 1979–1991. Univ. Press Colorado, 1993: vii–ix + 1–1082.

Smith, H. M., and W. W. Tanner, 1944: Description of a new snake from Mexico. Copeia, 1944 (3): 131–136.

Smith, H. M., and E. H. Taylor, 1941: A review of the snakes of the genus *Ficimia*. J. Washington Acad. Sci., 31: 356–368.

Smith, H. M., and E. H. Taylor, 1942: The snake genus *Conopsis* and *Toluca*. Univ. Kansas Sci. Bull., 28 (15): 325–363.

Smith, H. M., and E. H. Taylor, 1945: An annotated checklist and key to the snakes of Mexico. Bulletin 187, Washington, D.C., Smithsonian Institute, United States National Museum, United States Government Printing Office, 239 pp.

Smith, H. M., and E. H. Taylor, 1950: Type localities of Mexican reptiles and amphibians. Univ. Kansas Sci. Bull., 33 (8): 313–380.

Smith, H. M., and E. H. Taylor, 1966: Herpetology of Mexico: annotated checklist and keys to the amphibians & reptiles. Ashton, MD, Eric Lundberg.

Smith, H.M, and V. Wallach, and D. Chiszar, 1995): Observations on the snake genus *Pliocercus*, I. Bull. Maryland Herpetol. Soc., 31: 204–214.

Smith, H. M., and K. L. Williams, 1963: New and noteworthy amphibians and reptiles from southern Mexico. Herpetologica, 19: 22–27.

Smith, H. M., and K. L. Williams, 1966a: A new snake (*Tantilla*) from Las Islas de la Bahia, Honduras. Southwest. Naturalist, 11 (4): 483–487.

Smith, H. M., and K. L. Williams, 1966b: The ratsnake of the Bay Islands, Honduras. Nat. Hist. Misc. Chicago Acad. Sci., (185): 1–2.

Smith, H. M., K. L. Williams and G. Pérez-Higareda, 1986: The specific name for the Linnaean *Oxyrhopus*, or the calico false coral snake. Bull. Maryland Herpetol. Soc., 22: 10–13.

Smith, L.E, and F. K. Branom (ed.), 1937: National atlas of the world with new United States and world maps. Geographical Publishing Company, Chicago, 1–132 + i–xii.

Smith, M., 1926: Monograph of the seasnakes (Hydrophiidae). British Museum (Natural History), London.

Smith, P. W. A., 1956: The geographical distribution and constancy of the *semifasciata* pattern in the eastern garter snake. Herpetologica., 12 (2): 81–84.

Smith, P. W. A., and D. M. Darling, 1952: Results of a herpetological collection from eastern central Mexico. Herpetologica, 8 (3): 81–86.

Smith, R., 1977: Southern Barasano sentence structure. pp 175–205. *In* Longacre, R. E., ed., 1977: Discourse grammar—Studies in indigenous languages of Colombia, Panama, and Ecuador Part III.

Smith, R., 1993: Crisis under the canapoy-tourism & other problems facing the present day Huaorani. ABYA-YACA, September 1993.

Smith, R., and G. Seeberan, 1979: Occurrence of the rare checkerbelly snake, *Siphlophis cervinus*, in the Nariva Swamp, Trinidad. Living World, J. Trinidad Tobago Field Naturalist's Club, 1978–1979: 11.

Smith, W. H., 1882: Report on the reptiles and amphibians of Ohio. Geol. Survey Ohio, 4 (1): 631–734.

Smithsonian Institute, 1912: Bureau of American ethnology: A dictionary of the Biloxi and Ofo languages. Dorsey & Swanton eds. Bureau of American Ethnology, Bulletin no. 47, U.S. Government Printing Office, Washington, D.C.

Sokolov, V. E. (Ed.), 1988: Dictionary of animal names in five languages: amphibians and reptiles-Latin, Russian, English, German, French-12126 names. Russky Yazyk Publishers, Moscow.

Soulé, M., and A. J. Sloan, 1966: Biogeography and distribution of the reptiles and amphibians on islands in the Gulf of California, Mexico. Trans. San Diego Soc. Nat. Hist., 14: 137–156.

Solis Alcala, E., 1950: Diccionario Español-Mayan. Editorial Yikal Maya Than, Mexico D.F.

Solórzano, A., 1994: Una nueva especie de serpiente venenosa terrestre del género *Porthidium* (Serpentes: Viperidae), del suroeste de Costa Rica. Rev. Biol. Trop., 42: 695–701.

Solórzano, A., and L. Cerdas, 1986: A new subspecies of the bushmaster, *Lachesis muta*, from sw Costa Rica. J. Herpetol., 20: 463–466.

Solórzano, A., L. D. Gómez; J. Monge-Nájera; and B. I. Crother, 1998: Redescription and validation of *Bothriechis supraciliaris* (Serpentes: Viperidae). Revista de Biologia Tropical, 46: 453–462.

Solórzano, A., J. M. Gutiérrez and L. Cerdas, 1988: *Bothrops ophryomegas* Bocourt (Serpentes: Viperidae) en Costa Rica: distribución, lepidosis, variación sexual y cariotipo. Rev. Biol. Trop., 36 (2A): 187–190.

Solórzano, A., J. M. Gutiérrez and L. Cerdas, 1989: Notes on the natural history and karyotype of the colubrid snake, *Hydromorphus concolor* Peters, from Costa Rica. J. Herpetol., 23 (3): 314–315.

Sonnini, C. S., and P. A. Latreille, 1801–1802: Histoire naturelle des reptiles. J. B. Bailliére et Filo., Paris, 4 volumes.

Sosa, R. A., R. W. Henderson and R. Powell, 1995: *Hyspirhynchus* Günther. Cat. Amer. Amph. Rept., (617): 1–4.

Sothern, J. M., 1977: Cajun Dictionary-a collection of some commonly used words and phrases by the people of South Louisiana. Rachel Wells, Houma, Louisiana.

Spiteri, D. E., 1987 [1988]: The geographic variability of the species *Lichanura trivirgata* and a description of a new subspecies. *In* H. F. De Lisle, P. R. Brown, B. Kaufmann, and Brian M. McGurty (eds.), Proceedings of the Conference on California Herpetology-1987. Southwest. Herpetol. Soc. Spec. Pub., (4): 113–130.

Spiteri, D. E., 1991: The subspecies of *Lichanura trivirgata*. Why the confusion? Bull. ChicagocHerpetol. Soc., 26: 153–156.

Spiteri, D. E., 1993: The current taxonomy and captive breeding of the rosy boa (*Lichanura trivirgata*). Vivarium., 5 (3).

Spix, J. B. von, and J. G. Wagler, 1824: Herpetology of Brazil. With introduction by P. E. Vanzolini, reprinted under the patronage of The Government of João Baptista de Oliveira Figueiredo, President of the Federative Republic of Brazil, Society for the Study of Amphibians and Reptiles, 1981. Containing Introduction and three volumes.

Sprackland, R. G., 2000: Pronunciation of scientific names. Reptile & Amphibian Hobbyist, April, 2000, 5 (8): 44–48.

Stafford, P. J., 1986: Pythons and boas. T.F.H. Pub., Inc. Ltd. Neptune City, New Jersey.

Stafford, P., 1994.: Belize—amazing reptiles-Part I. Aquarist & Pondkeeper. September, 1994. Ashford, Kent, UK., 1994: 84–86.

Stafford, P. J., 1996: Geographic distribution: *Stenorrhina degenhardtii* (Degenhardt's Scorpion-eating Snake). Herpetol. Rev., 27 (4): 214–215.

Stafford, P. J., 1999: *Urotheca* (*Pliocercus*) *elapoides* (false coral snake): reproduction and diel activity. Herpetol. Rev., 30: 48.

Stafford, P. J., 2000: Snakes. Nat. Hist. Mus., London.

Stafford, P. J., and R. W. Henderson, 1996: Kaleidoscopic tree boas-The genus *Corallus* of tropical America. Krieger Publishing Co., Malabar. Florida.

Stafford, P. J., and J. R. Meyer, 2000: A guide to the reptiles of Belize. Academic Press, London and San Diego.

Stahnke, H. L., 1966: The treatment of venomous bites & stings. Revised Edition, Arizona State Univ., Tempe, Arizona.

Starace, F., 1995: Contribution to the study of the ophidians of French Guyana. Presence of a new Colubrid in French Guyana, *Xenopholis scalaris* (Wucherer, 1861). Litteratura Serpentium, English Edition, Dec. 1995, 15 (6): 153–157.

Starace, F., 1997: Contribution to the study of the snakes of French Guyana 2. Litteratura Serpentium, English Edition, 17 (5): 97–101.

Starace, F., 1998: Guide des serpents et amphisbènes de Guyane [Guide des serpents et amphisbènes de Guyane Française].(in French) Ibis Rouge Editions, Guadeloupe-Guyane.

Staszko, R., and J. G. Walls, 1994: Rat snakes: A hobbyist's guide to *Elaphe* and kin. T.F.H. Pub., Inc., Neptune City, New Jersey.

Stebbins, Robert C., 1966: A field guide to western reptiles & amphibians. 2nd Printing, Boston, Houghton Mifflin Co.

Stebbins, Robert C., 1972: Amphibians and reptiles of California. California Natural History Guides: 31. Univ. California Press, Berkley, Los Angeles and London.

Stebbins, Robert C., 1985: A field guide to western reptiles and amphibians. 2nd ed. Rev. Boston, Houghton Mifflin Co.

Steindachner, F., 1867a: Herpetologische Notizen. Sitzungsber. Akad. Wiss. Wien., 55 (1): 265–273.

Steindachner, F., 1867b: Reise der Österreichischen Fregatte Novara um die Erde in den Jahren 1857, 1858, 1859 unter dem Befehl des Commodore B. Von Wüllerstorf—Urbair. Zoologischer Teil. I. Band (Wirbeltiere. Reptilien. I. Kaiser. Hof—und Staatsdruckerie, Wien, 3 Reptilien.

Steindachner, F., 1891: Über einige neue und seltenere Reptilien und Amphibien. Anz. K. Akad. Wiss. Wein,. 14: 141–144.

Steindachner, F., 1901: Herpetologische und ichthyologische Ergebnisse einer Reise nach Südamerika mit einer Einleitung von Therese Prinzessin von Baiern. (Abstract). Anz. K. Akad. Wiss. Wein., 38: 194–196.

Steindachner, F., 1902: Herpetologische und ichthyologische Ergebnisse einer Reise nach Südamerika mit einer Einleitung von Therese Prinzessin von Bayern. Denkschr. Anz. K. Akad. Wiss. Wein., 72: 89–148.

Stejneger, L. H., 1890: Annotated list of reptiles and batrachians collected by D. C. Hart Merriam and Vernon Bailey on the San Francisco Mountain Plateau and desert of the Little Colorado, Arizona, with descriptions of new species. North American Fauna, 3: 103–118.

Stejneger, L. H., 1891: Notes on some North American snakes. Proc. U.S. Natl. Mus., 14: 501–505.

Stejneger, L. H., 1899: Reptiles of the Tres Marias and Isabel Islands. North American Fauna, (14): 63–71

Stejneger, L. H., 1901: A new systematic name for the yellow boa of Jamaica. Proc. U.S. Natl. Mus., 23 (1218): 468–470.

Stejneger, L. H., 1902a: An annotated list of batrachians and reptiles collected in the vicinity of La Guaira, Venezuela, with descriptions of two new species of snakes. Proc. U.S. Natl. Mus. 24 (1248): 179–192.

Stejneger, L. H., 1902b: The herpetology of Puerto Rico. Reprint of the U.S. Natl. Mus. Washington, D.C., 1902: 549–724.

Stejneger, L. H., 1903: The reptiles of the Huachuca Mountains,—Arizona. Proc. U.S. Natl. Mus., 25 (1282): 149–158.

Stejneger, L. H., 1904: The herpetology of Puerto Rico. Report U.S. Natl. Mus., 129: 549–724.

Stejneger, L. H., 1907: Herpetology of Japan and adjacent territory. Bull. U.S. Natl. Mus., 58: I-XX, 1-577, 35 pls.

Stejneger, L. H., 1909: Batrachians and Reptiles. pp. 211-224 In W. B. Scott (ed.), 1905-1911: Reports of the Princeton University Expedition to Patagonia to Patagonia, 1896-1899. Vol. III., 1, Zool., Princeton: The Univ.; Stuttgart: Schweizerbat'sche Verlagshandlung (E. Nägele & Dr. Sproesser), 1909: xii-374 pp.

Stejneger, L. H., 1913: Results of the Yale Peruvian Expedition of 1911. Batrachians and reptiles. Proc. U.S. Natl. Mus., 45: 541-547.

Stejneger, L. H., 1917: Cuban amphibians and reptiles collected for the United States National Museum from 1899 to 1902. Proc. U.S. Natl. Mus., 53 (2205): 259-291.

Stejneger, L. H., 1933: Studies on Neotropical Colubrinae. II. Some new species and subspecies of *Eudryas* Fitzinger, with annotated list of the forms of *Eudryas boddarti* (Sentzen). Occ. Pap. Mus. Zool. Univ. Michigan, (254): 1-10.

Stejneger, L. H., 1938: Studies on Neotropical Colubrinae. IV. A new species of *Eudryas* from South America. Copeia, 1938 (1): 7-8.

Stejneger, L. H., 1939: A new name for the genus *Eudryas* Fitzinger, 1943. Copeia, 1939 (1): 55.

Stejneger, L. H., 1941: Studies of Neotropical Colubrinae VII. A revision of the genus *Dryadophis* Stuart, 1939,. Misc. Pub. Mus. Zool. Univ. Michigan, (49): 1-106.

Stejneger, L. H., and T. Barbour, 1917: A checklist of North American amphibians and reptiles. Cambridge, Harvard Union Press, iv: 5-125.

Stewart, G. R., 1977: *Charina, c. bottae*. Cat. Amer. Amph. Rept., (205): 1-2.

Stewart, G. R., 1987 [1988]: The rubber boa (*Charina bottae*) in California, with particular references to the southern subspecies. *In* H. F. De Lisle, P. R. Brown, B. Kaufman, and B. M. McGurty: Proc. Conference Calif. Herpetol.-1987. Southwest. Herpetol. Soc., Special Pub., November 1, 1988, (4): 1-143.

Stickel, W. H., 1951: Distinction between the snake genera *Contia* and *Eirenis*. Herpetologica, 7 (part 3): 125-132.

Strecker, J. K., 1915: Reptiles and amphibians of Texas. Baylor Univ. Bull., 18 (4): 82.

Stickel, W. H., 1938: The snakes of the genus *Sonora* in the United States and Lower California. Copeia, (4): 182-190.

Stickel, W. H., 1943: The Mexican snakes of the genera *Sonora* and *Chionactis* with notes on the status of other colubrid genera. Proc. Biol. Soc. Washington, 56: 109-218.

Stimson, A. C., 1982: The snake's advocate. Eakin Press, Austin, Texas.

Stimson, A. F., 1969: Liste der rezenten Amphibien und Reptilien: Boidae. Das Tierreich, Walter de Gruyter, Berlin, 89:1-49.

Stoll, O., 1958: Etnografía de Guatemala. Editorial de Ministerio de Educacción Publica. Guatemala City.

Stoops, E. D., and A. Wright, 1992: Snakes and other reptiles of the Southwest-a user-friendly guide to snakes, lizards, and turtles.

Strecker, J. K. Jr., 1902: Reptiles and batrachians of McLennon County, Texas. Proc. Trans. Texas Acad. Sci., 4, Pt. 2 (5): 95-101.

Strecker, J. K. Jr., 1908a: The reptiles and batrachians of McLennan County, Texas. Proc. Biol. Soc. Washington, 21: 165-170.

Strecker, J. K. Jr., 1908b: The reptiles and batrachians of Victoria and Refugio counties, Texas. Proc. Biol. Soc. Washington, 21: 47-52.

Strecker, J. K. Jr., 1909a: Notes on the herpetology of Burnet County, Texas. Baylor Univ. Bull., 12 (1): 1-9.

Strecker, J. K. Jr., 1909b: Reptiles and amphibians collected in Brewster County, Texas. Baylor Univ. Bull., 12 (1): 11-16.

Strecker, J. K. Jr., 1915: Reptiles and amphibians of Texas. Baylor Univ. Bull., Baylor Univ. Press, Waco, Texas, August, 18 (4): 1-82.

Strecker, J. K. Jr., 1928: Common English and folk names for Texas amphibians and reptiles. Contrib. Baylor Univ. Mus., 16: 1-21.

Strecker, J. K. Jr., and L. S. Frierson, Jr., 1926: The herpetology of Caddo and DeSoto parishes, Louisiana. Cont. Baylor Univ. Mus., 5: 1-10.

Strecker, J. K. Jr., and W. J. Williams, 1927: Herpetological records from the vicinity of San Marcos, Texas, with distributional data on the amphibians and reptiles of the Edwards Plateau region and central Texas. Baylor Univ. Mus. Contrib., 12: 1-16.

Strimple, P. D., 1994: Husbandry of juvenile green anacondas (*Eunectes murinus*) in Captivity. The Vivarium, 6 (1): 52-54.

Strimple, P. D., 1995: Mexico's cross-banded mountain rattlesnake. Reptile & Amphibian Magazine, May/June 1995: 46-51.

Strimple, P. D., G. Puorto, W. E. Holmstrom, R. W. Henderson and Roger Conant, 1997: On the status of the anaconda *Eunectes barbouri* Dunn & Conant. J. Herpetol., 31 (4): 607-609.

Storer, D. H., 1839: Reptiles of Massachusetts. *In* Commissioners on the zoological and botanical survey of the State. Reports on Fisheries, Reptiles, and Birds of Massachusetts, Dalton and Wentworth, Boston, Mass., 1839: 203-253.

Strussman, C., and M. A. de Carvalho, 1998: New herpetological records for the state of Mato Grosso, western Brazil. Herpetol. Rev., 29 (3): 183-185.

Strüssmann, C., and I. Sazima, 1993: The snake assemblage of the Pantanal at Poconé, western Brazil: Faunal composition and ecology summary. Stud. Neotropical, Fauna and Environ, 28: 157-168.

Strüssmann, C., and I. Sazima, 1997: Hábitos alimentares da sucuri-amarela, *Eunectes notaeus* Cope, 1862, no Pantanal Matogrossense. Biociências, Porto Alegre, 5: 35-52.

Stuart, G. E., and G. S. Stuart, 1977: The mysterious Maya. National Geographic Soc., Washington, D.C.

Stuart, L. C., 1932: Studies of Neotropical Colubrinae: I. The taxonomic status of the genus *Drymobius* Fitzinger. Occ. Pap. Mus. Zool. Univ. Michigan, (236): 1-16.

Stuart, L. C., 1933: Studies of Neotropical Colubrinae: II. Some new species and subspecies of *Eudryas* Fitzinger, with an annotated list of the frogs of *Eudryas boddaertii* (Sentzen). Occ. Pap. Mus. Zool. Univ. Michigan, (254): 1-10.

Stuart, L. C., 1935a: A contribution to a knowledge of the herpetology of a portion of the Savanna Region of Central Petén, Guatemala. Mus. Zool. Misc. Pub. Univ. Michigan, (29): 1-56.

Stuart, L. C., 1935b: The king snake of the Tres Marías Islands. Occ. Paps Mus. Zool. Univ. Michigan, (323): 1-3.

Bibliography

Stuart, L. C., 1937: Some further notes on the amphibians and reptiles of the Peten Forest of northern Guatemala. Copeia, 1937 (1): 67-70.

Stuart, L. C., 1939: A new name for the genus *Eudryas* Fitzinger, 1843. Copeia, 1939 (1): 55.

Stuart, L. C., 1941a: Some new snakes from Guatemala. Occ. Paps. Mus. Zool. Univ. Michigan, (452): 1-7.

Stuart, L. C., 1941b: Studies of Neotropical Colubrinae: VIII. A revision of the genus, *Dryadophis* Stuart, 1939. Misc. Pub. Mus. Zool. Univ. Michigan, (49): 1-106.

Stuart, L. C., 1948: The amphibians and reptiles of Alta Verapaz, Guatemala. Misc. Pub. Mus. Zool. Univ. Michigan, 69: 1-109.

Stuart, L. C., 1950: A geographic study of the herpetofauna of Alta Verapaz, Guatemala. Contrib. Lab. Vert. Biol. Univ. Michigan, (45): 1-77.

Stuart, L. C., 1963: A checklist of the herpetofauna of Guatemala. Misc. Pub. Mus. Zool. Univ. Michigan, (122): 1-150.

Stuart, L. C., and J. R. Bailey, 1941: Three new species of the genus *Rhadinaea* from Guatemala. Occ. Pap. Mus. Zool. Univ. Michigan, (442): 1-11.

Stull, O. G., 1928: A revision of the genus *Tropidophis*. Occ. Pap. Mus. Zool. Univ. Michigan, 195: 1-49.

Stull, O. G., 1935: A checklist of the family boidae. Proc. Boston Soc. Nat. Hist., 40 (8): 387-408.

Stull, O. G., 1938: Three new subspecies of the family boidae. Occ. Pap. Boston Soc. Nat. Hist., 8: 297-300.

Stull, O. G., 1940: Variations and relationships in the snakes of the genus *Pituophis*. Bull. U.S. Natl. Mus., (175): 1-225.

Stumpel, A. H. P., 1995: Natural history notes. *Masticophis taeniatus taeniatus*. Herpetol. Rev., 26: 102.

Swanton, J. R., 1919: A structural lexical comparison of the Tunica, Chitimacha, and Atakapa languages. Smithsonian Inst. Bureau of American Ethnology No. 68.

Swash, A., and R. Still, 2000: Birds, mammals, & reptiles of the Galápagos Islands. Yale Univ. Press, New Haven and London.

Sweet, S. S., and W. S. Parker, 1990: *Pituophis melanoleucus*. Cat. Amer. Amph. Rept., (474): 1-8.

Sylestine, C., H. K. Hardy and T. Montler, 1993: Dictionary of the Alabama language. Univ. Texas Press, Austin.

Tanner, Vasco M., 1938: A new subspecies of worm snake from Utah. Proc. Utah Acad. Sci., 15: 149-150.

Tanner, W. W., 1941: The reptiles and amphibians of Idaho. No.1. Great Basin Naturalist, 2: 87-97.

Tanner, W. W., 1944 [1946]: A taxonomic study of the genus *Hypsiglena*. Great Basin Naturalist, 5 (3-4): 25-92.

Tanner, W. W., 1954a: Additional note on the genus *Hypsiglena* with a description of a new subspecies. Herpetologica, 10: 54-56.

Tanner, W. W., 1954b: Herpetological notes concerning some reptiles of Utah and Arizona. Herpetologica, 10: 92-96.

Tanner, W. W., 1959: A new *Thamnophis* from western Chihuahua with notes on four other species. Herpetologica, 15: 165-172.

Tanner, W. W., 1961a: A new subspecies of *Conopsis nasus* from Chihuahua, Mexico. Herpetologica, 17 (1): 67-70.

Tanner, W. W., 1961b: Notes on a collection of amphibians and reptiles for southern Mexico, with a description of a new *Hyla*. Great Basin Naturalist, 17 (1-2): 53-56.

Tanner, W. W., 1966a: A new rattlesnake from western Mexico. Herpetologica, 22 (4): 298-302.

Tanner, W. W., 1966b: A reevaluation of the genus *Tantilla* in the southwestern United States and northwestern Mexico. Herpetologica, 22 (2): 134-152, June 30.

Tanner, W. W., 1966c: The night snakes of Baja California. Transactions of the San Diego Society of Natural History, 14 (15): 189-196.

Tanner, W. W., 1969: New records and distribution notes for reptiles of the Nevada test site. Great Basin Naturalist 29: 31-34.

Tanner, W. W., 1970: A catalogue of the fish, amphibian, and reptiles types in the Brigham Young University Museum of Natural History. Great Basin Naturalist, 30 (4): 219-226.

Tanner, W. W., 1981: A new *Hypsiglena* from Tiburon Island, Sonora, Mexico. Great Basin Naturalist, 41 (1): 139-142.

Tanner, W. W., 1983: *Lampropeltis pyromelana*. Cat. Amer. Amph. Rept., (342): 1-2.

Tanner, W. W. [1985] 1986: Snakes of western Chihuahua. Great Basin Naturalist, 45 (4): 615-676.

Tanner, W. W., 1988: Status of *Thamnophis sirtalis* in Chihuahua, Mexico (Reptilia: Colubridae). Great Basin Naturalist, 48 (4): 499-507.

Tanner, W. W., 1990: *Thamnophis rufipunctatus* (Cope)—Narrow-headed Garter Snake. Cat. Amer. Amph. Rept., (505): 1-2.

Tanner, W. W. and B. H. Banta, 1962: Description of a new *Hypsiglena* from San Martin Island, Mexico, with a resumé of the reptile fauna of the Island, Herpetologica, 18 (1): 21-25.

Tanner, W. W. and B. H. Banta, 1966: A systematic review of the Great Basin reptiles in the collections of Brigham Young University and the University of Utah. Great Basin Naturalist, 26: 87-135.

Tanner, W. W., J. R. Dixon and H. S. Harris, Jr., 1972: A new subspecies of *Crotalus lepidus* for western Mexico. Great Basin Naturalist, 32 (1): 16-24.

Tanner, W. W., and A. E. Leviton, 1960: The generic allocations of *Hypsiglena slevini* Tanner (Serpentes: Colubridae). Occ. Pap. California Acad. Sci., (27): 1-7.

Tanner, W. W., and C. H. Lowe, Jr., 1989: Variations in *Thamnophis elegans* with descriptions of new subspecies. Great Basin Naturalist, 49 (4): 511-516.

Tanner, W. W., and W. G. Robison, Jr., 1960a: Herpetology notes for northwestern Jalisco, Mexico. Herpetologica, 16: 59-62.

Tanner, W. W., and W. G. Robison, Jr., 1960b: New and unusual serpents from Chihuahua, Mexico. Herpetologica, 16: 67-70: 13-18.

Taylor, D., 1977: Languages of the West Indies. Johns Hopkins Univ. Press, Baltimore, 278 pp.

Taylor, E. H., 1931: Notes on two specimens of the rare snake *Ficimia cana* and the description of a new species of *Ficimia* from Texas. Copeia, 1931 (3): 4-7.

Taylor, E. H., 1936: Notes and comments on certain American and Mexican snakes of the genus *Tantilla* with descriptions of new species. Trans. Kansas Acad. Sci., 39: 335-348.

Taylor, E. H., 1938: Notes on Mexican snakes of the genus *Leptodeira*, with a proposal of a new snake genus *Pseudoleptodeira*. Univ. Kansas Sci. Bull., 25 (15) : 315–344.

Taylor, E. H., 1938 [1939]: On Mexican snakes of the general *Trimorphodon* and *Hypsiglena*. Univ. Kansas Sci. Bull., 25 (16): 357–383.

Taylor, E. H., 1939a: On North American Snakes of the Genus *Leptotyphlops*. Copeia, 1939 (1): 1–7.

Taylor, E. H., 1939b: Some Mexican serpents. Univ. Kansas Sci. Bull., 26 (13): 445–487.

Taylor, E. H., 1939c: Two new species of the genus *Anomalepis* Jan, with a proposal of a new family of snakes. Proc. New England Zool. Club, 17: 117–187.

Taylor, E. H., 1942: Mexican snakes of the genus *Adelophis* and *Storeria*. Herpetologica, 2 (4): 75–79.

Taylor, E. H., 1949: A preliminary account of the herpetology of the snakes of San Luis Potosí, Mexico. Univ. Kansas Sci. Bull., 33 (2): 169–215.

Taylor, E. H., 1950: Second contribution to the herpetology of San Luis Potosí. Univ. Kansas Sci. Bull., 33—Pt.II (11): 441–457.

Taylor, E. H., 1951: A brief review of the snakes of Costa Rica. Univ. Kansas Sci. Bull., 34-pt.I, (1): 3–188.

Taylor, E. H., 1952: Third contribution to the herpetology of San Luis Potosí. Univ. Kansas Sci. Bull., 34-Pt.II (13): 793–815.

Taylor, E. H., 1953a: Early records of the seasnake in Latin America. Copeia, 1953 (2): 124.

Taylor, E. H., 1953b: Fourth contribution to the herpetology of San Luis Potosí. Univ. Kansas Sci. Bull., 35-Pt.II (13): 1587–1614.

Taylor, E. H., 1954: Further studies on the serpents of Costa Rica. Univ. Kansas Sci. Bull., 36: 673–800.

Taylor, E. H., 1960: Mexican snakes of the genus *Typhlops*. Kansas Univ. Sci. Bull., 26: 441–444.

Taylor, E. H., A. Flores and R. Bolaños, 1974: Geographical distribution of viperidae, elapidae and hydrophidae in Costa Rica. Rev. Biol. Trop., 21: 383–397.

Taylor, E. H., and H. M. Smith, [1938] 1939: Miscellaneous notes on Mexican snakes. Univ. Kansas Sci. Bull., 25 (13): 239–258.

Taylor, E. H., and H. M. Smith, 1942: The snake genus *Conopsis* and *Toluca*. Univ. Kansas Sci. Bull., 28 (15): 325–363.

Taylor, E. H., and H. M. Smith, 1943: A review of American Sibynophine snakes, with a proposal of a new genus. Univ. Kansas Sci. Bull., 29 (6): 301–336.

Taylor, E. H., and D. Weyer 1958: Report on a collection of amphibians and reptiles from Harbel, Republic of Liberia. Univ. Kansas Sci. Bull., 38-Pt.II (14): 1191–1229.

Taylor, R. C., 1835: On the geology and natural history of the north-eastern extremity of the Allegheny Mountain Range, in Pennsylvania, United States. Mag. Nat. Hist., 8 (54): 529–541.

Taylor, W. E., 1892a: Catalogue of the snakes of Nebraska with notes on their habits and distribution. Amer. Naturalist, 26 (309): 742–752.

Taylor, W. E., 1892b: The ophidia of Nebraska. Ann. Report Nebraska State Boards Agric. For 1891, 1892: 310–357.

Teixeira, D. M., and M. Porto; 1991: Life history notes: *Leptophis ahaetulla*, feeding behavior. Herpetol. Rev., 22 (4): 132.

Telford, S. R., 1966: Variation among the southeastern crowned snakes, genus *Tantilla*. Bull. Florida State Mus. Biol. Sci., 10: 261–304.

Telford, S. R., 1980a: *Tantilla oolitica*. Cat. Amer. Amph. Rept., (256): 1.

Telford, S. R., 1980b: *Tantilla relicta*. Cat. Amer. Amph. Rept., (257): 1–2.

Telford, S. R., 1982: *Tantilla coronata*. Cat. Amer. Amph. Rept., (308): 1–2.

Tennant, A., 1984: The snakes of Texas. Austin, Texas, Texas Monthly Press.

Tennant, A., 1997a: A field guide to snakes of Florida. Gulf Publishing Co., Houston, Texas.

Tennant, A., 1985: A field guide to Texas snakes. Austin, Texas, Texas Monthly Press.

Tennant, A., 1998: A field guide to Texas snakes. Second Edition, Field Guide Series, Gulf Publishing Company, Houston, Texas.

Tennant, A., and R. D. Bartlett, 2000: Snakes of North America-eastern and central regions. Gulf Publishing Company, Houston, Texas.

Terent'ev, P. V., 1965: Herpetology-a manual on amphibians and reptiles. Translated from Russian by Israel Program for Scientific Translation, Jerusalem.

Texas Parks & Wildlife Dept., 1989: Snakebite, it could happen to you. Publication PWD-LF-9200-025-12/89, Austin, Texas.

Theodore, C., 291 pp.: Haitian Creole-English/English-Haitian Creole dictionary. Hippocrene Concise Dictionary, New York.

Thomas, J. P. R., 1976: Systematics of Antillean Blind Snakes of the genera *Typhlops* (Serpentes: Typhlopidae). Ph.D. Dissertation, Louisiana State University & Agricultural & Mechanical College, August 1976: i–xvi + 288 pp.

Thomas, R., 1965a: A new species of *Typhlops* from the Barahona Peninsula of Hispaniola. Copeia, 1965 (4): 436–439,

Thomas, R., 1965b: A reassessment of the Virgin Islands *Typhlops* with the description of two new subspecies. Rev. Biol. Trop., 13 (2): 187–201.

Thomas, R., 1965c: The genus *Leptotyphlops* in the West Indies, with description of a new species from Hispaniola (Serpentes, Leptotyphlopidae). Breviora., (222): 1–12.

Thomas, R., 1966a: A reassessment of the herpetofauna of Navassa Island. J. Ohio Herpetol. Soc., 5 (3): 73–89.

Thomas, R., 1966b: Leeward Islands *Typhlops* (Reptilia, Serpentes). Proc. Biol. Soc. Washington, 79: 255–266.

Thomas, R., 1968: The *Typhlops biminensis* Group of Antillean blind snakes. Copeia, 1968 (4): 713–722.

Thomas, R., 1974a: A new species of Lesser Antillian *Typhlops* (Serpentes: Typhlopidae). Occ. Pap. Mus. Zool. Louisiana State Univ., (6): 1–5.

Thomas, R., 1974b: A new species of *Typhlops* (Serpentes: Typhlopidae) from Hispaniola. Proc. Bio. Soc. of Washington, 87 (2): 11–18.

Thomas, R., 1989: The relationships of Antillean *Typhlops* (Serpentes: Typhlopidae) and the descriptions of three new Hispaniolan species. pp 409–432. *In* C. A. Woods (ed.) Biogeography of the West Indies: Past, Present, and Future. Sandhill Crane Press, Gainsville, FL.

Thomas, R., 1999: The Puerto Rican Area. pp 169–179. *in* B. I. Crother, Caribbean amphibians and reptiles. Academic Press, New York.

Thomas, R., R. W. McDiarmid and F. G. Thompson, 1985: Three new species of thread snakes (Serpentes: Leptotyphlopidae) from Hispaniola. Proc. Biol. Soc. of Washington, 98 (1): 204–220.

Thomas, R., and R. Powell, 1992: *Typhlops hectus*. Cat. Amer. Amph. Rept., (550): 1-2.

Thomas, R., and R. Powell, 1994a: *Typhlops schwartzi*—Thomas. Cat. Amer. Amph. Rept., (596): 1-2.

Thomas, R., and R. Powell, 1994b: *Typhlops sulcatus*-Thomas. Cat. Amer. Amph. Rept., (597): 1-2.

Thomas, R., and R. Powell, 1994c: *Typhlops tetrathyreus* -Thomas. Cat. Amer. Amph. Rept., (598): 1-2.

Thomas, R., and R. Powell, 1995a: *Typhlops capitulatus*. Cat. Amer. Amph. Rept., (618): 1-2.

Thomas, R., and R. Powell, 1995b: *Typhlops gonavensis*. Cat. Amer. Amph. Rept., (619): 1-2.

Thomas, R., and R. Powell, 1995c: *Typhlops titanus*. Catalogue of American Amphibians and Reptiles, 620: 1-2.

Thomas, R., and A. Schwartz, 1965: Hispanolian snakes of the genus *Dromicus* (Colubridae). Rev. Biol. Trop., 13 (1): 59-83.

Thomas, R. A., 1974: Geographic variation in *Elaphe guttata* (Linnaeus) (Serpentes: Colubridae). M. S. Thesis, Texas A&M University, College Station, TX.

Thomas, R. A., 1974 [revised 1976]: A checklist of Texas amphibians and reptiles. Texas Parks and Wildlife Dept. Tech. Ser., (17): a-b +1-15.

Thomas, R. A., 1975: Taxonomic chaos: *Elaphe guttata*, a case in point. Bull. Maryland Herpetol. Soc., 11 (4): 171-176.

Thomas, R. A., 1976: A revision of the South American colubrid snake genus *Philodryas* Wagler, 1930. Ph.D. Dissertation, Texas A&M University, August 1976.

Thomas, R. A., 1977a: A new generic arrangement for *Incaspis* and mainland South American *Alsophis* and the status of two additional Peruvian species. Copeia, 1977 (4): 648-652.

Thomas, R. A., 1977b: Generic relationships of *Philodryas elegans* (Tschudi). Abst. VII, Congr. Latin American Zool. (Tucuman): 120.

Thomas, R. A., 1977: *Philodryas nattereri* Steindachner, 1870 (Reptilia: Serpentes): proposed conservation. Z. N.(S.) 2166. Bull. Zool. Nomencl., 33 (3/4): 216-217, March 1977.

Thomas, R. A., 1994: Geographic Distribution: *Rhamphotyphlops braminus* (Brahaminy Blind Snake). Herpetol. Rev., 25 (1): 34.

Thomas, R. A., 1997: Galápagos terrestrial snakes. Biogeography and systematics. Herpetol. Nat. Hist., 5 (1): 19-40.

Thomas, R. A., [in preparation]: *Philodryas* revisited. Paper presented at Annual Fall Meeting of the Texas Herpetological Society at Texas A&M University, Saturday 23, October, 1999 [details to be published].

Thomas, R. A., A., Barrio and R. F. Laurent, 1977: *Philodryas borellii* Peracca (Serpentes: Colubridae); a distinct species. Herpetologica, 33 (1): 82-86.

Thomas, R. A., and J. A. Davis, 1983: Poisonous snakes of the southeastern U.S.A. (A full color poster). Miss. Gulf Coast Emerg. Med. Serv., Gulfport, MS.

Thomas, R. A., B. J. Davis and M. R. Culbertson, 1976: Notes on variation and range of the Louisiana Pine Snake *Pituophis melanoleucus ruthveni* Stull (Reptilia, Serpentes, Colubridae). J. Herpetol., 10: 252-254.

Thomas, R. A., and J. R. Dixon, 1975: *Philodryas olfersii* (Lichenstein) new to Colombia and Venezuela. Herpetol. Rev., 6 (4): 108-109.

Thomas, R. A., and J. R. Dixon, 1977: A new systemic arrangement for *Philodryas serra* (Schlegel) and *Philodryas pseudoserra* Amaral (Serpentes: Colubridae). The Pearce-Sellardo Series, July 1977, (27): 1-20.

Thomas, R. A., and R. Fernandes, 1996: The systematic status of *Platynion lividum* Amaral, 1923 (Serpentes: Colubridae: Xenodentinae). Herpetologica, 52 (2): 271-275.

Thomas, R. A., and V. Friloux, 1983: Life history notes: *Elaphe obsoleta lindheimeri* (Texas Rat Snake). Size. Herpetol. Rev., 14 (1): 19-20.

Thomas, R. A., and I. Ineich, 1999: The identity of the syntypes of *Dryophylax freminvillei* with comments on their locality data. J. Herpetol., 33 (1): 152-153.

Thomas, R. A., and J. D. Johnson, 1984: *Philodryas varius* (Jan, 1863) a senior synonym of *Philodryas borellii* Peracca (Serpentes, Colubridae). J. Herpetol., 18 (1): 80.

Thomas, R. A., R. F. Laurent and Avelino Barrio, 1977: *Philodryas borellii* Peracca (Serpentes: Colubridae), a distinct species. Herpetologica, 33 (1): 82-86.

Thomas, R. A., Julio Cesar de Moura-Leite, S. A. Abrahão Morato, and R. Silveira Bérnils, in press: *Ditaxodon taeniatus* (Hensel, 1868), its variation and distribution (Serpentes, Colubridae, Xenodontinae).

Thompson, F. G., 1957: A new Mexican gartersnake (genus *Thamnophis*) with notes on related forms. Occ. Pap. Mus. Zool. Univ. Michigan, 584: 1-10.

Thompson, F. G., 1997: Galápagos terrestrial snakes: Biogeography & Systematics. Herpetol. Nat. Hist., 5 (1): 19-40.

Thord-Gray, I., 1955: Tarahumara-English, English-Tarahumara dictionary and an introduction to Tarahumara grammar. Univ. Miami Press, Coral Gables, FL.

Thorn, S., 1995: Honduran protected areas are a powerhouse of biodiversity. Honduras This Week, Tegucigalpa, Honduras, June 10, 1995, 8 (22) [3388]: 16.

Tinoco V., R. A., 1978: Las serpientes de Colombia. Ciencia, mitos y leyendas. Ediciones Editorial Mejoras, Universidad del Atlantico, Colombia.

Tinkle, D. W., and R. Conant, 1961: The rediscovery of the water snake, *Natrix harteri*, in western Texas, with the description of a new subspecies. Southwest. Naturalist, 6 (1): 33-44.

Tolson, P. J., and Henderson, R. W., 1993: The natural history of West Indian Boas. Tauton, Somerset, England, R & A Publishing Limited.

Torn, N., with M. Prehtel, 1997: Scandals in the house of birds, shamans and priests on Lake Atitlán. Marsillo Publisher, New York.

Touzet, J. M., 1997: *Bothriopsis bilineatus smaragdinus*: Ecuador's eye-grabbing emerald. Reptiles, August, 1997, 5 (8): 40-46.

Tozzer, A. M., 1977: A Maya grammar. Dover Publications, Inc., New York.

Trapido, H., 1941: A new species of *Natrix* from Texas. Amer. Midl, Naturalist, 25 (3): 673-680.

Trapido, H., 1944: The snakes of the genus *Storeria*. Amer. Midl. Naturalist, 31: 1-84.

Troschel, F. H., 1848: Amphibien. *In* Versuch einer Zusammenstellung der Fauna und Flora von British-Guiana. Vol. 3, Richard Schomburgk, Leipzig, 3: 645-661.

True, F. W., 1883: On the bite of the North American coral snakes (genus Elaps). Amer. Naturalist, 17: 26-31.

Trutnau, L., 1981: Nonvenomous snakes. 1st English Series, Hauppauge, New York, Barron's Educational Series, Inc.

Trutnau, L., 1982: Einige Bemerkungen über die neuwettlichen Riesenschlangen der Gattung *Eunectes* Wagler, 1830. Herpetofauna, 17: 14–21.

Tschudi, J. J. von, 1845: Reptilium conspectus quae in Republica Peruana reperiuntur et pleraque observata vel collecta sunt in itinere. Arch. für Naturg., Berlin, 11 (1): 150–170.

Tschudi, J. J. von, 1846: Untersuchungen über die Fauna Peruana. Herpetologie. Zeiter und Zollikofer.

Tucker, J. K., 1977: Notes on the food habits of Kirtland's Water Snake *Clonophis kirtlandi*. Bull. Maryland Herpetol. Soc., 13 (3): 193–195.

Tucker, J. K., and G. L. Paukstis, 1995: Geographic distribution: *Coluber constrictor foxii*. Herpetol. Rev., 26 (1): 46.

Tumlison, R., M. Karnes and M. Clark, 1992: New records of vertebrates in southwestern Arkansas. Arkansas Acad. Sci. Proc., 46: 109–111.

Turner, F. B., and R. H. Wauer, 1963: A survey of the herpetofauna of the Death Valley area. Great Basin Naturalist, 23: 119–128.

Turner, L. D., 1969: Africanisms in the Gullah dialect. Arno Press and The New York Times, New York.

Underwood, G., 1962: Reptiles of the eastern Caribbean—No. 1. Caribbean Affairs (New Series) Department of Extra-Mural Studies—Univ. West Indies, (1): 1–192.

Underwood, G. L., 1967: A contribution to the classification of snakes. Trustees of the British Museum (Natural History), London, (653): 1–179.

Underwood, G. L., 1979: Classification and distribution of venomous snakes of the world. *In* Lee: Snake Venoms, Handbook Exp. Pharmacol, 52: 15–40.

Underwood, G. L., 1993: A new snake from St. Lucia, West Indies. Bull. Nat. Hist. Museum (Zoology) 59: 1–9.

U.S. Government Printing Office, 1988: 1988 0-220-677, 50 CFR 23 (Ref.6/1/88), ENF 4-REG-23.

U.S. Government Printing Office, 1989: 1989 242-386/05952, 50 CFR 17 (Rev.1/1/89), ENF 4-REG 17.

Utiger, V., N. Helfenberger, B. Schätti, C. Schmidt, M. Ruf, and V. Ziswiler, 2002: Molecular systematics and phylogeny of Old and New World ratsnakes, *Elaphe* Auct., and related genera (Reptilia, Squamata, Coubridae). Russian J. Herpetology, 9(2): 105-124.

Uzzell, T. M. Jr., and P. Starrett, 1958: Snakes from El Salvador. Copeia, 1958: (4): 339–342.

Valardi, Francisco, 1962: Natura viva-enciclopedia sistematica del reino animal. Vol II. Aves, Reptiles, Anfibios. Editorial Exito S.A., Barcelona, Spain.

Valdman, A., 1977: Pidgin and Creole linguistics. Indiana Univ. Press, Bloomington, Ind.

Valdujo, P. H., and C. Nogueira, 1999: Geographic distribution—*Philodryas livida*. Herpetol. Rev., 30 (1): 55.

Valdujo, P. H., and C. Nogueira, 2000: *Bothrops neuwiedi pauloensis* (Jararaca-Rabo-de-Osso). Herpetol. Rev., 31 (1): 45.

Van Denburgh, J., 1895: A review of the herpetology of lower California. Part I. Reptiles. Proc. California Acad. Sci., Ser. 2, (5): 77-162 + 11 pls. *In* Herpetology of Lower California, collected papers by John Van Denburgh. SSAR Fascimile Reprints in Herpetol., 1978.

Van Denburgh, J., 1897a: Reptiles from Sonora, Sinaloa And Jalisco, Mexico, with a description of a new species of *Sceloporus*. Proc. Acad. Nat. Sci. Philadelphia, 1897: 460–464.

Van Denburgh, J., 1897b: The reptiles of the Pacific Coast and Great Basin. Occ. Pap. California Acad. Sci., 5: 1–236.

Van Denburgh, J., 1898: Herpetological notes, April 1, 1898. On the time of laying of the Western Gopher Snake in Central California. Proc. Amer. Philos. Soc., 37 (157): 139–141.

Van Denburgh, J., 1912a: Expedition of the California Academy of Sciences to the Galapagos Islands, 1905-1906-IV The snakes of the Galapagos Islands. Proc. California Acad. Sci., 4th Ser., 1: 323–374.

Van Denburgh, J., 1912b: Notes on some reptiles and amphibians from Oregon, Idaho and Utah. Proc. California Acad. Sci., 4th Ser., 3: 155–160.

Van Denburgh, J., 1912c: Notes on the collection of reptiles from Southern California and Arizona. Proc. California Acad. Sci., 4th Ser., 3: 147–154.

Van Denburgh, J., 1920: II. Description of a new species of rattlesnake (*Crotalus lucasensis*) from Lower California. Proc. California Acad. Sci., 10 (2): 29–30.

Van Denburgh, J., 1920a: III. Description of a new species of boa (*Charina bottae utahensis*) from Utah. Proc. California Acad. Sci., 10 (3): 31–32.

Van Denburgh, J., 1920b: I. Further study of variation in the gopher snake of Western North America. Proc. California Acad. Sci., 10 (1): 1–27.

Van Denburgh, J., 1922: The reptiles of western North America. Occ. Pap. California Acad. Sci., 10: 623–1028.

Van Denburgh, J., 1924: Notes on the herpetology of New Mexico with a list of species known from that State. Proc. California Acad. Sci., 4th Ser., 13 (12): 189–230.

Van Denburgh, J., 1978: Herpetology of Lower California-Collected Papers. SSAR, 1978: 77–162.

Van Denburgh, J., and J. R. Slevin, 1913: A list of the amphibians and reptiles of Arizona, with notes on the species in the collection of the Academy. Proc. California Acad. Sci., Fourth Series, 3: 391–454.

Van Denburgh, J., and J. R. Slevin, 1914: V. A list of the amphibians and reptiles of the Island of the West Coast of North America. Proc. California Acad. Sci., Fourth Series, 4: 129–152.

Van Denburgh, J., and J. R. Slevin, 1915: IV. A list of the amphibians and reptiles of Utah, with notes on the species in the collection of the Academy. Proc. California Acad. Sci., Fourth Series, 5: 99–110.

Van Denburgh, J., and J. R. Slevin, 1918: VI. The garter-snakes of Western North America. Proc. California Acad. Sci., Fourth Series, 8 (6): 181–270.

Van Denburgh, J., and J. R. Slevin, 1919: The gopher-snakes of Western North America. Proc. California Acad. Sci., Fourth Series, 9 (6): 197–220.

Van Denburgh, J., and J. R. Slevin, 1921a: II. A list of the amphibians and reptiles of Nevada, with notes on the species in the collection of the Academy. Proc. California Acad. Sci., Fourth Series, 11 (2): 27–37.

Van Denburgh, J., and J. R. Slevin, 1921b: III. A list of the amphibians and reptiles of Idaho, with notes on the species

Bibliography

in the collection of the Academy. Proc. California Acad. Sci., Fourth Series, 11 (3): 38-48.

Van Denburgh, J., and J. R. Slevin, 1921c: IV. A list of the amphibians and reptiles of the peninsula of Lower California, with notes on the species in the Collection of the Academy. Proc. California Acad. Sci., Fourth Series, 11 (4): 49-72.

Van Denburgh, J., and J. R. Slevin, 1921d: Preliminary diagnosis of more new species of reptiles from islands in the Gulf of California, Mexico. Proc. California Acad. Sci., Fourth Series, 11 (17): 395-398.

Van Denburgh, J., and J. R. Slevin, 1923: I. Preliminary diagnosis of four new snakes from Lower California, Mexico. Proc. California Acad. Sci., Fourth Series, 8 (1): 1-2.

Van Denburgh, J., and J. R. Slevin, 1926: VI. Preliminary diagnosis of new species of reptiles from Islands in the Gulf of California, Mexico. Proc. California Acad. Sci., Fourth Series, 11 (6): 95-98.

Van Devender, R. W., and C. J. Cole, 1977: Notes on a colubrid snake, *Tantilla vermiformis*, from Central America. Amer. Mus. Novitates, (2625): 1-12.

Van Devender, T. R., C. H. Lowe and Howard E. Lawler, 1994: Factors influencing the distribution of the Neotropical vine snake (*Oxybelis aeneus*) in Arizona and Sonora, Mexico. Herpetol. Nat. Hist., 2 (1): 25-42.

Van Hyling, O. C., 1933: Batrachia and reptiles of Alachua County, Florida. Copeia (1933): 3-7.

Vanzolini, P. E., 1947: Nota nomenclatural sôbre *Leimadophis almada* (Wagler, 1924) (= *Leimadophis almadensis* Auct.). Pap. Avuls. Dept. Zool., São Paulo, 8 (25): 285-286.

Vanzolini, P. E., 1948: Notas sôbre os ofídios e lagartos da Cachoeira de Emas, no muncíipo de Pirassununga, Estado de São Paulo. Rev. Brasil. Biol., 8 (3): 377-400.

Vanzolini, P. E., 1956/58: Notas sôbre a zoologia dos índios Canela. Rev. Mus. Paulista—Noua Série, 10: 155-171.

Vanzolini, P. E., 1972: *Typhlops brongersmai* spec. nov. from the coast of Bahia, Brazil (Serpentes, Typhlopidae). Zool. Mededelingen. Rijksmus. Natuurlijke, Leiden, 47: 27-29.

Vanzolini, P. E., 1976: *Typhlops brongersmianus*, a new name for *Typhlops brongersmai* Vanzolini, 1972, preoccupied (Serpentes, Typhlopidae). Pap. Avuls. Dept. Zool., Universidade de São Paulo, 29: 247.

Vanzolini, P. E., 1977: An annotated bibliography of the land and fresh-water reptiles of South America (1758-1975). Vol. I (1758-1900), Vol. II (1901-1975). Mus. Zool. Univ. São Paulo.

Vanzolini, P. E., 1978: An annotated bibliography of the land and fresh-water reptiles of South America (1758-1975). Vol. II (1901-1975). Mus. Zool. Univ. São Paulo.

Vanzolini, P. E., 1986: Addenda and corrigenda to the Catalogue of the Neotropical Squamata, Part I Snakes. Smithsonian Herpetol. Infor. Serv., (70): 1-25.

Vanzolini, P. E., 1981: Introduction: The scientific and political contexts of the Bavarian Expedition to Brasil. pp ix-xxix. *In* Spix, J. B., von and J. G. Wagler, 1824: Herpetology of Brazil. Reprinted under the patronage of The Government of João Baptista de Oliveira Figueiredo, President of the Federative Republic of Brazil, Society for the Study of Amphibians and Reptiles, 1981. Containing introduction and three volumes.

Vanzolini, P. E., A. M. Ramos-Costa and L. J. Vitt, 1980: Repteis das Caatingas. Acad. Brasileira Ci., Rio de Janiero.

Varkey, A., 1979: Comparative cranial myology of North American natricine snakes. Pub. Biol. Geol. Milwaukee Publ. Mus., (4): 1-70.

Vaughan, R. K., J. R. Dixon and R. A. Thomas, 1996: A reevaluation of populations of the Corn Snake *Elaphe guttata* (reptilia: serpentes: colubridae) in Texas. Texas J. Sci., 48 (3): 175-190.

Vázquez de Kartzow (M. D.), A. R., 1994: Mordedura de serpientes venenosas-guía práctica para la identificación, clasificación, diagnóstico y tratamiento del accidente ofídico. Colegio Mayor de Nuestra Señora del Rosario, Facultad de Medicina, Santafé de Bogotá.

Vásquez-Diaz, J., and G. Quintero-Díaz, 1997: Anfibios y reptiles de Aguascalientes. Centro de Investigaciones y Estudios Multidisciplinarios de Aguascalientes, Aguascalientes,

Vásquez-Diaz, J., and G. Quintero-Díaz, 1998: *Nerodia nasicus*. Herpetol. Rev., 29 (3): 177

Vásquez-Diaz, J., and G. Quintevo-Díaz, 1999a: Geographic distribution: *Geophis dugesi aquilonaris*. Herpetol. Rev., 30 (4): 235.

Vásquez-Diaz, J., and T. G. Quinntero-Díaz, 1999b: *Rhadinaea hesperia* (Western Graceful Brown Snake)—Geographic distribution. Herpetol. Rev., 30 (4): 236.

Vaz-Ferreia, R., F. Achával and J. González, 1980: Effectos de la formacion de un lago de represa sobre la fauna de anfibios y reptiles y sobre su comportamiento. VIII. Cong. Latinoam. De Zoológia, Mérida, Venezuela. Resúmes: 146-147.

Vaz-Ferreia, R., F. Achával and M. P. Meneghel, 1980: Relaciones entre progenitores y cria en reptiles de la Rep. O. Del Uruguay. Res. J. C. nat., Montevideo, Uruguay, 1: 121-122.

Vaz-Ferreia, R., L. Covelo de Zolessi and F. Achával, 1970: Oviposicion y desarrollo de ofidios y lacertilios en hormigueros de *Acromyromex*. Physis, Buenos Aires, Argentina, 29 (79): 431-459.

Vaz-Ferreia, R., L. Covelo de Zolessi and F. Achával, 1973: Oviposicion y desarrollo de ofidios y lacertilios en hormigueros de *Acromyromex*. II. Trab. V. Congr. Latinoam. Zool., Montevideo, Uruguay, 1: 232-244.

Vaz-Ferreia, R. L. Covelo de Zollessi and F. Achával, 1976: Incubacion de huevos de reptiles en la naturaleza y en cautividad. Primer Congr. Parques Zoolóios del Cono Sur. 17-18-V-1976, Durazno, 1976: 1-4.

Vega-López, A. A., and T. Alvarez S., 1992: La herpetofauna de los volcanes Popocatépetl e Iztaccíhuatl. Acta Zoologica Mexicana, Nueva Serie, (51): 1-120.

Vellard, J., 1943: Una nueva forma de "*Oxyrhopus,*" "*O. rhombifer septentrionalis.*" Acta Zool. Lillo., 1: 89-91.

Vences, M., M. Franzen, A. Flaschendräger, R. Schmidt and J. Regös, 1998: Beobechtungen Zur Herpetolfauna von Nicaragua: Kommentierte Artenliste der Reptilen. Salamandra, 34 (1): 17-42.

Vereau, C. M., 1992: Variedad de serpientes tiene el Perú. El Comercio. Seccion C, Lima, Peru, miércoles 23 de septiembre de 1992,

Vermersch, T. G., and R. E. Kuntz, 1986: Snakes of south-central Texas. Austin, Texas, Eakin Press.

Vial, J. L., and J. M. Jiménez-Porras, 1967: The ecogeography of the bushmaster, *Lachesis muta*, in Central America. Amer. Midl. Naturalist, 78: 182-187.

Vidal, N., J. C. de Massary and C. Marty, 1998: Nouvelles especes de serpents pour la Guyane française. Revue Française D'aquariologie Herpetologie, 25 (3-4): 131-134.

Villa R., J., 1962: Las serpientes venonosas de Nicaragua. Ed. Vovedades, Managua, D. N.

Villa, J., 1968: A new colubrid snake from the Corn Islands, Nicaragua. Rev. Biol. Trop., 15: 117-121.

Villa, J., 1969a: Notes on *Conophis nevermanni*, an addition to the herpetofauna of Nicaragua. J. Herpetol., 3: 169-171.

Villa, J., 1969b: Two new insular subspecies of the natricid snake *Tretanorhinus nigroluteus* Cope from Honduras and Nicaragua. J. Herpetol., 3: 145-150.

Villa, J. D., [1970] 1971: The Snake *Hydromorphus* in Nicaragua, with a description of it hemipenis. Caribbean J. Sci., 10: 119-121.

Villa, J. D., 1971a: *Crisantophis*, A new genus for *Conophis nevermanni* Dunn, 1937., J. Herpetol., 5 (3-4): 173-177.

Villa, J. D., 1971b: Geographic distribution. *Typhlops costaricensis* (Costa Rican Blind Snake). Herpetol. Rev., 9 (2): 62.

Villa, J. D., 1971c: Notes on Nicaraguan reptiles. J. Herpetol., 5 (1-2): 45-48.

Villa, J. D., 1972a: Snakes of the Corn Islands, Caribbean Nicaragua. Brenesia, 1: 14-18.

Villa, J., 1972b: Un coral (*Micrurus*) blanco y negro de Costa Rica. Brenesia, 1: 10-13.

Villa, J. D., 1978: Geographical distribution: *Typhlops costaricensis*—distribution Herpetol. Rev., 9: 62.

Villa, J., 1983: Peces, anfibios, y reptiles Nicaraguenses: lista y bibliografía. Nicaraguan fishes, amphibians & reptiles: a checklist and bibliography. Univ. Centroamericana, Managua, Nicaragua.

Villa, J. D., 1984: The venomous snakes of Nicaragua. Contrib. Biol. Geol. Milwaukee Publ. Mus., (59): 1-41.

Villa, J., 1988a: *Cristanophis, C. nevermanni*. Cat. Amer. Amph. Rept., (429): 1-2.

Villa, J. D., 1988b: *Typhlops costaricensis* Jiménez & Savage—Costa Rican Blind Snake. Cat. Amer. Amph. Rept., (435): 1-2.

Villa, J. D., 1990a: *Hydromorphus* Peters, *H. concolor, H. dunni*. Cat. Amer. Amph. Rept., (472): 1-2.

Villa, J. D., 1990b: *Leptotyphlops nasalis*. Cat. Amer. Amph. Rept., (473): 1.

Villa, J. D. and J. R. McCranie, 1995: *Oxybelis wilsoni*, a new species of vine snake from Isla Roatán, Honduras (Serpentes; Colubridae). Rev. Biol. Trop., 43 (1-3): 297-305.

Villa, J. D. and L. D. Wilson, 1990: *Ungaliophis* Müller—Central American Dwarf Boa, *U. continentalis, U. panamensis*. Cat. Amer. Amph. Rept., (480): 1-4.

Villa, J., L. D. Wilson, and J. P. Johnson, 1988: Middle America herpetology: A bibliographic checklist. Colombia, Missouri, Univ. Missouri Press.

Villarejo, A., 1988: Asi es la selva. 4th Ed., Centro de Estudios Teológicas de la Amazonía, Iquitos, Peru.

Visinoni, A., 1995 [1996]: *Leptomicrurus narduccii narduccii* endemic to the Bolivian orient [in Italian]. Atti della societa Italiana di Scienze naturali del Museo Civico di Storia Naturale di Milano, 136 (1): 86-93.

Vitt, L. J., 1996: Book review-revision of the Neotropical snakes genus *Chironius* Fitzinger (Serpentes: Colubridae) J. R. Dixon, J. A. Wiest, Jr. and J. M. Cei (1993). Herpetol. Rev., 27 (2): 95-97.

Voegelin, C. F., F. M. Voegelin and K. L. Hale, 1962: Typological and comparative grammar of Uto-Aztecan: I (Phonology). IJAL Memoir 17,

Vogt, R. C., 1981: Natural history of amphibians and reptiles of Wisconsin. Milwaukee Publ. Mus., Milwaukee, Wisconsin.

Vorhies, C. T., 1917: Poisonous animals of the desert. Univ. Arizona Agric. Exp. Station, Bulletin 83, (83): 356-392.

Vosjoli, P. de, with photos by D. Northcot and P. de Vosjoli, 1998: A new look at boa constrictors. The Vivarium, 9 (1): 37-43, December 1997/January 1998.

Vúletin, A., 1960: Zoonimia Andina (Nomenclador Zoologica). Universidad Nacional de Tucumán, Santiago de Estero, Argentina.

Vuoto, J. A., 1996: Ampliación del area de distribución de *Waglerophis merremi* (Wagler, 1824) (Serpentes: Colubridae) sobre las provincias de Entre Rios, Santa Fe y Buenos Aires, Argentina. Cuadernos de Herpetologia, 10 (1-2): 59-70.

Wagler, J. G., 1824: Naturliches system der Amphibien mit voranghender classification der Saugetheire und Vogel Ein Beitrag zur vergleichender Zoologie. Munich.

Wagler, J. G., 1824: Serpentum Brasilensium species novae ou historie naturelle des espès nouvelles de serpens, recueillies et observées pendant le voyage dans l'intérieur du Brasil dans les annees 1817, 1818, 1819, 1820, exécuté par order de sa Majesté le Roide Baviére, publiée par Jean de Spix, ecrite d'après les notes du voyageur par Jean Wagler. Franc. Serraph. Jübschmann, Monachii.

Wagler, J. C., 1830: Natürlishes System der Amphibien mit Vorangehender Classification dor Säugthiere und Vögel fin Beitrag zur vergleichenden Zoologie. J. G. Cotta'schen Buchhandlung, Munich.

Wahlgren, R., 2000: Encyclopedia Londiensis (1796-1829) and a genuine and universal system of natural history, vol. 12 (1809 or 1810)—two little known contributions in the history of herpetology. Inter. Soc. Hist. Biblio. Herpetol.-Newsletter and Bull., 2 (1): 20-26.

Walker, W., 1945: A study of the snake *Tachymenis peruviana* Wiegmann and it's allies. Bull. Mus. Comp. Zool., 96 (1): 4-55.

Wallach, V., 1995: Revalidation of the genus *Tropidodipsas* Günther, with notes on the dipsadini and nothopisini (Serpente: Colubridae). J. Herpetol., 29 (3): 476-481.

Wallach, V., 1996: Note and corrections on two scolecophidians: *Ramphotyphlops albiceps* and *Leptotyphlops brasilensis*. Herpetol. Rev., 27 (1): 10.

Wallach, V., 1998: The visceral anatomy of blindsnakes and wormsnakes and its systematic implications (serpentes, anomalepididae, typhlopidae, lepetotyphlopidae). PhD Thesis for Department of Biology, Northeastern University, August 1998.

Wallach, V., 1999: Geographic distribution: *Ramphotyphlops braminus*. Herpetol. Rev., 30 (4): 236.

Wallach, V., and R. Guenther, 1997: Typhlopidae vs. Anomalepididae: The identity of *Typhlops mutilatus* Werner (Reptilia: Serpentes). Mitt. Zool. Mus. Berl., 73: 233-342.

Bibliography

Wallach, V., and H. M. Smith, 1992: *Boella tenella* is *Epicrates inornatus* (Reptilia: Serpentes). Bull. Maryland Herpetol. Soc., 28 (4): 162-170.

Waller, T., P. A. Micucci and E. Buongermini Palumbo, 1995: Distribución y conservación de la familia boidae en el Paraguay. Autoridad Científica CITES del Paraguay. Asunción, Paraguay, Julio 1995.

Walley, H. D., 1999: *Rhadinaea flavilata* Cope—Pine Woods Snake. Cat. Amer. Amph. Rept., (699): 1-5.

Walley, H. D., and C. M. Eckerman, 1999: *Heterodon nasicus* Baird and Girard-Western Hognose Snake. Cat. Amer. Amph. Rept., (698): 1-10.

Walley, H. D., and M. V. Plummer, 2000: *Opheodrys aestivus* (Linnaeus)—Rough Green Snake. Cat. Amer. Amph. Rept., (718): 1-14.

Walls, J. G., 1994: All rat snakes are not created equal. Reptile and Amphibian Magazine, May-June, 1994: 69-75.

Walsh, J., with R. Gannor, 1967: Time is short and the water rises. E. P. Dutton & Co., Inc., New York.

Waltz, N. E., 1976: Discourse functions of Guanano sentence paragraph. pp. 21-145. *In* Longacre, R. E., (Ed.), 1976: Discourse grammar—Studies in indigenous languages of Colombia, Panama, and Ecuador Part I.

Webb, R. G., 1961: A new kingsnake from Mexico, with remarks on the *mexicana* group of the genus *Lampropeltis*. Copeia, 1961 (3): 326-333.

Webb, R. G., 1966: Resurrected names for Mexican populations of black-necked garter snakes, *Thamnophis cyrtopsis* (Kennicott). Tulane Studies Zool., 13: 1-70.

Webb, R. G., 1970: Reptiles of Oklahoma. Univ. Oklahoma Press, Norman, Oklahoma, 370 pp.

Webb, R. G., 1976: A review of the garter snake *Thamnophis elegans* in Mexico. Nat. Hist. Mus. Los Angeles County Contrib. Sci., (284): 1-13.

Webb, R. G., 1977: Comments on snakes of the genus *Geophis* (Colubridae) from the Mexican states of Durango and Sinaloa. Southwest. Naturalist, 21 (4): 548-551.

Webb, R. G., 1978: A review of the Mexican garter snake *Thamnophis cyrtopsis postremus* Smith with comments on *Thamnophis vicinus* Smith. Contrib. Biol. Geol. Milwaukee Publ. Mus., (19): 1-13.

Webb, R. G., 1980: *Thamnophis cyrtopsis*, Cat. Amer. Amph. Rept., (245): 1-4.

Webb, R. G., 1982a: Distributional records for Mexican reptiles. Herpetol. Rev., 13: 132.

Webb, R.G, 1982b.: Taxonomic status of some Neotropical garter snakes (Genus *Thamnophis*). Bull. Southern California Acad. Sci., 81 (1): 26-40.

Webb, R. G., 1990: Description of a new subspecies of *Bogertophis subocularis* (Brown) from northern Mexico (Serpientes: Colubridae). Texas J. Sci., 42 (3): 227-243. Merriam-Webster, 1986: Webster's Ninth New Collegiate Dictionary. Merriam-Webster Inc., publishers, Springfield, Massachusetts.

Wehkind, L., 1955: Notes on the foods of the Trinidad snakes. British. J. Herpetol., 2: 9-13.

Wehekind, L., 1960: Trinidad snakes. J. British Guiana Mus. Zool., 27: 71-76.

Weir, J., 1993: The taxonomic status of *Elaphe guttata intermontana*—The intermountain ratsnake (Woodbury & Woodbury, 1942). Herptile, 18 (4): 167-179, figs. 7-9.

Weldon, P. J., R. Ortiz and T. R. Sharp, 1992: The chemical ecology of crotaline snakes. pp 309-319 *in* J. A. Campbell and E. D. Brodie, Jr.: Biology of the pitvipers. Selva, Tyler, Texas.

Wellman, J., 1959: Notes on the variation in and distribution of the Mexican colubrid snake *Coniophanes lateritius*. Herpetologica, 15 (3): 127-128.

Wellman, J., 1963: A revision of the snakes of the genus *Conophis* (family Colubridae). Univ. Kansas Pub. Mus. Nat. Hist., 15 (6): 251-295.

Werler, J. E., 1948: *Natrix cyclopion cyclopion* in Texas. Herpetologica, 4: 148.

Werler, J. E., 1970: Poisonous snakes in Texas and first aid treatment of their bites. Bulletin No. 31, Austin, Texas, Texas Parks & Wildlife Dept., (31): 1-62.

Werler, J. E., and J. R. Dixon, 2000: Texas snakes-identification, distribution, and natural history. Univeristy of Texas Press, Austin, Texas.

Werler, J. E., and H. M. Smith, 1952: Notes on a collection of reptiles and amphibians from Mexico, 1951-1952. Texas J. Sci., 14: 551-573.

Werman, S. D., 1984: Taxonomic comments on the Costa Rican pit viper *Bothrops picadoi*. J. Herpetol., 18: 207-210.

Wermen, S. D., 1984: The taxonomic status of *Bothrops supraciliaris* Taylor. J. Herpetol., 18: 484-486.

Werman, S. D., 1992: Phylogenetic relationships of Central and South American pitvipers of the genus *Bothrops* (*sensu latu*): Cladistic analyses of biochemical and anatomical characters. pp. 21-40. *In* Campbell and Brodie, Jr., (eds.). Biology of the pitvipers. Selva, Tyler, Texas.

Werner, J., 1896: Beiträge zur Kenntnis der Reptilien und Batrachier von Zentralamerika und Chile sowie einiger seltenerer Schlangenarten. Ver. K.K. Zool. Bot. Gesell. Wien. 1896: 244-278.

Werner, F., 1898: Die Reptilien und Batrachier der Sammlung Plate. Zool. Jahrb. Abt. Syst. Oekol. Geogr. Tiere (Suppl. 4): 244-278.

Werner, F., 1899: Ueber Reptilien und Batrachier aus Colimbien und Trinidad.Verhandl. Zool. Bot. Ges. Wien, 49: 470-484.

Werner, F., 1900: Ueber Reptilien und Batrachier aus Colimbien und Trinidad, II. Verhandl. Zool. Bot. Ges. Wien, 50: 262-272.

Werner, F., 1901a: Über Reptilien und Batrachier aus Ecuador und Neu-Guinea. Verhandl. Zool. Bot. Ges. Wien, 51: 593-614.

Werner, F., 1901b: Reptilien und Batrachier aus Peru and Boliven. Abhandl. Beri. Königl. Zool. Anthro.-Ethnogra. Mus. Dresden, 9 (2): 1-13.

Werner, F., 1903: Neue Reptilien und Batrachier aus dem naturhistorischen Museum in Brussel. Zool. Anz., 26 (693): 246-253.

Werner, F., 1904: Über Reptilien und Batrachier aus Guatemala und China in der zoologischen Staats-Sammlung in München, nebst einem Anhang über seltene Formen aus anderen Gegenden. Abhandl. Bayerische Akad. Wiss., 22 (2): 343-384.

Werner, F. *in* Burger, 1907: Estudios sobre reptiles chilenos. Anales Univ. Chile, 121: 149-155 + pl.I-II.

Werner, F., 1908: Neue Reptilier und Barachier aus dem naturhistorischen Museum in Brüssel. Zool. Anz., 26: 246-253.

Werner, F., 1909a: Beschreibung neuer Reptilien dem Kgl. Naturalien-Kabinett

Werner, F., 1909a: ... in Stuttgart, Jahrb. Ver. Naturk. Stuttgart, 65: 55–63.

Werner, F., 1909b: Neue oder seltene Reptilien des Musee Royal d'Histoire naturelle de Belgique in Brüssel. Zool. Jahrb., 28 (3): 263–288.

Werner, F., 1909c: Neue oder seltene Reptilien des Naturhistorischen Museums in Hamburg. Mitt. Naturhist. Mus., Hamburg, 26: 205–247.

Werner, F., 1909d: Über neue oder seltene Reptilien des Naturhistorischen Museums in Hamburg. I. Schlangen. Mitt. Naturhist. Mus., Hamburg, 26: 205–247.

Werner, F., 1910: Über neue oder seltenere Reptilien des Naturhistorischen Museums Hamburg. 2 Eidechsen. Mitt. Naturhist. Mus., 27: 1–90.

Werner, F., 1913: Neue oder seltene Reptilien und Frösche des naturhistorischen Museums Hamburg. Mitt. Naturh. Mus. Hamburg., 30: 1–51.

Werner, F., 1916: Bemerkungen über einige niedere Wirbeltiere der Andsn von Kolumbien mit Beschroibungen neuer Arten. Zool. Anz. Leipzig, 47: 301–304, 305–311.

Werner, F., 1917: Versuch einer Synopsis der Schlangenfamilie der Glauconidae. Mitt. Zool. Mus. Hamburg, 34: 191–208.

Werner, F., 1923: Übersicht der Gattungen und Arten der Schlangen der Familie Colubridae. I. Teil. Arch. Natur., Jahrg. 89, A (8): 138–199.

Werner, F., 1923: Separat-Abdruk aus dem XXXVI. Bande der Ann. Naturh. Mus. Wien, 36: 160–163.

Werner, F., 1924a: Neue oder wenig bekannte Schlangen aus dem Naturhistorischen Staatsmuseum. Sitzungsber. Akak. Wiss. Wien (1) 133: 29–56.

Werner, F., 1924b: Übersicht der Gattungen und Arten der Schlangen der Familie Colubridae. II. Teil. Dipsadomorphinae und Hydrophiinae. Arch. Naturg., 12: 108–166.

Werner, F., 1927: Neue oder wenig bekannte Schlangen aus dem Weiner naturhistorischen Staatsmuseum. (III Teil). Sitzber. Akad. Wiss, 135: 243–257.

Werner, F., 1929: Übersicht der Gattungen und Arten der Schlangen aus der Familie Colubridae. II. Teil. Colubrinae. Zool. Jahrb. Syst., 57: 1–196.

West, B., 1977: Results of a Tucan syntax questionnaire. pp 339–377. *In* Longacre, R. E., ed., 1977: Discourse grammar—Studies in indigenous languages of Colombia, Panama, and Ecuador Part III.

Weyer, D., 1995: Snakes of Belize. Belize Audubon Society, Sponsored by World Wildlife Fund, March 1995.

Wheeler, G. C., and J. Wheeler, 1966: The amphibian and reptiles of North Dakota. Univ. North Dakota Press, Grand Forks.

White, L. R., J. S. Parmerlee and R. Powell, Jr., 1992: *Typhlops syntherus*. Cat. Amer. Amph. Rept., (551): 1–2.

Wiebe, N., 1977: The structure of events and participants in Cayapo narrative discourse. pp. 191–227. *In* Longacre, R. E., ed., 1977: Discourse grammar—Studies in indigenous languages of Colombia, Panama, and Ecuador Part II.

Wickler, W., 1968: Mimicry in plants and animals. World Univ. Library, 256 pp.

Wied-Neuwied, M. Prinzen von, 1820: Reise nach Brasilien in den Jahren 1815 bis 1817. 2 vol, Frankfurt a.M.: Henrich L. Bronner.

Wied-Neuwied, M. Prinzen von, 1824: Abbildungen zur Naturgeschichte Brasilenss. Weimar, 1824: 4–8.

Wied-Neuwied, M. Prinzen von zu, 1825: Beitrange zur Naturgeschichte von Brasilien von Maximilian Prinzen (zu Wied). Weimar: im Verlage des Gr. H. S. priv. Landes-Industrie-Comptoirs (1825–1833). I: 495–498.

Wied-Neuwied, M. Prinzen von au, 1829: Abbildungen zur Nat. Brasiliens. Driezehnte Lieferung. Weiner, im Verlage des Drofsherzogl. sächs. Priv. Landes-Industrie-Comptoirs.

Wied-Neuwied, M. Prinzen von au, 1830: Abbildungen zur Naturgeschichte Brasiliens, Verzehnte Lieferung. Weimer: Im Verlage des Grofsherzogl. sächs. priv. Landes-Industrie-Comptoirs.

Wiegmann, A. F. A., 1834: Beiträge zur Zoologie gesammelt auf einer Reise um die Erde, von Dr. F. J. F. Meyen. Siebente Abhandlung. Amphibien. Nova Acta Acad. Caes. Leop.—Carol, 17 (1): 185–268 + pls. 13–22.

Wiegmann, A. F. A., 1834: Herpetologia Mexicana seu Descriptio Amphibiorum Novae Hispaniae quae Itinerbus Comitis de Sack, Ferdinandi Deppe et. Chr. Guil. Schiede in Museum Zoologicum Berolinense Pervenerunt. Pars prima, Sasurorum Species Amplectans, Adiecto Systematis Saurorum Prodromao, Additisque Multis in Hunc Amphibriorum Ordinem observationbus. Berlin, C. G. Lëderitz.

Wiest, J. A. Jr., 1978: Revision of the Neotropical snake genus *Chironius* Fitzinger (Serpentes, Colubridae). Ph.D. Dissertation, Texas A&M University, December 1978: xv + 370.

Wilderness Southeast: Experience the Okefenokee. 711-VC Sandtown Road, Savannah, Georgia.

Willett, T. L., 1991: A reference grammar of Southeastern Tepehuan. Publication 100-Summer Institute of Linguistics and The University Texas at Arlington Publications in Linguistics.

Williams, J. D., and G. A. Couturier, 1984: Primera cita del género *Hydrops* Wagler, 1830 para la República Argentina (Serpentes: Colubridae). Hist. Nat., Corrientes, Argentina, 4 (7): 61–66.

Williams, J. D., and F. Francini, 1991: A check-list of the Argentine snakes. Boll. Mus. reg. Sci. nat. Torino, 9 (1): 55–90.

Williams, J. D., and E. Gudynas, 1991: Revalidation and redescription of *Atractus taeniatus* Griffin 1916 (Serpentes, Colubridae). Cont. Biol., Montevideo, 15: 1–8.

Williams, J. D., and G. J. Scrocchi, 1994: Fauna de agua dulce de la República Argentina. Vol. 42, Fasc. 3—Ophidia, Lepidosauria. PROFADU (CONICET) La Plata, 1994: 1–55.

Williams, K. L., 1978: Systematics and natural history of the American milk snake, *Lampropeltis triangulum*. Milwaukee Pub. Mus. Publ. Biol. Geol., (2): 1–258.

Williams, K. L., 1985: *Cemophora, C. coccinea*. Cat. Amer. Amph. Rept., (374): 1–4.

Williams, K. L., 1988: Systematics and natural history of the American Milk Snake, *Lampropeltis triangulum*. Milwaukee Publ. Mus.

Williams, K. L., 1994: *Lampropeltis triangulum* (Lacépède)—Milk Snake. Cat. Amer. Amph. Rept., (594): 1–10.

Williams, K. L., B.C. Brown and L. D. Wilson, 1966: A new subspecies of the colubrid snake *Cemophora coccinea* (Blumenbach) from southern Texas. Texas J. Sci., 18 (1): 85–88.

Williams, K. L., and L. D. Wilson, 1967: A review of the colubrid snake genus *Cemophora*. Tulane Studies Zool. Bot., 13 (4): 103–124.

Bibliography

Williams, K. L., and V. Wallach, 1989: Snakes of the world-Volume I-synopsis of snake generic names. Krieger Publishing Company, Malabar, FL.

Williams, K. L., and L. D. Wilson, 1967: A review of the colubrid snake genus, *Cemophora* Cope. Tulane Stud. Zool., 13 (4): 103-124.

Williamson, J. P., 1970: An English-Dakota dictionary. Ross & Haines, Inc., Minneapolis, 264 pp.

Williamson, M. A. Jr., and N.J. Scott, 1982: Geographic distribution. *Masticophis taeniatus*. Herpetol. Rev., 13: 25.

Wilson, L. D., 1966a: *Dendrophidion vinitor*: an addition to the snake fauna of British Honduras. J. Ohio Herpetol. Soc., 5: 103.

Wilson, L. D., 1966b: The range of the Rio Grande racer in México and the status of *Coluber oaxaca* (Jan). Herpetologica, 22: 42-47.

Wilson, L. D., 1968: *Leptotyphlops phenops* (Cope) in Honduras. J. Herpetol., 2: 166-167.

Wilson, L. D., 1970a: *Amastridium*. In J. A. Peters and B. Orejas-Miranda. Catalogue of the Neotropical Squamata: Part I. Snakes. Bull. U.S. Natl. Mus., (297): 18.

Wilson, L. D., 1970b: A Review of the *chloroticus* Group of the Colubrid Snake Genus *Drymobius*, with notes on a twin-striped form of *D. chloroticus* (Cope) from Southern Mexico. J. Herpetol., 4 (3-4): 155-163.

Wilson, L. D., 1970c: *Tantilla brevicauda*: an addition to the snake fauna of Guatemala, with comments on its relationships. Bull. Southern California Acad. Sci., 69: 118-120.

Wilson, L. D., 1970d: The coachwhip snake, *Masticophis flagellum* (Shaw): Taxonomy and distribution. Tulane Studies Zool. Bot., 16: 31-99.

Wilson, L. D., 1970e: The racer *Coluber constrictor* (Serpentes: Colubridae) in Louisiana and eastern Texas. Texas J. Sci., 22: 67-99.

Wilson, L. D., 1972: A review of the colubrid snakes of the genus *Tantilla* in Central America. Milwaukee Publ. Mus. Contrib. Biol. Geol., (52): 1-77.

Wilson, L. D., 1973a: *Masticophis*. Cat. Amer. Amph. Rept., (144): 1-2.

Wilson, L. D., 1973b: *Masticophis flagellum*. Cat. Amer. Amph. Rept., (145): 1-4.

Wilson, L. D., 1974a: *Drymobius margaritiferus* (Schlegel)-Central American speckled racer. Cat. Amer. Amph. Rept., (172): 1-2.

Wilson, L. D., 1974b: *Tantilla taeniata* (Bocourt): An addition to the snake fauna of El Salvador. Bull. Southern California Acad. Sci., 72: 53-54.

Wilson, L. D., 1975a: *Drymobius*. Cat. Amer. Amph. Rept., (170): 1-2.

Wilson, L. D., 1975b: *Drymobius chloroticus*. Cat. Amer. Amph. Rept., (171): 1-2.

Wilson, L. D., 1975c: *Drymobius margaritiferus*. Cat. Amer. Amph. Rept., (172): 1-2.

Wilson, L. D., 1975d: *Drymobius melanotropis*. Cat. Amer. Amph. Rept., (173): 1-2.

Wilson, L. D., 1976: Variation in the colubrid snake *Tantilla semicincta* (Duméril, Bibron and Duméril,), with comments on pattern dimorphism, Bull. Southern California Acad. Sci., 75: 42-48.

Wilson, L. D., 1978: *Coluber constrictor*. Cat. Amer. Amph. Rept., (218): 1-4.

Wilson, L. D., 1979: A new snake of the genus *Tantilla* from Ecuador. Herpetologica, 35 (3): 274-276.

Wilson, L. D., 1982a: A Review of the Colubrid snakes of the genus *Tantilla* of Central America. Contrib. Biol. Geol. Milwaukee Publ. Mus., (52): 1-77.

Wilson, L. D., 1982b: *Tantilla*. Cat. Amer. Amph. Rept., (307): 1-4.

Wilson, L. D., 1983a: A new species of *Tantilla* (Serpentes: Colubridae) of the *taeniata* group from Chiapas, México. J. Herpetol., 17: 54-59.

Wilson, L. D., 1983b: *Tantilla taeniata*. Cat. Amer. Amph. Rept., (344): 1-2.

Wilson, L. D., 1984a: Additional notes on Colubrid snakes of the genus *Tantilla* from tropical America. Herpetol. Rev., 15: 8-10.

Wilson, L. D., 1984b: The status of *Micrurus ruatanus* (Günther), a coral snake endemic to the Bay Islands of Honduras. Herpetol. Rev., 15: 67-68.

Wilson, L. D., 1985a: Rediscovery of *Tantilla bairdi* Stuart and a definite Guatemalan locality for *Tantilla taeniata* (Bocourt). Herpetol. Rev., 16: 105.

Wilson, L. D., 1985b: *Tantilla albiceps*. Cat. Amer. Amph. Rept., (377): 1.

Wilson, L. D., 1985c: *Tantilla andinista*. Cat. Amer. Amph. Rept., (378): 1.

Wilson, L. D., 1985d: *Tantilla annulata*. Cat. Amer. Amph. Rept., (379): 1.

Wilson, L. D., 1985e: *Tantilla bairdi*. Cat. Amer. Amph. Rept., (380): 1.

Wilson, L. D., 1985f: *Tantilla briggsi*. Cat. Amer. Amph. Rept., (365): 1.

Wilson, L. D., 1985g: *Tantilla cuesta*. Cat. Amer. Amph. Rept., (366): 1.

Wilson, L. D., 1985h: *Tantilla cuniculator*. Cat. Amer. Amph. Rept., (367): 1

Wilson, L. D., 1985i: *Tantilla jani*. Cat. Amer. Amph. Rept., (369): 1.

Wilson, L. D., 1985j: *Tantilla flavilineata*. Cat. Amer. Amph. Rept., (368): 1.

Wilson, L. D., 1985k: *Tantilla reticulata*. Cat. Amer. Amph. Rept., (370): 1.

Wilson, L. D., 1986a: *Tantilla alticola*. Cat. Amer. Amph. Rept., (400): 1.

Wilson, L. D., 1986b: The status of *Homalocranium breve* Günther, 1895 (Serpentes: Colubridae). Southwest. Naturalist, 31: 61-62.

Wilson, L. D., 1987a: A résumé of the Colubrid snakes of the genus *Tantilla* of South America. Contrib. Biolo. Geol. Milwaukee Publ. Mus., (68): 1-35.

Wilson, L. D., 1987b: *Geagras, G. redimitus*. Cat. Amer. Amph. Rept., (430): 1-2.

Wilson, L. D., 1987c: *Tantilla schistosa*. Cat. Amer. Amph. Rept., (409): 1-2.

Wilson, L. D., 1987d: *Tantilla vermiformis*. Cat. Amer. Amph. Rept., (410): 1.

Wilson, L. D., 1988a: *Amastridium, A. veliferum*. Cat. Amer. Amph. Rept., (449): 1-3.

Wilson, L. D., 1988b: *Tantilla brevicauda*. Cat. Amer. Amph. Rept., (432): 1.

Wilson, L. D., 1988c: *Tantilla calamarina*. Cat. Amer. Amph. Rept., (433): 1-2.

Wilson, L. D., 1988d: *Tantilla canula*. Cat. Amer. Amph. Rept., (434): 1.

Wilson, L. D., 1988e: *Tantilla cascadae*. Cat. Amer. Amph. Rept., (451): 1.

Wilson, L. D., 1988f: *Tantilla deppei*. Cat. Amer. Amph. Rept., (452): 1.

Wilson, L. D., 1988g: *Tantilla equatoriana*. Cat. Amer. Amph. Rept., (453): 1.

Wilson, L. D., 1988h: *Tantilla moesta*. Cat. Amer. Amph. Rept., (454): 1.

Wilson, L. D., 1988i: *Tantillita, T. brevissima and T. lintoni*. Cat. Amer. Amph. Rept., (455): 1-2.

Wilson, L. D., 1988j: The status of *Tantilla excubitor* Wilson. J. Herpetol., 22: 469-470.

Wilson, L. D., 1990a: *Tantilla capistrata*. Cat. Amer. Amph. Rept., (475): 1.

Wilson, L. D., 1990b: *Tantilla coronadoi*. Cat. Amer. Amph. Rept., (501): 1.

Wilson, L. D., 1990c: *Tantilla insulmontana*. Cat. Amer. Amph. Rept., (502): 1.
Wilson, L. D., 1990d: *Tantilla lempira*. Cat. Amer. Amph. Rept., (476): 1.
Wilson, L. D., 1990e: *Tantilla miyatai*. Cat. Amer. Amph. Rept., (477): 1.
Wilson, L. D., 1990f: *Tantilla oaxacae*. Cat. Amer. Amph. Rept., (503): 1.
Wilson, L. D., 1990g: *Tantilla semicincta*. Cat. Amer. Amph. Rept., (478): 1.
Wilson, L. D., 1990h: *Tantilla striata*. Cat. Amer. Amph. Rept., (504): 1.
Wilson, L. D., 1990i: *Tantilla tayrae*. Cat. Amer. Amph. Rept., (479): 1.
Wilson, L. D., 1991a: *Tantilla shawi*. Cat. Amer. Amph. Rept., (528): 1
Wilson, L. D., 1991b: *Tantilla petersi*. Cat. Amer. Amph. Rept., (527): 1.
Wilson, L. D., 1992a: *Tantilla melanocephala*. Cat. Amer. Amph. Rept., (54) 7: 1.
Wilson, L. D., 1992b: *Tantilla nigra*. Cat. Amer. Amph. Rept., (548): 1.
Wilson, L. D., 1999: Checklist and key to the species of the genus Tantilla (Serpentes: Colubridae), with some commentary on distribution. Smithsonian Herpetol. Information Serv., (122): 1–34 + table.
Wilson, L. D., and J. A. Campbell, 2000: A new species of *calamaria* group of the colubrid snake genus *Tantilla* (Reptilia: Squamata) from Guerrero, Mexico, with a review of and key to members of the group. Proc. Biol. Soc. Washington, 113 (3): 820–827.
Wilson, L. D., and G. A. Cruz Díaz, 1986: Two additions to the herpetofauna of Honduras: *Smilisca sordida* and *Drymobius melanotropis*. Southwest. Naturalist, 31 (2): 66.
Wilson, L. D., G. A. Cruz Diaz, E. Villeda, and S. Flores, 1988: *Typhlops costaricensis* Jimenez and Savage: An addition to the snake fauna of Honduras. Southwest. Naturalist, 33 (4): 499–500.
Wilson, L. D., and D. D. Dugas, 1972: *Pliocercus euryzonus* Cope: An addition to the snake fauna of Honduras. Bull. Southern California Acad. Sci., 71: 159.
Wilson, L. D., and D. E. Hahn, 1973: The herpetofauna of the Islas de la Bahía, Honduras. Bull. Florida State Mus. Biol. Ser., 17: 93–150.
Wilson, L. D., J. R. McCranie, 1979: Notes on the herpetofauna of two mountain ranges in Mexico (Sierra Fria, Aguascalientes, and Sierra Morones, Zacatecas). J. Herpetol., 13 (3): 271–278.
Wilson, L. D., and J. R. McCranie, 1992: *Bothriechis marchi*. Cat. Amer. Amph. Rept., (544): 1–2.
Wilson, L. D., and J. R. McCranie, 1997: Publication in non-peer-reviewed outlets: The case of Smith and Chiszar's "Species-group taxa of the false coral snake genus *Pliocercus*". Herpetol. Rev., 28 (1): 18–21.
Wilson, L. D., and J. R. McCranie, 1999: The systematic status of Honduran populations of the *Tantilla taeniata* group (Serpentes: Colubridae), with notes on other populations. Amphibia-Reptilia, 20: 326–329.
Wilson, L. D., and J. R. McCranie, 2002: Update on the list of reptiles known from Honduras. Herp. Review, 32 (2): 90–94
Wilson, L. D., J. R. McCranie, and M. R. Espinal, 1996: Coral snake mimics of the genus *Pliocercus* (family Colubridae) in Honduras and their mimetic relationships with *Micrurus* (family Elapidae). Herpetol. Nat. Hist., 4: 57–63.
Wilson, L. D., J. R. McCranie, and L. Porras, 1977: Taxonomic notes on *Tantilla* (Serpentes: Colubridae) from tropical America. Bull. Southern California Acad. Sci., 76: 49–56.
Wilson, L. D., J. R. McCranie, and L. Porras, 1978: Two snakes, *Leptophis modestus* and *Pelamis platurus*, new to the herpetofauna of Honduras. Herpetol. Rev., 9: 63–64.
Wilson, L. D., J. R. McCranie, and L. Porras, 1979: *Rhadinaea montecristi* Mertens: An addition to the snake fauna of Honduras. Herpetol. Rev., 10 (2): 62.
Wilson, L. D., J. R. McCranie, and J. B. Slowinski, 1992: *Micrurus ruatanus* (Günther)—Babaspul (local dialect), Coral (Spanish), Roatan Coral Snake (English). Cat. Amer. Amph. Rept., (545): 1–2.
Wilson, L. D., J. R. McCranie and K. L. Williams, 1998: A new species of *Geophis* of the *sieboldi* group (Reptilia: Serpentes: Colubridae) from northern Honduras, Proc. Biol. Soc. Washington, 111 (2): 410–417.
Wilson, L. D., and C. E. Mena, 1980: Systematics of the *melanocephala* group of the colubrid snake genus *Tantilla*. Mem. San Diego Soc. Nat. Hist., 11: 1–58.
Wilson, L. D., and J. R. Meyer, 1969: A review of the Colubrid snake genus *Amastridium*. Bull. Southern California Acad. Sci., 68: 146–160.
Wilson, L. D., and J. R. Meyer, 1971: A revision of the *taeniata* group of the Colubrid snake genus *Tantilla*. Herpetologica, 27: 11–40.
Wilson, L. D., and J. R. Meyer, 1972a: (Review of) anfibios de Nicaragua: Introducción a su sistemática, vida y costumbres by Jaime Villa. Herpetol. Rev., 4: 219.
Wilson, L. D., and J. R. Meyer, 1972b: *Rhadinaea godmani*, an addition to the snake fauna of Honduras. Bull. Southern California Acad. Sci., 71: 50–52.
Wilson, L. D., and J. R. Meyer, 1972c: The coral snake *Micrurus nigrocinctus* in Honduras. Bull. Southern California Acad. Sci., 71: 139–145.
Wilson, L. D., and J. R. Meyer, 1981: Systematics of the *calamarina* group of the Colubrid snake genus *Tantilla*. Contrib. Biol. Geol. Milwaukee Publ. Mus., (42): 1–25.
Wilson, L. D., and J. R. Meyer, 1982: The snakes of Honduras. Contrib. Biol. Geol. Milwaukee Publ. Mus., (6): 1–159.
Wilson, L. D., and J. R. Meyer, 1985: The snakes of Honduras. 2nd. ed., Contrib. Biol. Geol. Milwaukee Publ. Mus., (6): 1–150.
Wilson, L. D., and L. Porras, 1983: The ecological impact of man on the south Florida herpetofauna. Univ. Kansas Mus. Nat. Hist. Spec. Pub., (9): vi–89.
Wilson, L. D., L. Porras and J. R. McCranie, 1986: Distributional and taxonomic comments on some members of the Honduran herpetofauna. Milwaukee Publ. Mus. Contrib. Biol. Geol., (66): 1–18.
Wilson, L. D., and D. C. Robinson, [1970] 1971: Additional specimens of the colubrid snake *Amastridium veliferum* (Colubridae) from Costa Rica, with comments on a pseudohermaphrodite. Bull. Southern California Acad. Sci., 70: 53–54.
Wilson, L. D., R. K. Vaughan and J. R. Dixon, 1999: Another new species of *Tantilla* of the *taeniatus* group from Chiapas, Mexico. J. Herpetol., 33 (1): 1–5.
Wilson, L. D., R. K. Vaughan and J. R. Dixon, 2000a: *Tantilla cucullata* Minton —Trans-Pecos black-headed Snake. Cat. Amer. Amph. Rept., (719): 1–2.

Wilson, L. D., R. K. Vaughan and J. R. Dixon, 2000b: *Tantilla rubra* Cope—La Rojilla (Red black-headed snake). Cat. Amer. Amph. Rept., (720): 1–3.

Wilson, L. D., and J. Villa, 1973: Colubrid snakes of the genu *Tantilla* from Nicaragua. Bull. Southern California Acad. Sci., 72: 93–96.

Wirz, S., 2001: Erfahrungen mit selten gepflegten *Corallus*-Arten—*Corallus annulatus annulatus, Corallus hortulanus hortulanus* und *Corallus hortulanus ruschenbergeri*. Draco, Berlin, Germany, 2 (5): 26–31.

Wistrand-Robinson, L., 1991: Uto-Aztecan affinities with Panoan of Peru I: Correspondences. pp. 243–276. *In* Mary Ritchie Kay (Ed.): Language changes in South American Indian Languages. Univ. Pennsylvania Press, Philadelphia.

Wistrand-Robinson, L., and J. Armagost, 1990: Comanche dictionary and grammar. A Publication of the Summer Institute of Linguistics and the Univ. Texas Arlington. Publication (92): 1–338.

Wolff, B., and J. Ronne, 1997: Boas-an essay on current taxonomy. Reptile & Amphibian Mag., September/October 1997 (50): 24–31.

Wood, W., 1634: New Englands prospect. London, 1634: viii + 98 + [v]. [Boston reprint, pp. (vi) + 103 + (vi), 1898].

Woodbury, A. M., 1952: Amphibians and reptiles of the Great Salt Lake Valley. Herpetologica, 8 (2): 42–49.

Woodin, W. H., 1952: Notes on some reptiles from the Huachuca area of southeastern Arizona. Bull. Chicago Acad. Sci., 9: 285–296.

Woodbury, A. M., and D. M. Woodbury, 1942: Studies of the rat snake, *Elaphe laeta*, with description of a new subspecies. Proc. Biol. Soc. Washington, 55: 133–142.

Worthington, R. D., 1980: *Elaphe subocularis*. Cat. Amer. Amph. Rept., (268): 1–2.

Wright, A. H., and A. A. Wright, 1962: Handbook of snakes of the United States and Canada. Vol. III. Bibliography. (Privately Printed), Ithaca, New York, reprinted by Study of Amphibians and Reptiles Fascimile Reprints in Herpetology, 1979.

Wright, A. H., and A. A. Wright, 1957 [1994 printing]: Handbook of snakes of the United States and Canada. Comstock Publishing Associates, A Division of Cornell Univ. Press, Ithaca and London, Volume I and Volume II, 1957, forward by Campbell (1994).

Wright, N. P., 1970: A guide to Mexican mammals and reptiles. Personal Adventure Series, Editorial Minutiae Mexicana, S.A. de C.V., Mexico, D.F.

Wucherer, O., 1861: On the ophidians of the province of Bahia, Brazil. Proc. Zool. Soc., London, 8: 113–116 + 19 pls.

Wüster, W., 1997: Jararaca! An introduction to Brazilian pitvipers of the genus *Bothrops*. Reptile & Amphibian Mag., (51): 12–21.

Yanosky, A. A., 1989: Approche de l'herpétofaune de la resrve écologiique El Bagual (Formosa, Argentina). I. Anoures et Ophidiens. Reveue fr. Aquariol., 16 (1989): 57–62, 30 November, 1989.

Yanosky, A. A., 1998: Las reservas naturales privadas del Paraguay: Asistencia al maintenimiento de la diversidad biologica paraguaya. Master's Thesis, Universidad Nacional de Entre Rios Facultadad de Ciencias Economica, Feb. 1998.

Yanosky, A. A., and J. M. Chani, 1986: *Lystrophis dorbignyi* (Duméril, Bibron et Duméril). Eds. B. Condé & D. Terver. Supplement to the Revue française d'Aquariologie, Sté Nlle, Imp. Pagel Vandoeuvre (3/85): 297–298.

Yanosky, A. A., and J. M. Chani, 1988: Possible dual mimicry of *Bothrops* and *Micrurus* by the colubrid *Lystrophis dorbignyi*. J. Herpetol., 22 (2): 222–224.

Yanosky, A. A., J. R. Dixon, and C. Mercolli, 1992: First specific locality record for *Liophis dilepis* (Cope, 1862) in Argentina. Rev. Esp. Herpetol, Spain, 7: 33–35.

Yanosky, A. A., J. R. Dixon, and C. Mercolli, 1993: The herpetofauna of El Bagual Ecological Reserve (Formosa, Argentina) with comments on its herpetological collection. Bull. Maryland Herpetol. Soc., 29 (4): 160–171.

Yarlequé Chocas, A., 2000: Las serpientes peruanas y sus venonos. Universidad Nacional, Peru, Depósito Legal de Acuerdo al Art. 23 del D.S. No. 017-98-Ed de la Ley de Depósito Legal vigente, 150102000-1091, printed by: Centro de Producción Editorial de la Universidad Nacional Mayor de San Marcos. Abril, 2000.

Yarrow, H., 1875: Report upon the collections of batrachians and reptiles made in portions of Nevada, Utah, California, Colorado, New Mexico, and Arizona, during the years 1871, 1872, 1873, 1874: pp. 509–584. *In* G. M. Wheeler, Report upon geographical and geological explorations and surveys west of the one hundreth meridian. Vol. V, Zoology, Engineer Dept., United States Army, Washington, D.C.

Yarrow, H., 1882: Checklist of North American reptilia and batrachia with catalogue of specimens in the U.S. National Museum. Bull. U.S. Natl. Mus., 24: i -vi +1–249.

Yarrow, H., 1883a: Check list of North American reptiles and batrachia, based on specimens contained in the United States National Museum. Smithsonian Misc. Coll., (517): 1–28.

Yarrow, H., 1883b: Description of new species of reptiles in the United States National Museum. Proc. U.S. Natl. Mus., 6: 152–154.

Yingling, P. R., 1982: *Lichanura, L. trivirgata*. Cat. Amer. Amph. Rept., (294): 1–2.

Yoon, C. K., 1992: Hawaii squirms over threat of Guam brown tree snake. Denver Post.

Young, R., and W. Morgan, 1980: The Navajo language—a grammar and colloquial dictionary. Univ. New Mexico Press, Albuquerque.

Young. R., and W. Morgan, 1992: Analytical lexicon of Navajo. Univ. New Mexico Press, Albuquerque.

Yuki, R. N., 1994: Sobre *Helicops danieli* Amaral, 1937, com a descrição do hemipênis (Serpentes, Colubridae, Xenodontinae). Boll. Mus. Paraense Emilio Goeldi, Ser. Zool., 10 (2): 203–209.

Yuki, R. N., 1997a: Geographic distribution —*Bothrops neuwiedi* (Neuwied's Lancehead). Herpetol. Rev., 28 (3): 158.

Yuki, R. N., 1997b: Geographic distribution —*Hydrops triangularis* (Triangle Water Snake). Herpetol. Rev., 28 (1): 52.

Yuki, R. N., 1999: Geographic distribution —*Taenophallus brevirostris*. Herpetol. Rev., 30 (1): 55.

Yuki, R. N., and O. V. Castano, [1998] 1999: Geographic distribution note of the water-snake *Helicops danieli* Amaral, 1937 (Colubridae: Xenodontinae). The Snake, [1998] 28 (1–2): 90–92.

Yuki, R. N., and R.M de Santos, (1996)

1998: Snakes from Marajo and the Mexicana Isls, Para State Brazil. Boll. Mus. Paraense Emilio Goeldi, Ser Zool., 12 (1): 41–53.

Yuki, V. L., Ferreira, 1993: Realocação genérica de *Xenodon werneri* Eiselt, 1963 (Serpentes: Colubridae). Com. Mus. Ci. PUCRS. Porto Alegre, (53): 39–47.

Zaher, H., 1996: A new genus of Pseudoboinae snake, with a revision of the genus of *Clelia* (Serpentes, Xenodontinae). Boll. Mus. Reg. Sci. Nat. (Turin), 14: 289–337.

Zaher, H., 1999: Hemipenial morphology of the South American xenodontine snakes, with a proposal for a monophyletic xenodontinal and a reappraisal of Colubrid hemipenes. Bull. Amer. Mus. Nat. Hist, (240): 1–168.

Zaher, H., and U. Caramaschi, 1992: Sur le statut taxinomique d'*Oxyrhopus trigeminus* et *O. guibei* (Serpentes, Xenodontidae). Bull. Mus. Natl. Hist., Paris, 4 (Sér. 14): 805–827.

Zaher, H., and U. Caramaschi, 1996: Geographic distribution—*Hydrops triangularis* (Triangle Water Snake), Herpetol. Rev., 27 (4): 212.

Zaher, H., and L. C. Prudente, 1999: Intraspecific variation of the hemipenis in *Siphlophis* and *Tripanurgos*. J. Herpetol., 33 (4): 698–702.

Zamprogno, C., 1997: Geographic distribution—*Uromancerina ricardinii* (Cobracipó, São Paulo Sharp Snake). Herpetol. Rev., 28 (4): 211.

Zamprogno, C., M. Das Graças F. Zamprogno, and T. de Lema, 1998: Contribuição ao conhecimento de *Apostolepis cearensis* Gomes, 1915, serpente fossorial do Brasil (colubridae: elapomorphinae). Acta Biol. Leopoldensia, 20 (2): 207–216.

Zamudio, K. R., and H. W. Greene, 1997: Phylogeography of the Bushmaster (*Lachesis muta*: Viperidae): implications for Neotropical biogeography, systematics and conservation. Biol. J. Linnaean Soc., 62 (3): 421–442.

Zippel, K. C., J. S. Parmerlee Jr., and R. Powell, 1994: *Ialtris dorsalis*. Cat. Amer. Amph. Rept., (592): 1.

Zug, G. R., 1977: Distribution and variation of *Leptotyphlops tricolor*. Copeia, 1977 (4): 744–745.

Zug, G. R., 1993: Herpetology. An introductory biology of amphibians and reptiles. Academic Press, San Diego, California, 1993: xv + 572 pp.

Zweifel, R. G., 1954: Notes on the distribution of some reptiles in western Mexico. Herpetologica, 10 (3): 145–149.

Zweifel, R. G., 1959a: Additions to the herpetofauna of Nayarit, Mexico. Amer. Mus. Novitates, 1953: 1–13.

Zweifel, R. G., 1959b: Snakes of the genus *Imantodes* in western Mexico. Amer. Mus. Novitates, (1961): 1–18.

Zweifel, R. G., 1959c: The provenance of reptiles and amphibians collected in western Mexico by J. J. Major. Amer. Mus. Novitates, (1949): 1–9.

Zweifel, R. G., 1973: Reptiles and amphibians. Amer. Mus. Novitates, 1973: 434–439.

Zweifel, R. G., 1974: *Lampropeltis zonata*. Cat. Amer. Amph. Rept., (174): 1–4.

Zweifel, R. G., and K. S. Norris, 1955: Contribution to the herpetology of Sonora, Mexico: descriptions of new subspecies of snakes (*Micrurus euryxanthus* and *Lampropeltis getulus*) and miscellaneous collecting notes. Amer. Midl. Natutalist, 54 (1): 230–249.

WEBSITE REFERENCES

Achaval, F., 1999: Reptiles del Uruguay. Mod. 20-XII-1999, Sección Zoología Vertebrados, zvert.feien.edu.uy//reptiles.html

Arassari Trek, 1999: arassari.co/5.h9dayl.html carivero@telcel.net.ve

Chapter IV-analysis of the National Park System, 1999: www.oas.org/usde/publications/Unit/oea51e/ch06/html

Cherno, K. B., 1999: http://www4.nesu.edu/~Kbcherno/index.html

Contemporary Herpetology http://research.calacademy.org/herpetology/herpdocs/ch/2000/2/index.htm

Finnish Museum of Natural History, 2001: Untitled—http://www.fmnh.helsinki.fi/users/haaromo/Metazoa/Deuterostoma/Chordata/Reptilia/lepidosauromorpha/pythonomorpha/Serpentesfamilies.htmpuuttuu!

Hardenbrook, D. B., 2000: ¿A que vicne todo ese "Cuchicheo" sobre los reptiles venosos de Nevada? http://www.extension.unr.edu/venomousreptiles/reptilesvenonosos.html

Hedges, S. B., 1998: Checklist of West Indian amphibians and reptiles. Caribherp Website, http//carib.bio.psu.edu/caribherp/lists/wi-list.html.

http://www.argenet.com.ar/~reserva/Fauna.html

http://www.cariari.ucr.ac.cr/~icpucr/sernoven.html

Collins, J., http://eagle.cc.ukans.edu/~cnaar/serpentes.html.

http://www.cucba.udg.mx/es/paginter/apeVanimales_de_jalisco_con_veneno.html: Animales de Jalisco con veneno mortal al hombre

Denpewolf, C., 11.10.1999: http://www.claus-dempewolf.de/Schlangenvereichnis

Jogler, R. L. and P. Burrowes, 1999: Lista de reptiles que habiten en Puerto Rico. Recinto de Río Piedras Dept. De Biología, UPR. Colegio Universitario de Cayey, home.coqui.net/emart/Espanol/reptil.html.

Johnson, H., 2000: Zolque:San Miguel Chimalapa Soke. http://www.albany/maldp/mig/html.

Klallam Classified Word List, 1999: http://www.ling.unt.edu/~montler/klallam/wordlist/intex/html

Lamar, W. W., 1999: A checklist with common names of the reptiles of the Peruvian lower Amazon. Amazon Reptile List Website, http://www.greentracks.com/RepList.html, 11/28/99.

Leenders, T., Oct. 18, 1999: Checklist of the amphibians and reptiles of Rara Avis, Costa Rica. http://www.rara-avis.com/herplist.html

Leman, W., 1999: Chyenne/English online dictionary. wleman@earthlink.net

Mastroscello, N., 2000: Salvar a la laguna de Roche. www.ambiente-ecologico.com/revist27/rocha27.html

McKinney, S., 1997: Potawatomi dictionary. In Potawatomi Web in Internet.

Meyer, E., 1999: The herpetofauna of Genesis II in the cloud forest of Costa Rica. wwww.hotels.co.cr/genesis/herpt.html

Nicolai, Renata, 2001: Vocabularios e dicionários de língues indíguenas brasileiras-vocabularies and dictionaries of Brazilian Indians language. Http://orbita.starmedias.com/~i.n.d.o.s/menu.html.

Ormsby L., Harold, 2000: Centro de investigaciones y estudios superiores en antropologia social (CIESAS). Mexico. ormsby@servidor.unam.mx

Petrucci, V. A., 2001: Indigenous Languages 2000. Brazil. http://www.geocities.com/indian languages_2000/equatorial.htm.

Bibliography

Revista Digital Universidad: http://www.revista.unam.htl

Saanich, 1999: http://www.cos.unt.edu~montler/Saanich/wordlist/index/html

Salazar, M. and S. Chapman, 1999: Paumarí-Portuguese Bilingual Dictionary: Diccionário Paumarí. http://www.sil.org/americas/brasil/dictgram/pmdicpor.htm.

San Diego Natural History Museum, July 2001: Checkliast-reptiles of Baja California and nearby islands in the Gulf of California and Pacific Ocean. Http://www.sdnhm.org/research/herpetology/bajarept.html.

Schwaller, J. R., 1999:Nahuatl gateway. http://www.umt.edu/history/Nahuatl

Sigala Rodríguez, J., 2000: http://www.geocities.com/jjsigala

Uetz, P., 2000: The EMBL Reptile Database. The EMBL = European Molecular biology Laboratory database maintained at the University of Washington by Peter Uetz, created Nov. 10, 1995, most recent update 17 March, 2000, htt;://www.embl-heidelberg.de/~uetz/LivingReptiles.html. Copyrighted by Peter Uetz & EMBL Heidelberg.

Walker, G., 2000: Indigenous languages website. Nativelit@earthlink.net

Wisdom, C., 1950: Chortí Maya Dictionary. *In* B. Stross website from microfilm of handwritten notes deposited at the University of Chicago by Charles Wisdom. The dictionary was transcribed and transliterated by Brian Stross, Department of Anthropology, University of Texas, Austin. http://www.utexas.edu/courses/stross/chorti/index/html.

www.naherpetology.org

Index

Abaco Island Boa, 44
abacura, 129
abnorma, 154
abyssus, 7, 332
Acapulco Coral Snake, 291
accedens, 125
acrantophis, 38, 39
acricus, 110
Acrochordoids, 37
acuta, 268
adamanteus, 320, 345
Adelophis, 58, 59
Adelophis copei, 59
Adelophis foxi, 59
Adelphicos, 59
Adelphicos daryi, 59
Adelphicos ibarrorum, 59
Adelphicos latifasciatum, 59
Adelphicos nigrilatum, 60
Adelphicos quadrivirgatum, 60
Adelphicos quadrivirgatum newmanorum, 60
Adelphicos quadrivirgatum quadrivirgatum, 60
Adelphicos quadrivirgatum sargii, 60
Adelphicos quadrivirgatum visoninum, 60
Adelphicos sargii, 60
Adelphicos veraepacis, 60
Adelphicos visoninus, 60
Adorned Graceful Brownsnake, 219
adspersus, 62
advanced snakes, 8, 37, 58
aemula, 222, 235
aeneus, 89, 193
aequalis, 281, 282
aequicinctus, 177
aequifasciatus, 196
Aesculapian False Coral Snake, 127
aesculapii, 127, 128
aestiva, 200
aestivus, 8, 10, 192

affinis, 20, 120, 145, 208, 218, 243, 244, 292
agalma, 156
agametus, 45
agassizii, 211
Agkistrodon, 306
Agkistrodon bilineatus, 6, 306
Agkistrodon bilineatus bilineatus, 306
Agkistrodon bilineatus howardgloydi, 306
Agkistrodon bilineatus lemasespinali, 6
Agkistrodon bilineatus russeolus, 306
Agkistrodon bilineatus taylori, 307
Agkistrodon contortrix, 306
Agkistrodon contortrix contortrix, 306
Agkistrodon contortrix laticinctus, 306
Agkistrodon contortrix mokasen, 306
Agkistrodon contortrix phaeogaster, 307
Agkistrodon contortrix pictigaster, 307
Agkistrodon piscivorus, 307
Agkistrodon piscivorus conanti, 307
Agkistrodon piscivorus leucostoma, 307
Agkistrodon piscivorus piscivorus, 307
Agkistrodon taylori, 307
aglaeope, 292
Agouti Snake, 93
Aguanaval Watersnake, 186
agyrtes, 147
Ahaetulla, 161
Ahaetulla modesta, 163
aidae, 314
ailurus, 46
ainictum, 73
Alameda Striped Racer, 180
alasukai, 78, 132
Alausí Centipede Snake, 246
albiceps, 246
albicinctus, 289
albifrons, 20, 21, 27, 112, 347
albipunctus, 21
albirostris, 17
albiventris, 135, 167

albocarinata, 311
albonuchalis, 227
albuquerquei, 75
Albuquerque Tellurian Snake, 75
Alethinophidia, 3, 5, 8, 35, 37, 58
alienus, 292
alleni, 67, 217, 289
Allen's Racer, 67
Almada Legion Snake, 165
almadensis, 165
almawebi, 312
alphonsehogei, 75
alpinus, 262
Alsophis, 61, 72, 165, 200
Alsophis anguistilineatus, 203
Alsophis anomalus, 61
Alsophis antiguae, 61
Alsophis antiguae antigua, 61
Alsophis antiguae sajdaki, 61
Alsophis antillensis, 61
Alsophis antillensis antillensis, 61
Alsophis antillensis danforthi, 61
Alsophis antillensis manselli, 62
Alsophis antillensis sanctorum, 62
Alsophis antillensis sibonius, 62
Alsophis ater, 62
Alsophis biserialis, 62
Alsophis biserialis biserialis, 62
Alsophis biserialis dorsalis, 62
Alsophis biserialis occidentalis, 62
Alsophis cantherigerus, 62
Alsophis cantherigerus adspersus, 62
Alsophis cantherigerus brooksi, 63
Alsophis cantherigerus cantherigerus, 63
Alsophis cantherigerus cayamus, 63
Alsophis cantherigerus fuscicaudus, 63
Alsophis cantherigerus pepei, 63
Alsophis cantherigerus ruttyi, 63
Alsophis cantherigerus schwartzi, 63
Alsophis elegans, 63

Alsophis elegans elegans, 63
Alsophis elegans rufidorsatus, 63
Alsophis inca, 203
Alsophis melanichnus, 64
Alsophis portoricensis, 64
Alsophis portoricensis anegadae, 64
Alsophis portoricensis aphantus, 64
Alsophis portoricensis nicholsi, 64
Alsophis portoricensis portoricensis, 64
Alsophis portoricensis prymnus, 64
Alsophis portoricensis richardi, 64
Alsophis portoricensis variegatus, 64
Alsophis rijersmai, 64
Alsophis rufiventris, 64
Alsophis sancticrucis, 65
Alsophis utowanae, 65
Alsophis vudii, 65
Alsophis vudii aterrimus, 65
Alsophis vudii picticeps, 65
Alsophis vudii raineyi, 65
Alsophis vudii utowanae, 65
Alsophis vudii vudii, 65
alta, 186
alterna, 150, 151, 153
alternatus, 184, 312
alticola, 246, 311
alticolus, 168
altirostris, 289, 295
alvarezi, 43, 100
amabilis, 109, 110, 333
amarali, 39, 182, 348
Amaral's Boa Constrictor, 39
Amaral's Legion Snake, 348
Amaral's Lizard-eater, 182
Amastridium, 65
Amastridium sapperi, 66
Amastridium veliferum, 65, 66
Amastridium veliferum sapperi, 66
Amastridium veliferum veliferum, 66
amaura, 154
Amazon Basin Blindsnake, 31
Amazon Basin Tellurian Snake, 82
Amazon Bird-eating Snake, 215
Amazon Coffee Snake, 190
Amazon Dwarf Boa, 57
Amazon Egg-eating Snake, 116
Amazon False Coral Snake, 196
Amazon False Pit Viper, 285
Amazonian Aquatic Coral Snake, 303
Amazonian Banded Tree Snake, 223
Amazonian Big-headed Snail-eater, 113
Amazonian Bushmaster, 334
Amazonian Coral Snake, 302
Amazonian Forest Pit Viper, 310
Amazonian Short-nosed Snake, 216
Amazonian Snail-eater, 113
Amazonian Threadsnake, 22

Amazonian Toadheaded Pit Viper, 312
amazonicus, 171
Amazon Legion Snake, 166
Amazon Moss Snake, 277
Amazon Scarletsnake, 211
Amazon Tree Boa, 42
Amazon Tree Snake, 150
Amazon Watersnake, 143
Amazon Wormsnake, 69
Ambergris Cay Dwarf Boa, 55
ambiniger, 68
American Burrowing Snake, 59
American Cat-eyed Snake, 157
American Copperhead, 306
American Coral Snake, 289
American Egg-eating Snake, 116
American Gartersnake, 258
American House Snake, 256
American Lancehead Pit Viper, 312
American Large-eyed Snake, 257
American Pipesnake, 38
American Ribbonsnake, 258
American Slug-eating Snake, 112
American Snail-eating Snake, 112
American Vinesnake, 193
ammodytoides, 313
amoenus, 88
amonena, 120
amphisticha, 268
amplinotus, 86
anachoreta, 218
Anaconda, 47
anahuacus, 332
Anchor Coral Snake, 289
ancoralis, 289
anctorum, 38
Andean Black-backed Coral Snake, 287
Andean Lancehead Pit Viper, 313
Andean Milksnake, 154
Andean Racer, 63
Andean Red-tailed Coral Snake, 299
Andean Tellurian Snake, 81
Andean Whipsnake, 94
andesiana, 154
andianus, 313
andinista, 246
andinus, 75, 166
andreae, 66
andrei, 347
andresensis, 100
andrewsi, 279, 281
androsi, 54
Andros Island Dwarf Boa, 54
anegadae, 64
Angel de la Guardia Rattlesnake, 328
angelensis, 6, 328
Angry Legion Snake, 173

Anguilla Bank Racer, 64
angulatus, 138
angulifer, 43
angustilineatus, 203
angustirostris, 263
Aniliidae, 10, 11, 37
Anilioid, 37
Anilius, 37, 38
Anilius scytale, 38
Anilius scytale phelpsorum, 38
Anilius scytale scytale, 38
annae, 29
annectens, 208, 265, 274, 346
annellatus, 279, 290
annulata, 92, 154, 157, 158, 252, 276, 344, 349
Annulated Coral Snake, 290
Annulated Tree Boa, 41
annulatus, 41, 227, 276
annulifera, 229, 275
anocularis, 132
anomala, 239
Anomalepididae, 15
anomalepis, 237
Anomalepis, 16
Anomalepis aspinosus, 16
Anomalepis colombia, 16
Anomalepis flavapices, 16
Anomalepis mexicanus, 16
anomalus, 61, 116, 166, 299
Anomochilus, 37
anops, 17, 18
anthicus, 99
anthonyi, 109, 110, 178
Anthony's Ring-necked Snake, 110
anthracinus, 21
anthracops, 230
Antigua Coral Snake, 291
antiguae, 61
Antiguan Racer, 61
Antillean Racer, 66
Antillean Snake, 66
antillensis, 61
Antillophis, 66, 165, 200
Antillophis andreae, 66
Antillophis andreae andreae, 66
Antillophis andreae melopyrrha, 66
Antillophis andreae morenoi, 66
Antillophis andreae nebulatus, 66
Antillophis andreae orientalis, 67
Antillophis andreae peninsulae, 67
Antillophis parvifrons, 67
Antillophis parvifrons alleni, 67
Antillophis parvifrons lincolni, 67
Antillophis parvifrons niger, 67
Antillophis parvifrons paraniger, 67
Antillophis parvifrons parvifrons, 67

Index

Antillophis parvifrons protenus, 67
Antillophis parvifrons rosamondae, 67
Antillophis parvifrons stygius, 67
Antillophis parvifrons tortuganus, 67
Antillophis slevini, 68
Antillophis slevini slevini, 68
Antillophis slevini steidachneri, 68
Antillophis steindachneri, 68
Antioquia False Rhadinaea, 226
Antioquian Coral Snake, 294
Antioquia Tellurian Snake, 75
antioquiensis, 226, 294
antoni, 224
aphantus, 64
apiatus, 292
Apostolepis, 68, 70, 124, 197
Apostolepis ambiniger, 68
Apostolepis arenarius, 68
Apostolepis assimilis, 68
Apostolepis barrioi, 69
Apostolepis borellii, 69
Apostolepis breviceps, 68
Apostolepis cearensis, 68
Apostolepis coronata, 70, 124
Apostolepis coronatus, 124
Apostolepis dimidiata, 68
Apostolepis dorbignyi, 69
Apostolepis flavotorquata, 69, 70
Apostolepis gaboi, 69
Apostolepis goiasensis, 69
Apostolepis goyasensis, 69
Apostolepis intermedia, 69
Apostolepis longicaudata, 69
Apostolepis multicincta, 69
Apostolepis niceforoi, 69
Apostolepis nigrolineata, 70
Apostolepis nigroterminata, 70
Apostolepis phillipsi, 70
Apostolepis polylepis, 70
Apostolepis pymi, 69, 70
Apostolepis quinquelineata, 70
Apostolepis quirogai, 71
Apostolepis rondoni, 70
Apostolepis sanctaeritae, 71
Apostolepis tenuis, 71
Apostolepis ventrimaculata, 69
Apostolepis villaricae, 69
Apostolepis vittata, 71
Aquatic Coral Snake, 303
Aquatic Gartersnake, 258
aquaticus, 258
aquilonaris, 134
aquilus, 320
arangoi, 75
Arboreal Bird Snake, 214
arcifera, 154
arcosae, 336

arenarius, 68
arenicola, 71, 206, 239
argaleus, 18
argenteus, 192, 285, 286
Argentine Boa Constrictor, 40
Argentine Coral Snake, 302
Argentine False Coral Snake, 196
Argentine False Pit Viper, 213
Argentine Hog-nosed Snake, 176
Argentine Large-eyed Snake, 256
Argentine Legion Snake, 173
Argentine Painted Lancehead Pit Viper, 317
Argentine Racer, 203
Argentine Rainbow Boa, 43
Argentine Tellurian Snake, 76
argentinus, 174
argus, 214, 230
argusiformis, 237
Arid Land Ribbonsnake, 262
arizona, 50
Arizona, 71
Arizona Black Rattlesnake, 333
Arizona Coral Snake, 288
arizonae, 259
Arizona elegans, 71
Arizona elegans arenicola, 71
Arizona elegans australis, 72
Arizona elegans candida, 71
Arizona elegans eburnata, 72
Arizona elegans elegans, 71
Arizona elegans expolita, 72
Arizona elegans noctivaga, 72
Arizona elegans occidentalis, 72
Arizona elegans pacta, 72
Arizona elegans philipi, 72
Arizona Glossy Snake, 72
Arizona Mountain Kingsnake, 153
Arizona occidentalis, 71
Arizona pacta, 71
Arizona Ridge-nosed Rattlesnake, 333
Arizona Rosy Boa, 50
Arizona Wandering Gartersnake, 259
armstrongi, 332
Armstrong's Dusky Rattlesnake, 332
arnaldoi, 201
Arnaldoi's Green Racer, 201
arnyi, 110, 111
Arrhyton, 61, 72, 165, 200
Arrhyton ainictum, 73
Arrhyton bivittatum, 74
Arrhyton callilaemum, 73
Arrhyton dolichurum, 73
Arrhyton exiguum, 73
Arrhyton exiguum exiguum, 73
Arrhyton exiguum stahli, 73
Arrhyton exiguum subspadix, 73

Arrhyton funereum, 72, 73
Arrhyton landoi, 73
Arrhyton polylepis, 72, 73
Arrhyton procerum, 73
Arrhyton quenselii, 116
Arrhyton supernum, 74
Arrhyton taeniatum, 74
Arrhyton tanyplectum, 74
Arrhyton vittatum, 74
Arrow-headed Coral Snake, 289
articulata, 112
Aruba Island Rattlesnake, 324
Aruban Cat-eyed Snake, 158
arubricus, 279, 281
asbolepis, 21
ashmeadi, 158
Ashmead's Cat-eyed Snake, 158
asper, 313, 314
aspinosus, 16
assimilis, 68
assisi, 43
Atacama Racer, 201
atahuallpae, 165, 226
Atahuallpa False Rhadinaea, 226
ater, 24, 62, 168
aterrimus, 65
Atlantic Forest Bushmaster, 334
Atlantic Forest False Coral Snake, 128
Atlantic Parrot Snake, 161
Atlantic Saltmarsh Snake, 186
Atlantic Snail-eater, 114
Atractaspis, 305
Atractus, 74
Atractus albuquerquei, 75
Atractus alphonsehogei, 75
Atractus andinus, 75
Atractus arangoi, 75
Atractus badius, 75, 78
Atractus balzani, 76
Atractus biseriatus, 76
Atractus bocki, 80
Atractus bocourti, 76
Atractus boettgeri, 76, 83
Atractus boulengeri, 76
Atractus canedii, 75, 76
Atractus carrioni, 76
Atractus clarki, 76
Atractus collaris, 76
Atractus crassicaudatus, 76, 344
Atractus darienensis, 77
Atractus depressiocellus, 77
Atractus duidensis, 77
Atractus dunni, 77
Atractus ecuadorensis, 77
Atractus elaps, 77
Atractus emigdioi, 77
Atractus emmeli, 77

Atractus erythromelas, 77
Atractus favae, 77, 344
Atractus flammingerus, 78
Atractus flammingerus flammingerus, 78
Atractus flammingerus snethlageae, 78, 83
Atractus fuliginosus, 78
Atractus gaigeae, 78
Atractus guentheri, 78
Atractus hostilitractus, 78
Atractus imperfectus, 78
Atractus indistinctus, 78
Atractus insipidus, 78
Atractus iridescens, 78
Atractus lancinii, 79
Atractus lasallei, 79
Atractus latifrons, 79
Atractus lehmanni, 79
Atractus limitaneus, 79
Atractus longimaculatus, 85
Atractus loveridgei, 79
Atractus maculatus, 79
Atractus major, 79
Atractus manizalesensis, 79, 82
Atractus mariselae, 79
Atractus melanogaster, 80
Atractus melas, 80
Atractus micheli, 75
Atractus microrhynchus, 75
Atractus modestus, 80
Atractus multicinctus, 80
Atractus nebularis, 80
Atractus nicefori, 80
Atractus nigricaudus, 80
Atractus nigriventris, 80
Atractus obesus, 80
Atractus obtusirostris, 81
Atractus occidentalis, 81
Atractus occipitoalbus, 81
Atractus oculotemporalis, 81
Atractus pamplonensis, 81
Atractus pantostictus, 81
Atractus paraguayensis, 81
Atractus paravertebralis, 81
Atractus paucidens, 81
Atractus pauciscutatus, 81
Atractus peruvianus, 81
Atractus poeppigi, 82
Atractus potschi, 82
Atractus punctiventris, 82
Atractus resplendens, 82
Atractus reticulatus, 82
Atractus reticulatus paraguayensis, 81, 82
Atractus riveroi, 82
Atractus roulei, 82
Atractus sanctaemartae, 82
Atractus sanguineus, 82
Atractus schach, 82

Atractus serranus, 83, 84, 343
Atractus snethlageae, 83
Atractus steyermarki, 83
Atractus subbcinctum, 75
Atractus taeniatus, 83
Atractus taphorni, 83
Atractus torquatus, 83
Atractus trihedrurus, 83, 84, 343
Atractus trilineatus, 84
Atractus trivittatus, 84
Atractus univittatus, 84
Atractus variegatus, 84
Atractus ventrimaculatus, 84
Atractus vertebralis, 84
Atractus vertebrolineatus, 84
Atractus vittatus, 74, 343
Atractus wagleri, 85
Atractus werneri, 85
Atractus zebrinus, 85
Atractus zidoki, 85
atrata, 189
atratus, 258
atraventer, 166
atricaudatus, 326
atriceps, 246
atrocinctus, 228
Atropoides, 308, 309, 310, 311, 312, 319, 335
Atropoides nummifer, 308
Atropoides nummifer mexicanus, 308
Atropoides nummifer nummifer, 308
Atropoides nummifer occiduus, 308
Atropoides olmec, 308
Atropoides picadoi, 308
atrox, 313, 314, 320, 321
attenuata, 243
aurata, 89
auratus, 89
aurifer, 309
aurigulus, 178
Australasian Blindsnake, 28
australis, 21, 72, 171, 181, 288
Autlán Rattlesnake, 327
averyi, 289
ayarzaguenai, 19
Ayarzagüena's Similar-scaled Snake, 19

babaspul, 300
Babaspul, 300
bachmanni, 196
Bachmann's False Coral Snake, 196
badius, 75
Bahamian Dwarf Boa, 54
Bahamian Threadsnake, 22
Bahia Lancehead Pit Viper, 318
Bahia Miner Snake, 204
Bahia Snail-eater, 114
Bahia Threadsnake, 22

Bahia Wormsnake, 68
baileyi, 195, 196, 220
bairdi, 121, 225, 246
Baird's Centipede Snake, 246
Baird's Patch-nosed Snake, 225
Baird's Ratsnake, 121
Baja California Cape Threadsnake, 24
Baja California Coachwhip, 179
Baja California Gophersnake, 210
Baja California Leaf-nosed Snake, 206
Baja California Lyresnake, 272
Baja California Mountain Kingsnake, 156
Baja California Nightsnake, 127
Baja California Patch-nosed Snake, 225
Baja California Ratsnake, 85
Baja California Rattlesnake, 324
Baja California Striped Whipsnake, 178
Baja California Threadsnake, 24
Baja California Whipsnake, 178
bakeri, 158
Baker's Cat-eyed Snake, 158
bakewelli, 24
Bakewell's Threadsnake, 24
baliocoryphus, 290
Ball Python, 5
Ball Snake, 137
Balsan Coral Snake, 297
balzani, 76, 290
Balzapote Snail-eater, 231
Banana Boa, 58
Banded Calico Snake, 196
Banded Cat-eyed Snake, 157
Banded Coffee Snake, 190
Banded False Coral Snake, 252
Banded Galápagos Antillophis, 68
Banded Hog-nosed Snake, 177
Banded Rock Rattlesnake, 327
Banded Sandsnake, 90
Banded Sea Snake, 304
Banded Snail-eater, 275
Banded Snake, 223
Banded Tree Snake, 223
Banded Watersnake, 187
Bandless Sandsnake, 91
Barahona Blindsnake, 32
Barahona Threadsnake, 26
Barba Amarilla, 313
Barbados Legion Snake, 172
barbouri, 44, 47, 54, 178, 270, 319, 338
Barbour's Centipede Snake, 246
Barbour's Montane Pit Viper, 319
Barbour's Tropical Groundsnake, 270
barnetti, 315
Barnett's Lancehead Pit Viper, 315
baroni, 201
Baron's Racer, 201
Barranquilla Pygmy Coral Snake, 293

Index

Barred Forest Racer, 109
barrioi, 69, 95
Basal Alethinophidia, 8, 37, 38, 58
Basal Macrostomatans, 38
basiliscus, 321
bassleri, 143
battersbyi, 53
Battersby's Dwarf Boa, 53
baueri, 145
Bauer's Nightsnake, 145
baupertuisii, 128
Bay Islands Threadsnake, 24
Beautiful Road Guarder, 104
Beh Belle Chemin, 170
Belize Coral Snake, 297
Bellavista Snail-eater, 233
bellus, 132
bernadi, 290
bertholdi, 284
Berthold's Neotropical Brownsnake, 281
betaniensis, 132
beui, 18, 245
bicarinatus, 93
bicinctus, 142
bicolor, 52, 97, 112, 132, 171, 242, 279, 281, 309
Bicolored Mussurana, 96
Bicolored Snail-eater, 112
Bicolored Watersnake, 242
bifossatus, 182
bifoveatus, 126
Big Bend Patch-nosed Snake, 225
Big-eyed Blindsnake, 33
Big-eyed Snail-eater, 230
Big-eyed Threadsnake, 26
Big-nosed Tellurian Snake, 81
Big-scaled Threadsnake, 26
bilineata, 22, 120, 310
bilineatus, 6, 9, 22, 178, 198, 244, 306, 310
bimaculatus, 168
bimaris, 207, 209, 210
Bimini Blindsnake, 29
Bimini Boa, 46
biminiensis, 29
Bimini Island Dwarf Boa, 54
Bimini Racer, 65
binghami, 270
Bingham's Cativo, 270
bipraeocularis, 168
bipunctatus, 100
Bird-eating Tree Snake, 214
biscutatus, 271
biserialis, 62, 105
biseriatus, 76, 100
bivittatum, 74, 108
bizonus, 128
Black and White Earthsnake, 135

Black and White Racer, 66
Black-backed Coral Snake, 287
Black-backed Tellurian Snake, 82
Black-banded Cat-eyed Snake, 159
Black-banded Earthsnake, 136
Black-banded Snake, 228
Black-bellied Centipede Snake, 250
Black-bellied Dwarf Boa, 56
Black-bellied Legion Snake, 166
Black-bellied Tellurian Snake, 80
Black Blindsnake, 29
Black-chinned Watersnake, 140
Black-collared Snake, 116
Black Crown Burrowing Snake, 124
Black-headed Amazonian Coral Snake, 302
Black-headed Bushmaster, 334
Black-headed Calico Snake, 195
Black-headed Coral Snake, 290
Black-headed Lyre Snake, 273
Black-headed Lyresnake, 273
Black-headed Snake, 245, 246
Black-headed Stripeless Snake, 102
Black Hills Red-bellied Snake, 241
Black Kingsnake, 152
Black-masked Racer, 99
Black-naped Forest Racer, 108
Black-necked Amazonian Coral Snake, 302
Black-necked Garter Snake, 259
Black-necked Gartersnake, 259
Black Pinesnake, 209
Black Ratsnake, 123
Black-ringed Threadsnake, 21
Black-skinned Parrot Snake, 161
Black-speckled Palm Pit Viper, 309
Black Speckled Racer, 119
Black-striped Snake, 100
Black Swampsnake, 228
Black-tailed Boa, 56
Black-tailed Cribo, 117
Black-tailed Golden Snake, 215
Black-tailed Horned Pit Viper, 335
Black-tailed Rattlesnake, 328
Black-tailed Tellurian Snake, 80
Black Threadsnake, 23
blairi, 151
blanchardi, 111, 132, 155, 164, 188
Blanchard's Earthsnake, 132
Blanchard's Milksnake, 155
Blindsnake, 3, 5, 8, 10, 15, 28, 29, 58
blombergi, 41
Blomberg's Annulated Tree Boa, 41
Blood Snake, 237, 238
Blotched Amazonian False Water Cobra, 142
Blotched Coral Snake, 290

Blotched Hook-nosed Snake, 130
Blotched Racer, 119
Blotched Tellurian Snake, 78
Blotched Watersnake, 187
Blue Parrot Snake, 161
Blue Racer, 99
Blue-striped Gartersnake, 266
Blue-striped Mexican Gartersnake, 260
Blue-striped Ribbonsnake, 264
Blunt-headed Tree Snake, 148
Boa, 6, 38, 39
Boa constrictor, 4, 39
Boa Constrictor, 39
Boa constrictor amarali, 39
Boa constrictor constrictor, 39
Boa constrictor imperator, 39, 40
Boa constrictor longicauda, 40
Boa constrictor melanogaster, 39, 40
Boa constrictor mexicana, 39
Boa constrictor nebulosus, 40
Boa constrictor occidentalis, 40
Boa constrictor orophias, 40
Boa constrictor ortonii, 40
Boa constrictor sabogae, 40
Boa diviniloquax var. *mexicana*, 39
Boa nebulosa, 40
Boa nebulosus, 40
Boas, 3, 4, 8, 9, 37, 38
bocki, 80
bocourti, 76, 161, 246, 290
Bocourt's Black-headed Snake, 246
Bocourt's Parrot Snake, 161
Bocourt's Red-backed Coffee Snake, 191
Bocourt's Snail-eater, 116
Bocourt's Tellurian Snake, 76
boddaerti, 183
Boddaert's Lizard-eater, 183
Boella tenella, 45
boettgeri, 24, 76, 83, 112
Boettger's Sipo, 94
Boettger's Snail-eater, 112
Boettger's Tellurian Snake, 76
bogerti, 186, 256, 290
Bogertophis, 85, 121, 228
Bogertophis rosaliae, 85
Bogertophis subocularis, 85
Bogertophis subocularis amplinotus, 86
Bogertophis subocularis subocularis, 85
bogertorum, 218
Bogert's Centipede Snake, 253
Bogert's Coral Snake, 290
Boid, 38, 52, 57
Boidae, 38
Boidae *auctorum*, 38
Boids, 10, 11
Boiga, 4, 6, 86
Boiga irregularis, 4, 6, 86

Boilla, 58
Boinae, 38
boipiranga, 246
Boiruna, 86, 96, 194, 211
Boiruna maculata, 87
Boiruna sertaneja, 87
Bolivar Legion Snake, 167
boliviana, 201, 243
Bolivian Aesculapian False Coral Snake, 128
Bolivian Amazonian Coral Snake, 302
Bolivian Black-backed Coral Snake, 287
Bolivian Boa Constrictor, 39
Bolivian Coral Snake, 290
Bolivian Eel Snake, 212
Bolivian Lancehead Pit Viper, 317
Bolivian Mudsnake, 144
Bolivian Parrot Snake, 161
Bolivian Racer, 201
Bolivian Slender Snake, 243
Bolivian Tellurian Snake, 76
Bolivian Triad Coral Snake, 295
bolivianus, 144, 161, 290, 317
Bolyeriidae, 38
Booid, 37
borapeliotes, 22
bordensis, 18
Border Tellurian Snake, 78
borealis, 164
borellii, 69, 203
Borelli's Racer, 203
borrichianus, 22
boshelli, 108, 345
bostici, 50
Bothriechis, 308, 309, 310, 311, 312, 319, 335
Bothriechis aurifer, 309
Bothriechis bicolor, 309
Bothriechis lateralis, 309
Bothriechis marchi, 309
Bothriechis nigroviridis, 309
Bothriechis rowleyi, 309
Bothriechis schlegeli, 309
Bothriechis supraciliaris, 309
Bothriechis thalassinus, 7
Bothriopsis, 308, 309, 310, 311, 312, 319, 335
Bothriopsis albocarinata, 311
Bothriopsis alticola, 311
Bothriopsis bilineata, 310
Bothriopsis bilineata bilineata, 310
Bothriopsis bilineata smaragdinae, 310
Bothriopsis mahnerti, 311
Bothriopsis medusa, 310
Bothriopsis oligolepis, 310
Bothriopsis osbornei, 311
Bothriopsis peruviana, 310
Bothriopsis pulchra, 310
Bothriopsis punctata, 311

Bothriopsis taeniata, 311
Bothriopsis taeniata lichenosa, 311
Bothriopsis taeniata taeniata, 311
Bothrocophias, 308, 309, 310, 311, 312, 335
Bothrocophias campbelli, 312
Bothrocophias hyoprora, 312
Bothrocophias microphthalmus, 312
Bothrocophias myersi, 312
Bothrops, 308, 309, 310, 311, 312, 319, 335, 349
Bothrops alternatus, 312
Bothrops ammodytoides, 313
Bothrops andianus, 313
Bothrops andinus asper, 313
Bothrops asper, 313, 314
Bothrops atrox, 313, 314
Bothrops atrox aidae, 314
Bothrops atrox asper, 313
Bothrops atrox atrox, 314
Bothrops atrox nacaritae, 315
Bothrops atrox xanthogrammus, 314
Bothrops barnetti, 315
Bothrops bilineatus, 310
Bothrops brazili, 315
Bothrops caribbaeus, 315
Bothrops castelnaudi, 311
Bothrops castelnaudi castelnaudi, 311
Bothrops castelnaudi lichenosus, 311
Bothrops colombianus, 315
Bothrops colombiensis, 314
Bothrops cotiara, 315
Bothrops erythromelas, 315
Bothrops fonsecai, 315
Bothrops iglesiasi, 316
Bothrops insularis, 316
Bothrops isabelae, 313, 314
Bothrops itapetinigae, 316
Bothrops jararaca, 311, 316
Bothrops jararacussu, 316
Bothrops jonathani, 316
Bothrops lanceolatus, 316
Bothrops lanceolatus aidae, 314
Bothrops lanceolatus lanceolatus, 316
Bothrops lanceolatus nacaritae, 315, 316
Bothrops lansbergii, 349
Bothrops leucurus, 316
Bothrops lojanus, 317
Bothrops marajoensis, 314, 317
Bothrops medusa, 310
Bothrops microphthalmus, 315
Bothrops moojeni, 317
Bothrops neuwiedi, 317
Bothrops neuwiedi bolivianus, 317
Bothrops neuwiedi diporus, 317
Bothrops neuwiedi goyazensis, 317
Bothrops neuwiedi lutzi, 317
Bothrops neuwiedi mattogrossensis, 317

Bothrops neuwiedi meridionalis, 318
Bothrops neuwiedi neuwiedi, 317
Bothrops neuwiedi paranaensis, 318
Bothrops neuwiedi pauloensis, 318
Bothrops neuwiedi piauhyensis, 318
Bothrops neuwiedi pubescens, 318
Bothrops neuwiedi urutu, 318
Bothrops nigroviridis, 349
Bothrops pictus, 318
Bothrops pifanoi, 318
Bothrops pirajai, 318
Bothrops pradoi, 314, 318
Bothrops roadingeri, 318
Bothrops sanctaecrucis, 318
Bothrops taeniatus, 311
Bothrops venezuelae, 319
Bothrops venezuelensis, 319
bottae, 48
boulengeri, 53, 76, 192, 286
Boulenger's Nightsnake, 145
Boulenger's Red-bellied Swamp Legion Snake, 169
Boulenger's Snail-eater, 113
Boulenger's Tellurian Snake, 76
Boulenger's Tree Snake, 233
boursieri, 226
Boursier's False Rhadinaea, 226
bovallii, 223, 261
Box-headed Snail-eater, 114
brachycephalus, 133, 346
brachystoma, 258
brachyurus, 175
Brahminy's Blindsnake, 28
Braided-back Watersnake, 242
braminus, 28
brasiliensis, 22, 137, 291
brazili, 119, 218, 315
Brazilian Ball Snake, 137
Brazilian Bird Snake, 218
Brazilian Boa, 42
Brazilian Burrowing Snake, 124
Brazilian Calico Snake, 196
Brazilian Cipó, 94
Brazilian Coral Snake, 291
Brazilian Dwarf Boa, 57
Brazilian False Water Cobra, 142
Brazilian Forest Snake, 236
Brazilian Giant Coral Snake, 295
Brazilian Green Legion Snake, 175
Brazilian Green Racer, 201
Brazilian Green Tree Snake, 201
Brazilian Lancehead Pit Viper, 316
Brazilian Liana Snake, 234
Brazilian Mussurana, 97
Brazilian Rainbow Boa, 43
Brazilian Ribbon Coral Snake, 297
Brazilian Short-tailed Coral Snake, 291

Brazilian Slug-eater, 232
Brazilian Speckled Racer, 201
Brazilian Striped Racer, 203
Brazilian Tropical Snake, 120
Brazilian Watersnake, 139
Brazilian Yellow-lined Snake, 216
Brazil's Glossy Racer, 119
Brazil's Lancehead Pit Viper, 315
Brazos River Watersnake, 139, 187
bressoni, 22, 160
Bresson's Splendid Cat-eyed Snake, 160
brevicauda, 247
breviceps, 68, 166
brevifacies, 112, 345
brevirostris, 120, 193, 215, 216, 218, 244
brevis, 230, 254, 347
brevissima, 254
brevissimus, 22
breviventris, 287
briggsi, 247
Briggs's Centipede Snake, 247
British Virgin Islands Racer, 64
British Virgins Blindsnake, 30
Broad-banded Copperhead, 306
Broad-banded Dwarf Boa, 54
Broad-banded Watersnake, 187
Broad-ringed Coral Snake, 297
Broad-striped Groundsnake, 74
Broken-collar Graceful Brownsnake, 222
Broken-ringed Earthsnake, 135
brongersmai, 29
Brongersma's Blindsnake, 29
brongersmianus, 29
Bronze-striped Parrot Snake, 163
brooksi, 63, 152
Brown-banded Watersnake, 138
Brown Bush Snake, 116
Brown Centipede Snake, 246
Brown-chinned Racer, 99
Brown Forest Racer, 108
Brown Hook-nosed Snake, 130
browni, 205, 291
Brown-lined Lizard-eater, 182
Brown Parrot Snake, 162
Brown Racer, 147
Brown's Coral Snake, 291
Brown-spotted Mock Viper, 269
Brown-spotted Night Viper, 269
Brown-striped Whipsnake, 94
Brown Tellurian Snake, 83
Brown Tree Snake, 86
Brown Vinesnake, 193
Brown Watersnake, 189
bruesi, 183
brunneum, 108
brunneus, 338
bucculentus, 56

bucephala, 114
Buck Island Racer, 64
Bullsnake, 5, 207, 208
burmeisteri, 203
Burrowing Night Snake, 211
Burrowing Snake, 15, 68
Bushmaster, 334
butleri, 258
Butler's Gartersnake, 258
Buttermilk Racer, 99

Caatinga Coral Snake, 297
Caatinga Lancehead Pit Viper, 315
Caatingas Green Legion Snake, 176
Caatingas Legion Snake, 176
Caatingas Rainbow Boa, 43
Caatinga Threadsnake, 22
Cabrit Island Ground Boa, 45
Caenophidia, 3, 8, 37, 38, 58
caesius, 172, 173
cahuilae, 24
Caissaca, 317
Calabar Burrowing Python, 48, 49
Calabaria, 8, 48, 49
Calamaria, 344
Calamaria favae, 344
calamarina, 247, 343
calamitus, 242
Calamodon, 87
Calamodontophis, 87
Calamodontophis paucidens, 87
Calico False Coral Snake, 195
Calico Snake, 194
california, 152
California Glossy Snake, 72
California Kingsnake, 152
California Lyresnake, 272
California Mountain Kingsnake, 156
California Nightsnake, 146
California Rosy Boa, 51
California Striped Racer, 180
californica, 348
caliginis, 7, 333
calligaster, 151, 219
callilaemum, 73
calliliaemus, 72
callostictus, 143
calypso, 22
campbelli, 155, 312
canaima, 166
Canasi Dwarf Boa, 54
cancellatus, 133
candida, 71
Canebreak Rattlesnake, 326
canedii, 75, 76
canellei, 18
canescens, 261

caninus, 5, 42
cantherigerus, 62
Cantil, 6, 306
canula, 254
canum, 138
canus, 53, 54
Cape Gartersnake, 267
Cape Gophersnake, 210
capensis, 346
Cape San Lucas Racer, 178
Cape Whipsnake, 178
capistra, 247
capistrata, 249
capitulatus, 29
Capuchin Coral Snake, 293
Caqueta Threadsnake, 22
Carabobo Common Lancehead Pit Viper, 315
caracasensis, 18
carajasensis, 166
carbonelli, 202
caribbaeus, 315
Caribbean Banded Coffee Snake, 190
Caribbean Dwarf Boa, 53
Caribbean Groundsnake, 72
Caribbean Ring-necked Coffee Snake, 190
Caribbean Watersnake, 269
Carib Coral Snake, 301
Carib False Coral Snake, 282
carinatus, 93, 192
carinicauda, 294
carinicaudus, 138, 139
carinosus, 133
Carmen Island Threadsnake, 25
carnicauda, 294
Carolina Pygmy Rattlesnake, 338
Carolina Swampsnake, 228
Carolina Watersnake, 189
Carphophis, 87, 88
Carphophis amoenus, 88
Carphophis amoenus amoenus, 88
Carphophis amoenus helenae, 88
Carphophis amoenus vermis, 88
Carphophis vermis, 88
carri, 230
carrioni, 76
Carr's Snail-eater, 230
carvalhoi, 297
Cascabel, 320
cascadae, 247
cascavella, 322
castaneus, 327
Castellana, 306
castelnaudi, 311
Castelnaud's Forest Pit Viper, 311
catalensis, 152
catalinae, 145

catalinensis, 152, 321
catamayensis, 291
Catamayo Coral Snake, 291
catapontus, 30, 32
Catastoma, 132
Catastomus, 132
catenatus, 337
catenifer, 207, 210
catesbyi, 112, 278
Catesby's Snail-Eater, 112
Cat-eyed Snake, 157
Cat Island Slender Boa, 46
Cativo, 269
Cauca Coral Snake, 300
Cauca False Rhadinaea, 226
cavalheiroi, 112
cayamus, 63
Cayemite Vinesnake, 278
Cayman Brac Blindsnake, 30
Cayman Brac Dwarf Boa, 54
Cayman Brac Racer, 63
caymanensis, 30, 54
Cayman Islands Dwarf Boa, 54
Cayman Watersnake, 270
Cayo Cantiles Racer, 66
cearensis, 68
Cecilla Threadsnake, 27
Cedros Island Diamond Rattlesnake, 325
Cedros Island Gophersnake, 208
Cedros Island Rosy Boa, 50
ceii, 166
Cei's Burrowing Snake, 198
Cei's Legion Snake, 166
celaeno, 266, 267
celaenops, 155
celata, 189
celatus, 281
celine, 54
Cemophora, 88
Cemophora coccinea, 88
Cemophora coccinea coccinea, 88
Cemophora coccinea copei, 88
Cemophora coccinea lineri, 88
cenchoa, 148
cenchria, 5, 43
Centipede-eating Snake, 246
Centipede Snake, 245, 246
Central Amazon Coral Snake, 302
Central American Banded Centipede Snake, 228
Central American Bird Snake, 214
Central American Boa Constrictor, 39
Central American Bushmaster, 335
Central American Cat-eyed Snake, 160
Central American Centipede Snake, 252
Central American Coral Snake, 300
Central American Cribo, 118

Central American Dawn Blindsnake, 16
Central American Dwarf Boa, 58
Central American Earthsnake, 132
Central American False Coral Snake, 128, 129
Central American Lizard-eater, 184
Central American Lyresnake, 272
Central American Milksnake, 156
Central American Neck-banded Snake, 227
Central American Racer, 118
Central American Ratsnake, 121
Central American Rattlesnake, 322
Central American Road Guarder, 103
Central American Slender Blindsnake, 17
Central American Speckled Racer, 118
Central American Tree Snake, 149
Central American Treesnake, 148
Central American Watersnake, 142
Central Baja California Gophersnake, 210
Central Brazilian Rattlesnake, 322
Central Brazil Tellurian Snake, 81
Central Chile Swift Andean Snake, 243
Central Florida Crowned Snake, 251
Central Galápagos Racer, 62
centralis, 180
Central Lined Snake, 274
Central Plains Milksnake, 155
Central Plateau Dusky Rattlesnake, 332
Central Plateau Pygmy Rattlesnake, 338
Central Sipo, 95
Central Texas Whipsnake, 182
Central Wormsnake, 69
cephalomaculata, 120
cephalostriata, 120
cerastes, 321
cerberus, 7
Cercal Snake, 89
cercobombus, 321
Cercophis, 89
Cercophis auratus, 89
cercostroha, 203
cereberus, 7
cereolineatus, 278
Cerrados Black Snake, 212
Cerrados Legion Snake, 171
cerralvensis, 6, 325
Cerralvo Island Long-nosed Snake, 224
Cerralvo Island Rattlesnake, 325
Cerralvo Island Sandsnake, 91
cerroensis, 190
Cerro Guaiquinima Legion Snake, 175
Cerrophidion, 308, 309, 310, 311, 312, 319, 335
Cerrophidion barbouri, 319
Cerrophidion godmani, 319
Cerrophidion petlalcalensis, 319

Cerrophidion tzolzilorum, 319
Cerro Saslaya Graceful Brownsnake, 222
Cerro Yavi Large-eyed Snake, 257
cervinus, 234
Chacoan Hog-nosed Snake, 176
Chaco Diminutive Snake, 216
chacoensis, 176
Chaco Forest Legion Snake, 175
Chaco Large-eyed Snake, 256
Chaco Legion Snake, 173
Chaco Tellurian Snake, 83
Chaco Whipsnake, 95
chalybeus, 133
chamissona, 201
chamissonis, 201
championi, 133
chaparensis, 113
Chapare Snail-eater, 113
Chapinophis, 89, 90
Chapinophis xanthocheilus, 90
Chaqueña False Coral Snake, 196
Chaqueña Miner Snake, 205
chaquensis, 256
Charina, 8, 48, 49
Charina bottae, 48
Charina bottae bottae, 48
Charina bottae umbratica, 49
Charina bottae utahensis, 49
Checkered Gartersnake, 261
Chersodromus, 90
Chersodromus liebmanni, 90
Chersodromus rubiventris, 90
Chiapan Earthsnake, 133
Chiapan Stripeless Snake, 100
Chiapas Highland Ribbonsnake, 262
Chicago Gartersnake, 266
Chiclín Threadsnake, 26
chihuahuaensis, 25, 262
Chihuahua Mexican Earthsnake, 106
Chihuahuan Black-bellied Gartersnake, 262
Chihuahuan Black-headed Snake, 253
Chihuahuan Earthsnake, 133
Chihuahuan Glossy Snake, 72
Chihuahuan Hook-nosed Snake, 138
Chihuahuan Ratsnake, 85
Chihuahua Threadsnake, 25
Chilean Racer, 201
Chilean Slender Snake, 243
Chilean Swift Andean Snake, 243
chilensis, 243
Chilomeniscus, 90
Chilomeniscus cinctus, 90
Chilomeniscus fasciatus, 91
Chilomeniscus punctatissimus, 91
Chilomeniscus savagei, 91
Chilomeniscus stramineus, 91

Index

Chilomeniscus stramineus esterensis, 91
Chilomeniscus stramineus stramineus, 91
chimanta, 256
Chimanta Large-eyed Snake, 256
Chinilla, 337
Chionactis, 91, 92
Chionactis occipitalis, 92
Chionactis occipitalis annulata, 92, 344
Chionactis occipitalis klauberi, 92
Chionactis occipitalis occipitalis, 92
Chionactis occipitalis talpina, 92
Chionactis palarostris, 92
Chionactis palarostris organica, 92
Chionactis palarostris palarostris, 92
Chionactis saxatilis, 91, 92, 344
Chironius, 92, 93
Chironius barrioi, 95
Chironius bicarinatus, 93
Chironius carinatus, 93
Chironius carinatus carinatus, 93
Chironius carinatus flavopictus, 93
Chironius carinatus spixii, 93
Chironius cinnamomeus, 93, 348, 349
Chironius exoletus, 93
Chironius flavolineatus, 94
Chironius fuscus, 94
Chironius fuscus fuscus, 94
Chironius fuscus leucometapus, 94
Chironius grandisquamis, 94
Chironius laevicollis, 94
Chironius laurenti, 94
Chironius monticola, 94
Chironius multiventris, 94
Chironius multiventris cochrane, 95
Chironius multiventris foveatus, 95
Chironius multiventris multiventris, 95
Chironius multiventris septentrionalis, 95
Chironius quadricarinatus, 95
Chironius quadricarinatus maculoventris, 95
Chironius quadricarinatus quadricarinatus, 95
Chironius schlueteri, 94
Chironius scurrulus, 95, 349
Chironius sexcarinatus, 215
Chironius vicenti, 95
Chita Tellurian Snake, 80
chlorauges, 278
chlorophae, 145
chlorophaea, 145
chloroticus, 118
Chocan Blunt-headed Tree Snake, 150
Chocó Anchor Coral Snake, 290
Chocó Centipede Snake, 250
Chocó Parrot Snake, 162
Chocoan Forest Pit Viper, 311
Chocoan Toadheaded Pit Viper, 312
chocoensis, 161, 162
chrysobronchus, 215
chrysocephalus, 258
chrysogaster, 44
chrysostomus, 171
chui, 204
cinctus, 90
cinereus, 345, 349
cingulum, 179
cinnamomea, 93, 348
cinnamomeus, 93, 348, 349
circinalis, 291, 301
cisticeps, 114
Clarion Island Nightsnake, 146
Clarion Island Whipsnake, 178
clarki, 76, 108, 143, 186, 291
Clark's Coral Snake, 291
Clark's Tellurian Snake, 76
clarus, 224
clathratus, 194
clavatus, 102
Clear-banded Coral Snake, 293
clelia, 96, 97
Clelia, 86, 87, 96, 97, 194, 211
Clelia baileyi, 196
Clelia bicolor, 97
Clelia clelia, 96, 97
Clelia clelia clelia, 87, 97
Clelia clelia groomei, 97
Clelia clelia immaculata, 97
Clelia equatoriana, 97
Clelia errabunda, 96, 97
Clelia montana, 97
Clelia occipitolutea, 86, 87
Clelia plumbea, 96, 97
Clelia quimi, 96, 97
Clelia rustica, 97
Clelia scytalina, 97
cleofae, 118
cliftoni, 184
Clifton's Lizard-eater, 184
cloelia, 96
Cloelia, 96
Cloelia cloelia, 96
Clonophis, 98, 186
Clonophis kirtlandii, 98
Clouded Leaf-nosed Snake, 206
Cloud Forest Coral Snake, 300
Cloud Forest Parrot Snake, 163
Cloud Forest Snake, 107
Cloud Forest Tellurian Snake, 80
Cloudy Snail-Sucker, 231
Coachwhip, 178, 179
Coastal Coral Snake, 290
Coastal Dunes Crowned Snake, 251
Coastal Dwarf Blindsnake, 30
Coastal House Snake, 256
Coastal Plain Centipede Snake, 252
Coastal Rosy Boa, 51
Coastal Snail-sucker, 231
Coastal Whipsnake, 95
Coast Gartersnake, 259
Coast Mountain Kingsnake, 157
Coast Patch-nosed Snake, 226
Cobel Legion Snake, 166
cobellus, 166, 345
Cobra Bola, 137
coccinea, 88
Cochliophagus, 345
Cochliophagus isolepis, 345
cochranae, 95
Cochrane's Whipsnake, 95
coeruleodorsus, 161
Coffee Blindsnake, 33
Coffee Earthsnake, 135
Coffee Snake, 189
Coiba Coral Snake, 300
coibensis, 300
Colima Clear-banded Coral Snake, 293
Colima Gartersnake, 267
collaris, 22, 76, 171, 259, 272, 287
collilineatus, 322
colombia, 16
Colombia Long-tailed Snake, 125
Colombian Annulated Tree Boa, 42
Colombian Coral Snake, 303
Colombian Dawn Blindsnake, 16
Colombian Earthsnake, 132
Colombian Forest Racer, 108
Colombian Frog-eating Snake, 111
Colombian Lancehead Pit Viper, 315
Colombian Racer, 184
Colombian Rainbow Boa, 44
colombianus, 42, 294, 315
Colombian Watersnake, 139
Colombian Whipsnake, 180
colombiensis, 314
Colorado Desert Sidewinder, 321
Colorado Shovel-nosed Snake, 92
Coluber, 98, 264
Coluber cinereus, 345
Coluber constrictor, 98
Coluber constrictor anthicus, 99
Coluber constrictor constrictor, 98
Coluber constrictor etheridgei, 99
Coluber constrictor flaviventris, 99
Coluber constrictor foxi, 99
Coluber constrictor helvigularis, 99
Coluber constrictor latrunculus, 99
Coluber constrictor mormon, 98, 99
Coluber constrictor oaxaca, 99
Coluber constrictor paludicola, 99
Coluber constrictor priapus, 100
Coluber constrictor stejnegerianus, 99

Coluber minervae, 345
Coluber mormon, 98, 99
Coluber sirtalis, 264
Colubrid, 5, 9, 10, 11, 58, 286, 305
Colubridae, 2, 5, 58
colubrina, 304
colubrinus, 284
Colubroid, 37
columbi, 22
Common Aesculapian False Coral, 127
Common Annulated Coral Snake, 290
Common Bird-eating Snake, 215
Common Black-headed Snake, 249
Common Blindsnake, 28, 29
Common Blind Wormsnake, 29
Common Blunt-headed Tree Snake, 148
Common Boa Constrictor, 39
Common Brazilian Green Legion Snake, 176
Common Brown-lined Lizard-eater, 183
Common Calico Snake, 195
Common Cantil, 306
Common Capuchin Coral Snake, 293
Common Cat-eyed Snake, 158
Common Central American Coral Snake, 300
Common Clear-banded Coral Snake, 293
Common Cobel Legion Snake, 166
Common Coffee Snake, 189
Common Copeia Legion Snake, 173
Common Cribo, 117
Common Cuban Groundsnake, 74
Common Eel Snake, 212
Common False Coral Snake, 280
Common False Fer-de-Lance, 313
Common False Mussurana, 87
Common False Pit Viper, 284
Common Fire-bellied Legion Snake, 167
Common Forest Racer, 108
Common Gartersnake, 265
Common Glossy Racer, 119
Common Graceful Brownsnake, 221
Common Green Racer, 204
Common Haitian Vinesnake, 278
Common Hispaniolan Black Racer, 67
Common Hispaniolan Racer, 67
Common Hog-nosed Racer, 147
Common Jumping Pit Viper, 308
Common Kingsnake, 151
Common Lancehead Pit Viper, 313
Common Large-eyed Snake, 257
Common Liana Snake, 234
Common Lizard-eater, 184
Common Lyresnake, 271
Common Mexican Moccasin, 306
Common Miner Snake, 204
Common Mudsnake, 143

Common Mussurana, 96
Common Neotropical Whipsnake, 180
Common Nightsnake, 144
Common Olive Whipsnake, 94
Common Parrot Snake, 161
Common Puerto Rican Groundsnake, 73
Common Rainbow Snake, 130
Common Ratsnake, 121
Common Red-backed Coffee Snake, 191
Common Regal Legion Snake, 173
Common Ribbonsnake, 264
Common Ringed Tree Boa, 41
Common Savanna Racer, 244
Common Sierra Madre Coral Snake, 291
Common South American Swamp Legion Snake, 170, 171
Common South American Watersnake, 170
Common Southeastern Mussurana, 97
Common Southern Whipsnake, 93
Common Striped Racer, 203
Common Tellurian Snake, 82
Common Tiger Ratsnake, 237
Common Tree Boa, 42
Common Varigated Snake-eater, 115
Common Watersnake, 188
Common Wormsnake, 29
compressicauda, 186
compressus, 234, 235
conanti, 155, 192, 307
Conant's Milksnake, 155
Concho Watersnake, 187
concinnus, 265
concolor, 7, 104, 142, 198, 333
confluens, 187
confluentes, 332
conica, 268
Coniophanes, 100, 346
Coniophanes alvarezi, 100
Coniophanes andresensis, 100
Coniophanes bipunctatus, 100
Coniophanes bipunctatus bipunctatus, 100
Coniophanes bipunctatus biseriatus, 100
Coniophanes dromiciformis, 101
Coniophanes fissidens, 101
Coniophanes fissidens convergens, 101
Coniophanes fissidens dispersus, 101
Coniophanes fissidens fissidens, 101
Coniophanes fissidens obsoletus, 101
Coniophanes fissidens proterops, 101
Coniophanes fissidens punctigularis, 101
Coniophanes imperialis, 101
Coniophanes imperialis clavatus, 102
Coniophanes imperialis copei, 102
Coniophanes imperialis imperialis, 101
Coniophanes joanae, 102
Coniophanes lateritius, 102, 103

Coniophanes longinquus, 102
Coniophanes melanocephalus, 102
Coniophanes mentalis, 346
Coniophanes meridanus, 102
Coniophanes piceivittis, 102
Coniophanes piceivittis frangivirgatus, 102
Coniophanes piceivittis piceivittis, 102
Coniophanes piceivittis schmidti, 103
Coniophanes piceivittis taylori, 103
Coniophanes quinquevittatus, 103
Coniophanes sarae, 103
Coniophanes schmidti, 103
Conophis, 103
Conophis lineatus, 103, 104
Conophis lineatus concolor, 104
Conophis lineatus dunni, 104
Conophis lineatus lineatus, 104
Conophis nevermanni, 106
Conophis pulcher, 103, 104
Conophis pulcher pulcher, 104
Conophis pulcher similes, 104
Conophis taeniatus, 116
Conophis vittatus, 104
Conophis vittatus viduus, 104, 105
Conophis vittatus vittatus, 104
Conopsis, 105, 130, 137, 212, 267
Conopsis biserialis, 105
Conopsis nasus, 105
Conopsis nasus labialis, 106
Conopsis nasus nasus, 105
constrictor, 39, 98
Constrictor, 39
Constrictor constrictor, 39
Contia, 106
Contia tenuis, 106
continentalis, 58
contortrix, 306
convergens, 101
cooki, 42
Cook's Tree Boa, 42
copeae, 169
copei, 59, 88, 102, 113, 161
Copeia Legion Snake, 172
Cope's Black-striped Snake, 102
Cope's Blunt-headed Tree Snake, 148
Cope's Burrowing Snake, 60
Cope's Central American Parrot Snake, 162
Cope's False Coral Snake, 281
Cope's Gophersnake, 209
Cope's Graceful Brownsnake, 223
Cope's Graceful Neotropical Brownsnake, 282
Cope's Green Racer, 202
Cope's Legion Snake, 168
Cope's Lizard-eater, 185
Cope's Mountain Meadow Snake, 59

Index

Cope's Neotropical Brownsnake, 281
Cope's Parrot Snake, 161
Cope's Rainforest Cat-eyed Snake, 158
Cope's Savanna Racer, 245
Cope's Slender Blindsnake, 18
Cope's Snail-eater, 113
Cope's Snail-sucker, 230
Cope's Stripeless Snake, 102
Cope's Tropical Groundsnake, 270, 271
Cope's Tropical Snake, 120
Cope's Vinesnake, 193
Cope's Wormsnake, 71
Cope's Yellow-bellied Snake, 101
Copper-bellied Watersnake, 187
corais, 117
Coral-backed Snail-eater, 230
Coral-bellied Ring-necked Snake, 110
Coral Earthsnake, 136
corallinus, 291
coralliventris, 169
Corallus, 41
Corallus annulatus, 41
Corallus annulatus annulatus, 41
Corallus annulatus blombergi, 41
Corallus annulatus colombianus, 42
Corallus caninus, 5, 42
Corallus cooki, 42
Corallus cropanii, 42
Corallus enydris, 42
Corallus grenadensis, 42
Corallus hortulanus, 42
Corallus hortulanus cooki, 42
Corallus hortulanus grenadensis, 42
Corallus hortulanus hortulanus, 42
Corallus hortulanus ruschenbergeri, 42
Corallus ruschenbergeri, 42
Coral Mudsnake, 143
Coral Pipesnake, 38
Corals, 286
Coral Snake, 8, 11, 122, 286
cordata, 200, 201
Cornsnake, 122
corocoroensis, 256
Corocoro Large-eyed Snake, 256
coronadoi, 247
Coronado Island Gophersnake, 208
Coronado Island Rattlesnake, 333
coronalis, 208
coronata, 70, 124, 211, 247
coronatus, 124
coronellina, 243
Costa Rican Banded Tree Snake, 223
Costa Rican Bird Snake, 215
Costa Rican Blindsnake, 30
Costa Rican Cat-eyed Snake, 159
Costa Rican Coffee Snake, 189
Costa Rican Lizard-eater, 185

Costa Rican Many-banded Coral Snake, 299
Costa Rican Palm Pit Viper, 309
Costa Rican Watersnake, 143
costaricensis, 30
cotiara, 315
Cotiara, 315
Cotiarinha, 316
Cottonmouth, 307
couchi, 258
couperi, 117
crassicaudatum, 344
crassicaudatus, 76, 344
crassus, 44
Crayfish Snake, 216
Cribo, 5, 117
Crisanta's Snake, 106
Crisantophis, 106
Crisantophis nevermanni, 106
Crooked Island Racer, 65
cropanii, 42
Cropani's Boa, 42
Cross-banded Mountain Rattlesnake, 331
Crossed Lancehead Pit Viper, 313
Crotalinae, 10, 11, 58, 305
Crotalus, 319, 320, 337
Crotalus abyssus, 7
Crotalus adamanteus, 320, 345
Crotalus angelensis, 6
Crotalus aquilus, 320
Crotalus atrox, 320, 321
Crotalus atrox atrox, 321
Crotalus atrox tortugensis, 321, 331
Crotalus basiliscus, 321
Crotalus basiliscus oaxacus, 329
Crotalus caliginis, 7
Crotalus catalinensis, 321
Crotalus cerastes, 321
Crotalus cerastes cerastes, 321
Crotalus cerastes cercobombus, 321
Crotalus cerastes laterorepens, 321
Crotalus cereberus, 7
Crotalus cerralvensis, 6
Crotalus concolor, 7
Crotalus confluentes, 332
Crotalus durissus, 321, 322, 323, 345
Crotalus durissus cascavella, 322
Crotalus durissus collilineatus, 322
Crotalus durissus culminatus, 322
Crotalus durissus cumanensis, 322
Crotalus durissus dryinas, 323
Crotalus durissus durissus, 322
Crotalus durissus marajoensis, 323
Crotalus durissus neoleonensis, 324
Crotalus durissus ruruima, 323, 324
Crotalus durissus terrificus, 323
Crotalus durissus totonacus, 323

Crotalus durissus trigonicus, 324
Crotalus durissus tzabcan, 324
Crotalus durissus unicolor, 324
Crotalus durissus vegrandis, 324
Crotalus enyo, 6, 324
Crotalus enyo cerralvensis, 6, 325
Crotalus enyo enyo, 325
Crotalus enyo furvus, 325
Crotalus estebanensis, 6
Crotalus exsul, 325, 330
Crotalus exsul exsul, 330
Crotalus exsul rubber, 325, 330
Crotalus helleri, 7
Crotalus horridus, 326
Crotalus horridus atricaudatus, 326
Crotalus horridus horridus, 326
Crotalus intermedius, 326
Crotalus intermedius gloydi, 326
Crotalus intermedius intermedius, 326
Crotalus intermedius omiltemanus, 327
Crotalus lannomi, 327
Crotalus lepidus, 327
Crotalus lepidus castaneus, 327
Crotalus lepidus klauberi, 327
Crotalus lepidus lepidus, 327
Crotalus lepidus maculosus, 327
Crotalus lepidus morulus, 327
Crotalus lorenzoensis, 6, 334
Crotalus lutosus, 7
Crotalus maricelae, 322
Crotalus mitchelli, 328
Crotalus mitchelli angelensis, 6, 328
Crotalus mitchelli mitchelli, 328
Crotalus mitchelli muertensis, 6, 328
Crotalus mitchelli pyrrhus, 328
Crotalus mitchelli stephensi, 328
Crotalus molossus, 321, 328
Crotalus molossus estebanensis, 6, 329
Crotalus molossus molossus, 328
Crotalus molossus nigrescens, 329
Crotalus molossus oaxacus, 329
Crotalus muertensis, 6
Crotalus oaxacus, 320
Crotalus oreganos, 7
Crotalus pifanorum, 322
Crotalus polystictus, 329
Crotalus pricei, 329
Crotalus pricei miquihuanus, 329
Crotalus pricei pricei, 329
Crotalus pusillus, 329
Crotalus ruber, 325, 330
Crotalus ruber elegans, 330
Crotalus ruber lorenzoensis, 6, 331
Crotalus ruber lucasensis, 330
Crotalus ruber ruber, 330
Crotalus scutulatus, 331
Crotalus scutulatus salvini, 331

Crotalus scutulatus scutulatus, 331
Crotalus stejnegeri, 331
Crotalus tesselatus, 345
Crotalus tigris, 331
Crotalus tortugensis, 331
Crotalus transversus, 331
Crotalus triseriatus, 332, 345
Crotalus triseriatus anahuacus, 332
Crotalus triseriatus aquilus, 320
Crotalus triseriatus armstrongi, 332
Crotalus triseriatus quadrangularis, 320
Crotalus triseriatus triseriatus, 332
Crotalus unicolor, 324
Crotalus vegrandis, 324
Crotalus viridis, 6, 7, 332
Crotalus viridis abyssus, 7, 332
Crotalus viridis caliginis, 333
Crotalus viridis cerberus, 7, 333
Crotalus viridis concolor, 333
Crotalus viridis helleri, 7, 333
Crotalus viridis lutosus, 7, 333
Crotalus viridis nuntius, 333
Crotalus viridis oreganos, 7, 333
Crotalus viridis viridis, 332
Crotalus willardi, 333
Crotalus willardi amabilis, 333
Crotalus willardi meridionalis, 334
Crotalus willardi obscurus, 334
Crotalus willardi silus, 334
Crotalus willardi willardi, 333
Crowned Graceful Brownsnake, 221
Crowned Snake, 245, 246
Cryophis, 106, 107
Cryophis hallbergi, 107
Cuban Black-tailed Dwarf Boa, 56
Cuban Boa, 43
Cuban Cativo, 270
Cuban Dusky Dwarf Boa, 54
Cuban Dwarf Boa, 56
Cuban Racer, 62
Cuban Slender Groundsnake, 73
Cuban White-necked Dwarf, 57
cucullata, 247
cucutae, 18
cuesta, 249
Cul-de-Sac Blind Snake, 33
culminatus, 322
cumanensis, 322
cuneata, 219
cuniculator, 247
cupinensis, 22
cupreus, 162
Curacao Legion Snake, 175
cursor, 167
curtus, 54
cussiliris, 158
cuyanus, 198

Cuzco Calico Snake, 194
cyanopleura, 120
Cyclagras, 141, 142
Cyclagras gigas, 142
cyclas, 228
cyclopion, 186
Cylindrophis, 37
cyrtopsis, 259

Dainty Blindsnake, 22
damiani, 133
Damian's Earthsnake, 133
danforthi, 61
Danforth's Racer, 61
danieli, 58, 139, 184
darienensis, 77
Darien Tellurian Snake, 77
Dark-bellied Trope, 56
Dark Slender Blindsnake, 18
Dark-spotted Anaconda, 47
Dark Swift Andean Snake, 243
Dark Tellurian Snake, 78
Dark Water Mapanare, 140
Darlingtonia, 72, 107
Darlingtonia haetiana, 107
Darlingtonia haetiana haetiana, 107
Darlingtonia haetiana perfector, 107
Darlingtonia haetiana vaticinata, 107
daryi, 59
Dawn Blindsnake, 16
decipiens, 280
deckerti, 123
decorata, 219
Decorated Coral Snake, 291
decoratus, 291
decurtatus, 206
decussatus, 299
degener, 268
degenhardti, 238
Degenhardt's Scorpion-eater, 238
dekayi, 239
DeKay's Brownsnake, 239
DeKay's Snake, 239
Del Nido Ridge-nosed Rattlesnake, 333
Delta Crayfish Snake, 217
deltae, 217
Dendrophidion, 107, 108, 345
Dendrophidion bivittatum, 108
Dendrophidion boshelli, 108, 345
Dendrophidion brunneum, 108
Dendrophidion clarki, 108
Dendrophidion dendrophis, 108
Dendrophidion nuchale, 108
Dendrophidion paucicarinatum, 108
Dendrophidion percarinatum, 108, 345
Dendrophidion vinitor, 109
dendrophis, 108

Dendrophis, 89
Dendrophis aurata, 89
deppei, 207, 209, 210, 247, 281
Deppe's Centipede Snake, 248
depressa, 132
depressiocellus, 77
depressirostris, 162
deschauenseei, 47
Desert Coral Snake, 304
Desert Glossy Snake, 72
Desert Ground Snake, 235
Desert Hook-nosed Snake, 137
deserticola, 145, 225
deserticolus, 208
Desert Kingsnake, 152
Desert Lancehead Pit Viper, 318
Desert Massasauga, 337
Desert Nightsnake, 145
desertorum, 137, 138
Desert Patch-nosed Snake, 225
Desert Ratsnake, 85
Desert Rosy Boa, 50
Desert Striped Whipsnake, 181
Desert Threadsnake, 24
deviatrix, 246
Diablo Range Gartersnake, 258
diabola, 247, 248
diabolicus, 262
diademata, 189
Diadem Snake, 124, 197
Diadophis, 5, 109
Diadophis amabilis, 109
Diadophis amabilis amabilis, 110
Diadophis amabilis modestus, 110
Diadophis amabilis occidentalis, 110
Diadophis amabilis pulchellus, 110
Diadophis amabilis vandenburghi, 111
Diadophis elinorae, 109
Diadophis punctatus, 5, 109
Diadophis punctatus acricus, 110
Diadophis punctatus amabilis, 110
Diadophis punctatus anthonyi, 110
Diadophis punctatus arnyi, 109, 110
Diadophis punctatus blanchardi, 111
Diadophis punctatus dugesi, 110
Diadophis punctatus edwardsi, 110
Diadophis punctatus modestus, 110
Diadophis punctatus occidentalis, 110
Diadophis punctatus pulchellus, 110
Diadophis punctatus punctatus, 109
Diadophis punctatus regalis, 110, 111
Diadophis punctatus similes, 111
Diadophis punctatus stictogenys, 111
Diadophis punctatus vandenburghi, 111
Diadophis regalis, 109, 110
Diadophis regalis regalis, 110
Diamond-backed Watersnake, 188

Index

diana, 292
Diana's Coral Snake, 292
Diaphololepis, 125
Diaphorolepis, 111, 242
Diaphorolepis laevis, 111
Diaphorolepis wagneri, 111
diaplocius, 22
diastema, 292
Diastema Coral Snake, 292
Di-Bernardo's Tropical Snake, 120
dichrous, 119
dictyodes, 171
digitalis, 195, 196
digueti, 261
dilepis, 167
dimidiata, 68, 230
dimidiatus, 22, 69, 198, 199, 279, 280, 281
diperkini, 215
diplotrophis, 162
diporus, 317
Dipsas, 112, 229, 232, 275, 345
Dipsas albifrons, 112
Dipsas albifrons albifrons, 112
Dipsas albifrons cavalheiroi, 112
Dipsas articulate, 112
Dipsas bicolor, 112
Dipsas boettgeri, 112
Dipsas brevifacies, 112, 345
Dipsas catesbyi, 112
Dipsas chaparensis, 113
Dipsas copei, 113
Dipsas elegans, 113
Dipsas ellipsifera, 113
Dipsas gaigeae, 113
Dipsas gracilis, 113
Dipsas incerta, 113
Dipsas indica, 113
Dipsas indica bucephala, 114
Dipsas indica cisticeps, 114
Dipsas indica ecuadorensis, 114
Dipsas indica indica, 113
Dipsas indica petersi, 114
Dipsas infrenalis, 345
Dipsas latifasciata, 114
Dipsas latifrontalis, 114
Dipsas maxillaris, 114
Dipsas neivai, 112, 114
Dipsas niceforoi, 115
Dipsas nicholsi, 349
Dipsas oreas, 114
Dipsas oreas elegans, 113
Dipsas oreas ellipsifera, 113
Dipsas pavonina, 114
Dipsas perijanensis, 114
Dipsas peruana, 115
Dipsas poecilolepis, 233
Dipsas polylepis, 115
Dipsas pratti, 115
Dipsas sanctijoannis, 115
Dipsas schunkii, 115
Dipsas subaequalis, 346
Dipsas temporalis, 115
Dipsas tenuissima, 115
Dipsas tolimensis, 115
Dipsas variegata, 115
Dipsas variegata nicholsi, 115
Dipsas variegata trinitatis, 115
Dipsas variegata variegata, 115
Dipsas vermiculata, 115
Dipsas viguieri, 116
disastemus, 281
discolor, 245
dispersus, 101
Disphorolepis, 242
Disphorolepis lasallei, 242
dissectus, 23
dissoleucus, 292, 293
distans, 293
Ditaxodon, 16
Ditaxodon taeniatus, 116
diutius, 298
divaricatus, 300
diviniloquax, 39
divittatus, 198
dixoni, 155
Dixon's Milksnake, 155
doliata, 154
doliatus, 194
dolichurum, 73
Dominica Blindsnake, 30
Dominican Boa Constrictor, 40
Dominican Republic Dwarf Boa, 55
Dominican Republic Ground Boa, 45
Dominican Republic Racer, 67
Dominican Republic Vinesnake, 278
dominicanus, 30
Dominica Racer, 62
donosoi, 301
dorbignyi, 69, 176
Dorbigny's Wormsnake, 69
dorsalis, 62, 147, 169, 184, 265, 278
dorsatus, 269
Dotted Brownsnake, 236
Dotted Burrowing Snake, 199
Double-banded False Coral Snake, 128
Double-banded Watersnake, 142
Double-black Coral Snake, 294
Double Collar Coral Snake, 297
Double-keeled-scale Snake, 111
downsi, 133
Downs's Earthsnake, 134
Drab Legion Snake, 168
Drepanodon, 116
Drepanoides, 116
Drepanoides anomalus, 116
dromiciformis, 101
Dromicus, 61, 165, 200
Dromicus angustilineatus, 203
Dromicus biserialis, 62
Dromicus biserialis dorsalis, 62
Dromicus biserialis eibli, 62
Dromicus biserialis helleri, 62
Dromicus capensis, 346
Dromicus chamissonis, 201
Dromicus tachymenoides, 203
Dryadophis [as *Mastigodryas*], 8, 182, 192
Dryadophis amarali [as Mastigodryas], 182
Dryadophis bifossatus [as *Mastigodryas*], 182
Dryadophis bifossatus bifossatus [as *Mastigodryas*], 183
Dryadophis bifossatus lacerdai [as *Mastigodryas*], 183
Dryadophis bifossatus striatus [as *Mastigodryas*], 183
Dryadophis bifossatus triseriatus [as *Mastigodryas*], 183
Dryadophis bifossatus villelai [as *Mastigodryas*], 183
Dryadophis boddaerti [as *Mastigodryas*], 183
Dryadophis boddaerti boddaerti [as Mastigodryas], 183
Dryadophis boddaerti dunni [as *Mastigodryas*], 183
Dryadophis boddaerti ruthveni [as *Mastigodryas*], 183
Dryadophis bruesi [as *Mastigodryas*], 183
Dryadophis cliftoni [as *Mastigodryas*], 184
Dryadophis dorsalis [as *Mastigodryas*], 184
Dryadophis fasciatus, 184
Dryadophis heathi [as *Mastigodryas*], 184
Dryadophis melanolomus [as *Mastigodryas*], 184
Dryadophis melanolomus alternatus [as *Mastigodryas*], 184
Dryadophis melanolomus laevis [as *Mastigodryas*], 184
Dryadophis melanolomus melanolomus [as *Mastigodryas*], 184
Dryadophis melanolomus slevini [as *Mastigodryas*], 184
Dryadophis melanolomus stuarti [as *Mastigodryas*], 185
Dryadophis melanolomus tehuanae [as *Mastigodryas*], 185
Dryadophis melanolomus veraecrucis [as *Mastigodryas*], 185
Dryadophis pleei [as *Mastigodryas*], 185
Dryadophis pulchriceps [as *Mastigodryas*], 185
Dryadophis sanguiventris [as *Mastigodryas*], 185

Dry Forest Legion Snake, 167
dryinas, 323
Dryiophis, 89
Dryiophis auratus, 89
Drymarchon, 5, 116, 117
Drymarchon corais, 117
Drymarchon corais cleofae, 118
Drymarchon corais corais, 117
Drymarchon corais couperi, 117
Drymarchon corais erebennus, 117
Drymarchon corais margaritae, 117
Drymarchon corais melanurus, 117
Drymarchon corais orizabensis, 117
Drymarchon corais rubidus, 117, 118
Drymarchon corais unicolor, 118
Drymobius, 118
Drymobius chloroticus, 118
Drymobius margaritiferus, 118
Drymobius margaritiferus fistulosus, 118
Drymobius margaritiferus margaritiferus, 118
Drymobius margaritiferus maydis, 119
Drymobius margaritiferus occidentalis, 119
Drymobius melanotropis, 119
Drymobius rhombifer, 119
Drymoluber, 119
Drymoluber brazili, 119
Drymoluber dichrous, 119
dubius, 133
duellmani, 133
Duellman's Cat-eyed Snake, 158
dugandi, 23
Dugès Brownsnake, 239
Dugès Earthsnake, 133, 134
dugesi, 25, 110, 133, 134
duida, 256
Duida Large-eyed Snake, 256
Duida Tellurian Snake, 77
duidensis, 77
dulcis, 23
dumerili, 219, 281, 293
Dumeril's Burrowing Snake, 198
Dumeril's Neotropical Brownsnake, 281
Dumeril's Coral Snake, 293
Dumeril's Graceful Brownsnake, 219
Dunes Hog-nosed Snake, 177
dunkeli, 145
dunni, 77, 104, 134, 143, 183, 230, 293, 336
Dunn's Checkered Gartersnake, 261
Dunn's Forest Racer, 108
Dunn's Hog-nosed Pit Viper, 336
Dunn's Legion Snake, 168
Dunn's Lizard-eater, 183
Dunn's Road Guarder, 104, 106
Dunn's Snail-eater, 230
Dunn's Tellurian Snake, 77
Dunn's Watersnake, 143
Durangan Rock Rattlesnake, 327

Durango Ratsnake, 86
durissus, 321, 322, 323, 345
Dusky Forest Pit Viper, 310
Dusky Pygmy Rattlesnake, 338
Dusky Rattlesnake, 332
Dwarf Boas, 52
Dwarf Short-tailed Snake, 254
dysodes, 56
dyticus, 167

Earth Runner, 90
Earthsnake, 29
Earthworm Blindsnake, 31
Eastern Black Kingsnake, 152
Eastern Black-necked Gartersnake, 259
Eastern Calico Snake, 195
Eastern Coachwhip, 179
Eastern Coral Snake, 296
Eastern Cottonmouth, 307
Eastern Cuba Racer, 62
Eastern Diamond-backed Rattlesnake, 320
Eastern Double Collar Coral Snake, 297
Eastern Foxsnake, 122
Eastern Galápagos Racer, 62
Eastern Gartersnake, 265
Eastern Hemprich's Coral Snake, 296
Eastern Hog-nosed Snake, 141
Eastern Hook-nosed Snake, 130
Eastern Indigo Snake, 117
Eastern Kingsnake, 151
Eastern Legion Snake, 167
Eastern Massasauga, 337
Eastern Mexican Bullsnake, 209
Eastern Milksnake, 154
Eastern Mudsnake, 129
Eastern Patch-nosed Snake, 225
Eastern Racer, 98
Eastern Ratsnake, 123
Eastern Ribbonsnake, 264
Eastern Short-tailed Blindsnake, 32
Eastern Smooth Earthsnake, 283
Eastern Smooth Green Snake, 164
Eastern Snail-eater, 232
Eastern Striped Forest Pit Viper, 310
Eastern Twin-spotted Rattlesnake, 329
Eastern Worm-eating Coral Snake, 296
Eastern Wormsnake, 88
Eastern Yellow-bellied Racer, 99
eburnata, 72
Echinanthera, 119, 120, 218, 244
Echinanthera affinis, 120, 244
Echinanthera amonena, 120
Echinanthera bilineata, 120
Echinanthera bilineatus, 244
Echinanthera brevirostris, 120, 218
Echinanthera cephalomaculata, 120
Echinanthera cephalostriata, 120

Echinanthera cyanopleura, 120
Echinanthera melanostigma, 120
Echinanthera persimilis, 245
Echinanthera undulata, 120
Echynanthera, 244, 245
Echynanthera affinis, 244
Echynanthera occipitalis, 245
ecuadorensis, 77, 114, 168, 212
Ecuadorian Amazon Snail-eater, 114
Ecuadorian Anchor Coral Snake, 289
Ecuadorian Boa Constrictor, 40
Ecuadorian Coral Snake, 290
Ecuadorian Dawn Blindsnake, 16
Ecuadorian Frog-eating Snake, 111
Ecuadorian Legion Snake, 167
Ecuadorian Lowland Threadsnake, 21
Ecuadorian Milksnake, 155
Ecuadorian Montane Snail-eater, 114
Ecuadorian Snail-eater, 232
Ecuadorian Tellurian Snake, 77
Ecuadorian Tiger Ratsnake, 237
Ecuadorian Toadheaded Pit Viper, 312
Ecuadorian Whipsnake, 94
edwardsi, 110, 337
Eel Snake, 212
Egg-eater, 116
eibli, 62
elaeoides, 175
Elaphe, 5, 6, 85, 120, 121, 228
Elaphe bairdi, 121
Elaphe flavirufa, 121, 122
Elaphe flavirufa flavirufa, 121, 122
Elaphe flavirufa matudai, 121
Elaphe flavirufa pardalina, 121
Elaphe flavirufa phaescens, 121
Elaphe flavirufa polysticha, 122
Elaphe gloydi, 122
Elaphe guttata, 5, 121, 122
Elaphe guttata emoryi, 122
Elaphe guttata guttata, 122
Elaphe guttata intermontana, 122
Elaphe guttata meahllmorum, 122, 123
Elaphe guttata rosaeca, 122
Elaphe laetus, 123
Elaphe obsoleta, 121, 123
Elaphe obsoleta deckerti, 123
Elaphe obsoleta lindheimeri, 123
Elaphe obsoleta obsoleta, 123
Elaphe obsoleta quadrivittata, 123
Elaphe obsoleta rossalleni, 121, 123
Elaphe obsoleta spiloides, 123
Elaphe phaescens, 121
Elaphe triaspis, 228, 229
Elaphe triaspis mutabilis, 229
Elaphe triaspis triaspis, 229
Elaphe vulpina, 121, 124
Elapid, 5, 9, 58

Index

Elapidae, 10, 11, 58, 286
elapoides, 279, 280, 281
Elapomojus, 69
Elapomojus dimidiatus, 69
Elapomorphus, 68, 124, 125, 197
Elapomorphus accedens, 125
Elapomorphus coronatus, 124
Elapomorphus lepidus, 124, 197
Elapomorphus quinquelineatus, 124, 125, 197
Elapomorphus wuchereri, 125
Elaps, 77
Elaps mipartitus, 299
elapsoides, 155
elegans, 63, 71, 113, 259, 285, 294, 330
Elegant Coral Snake, 294
Elegant Green Racer, 63
elegantissimus, 167
Elegant Snail-eater, 113
elinorae, 109
ellipsifera, 113
El Muerto Island Rattlesnake, 328
elongata, 63
Eluthera Island Dwarf Boa, 54
Emerald Tree Boa, 42
emigdioi, 77
Emigdio's Tellurian Snake, 77
emmeli, 77
Emmel's Tellurian Snake, 77
Emmochliophis, 125, 242
Emmochliophis fugleri, 125
Emmochliophis miops, 125
emoryi, 122
engelsi, 189
Enicagnethus, 346
Enicagnethus melanoauchen, 346
Enuliophis, 125
Enuliophis sclateri, 125
Enulius, 125, 126
Enulius bifoveatus, 126
Enulius flavitorques, 126
Enulius flavitorques flavitorques, 126
Enulius flavitorques sumichrasti, 126
Enulius flavitorques unicolor, 126
Enulius oligostichus, 126
Enulius roatanensis, 126
enydris, 42
enyo, 6, 324, 325
Epacta, 21
Epacta undecimstriata, 21
epactius, 30
ephippiata, 160
ephippifer, 294
Epicrates, 43
Epicrates angulifer, 43
Epicrates cenchria, 5, 43
Epicrates cenchria alvarezi, 43
Epicrates cenchria assisi, 43

Epicrates cenchria barbouri, 44
Epicrates cenchria cenchria, 43
Epicrates cenchria crassus, 44
Epicrates cenchria gaigei, 44
Epicrates cenchria hygrophilus, 44
Epicrates cenchria maurus, 44
Epicrates cenchria polylepis, 44
Epicrates cenchria xerophilus, 43
Epicrates chrysogaster, 44
Epicrates chrysogaster chrysogaster, 44
Epicrates chrysogaster relicquus, 44
Epicrates chrysogaster schwartzi, 44
Epicrates exsul, 44
Epicrates fordi, 45
Epicrates fordi agametus, 45
Epicrates fordi fordi, 45
Epicrates fordi manototus, 45
Epicrates gracilis, 45
Epicrates gracilis gracilis, 45
Epicrates gracilis hapalus, 45
Epicrates inornatus, 45
Epicrates monensis, 45
Epicrates monensis granti, 46
Epicrates monensis monensis, 46
Epicrates striatus, 46
Epicrates striatus ailurus, 46
Epicrates striatus exagistus, 46
Epicrates striatus fosteri, 46
Epicrates striatus fowleri, 46
Epicrates striatus mccraniei, 46
Epicrates striatus striatus, 46
Epicrates striatus strigilatus, 46
Epicrates striatus warreni, 46
Epicrates subflavus, 47
Epicta, 347
Epicta undecimstriata, 21, 347
epinephelus, 167, 168
Equal-banded Coral Snake, 297
Equatorial Mussurana, 97
Equatorial Watersnake, 140
equatoriana, 97, 247
Equator Tree Snake, 232
eques, 260, 263
erebennus, 117
eremicola, 201
ericksoni, 56
Eridiphas, 126, 127, 144, 213, 245
Eridiphas marcosensis, 127
Eridiphas slevini, 127
Eridiphas slevini marcosensis, 127
Eridiphas slevini slevini, 127
errabunda, 96, 97
errans, 260
Erycinae, 48
erythrogaster, 186
erythrogramma, 129, 130
Erythrolamprus, 127

Erythrolamprus aesculapii, 127
Erythrolamprus aesculapii aesculapii, 127
Erythrolamprus aesculapii monzonus, 128
Erythrolamprus aesculapii ocellatus, 128
Erythrolamprus aesculapii tetrazonus, 128
Erythrolamprus aesculapii venustissimus, 128
Erythrolamprus baupertuisii, 128
Erythrolamprus bizonus, 128
Erythrolamprus guentheri, 128
Erythrolamprus mentalis, 346
Erythrolamprus mimus, 128
Erythrolamprus mimus impar, 129
Erythrolamprus mimus micrurus, 129
Erythrolamprus mimus mimus, 129
Erythrolamprus ocellatus, 128
Erythrolamprus pseudocorallus, 129
erythromelas, 77, 315
Españolan Galapágos Racer, 202
espinali, 190
Espirito Santo Tropical Snake, 120
Espiritu Santo Island Sandsnake, 91
Espiritu Santo Striped Whipsnake, 178
estebanensis, 6, 329
esterensis, 91
Estero Salina Sandsnake, 91
etheridgei, 99, 224
Eudryas, 182
Eunectes, 47
Eunectes barbouri, 47
Eunectes deschauenseei, 47
Eunectes murinus, 47
Eunectes murinus gigas, 47
Eunectes murinus murinus, 47
Eunectes notaeus, 47
euryxanthus, 180, 288
euryzona, 281
euryzonus, 279, 281
Everglades Racer, 99
Everglades Ratsnake, 123
exagistus, 46
excubitor, 254
exedrus, 147
Exelencophis, 275
Exelencophis nelsoni, 275
exiguum, 73
exiguous, 338, 339
Exiled Gartersnake, 260
Exiliboa, 52
Exiliboa placata, 53
exoletus, 93
expolita, 72
exsul, 44, 260, 325, 330
extenuatum, 239
Eyelash Boa, 52
Eyelash Dwarf Boa, 53
Eyelash Palm Pit Viper, 309

Faded Black-striped Snake, 103
Falcón Blindsnake, 31
False Boa, 211
False Cat-eyed Snake, 213
False Coral Mudsnake, 143
False Coral Snake, 37, 127, 194
False Fer-de-Lance, 284
False Ficimia, 213
False Harmless Snake, 211
False Middle American Snail-eater, 275
False Mussurana, 87
False Pit Viper, 284
False Regal Legion Snake, 171
False Rhadinaea, 226
False Sao Paulo Tellurian Snake, 83
False Tomodon Snake, 213, 214
False Triad Coral Snake, 290
False Yarará Snail-eater, 114
False Yarará, 213, 214
Fanged Snake, 147
Fangless Nightsnake, 144
Farancia, 129
Farancia abacura, 129
Farancia abacura abacura, 129
Farancia abacura reinwardti, 129
Farancia erythrogramma, 129
Farancia erythrogramma erythrogramma, 130
Farancia erythrogramma seminola, 130
fasciata, 187, 229, 275
fasciatus, 91, 144, 184
fasciolatus, 272
favae, 77, 344
feicki, 54
Feick's Dwarf Boa, 54
Fer-de-Lance, 313
Fernandes's Atlantic Forest Tellurian Snake, 82
ferox, 147
Ferrarezzi's Burrowing Snake, 198
festae, 168
Few-ringed Coral Snake, 300
Ficimia, 105, 130, 137, 212, 267
Ficimia hardyi, 130
Ficimia olivacea, 130
Ficimia publia, 130, 131
Ficimia publia publia, 131
Ficimia publia taylori, 131
Ficimia publia wolffsohni, 131
Ficimia ramirezi, 131
Ficimia ruspator, 131
Ficimia streckeri, 131
Ficimia variegata, 131
File-tailed Groundsnake, 235
filiformis, 295
Fire-bellied Legion Snake, 167
fischeri, 229, 276

Fischer's Long-tailed Snake, 126
Fischer's Snail-eater, 276
Fishing Snake, 242
fissidens, 101
fistulosus, 118
fitchi, 265
fitzingeri, 194, 296
Fitzinger's False Coral Snake, 194
Five-lined Burrowing Snake, 125
Five-striped Snake, 103
flagellum, 179
Flame Snake, 195
flammingerus, 78
Flat-headed Snake, 246, 248
Flathead Snake, 245
Flat-tailed Sea Snake, 304
flavapices, 16
flavifrenatus, 168
flavigaster, 187
flavigularis, 180
flavilata, 219
flavilineata, 247
flavirufa, 121, 122
flavitorques, 126
flaviventris, 99
flavolineatus, 94
flavopictus, 93
flavoterminatus, 17
flavotorquata, 69, 70
flavovirgatus, 216
Florida Brownsnake, 240
Florida Cottonmouth, 307
Florida Crowned Snake, 251
Florida Green Watersnake, 187
Florida Kingsnake, 152
floridana, 152, 187
Florida Pinesnake, 209
Florida Red-bellied Snake, 241
Florida Scarletsnake, 88
Florida Watersnake, 187
fonsecai, 315
Fonseca's Lancehead Pit Viper, 316
forbesi, 219, 272
Forbes's Graceful Brownsnake, 219
fordi, 45
Ford's Boa, 45
Forest Flame Snake, 195
Forest Pit Viper, 310
Forest Queen Legion Snake, 175
Forest Racer, 108
Forest Watersnake, 140
Forest Whipsnake, 95
formosus, 194
forreri, 162
Forrer's Parrot Snake, 162
forsteri, 172
fortitus, 205

fosteri, 46
Four-lined Burrowing Snake, 60
foveatus, 95
fowleri, 46
Fowler's Boa, 46
foxi, 59, 99
Fox's Mountain Meadow Snake, 59
Fox Snake, 122, 124
frangivirgatus, 102
fraseri, 168, 249
Fraser's Legion Snake, 168
freminvilli, 238
frenata, 158
frenatus, 168, 179, 278
Fresh-water Watersnake, 143
frizzelli, 194
Frog-eating Snake, 111
frondicolor, 278
frontalis, 17, 217, 289, 295
frontifasciatus, 295
fugleri, 125
fulgidus, 193
fulginatus, 208
fulginosus, 179
fuliginosus, 78
fulviceps, 281
fulvius, 295, 296
fulvivittis, 219
fulvoguttatus, 134
fulvus, 260
funereum, 72, 73
furvus, 325
fusca, 171
fuscicaudus, 63
fuscus, 54, 94, 171

gaboi, 69
gadowi, 347
gaigae, 155
gaigeae, 78, 113, 219
gaigei, 44
Gaige's Pine Forest Snake, 219
Gaige's Snail-eater, 113
Gaige's Tellurian Snake, 78
galacelidus, 57
Galápagos Racer, 62
Gambote Large-eyed Snake, 256
gambotensis, 256
Geagras, 131, 132
Geagras redimitus, 132
Geatractus, 275
Geatractus tecpanecus, 275
geminatus, 235
gemmistratus, 149, 150, 347
genimaculata, 218
genimaculatus, 216
gentilis, 155

Index

Geophis, 74, 132
Geophis alasukai, 78, 132
Geophis anocularis, 132
Geophis aquilonaris, 134
Geophis bellus, 132
Geophis betaniensis, 132
Geophis bicolor, 132
Geophis blanchardi, 132
Geophis brachycephalus, 133, 346
Geophis brachycephalus auctorum, 346
Geophis cancellatus, 133
Geophis carinosus, 133
Geophis chalybeus, 133
Geophis championi, 133
Geophis damiani, 133
Geophis downsi, 133
Geophis dubius, 133
Geophis duellmani, 133
Geophis dugesi, 133
Geophis dugesi aquilonaris, 134
Geophis dugesi dugesi, 134
Geophis dunni, 134
Geophis fulvoguttatus, 134
Geophis gertschi, 247
Geophis godmani, 134
Geophis hoffmanni, 134
Geophis immaculatus, 134
Geophis incomptus, 134
Geophis isthmicus, 134
Geophis juliai, 134
Geophis laticinctus, 135
Geophis laticinctus albiventris, 135
Geophis laticinctus laticinctus, 135
Geophis laticollaris, 135
Geophis latifrontalis, 135
Geophis latifrontalis latifrontalis, 135
Geophis latifrontalis semmiannulatus, 135
Geophis maculiferus, 135
Geophis mutitorques, 135
Geophis nasalis, 135
Geophis nigroalbus, 135
Geophis nigrocinctus, 136
Geophis omiltemanus, 136
Geophis petersi, 136
Geophis pyburni, 136
Geophis rhodogaster, 136
Geophis rostralis, 136
Geophis russatus, 136
Geophis ruthveni, 136
Geophis sallaei, 135, 136
Geophis sallei, 136
Geophis semidoliatas, 136
Geophis sieboldi, 136, 137
Geophis, species inquirenda, 346
Geophis talamancae, 137
Geophis tarascae, 137
Geophis zeledoni, 137

geotomus, 31
gertschi, 247
getula, 151
Ghost Blunt-headed Tree Snake, 150
Giant Bird Snake, 215
Giant Blindsnake, 32
Giant Central American Milksnake, 155
Giant False Water Cobra, 142
Giant Gartersnake, 260
Giant Pine-Oak Snake, 222
Giant Tellurian Snake, 79
gibsoni, 209
Gibson's Gophersnake, 209
gigas, 47, 142, 260
girardi, 182
Glauconia, 347
Glauconia albifons, 21, 347
Glossy Crayfish Snake, 217
Glossy Racer, 119
Glossy Snake, 71
gloydi, 122, 141, 326
Gloyd's Cantil, 306
godmani, 134, 219, 260, 319
Godman's Earthsnake, 134
Godman's Gartersnake, 260
Godman's Graceful Brownsnake, 219
Godman's Montane Pit Viper, 319
goiasensis, 69
Goiás Legion Snake, 172
Goiás Wormsnake, 69
Goiaz Lancehead Pit Viper, 317
goini, 152
Golden-headed Gartersnake, 258
Golden Lancehead Pit Viper, 316
Golden Snake, 89
gomesi, 139
Gomesophis, 137
Gomesophis brasiliensis, 137
Gomes's Burrowing Snake, 199
Gomes's Miner Snake, 204
Gomes's Snake, 137
Gomes's Wormsnake, 68
Gomes's Watersnake, 139
Gonâve Blindsnake, 30
Gonâve Hog-nosed Racer, 147
gonavensis, 30
Gonâve Vinesnake, 278
Gophersnake, 5, 207
goudoti, 23, 24
Goudoti's Threadsnake, 24
goyasensis, 69
goyazensis, 317
Graceful Brown Snake, 218
Graceful Brownsnake, 218
Graceful Mountain Snake, 223
Graceful Snail-eater, 113
gracia, 50, 51

gracile, 270
Gracile Tropical Groundsnake, 270
gracilis, 45, 113, 247
gracillimus, 149
grahamae, 225
grahami, 217
grahamiae, 225
Graham's Crayfish Snake, 217
Grand Bahamas Racer, 65
Grand Canyon Rattlesnake, 332
Grand Cayman Blindsnake, 30
Grand Cayman Dwarf Boa, 54
Grand Cayman Racer, 63
grandisquamis, 94
grandocula, 230
granti, 30, 46
Grant's Blindsnake, 30
graphicus, 173
Grass Snake, 164
Gray-banded Kingsnake, 151
Gray-bellied Lizard-eater, 183
Gray Black-bellied Gartersnake, 261
Gray Earthsnake, 133
Gray Ratsnake, 123
Gray Tellurian Snake, 81
Great Basin Gophersnake, 208
Great Basin Rattlesnake, 333
Great Bird Island Racer, 61
Great Corn Island Coral Snake, 300
Great Corn Island Racer, 119
Greater Blindsnake, 17
Greater Blunt-headed Tree Snake, 149
Great Inagua Dwarf Boa, 53, 54
Great Inauga Island Blindsnake, 29
Great Inaugua Island Boa, 44
Great Inagua Racer, 65
Great Plains Ratsnake, 122
Green Anaconda, 47
Green-bellied Palm Snake, 202
Green-headed Parrot Snake, 163
Green-headed Racer, 160
Green Highland Racer, 118
Green Horse-Whip Snake, 161
Green Littersnake, 219
Green Parrot Snake, 161
Green Racer, 200
Green Ratsnake, 229
Green-striped Vinesnake, 285
Green Tree Snake, 162
Green Vinesnake, 193
Green Watersnake, 186
greenwayi, 54, 55
greeri, 153
Grenada Blindsnake, 33
Grenada Tree Boa, 42
grenadensis, 42
groomei, 97

Groundsnake, 74
guadeloupe, 30
Guadeloupe Blindsnake, 30
Guadeloupe Racer, 61
Guanaja Centipede Snake, 253
Guanaja Long-tailed Snake, 126
Guanajuato Coral Snake, 296
Guaniguanico Groundsnake, 74
Guatemalan Centipede Snake, 248
Guatemalan Lizard-eater, 184
Guatemalan Milksnake, 154
Guatemalan Neotropical Ratsnake, 229
Guatemalan Palm Pit Viper, 309
Guatemalan Snake, 90
guayaquilensis, 24
Guayaquil Racer, 63
Guayaquil Threadsnake, 24
guentheri, 19, 78, 128, 169, 281, 284
Guenther's False Pit Viper, 284
Guenther's Legion Snake, 169
Guenther's Neotropical Brownsnake, 281
Guenther's Splendid Cat-eyed Snake, 160
Guenther's Two-spotted Snake, 100
guerini, 204
Guerreran Centipede Snake, 247
Guerreran Earthsnake, 136
Guerreran Hook-nosed Snake, 131
Guerreran Pine Woods Snake, 221
Guerreran Pygmy Rattlesnake, 339
Guerreran Snail-eater, 275
guerreroensis, 275
Guerrero Threadsnake, 26
Guiacurus Blue Racer, 202
Guianan Bird Snake, 215
Guianan Black-backed Coral Snake, 287
Guianan Miner Snake, 204
Guianan Mudsnake, 144
Guianan Savanna Racer, 244
Guiana's Ribbon Coral Snake, 297
Guianas Threadsnake, 22
guianensis, 204
Guianian Rattlesnake, 323
guibei, 194, 195
Guibe's False Coral Snake, 194
gularis, 53, 145
Gulf Coast Ribbonsnake, 263
Gulf Crayfish Snake, 217
Gulf Saltmarsh Snake, 186
Gunanbara Liana Snake, 235
Günther's False Coral Snake, 128
Günther's Liana Snake, 234
Günther's Tellurian Snake, 78
guttata, 5, 121
Guyana Wormsnake, 70
Gyalopion, 105, 130, 137, 212, 267
Gyalopion canum, 138
Gyalopion desertorum, 137, 138

Gyalopion quadrangulae, 138
Gyalopion quadrangulare desertorum, 138
Gyalopion quadrangulare quadrangulare, 138

haasi, 211
Haas's False Mussurana, 211
Haenophidia, 3, 5, 37, 58
Haenophidian, 37
haetiana, 107
haetianus, 55
hagmanni, 139
Hagmann's Squint-eyed Watersnake, 139
Haitian Boa, 46
Haitian Border Threadsnake, 25
Haitian Dwarf Boa, 55
Haitian Gracile Boa, 45
Haitian Ground Boa, 45
Haitian Groundsnake, 107
Haitian Long-tailed Snake, 278
Haitian Pale-lipped Blindsnake, 29
Haitian Southern Border Vinesnake, 278
Haitian Vinesnake, 278
Haldea, 282
Half-banded Blunt-headed Tree Snake, 149
hallbergi, 107
Hallberg's Cloud Forest Snake, 107
Halloween Snake, 280
Hallowell's Centipede Snake, 253
Hallowell's Coffee Snake, 189
hammondi, 261
hannsteini, 220
Hannstein's Spot-lipped Snake, 220
hapalus, 45
hardyi, 56, 130
Hardy's Dwarf Boa, 56
hariolatus, 278
Harlequin Coral Snake, 295
harteri, 187
hartwegi, 231
Hartweg's Snail-sucker, 231
Havana Groundsnake, 73
heathi, 184
Heath's Lizard-eater, 184
hectus, 30
helenae, 88
Helicops, 138
Helicops angulatus, 138
Helicops carinicaudus, 138, 139
Helicops danieli, 139
Helicops gomesi, 139
Helicops hagmanni, 139
Helicops hogei, 139
Helicops infrataeniatus, 139
Helicops leopardinus, 139
Helicops leprieuri, 346
Helicops leprieuri moesta, 346
Helicops modestus, 139

Helicops pastazae, 139
Helicops petersi, 140
Helicops pictiventris, 140
Helicops polylepis, 139, 140
Helicops scalaris, 140
Helicops schistosus, 346
Helicops septemvittatus, 139, 346
Helicops trivittatus, 140
Helicops yacu, 140
helleri, 7, 62, 298, 333
Helmintophis, 16, 17
Helmintophis flavoterminatus, 17
Helmintophis frontalis, 17
Helmintophis praeocularis, 17
helvigularis, 99
hemprichi, 296
Hemprich's Coral Snake, 296
hempsteadae, 220
Hempstead's Graceful Brownsnake, 220
hendersoni, 55
Henderson's Dwarf Boa, 55
Hensel's Snake, 116
herbea, 202
hermerus, 55
Herpetodryas, 346
Herpetodryas annectens, 346
herrerae, 157
hertwigi, 299
hespere, 336
hesperia, 220
hesperiodes, 220
Heterodon, 140
Heterodon gloydi, 141
Heterodon nasicus, 140
Heterodon nasicus gloydi, 141
Heterodon nasicus kennerlyi, 141
Heterodon nasicus nasicus, 140
Heterodon platirhinos, 141
Heterodon platyrhinos, 141
Heterodon simus, 141
Heterorhachis, 233
Heterorhachis poecilolepis, 233
hexalepis, 225
Hidalgo Brownsnake, 240
hidalgoensis, 240
Hidalgo Hook-nosed Snake, 130
Hidalgo Ring-necked Coffee Snake, 190
Higher Montane Groundsnake, 107
Highland Coffee Snake, 190
Highland Mussurana, 97
Highlands Burrowing Snake, 59
Highlands Centipede Snake, 246
Highlands Earthsnake, 135
Highlands Gartersnake, 260
Highland Swift Snake, 243
Highland Tellurian Snake, 79
Highly Variable Tellurian Snake, 78

Index

Hillis's Watersnake, 242
hiltoni, 213
Himantodes, 148
hippocrepis, 297
Hispaniolan Black Racer, 67
Hispaniolan Boa, 45
Hispaniolan Brown Racer, 61
Hispaniolan Common Blindsnake, 32
Hispaniolan Gracile Boa, 45
Hispaniolan Gracile Ground Boa, 45
Hispaniolan Groundsnake, 107
Hispaniolan Hog-nosed Racer, 147
Hispaniolan Racer, 147
histricus, 177
hobartsmithi, 247, 281
hoeversi, 163
hoffmanni, 134
Hoffman's Earthsnake, 134
hogei, 139
Hoge's Water Mapanare, 139
Hog-nosed False Coral Snake, 177
Hog-nosed Pit Viper, 336
Holartic ratsnakes, 6
holbrooki, 152
Honduran Centipede Snake, 249
Honduran Coffee Snake, 190
Honduran Milksnake, 155
Honduran Neotropical Ratsnake, 229
Honduran Nightsnake, 121
Honduran Ratsnake, 121
Honduras Coral Snake, 300
hondurensis, 155
hoodensis, 202
Hopi Rattlesnake, 333
horridus, 326
hortulanus, 42
hostilitractus, 78
houtmann, 336
Houtmann's Hog-nosed Pit Viper, 336
howardgloydi, 306
Huachuca Black-headed Snake, 253
Huachuca Mountain Kingsnake, 154
Huamantlan Rattlesnake, 331
Huancabamba Snail-eater, 233
Huatupilla Threadsnake, 26
hudsoni, 190
hueyi, 259
humilis, 24
Hydrodynastes, 61, 72, 141, 142
Hydrodynastes bicinctus, 142
Hydrodynastes bicinctus bicinctus, 142
Hydrodynastes bicinctus schultzi, 142
Hydrodynastes gigas, 142
Hydromorphus, 142
Hydromorphus clarki, 143
Hydromorphus concolor, 142
Hydromorphus dunni, 143

Hydrophiidae, 10, 11, 304
Hydrophiinae, 305
hydrophilus, 258
Hydrops, 143
Hydrops martii, 143
Hydrops martii callostictus, 143
Hydrops martii martii, 143
Hydrops triangularis, 143
Hydrops triangularis bassleri, 143
Hydrops triangularis bolivianus, 144
Hydrops triangularis fasciatus, 144
Hydrops triangularis neglectus, 144
Hydrops triangularis triangularis, 144
Hydrops triangularis venezuelensis, 144
hygrophilus, 44
hyoprora, 312
hypoconia, 256
hypomethes, 30
Hypsiglena, 126, 144, 213, 245
Hypsiglena gularis, 145
Hypsiglena tanzeri, 144
Hypsiglena torquata, 144, 145, 146
Hypsiglena torquata affinis, 145
Hypsiglena torquata baueri, 145
Hypsiglena torquata catalinae, 145
Hypsiglena torquata chlorophaea, 145
Hypsiglena torquata deserticola, 145
Hypsiglena torquata dunkeli, 145
Hypsiglena torquata gularis, 145
Hypsiglena torquata jani, 144, 145, 146
Hypsiglena torquata klauberi, 145
Hypsiglena torquata loreala, 145
Hypsiglena torquata martinensis, 146
Hypsiglena torquata nuchalata, 146
Hypsiglena torquata ochrorhyncha, 146
Hypsiglena torquata texana, 145
Hypsiglena torquata tiburonensis, 146
Hypsiglena torquata torquata, 144
Hypsiglena torquata tortugaensis, 146
Hypsiglena torquata unaocularis, 146
Hypsiglena torquata venusta, 146
Hypsirhynchus, 146, 147
Hypsirhynchus ferox, 147
Hypsirhynchus ferox exedrus, 147
Hypsirhynchus ferox ferox, 147
Hypsirhynchus ferox paracrousis, 147
Hypsirhynchus ferox scalaris, 147

Ialtris, 147
Ialtris agyrtes, 147
Ialtris dorsalis, 147
Ialtris parishi, 148
Ibarra's Burrowing Snake, 59
ibarrorum, 59
ibiboboca, 297
Ibiboboca, 297
iglesiasi, 204, 316

iheringi, 165, 198
Ihering's Burrowing Snake, 198
Ihering's Snake, 165
Île-à-Vache Racer, 67
Îles des Saintes Racer, 62
Ilysiidae, 37
Imantodes, 148
Imantodes cenchoa, 148
Imantodes cenchoa cenchoa, 148
Imantodes cenchoa leucomelas, 148
Imantodes cenchoa semifasciatus, 149
Imantodes gemmistratus, 149, 150, 347
Imantodes gemmistratus gemmistratus, 149
Imantodes gemmistratus gracillimus, 149
Imantodes gemmistratus latistratus, 149
Imantodes gemmistratus luciodorsus, 149
Imantodes gemmistratus oliveri, 149, 150, 347
Imantodes gemmistratus reticulatus, 150, 347
Imantodes gemmistratus splendidus, 150
Imantodes inornatus, 150
Imantodes lentiferus, 150
Imantodes phantasma, 150
Imantodes semifasciatus, 148
Imantodes splendidus oliveri, 149
Imantodes tenuissimus, 150
immaculata, 97, 191
Immaculate Red-backed Coffee Snake, 191
immaculatus, 134
impar, 129
impensa, 247
imperator, 39, 40
Imperfect Tellurian Snake, 78
imperfectus, 78
imperialis, 100, 101
importunus, 291
inaequifasciatus, 196, 232
inca, 203
Incachaca Legion Snake, 166
Inca Forest Pit Viper, 310
Incaspis, 203
Incapis cercostroha, 203
incerta, 113
incertae sedis amarali, 348
incertae sedis melanoauchen, 346
Incertae Sedis, 5, 9, 10, 11, 341, 343
incertus, 18
incháusteguii, 278
incomptus, 134
indica, 113
Indigo Snake, 5, 117
Indistinct Tellurian Snake, 78
indistinctus, 78
infernalis, 265, 266
infralabialis, 154
infrataeniatus, 139

infrenalis, 345
ingeri, 167
Inglesia's Lancehead Pit Viper, 316
inornatus, 150
insignissimus, 245
insipidus, 78
insulaepinorum, 270
insulaevaccarum, 278
insulamontana, 249
insulanus, 208
insularis, 316
insularum, 188
Inter-Andean Blindsnake, 17
Interandean Forest Racer, 108
intermedia, 50, 69, 226, 229
Intermediate Capuchin Coral Snake, 294
Intermediate Montane Groundsnake, 107
intermedius, 5, 170, 172, 174, 326
intermontana, 122
Intermontane Valley Tellurian Snake, 76
Intermountain Wandering Gartersnake, 259
iridescens, 78
Iridescent Tellurian Snake, 78
Irregular Coral Snake, 292
irregularis, 86
isabelae, 313, 314
isabelleae, 267
Isla de la Juventud Cativo, 270
Isla de la Juventud Racer, 66
Island Burrowing Sandsnake, 91
Island Racer, 72
Islands Nightsnake, 146
Isla Saóna Racer, 67
isolepis, 175, 345
isozonus, 297
Isthmian Dwarf Boa, 58
Isthmian Earthsnake, 134
Isthmican White-Lipped Snake, 241
isthmicus, 134
itapetinigae, 316
Itá Threadsnake, 26
iversoni, 23

jaegeri, 169
Jaeger's Swamp Legion Snake, 169
Jalisco Milksnake, 154
Jamaican Black Groundsnake, 73
Jamaican Blindsnake, 31
Jamaican Boa, 47
Jamaican Dwarf Boa, 55
Jamaican Long-tailed Groundsnake, 73
Jamaican Racer, 62
Jamaican Red Groundsnake, 73
Jamaican Tree Snake, 62
jamaicensis, 31, 55
janaleeae, 169

Janalee's Legion Snake, 169
jani, 144, 145, 146, 209, 249, 290
Jan's Atlantic Forest Tellurian Snake, 85
Jan's Burrowing Snake, 60
Jan's Centipede Snake, 249
Jan's Diadem Snake, 198
Jan's Hog-nosed Snake, 177
Jan's Racer, 203
Jan's Snail-eater, 113
Jan's Wormsnake, 68
jararaca, 311, 316
Jararacá, 316
Jararacuçu, 316
jararacussu, 316
Jararacusu, 316
Jensen's Groundsnake, 285
Jericoa Snake, 165
Jericó Tellurian Snake, 81
Jericó Threadsnake, 25
joanae, 102
Joan's Highland Snake, 102
joberti, 216, 218
Jobert's Diminutive Snake, 216
johnsoni, 249
Johnson's Centipede Snake, 249
jonathani, 316
Jonathan's Lancehead Pit Viper, 316
joshuai, 25
juliae, 169
juliai, 134
Jumping Pit Viper, 308
Junamina Threadsnake, 23
juvenalis, 168

kanalchutchan, 220
Kanalchutchan, 220
Kansas Glossy Snake, 71
karlschmidti, 287
Keelback, 138
Keeled Earthsnake, 133
Keeled Whipsnake, 93
kennerlyi, 141
Key Ring-necked Snake, 110
kidderi, 276
Kidder's Snail-eater, 276
Kingsnake, 5, 151
kinkelini, 220
Kinkelini's Graceful Brownsnake, 220
kirtlandii, 98
Kirtland's snake, 98
klauberi, 92, 145, 205, 225, 327
Klauber's Leaf-nosed Snake, 205
Klauber's Threadsnake, 27
knoblocki, 153
Kogi Cloud Forest Legion Snake, 168
kogiorum, 168
koppesi, 25

labialis, 106
labiosa, 189
lacerdai, 183
Lacerda's Lizard-eater, 183
La Chatilla, 52
Lachesis, 334
Lachesis melanocephala, 334
Lachesis muta, 334
Lachesis muta muta, 334
Lachesis muta noctivaga, 335
Lachesis muta rhombeata, 334
Lachesis stenophrys, 335
lachrymans, 220
laetus, 123
laevicollis, 94
laevis, 111, 184
La Hotte Blind Snake, 30
Lake Erie Watersnake, 188
lamari, 170
Lamar's Swamp Legion Snake, 170
lambda, 272, 273
lamonae, 168
Lamon Legion Snake, 168
Lampropeltis, 5, 150, 151
Lampropeltis alterna, 150, 151, 153
Lampropeltis alterna alterna, 151
Lampropeltis alterna blairi, 151
Lampropeltis blairi, 151
Lampropeltis blairi alterna, 151
Lampropeltis blairi "alterna" form, 151
Lampropeltis blairi "blairi" phase, 151
Lampropeltis calligaster, 151
Lampropeltis calligaster calligaster, 151
Lampropeltis calligaster occipitolineata, 151
Lampropeltis catalinensis, 152
Lampropeltis doliata, 154
Lampropeltis getula, 151
Lampropeltis getula brooksi, 152
Lampropeltis getula californiae, 152
Lampropeltis getula catalensis, 152
Lampropeltis getula floridana, 152
Lampropeltis getula getula, 151
Lampropeltis getula goini, 152
Lampropeltis getula holbrooki, 152
Lampropeltis getula niger, 152
Lampropeltis getula nigra, 152
Lampropeltis getula nigrits, 152
Lampropeltis getula splendida, 152
Lampropeltis getula sticticeps, 153
Lampropeltis getula yumensis, 152
Lampropeltis getulus catalinensis, 152
Lampropeltis getulus floridana, 152
Lampropeltis mexicana, 153
Lampropeltis mexicana alterna, 151
Lampropeltis mexicana blairi, 151
Lampropeltis mexicana "greeri", 153
Lampropeltis mexicana "leonis", 153

Index

Lampropeltis mexicana "thayeri", 153
Lampropeltis pyromelana, 153
Lampropeltis pyromelana infralabialis, 154
Lampropeltis pyromelana knoblocki, 153
Lampropeltis pyromelana pyromelana, 153
Lampropeltis pyromelana woodini, 154
Lampropeltis rhombomaculata, 151
Lampropeltis ruthveni, 150, 154
Lampropeltis triangulum, 154
Lampropeltis triangulum abnorma, 154
Lampropeltis triangulum amaura, 154
Lampropeltis triangulum andesiana, 154
Lampropeltis triangulum annulata, 154
Lampropeltis triangulum arcifera, 154
Lampropeltis triangulum blanchardi, 155
Lampropeltis triangulum campbelli, 155
Lampropeltis triangulum celaenops, 155
Lampropeltis triangulum conanti, 155
Lampropeltis triangulum dixoni, 155
Lampropeltis triangulum elapsoides, 155
Lampropeltis triangulum gaigae, 155
Lampropeltis triangulum gentilis, 155
Lampropeltis triangulum hondurensis, 155
Lampropeltis triangulum micropholis, 155
Lampropeltis triangulum multistrata, 155
Lampropeltis triangulum nelsoni, 155
Lampropeltis triangulum oligozona, 156
Lampropeltis triangulum polyzona, 156
Lampropeltis triangulum schmidti, 156
Lampropeltis triangulum sinaloae, 156
Lampropeltis triangulum smithi, 156
Lampropeltis triangulum stuarti, 156
Lampropeltis triangulum syspila, 156
Lampropeltis triangulum taylori, 156
Lampropeltis triangulum triangulum, 154
Lampropeltis zonata, 156
Lampropeltis zonata agalma, 156
Lampropeltis zonata herrerae, 157
Lampropeltis zonata multicincta, 157
Lampropeltis zonata multifasciata, 157
Lampropeltis zonata parvirubra, 157
Lampropeltis zonata pulchra, 157
Lampropeltis zonata zonata, 5, 156
lanceolatus, 316
lancinii, 79
Lancini's Tellurian Snake, 79
landoi, 73
langsdorffi, 297
Langsdorff's Coral Snake, 297
lannomi, 327
lansbergi, 336
lansbergii, 349
Lansberg's Hog-nosed Pit Viper, 336
lanthanus, 55
La Pedrera Tellurian Snake, 79
larcorum, 159
Large-blotched Tolucan Groundsnake, 268

Large-headed Snail-eater, 114
Large-nose Mexican Earthsnake, 105
La Rojilla, 251
lasallei, 79, 242
Las Tunas Groundsnake, 73
lateralis, 180, 269, 309
lateristriga, 281
lateritius, 102, 103
laterorepens, 321
laterstriga, 218, 279
Laticauda, 304, 305
Laticauda colubrina, 304
Laticaudinae, 304
laticeps, 204
laticinctus, 135, 306
laticollaris, 135, 281, 297
latifascia, 273
latifasciata, 114, 213
latifasciatum, 59
latifasciatus, 297
latifrons, 79
latifrontalis, 114, 135
Latin American Centipede-eating Snake, 237
Latin American Earthsnake, 132
Latin American Spider-eating Snake, 237
latirostris, 202
latistratus, 149
lativittata, 203
lativittatus, 198
latrunculus, 99
laureata, 220
laurenti, 94
Laurent's Whipsnake, 94
La Vega Racer, 64
lavillai, 232
La Villas Province Dwarf Boa, 57
La Villa's Tree Snake, 232
Leaf-nosed Snake, 205
lecontei, 224
Leeward Blindsnake, 31
Leeward Legion Snake, 169
Leeward Racer, 61
Legion Snake, 165
lehmanni, 79
Lehmann's Tellurian Snake, 79
lehneri, 31
Leimadophis, 119, 165, 172, 200
Leimadophis atahuallpae, 165
Leimadophis epinephelus ecuadorensis, 168
Leimadophis melanostigma, 218
Leimadophis oligolepis, 174
Leimadophis poecilogryus, 172, 348
Leimadophis poecilogyrus intermedius, 172
Leimadophis poecilotyrus californica incertae sedis, 348
Leimadophis pygmaeus, 165

Leimadophis simonsii, 165
Leimadophis typhlus typhlus, 175
Lely Mountains Threadsnake, 22
Lema's Burrowing Snake, 198
Lema's Snake, 255
lemniscata, 226
lemniscatus, 198, 297
lemosespinali, 6
lempira, 249
lentiferus, 150
lentiginosum, 223
leonis, 153
Leopard Dwarf Boa, 57
leopardinus, 139
lepidus, 124, 197, 327
leprieuri, 346
leptepileptus, 25
Leptocalamus limitaneus, 79
Leptodeira, 157, 213, 245
Leptodeira annulata, 157
Leptodeira annulata annulata, 158
Leptodeira annulata ashmeadi, 158
Leptodeira annulata cussiliris, 158
Leptodeira annulata pulchriceps, 158
Leptodeira annulata rhombifera, 158, 159
Leptodeira bakeri, 158
Leptodeira (?) discolor, 245
Leptodeira frenata, 158
Leptodeira frenata frenata, 158
Leptodeira frenata malleisi, 158
Leptodeira frenata yucatanensis, 159
Leptodeira maculata, 159
Leptodeira nigrofasciata, 159
Leptodeira nigrofasciata mystacina, 159
Leptodeira nigrofasciata nigrofasciata, 159
Leptodeira polysticta, 160
Leptodeira punctata, 159
Leptodeira rubricata, 159
Leptodeira septentrionalis, 159
Leptodeira septentrionalis larcorum, 159
Leptodeira septentrionalis ornata, 160
Leptodeira septentrionalis polysticta, 160
Leptodeira septentrionalis septentrionalis, 159
Leptodeira smithi, 159
Leptodeira splendida, 160
Leptodeira splendida bressoni, 160
Leptodeira splendida ephippiata, 160
Leptodeira splendida splendida, 160
Leptodrymus, 160
Leptodrymus pulcherrimus, 160
Leptognathus, 347
Leptognathus andrei, 347
Leptognatus brevis, 347
Leptomicrurus, 10, 286, 287, 289
Leptomicrurus collaris, 287
Leptomicrurus collaris breviventris, 287
Leptomicrurus collaris collaris, 287

Leptomicrurus narduccii, 287
Leptomicrurus narduccii melanotus, 287
Leptomicrurus narduccii narduccii, 287
Leptomicrurus narducci schmidti, 288
Leptomicrurus schmidti, 287, 288
Leptomicrurus scutiventris, 287, 288
Leptophis, 160, 161
Leptophis ahaetulla, 161
Leptophis ahaetulla ahaetulla, 161
Leptophis ahaetulla bocourti, 161
Leptophis ahaetulla bolivianus, 161
Leptophis ahaetulla chocoensis, 161, 162
Leptophis ahaetulla coeruleodorsus, 161
Leptophis ahaetulla copei, 161
Leptophis ahaetulla liocercus, 161
Leptophis ahaetulla marginatus, 161
Leptophis ahaetulla nigromarginatus, 161
Leptophis ahaetulla occidentalis, 162
Leptophis ahaetulla ortoni, 162
Leptophis ahaetulla praestans, 162
Leptophis ahaetulla urostictus, 161, 162
Leptophis cupreus, 162
Leptophis depressirostris, 162
Leptophis diplotrophis, 162
Leptophis diplotrophis diplotrophis, 162
Leptophis diplotrophis forreri, 162
Leptophis mexicanus, 162
Leptophis mexicanus hoeversi, 163
Leptophis mexicanus mexicanus, 163
Leptophis mexicanus septentrionalis, 163
Leptophis mexicanus yucatanensis, 163
Leptophis modestus, 161, 163
Leptophis nebulosus, 163
Leptophis riveti, 163
Leptophis santamartensis, 163
Leptophis stimsoni, 163
Leptotyphlopidae, 15, 19
Leptotyphlops, 20
Leptotyphlops affinis, 20
Leptotyphlops albifrons, 20, 21, 27, 347
Leptotyphlops albifrons albifrons, 21
Leptotyphlops albifrons margaritae, 21
Leptotyphlops albipunctus, 21
Leptotyphlops anthracinus, 21
Leptotyphlops asbolepis, 21
Leptotyphlops australis, 21
Leptotyphlops bilineatus, 22
Leptotyphlops borapeliotes, 22
Leptotyphlops borrichianus, 22
Leptotyphlops brasiliensis, 22
Leptotyphlops bressoni, 22
Leptotyphlops brevissimus, 22
Leptotyphlops calypso, 22
Leptotyphlops collaris, 22
Leptotyphlops columbi, 22
Leptotyphlops cupinensis, 22
Leptotyphlops diaplocius, 22

Leptotyphlops dimidiatus, 22
Leptotyphlops dugandi, 23
Leptotyphlops dulcis, 23
Leptotyphlops dulcis dissectus, 23
Leptotyphlops dulcis dulcis, 23
Leptotyphlops dulcis iversoni, 23
Leptotyphlops dulcis myopicus, 23
Leptotyphlops dulcis supraocularis, 23
Leptotyphlops gadowi, 347
Leptotyphlops goudoti, 23
Leptotyphlops goudoti ater, 24
Leptotyphlops goudoti bakewelli, 24
Leptotyphlops goudoti goudoti, 23
Leptotyphlops goudoti magnamaculata, 24
Leptotyphlops goudoti phenops, 24
Leptotyphlops guayaquilensis, 24
Leptotyphlops humilis, 24, 25
Leptotyphlops humilis boettgeri, 24
Leptotyphlops humilis cahuilae, 24
Leptotyphlops humilis chihuahuaensis, 25
Leptotyphlops humilis dugesi, 25
Leptotyphlops humilis humilis, 24
Leptotyphlops humilis levitoni, 25
Leptotyphlops humilis lindsayi, 25
Leptotyphlops humilis segregus, 25
Leptotyphlops humilis slevini, 24
Leptotyphlops humilis tenuiculus, 25
Leptotyphlops humilis utahensis, 25
Leptotyphlops joshuai, 25
Leptotyphlops koppesi, 25
Leptotyphlops leptepileptus, 25
Leptotyphlops macrolepis, 26
Leptotyphlops maximus, 26
Leptotyphlops melanotermus, 26
Leptotyphlops melanurus, 26
Leptotyphlops munoai, 26
Leptotyphlops myopicus, 23
Leptotyphlops myopicus dissectus, 23
Leptotyphlops myopicus myopicus, 23
Leptotyphlops nasalis, 26
Leptotyphlops nicefori, 26
Leptotyphlops peruvianus, 26
Leptotyphlops pyrites, 26
Leptotyphlops rubrolineatus, 26
Leptotyphlops rufidorsus, 26
Leptotyphlops salgueiroi, 26
Leptotyphlops septemstriatus, 27
Leptotyphlops signatus, 27
Leptotyphlops striatulus, 27
Leptotyphlops subcrotillus, 27
Leptotyphlops teaguei, 27
Leptotyphlops tenella, 20
Leptotyphlops tenellus, 27
Leptotyphlops tesselatus, 27
Leptotyphlops tricolor, 27
Leptotyphlops unguirostris, 28
Leptotyphlops undecimstriata, 21, 347

Leptotyphlops undecimstriatus, 20, 21
Leptotyphlops vellardi, 28
Leptotyphlops weyrauchi, 28
Lesser Blindsnake, 17
Lesser Centipede-eater, 251
leucocephalus, 234
leucogaster, 165
leucomelas, 148, 195, 231
leucometapus, 94
leucostoma, 307
leucostomus, 241
leucurus, 316
levitoni, 25
lewisi, 270
Liana Snake, 234
Lichanura, 8, 48, 49
Lichanura gracia, 51
Lichanura myriolepis, 51
Lichanura roseofusca, 49, 51
Lichanura saslowi, 51
Lichanura trivirgata, 49, 50, 51
Lichanura trivirgata arizona, 50
Lichanura trivirgata bostici, 50
Lichanura trivirgata gracia, 50, 51
Lichanura trivirgata intermedia, 50
Lichanura trivirgata myriolepis, 49, 50, 51
Lichanura trivirgata roseofusca, 50, 51
Lichanura trivirgata saskelli, 51
Lichanura trivirgata saslowi, 51
Lichanura trivirgata trivirgata, 50
Lichen-like Forest Pit Viper, 311
lichenosa, 311
lichenosus, 311
Lichtenstein's Green Racer, 202
liebmanni, 90
Liebmann's Earth Runner, 90
limbatus, 298
limitaneus, 79
limnetes, 240
lincolni, 67
Lincoln's Racer, 67
lindheimeri, 123
Lindheimer's Ratsnake, 123
lindsayi, 25
lineapiatus, 217
linearis, 231, 262
lineata, 169, 225, 268
lineaticollis, 207, 209
lineatulus, 179
lineatum, 274
lineatus, 103, 104, 169, 180, 345
Lined Black-bellied Gartersnake, 262
Lined Blindsnake, 33
Lined Coachwhip, 179
Lined Legion Snake, 169
Lined Road Guarder, 104
Lined Snake, 274

Index

Lined Tolucan Groundsnake, 268
Lineless Tolucan Groundsnake, 268
lineolatus, 9, 179
lineri, 88
lintoni, 254, 255
Linton's Dwarf Short-tailed Snake, 255
liocercus, 161
Liochlorophis, 163, 164, 192
Liochlorophis vernalis, 164
Liochlorophis vernalis blanchardi, 164
Liochlorophis vernalis borealis, 164
Liochlorophis vernalis vernalis, 164
Liodytes, 216
Lioheterophis, 164, 165
Lioheterophis ineringi, 165
Liophis, 61, 72, 119, 165, 200, 218
Liophis almadensis, 165
Liophis alticolus, 168
Liophis amarali, 348
Liophis andinus, 166
Liophis anomalus, 166
Liophis argentinus, 174
Liophis ater, 168
Liophis atraventer, 166
Liophis australis, 171
Liophis bicolor, 171
Liophis bipraeocularis, 168
Liophis breviceps, 166
Liophis breviceps breviceps, 166
Liophis breviceps canaima, 166
Liophis californica, 348
Liophis carajasensis, 166
Liophis ceii, 166
Liophis cobellus, 166, 345
Liophis cobellus cobellus, 166
Liophis cobellus dyticus, 167
Liophis cobellus ingeri, 167
Liophis cobellus taeniogaster, 167
Liophis cobellus trebbaui, 167
Liophis collaris, 171
Liophis cursor, 167
Liophis dictyodes, 171
Liophis dilepis, 167
Liophis dorsalis, 169
Liophis elegantissimus, 167
Liophis epinephelus, 167
Liophis epinephelus albiventris, 167
Liophis epinephelus bimaculatus, 168
Liophis epinephelus ecuadorensis, 168
Liophis epinephelus epinephelus, 167
Liophis epinephelus fraseri, 168
Liophis epinephelus juvenalis, 168
Liophis epinephelus kogiorum, 168
Liophis epinephelus lamonae, 168
Liophis epinephelus opisthotaenius, 168
Liophis epinephelus pseudocobellus, 168
Liophis festae, 168

Liophis flavifrenatus, 168
Liophis forsteri, 172
Liophis frenatus, 168
Liophis fuscus, 171
Liophis graphicus, 173
Liophis guentheri, 169
Liophis ingeri, 167
Liophis isolepis, 175
Liophis jaegeri, 169
Liophis jaegeri coralliventris, 169
Liophis jaegeri jaegeri, 169
Liophis janaleeae, 169
Liophis joberti, 216
Liophis juliae, 169
Liophis juliae copeae, 169
Liophis juliae juliae, 169
Liophis juliae mariae, 169
Liophis leucogaster, 165
Liophis lineata, 169
Liophis lineatus, 169, 345
Liophis longiventris, 169
Liophis macrops, 175
Liophis mariahellanae, 170
Liophis maryellenae, 170
Liophis melanotus, 170
Liophis melanotus lamari, 170
Liophis melanotus nesos, 170
Liophis melaontus melanotus, 170
Liophis meridionalis, 170
Liophis miliaris, 170, 174
Liophis miliaris amazonicus, 171
Liophis miliaris chrysostomus, 171
Liophis miliaris intermedius, 5, 170, 172, 174
Liophis miliaris merremi, 171
Liophis miliaris miliaris, 171
Liophis miliaris mossoroensis, 171
Liophis miliaris orinus, 171
Liophis miliaris semiaureus, 171
Liophis m-nigrum, 172
Liophis natricoides, 171
Liophis obtusus, 216
Liophis oligolepis, 171, 174
Liophis orientalis, 171
Liophis ornata, 171
Liophis ornata var. semiaureus, 171
Liophis ornatus, 172
Liophis paucidens, 172
Liophis perfuscus, 172
Liophis poecilogyrus, 172
Liophis poecilogyrus caesius, 172, 173
Liophis poecilogyrus intermedius, 170, 172
Liophis poecilogyrus pictostriatus, 173
Liophis poecilogyrus platensis, 173
Liophis poecilogyrus poecilogyrus, 172, 173
Liophis poecilogyrus schotti, 172, 173
Liophis poecilogyrus subfasciatus, 173

Liophis poecilogyrus sublineatus, 172, 173
Liophis poecilogyrus xerophilus, 172
Liophis problematicus, 173
Liophis pulcher, 174
Liophis purpurans, 166, 171
Liophis quadrilineatus, 168
Liophis reginae, 170, 173
Liophis reginae macrosomus, 174
Liophis reginae maculatus, 348
Liophis reginae maculicaudus, 174
Liophis reginae reginae, 173
Liophis reginae semilineatus, 174
Liophis reginae zweifeli, 174
Liophis sagittifer, 174
Liophis sagittifer modestus, 174
Liophis sagittifer sagittifer, 174
Liophis serpentinus, 167
Liophis taeniurus, 175
Liophis torrenicolus, 175
Liophis trifasciatus, 174
Liophis triscalis, 175
Liophis typhlus, 175
Liophis typhlus brachyurus, 175
Liophis typhlus elaeoides, 175
Liophis typhlus typhlus, 175
Liophis undulatus, 175, 218
Liophis vanzolinii, 175
Liophis violaceus, 173
Liophis viridis, 175
Liophis viridis prasinus, 176
Liophis viridis viridis, 176
Liophis vitti, 176
Liophis williamsi, 176
Liotyphlops, 16, 17, 19
Liotyphlops albirostris, 17
Liotyphlops anops, 17, 18
Liotyphlops argaleus, 18
Liotyphlops beui, 18
Liotyphlops bordensis, 18
Liotyphlops canellei, 18
Liotyphlops caracasensis, 18
Liotyphlops cucutae, 18
Liotyphlops guentheri, 19
Liotyphlops incertus, 18, 19
Liotyphlops metae, 18
Liotyphlops petersi, 17, 18
Liotyphlops rowani, 17, 18
Liotyphlops schubarti, 18
Liotyphlops ternetzii, 18
Liotyphlops wilderi, 18, 19
lippiens, 242
Littersnake, 218
Little Black Coral Snake, 287
Little Cayman Dwarf Boa, 54
Little Cayman Racer, 63
Little-eyed Tellurian Snake, 77
livida, 202

lividum, 202
Lizard-eater, 182
lodingi, 209
Lojan Lancehead Pit Viper, 317
lojanus, 317
Loja Tellurian Snake, 76
Long-banded Coral Snake, 297
Long Black-backed Coral Snake, 287
longicauda, 40
longicaudata, 69
longicaudatus, 234
longifrenis, 231
longifrontalis, 249
longimaculatus, 85
longinquus, 102
longissimus, 349
longiventris, 169
Long-nosed Snake, 224
Longtail Blindsnake, 28
Long-tailed blindsnake, 28
Long-tailed Rattlesnake, 331
Long-tailed Whipsnake, 95
loreala, 145
lorenzoensis, 6, 331
Los Mangos Tellurian Snake, 80
Lost Centipede-eater, 253
Lost Snail-eater, 230
Los Tuxtlas Rainforest Earthsnake, 134
Louisiana Milksnake, 154
Louisiana Pinesnake, 210
loveridgei, 79
Loveridge's Tellurian Snake, 79
lowei, 265
Lower California Rattlesnake, 325
Lower Central American Neotropical Brownsnake, 280
Lower Central American Snail-eater, 112
Lower Montane Groundsnake, 107
Lowland Watersnake, 242
Loxocemidae, 38, 52
Loxocemus, 52
Loxocemus bicolor, 52
lucasensis, 330
lucidus, 206
luciodorsus, 149
lumbricalis, 31
lutescens, 338
lutosus, 7, 333
lutzi, 317
Lutz's Lancehead Pit Viper, 317
Lygophis, 119, 165, 168, 169, 200
Lygophis coralliventris, 169
Lygophis flavifrenatus, 168
Lyresnake, 271
lyrophanes, 272, 273
Lystrophis, 176
Lystrophis dorbignyi, 176

Lystrophis dorbignyi chacoensis, 176
Lystrophis dorbignyi dorbignyi, 176
Lystrophis dorbignyi orientalis, 177
Lystrophis dorbignyi uruguayensis, 177
Lystrophis histricus, 177
Lystrophis matogrossensis, 177
Lystrophis nattereri, 177
Lystrophis pulcher, 177
Lystrophis semicinctus, 177
Lystrophis semicinctus aequicinctus, 177
Lystrophis semicinctus multicinctus, 177
Lystrophis semicinctus nigrocinctus, 177

mabila, 217
macdougalli, 221, 276, 292
MacDougall's Coral Snake, 292
MacDougall's Graceful Brownsnake, 221
MacDougall's Snail-eater, 276
macrolepis, 26
macrops, 175
macrosomus, 174
Macrostomata, 8, 38
maculata, 77, 87, 159, 190
maculatas, 296
maculatum, 77
maculatus, 56, 79, 87, 237, 348
maculicaudus, 174
maculiferus, 135
maculirostris, 297
maculosus, 327
maculoventris, 95
Madre de Dios Tellurian Snake, 81
magnamaculata, 24
Magnotes Threadsnake, 26
mahnerti, 311
Majá, 43
majalis, 192
major, 79
malleisi, 158
Malleis's Cat-eyed Snake, 158
Manabí Hog-nosed Pit Viper, 336
Managua Threadsnake, 26
manegarzoni, 201
Mangrove Saltmarsh Snake, 186
Mangrove Snake, 86
manizalesensis, 79, 82
Mano de Metate, 308
Manolepis, 177
Manolepis putnami, 178
manototus, 45
manselli, 62
Many-banded Coral Snake, 299
Many-banded Tellurian Snake, 80
Many-lined Neotropical Brownsnake, 282
Many-scaled Snail-eater, 115
Manzinaies Tellurian Snake, 76
Maracaibo False Coral Snake, 129

Marahuaca Large-eyed Snake, 256
marahuaquensis, 256
Marajoan Rattlesnake, 323
marajoensis, 323
marajoensis, 314, 317
Marajó Island Anaconda, 47
Marajó Island Rainbow Boa, 44
Marajó Lancehead Pit Viper, 317
Maranhao Slug-eater, 232
Mara's False Coral Snake, 282
marcapatae, 95
marcellae, 221
Marcella's Graceful Brownsnake, 221
marchi, 309
March's Palm Pit Viper, 309
marcianus, 261
marcosensis, 127
Marcy's Checkered Gartersnake, 261
margaritae, 21, 117
Margarita Island Cribo, 117
margaritiferus, 118, 298
marginatus, 161
mariae, 169
mariahellanae, 170
maricelae, 322
Maricopa Leaf-nosed Snake, 206
mariselae, 79
Marisela's Tellurian Snake, 79
Maritime Gartersnake, 266
Marsh Brownsnake, 240
Marsh Burrowing Snake, 199
martii, 143
martindelcampo, 247
martinensis, 146
Martin García Threadsnake, 21
Martinique Lancehead Pit Viper, 316
Martinique Legion Snake, 167
Martinique Threadsnake, 22
martiusi, 302
maryellenae, 170
Mary Ellen's Legion Snake, 170
Massasauga, 5, 337
Masticophis, 9, 178
Masticophis anthonyi, 178
Masticophis aurigulus, 178
Masticophis aurigulus aurigulus, 178
Masticophis aurigulus barbouri, 178
Masticophis bilineatus, 9, 178
Masticophis bilineatus bilineatus, 9, 179
Masticophis bilineatus lineolatus, 9, 179
Masticophis bilineatus slevini, 9, 179
Masticophis flagellum, 179
Masticophis flagellum cingulum, 179
Masticophis flagellum flagellum, 179
Masticophis flagellum flavigularis, 180
Masticophis flagellum frenatus, 179
Masticophis flagellum fulginosus, 179

Index

Masticophis flagellum lineatulus, 179
Masticophis flagellum lineatus, 180
Masticophis flagellum piceus, 179
Masticophis flagellum ruddocki, 179
Masticophis flagellum stirolatus, 180
Masticophis flagellum testaceus, 180
Masticophis lateralis, 180
Masticophis lateralis euryxanthus, 180
Masticophis lateralis lateralis, 180
Masticophis lineolatus, 179
Masticophis menotvarius, 180
Masticophis menotvarius centralis, 180
Masticophis menotvarius mentovarius, 180
Masticophis menotvarius striolatus, 180
Masticophis menotvarius suborbitalis, 181
Masticophis menotvarius variolosus, 181
Masticophis ruthveni, 181
Masticophis schotti, 181
Masticophis schotti ruthveni, 181
Masticophis schotti schotti, 181
Masticophis sleveni, 181
Masticophis taeniatus, 181
Masticophis taeniatus australis, 181
Masticophis taeniatus girardi, 182
Masticophis taeniatus ornatus, 182
Masticophis taeniatus ruthveni, 181
Masticophis taeniatus schotti, 181
Masticophis taeniatus taeniatus, 181
Mastigodryas, 8, 182
Mastigodryas amarali, 182
Mastigodryas bifossatus, 182
Mastigodryas bifossatus bifossatus, 183
Mastigodryas bifossatus lacerdai, 183
Mastigodryas bifossatus striatus, 183
Mastigodryas bifossatus triseriatus, 183
Mastigodryas bifossatus villelai, 183
Mastigodryas boddaerti, 183
Mastigodryas boddaerti boddaerti, 183
Mastigodryas boddaerti dunni, 183
Mastigodryas boddaerti ruthveni, 183
Mastigodryas bruesi, 183
Mastigodryas cliftoni, 184
Mastigodryas danieli, 184
Mastigodryas dorsalis, 184
Mastigodryas heathi, 184
Mastigodryas melanolomus, 184
Mastigodryas melanolomus alternatus, 184
Mastigodryas melanolomus dorsalis, 184
Mastigodryas melanolomus laevis, 184
Mastigodryas melanolomus melanolomus, 184
Mastigodryas melanolomus slevini, 184
Mastigodryas melanolomus stuarti, 185
Mastigodryas melanolomus tehuanae, 185
Mastigodryas melanolomus veraecrucis, 185
Mastigodryas pleei, 185
Mastigodryas pulchriceps, 185

Mastigodryas sanguiventris, 184, 185
Matagalpa Earthsnake, 134
matogrossensis, 177
Mato Grosso Hog-nosed Snake, 177
Mato Grosso Lancehead Pit Viper, 318
Mato Grosso Racer, 202
Mato Grosso Wormsnake, 69
mattogrossensis, 202, 317
matudai, 121
Matuda's Nightsnake, 121
Matuda's Ratsnake, 121
maurus, 44
maxillaris, 114
Maximilian's Lancehead Pit Viper, 317
maximus, 26
Maya Coral Snake, 297
mayae, 241
Mayan Black-headed Centipede-eater, 248
Mayan Golden-backed Snake, 241
maydis, 119
mccraniei, 46
meahllmorum, 122, 123
medemi, 298, 301
medusa, 310
megalodon, 268
megalolepis, 237
megalops, 260
melanichnus, 64
melanoauchen, 346
melanocephala, 249, 334
melanocephalus, 102
melanogaster, 39, 40, 80, 261, 263
melanogenys, 195, 293
melanoleucus, 207, 209
melanolomus, 184
melanostigma, 120, 218
melanotermus, 26
melanotropis, 119
melanotus, 170, 287
melanura, 243
melanurus, 26, 56, 117, 335
melas, 80
melopyrrha, 66
Mena's Centipede Snake, 249
mendax, 262
menotvarius, 180
mentalis, 346
meridanus, 102
Mérida Pygmy Coral Snake, 298
meridensis, 298
meridionalis, 170, 318, 334
merremi, 171, 283
Merrem's False Pit Viper, 283
Merrem's Legion Snake, 171
mertensi, 199, 269, 274, 277, 298
Mertens's Burrowing Snake, 199
Mertens's Coral Snake, 298

Mertens's Earthsnake, 134
Mertens's Moss Snake, 277
Mesa Central Blotched Gartersnake, 264
Mesa Central Earthsnake, 135
Mesa del Sur Earthsnake, 133
Mesa Verde Nightsnake, 145
Meso-American Gartersnake, 260
Mesopotamian Coral Snake, 290
mesopotamicus, 290
metae, 18
Meta Moss Snake, 277
Metapil, 307
Meta Slender Blindsnake, 18
mexicana, 39, 153, 226, 238, 250
Mexican Alpine Blotched Gartersnake, 264
Mexican Banded Burrowing Snake, 242
Mexican Black-bellied Gartersnake, 261
Mexican Black-headed Snake, 246
Mexican Black-tailed Rattlesnake, 329
Mexican Brownsnake, 241
Mexican Bullsnake, 209
Mexican Burrowing Python, 52
Mexican Coast Ribbonsnake, 263
Mexican Cornsnake, 121
Mexican Dwarf Boa, 52
Mexican Earthsnake, 105
Mexican False Pit Viper, 284
Mexican Gartersnake, 260
Mexican Groundsnake, 235
Mexican Hog-nosed Snake, 141
Mexican Horned Pit Viper, 335
Mexican Jumping Pit Viper, 308
Mexican Kingsnake, 153
Mexican Lance-headed Rattlesnake, 329
Mexican Long-nosed Snake, 224
Mexican Long-tailed Snake, 126
Mexican Lyresnake, 272
Mexican Milksnake, 154
Mexican Moccasin, 306
Mexican Mussurana, 97
Mexican Nightsnake, 121
Mexican Pacific Lowlands Gartersnake, 266
Mexican Parrot Snake, 162
Mexican Patch-nosed Snake, 226
Mexican Plateau Earthsnake, 132
Mexican Pygmy Rattlesnake, 337, 338
Mexican Python, 52
Mexican Racer, 99
Mexican Ring-necked Snake, 110
Mexican Rosy Boa, 50
Mexican Scorpion-eater, 238
Mexican Short-tailed Snake, 241, 242
Mexican Small-headed Rattlesnake, 326
Mexican Snake Eater, 97
Mexican Threadsnake, 25
Mexican Tiger Ratsnake, 237

mexicanus, 16, 162, 163, 237, 308
Mexican Wandering Gartersnake, 260
Mexican West Coast Cribo, 117
Mexican West Coast Rattlesnake, 321
Mexican Whipsnake, 180
Mexican Yellow-bellied Brownsnake, 240
meyeri, 216
micheli, 75
Michoacán Centipede Snake, 247
Michoacán Clear-banded Coral Snake, 293
Michoacán Earthsnake, 135
michoacanensis, 235, 293
Michoacán Groundsnake, 235
Michoacán Stripeless Snake, 103
Michoacán Threadsnake, 22
microgalbineus, 296
micropholis, 155
microphthalmus, 312, 315
microrhynchus, 75
microstomus, 31
Micrurinae, 286
Micruroides, 288
Micruroides euryxanthus, 288
Micruroides euryxanthus australis, 288
Micruroides euryxanthus euryxanthus, 288
Micruroides euryxanthus neglectus, 288
micrurus, 129
Micrurus, 286, 287, 288, 289
Micrurus albicinctus, 289
Micrurus alleni, 289
Micrurus alleni alleni, 289
Micrurus alleni yatesi, 289
Micrurus altirostris, 289
Micrurus ancoralis, 289
Micrurus ancoralis ancoralis, 289
Micrurus ancoralis jani, 290
Micrurus annellatus, 290
Micrurus annellatus annellatus, 290
Micrurus annellatus balzani, 290
Micrurus annellatus bolivianus, 290
Micrurus annellatus montanus, 290
Micrurus averyi, 289
Micrurus baliocoryphus, 290
Micrurus bernadi, 290
Micrurus bocourti, 290
Micrurus bogerti, 290
Micrurus brasiliensis, 291
Micrurus browni, 291
Micrurus browni browni, 291
Micrurus browni importunus, 291
Micrurus browni taylori, 291
Micrurus carnicauda, 294
Micrurus catamayensis, 291
Micrurus circinalis, 291
Micrurus clarki, 291
Micrurus collaris, 287
Micrurus collaris breviventris, 287

Micrurus collaris collaris, 287
Micrurus corallinus, 291
Micrurus decoratus, 291
Micrurus diana, 292
Micrurus diastema, 292
Micrurus diastema affinis, 292
Micrurus diastema aglaeope, 292
Micrurus diastema alienus, 292
Micrurus diastema apiatus, 292
Micrurus diastema diastema, 292
Micrurus diastema macdougalli, 292
Micrurus diastema sapperi, 292
Micrurus dissoleucus, 292
Micrurus dissoleucus dissoleucus, 293
Micrurus dissoleucus dunni, 293
Micrurus dissoleucus melanogenys, 293
Micrurus dissoleucus nigrirostris, 293
Micrurus distans, 293
Micrurus distans distans, 293
Micrurus distans michoacanensis, 293
Micrurus distans oliveri, 293
Micrurus distans zweifeli, 293
Micrurus donosoi, 301
Micrurus dumerili, 293
Micrurus dumerili antioquiensis, 294
Micrurus dumerili carinicauda, 294
Micrurus dumerili colombianus, 294
Micrurus dumerili dumerili, 293
Micrurus dumerili transandinus, 294
Micrurus dumerili venezuelensis, 294
Micrurus elegans, 294
Micrurus elegans elegans, 294
Micrurus elegans veraepacis, 294
Micrurus ephippifer, 294
Micrurus ephippifer ephippifer, 294
Micrurus ephippifer zapotecus, 294
Micrurus filiformis, 295
Micrurus frontalis, 289, 295
Micrurus frontalis altirostris, 289, 295
Micrurus frontalis baliocoryphus, 290
Micrurus frontalis brasiliensis, 291
Micrurus frontalis diana, 292
Micrurus frontalis frontalis, 295
Micrurus frontalis mesopotamicus, 290
Micrurus frontalis multicinctus, 295
Micrurus frontalis pyrrhocryptus, 302
Micrurus frontalis tricolor, 302, 303
Micrurus frontifasciatus, 295
Micrurus fulvius, 295, 296
Micrurus fulvius fitzingeri, 296
Micrurus fulvius fulvius, 296
Micrurus fulvius maculatus, 296
Micrurus fulvius microgalbineus, 296
Micrurus fulvius tener, 296
Micrurus hemprichi, 296
Micrurus hemprichi hemprichi, 296
Micrurus hemprichi ortoni, 296

Micrurus hemprichi rondonianus, 296
Micrurus hippocrepis, 297
Micrurus ibiboboca, 297
Micrurus isozonus, 297
Micrurus karlschmidti, 287
Micrurus langsdorffi, 297
Micrurus langsdorffi ornatissimus, 289
Micrurus laticollaris, 297
Micrurus laticollaris laticollaris, 297
Micrurus laticollaris maculirostris, 297
Micrurus latifasciatus, 297
Micrurus lemniscatus, 297
Micrurus lemniscatus carvalhoi, 297
Micrurus lemniscatus diutius, 298
Micrurus lemniscatus frontifasciatus, 295
Micrurus lemniscatus helleri, 298
Micrurus lemniscatus lemniscatus, 297
Micrurus limbatus, 298
Micrurus limbatus limbatus, 298
Micrurus limbatus spilosomus, 298
Micrurus margaritiferus, 298
Micrurus medemi, 298
Micrurus meridensis, 298
Micrurus mertensi, 298
Micrurus mipartitus, 298
Micrurus mipartitus anomalus, 299
Micrurus mipartitus decussatus, 299
Micrurus mipartitus mipartitus, 299
Micrurus mipartitus popayanensis, 299
Micrurus mipartitus rozei, 299
Micrurus mipartitus semipartitus, 299
Micrurus multifasciatus, 299
Micrurus multifasciatus hertwigi, 299
Micrurus multifasciatus multifasciatus, 299
Micrurus multiscutatus, 300
Micrurus narduccii, 287
Micrurus nebularis, 300
Micrurus nigrocinctus, 300
Micrurus nigrocinctus babaspul, 300
Micrurus nigrocinctus coibensis, 300
Micrurus nigrocinctus divaricatus, 300
Micrurus nigrocinctus mosquitensis, 300
Micrurus nigrocinctus nigrocinctus, 300
Micrurus nigrocinctus ovandoensis, 300
Micrurus nigrocinctus wagneri, 300
Micrurus nigrocinctus zunilensis, 300
Micrurus oligoanellatus, 300
Micrurus ornatissimus, 289, 301
Micrurus pachecogili, 301
Micrurus paraensis, 301
Micrurus peruvianus, 301
Micrurus petersi, 301
Micrurus proximans, 301
Micrurus psyches, 301
Micrurus psyches circinalis, 291, 301
Micrurus psyches donosoi, 301
Micrurus psyches medemi, 301

Index

Micrurus psyches paraensis, 301
Micrurus psyches psyches, 301
Micrurus psyches remotus, 301, 302
Micrurus putumayensis, 301
Micrurus pyrrhocryptus, 302
Micrurus pyrrhocryptus tricolor, 302, 303
Micrurus remotus, 302
Micrurus ruatanus, 302
Micrurus sangilensis, 302
Micrurus spixi, 302
Micrurus spixi martiusi, 302
Micrurus spixi obscurus, 302
Micrurus spixi princeps, 302
Micrurus spixi spixi, 302
Micrurus spurelli, 303
Micrurus steindachneri, 303
Micrurus steindachneri orcesi, 303
Micrurus steindachneri steindachneri, 303
Micrurus stewarti, 303
Micrurus stuarti, 303
Micrurus surinamensis, 303
Micrurus surinamensis nattereri, 303
Micrurus surinamensis surinamensis, 303
Micrurus tener, 296
Micrurus tricolor, 303
Micrurus tschudii, 304
Micrurus tschudii olssoni, 304
Micrurus tschudii tschudii, 304
Mid-Baja California Rosy Boa, 51
Middle American Burrowing Snake, 59
Middle American Earthsnake, 132
Middle American Gophersnake, 209
Middle American Long-tailed Snake, 126
Middle American Scorpion-eating Snake, 237
Middle American Snail-eating Snake, 230
Middle American Snail-sucker, 230
Middle American Swampsnake, 269
Midget Faded Rattlesnake, 333
Midland Brownsnake, 240
Midland Watersnake, 188
Midwestern Wormsnake, 88
mikani, 232
miliaris, 170, 171, 174, 338
Milksnake, 5, 151, 154
mimeticus, 212
Mimic False Coral Snake, 128, 129
Mimic Pine-Oak Snake, 222
mimus, 128, 129
Minas Geraís Burrowing Snake, 198
Miner, 204
Miner Snake, 5, 204
minervae, 345
miniata, 251
minuisquamus, 31
Minute Snakes, 15
miops, 125

mipartitus, 298, 299
miquihuanus, 329
Misiones Mosetenes Tellurian Snake, 76
Misiones Wormsnake, 71
Mississippi Green Watersnake, 186
Mississippi Ring-necked Snake, 111
mitchelli, 328
miyatai, 250
Miyata's Centipede Snake, 250
m-nigrum, 172
Moccasin, 306
mocquardi, 269
Modern Group, 35, 37
modern snakes, 37, 58
Modest Tellurian Snake, 80
modestus, 80, 109, 110, 139, 161, 163, 174
moesta, 250, 346
Mogotes Threadsnake, 26
Mojave Desert Sidewinder, 321
Mojave Glossy Snake, 71
Mojave Green Rattlesnake, 331
mojavensis, 226
Mojave Patch-nosed Snake, 226
Mojave Rattlesnake, 331
Mojave Shovel-nosed Snake, 92
mokasen, 306
mokeson, 307
Mole Kingsnake, 151
Molesnake, 151
molossus, 321, 328
Mona Island Blindsnake, 31
Mona Island Racer, 64
Mona Island Slender Boa, 45, 46
monastus, 31
monensis, 31, 45, 46
Mongrel Snake, 191, 192
Monserrat Blindsnake, 31
Monserrate Leaf-nosed Snake, 206
montana, 97, 221
Montane Groundsnake, 107
Montane Pit Viper, 319
montanus, 290
Monte Cristi Graceful Brownsnake, 221
montecristi, 221
Monterey Ring-necked Snake, 111
monticola, 94, 223
Montserrat Racer, 62
monzonus, 128
moojeni, 317
morenoi, 56, 66
Moreno's Dwarf Boa, 56
Moreno's Racer, 66
morgani, 251
morleyi, 191
mormon, 98, 99
morulus, 327
mosquitensis, 300

Mosquito Coral Snake, 300
mossoroensis, 171
Moss Snake, 277
Mottled Rock Rattlesnake, 327
Mottled Snail-Eater, 231
Mountain Coral Snake, 301
Mountain Earthsnake, 135
Mountain Gartersnake, 259
Mountain Meadow Snake, 59
Mountain Patch-nosed Snake, 225
Mountain Ratsnake, 229
Mountain Shovel-nosed Snake, 91, 344
Mountain Shovel-nose Snake, 91, 344
Mountain Smooth Earthsnake, 283
Mt. Aripo Parrot Snake, 163
Mt. Roraima Rattlesnake, 323
Mudsnake, 129
muertensis, 6, 328
mugitus, 209
multicincta, 69, 157
multicinctus, 80, 177, 295
multifasciata, 157
multifasciatus, 299
multilineata, 282
multilineatus, 29
multipunctatus, 199
multiscutatus, 300
multistictum, 239
multistrata, 155
multiventris, 94
munoai, 26
murinus, 47
Mussurana, 96
muta, 334
mutabilis, 229, 235
mutitorques, 135
myersi, 221, 282, 312
Myers's Graceful Brownsnake, 221
Myers's Neotropical Brownsnake, 282
myopicus, 23
myriolepis, 49, 50, 51
mystacina, 159

nacaritae, 315
narduccii, 287
Narrow-headed Gartersnake, 263
Narrow-headed Serra Snake, 277
nasalis, 26, 135
nasicus, 140
Nassau Boa, 46
nasus, 105
nasutum, 336
nasutus, 199
natricoides, 171
Natrix, 10, 93, 185, 216, 266
Natrix cinnamomea, 93, 348
Natrix melanostigma, 120

Natrix sipedon, 5
Natrix valida, 266
Natrix valida celano, 266
nattereri, 177, 202, 255, 303
naugus, 30
Navassa Island Dwarf Boa, 56
Nayarit Coral Snake, 301
Nazas Watersnake, 186
nebularis, 80, 300
nebulata, 231, 347
nebulatus, 66
nebulosa, 40
nebulosus, 40, 163
Neck-banded Snake, 227
neglecta, 187
neglectus, 144, 288
neilli, 231, 251
Neill's Pygmy Snail-eater, 231
neivai, 112, 114
nelsoni, 155, 275
Nelson's Milksnake, 155
neoleonensis, 324
Neotropical Bird Snake, 214
Neotropical Brownsnake, 280
Neotropical Lizard-eater, 183
Neotropical Lizard-eating Snake, 182
Neotropical Mudsnake, 143
Neotropical Pit Viper, 312
Neotropical Python, 52
Neotropical Racer, 61, 63, 118
Neotropical Ratsnake, 229
Neotropical Rattlesnake, 321
Neotropical Snail-eating Snake, 112
Neotropical Speckled Racer, 118
Neotropical Threadsnake, 27
Neotropical Tree Boa, 41
Neotropical Vinesnake, 193
Neotropical Watersnake, 138
Neotropical Whipsnake, 93, 180
Nerodia, 5, 6, 10, 185, 186, 257, 266, 267
Nerodia clarki, 186
Nerodia clarki clarki, 186
Nerodia clarki compressicauda, 186
Nerodia clarki taeniata, 186
Nerodia cyclopion, 186
Nerodia cyclopion floridana, 187
Nerodia erythrogaster, 186
Nerodia erythrogaster alta, 186
Nerodia erythrogaster bogerti, 186
Nerodia erythrogaster erythrogaster, 186
Nerodia erythrogaster flavigaster, 187
Nerodia erythrogaster neglecta, 187
Nerodia erythrogaster transversa, 187
Nerodia fasciata, 187
Nerodia fasciata confluens, 187
Nerodia fasciata fasciata, 187
Nerodia fasciata pictiventris, 187

Nerodia floridana, 187
Nerodia harteri, 187
Nerodia harteri paucimaculata, 188
Nerodia melanogaster, 261
Nerodia paucimaculata, 187
Nerodia rhombifer, 188
Nerodia rhombifer blanchardi, 188
Nerodia rhombifer rhombifer, 188
Nerodia rhombifer werleri, 188
Nerodia sipedon, 5, 188
Nerodia sipedon engelsi, 189
Nerodia sipedon insularum, 188
Nerodia sipedon pleuralis, 188
Nerodia sipedon sipedon, 188
Nerodia sipedon williamengelsi, 189
Nerodia taxispilota, 189
Nerodia valida, 267
Nerodia valida celaeno, 266
nesos, 170
neuwiedi, 211, 217, 232, 284
Neuwied's False Boa, 211
Neuwied's False Pit Viper, 284
Neuwied's Lancehead Pit Viper, 317
Neuwied's Snail-eater, 232
Nevada Shovel-nosed Snake, 92
nevermanni, 106
newmanorum, 60
Newman's Burrowing Snake, 60
New Mexico Gartersnake, 265
New Mexico Lined Snake, 274
New Mexico Milksnake, 155
New Mexico Ridge-nosed Rattlesnake, 334
New Mexico Threadsnake, 23
New World Coral Snake, 286
nicagus, 120, 244
nicefori, 26, 80
Niceforo's Tellurian Snake, 80
nicholsi, 64, 69, 115,
Nichols's Snail-eater, 115
nietoi, 190
niger, 67, 152
Nightsnake, 144
nigra, 152, 212, 250
nigrescens, 260, 329
nigricaudus, 80
nigriceps, 250
nigrilatum, 60
nigrilatus, 199
nigrirostris, 293
nigrita, 152
nigriventris, 56, 80
nigroalbus, 135
nigrocinctus, 136, 177, 300
nigrofasciata, 159
nigroleteus, 269
nigrolineata, 70
nigroluteus, 269

nigromarginatus, 161
nigronuchalis, 262, 264
nigroterminata, 70
nigroviridis, 309, 349
Ninia, 189
Ninia atrata, 189
Ninia celata, 189
Ninia cerroensis, 190
Ninia diademata, 189
Ninia diademata diademata, 189
Ninia diademata labiosa, 189
Ninia diademata nietoi, 190
Ninia diademata plorator, 190
Ninia espinali, 190
Ninia hudsoni, 190
Ninia maculata, 190
Ninia maculata maculata, 190
Ninia maculata tessellata, 190
Ninia oxynota, 190
Ninia pavimentata, 190
Ninia psephota, 190
Ninia sebae, 190, 191
Ninia sebae immaculata, 191
Ninia sebae morleyi, 191
Ninia sebae punctulata, 191
Ninia sebae sebae, 191
nitae, 264
noctivaga, 72, 335
norrisi, 206
Norris's Leaf-nosed Snake, 206
North American Brownsnake, 239
North American Coachwhip, 178
North American Coral Snake, 295
North American Earthsnake, 282
North American Groundsnake, 235
North American Hog-nosed Snake, 140
North American Long-nosed Snake, 224
North American Mudsnake, 129
North American Queen Snake, 217
North American Racer, 98
North American Rainbow Snake, 129
North American Sandsnake, 90
North American Scarletsnake, 88
North American Smooth Grass Snake, 164
North American Smooth Green Snake, 164
North American Swampsnake, 228
North American Watersnake, 186
North American Whipsnake, 178
North American Wormsnake, 88
Northeastern Brazilian Rattlesnake, 322
Northern Amazon Threadsnake, 27
Northern Black Racer, 98
Northern Black-tailed Rattlesnake, 328
Northern Brownsnake, 239
Northern Cat-eyed Snake, 159
Northern Copperhead, 306
Northern Cuba Racer, 63

Index

Northern Desert Coral Snake, 304
Northern Diamond-backed Water-snake, 188
Northern Eyelash Boa, 53
Northern Florida Swampsnake, 228
Northern Green Frogger, 162
Northern Green Tree Snake, 201
Northern Haiti Vinesnake, 278
Northern Large-eyed Snake, 257
Northern Lined Snake, 274
Northern Mexican Gartersnake, 260
Northern Mexican Short-tailed Snake, 242
Northern Neuwied's Lancehead Pit Viper, 318
Northern Pacific Rattlesnake, 333
Northern Pinesnake, 209
Northern Puerto Rican Groundsnake, 73
Northern Red-bellied Snake, 240
Northern Ribbonsnake, 264
Northern Ring-necked Snake, 110
Northern Rubber Boa, 48
Northern Scarletsnake, 88
Northern Smooth Green Snake, 164
Northern South American Blindsnake, 17
Northern Speckled Racer, 118
Northern Tiger Ratsnake, 237
Northern Watersnake, 188
Northwestern Gartersnake, 262
Northwestern Neotropical Rattlesnake, 322
Northwestern Ring-necked Snake, 110
notaeus, 47
Nothopis, 191
Nothopis rugosus, 192
nubilis, 206
nuchalata, 146
nuchale, 108
Nuevo León Graceful Brownsnake, 221
nummifer, 308
nuntius, 7, 333

oaxaca, 99
oaxacae, 250
Oaxacan Black-tailed Rattlesnake, 329
Oaxacan Burrowing Snake, 59
Oaxacan Cat-eyed Snake, 245
Oaxacan Centipede Snake, 250
Oaxacan Dwarf Boa, 53
Oaxacan Graceful Brownsnake, 218
Oaxacan Patch-nosed Snake, 226
Oaxacan Pygmy Rattlesnake, 339
Oaxacan Small-headed Rattlesnake, 326
Oaxacan Speckled Coral Snake, 292
oaxacus, 320, 329
obesus, 80
obscura, 241
obscurus, 302, 334
obsoleta, 121, 123

obsoletus, 101
obtusirostris, 81
obtusus, 216, 218
Ocaña Tellurian Snake, 84
occidentalis, 40, 71, 72, 81, 109, 110, 119, 162, 281
occiduus, 308
occipitalis, 92, 120, 195, 244, 245
occipitoalbus, 81
occipitolineata, 151
occipitolutea, 86, 87
occipitomaculata, 240
ocellata, 238
Ocellated Mock Viper, 269
Ocellated Night Viper, 269
ocellatus, 128, 214, 268, 259, 269
ochrorhyncha, 146
Octera (la), 60
oculotemporalis, 81
Old World Cat-eyed Snake, 86
olfersii, 202
oligoannelatus, 300
oligolepis, 171, 174, 310
oligostichus, 126
oligozona, 156
oligozonatus, 232
olivacea, 130
Olive Forest Racer, 108
Olive Lizard-eater, 185
Olive Watersnake, 139
Olive Whipsnake, 94
oliveri, 149, 150, 293, 347
Oliver's Blunt-headed Tree Snake, 149
olmec, 308
Olmecan Pit Viper, 308
olssoni, 304
omiltemana, 221
Omilteman Small-headed Rattlesnake, 327
omiltemanus, 136, 327
One-banded Tellurian Snake, 84
oneilli, 232
O'Neill's Snail-eater, 232
oolitica, 250
Opheodrys, 8, 10, 163, 164, 192
Opheodrys aestivus, 8, 10, 192
Opheodrys aestivus aestivus, 192
Opheodrys aestivus carinatus, 192
Opheodrys aestivus conanti, 192
Opheodrys aestivus majalis, 192
Opheodrys mayae, 241
Opheodrys vernalis, 192
Ophis, 283
Ophryacus, 308, 309, 310, 311, 312, 319, 335
Ophryacus melanurus, 335
Ophryacus undulatus, 335
ophryomegas, 336
Opisthoplus, 268

Opisthoplus degener, 268
opisthotaenius, 168
Orange-bellied Lizard-eater, 184
Orange-bellied Swampsnake, 269
Orange-striped Ribbonsnake, 262
orarius, 263
orcesi, 303
ordinoides, 262
oreas, 114
oreganus, 7, 333
Oregon Gartersnake, 258
organica, 92
Organ Pipe Shovel-nosed Snake, 92
orientalis, 67, 171, 177, 195
Oriente Black Groundsnake, 74
Oriente Brown-capped Groundsnake, 73
Oriente Racer, 67
orinus, 171
Orizaba Cribo, 117
orizabensis, 117
ornata, 160, 171, 196, 265
Ornate Cat-eyed Snake, 160
Ornate Coral Snake, 301
Ornate Tellurian Snake, 77
ornatissimus, 289, 301
ornatus, 172, 182
orophias, 40
ortonii, 40
Orton's Parrot Snake, 162, 296
Osage Copperhead, 307
osbornei, 311
Outer Banks Kingsnake, 153
ovandoensis, 300
Oxybelis, 192, 193, 285
Oxybelis aeneus, 89, 193
Oxybelis argenteus, 192
Oxybelis boulengeri, 192
Oxybelis brevirostris, 193
Oxybelis fulgidus, 193
Oxybelis wilsoni, 193
oxynota, 190
Oxyrhopus, 86, 96, 193, 211
Oxyrhopus baileyi, 195
Oxyrhopus clathratus, 194
Oxyrhopus doliatus, 194
Oxyrhopus fitzingeri, 194
Oxyrhopus fitzingeri fitzingeri, 194
Oxyrhopus fitzingeri frizzelli, 194
Oxyrhopus formosus, 194
Oxyrhopus guibei, 194
Oxyrhopus leucomelas, 195
Oxyrhopus maculatus, 87
Oxyrhopus marcapatae, 195
Oxyrhopus melanogenys, 195
Oxyrhopus melanogenys melanogenys, 195
Oxyrhopus melanogenys orientalis, 195
Oxyrhopus occipitalis, 195

Oxyrhopus petola, 195
Oxyrhopus petola aequifasciatus, 196
Oxyrhopus petola baileyi, 196
Oxyrhopus petola digitalis, 195, 196
Oxyrhopus petola petola, 195
Oxyrhopus petola sebae, 196
Oxyrhopus petola semifasciatus, 195, 196
Oxyrhopus rhombifer, 196
Oxyrhopus rhombifer bachmanni, 196
Oxyrhopus rhombifer inaequifasciatus, 196
Oxyrhopus rhombifer rhombifer, 196
Oxyrhopus rhombifer septentrionalis, 196
Oxyrhopus trigeminus, 196
Oxyrhopus trigeminus guibei, 194, 195
Oxyrhopus trigeminus trigeminus, 194, 197
Oxyrhopus venezuelanus, 197
oxyrhynchus, 279

pachecogili, 301
pachyura, 282
Pacific Banded Coffee Snake, 190
Pacific Cativo, 270
Pacific Coast Centipede Snake, 247
Pacific Coast Parrot Snake, 162
Pacific Gartersnake, 267
Pacific Gophersnake, 208
Pacific Long-tailed Snake, 126
Pacific Lowland False Coral Snake, 194
Pacific Patch-nosed Snake, 226
Pacific Red-tailed Coral Snake, 299
Pacific Ring-necked Coffee Snake, 190
Pacific Ring-necked Snake, 110
pacificus, 281
pacta, 71, 72
pahasapae, 241
Painted Coral Snake, 291
Painted Desert Glossy Snake, 72
Painted Lancehead Pit Viper, 318
Painted Steppe Legion Snake, 174
Painted Watersnake, 139
palarostris, 92
Pale Milksnake, 155
pallidulus, 266
pallidus, 257
Palm Pit Viper, 309
paludicola, 99
paludis, 228
pamlica, 251
Pampa Burrowing Snake, 198
Pampa Grande Wormsnake, 69
Pampa Savanna Racer, 245
Pampa Smooth Legion Snake, 171
Pampas Snake, 204
pampineus, 278
Pamplona Tellurian Snake, 81
pamplonensis, 81
Panama Many-banded Coral Snake, 299

Panamanian Coral Snake, 303
Panamanian Dwarf Boa, 58
Panamanian Highland Snake, 102
Panama Pygmy Coral Snake, 293
Panamá Swampsnake, 269
panamensis, 58
Panamint Rattlesnake, 328
Pantanal Coral Snake, 303
Pantherophis, 6
pantostictus, 81
Pará Coral Snake, 301
paracrousis, 147
paradoxus, 29
paraensis, 301
Paraguayan Burrowing Snake, 199
Paraguayan Racer, 202
Paraguayan Rainbow Boa, 44
Paraguayan Tellurian Snake, 81
Paraguay False Coral Snake, 234
paraguayensis, 81, 82
paranaensis, 318
Paraná Lancehead Pit Viper, 318
paraniger, 67
Parapostolepis, 70
Parapostolepis polylepis, 70
Paraptychophis meyeri, 216
paravertebralis, 81
pardalina, 121
pardalis, 57
parietalis, 266
parishi, 148
parkeri, 54
Paroxyrhopus, 86, 285
Paroxyrhopus reticulatus, 285
Parrot Snake, 161
Partida Norte Nightsnake, 145
parvifrons, 67
parvirubra, 157
pastazae, 139
Pastaza Piedmont Coral Snake, 303
Pastaza Watersnake, 139
Patagonian Lancehead Pit Viper, 313
Patagonian Racer, 202
patagoniensis, 202
Patch-nosed Snake, 225
patoquilla, 349
paucicarinatum, 108
paucidens, 81, 87, 172
paucimaculata, 187, 188
paucimaculatus, 272
pauciscutatus, 81
paucisquamis, 57
paucisquamus, 31
pauloensis, 318
pavimentata, 190
pavonina, 114
Pelagic Sea Snake, 305

Pelamis, 304, 305
Pelamis platurus, 305
Peninsula Crowned Snake, 251
Peninsula de Samaná Racer, 67
peninsulae, 67
Peninsular Cat-eyed Snake, 159
Peninsular Glossy Snake, 72
Peninsula Ribbonsnake, 264
Peninsular Racer, 67
Peninsular Rat Snake, 229
Peninsular Road Guarder, 104
Peninsular Stripeless Snake, 102
pepei, 63
Peracca's Legion Snake, 168
Peramba Snake, 125
percarinatum, 108, 345
perfector, 107
perfuscus, 172
perijanensis, 114
Perija Snail-eater, 114
perkinsi, 206
Pernambuco Blindsnake, 31
persimilis, 120, 245
peruana, 115
peruviana, 243, 244, 310
Peruvian Boa Constrictor, 40
Peruvian Centipede Snake, 247
Peruvian Coral Snake, 301
Peruvian Dawn Blindsnake, 16
Peruvian Desert Coral Snake, 298
Peruvian Dwarf Boa, 57
Peruvian Forest Pit Viper, 310
Peruvian Highland False Coral Snake, 195
Peruvian Long-tailed Boa Constrictor, 40
Peruvian Mudsnake, 143
Peruvian Rainbow Boa, 44
Peruvian Running Snake, 102
Peruvian Slender Snake, 243
Peruvian Snail-eater, 115
Peruvian Speckled Coral Snake, 298
Peruvian Tellurian Snake, 81
Peruvian Threadsnake, 26
peruvianus, 26, 81, 301
Peruvian Whipsnake, 94
Peruvian Wormsnake, 70
petersi, 17, 18, 114, 136, 140, 233, 250, 301
Peters's Atlantic Snail-sucker, 114
Peters's Black-striped Snake, 102
Peters's Centipede Snake, 250
Peters's Coral Snake, 301
Peters's Earthsnake, 136
Peters's Running Snake, 101
Peters's Snail-eater, 233
Peters's Watersnake, 140
petlalcalensis, 319
petola, 195
Peugeot Sound Gartersnake, 266

Index

phaeogaster, 307
phaescens, 121, 122
Phalotris, 68, 124, 197
Phalotris bilineatus, 198
Phalotris concolor, 198
Phalotris cuyanus, 198
Phalotris dimidiatus, 198, 199
Phalotris lativittatus, 198
Phalotris lemniscatus, 198
Phalotris lemniscatus divittatus, 198
Phalotris lemniscatus iheringi, 198
Phalotris lemniscatus lemniscatus, 198
Phalotris lemniscatus trilineatus, 199
Phalotris mertensi, 199
Phalotris multipunctatus, 199
Phalotris nasutus, 199
Phalotris nigrilatus, 199
Phalotris punctatus, 199
Phalotris spegazzinii, 198, 199
Phalotris spegazzinii spegazzinii, 199
Phalotris spegazzinii suspectus, 199
Phalotris tricolor, 199
phantasma, 150
phelpsorum, 38
Phelps's Coral Pipesnake, 38
phenops, 24
philipi, 72
philippi, 229, 276
Philipp's Snail-eater, 276
phillipsi, 70
Phillips's Wormsnake, 70
Philodryas, 61, 72, 165, 200, 277
Philodryas aestiva, 200
Philodryas aestiva aestiva, 201
Philodryas aestiva manegarzoni, 201
Philodryas aestiva subcarinata, 201
Philodryas arnaldoi, 201
Philodryas baroni, 201
Philodryas boliviana, 201
Philodryas borellii, 203
Philodryas burmeisteri, 203
Philodryas carbonelli, 202
Philodryas chamissona, 201
Philodryas chamissona chamissana, 201
Philodryas chamissona eremicola, 201
Philodryas cordata, 200, 201
Philodryas elegans, 63
Philodryas elegans rufidorsatus, 63
Philodryas elegans rufodorsatus, 63
Philodryas hoodensis, 202
Philodryas livida, 202
Philodryas mattogrossensis, 202
Philodryas nattereri, 202
Philodryas olfersii, 202
Philodryas olfersii herbea, 202
Philodryas olfersii latirostris, 202
Philodryas olfersii olfersii, 202

Philodryas patagoniensis, 202
Philodryas psammophidea, 203
Philodryas psammophidea lativittata, 203
Philodryas psammophidea psammophidea, 203
Philodryas psammophideus, 201
Philodryas simonsi, 203, 203
Philodryas simonsii, 203
Philodryas tachymenoides, 203
Philodryas trilinaeatus, 203
Philodryas trilineata, 203
Philodryas varia, 203
Philodryas viridissima, 204
Philodryas viridissima laticeps, 204
Philodryas viridissima viridissima, 204
Phimophis, 204
Phimophis chui, 204
Phimophis guerini, 204
Phimophis guianensis, 204
Phimophis iglesiasi, 204
Phimophis scriptorcibatus, 204
Phimophis vittatus, 205
phrenitica, 251
Phyllorhynchus, 205
Phyllorhynchus browni, 205
Phyllorhynchus browni browni, 205
Phyllorhynchus browni fortitus, 205
Phyllorhynchus browni klauberi, 205
Phyllorhynchus browni lucidus, 206
Phyllorhynchus decurtatus, 206
Phyllorhynchus decurtatus arenicola, 206
Phyllorhynchus decurtatus decurtatus, 206
Phyllorhynchus decurtatus norrisi, 206
Phyllorhynchus decurtatus nubilis, 206
Phyllorhynchus decurtatus perkinsi, 206
piauhyensis, 318
picadoi, 308
Picado's Pit Viper, 308
piceivittis, 102
piceus, 179
Pichincha Centipede Snake, 250
Pichincha Snake, 125
pickeringi, 266
picticeps, 65
pictigaster, 307
pictiventris, 140, 187
pictostriatus, 173
pictus, 318
Piedmont Coral Snake, 303
pifanoi, 318
pifanorum, 322
pilonarum, 221
pilsbryi, 57
Pilsbry's Dwarf Boa, 57
Pima Leaf-nosed Snake, 205
Pinar del Rio Dwarf Boa, 56
Pine Island Dwarf Boa, 56

Pine-Oak Snake, 222
Pinesnake, 5, 207, 209
Pine Woods Littersnake, 219
pinicola, 220
Pink-headed Blindsnake, 17
Pink-naped False Coral Snake, 128
Pipe Snakes, 37
Pipesnakes, 8, 37
pirajai, 318
Piraja's Lancehead Pit Viper, 318
piscivorus, 307
Pituophis, 5, 207
Pituophis catenifer, 207, 210
Pituophis catenifer affinis, 208
Pituophis catenifer annectens, 208
Pituophis catenifer bimaris, 209
Pituophis catenifer catenifer, 207
Pituophis catenifer coronalis, 208
Pituophis catenifer deppei, 210
Pituophis catenifer deserticolus, 208
Pituophis catenifer fulginatus, 208
Pituophis catenifer insulanus, 208
Pituophis catenifer pumilus, 208
Pituophis catenifer sayi, 208
Pituophis catenifer vertebralis, 210
Pituophis deppei, 207, 209, 210
Pituophis deppei deppei, 209
Pituophis deppei jani, 209
Pituophis lineaticollis, 207, 209
Pituophis lineaticollis gibsoni, 209
Pituophis lineaticollis lineaticollis, 209
Pituophis melanoleucus, 207, 209
Pituophis melanoleucus lodingi, 209
Pituophis melanoleucus melanoleucus, 209
Pituophis melanoleucus mugitus, 209
Pituophis melanoleucus ruthveni, 210
Pituophis ruthveni, 207, 210
Pituophis vertebralis, 207, 210
Pituophis vertebralis bimaris, 210
Pituophis vertebralis vertebralis, 210
pit viper, 5, 9, 10, 11, 58, 305
Piuaí Wormsnake, 70
placata, 53
Plain-bellied Watersnake, 186
planiceps, 250
Plains Black-headed Snake, 250
Plains Gartersnake, 263
Plains Hog-nosed Snake, 140
Plains Threadsnake, 23
Plateau Mexican Earthsnake, 105
platensis, 173
platirhinos, 141
platurus, 305
platycephalus, 32
Platyinion, 200
Platyinion lividum, 202
platyrhinos, 141

plectovertebralis, 242
pleei, 185
Plee's Lizard-eater, 185
pleuralis, 188
plicatilis, 212
Pliocercus, 8, 279, 280
Pliocercus andrewsi, 279, 281
Pliocercus annellatus, 279
Pliocercus arubricus, 279, 281
Pliocercus bicolor, 279
Pliocercus celatus, 281
Pliocercus deppei, 281
Pliocercus dimidiatus, 279, 281
Pliocercus disastemus, 281
Pliocercus elapoides, 279, 280, 281
Pliocercus elapoides aequalis, 282
Pliocercus elapoides wilmarai, 282
Pliocercus euryzonus, 279, 281
Pliocercus hobartsmithi, 281
Pliocercus laticollaris, 281
Pliocercus occidentalis, 281
Pliocercus pacificus, 281
Pliocercus psycoides, 280
Pliocercus salvadorensis, 281
Pliocercus salvinii, 281
Pliocercus semicinctus, 281
Pliocercus tricinctus, 281
Pliocercus wilmarai, 280
pliolepis, 271
plorator, 190
plumbea, 96, 97
poecilogyrus, 172, 173, *348*
poecilolepis, 233
poecilonotus, 214
poecilopogon, 245
poecilostomus, 215
poeppigi, 82
Pointed Snake, 278
polylepis, 44, 70, 72, 73, 115, 139, 140, 215
polysticha, 122
polysticta, 160
polystictus, 329
polyzona, 156
Pond Snake, 212
popayanensis, 231, 299
Popayán Red-tailed Coral Snake, 299
Popayán Snail-sucker, 231
Porthidium, 308, 309, 310, 311, 312, 319, 335, 336
Porthidium almawebi, 312
Porthidium dunni, 336
Porthidium hespere, 336
Porthidium lansbergi, 336
Porthidium lansbergi arcosae, 336
Porthidium lansbergi houtmanni, 336
Porthidium lansbergi lansbergi, 336
Porthidium lansbergi rozei, 336

Porthidium nasutum, 336
Porthidium ophryomegas, 336
Porthidium volcanicum, 337
Porthidium yucatanicum, 337
Portland Ridge Dwarf Boa, 55
portoricensis, 64
Posadas's Graceful Brownsnake, 221
posadosi, 221
postremus, 262
Potosí Centipede Snake, 252
Potosí Earthsnake, 135
Potosí Threadsnake, 25
potschi, 82
pradoi, 314, 318
Prado's Lancehead Pit Viper, 318
praeocularis, 17, 261
praestans, 162
Prairie Kingsnake, 151
Prairie Rattlesnake, 332
Prairie Ring-necked Snake, 110
prasinus, 176
pratti, 115
Pratt's Snail-eater, 115
Pretty Earthsnake, 132
priapus, 100
pricei, 329
Primitive Alethinophidia, 37
Primitive Group, 15, 37
Primitive Snakes, 37
princeps, 302
Problematic Legion Snake, 173
problematicus, 173
procerum, 73
Procteria, 255
Procteria viridis, 255
protenus, 67
proterops, 101
proximans, 301
proximus, 262
prymnus, 64
psammophidea, 203
psammophideus, 201
psephota, 190
Pseudablabes, 210, 211
Pseudablabes agassizii, 211
Pseudelaphe, 6
Pseudoboa, 96, 194, 211
Pseudoboa cloelia, 96
Pseudoboa coronata, 211
Pseudoboa haasi, 211
Pseudoboa neuwiedi, 211
Pseudoboa nigra, 212
Pseudoboa ornata, 196
Pseudoboa rustica, 97
Pseudoboa serrana, 212
pseudocobellus, 168
pseudocorallus, 129

Pseudoeryx, 212
Pseudoeryx plicatilis, 212
Pseudoeryx plicatilis ecuadorensis, 212
Pseudoeryx plicatilis mimeticus, 212
Pseudoeryx plicatilis plicatilis, 212
Pseudoficimia, 105, 130, 137, 212, 213, 267
Pseudoficimia frontalis, 213
Pseudoficimia frontalis frontalis, 213
Pseudoficimia frontalis hiltoni, 213
Pseudoleptodeira, 126, 144, 157, 213, 245
Pseudoleptodeira latifasciata, 213
Pseudoleptodeira uribei, 213
Pseudotomodon, 213, 214, 268
Pseudotomodon trigonatus, 214
Pseustes, 214, 348
Pseustes poecilonotus, 214
Pseustes poecilonotus argus, 214
Pseustes poecilonotus chrysobronchus, 215
Pseustes poecilonotus poecilonotus, 214
Pseustes poecilonotus polylepis, 215
Pseustes sexcarinatus, 215
Pseustes shropshirei, 215
Pseustes sulphurerus, 215
Pseustes sulphurerus diperkini, 215
Pseustes sulphurerus poecilostomus, 215
Pseustes sulphurerus sulphureus, 215
Psomophis, 215, 216, 218
Psomophis brevirostris, 216
Psomophis genimaculatus, 216
Psomophis joberti, 216
Psomophis obtusus, 216
psyches, 301
psycoides, 280, 282
Ptychophis, 216
Ptychophis flavovirgatus, 216
pubescens, 318
publia, 130, 131
Pueblan Graceful Brownsnake, 221
Pueblan High Desert Coral Snake, 301
Pueblan Milksnake, 155
Puerto Asís Earth Snake, 75
Puerto Asís Tellurian Snake, 75
Puerto Rican Boa, 45
Puerto Rican Brown Blindsnake, 32
Puerto Rican Forest Blindsnake, 32
Puerto Rican Groundsnake, 73
Puerto Rican Racer, 64
Puffing Snake, 214
pulchellus, 109, 110
pulcher, 103, 104, 174, 177, 235
pulcherrimus, 160
pulchra, 157, 283, 310
pulchriceps, 158, 185
pulchrilatus, 263
pullatus, 237
pulveriventris, 221
pumilus, 208

Index

punctata, 159, 236, 311
punctatissimus, 91
punctatus, 5, 109, 111, 199
punctigularis, 101
punctiventris, 82
punctulata, 191
Puntarenas Yellow-bellied Snake, 101
purpurans, 166, 171
pusillus, 32, 329
putnami, 178
putumayensis, 301
Putumayo Coral Snake, 301
pyburni, 136, 277
Pyburn's Earthsnake, 136
pygaea, 228
pygmaea, 277
pygmaeus, 165
Pygmy Coral Snake, 292
Pygmy Moss Snake, 277
Pygmy Rattlesnake, 5, 337, 338
Pygmy Snail-eater, 231
pymi, 69, 70
pyrites, 26
pyromelana, 153
pyrrhocryptus, 302
pyrrhus, 328
Python, 5, 6
Pythonidae, 38
Python regis, 5, 6
Pythons, 3, 8, 9, 37, 38

quadrangulare, 138
quadrangularis, 320
quadricarinatus, 95
quadrilineatus, 168
quadrivirgatum, 60
quadrivittata, 123
quadruplex, 272
Queen Legion Snake, 175
quenselii, 116
Queretaran Dusky Rattlesnake, 320
Queretaro Kingsnake, 154
quimi, 96, 97
quinquelineata, 70, 221
quinquelineatus, 124, 125, 197
quinquevittatus, 103
quirogai, 71

rabdocephalus, 284
Rabdosoma maculatum, 77
radix, 263
Ragged Island Boa, 46
Rainbow Boa, 43
Rainbow Snake, 129
raineyi, 65
Rainforest Cat-eyed Snake, 158
Rainforest Hog-nosed Pit Viper, 336

Rainforest Tellurian Snake, 81
Rainforest Wormsnake, 70
ramirezi, 131
Ramirez's Hook-nosed Snake, 131
Ramphotyphlops, 4, 6, 28
Ramphotyphlops braminus, 4, 28
Rattlesnake, 320
ravus, 338
Rayed Hog-nosed Snake, 177
Reclusive Forest Snake, 218
rectilimbus, 242
Red and Black-banded False Coral
    Mudsnake, 143
Red-backed Coffee Snake, 191
Red-bellied Black-striped Snake, 102
Red-bellied Coffee Snake, 190
Red-bellied Diminutive Snake, 216
Red-bellied Earth Runner, 90
Red-bellied Legion Snake, 166
Red-bellied Mudsnake, 129
Red-bellied Snake, 239, 240
Red-bellied Watersnake, 186
Red Black-headed Snake, 251
Red-black Tellurian Snake, 77
Red Blunt-headed Tree Snake, 149
Red Coachwhip, 179
Red Diamond Rattlesnake, 330
Red Earth Centipede-eater, 251
Red Earthsnake, 136
Red-eyed Tree Snake, 234
redimitus, 132
Red-lined Threadsnake, 26
Red Milksnake, 156
Red Racer, 179
Red-sided Gartersnake, 266
Red-sided Mudsnake, 144
Red-sided Watersnake, 144
Red-spotted Gartersnake, 265
Red-striped Ribbonsnake, 263
Red-striped Threadsnake, 26
Red-tailed Boa Constrictor, 39
Red-tailed Coral Snake, 298
Red-tailed Legion Snake, 166
Red Tellurian Snake, 83
Regal Black-striped Snake, 101
regalis, 110, 111
Regal Legion Snake, 173
Regal Ring-necked Snake, 110
Regina, 186, 216
Regina alleni, 217
Regina alleni alleni, 217
Regina alleni lineapiatus, 217
reginae, 170, 173
Regina grahami, 217
Regina rigida, 217
Regina rigida deltae, 217
Regina rigida rigida, 217

Regina rigida sinicola, 217
Regina septemvittata, 217
Regina septemvittata mabila, 217
Regina septemvittata septemvittata, 217
Reinhardt's Wormsnake, 68
reinwardti, 129
relicquus, 44
relicta, 251
Remote Black-striped Snake, 102
Remote Coral Snake, 302
remotus, 301, 302
Renita Legion Snake, 168
resplendens, 82
Resplendent Tellurian Snake, 82
reticulata, 251
Reticulated Blindsnake, 32
Reticulated Blunt-headed Tree Snake, 150
Reticulated Centipede Snake, 251
Reticulated Regal Legion Snake, 174
reticulatus, 32, 82, 150, 285, 347
Reventazón Tropical Groundsnake, 271
Rhabdosoma crassicaudatum, 344
Rhachidelus, 217, 218
Rhachidelus brazili, 218
Rhadinaea, 4, 215, 218, 279, 280
Rhadinaea affinis, 218
Rhadinaea amarali, 348
Rhadinaea anachoreta, 218
Rhadinaea beui, 245
Rhadinaea bogertorum, 218
Rhadinaea brevirostris, 218, 244
Rhadinaea "brevirostris" group, 215,
    218, 244
Rhadinaea calligaster, 219
Rhadinaea cuneata, 219
Rhadinaea decorata, 219
Rhadinaea dumerali, 219
Rhadinaea flavilata, 219
Rhadinaea forbesi, 219
Rhadinaea fulviceps, 281
Rhadinaea fulvivittis, 219
Rhadinaea fusca, 171
Rhadinaea gaigeae, 219
Rhadinaea genimaculata, 218
Rhadinaea godmani, 219
Rhadinaea hannsteini, 220
Rhadinaea hempsteadae, 220
Rhadinaea hesperia, 220
Rhadinaea hesperia baileyi, 220
Rhadinaea hesperia hesperia, 220
Rhadinaea hesperia hesperiodes, 220
Rhadinaea insignissimus, 245
Rhadinaea joberti, 218
Rhadinaea kanalchutchan, 220
Rhadinaea kinkelini, 220
Rhadinaea lachrymans, 220
Rhadinaea laterstriga, 279

Rhadinaea "laterstriga" group, 218, 279
Rhadinaea laureata, 220
Rhadinaea macdougalli, 221
Rhadinaea marcellae, 221
Rhadinaea mentalis, 346
Rhadinaea montana, 221
Rhadinaea montecristi, 221
Rhadinaea multilineata, 282
Rhadinaea myersi, 221
Rhadinaea obtusus, 216, 218
Rhadinaea omiltemana, 221
Rhadinaea pilonarum, 221
Rhadinaea pinicola, 220
Rhadinaea posadosi, 221
Rhadinaea pulveriventris, 221
Rhadinaea quinquelineata, 221
Rhadinaea rogerromani, 222
Rhadinaea rubricollis, 282
Rhadinaea sargenti, 222
Rhadinaea schistosa, 222
Rhadinaea serperaster, 222
Rhadinaea stadelmani, 222
Rhadinaea steinbachi, 218, 349
Rhadinaea taeniata, 222
Rhadinaea taeniata aemula, 222
Rhadinaea taeniata taeniata, 222
Rhadinaea tolpanorum, 222
Rhadinaea undulatus, 119, 218
Rhadinaea "undulatus" group, 119, 218
Rhadinaea vermiculaticeps, 223
Rhadinella, 222
Rhadinella schitosa, 222
Rhadinophanes, 213, 223, 245
Rhadinophanes monticola, 223
Rhinobothryum, 223
Rhinobothryum bovallii, 223
Rhinobothryum lentiginosum, 223
Rhinocheilus, 224
Rhinocheilus etheridgei, 224
Rhinocheilus lecontei, 224
Rhinocheilus lecontei antoni, 224
Rhinocheilus lecontei clarus, 224
Rhinocheilus lecontei etheridgei, 224
Rhinocheilus lecontei lecontei, 224
Rhinocheilus lecontei tessellatus, 224
rhinostomus, 234
Rhinotyphlops, 17
rhodogaster, 136
rhombeata, 334
Rhombic Cat-eyed Snake, 158
rhombifer, 119, 188, 196
rhombifera, 158
rhombomaculata, 151
rhynosoma, 234
Ribbon Coral Snake, 297
Ribbon Graceful Brownsnake, 219
Ribbonsnake, 258

ricardinii, 279
richardi, 32, 64
Richard's Blindsnake, 32
Richard's Racer, 64
Ridge-headed Snake, 177, 178
Ridge-nosed Rattlesnake, 333
rigida, 217
rijersmai, 64
Rim Rock Crowned Snake, 250
Ringed Slug-eater, 276
Ringed Thirst Snake, 114
Ringed Tree Boa, 41
Ringed Tree Snake, 223
Ring-necked Coffee Snake, 189
Ring-necked Snake, 109, 110
Ring-necked Tellurian Snake, 76
Rio Amana Liana Snake, 235
Río Chotano Threadsnake, 27
Rio Grande do Sul Threadsnake, 26
Río Jubones Valley Centipede Snake, 249
Rio Manjura Legion Snake, 169
Río Marañon Valley Cat-eyed Snake, 159
Río Pandeiro Rainbow Boa, 44
riveroi, 82
Rivero's Tellurian Snake, 82
riveti, 163
Road Guarder, 103
roadingeri, 318
Roatán Coral Snake, 302
roatanensis, 126
Roatán Island Ratsnake, 122
Roatán Island Vinesnake, 193
Roatán Long-tailed Snake, 126
Rock Rattlesnake, 327
Rocky Mountain Rubber Boa, 49
Rodrigues's Miner Snake, 204
rogerromani, 222
rohdei, 234
rondoni, 70
Rondonia Coral Snake, 296
rondonianus, 296
rosaeca, 122
rosaliae, 85
rosamondae, 67
Rosario Rattlesnake, 325
Rose-bellied Earthsnake, 136
Rosen's Snail-eater, 345
roseofusca, 49, 50, 51
rossalleni, 121, 123
rossmani, 263
Rossman's Gartersnake, 263
rostellatus, 32
rostralis, 136
Rosy Boa, 48, 49, 50
Rough Coffee Snake, 191, 192
Rough Earthsnake, 282
Rough Grass Snake, 192

Rough Green Snake, 192
Rough-Scaled Boa, 53
Rough-tailed Watersnake, 138
roulei, 82
Roule's Tellurian Snake, 82
rowani, 17, 18
rowleyi, 309
Rowley's Palm Pit Viper, 309
rozei, 299, 336
rozellae, 255
Rozella's Dwarf Short-tailed Snake, 255
Roze's Hog-nosed Pit Viper, 336
Roze's Legion Snake, 166
ruatanus, 302
Rubber Boa, 48, 49
ruber, 325, 330
rubidus, 117, 118
rubiventris, 90
rubra, 247, 248, 251
rubricata, 159
rubricollis, 282
rubrilineatus, 263
rubrolineatus, 26
ruddocki, 179
ruficeps, 250
rufidorsatus, 63
rufidorsus, 26
rufipunctatus, 263, 264
rufiventris, 64
rufodorsatus, 63
rugosus, 192
Rupununi Savanna Rattlesnake, 324
Rupununi Savanna Threadsnake, 22
ruruima, 323, 324
ruschenbergeri, 42
Ruschenberger's Tree Boa, 42
ruspator, 131
russatus, 136
russeolus, 306
rustica, 97
Rusty-headed Snake, 65
Rusty Whipsnake, 95
ruthveni, 136, 150, 154, 181, 183, 207, 210
Ruthven's Earthsnake, 136
Ruthven's Lizard-eater, 183
Ruthven's Whipsnake, 181
Ruthven's Wormsnake, 71
rutilorus, 263
rutilus, 257
ruttyi, 63

Saba Racer, 64
sabogae, 40
Saboga Island Boa Constrictor, 40
sackeni, 264
Saddled Cat-eyed Snake, 160
Saddled Coral Snake, 290

Index

Saddled Leaf-nosed Snake, 205
sagittifer, 174
Saint Joann Snail-eater, 115
Saint Lucia Mussurana, 97
Saint Vincent Racer, 95
sajdaki, 61
salgueiroi, 26
sallaei, 135, 136
sallei, 136
Salle's Earthsnake, 136
Saltmarsh Snake, 186
Salvadora, 224, 225
Salvadora bairdi, 225
Salvadora deserticola, 225
Salvadora grahamae, 225
Salvadora grahamiae, 225
Salvadora grahamiae grahamiae, 225
Salvadora grahamiae lineata, 225
Salvadora hexalepis, 225
Salvadora hexalepis deserticola, 225
Salvadora hexalepis hexalepis, 225
Salvadora hexalepis klauberi, 225
Salvadora hexalepis mojavensis, 226
Salvadora hexalepis virgultea, 226
Salvadora intermedia, 226
Salvadora lemniscata, 226
Salvadora mexicana, 226
salvadorensis, 281
salvini, 331
salvinii, 281
Samana Threadsnake, 22
San Andres Groundsnake, 100
San Bernardino Mountain Kingsnake, 157
San Bernardino Ring-necked Snake, 110
sanctaecrucis, 318
sanctaemartae, 82
sanctaeritae, 71
sancticrucis, 65
sanctijoannis, 115
Sancti Spíritus Dwarf Boa, 57
sanctorum, 62
Sand Boas, 48
San Diego Gophersnake, 208
San Diego Mountain Kingsnake, 157
San Diego Nightsnake, 145
San Diego Ring-necked Snake, 111
Sandner-Montilla's Common Lancehead Pit Viper, 314
San Esteban Island Rattlesnake, 329
San Felipe Tolucan Groundsnake, 268
San Francisco Blindsnake, 33
San Francisco Gartersnake, 265
San Francisco Wormsnake, 69
San Gil Coral Snake, 302
sangilensis, 302
Sangrita Scorpion-eater, 238
sanguineus, 82

San Joaquin Coachwhip, 179
San Lorenzo Centipede Snake, 248
San Lorenzo Island Rattlesnake, 331
San Lucan Speckled Rattlesnake, 328
San Luis Potosí Kingsnake, 153
San Marcos Island Nightsnake, 127
San Martin Gophersnake, 208
San Martin Nightsnake, 127, 146
San Pedro Martir Gartersnake, 259
San Pedro Mountain Kingsnake, 156
sanguiventris, 184, 185
sanniola, 231
Santa Barbara Tellurian Snake, 80
Santa Catalina Island Kingsnake, 152
Santa Catalina Island Rattlesnake, 321
Santa Catalina Island Threadsnake, 25
Santa Catalina Nightsnake, 145
Santa Cruz Gartersnake, 258
Santa Cruz Gophersnake, 208
Santa Cruz Lancehead Pit Viper, 318
Santa Marta Capuchin Coral Snake, 294
Santa Marta Parrot Snake, 163
Santa Marta Pygmy Coral Snake, 293
Santa Marta Red-tailed Coral Snake, 299
santamartensis, 163
Santander Coral Snake, 302
Santa Philomena Wormsnake, 69
Santa Rita Groundsnake, 236
Santa Rita Wormsnake, 71
Santa Rosalia Ratsnake, 85
Santiago Piedmont Coral Snake, 303
Sanzinia, 38, 39
Saona Island Hog-nosed Racer, 147
Saona Vinesnake, 278
São Paulo Brown Whipsnake, 94
São Paulo False Coral Snake, 234
São Paulo Lancehead Pit Viper, 318
São Paulo Large-eyed Snake, 257
São Paulo Sharp Snake, 279
São Paulo Sharp Snake, 279
São Paulo Slender Blindsnake, 18
São Paulo Tellurian Snake, 83
São Paulo Watersnake, 139
São Paulo Whipsnake, 94
Saphenophis, 165, 226
Saphenophis antioquiensis, 226
Saphenophis atahuallpae, 226
Saphenophis boursieri, 226
Saphenophis Snake, 226
Saphenophis sneiderni, 226
Saphenophis tristriatus, 227
sapperi, 66, 292
sarae, 103
sargenti, 222
Sargent's Graceful Brownsnake, 222
sargii, 60
Sargi's Burrowing Snake, 60

sartorii, 229, 276
Sartorius' Snail-eater, 276
saskelli, 51
saslowi, 51
sauritus, 264
Sauvage's Snail-eater, 112
savagei, 91
Savage's Earthsnake, 133
Savanna Racer, 244
Savanna Striped Lizard-eater, 183
saxatilis, 91, 92, 344
sayi, 208
scalaris, 140, 147, 264, 285
Scaled Groundsnake, 285
scaliger, 264
Scaphiodontophis, 227
Scaphiodontophis albonuchalis, 227
Scaphiodontophis annulatus, 227
Scarlet Kingsnake, 155
Scarlet Pipesnake, 38
Scarletsnake, 88
schach, 82
Schach's Tellurian Snake, 82
schistosa, 222, 251
schistosus, 346
schlegeli, 309
schlueteri, 94
schmidti, 103, 156, 287, 288
Schmidt's Black-striped Snake, 103
schotti, 172, 173, 181
Schott's Legion Snake, 173
Schott's Whipsnake, 181
schubarti, 18
schultzi, 142
Schultz's False Water Cobra, 142
schunkii, 115
Schunk's Snail-eater, 115
schwartzi, 32, 44, 54, 63
Schwartz's Boa, 44
Schwartz's Racer, 63
sclateri, 125
Scolecophidia, 5, 8, 10, 11, 13, 15
Scolecophidian, 3, 5, 8, 15
Scolecophis, 227, 228
Scolecophis atrocinctus, 228
Scotophis laetus, 123
scriptorcibatus, 204
scurrulus, 95, 349
scutiventris, 287, 288
scutulatus, 331
scytale, 38
scytalina, 97
Sea Krait, 304
Sea Snakes, 5, 9, 10, 11, 304
sebae, 190, 191, 196
Seba's False Coral Snake, 196
segregus, 25

semiannulata, 235, 236
semiaureus, 171
semicincta, 252
semicinctus, 57, 177, 281
semidoliatas, 136
semifasciatus, 148, 149, 195, 196, 266
semilineatus, 174
Seminatrix, 228
Seminatrix pygaea, 228
Seminatrix pygaea cyclas, 228
Seminatrix pygaea paludis, 228
Seminatrix pygaea pgyaea, 228
seminola, 130
semipartitus, 299
semirutus, 271
semmiannulatus, 135
Senticolis, 6, 85, 121, 228, 229
Senticolis triaspis, 229
Senticolis triaspis intermedia, 229
Senticolis triaspis mutabilis, 229
Senticolis triaspis triaspis, 229
septemstriatus, 27
septemtrionalis, 159
septemvittata, 217
septemvittatus, 139, 346
septentrionalis, 95, 163, 196, 232, 264
serpentinus, 167
serperaste, 222
serra, 277
serrana, 212
Serrana False Boa, 212
Serrana False Coral Snake, 196
Serrana Slug-eater, 232
Serrana Smooth Legion Snake, 171
Serrana Tellurian Snake, 83
Serra Norte Legion Snake, 166
serranus, 83, 84, 343
Serra Snake, 277
sertaneja, 87
Sertaneja False Mussurana, 87
sertula, 252
Seven-rayed Threadsnake, 27
Seven-striped Threadsnake, 27
severus, 285
sexcarinatus, 95, 215
Sharpnose Bush Snake, 279
Sharp-nosed Snake, 285
Sharp Snake, 279
Sharp-tailed Snake, 106
shawi, 252
Shiny Threadsnake, 24
Short Black-backed Coral Snake, 287
Short-faced Snail-sucker, 345
Short-headed Gartersnake, 258
Short-headed Wormsnake, 68
Short-nosed Groundsnake, 244
Short-nosed Vinesnake, 193

Short-tailed Centipede Snake, 247
Short-tailed Coral Snake, 295
Short-tailed Snake, 239
Shovel-nosed Snake, 92
Shovel-toothed Snake, 227
shropshirei, 215
Shropshire's Puffing Snake, 215
Sibon, 112, 229, 230, 232, 274, 275
Sibon annulata, 276, 349
Sibon anthracops, 230
Sibon argus, 230
Sibon brevis, 230, 347
Sibon carri, 230
Sibon dimidiata, 230
Sibon dimidiata dimidiata, 230
Sibon dimidiata grandocula, 230
Sibon dunni, 230
sibonius, 62
Sibon linearis, 231
Sibon longifrenis, 231
Sibon nebulata, 231, 347
Sibon nebulata hartwegi, 231
Sibon nebulata leucomelas, 231
Sibon nebulata nebulata, 231, 347
Sibon nebulata popayanensis, 231
Sibon sanniola, 231
Sibon sanniola neilli, 231
Sibon sanniola sanniola, 231
Sibynomorphus, 112, 229, 232, 275, 345
Sibynomorphus inaequifasciatus, 232
Sibynomorphus lavillai, 232
Sibynomorphus mikani, 232
Sibynomorphus mikani mikani, 232
Sibynomorphus mikani septentrionalis, 232
Sibynomorphus neuwiedi, 232
Sibynomorphus oligozonatus, 232
Sibynomorphus oneilli, 232
Sibynomorphus petersi, 233
Sibynomorphus turgidus, 233
Sibynomorphus vagrans, 233
Sibynomorphus vagus, 233
Sibynomorphus ventrimaculatus, 233
Sibynomorphus williamsi, 233
Side-striped Palm Pit Viper, 309
Sidewinder, 321
sieboldi, 136, 137
Siebold's Earthsnake, 136
Sierra Coalcomán Earthsnake, 134
Sierra de la Ventana Legion Snake, 167
Sierra Gartersnake, 258
Sierra Juarez Earthsnake, 133
Sierra Madre Earthsnake, 136
Sierra Mije Earthsnake, 132
Sierra Mountain Kingsnake, 157
Sierra Nevada de Santa Marta Tellurian Snake, 82
signatus, 27

silus, 334
Similar-scaled Snake, 19
simile, 271
similis, 109, 111, 266
similus, 104
simonsi, 203, 203
simonsii, 165, 203
Simons's Racer, 203
Simophis, 233
Simophis rhinostomus, 234
Simophis rhynostoma, 234
Simophis rohdei, 234
simus, 141
sinaloae, 156
Sinaloan Coral Snake, 288
Sinaloan Milksnake, 156
sinaloensis, 338
sinicola, 217
sipedon, 5, 188
Siphlophis, 234
Siphlophis cervinus, 234
Siphlophis cervinus pulcher, 235
Siphlophis compressus, 234
Siphlophis geminatus, 235
Siphlophis leucocephalus, 234
Siphlophis longicaudatus, 234
Siphlophis pulcher, 235
Siphlophis worontzowi, 235
Sipo, 93
Sipurio Snail-eater, 230
sirtalis, 5, 264, 265
Sistrurus, 5, 320, 337
Sistrurus catenatus, 337
Sistrurus catenatus catenatus, 337
Sistrurus catenatus edwardsi, 337
Sistrurus catenatus tergeminus, 337
Sistrurus miliaris, 338
Sistrurus miliaris barbouri, 338
Sistrurus miliaris miliaris, 338
Sistrurus miliaris streckeri, 338
Sistrurus ravus, 338
Sistrurus ravus brunneus, 338
Sistrurus ravus exiguous, 339
Sistrurus ravus lutescens, 338
Sistrurus ravus ravus, 338
Sistrurus ravus sinaloensis, 338
Slaven's Centipede Snake, 252
slavensi, 252
Slender Black-backed Coral Snake, 287
Slender Blindsnake, 17
Slender Blunt-headed Tree Snake, 149
Slender Boa, 43
Slender Coral Snake, 295
Slender False Coral Snake, 128
Slender Hog-nosed Pit Viper, 336
Slender Snail-eater, 230
Slender Snake, 243

Index

Slender Vinesnake, 278
Slender Wormsnake, 19, 20
sleveni, 181
Slevin's Tropical Groundsnake, 271
slevini, 9, 24, 68, 127, 179, 184, 271
Slevin's Lizard-eater, 184
Slevin's Nightsnake, 127
Small-eyed Toadheaded Pit Viper, 312
Small-scaled Tellurian Snake, 81
smaragdinae, 310
smithi, 156, 159
Smith's Black-headed Snake, 248
Smith's Milksnake, 156
Smith's Two-spotted Snake, 100
Smith's Yellow-bellied Snake, 101
Smooth Brown Mussurana, 97
Smooth Earthsnake, 282
Smooth Green Snake, 164
Snail-eater, 112
Snail-sucker, 112
sneiderni, 226
snethlageae, 78, 83
Snethlage's Tellurian Snake, 83
Snouted Snake, 233
Solenoglyphic Snakes, 305
Sonora, 235
Sonora aemula, 235
Sonora michoacanensis, 235
Sonora michoacanensis michoacanensis, 235
Sonora michoacanensis mutabilis, 235
Sonora Mountain Kingsnake, 153
Sonoran Coachwhip, 179
Sonoran Coral Snake, 288
Sonoran Gophersnake, 208
Sonoran Groundsnake, 235
Sonoran Leaf-nosed Snake, 205
Sonoran Lyresnake, 272
Sonoran Nightsnake, 145
Sonoran Sidewinder, 321
Sonoran Whipsnake, 178
Sonora semiannulata, 235
Sonora semiannulata semiannulata, 236
Sonora semiannulata taylori, 236
Sonora Shovel-nosed Snake, 92
Sordellina, 236
Sordellina punctata, 236
South American Black Snake, 218
South American Burrowing Snake, 197
South American Bushmaster, 334
South American Coral Snake, 297
South American False Boa, 211
South American False Coral Snake, 127
South American False Water Cobra, 142
South American Flat-headed Snake, 285
South American Groundsnake, 244
South American Hog-nosed Snake, 176

South American Large-eyed Snake, 256
South American Legion Snake, 167
South American Mock Pit Viper, 269
South American Mock Viper, 255
South American Mudsnake, 143
South American Night Viper, 269
South American Pond Snake, 212
South American Racer, 200
South American Rattlesnake, 323
South American Scarletsnake, 211
South American Scorpion Snake, 211
South American Sipo, 94
South American Snail-eating Snake, 232
South American Striped Racer, 203
South American Threadsnake, 26
South American Tree Boa, 41
South American Tree Snake, 232
South American Watersnake, 138
South American Wormsnake, 68
Southeastern Crowned Snake, 247
Southeastern Green Racer, 202
South Florida Mole Kingsnake, 151
Southern Aesculapian False Coral Snake, 128
Southern Amazon Legion Snake, 171
Southern Anaconda, 47
Southern Banded Cat-eyed Snake, 158
Southern Bicolored Blindsnake, 32
Southern Black Racer, 100
Southern Black-striped Snake, 102
Southern Blunt-headed Tree Snake, 149
Southern Burrowing Snake, 199
Southern Copperhead, 306
Southern Coral Snake, 302
Southern Desert Coral Snake, 304
Southern Desert Racer, 147
Southern Durango Spotted Gartersnake, 262
Southern Eyelash Boa, 53
Southern Florida Rainbow Snake, 130
Southern Florida Rainbow Snake, 130
Southern Florida Swamp Snake, 228
Southern Florida Swampsnake, 228
Southern Graceful Brownsnake, 222
Southern Green Racer, 202
Southern Green Tree Snake, 201
Southern Hog-nosed Snake, 141
Southern Legion Snake, 170
Southern Many-banded Coral Snake, 295
Southern Mexican Gartersnake, 260
Southern Mexican Snail-sucker, 276
Southern Neuwied's Lancehead Pit Viper, 318
Southern Pacific Rattlesnake, 333
Southern Puerto Rico Racer, 64
Southern Regal Legion Snake, 174
Southern Ridge-nosed Rattlesnake, 334

Southern Ring-necked Snake, 109
Southern Rubber Boa, 49
Southern Sharp-nosed Snake, 286
Southern South American Wormsnake, 68
Southern Threadsnake, 28
Southern Tiger Ratsnake, 237
Southern Watersnake, 187
Southern Whipsnake, 93
South Texas Groundsnake, 236
South Todos Santos Island Kingsnake, 157
South Todos Santos Island Ring-necked Snake, 110
Southwestern Cat-eyed Snake, 159
Southwestern Puerto Rican Groundsnake, 73
Southwestern Ratsnake, 123
Southwestern Speckled Rattlesnake, 328
Southwestern Threadsnake, 24
Speckled Coral Snake, 292, 298
Speckled Dwarf Short-tailed Snake, 254
Speckled Forest Pit Viper, 311
Speckled Kingsnake, 152
Speckled Racer, 118
Speckled Rattlesnake, 328
Speckled Tellurian Snake, 84
spegazzinii, 199
Spegazzini's Burrowing Snake, 199
spiloides, 123
spilosomus, 298
Spilotes, 236, 237
Spilotes megalolepis, 237
Spilotes pullatus, 237
Spilotes pullatus anomalepis, 237
Spilotes pullatus argusiformis, 237
Spilotes pullatus maculatus, 237
Spilotes pullatus mexicanus, 237
Spilotes pullatus pullatus, 237
Spiny Boa, 53
Spirit Diminutive Snake, 216
Spirit Snake, 216
spiritus, 57
spixi, 302
spixii, 93
Spix's Whipsnake, 93
splendida, 152, 160
Splendid Blunt-headed Tree Snake, 150
Splendid Cat-eyed Snake, 160
Splendid Coral Snake, 292
splendidus, 149, 150
Spotted Black-striped Snake, 102
Spotted Burrowing Snake, 199
Spotted Coral Snake, 296
Spotted Dwarf Boa, 55
Spotted Leaf-nosed Snake, 206
Spotted Night Snake, 234
Spotted Nightsnake, 146
Spotted-nose Coral Snake, 292

Spotted Sandsnake, 91
Spotted Savanna Racer, 244
Spotted Steppe Legion Snake, 174
Spotted Tellurian Snake, 79
Spotted-throat Lizard Snake, 101
Spotted Tolucan Groundsnake, 268
Spotted Watersnake, 139
spurelli, 303
Squamos Similar-scaled Snake, 19
squamosus, 19
sticticeps, 153
stictogenys, 111
Stiff Snake, 125
Stilosoma, 238, 239
Stilosoma extenuatum, 239
Stilosoma extenuatum arenicola, 239
Stilosoma extenuatum extenuatum, 239
Stilosoma extenuatum multistictum, 239
stimsoni, 163
stirolatus, 180
stadelmani, 32, 33, 222
Stadelman's Blindsnake, 32
Stadelman's Graceful Brownsnake, 222
stahli, 73
Star Snake, 93
St. Barts Blindsnake, 29
St. Croix Racer, 65
steinbachi, 218, 349
steindachneri, 68, 303
Steindachner's Coral Snake, 303
stejnegeri, 55, 331
stejnegerianus, 99
Stejneger's Dwarf Boa, 55
stenophrys, 335
Stenorrhina, 237
Stenorrhina degenhardti, 238
Stenorrhina degenhardti degenhardti, 238
Stenorrhina degenhardti mexicana, 238
Stenorrhina degenhardti ocellata, 238
Stenorrhina freminvilli, 238
Stenostoma, 20
stephensi, 328
stewarti, 303
steyermarki, 83
Steyermark's Tellurian Snake, 83
St. Helena Mountain Milksnake, 156
St. Lucia Boa Constrictor, 40
St. Lucia Lancehead Pit Viper, 315
St. Lucia Legion Snake, 172
Storeria, 239
Storeria dekayi, 239
Storeria dekayi anomala, 239
Storeria dekayi dekayi, 239
Storeria dekayi limnetes, 240
Storeria dekayi temporalineata, 240
Storeria dekayi texana, 240
Storeria dekayi tropica, 240

Storeria dekayi victa, 240
Storeria dekayi wrightorum, 240
Storeria hidalgoensis, 240
Storeria occipitomaculata, 240
Storeria occipitomaculata obscura, 241
Storeria occipitomaculata occipitomaculata, 240
Storeria occipitomaculata pahasapae, 241
Storeria storerioides, 241
Storeria victa, 240
storerioides, 241
stramineus, 91
streckeri, 131, 338
striata, 252
striaticeps, 277
striatula, 282
striatulus, 27
striatus, 46, 183
strigatus, 257
strigilatus, 46
strigilis, 257
striolatus, 180
Striped Centipede Snake, 252
Striped Crayfish Snake, 217
Striped False Rhadinaea, 227
Striped Galápagos Antillophis, 68
Striped Lizard-eater, 184
Striped Lowland Snake, 160
Striped Parrot Snake, 163
Striped Racer, 180
Striped Road Guarder, 104
Striped Snail-eater, 114
Striped Swampsnake, 269
Striped Tellurian Snake, 75
Striped Whipsnake, 181
Stripeless Steppe Legion Snake, 174
Stripe-tailed Regal Legion Snake, 174
stuarti, 156, 185, 303
Stuart's Graceful Brownsnake, 221
Stuart's Lizard-eater, 185
Stuart's Milksnake, 156
stullae, 55
stygius, 67
subaequalis, 346
subannulata, 276
subbcinctum, 75
subcarinata, 201
subcrotillus, 27
subfasciatus, 173
subflavus, 47
sublineatus, 172, 173
subocularis, 85
suborbitalis, 181
subspadix, 73
sulcatus, 32
sulphurerus, 215
sumichrasti, 126, 266

Sumichrast's Gartersnake, 266
Sumichrast's Long-tailed Snake, 126
supernum, 74
supraciliaris, 309
supracincta, 252
supraocularis, 23
surinamensis, 63, 303
Suriname Tellurian Snake, 77
suspectus, 199, 284
Swamp Legion Snake, 168
Swan Island Racer, 63
Swift Andean Snake, 243
Swift Snake, 243
Symphimus, 241
Symphimus leucostomus, 241
Symphimus mayae, 241
Sympholis, 241
Sympholis lippiens, 242
Sympholis lippiens lippiens, 242
Sympholis lippiens rectilimbus, 242
Synophis, 125, 242
Synophis bicolor, 242
Synophis calamitus, 242
Synophis lasallei, 242
Synophis plectovertebralis, 242
syntherus, 32
Syphlophis, 235
Syphlophis pulcher, 235
syspila, 156

Tabasco Diamond-backed Watersnake, 188
Tachimenis, 242
Tachymenis, 242, 243
Tachymenis affinis, 243
Tachymenis attenuata, 243
Tachymenis attenuata attenuata, 243
Tachymenis attenuata boliviana, 243
Tachymenis chilensis, 243
Tachymenis chilensis chilensis, 243
Tachymenis chilensis coronellina, 243
Tachymenis chilensis melanura, 243
Tachymenis elongata, 63
Tachymenis peruviana, 243
Tachymenis peruviana peruviana, 244
Tachymenis peruviana yutoensis, 244
Tachymenis surinamensis, 63
Tachymenis tarmensis, 244
tachymenoides, 203
Tachymensis, 242, 256
Tachymensis hypoconia, 256
taczanowskyi, 57
taeniata, 186, 222, 252, 311
taeniatum, 74
taeniatus, 83, 116, 181, 270, 311
taeniogaster, 167
Taeniophallus, 120, 218, 244
Taeniophallus affinis, 120, 244

Index

Taeniophallus bilineatus, 244
Taeniophallus brevirostris, 120, 244
Taeniophallus nicagus, 120, 244
Taeniophallus occipitalis, 120, 244
Taeniophallus persimilis, 120, 245
Taeniophallus poecilopogon, 245
taeniurus, 175
talamancae, 137
Talamanca Earthsnake, 137
Talamanca Forest Racer, 108
talpina, 92
Tamaulipan Black-striped Snake, 101
Tamaulipan Hook-nosed Snake, 131
Tamaulipan Montane Gartersnake, 262
Tamaulipan Parrot Snake, 163
Tamaulipan Threadsnake, 23
Tamaulipian Milksnake, 154
Tamaulipian Nightsnake, 121
Tamaulipian Ratsnake, 121
Tamaulipian Rock Rattlesnake, 327
Tampico Coral Snake, 296
Tampico Diamond-backed Watersnake, 188
Tampico Threadsnake, 23
Tancitaran Dusky Rattlesnake, 329
Tan Lizard-eater, 183
Tanner's Threadsnake, 23
Tan Racer, 99
Tantalophis, 157, 213, 222, 245
Tantalophis discolor, 245
Tantilla, 245, 246, 254
Tantilla albiceps, 246
Tantilla alticola, 246
Tantilla andinista, 246
Tantilla annulata, 252
Tantilla atriceps, 246
Tantilla bairdi, 246
Tantilla bocourti, 246
Tantilla bocourti bocourti, 246
Tantilla bocourti deviatrix, 246
Tantilla boipiranga, 246
Tantilla brevicauda, 247
Tantilla brevis, 254
Tantilla briggsi, 247
Tantilla calamarina, 247
Tantilla canula, 254
Tantilla canula brevis, 254
Tantilla capistra, 247
Tantilla cascadae, 247
Tantilla coronadoi, 247
Tantilla coronata, 247
Tantilla cucullata, 247, 248
Tantilla cuesta, 249
Tantilla cuniculator, 247
Tantilla deppei, 247
Tantilla depressa, 132
Tantilla deviatrix, 246
Tantilla diabola, 248
Tantilla equatoriana, 247
Tantilla excubitor, 254
Tantilla flavilineata, 247
Tantilla fraseri, 249
Tantilla gracilis, 247
Tantilla hobartsmithi, 247
Tantilla impensa, 247
Tantilla insulamontana, 249
Tantilla jani, 249
Tantilla johnsoni, 249
Tantilla lempira, 249
Tantilla longifrontalis, 249
Tantilla martindelcampo, 247
Tantilla melanocephala, 249
Tantilla melanocephala capistrata, 249
Tantilla mexicana, 250
Tantilla miniata, 251
Tantilla miyatai, 250
Tantilla moesta, 250
Tantilla morgani, 251
Tantilla nigra, 250
Tantilla nigriceps, 250
Tantilla oaxacae, 250
Tantilla oolitica, 250
Tantilla petersi, 250
Tantilla phrenitica, 251
Tantilla phrenitica schistose, 251
Tantilla planiceps, 250
Tantilla planiceps utahensis, 248
Tantilla relicta, 251
Tantilla relicta neilli, 251
Tantilla relicta pamlica, 251
Tantilla relicta relicta, 251
Tantilla reticulata, 251
Tantilla rubra, 247, 248, 251
Tantilla rubra cucullata, 247
Tantilla rubra diabola, 247
Tantilla ruficeps, 250
Tantilla schistosa, 251
Tantilla semicincta, 252
Tantilla sertula, 252
Tantilla shawi, 252
Tantilla slavensi, 252
Tantilla striata, 252
Tantilla supracincta, 252
Tantilla taeniata, 252
Tantilla tayrae, 252
Tantilla tecta, 252
Tantilla trilineata, 253
Tantilla triseriata, 253
Tantilla tritaeniata, 253
Tantilla utahensis, 248
Tantilla vermiformis, 253
Tantilla vulcani, 253
Tantilla wilcoxi, 253
Tantilla wilcoxi wilcoxi, 253
Tantilla yaquia, 253
Tantilla yaquia bogerti, 253
Tantilla yaquia yaquia, 253
Tantillita, 254
Tantillita brevis, 254
Tantillita brevissima, 254
Tantillita canula, 254
Tantillita canula brevis, 254
Tantillita canula canula, 254
Tantillita lintoni, 254, 255
Tantillita lintoni lintoni, 255
Tantillita lintoni rozellae, 255
tanyplectum, 74
tanzeri, 144
Tanzer's Nightsnake, 144
taphorni, 83
Taphorn's Tellurian Snake, 83
tarascae, 137
Tarasca Earthsnake, 137
Tarma Swift Snake, 244
tarmensis, 244
tasymicris, 33
tau, 272, 273
taxispilota, 189
taylori, 103, 131, 156, 236, 291, 307
Taylor's Black-striped Snake, 103
Taylor's Cantil, 307
Taylor's Dawn Blindsnake, 16
Taylor's Lyresnake, 272
Taylor's Peru Blindsnake, 16
Taylor's Snail-eater, 115
Taylor's Threadsnake, 24
tayrae, 252
teaguei, 27
Tearful Pine-Oak Snake, 220
tecpanecus, 275
tecta, 252
tehuanae, 185
Tehuana Lizard-eater, 185
Tehuantepec Coral Snake, 294
Tehuantepec Hook-nosed Snake, 131
Tehuantepec Striped Snake, 131, 132
Tellurian Snake, 74
temporalineata, 240
temporalis, 115
Temporal Snail-eater, 115
tenella, 20, 45
tenellus, 27
tener, 296
tenuiculus, 25
tenuis, 33, 71, 106
tenuissima, 115
tenuissimus, 150
Tepalcatepec Valley Gartersnake, 262
Tepic Gartersnake, 267
Terenos Threadsnake, 25
tergeminus, 337

ternetzii, 18
Ternetz's Slender Blindsnake, 18
Terrestrial Gartersnake, 259
Terrestrial Snail-eating Snake, 276
terrestris, 259
terrificus, 323
tessellata, 190
tesselatus, 27, 345
tessellatus, 224
testaceus, 180
tetrataenia, 265
tetrathyreus, 33
tetrazonus, 128
texana, 145, 240
texanum, 274
Texas Brownsnake, 240
Texas Coral Snake, 296
Texas Gartersnake, 265
Texas Glossy Snake, 71
Texas Lined Snake, 274
Texas Long-nosed Snake, 224
Texas Lyresnake, 273
Texas Nightsnake, 145
Texas Patch-nosed Snake, 225
Texas Ratsnake, 123
Texas Scarletsnake, 88
Texas Threadsnake, 23
Texiguat Graceful Brownsnake, 222
thalassinus, 7
Thalerophis, 161
Thalesius, 255, 283, 284
Thalesius viridis, 255
Thamnobius, 214
Thamnodynastes, 255, 256
Thamnodynastes chaquensis, 256
Thamnodynastes chimanta, 256
Thamnodynastes corocoroensis, 256
Thamnodynastes duida, 256
Thamnodynastes gambotensis, 256
Thamnodynastes hypoconia, 256
Thamnodynastes marahuaquensis, 256
Thamnodynastes nattereri, 255
Thamnodynastes pallidus, 257
Thamnodynastes rutilus, 257
Thamnodynastes strigatus, 257
Thamnodynastes strigilis, 257
Thamnodynastes yavi, 257
Thamnophis, 165, 185, 257, 258, 266, 267
Thamnophis angustirostris, 263
Thamnophis atratus, 258
Thamnophis atratus atratus, 258
Thamnophis atratus atratus X *Thamnophis atratus hydrophilus*, 258
Thamnophis atratus hydrophilus, 258
Thamnophis atratus zanthus, 258
Thamnophis brachystoma, 258
Thamnophis butleri, 258

Thamnophis chrysocephalus, 258
Thamnophis couchi, 258
Thamnophis cyrtopsis, 259
Thamnophis cyrtopsis collaris, 259
Thamnophis cyrtopsis cyrtopsis, 259
Thamnophis cyrtopsis ocellatus, 259
Thamnophis digueti, 261
Thamnophis elegans, 259
Thamnophis elegans aquaticus, 258
Thamnophis elegans arizonae, 259
Thamnophis elegans elegans, 259
Thamnophis elegans hueyi, 259
Thamnophis elegans nigrescens, 260
Thamnophis elegans terrestris, 259
Thamnophis elegans vagrans, 259, 260
Thamnophis elegans vascotanneri, 260
Thamnophis eques, 260, 263
Thamnophis eques eques, 260
Thamnophis eques megalops, 260
Thamnophis eques virgatenuis, 260
Thamnophis errans, 260
Thamnophis exsul, 260
Thamnophis fulvus, 260
Thamnophis gigas, 260
Thamnophis godmani, 260
Thamnophis hammondi, 261
Thamnophis marcianus, 261
Thamnophis marcianus bovallii, 261
Thamnophis marcianus marcianus, 261
Thamnophis marcianus praeocularis, 261
Thamnophis melanogaster, 261, 263
Thamnophis melanogaster canescens, 261
Thamnophis melanogaster chihuahuaensis, 262
Thamnophis melanogaster linearis, 262
Thamnophis melanogaster melanogaster, 261
Thamnophis mendax, 262
Thamnophis nigronuchalis, 262
Thamnophis ordinoides, 262
Thamnophis postremus, 262
Thamnophis proximus, 262
Thamnophis proximus alpinus, 262
Thamnophis proximus diabolicus, 262
Thamnophis proximus orarius, 263
Thamnophis proximus proximus, 262
Thamnophis proximus rubrilineatus, 263
Thamnophis proximus rutilorus, 263
Thamnophis pulchrilatus, 263
Thamnophis radix, 263
Thamnophis rossmani, 263
Thamnophis rufipunctatus, 263, 264
Thamnophis rufipunctatus nigronuchalis, 264
Thamnophis rufipunctatus rufipunctatus, 264
Thamnophis rufipunctatus unilabialis, 264
Thamnophis sauritus, 264
Thamnophis sauritus nitae, 264

Thamnophis sauritus sackeni, 264
Thamnophis sauritus sauritus, 264
Thamnophis sauritus septentrionalis, 264
Thamnophis scalaris, 264
Thamnophis scaliger, 264
Thamnophis sirtalis, 5, 265
Thamnophis sirtalis annectens, 265
Thamnophis sirtalis concinnus, 265
Thamnophis sirtalis dorsalis, 265
Thamnophis sirtalis fitchi, 265
Thamnophis sirtalis infernalis, 265, 266
Thamnophis sirtalis lowei, 265
Thamnophis sirtalis ornata, 265
Thamnophis sirtalis pallidulus, 266
Thamnophis sirtalis parietalis, 266
Thamnophis sirtalis pickeringi, 266
Thamnophis sirtalis semifasciatus, 266
Thamnophis sirtalis similes, 266
Thamnophis sirtalis sirtalis, 265
Thamnophis sirtalis tetrataenia, 265
Thamnophis sumichrasti, 266
Thamnophis validus, 266
Thamnophis validus celaeno, 267
Thamnophis validus isabelleae, 267
Thamnophis validus thamnophisoides, 267
Thamnophis validus validus, 267
Thamnophis vicinus, 259
thamnophisoides, 267
Thanatophis, 349
Thanatophis patoquilla, 349
thayeri, 153
Thick-tailed Tellurian Snake, 76
Thin Legion Snake, 175
Thirst Snake, 112
Thornscrub Hook-nosed Snake, 138
Thread Coral Snake, 295
Threadsnake, 15, 19, 20
Thread Snakes, 15
Three-banded Tellurian Snake, 84
Three-lined Burrowing Snake, 199
Three-lined Centipede Snake, 253
Three-lined Tellurian Snake, 84
Tiburon Banded Racer, 148
Tiburon Dwarf Boa, 55
tiburonensis, 55, 146
Tiburon Hog-nosed Racer, 147
Tiburon Island Boa, 46
Tiburon Island Nightsnake, 146
Tiger Ratsnake, 237
Tiger Rattlesnake, 331
Tigrada Snail-eater, 233
Tigra Mariposa, 310
tigris, 331
Tigrita, 75
Tigrita Tellurian Snake, 75
Timber Rattlesnake, 326
titanops, 33

Index

Toadheaded Pit Viper, 312
Tobago False Coral Snake, 128
Todos Santos Island Kingsnake, 157
tolimensis, 115
tolpanorum, 222
Toluca, 105, 130, 137, 212, 267
Toluca amphisticha, 268
Toluca conica, 268
Toluca lineata, 268
Toluca lineata acuta, 268
Toluca lineata lineata, 268
Toluca lineata varians, 268
Toluca lineata wetmorei, 268
Toluca megalodon, 268
Tolucan Ground Snake, 267
Tolucan Groundsnake, 267
Tomodon, 268, 269
Tomodon dorsatus, 269
Tomodon ocellatus, 214, 268, 269
Torito, 335
torquata, 144, 145, 146
torquatus, 83
torrenicolus, 175
Tortrix, 38
tortugaensis, 146
Tortuga Island Diamond Rattlesnake, 331
Tortuga Island Nightsnake, 146
Tortuga Island Racer, 67
tortuganus, 67
tortugensis, 321, 331
Totalcan Small-headed Rattlesnake, 326
Totonacan Rattlesnake, 323
totonacus, 323
Trachyboa, 53
Trachyboa boulengeri, 53
Trachyboa gularis, 53
Transandean Capuchin Coral Snake, 294
Trans-Andean False Coral Snake, 129
transandinus, 294
Trans-Pecos Black-headed Snake, 247
Trans-Pecos Copperhead, 307
Trans-Pecos Ratsnake, 85
Trans-Pecos Threadsnake, 25
transversa, 187
transversus, 331
Trapido's Brownsnake, 240
trebbaui, 167
Tres Marias Islands Whipsnake, 181
Tretanorhinus, 269
Tretanorhinus mocquardi, 269
Tretanorhinus nigroluteus, 269
Tretanorhinus nigroluteus lateralis, 269
Tretanorhinus nigroluteus mertensi, 269
Tretanorhinus nigroluteus nigroleteus, 269
Tretanorhinus taeniatus, 270
Tretanorhinus variablis, 270
Tretanorhinus variablis binghami, 270

Tretanorhinus variablis insulaepinorum, 270
Tretanorhinus variablis lewisi, 270
Tretanorhinus variablis variablis, 270
Tretanorhinus variablis wagleri, 270
Triangle Watersnake, 143
triangularis, 143
triangulum, 154
triaspis, 228, 229
tricinctus, 281
tricolor, 27, 199, 302, 303
Tricolor Burrowing Snake, 199
Tri-colored False Coral Snake, 233
Tricolor Hog-nosed Snake, 177
Tricolor Threadsnake, 27
trifasciatus, 174
trigeminus, 196
trigonatus, 214
trigonicus, 324
Trigonocephalus, 313
Trigonocephalus xanthogrammus, 313
trihedrurus 83, 84, 343
trilinaeatus, 203
trilineata, 203, 252
trilineatus, 84, 199
Trimeresurus, 312
Trimetopon, 220, 270
Trimetopon barbouri, 270
Trimetopon gracile, 270
Trimetopon hannsteini, 220
Trimetopon pliolepis, 271
Trimetopon simile, 271
Trimetopon slevini, 271
Trimetopon viquezi, 271
Trimorphodon, 271
Trimorphodon biscutatus, 271
Trimorphodon biscutatus biscutatus, 271
Trimorphodon biscutatus lambda, 272
Trimorphodon biscutatus lyrophanes, 272, 273
Trimorphodon biscutatus paucimaculatus, 272
Trimorphodon biscutatus quadruplex, 272
Trimorphodon biscutatus semirutus, 271
Trimorphodon biscutatus vandenburghi, 272
Trimorphodon biscutatus vilkinsoni, 273
Trimorphodon collaris, 272
Trimorphodon fasciolatus, 272
Trimorphodon forbesi, 272
Trimorphodon lambda, 273
Trimorphodon lambda vilkinsoni, 273
Trimorphodon tau, 272, 273
Trimorphodon tau latifascia, 273
Trimorphodon tau tau, 272, 273
Trimorphodon tau upsilon, 273
Trimorphodon vilkinsoni, 273
Trinidad Blindsnake, 33
Trinidad Northern Coral Snake, 291

Trinidad Ribbon Coral Snake, 298
Trinidad Snail-eater, 115
Trinidad Thirst Snake, 115
trinitatis, 115
trinitatus, 33
Tripanurgos, 234, 235
Tripanurgos compressus, 234, 235
triscalis, 175
triseriata, 253
triseriatus, 183, 227, 332, 345
tritaeniata, 253
trivirgata, 49, 50, 51
trivittatus, 84, 140
Trope, 53
tropica, 240
Tropical Black-necked Gartersnake, 259
Tropical Brownsnake, 240
Tropical Forest Snake, 87
Tropical Groundsnake, 270
Tropical Watersnake, 142
Tropidoclonion, 273, 274
Tropidoclonion lineatum, 274
Tropidoclonion lineatum annectens, 274
Tropidoclonion lineatum lineatum, 274
Tropidoclonion lineatum mertensi, 274
Tropidoclonion lineatum texanum, 274
Tropidodipsas, 112, 229, 232, 274, 275
Tropidodipsas annulifera, 229, 275
Tropidodipsas fasciata, 229, 275
Tropidodipsas fasciata fasciata, 275
Tropidodipsas fasciata guerreroensis, 275
Tropidodipsas fasciata subannulata, 276
Tropidodipsas fischeri, 229, 276
Tropidodipsas fischeri fischeri, 276
Tropidodipsas fischeri kidderi, 276
Tropidodipsas philippi, 229, 276
Tropidodipsas sartorii, 229, 276
Tropidodipsas sartorii annulatus, 276
Tropidodipsas sartorii macdougalli, 276
Tropidodipsas sartorii sartorii, 276
Tropidodipsas zweifeli, 229, 276
Tropidodryas, 277
Tropidodryas serra, 277
Tropidodryas striaticeps, 277
Tropidophidae, 38
Tropidophiidae, 38, 52
Tropidophis, 53
Tropidophis battersbyi, 53
Tropidophis canus, 53
Tropidophis canus androsi, 54
Tropidophis canus barbouri, 54
Tropidophis canus canus, 54
Tropidophis canus curtus, 54
Tropidophis caymanensis, 54
Tropidophis caymanensis caymanensis, 54
Tropidophis caymanensis parkeri, 54
Tropidophis caymanensis schwartzi, 54

Tropidophis celine, 54
Tropidophis feicki, 54
Tropidophis fuscus, 54
Tropidophis greenwayi, 54
Tropidophis greenwayi greenwayi, 55
Tropidophis greenwayi lanthanus, 55
Tropidophis haetianus, 55
Tropidophis haetianus haetianus, 55
Tropidophis haetianus hermerus, 55
Tropidophis haetianus jamaicensis, 55
Tropidophis haetianus stejnegeri, 55
Tropidophis haetianus stullae, 55
Tropidophis haetianus tiburonensis, 55
Tropidophis hendersoni, 55
Tropidophis jamaicensis, 55
Tropidophis maculatus, 56
Tropidophis melanurus, 56
Tropidophis melanurus bucculentus, 56
Tropidophis melanurus dysodes, 56
Tropidophis melanurus ericksoni, 56
Tropidophis melanurus melanurus, 56
Tropidophis morenoi, 56
Tropidophis nigriventris, 56
Tropidophis nigriventris hardyi, 56
Tropidophis nigriventris nigriventris, 56
Tropidophis pardalis, 57
Tropidophis paucisquamis, 57
Tropidophis pilsbryi, 57
Tropidophis pilsbryi galacelidus, 57
Tropidophis pilsbryi pilsbryi, 57
Tropidophis semicinctus, 57
Tropidophis spiritus, 57
Tropidophis stejnegeri, 55
Tropidophis stullae, 55
Tropidophis taczanowskyi, 57
Tropidophis wrighti, 57
Trypanurgos, 234
Trypanurgos compressus, 234
tschudii, 304
Tschudi's False Coral Snake, 195
Tucson Shovel-nosed Snake, 92
Tucumán Threadsnake, 21
turgidus, 233
Turks and Caicos Island Boa, 44
Turks & Caicos Islands Dwarf Boa, 55
Turneffe Islands Parrot Snake, 163
Turquoise Parrot Snake, 163
Turtle Island Boa, 46
Tuxtlan Banded Coral Snake, 298
Tuxtlan Coral Snake, 298
Tuxtlan Spotted Coral Snake, 298
Twin-spotted Legion Snake, 168
Twin-spotted Racer, 203
Twin-spotted Rattlesnake, 329
Twin-spotted Tolucan Groundsnake, 268
Two-colored Racer, 202
Two-keeled Whipsnake, 93

Two-lined Burrowing Snake, 198
Two-lined Mexican Earthsnake, 105
Two-lined Savanna Racer, 244
Two-lined Tellurian Snake, 76
Two-spotted Snake, 100
Two-striped Forest Pit Viper, 310
Two-striped Gartersnake, 261
Tychlina, 28
Typhlina, 28
Typhlophidae, 19
Typhlophis, 16, 19
Typhlophis ayarzaguenai, 19
Typhlophis squamosus, 19
Typhlopidae, 15, 28
Typhlopoidae, 15
Typhlopoidea, 15
Typhlops, 29
Typhlops annae, 29
Typhlops biminiensis, 29
Typhlops biminiensis biminiensis, 29
Typhlops biminiensis paradoxus, 29
Typhlops brongersmai, 29
Typhlops brongersmianus, 29
Typhlops capitulatus, 29
Typhlops catapontus, 30, 32
Typhlops caymanensis, 30
Typhlops cinereus, 349
Typhlops costaricensis, 30
Typhlops diardi group, 33
Typhlops dominicanus, 30
Typhlops dominicanus dominicanus, 30
Typhlops dominicanus guadeloupe, 30
Typhlops epactius, 30
Typhlops gonavensis, 30
Typhlops granti, 30
Typhlops hectus, 30
Typhlops hypomethes, 30
Typhlops jamaicensis, 31
Typhlops lehneri, 31
Typhlops longissimus, 349
Typhlops lumbricalis, 31
Typhlops microstomus, 31
Typhlops minuisquamus, 31
Typhlops monastus, 31
Typhlops monastus geotomus, 31
Typhlops monastus monastus, 31
Typhlops monensis, 31
Typhlops multilineatus, 29
Typhlops paucisquamus, 31
Typhlops platycephalus, 32
Typhlops pusillus, 32
Typhlops reticulatus, 32
Typhlops richardi, 32
Typhlops richardi naugus, 30
Typhlops rostellatus, 32
Typhlops schwartzi, 32
Typhlops stadelmani, 32, 33

Typhlops sulcatus, 32
Typhlops syntherus, 32
Typhlops tasymicris, 33
Typhlops tenuis, 33
Typhlops tetrathyreus, 33
Typhlops titanops, 33
Typhlops trinitatus, 33
Typhlops unilineatus, 33
Typhlops yonenagae, 33
typhlus, 175
tzabcan, 324
tzolzilorum, 319
Tzotzil Montane Pit Viper, 319

umbratica, 49
Umbrivaga, 277
Umbrivaga mertensi, 277
Umbrivaga pyburni, 277
Umbrivaga pygmaea, 277
unaocularis, 146
undecimstriata, 21, 347
undecimstriatus, 20, 21
undulata, 120
undulatus, 119, 218, 285, 335
Ungaliophis, 57, 58
Ungaliophis continentalis, 58
Ungaliophis danieli, 58
Ungaliophis panamensis, 58
Ungualiidae, 52
unguirostris, 28
unicolor, 118, 126, 324
unilabialis, 264
unilineatus, 33
univittatus, 84
Upper Amazonian Basin Watersnake, 242
Upper Basin Gartersnake, 260
upsilon, 273
uribei, 213
Uribe's False Cat-eyed Snake, 213
Uromacer, 277, 278
Uromacer catesbyi, 278
Uromacer catesbyi catesbyi, 278
Uromacer catesbyi cereolineatus, 278
Uromacer catesbyi frondicolor, 278
Uromacer catesbyi hariolatus, 278
Uromacer catesbyi incháusteguii, 278
Uromacer catesbyi insulaevaccarum, 278
Uromacer catesbyi pampineus, 278
Uromacer frenatus, 278
Uromacer frenatus chlorauges, 278
Uromacer frenatus dorsalis, 278
Uromacer frenatus frenatus, 278
Uromacer frenatus wetmorei, 279
Uromacer oxyrhynchus, 279
Uromacerina, 279
Uromacerina ricardinii, 279
Uropeltis, 37

Index

Urorocan Rattlesnake, 324
urostictus, 162
urosticutus, 161
Urotheca, 8, 165, 218, 279, 280, 281
Urotheca aequalis, 281
Urotheca andrewsi, 281
Urotheca bicolor, 281
Urotheca celatus, 281
Urotheca decipiens, 280
Urotheca deppei, 281
Urotheca disastemus, 281
Urotheca dumerili, 281
Urotheca elapoides, 280, 281
Urotheca euryzona, 281
Urotheca fulviceps, 281
Urotheca guentheri, 281
Urotheca hobartsmithi, 281
Urotheca lateristriga, 281
Urotheca laticollaris, 281
Urotheca multilineata, 282
Urotheca myersi, 282
Urotheca occidentals, 281
Urotheca pachyura, 282
Urotheca pacificus, 281
Urotheca psycoides, 280, 282
Urotheca salvadorensis, 281
Urotheca salvinii, 281
Urotheca tricinctus, 281
Urotheca wilmarai, 280, 282
Uruguayan Coral Snake, 289
Uruguayan Hog-nosed Snake, 177
Uruguayan Painted Lancehead Pit Viper, 318
uruguayensis, 177
urutu, 318
Urutú, 313
utahensis, 25, 49, 248
Utah Milksnake, 156
Utah Mountain Kingsnake, 154
Utah Threadsnake, 25
utowanae, 65

Váche Vinesnake, 278
vagrans, 233, 259, 260
vagus, 233
valeriae, 282
valida, 266, 267
validus, 266, 267
Valley Gartersnake, 265
vandenburghi, 109, 111, 272
vanzolinii, 175
Vanzolini's, 175
varia, 203
Variable Coral Snake, 292
Variable Groundsnake, 235
variablis, 270
varians, 268

variegata, 115, 131
Variegated False Coral Snake, 281
Variegated Tellurian Snake, 84
Varigated Snail-eater, 115
variegatus, 64, 84
variolosus, 181
vascotanneri, 260
vaticinata, 107
vegrandis, 324
veliferum, 65, 66
vellardi, 28
Vellard's Threadsnake, 28
Velvety Swamp Legion Snake, 175
venezuelae, 319
Venezuelan Aquatic Coral Snake, 303
Venezuelan Calico Snake, 197
Venezuelan Capuchin Coral Snake, 294
Venezuelan Coral Snake, 297
Venezuelan False Coral Mudsnake, 144
Venezuelan False Coral Watersnake, 144
Venezuelan Forest Pit Viper, 310
Venezuelan Lancehead Pit Viper, 319
Venezuelan Pygmy Coral Snake, 293
Venezuelan Racer, 201
Venezuelan Rattlesnake, 322
Venezuelan Red-tailed Coral Snake, 299
Venezuelan Snail-eater, 114
Venezuelan Striped Tellurian Snake, 343
Venezuelan Tellurian Snake, 77, 343
Venezuelan Threadsnake, 20
venezuelanus, 197
Venezuelan Whipsnake, 181
venezuelensis, 144, 294, 319
ventrimaculata, 69
ventrimaculatus, 84, 233
venusta, 146
venustissimus, 128
Veracruz Coral Snake, 292
Veracruz Earthsnake, 133
Veracruz Graceful Brownsnake, 219
Veracruz Milksnake, 156
Veracruz Montane Pit Viper, 319
Veracruz Ring-necked Coffee Snake, 189
Veracruz Yellow-bellied Snake, 101
veraecrucis, 185
veraepacis, 60, 294
Verapaz Burrowing Snake, 60
Verapaz Elegant Coral Snake, 294
vermiculata, 115
Vermiculate Blindsnake, 17
Vermiculate Snail-eater, 115
vermiculaticeps, 223
vermiformis, 253
vermis, 88
vernalis, 164, 192
Vertebral Tellurian Snake, 84
vertebralis, 84, 207, 210

vertebrolineatus, 84
Víbora de Pico, 32
vicenti, 95
vicinus, 259
victa, 240
Vieques Island Racer, 64
viguieri, 116
vilkinsoni, 273
Villa María Tellurian Snake, 79
villaricae, 69
Villavicencio Coral Snake, 298
Villavicencio Tellurian Snake, 82
villelai, 183
Villela's Lizard-eater, 183
vinitor, 109
violaceus, 173
Viperidae, 10, 11, 58, 305
Viperids, 305
Vipers, 58, 305
viquezi, 271
Viquez's Tropical Groundsnake, 271
virgatenuis, 260
Virgin Islands Boa, 46
Virginia, 282
Virginia striatula, 282
Virginia valeriae, 282
Virginia valeriae elegans, 283
Virginia valeriae pulchra, 283
Virginia valeriae valeriae, 283
virgultea, 226
viridis, 175, 176, 255, 332
viridissima, 204
visoninum, 60
visoninus, 60
vittata, 71
vittatum, 74
vittatus, 74, 104, 205, 343
vitti, 176
Vitt's Legion Snake, 176
Volcán Chiriqui Earthsnake, 133
Volcán Earthsnake, 133
volcanicum, 337
Volcano Centipede Snake, 253
Volcano Coral Snake, 303
Volcano Hog-nosed Pit Viper, 337
Volcán Poás Earthsnake, 137
Volcán Tacaná Centipede Snake, 252
vudii, 65
vulcani, 253
vulpina, 121, 124

wagleri, 85, 270
Waglerophis, 255, 283, 284
Waglerophis merremi, 283
Wagler's Cativo, 270
Wagler's False Pit Viper, 283
Wagler's Snake, 283

Wagler's Tellurian Snake, 85
Wagler's Tropical Snake, 120
Wagler's Whipsnake, 95
wagneri, 111, 300
Wagner's Double-keeled Scale Snake, 111
Walker's Scrub Snake, 243
Walker's Swift Snake, 243
warreni, 46
Water Cobra, 142
Water Mapanare, 138
Wedge-tailed Tellurian Snake, 79
werleri, 188
werneri, 85, 255
Werner's Calico Snake, 195
Werner's Sipo, 93
Werner's Snail-eater, 114
Werner's Tellurian Snake, 85
Werner's Watersnake, 140
West Chihuahuan Ridge-nosed Rattlesnake, 334
Western Argentina Threadsnake, 22
Western Black-headed Snake, 250
Western Black-necked Gartersnake, 259
Western Calico Snake, 195
Western Cat-eyed Snake, 159
Western Coachwhip, 180
Western Coral Snake, 288
Western Cottonmouth, 307
Western Cuba Racer, 62
Western Desert Kingsnake, 152
Western Diamond-backed Rattlesnake, 320
Western Double Collar Coral Snake, 297
Western Elegant Coral Snake, 294
Western Foxsnake, 124
Western Galápagos Racer, 62
Western Graceful Brownsnake, 220
Western Hemprich's Coral Snake, 296
Western Hog-nosed Pit Viper, 336
Western Hog-nosed Snake, 140
Western Hook-nosed Snake, 137
Western Indigo Snake, 117
Western Jumping Pit Viper, 308
Western Leaf-nosed Snake, 206
Western Long-nosed Snake, 224
Western Lyresnake, 271
Western Massasauga, 337
Western Mexican Bullsnake, 209
Western Mudsnake, 129
Western Parrot Snake, 162
Western Patch-nosed Snake, 225
Western Pygmy Rattlesnake, 338
Western Rattlesnake, 332
Western Ribbon Coral Snake, 298
Western Ribbonsnake, 262
Western Shovel-nosed Snake, 92
Western Smooth Earthsnake, 283
Western Smooth Green Snake, 164

Western Snail-eating Snake, 275
Western South American Neotropical Racer, 63
Western Speckled Racer, 119
Western Striped Forest Pit Viper, 310
Western Threadsnake, 24
Western Tree Snake, 150
Western Twin-spottedRattlesnake, 329
Western Worm-eating Coral Snake, 296
Western Wormsnake, 88
Western Yellow-bellied Racer, 99
West Indies Racer, 65
West Mexican Coral Snake, 293
Wetlands Rainbow Boa, 44
wetmorei, 268, 279
Wetmore's Tolucan Groundsnake, 268
Wetmore's Vinesnake, 279
weyrauchi, 28
Weyrauch's Threadsnake, 28
W-headed Racer, 147
White-banded Coral Snake, 289
White-fronted Threadsnake, 20
White-headed Snake, 125
White-lipped Legion Snake, 171
White-lipped Snake, 241
White-nosed Blindsnake, 31
White-striped Centipede-eater, 252
White-tailed Lancehead Pit Viper, 317
Wide Collar Earthsnake, 135
Wied's Tropical Snake, 120
wilcoxi, 253
wilderi, 18, 19
Wilder's Slender Blindsnake, 19
willardi, 333
williamengelsi, 189
williamsi, 176, 233
Williams Legion Snake, 176
Williams's Snail-eater, 233
wilmarai, 280, 282
wilsoni, 193
Windward Tree Racer, 183
wolffsohni, 131
woodini, 154
Woodland Racer, 119
Wood Snake, 53
Worm-eating Coral Snake, 296
Wormsnake, 8, 10, 15, 19, 29
worontzowi, 235
Wreathed Centipede Snake, 252
wrighti, 57
wrightorum, 240
Wright's Dwarf Boa, 57
wuchereri, 125
Wucherer's Groundsnake, 285

Xadrez Snake, 87
Xaenophidia, 8

xanthocheilus, 90
xanthogrammus, 313, 314
Xenodon, 255, 283, 284
Xenodon bertholdi, 284
Xenodon colubrinus, 284
Xenodon guentheri, 284
Xenodon merremi, 283
Xenodon neuwiedi, 284
Xenodon rabdocephalus, 284
Xenodon rabdocephalus mexicanus, 284
Xenodon rabdocephalus rabdocephalus, 284
Xenodon severus, 285
Xenodon suspectus, 284
Xenodon werneri, 255
Xenopeltidae, 52
Xenopeltis, 37, 52
Xenopholis, 285
Xenopholis scalaris, 285
Xenopholis undulatus, 285
Xenoxybelis, 193, 285
Xenoxybelis argenteus, 285, 286
Xenoxybelis boulengeri, 286
xerophilus, 43, 172

yacu, 140
yaquia, 253
Yaqui Black-headed Snake, 253
Yarumal Tellurian Snake, 82
yatesi, 289
yavi, 257
Yellow Anaconda, 47
Yellow-banded Dwarf Boa, 57
Yellow-bellied Kingsnake, 151
Yellow-bellied Sea Snake, 305
Yellow-bellied Snake, 101
Yellow-bellied Watersnake, 187
Yellow-blotched Palm Pit Viper, 309
Yellow-headed Calico Snake, 194
Yellow-headed Threadsnake, 27
Yellow-lined Centipede Snake, 248
Yellow-lipped Sea Krait, 304
Yellow Queen Legion Snake, 170
Yellow Ratsnake, 123
Yellow-tailed Blindsnake, 17
Yellow-throated Gartersnake, 263
Yellow-throated Puffing Snake, 215
Yellow Tiger Ratsnake, 237
yonenagae, 33
Yucatán Blindsnake, 31
Yucatán Blunt-headed Tree Snake, 150
Yucatán Coral Snake, 292
Yucatán Dwarf Short-tailed Snake, 254
yucatanensis, 159, 163
Yucatán Hog-nosed Pit Viper, 337
yucatanicum, 337
Yucatán Nightsnake, 121

Index

Yucatán Parrot Snake, 163
Yucatán Ratsnake, 121
Yucatán Rattlesnake, 324
Yucatán Red-backed Coffee Snake, 191
Yucatán Snail-eater, 231
Yucatecan Cantil, 306
Yucatecan Checkered Gartersnake, 261
yumensis, 152
Yungas Coral Snake, 290

Yungas Threadsnake, 27
yutoensis, 244

Zacatera, 60
Zacatera Roja, 60
zanthus, 258
Zapotec Coral Snake, 294
zapotecus, 294
zebrinus, 85

zeledoni, 137
zidoki, 85
Zidok's Tellurian Snake, 85
zonata, 156
Zunil Coral Snake, 300
zunilensis, 300
zweifeli, 174, 229, 276, 293
Zweifel's Coral Snake, 293
Zweifel's Snail-eating Snake, 276